Physics of Condensed Matter

Physics of Condensed Matter

Prasanta K. Misra
Department of Physics
University of Houston

ELSEVIER

AMSTERDAM • BOSTON • HEIDELBERG • LONDON
NEW YORK • OXFORD • PARIS • SAN DIEGO
SAN FRANCISCO • SINGAPORE • SYDNEY • TOKYO
Academic Press is an imprint of Elsevier

Academic Press is an imprint of Elsevier
30 Corporate Drive, Suite 400, Burlington, MA 01803, USA
525 B Street, Suite 1800, San Diego, CA 92101-4495, USA
84 Theobald's Road, London WC1X 8RR, UK

Library of Congress Cataloging-in-Publication Data
Misra, Prasanta K. (Prasanta Kumar)
Physics of condensed matter / Prasanta K. Misra.
p. cm.
Includes bibliographical references and index.
ISBN 978-0-12-384954-0
1. Condensed matter. I. Title.
QC173.454.M57 2011
530.4'1–dc22 2010035448

British Library Cataloguing-in-Publication Data
A catalogue record for this book is available from the British Library.

For information on all Academic Press publications visit our Web site at *www.elsevierdirect.com*

Typeset by: diacriTech, Chennai, India

For Swayamprava, Debasis, Moushumi, Sandeep,
and Annika, Millan, Kishen, Nirvaan

Contents

Preface.......... xxi

CHAPTER 1 Basic Properties of Crystals.......... 1

1.1 Crystal Lattices.......... 2
- 1.1.1 Primitive Cell.......... 3
- 1.1.2 Unit Cell.......... 3
- 1.1.3 Wigner–Seitz Cell.......... 3
- 1.1.4 Lattice Point Group.......... 3

1.2 Bravais Lattices in Two and Three Dimensions.......... 4
- 1.2.1 Simple Cubic (sc) Lattice.......... 4
- 1.2.2 Lattice Constants.......... 5
- 1.2.3 Coordination Numbers.......... 5
- 1.2.4 Body-Centered Cubic (bcc) Lattice.......... 5
- 1.2.5 Face-Centered Cubic (fcc) Lattice.......... 7
- 1.2.6 Other Bravais Lattices.......... 9

1.3 Lattice Planes and Miller Indices.......... 11

1.4 Bravais Lattices and Crystal Structures.......... 13
- 1.4.1 Crystal Structure.......... 13
- 1.4.2 Lattice with a Basis.......... 13
- 1.4.3 Packing Fraction.......... 14

1.5 Crystal Defects and Surface Effects.......... 14
- 1.5.1 Crystal Defects.......... 14
- 1.5.2 Surface Effects.......... 14

1.6 Some Simple Crystal Structures.......... 15
- 1.6.1 Sodium Chloride Structure.......... 15
- 1.6.2 Cesium Chloride Structure.......... 15
- 1.6.3 Diamond Structure.......... 16
- 1.6.4 Zincblende Structure.......... 17
- 1.6.5 Hexagonal Close-Packed (hcp) Structure.......... 17

1.7 Bragg Diffraction.......... 19

1.8 Laue Method.......... 20

1.9 Reciprocal Lattice.......... 21
- 1.9.1 Definition.......... 21
- 1.9.2 Properties of the Reciprocal Lattice.......... 22
- 1.9.3 Alternative Formulation of the Laue Condition.......... 25

1.10 Brillouin Zones 27
1.10.1 Definition 27
1.10.2 One-Dimensional Lattice 28
1.10.3 Two-Dimensional Square Lattice 28
1.10.4 bcc Lattice 29
1.10.5 fcc Lattice 30
1.11 Diffraction by a Crystal Lattice with a Basis 31
1.11.1 Theory 31
1.11.2 Geometrical Structure Factor 32
1.11.3 Application to bcc Lattice 32
1.11.4 Application to fcc Lattice 33
1.11.5 The Atomic Scattering Factor or Form Factor 33
Problems 34
References 35

CHAPTER 2 Phonons and Lattice Vibrations 37
2.1 Lattice Dynamics 37
2.1.1 Theory 37
2.1.2 Normal Modes of a One-Dimensional Monoatomic Lattice 41
2.1.3 Normal Modes of a One-Dimensional Chain with a Basis 44
2.2 Lattice Specific Heat 48
2.2.1 Theory 48
2.2.2 The Debye Model of Specific Heat 49
2.2.3 The Einstein Model of Specific Heat 52
2.3 Second Quantization 53
2.3.1 Occupation Number Representation 53
2.3.2 Creation and Annihilation Operators 54
2.3.3 Field Operators and the Hamiltonian 58
2.4 Quantization of Lattice Waves 61
2.4.1 Formulation 61
2.4.2 Quantization of Lattice Waves 65
Problems 66
References 68

CHAPTER 3 Free Electron Model 71
3.1 The Classical (Drude) Model of a Metal 71
3.2 Sommerfeld Model 73
3.2.1 Introduction 73
3.2.2 Fermi Distribution Function 74
3.2.3 Density Operator 75
3.2.4 Free Electron Fermi Gas 77
3.2.5 Ground-State Energy of the Electron Gas 79
3.2.6 Density of Electron States 81

3.3 Fermi Energy and the Chemical Potential 82
3.4 Specific Heat of the Electron Gas 84
3.5 DC Electrical Conductivity 86
3.6 The Hall Effect 87
3.7 Failures of the Free Electron Model 89
Problems 90
References 93

CHAPTER 4 Nearly Free Electron Model 95
4.1 Electrons in a Weak Periodic Potential 96
4.1.1 Introduction 96
4.1.2 Plane Wave Solutions 97
4.2 Bloch Functions and Bloch Theorem 99
4.3 Reduced, Repeated (Periodic), and Extended Zone Schemes 99
4.3.1 Reduced Zone Scheme 100
4.3.2 Repeated Zone Scheme 100
4.3.3 Extended Zone Scheme 101
4.4 Band Index 101
4.5 Effective Hamiltonian 102
4.6 Proof of Bloch's Theorem from Translational Symmetry 103
4.7 Approximate Solution Near a Zone Boundary 105
4.8 Different Zone Schemes 109
4.8.1 Reduced Zone Scheme 109
4.8.2 Extended Zone Scheme 110
4.8.3 Periodic Zone Scheme 111
4.9 Elementary Band Theory of Solids 111
4.9.1 Introduction 111
4.9.2 Energy Bands in One Dimension 112
4.9.3 Number of States in a Band 112
4.10 Metals, Insulators, and Semiconductors 112
4.11 Brillouin Zones 117
4.12 Fermi Surface 119
4.12.1 Fermi Surface (in Two Dimensions) 119
4.12.2 Fermi Surface (in Three Dimensions) 121
4.12.3 Harrison's Method of Construction of the Fermi Surface 121
Problems 124
References 130

CHAPTER 5 Band-Structure Calculations 131
5.1 Introduction 131
5.2 Tight-Binding Approximation 131
5.3 LCAO Method 135

5.4 Wannier Functions 140
5.5 Cellular Method 142
5.6 Orthogonalized Plane-Wave (OPW) Method 145
5.7 Pseudopotentials 147
5.8 Muffin-Tin Potential 149
5.9 Augmented Plane-Wave (APW) Method 150
5.10 Green's Function (KKR) Method 152
5.11 Model Pseudopotentials 156
5.12 Empirical Pseudopotentials 157
5.13 First-Principles Pseudopotentials 158
Problems 160
References 163

CHAPTER 6 Static and Transport Properties of Solids 165
6.1 Band Picture 166
6.2 Bond Picture 167
6.3 Diamond Structure 168
6.4 Si and Ge 168
6.5 Zinc-Blende Semiconductors 170
6.6 Ionic Solids 172
6.7 Molecular Crystals 174
6.7.1 Molecular Solids 174
6.7.2 Hydrogen-Bonded Structures 174
6.8 Cohesion of Solids 174
6.8.1 Molecular Crystals: Noble Gases 174
6.8.2 Ionic Crystals 176
6.8.3 Covalent Crystals 177
6.8.4 Cohesion in Metals 178
6.9 The Semiclassical Model 179
6.10 Liouville's Theorem 182
6.11 Boltzmann Equation 183
6.12 Relaxation Time Approximation 184
6.13 Electrical Conductivity 186
6.14 Thermal Conductivity 187
6.15 Weak Scattering Theory of Conductivity 188
6.15.1 Relaxation Time and Scattering Probability 188
6.15.2 The Collision Term 188
6.15.3 Impurity Scattering 189
6.16 Resistivity Due to Scattering by Phonons 192
Problems 194
References 196

CHAPTER 7 Electron–Electron Interaction **199**
7.1 Introduction 199
7.2 Hartree Approximation 200
7.3 Hartree–Fock Approximation 203
7.3.1 General Formulation 203
7.3.2 Hartree–Fock Theory for Jellium 204
7.4 Effect of Screening 207
7.4.1 General Formulation 207
7.4.2 Thomas–Fermi Approximation 208
7.4.3 Lindhard Theory of Screening 209
7.5 Friedel Sum Rule and Oscillations 214
7.6 Frequency and Wave-Number-Dependent Dielectric Constant 217
7.7 Mott Transition 222
7.8 Density Functional Theory 223
7.8.1 General Formulation 223
7.8.2 Local Density Approximation 224
7.9 Fermi Liquid Theory 225
7.9.1 Quasiparticles 225
7.9.2 Energy Functional 227
7.9.3 Fermi Liquid Parameters 230
7.10 Green's Function Method 232
7.10.1 General Formulation 232
7.10.2 Finite-Temperature Green's Function Formalism for Interacting Bloch Electrons 233
7.10.3 Exchange Self-Energy in the Band Model 234
Problems 235
References 241

CHAPTER 8 Dynamics of Bloch Electrons **243**
8.1 Semiclassical Model 243
8.2 Velocity Operator 244
8.3 k · p Perturbation Theory 245
8.4 Quasiclassical Dynamics 246
8.5 Effective Mass 247
8.6 Bloch Electrons in External Fields 248
8.6.1 Time Evolution of Bloch Electrons in an Electric Field 250
8.6.2 Alternate Derivation for Bloch Functions in an External Electric and Magnetic Field 252
8.6.3 Motion in an Applied DC Field 253
8.7 Bloch Oscillations 254
8.8 Holes 255
8.9 Zener Breakdown (Approximate Method) 258

8.10 Rigorous Calculation of Zener Tunneling 261
8.11 Electron–Phonon Interaction 264
Problems 271
References 274

CHAPTER 9 Semiconductors 275
9.1 Introduction 275
9.2 Electrons and Holes 278
9.3 Electron and Hole Densities in Equilibrium 279
9.4 Intrinsic Semiconductors 283
9.5 Extrinsic Semiconductors 284
9.6 Doped Semiconductors 285
9.7 Statistics of Impurity Levels in Thermal Equilibrium 288
9.7.1 Donor Levels 288
9.7.2 Acceptor Levels 288
9.7.3 Doped Semiconductors 289
9.8 Diluted Magnetic Semiconductors 290
9.8.1 Introduction 290
9.8.2 Magnetization in Zero External Magnetic Field in DMS 291
9.8.3 Electron Paramagnetic Resonance Shift 291
9.8.4 $\vec{k} \cdot \vec{\pi}$ Model 295
9.9 Zinc Oxide 296
9.10 Amorphous Semiconductors 296
9.10.1 Introduction 296
9.10.2 Linear Combination of Hybrids Model for Tetrahedral Semiconductors 297
Problems 300
References 303

CHAPTER 10 Electronics 305
10.1 Introduction 305
10.2 p-n Junction 306
10.2.1 Introduction 306
10.2.2 p-n Junction in Equilibrium 307
10.3 Rectification by a p-n Junction 311
10.3.1 Equilibrium Case 311
10.3.2 Nonequilibrium Case ($V \neq 0$) 313
10.4 Transistors 318
10.4.1 Bipolar Transistors 318
10.4.2 Field-Effect Transistor 319
10.4.3 Single-Electron Transistor 321
10.5 Integrated Circuits 325

10.6 Optoelectronic Devices . . . 325
10.7 Graphene . . . 329
10.8 Graphene-Based Electronics . . . 332
Problems . . . 333
References . . . 336

CHAPTER 11 Spintronics . . . 339
11.1 Introduction . . . 339
11.2 Magnetoresistance . . . 340
11.3 Giant Magnetoresistance . . . 340
11.3.1 Metallic Multilayers . . . 340
11.4 Mott's Theory of Spin-Dependent Scattering of Electrons . . . 342
11.5 Camley–Barnas Model . . . 345
11.6 CPP-GMR . . . 348
11.6.1 Introduction . . . 348
11.6.2 Theory of CPP-GMR of Multilayered Nanowires . . . 350
11.7 MTJ, TMR, and MRAM . . . 352
11.8 Spin Transfer Torques and Magnetic Switching . . . 356
11.9 Spintronics with Semiconductors . . . 357
11.9.1 Introduction . . . 357
11.9.2 Theory of an FM-T-N Junction . . . 358
11.9.3 Injection Coefficient . . . 361
Problems . . . 364
References . . . 367

CHAPTER 12 Diamagnetism and Paramagnetism . . . 369
12.1 Introduction . . . 370
12.2 Atomic (or Ionic) Magnetic Susceptibilities . . . 371
12.2.1 General Formulation . . . 371
12.2.2 Larmor Diamagnetism . . . 372
12.2.3 Hund's Rules . . . 373
12.2.4 Van Vleck Paramagnetism . . . 374
12.2.5 Landé g Factor . . . 375
12.2.6 Curie's Law . . . 377
12.3 Magnetic Susceptibility of Free Electrons in Metals . . . 378
12.3.1 General Formulation . . . 378
12.3.2 Landau Diamagnetism and Pauli Paramagnetism . . . 380
12.3.3 De Haas–van Alphen Effect . . . 383
12.4 Many-Body Theory of Magnetic Susceptibility of Bloch Electrons in Solids . . . 388
12.4.1 Introduction . . . 388
12.4.2 Equation of Motion in the Bloch Representation . . . 388

12.4.3 Thermodynamic Potential ... 390
12.4.4 General Formula for χ ... 390
12.4.5 Exchange Self-Energy in the Band Model ... 393
12.4.6 Exchange Enhancement of χ_s ... 394
12.4.7 Exchange and Correlation Effects on χ_o ... 395
12.4.8 Exchange and Correlation Effects on χ_{so} ... 396
12.5 Quantum Hall Effect ... 396
12.5.1 Introduction ... 396
12.5.2 Two-Dimensional Electron Gas ... 396
12.5.3 Quantum Transport of a Two-Dimensional Electron Gas in a Strong Magnetic Field ... 397
12.5.4 Quantum Hall Effect from Gauge Invariance ... 400
12.6 Fractional Quantum Hall Effect ... 400
Problems ... 401
References ... 407

CHAPTER 13 Magnetic Ordering ... 409
13.1 Introduction ... 410
13.2 Magnetic Dipole Moments ... 411
13.3 Models for Ferromagnetism and Antiferromagnetism ... 412
13.3.1 Introduction ... 412
13.3.2 Heitler–London Approximation ... 412
13.3.3 Spin Hamiltonian ... 414
13.3.4 Heisenberg Model ... 416
13.3.5 Direct, Indirect, and Superexchange ... 416
13.3.6 Spin Waves in Ferromagnets: Magnons ... 417
13.3.7 Schwinger Representation ... 417
13.3.8 Application to the Heisenberg Hamiltonian ... 418
13.3.9 Spin Waves in Antiferromagnets ... 421
13.4 Ferromagnetism in Solids ... 422
13.4.1 Ferromagnetism Near the Curie Temperature ... 422
13.4.2 Comparison of Spin-Wave Theory with the Weiss Field Model ... 424
13.4.3 Ferromagnetic Domains ... 425
13.4.4 Hysteresis ... 426
13.4.5 Ising Model ... 427
13.5 Ferromagnetism in Transition Metals ... 427
13.5.1 Introduction ... 427
13.5.2 Stoner Model ... 428
13.5.3 Ferromagnetism in Fe, Co, and Ni from Stoner's Model and Kohn–Sham Equations ... 430
13.5.4 Free Electron Gas Model ... 431
13.5.5 Hubbard Model ... 433

13.6 Magnetization of Interacting Bloch Electrons 434
13.6.1 Introduction 434
13.6.2 Theory of Magnetization 434
13.6.3 The Quasiparticle Contribution to Magnetization 435
13.6.4 Contribution of Correlations to Magnetization 436
13.6.5 Single-Particle Spectrum and the Criteria for Ferromagnetic Ground State 437
13.7 The Kondo Effect 439
13.8 Anderson Model 439
13.9 The Magnetic Phase Transition 440
13.9.1 Introduction 440
13.9.2 The Order Parameter 441
13.9.3 Landau Theory of Second-Order Phase Transitions 441
Problems 443
References 448

CHAPTER 14 Superconductivity 451
14.1 Properties of Superconductors 452
14.1.1 Introduction 452
14.1.2 Type I and Type II Superconductors 453
14.1.3 Second-Order Phase Transition 454
14.1.4 Isotope Effect 454
14.1.5 Phase Diagram 454
14.2 Meissner–Ochsenfeld Effect 455
14.3 The London Equation 455
14.4 Ginzburg–Landau Theory 456
14.4.1 Order Parameter 456
14.4.2 Boundary Conditions 457
14.4.3 Coherence Length 457
14.4.4 London Penetration Depth 458
14.5 Flux Quantization 459
14.6 Josephson Effect 460
14.6.1 Two Superconductors Separated by an Oxide Layer 460
14.6.2 AC and DC Josephson Effects 462
14.7 Microscopic Theory of Superconductivity 462
14.7.1 Introduction 462
14.7.2 Quasi-Electrons 463
14.7.3 Cooper Pairs 464
14.7.4 BCS Theory 466
14.7.5 Ground State of the Superconducting Electron Gas 466
14.7.6 Excited States at $T=0$ 469
14.7.7 Excited States at $T\neq 0$ 470

14.8 Strong-Coupling Theory 472
14.8.1 Introduction 472
14.8.2 Upper Limit of the Critical Temperature, T_c 472
14.9 High-Temperature Superconductors 473
14.9.1 Introduction 473
14.9.2 Properties of Novel Superconductors (Cuprates) 474
14.9.3 Brief Review of *s*-, *p*-, and *d*-wave Pairing 474
14.9.4 Experimental Confirmation of *d*-wave Pairing 476
14.9.5 Search for a Theoretical Mechanism of High T_c Superconductors 481
Problems 481
References 485

CHAPTER 15 Heavy Fermions 487
15.1 Introduction 488
15.2 Kondo-Lattice, Mixed-Valence, and Heavy Fermions 490
15.2.1 Periodic Anderson and Kondo-Lattice Models 490
15.2.2 Mixed-Valence Compounds 492
15.2.3 Slave Boson Method 493
15.2.4 Cluster Calculations 494
15.3 Mean-Field Theories 498
15.3.1 The Local Impurity Self-Consistent Approximation 498
15.3.2 Application of LISA to Periodic Anderson Model 499
15.3.3 RKKY Interaction 500
15.3.4 Extended Dynamical Mean-field Theory 501
15.4 Fermi-Liquid Models 502
15.4.1 Heavy Fermi Liquids 502
15.4.2 Fractionalized Fermi Liquids 505
15.5 Metamagnetism in Heavy Fermions 506
15.6 Ce- and U-Based Superconducting Compounds 508
15.6.1 Ce-Based Compounds 508
15.6.2 U-Based Superconducting Compounds 509
15.7 Other Heavy-Fermion Superconductors 513
15.7.1 $PrOs_4Sb_{12}$ 513
15.7.2 $PuCoGa_5$ 513
15.7.3 $PuRhGa_5$ 515
15.7.4 Comparison between Cu and Pu Containing High-T_c Superconductors 516
15.8 Theories of Heavy-Fermion Superconductivity 516
15.9 Kondo Insulators 516
15.9.1 Brief Review 516
15.9.2 Theory of Kondo Insulators 517
Problems 519
References 524

CHAPTER 16 Metallic Nanoclusters 527

16.1 Introduction 528
16.1.1 Nanoscience and Nanoclusters 528
16.1.2 Liquid Drop Model 528
16.1.3 Size and Surface/Volume Ratio 528
16.1.4 Geometric and Electronic Shell Structures 530

16.2 Electronic Shell Structure 531
16.2.1 Spherical Jellium Model (*Phenomenological*) 531
16.2.2 Self-Consistent Spherical Jellium Model 532
16.2.3 Ellipsoidal Shell Model 535
16.2.4 Nonalkali Clusters 535
16.2.5 Large Clusters 535

16.3 Geometric Shell Structure 537
16.3.1 Close-Packing 537
16.3.2 Wulff Construction 537
16.3.3 Polyhedra 538
16.3.4 Filling between Complete Shells 540

16.4 Cluster Growth on Surfaces 540
16.4.1 Monte Carlo Simulations 540
16.4.2 Mean-Field Rate Equations 541

16.5 Structure of Isolated Clusters 542
16.5.1 Theoretical Models 542
16.5.2 Structure of Some Isolated Clusters 546

16.6 Magnetism in Clusters 547
16.6.1 Magnetism in Isolated Clusters 547
16.6.2 Experimental Techniques for Studying Cluster Magnetism 549
16.6.3 Magnetism in Embedded Clusters 553
16.6.4 Graphite Surfaces 555
16.6.5 Study of Clusters by Scanning Tunneling Microscope 555
16.6.6 Clusters Embedded in a Matrix 557

16.7 Superconducting State of Nanoclusters 558
16.7.1 Qualitative Analysis 558
16.7.2 Thermodynamic Green's Function Formalism for Nanoclusters 559

Problems 562
References 565

CHAPTER 17 Complex Structures 567

17.1 Liquids 568
17.1.1 Introduction 568
17.1.2 Phase Diagram 568
17.1.3 Van Hove Pair Correlation Function 569
17.1.4 Correlation Function for Liquids 570

17.2 Superfluid ^{4}He........ 570
17.2.1 Introduction........ 570
17.2.2 Phase Transition in ^{4}He........ 570
17.2.3 Two-Fluid Model for Liquid ^{4}He........ 571
17.2.4 Theory of Superfluidity in Liquid ^{4}He........ 571
17.3 Liquid ^{3}He........ 573
17.3.1 Introduction........ 573
17.3.2 Possibility of Superfluidity in Liquid ^{3}He........ 574
17.3.3 Fermi Liquid Theory........ 574
17.3.4 Experimental Results of Superfluidity in Liquid ^{3}He........ 575
17.3.5 Theoretical Model for the A and A_1 Phases........ 575
17.3.6 Theoretical Model for the B Phase........ 577
17.4 Liquid Crystals........ 578
17.4.1 Introduction........ 578
17.4.2 Three Classes of Liquid Crystals........ 578
17.4.3 The Order Parameter........ 580
17.4.4 Curvature Strains........ 581
17.4.5 Optical Properties of Cholesteric Liquid Crystals........ 581
17.5 Quasicrystals........ 583
17.5.1 Introduction........ 583
17.5.2 Penrose Tiles........ 583
17.5.3 Discovery of Quasicrystals........ 584
17.5.4 Quasiperiodic Lattice........ 584
17.5.5 Phonon and Phason Degrees of Freedom........ 586
17.5.6 Dislocation in the Penrose Lattice........ 589
17.5.7 Icosahedral Quasicrystals........ 589
17.6 Amorphous Solids........ 590
17.6.1 Introduction........ 590
17.6.2 Energy Bands in One-Dimensional Aperiodic Potentials........ 591
17.6.3 Density of States........ 593
17.6.4 Amorphous Semiconductors........ 593
Problems........ 594
References........ 597

CHAPTER 18 Novel Materials........ 599
18.1 Graphene........ 600
18.1.1 Introduction........ 600
18.1.2 Graphene Lattice........ 601
18.1.3 Tight-Binding Approximation........ 602
18.1.4 Dirac Fermions........ 606
18.1.5 Comprehensive View of Graphene........ 608

18.2 Fullerenes 608
18.2.1 Introduction 608
18.2.2 Discovery of C_{60} 609
18.3 Fullerenes and Tubules 613
18.3.1 Introduction 613
18.3.2 Carbon Nanotubeles 614
18.3.3 Three Types of Carbon Nanotubes 614
18.3.4 Symmetry Properties of Carbon Nanotubes 616
18.3.5 Band Structure of a Fullerene Nanotube 617
18.4 Polymers 617
18.4.1 Introduction 617
18.4.2 Saturated and Conjugated Polymers 618
18.4.3 Transparent Metallic Polymers 621
18.4.4 Electronic Polymers 621
18.5 Solitons in Conducting Polymers 622
18.5.1 Introduction 622
18.5.2 Electronic Structure 623
18.5.3 Tight-Binding Model 623
18.5.4 Soliton Excitations 624
18.5.5 Solitons, Polarons, and Polaron Excitations 626
18.6.6 Polarons and Bipolarons 626
18.6 Photoinduced Electron Transfer 627
Problems 627
References 630

APPENDIX A Elements of Group Theory 633
A.1 Symmetry and Its Consequences 633
A.1.1 Symmetry of Crystals 633
A.1.2 Definition of a Group 633
A.1.3 Symmetry Operations in Crystal Lattices 634
A.2 Space Groups 634
A.2.1 Introduction 634
A.2.2 Space Group Operations 634
A.3 Point Group Operations 636
A.3.1 Introduction 636
A.3.2 Description of Point Groups 636
A.3.3 The Cubic Group O_h 638

APPENDIX B Mossbauer Effect 641
B.1 Introduction 641
B.2 Recoilless Fraction 642
B.3 Average Transferred Energy 643
Reference 644

APPENDIX C Introduction to Renormalization Group Approach **645**
C.1 Critical Behavior 645
C.2 Theory for Scaling 646
C.3 Renormalization Group Approach 648
References 649

Index **651**

Preface

This textbook is designed for a one-year (two semesters) graduate course on condensed matter physics for students in physics, materials science, solid state chemistry, and electrical engineering. It can also be used as a one-semester course for advanced undergraduate majors in physics, materials science, chemistry, and electrical engineering, and another one-semester course for graduate students in these areas. The book assumes a working knowledge of quantum mechanics, statistical mechanics, electricity and magnetism, and Green's function formalism (for the second-semester curriculum). The book is written as a two-semester graduate-level textbook, but it can also be used as a reference book by faculty and other researchers actively engaged in research in condensed matter physics. With judicious choice of topics, the book can be divided into two parts: "Fundamental Concepts" designed to be taught in the first semester, and "Research Applications" to be taught in the second semester. Obviously, the first part can be taught to advanced undergraduate majors as an introductory course.

The later chapters are self-contained. Each research topic has a brief introduction, a review, and a summary of basic foundations for advanced research. This is done with the belief that the students will develop the skills and will be sufficiently prepared to develop an interest in one of the vast areas of the topics covered under the umbrella of "condensed matter physics." In fact, this wide diversity of topics, the research on which has been increasing exponentially during the past decade, makes it nearly impossible to write a two-semester textbook for graduate students. Probably that is the reason for a dearth of graduate-level textbooks in condensed matter physics. This has led to an increasingly difficult task for the instructor because he or she has to prepare notes from a variety of textbooks, reference books, and review articles, especially to teach in the second-semester graduate level.

There has been slow but steady growth in the area of solid state physics after it was recognized as a separate branch of physics around 1940, probably after the publication of the book *The Modern Theory of Solids* by Seitz. The main reason for this growth is solid state physics is essentially the applied branch of physics with a variety of technological applications and has attracted students from other disciplines. The slow but steady growth accelerated in the 1960s because of extensive research funding due to the space program, and eventually solid state physics became the major branch of physics attracting the maximum number of faculty and students. The American Physical Society officially changed the name of its largest group from "Solid State Physics" to "Condensed Matter Physics," thereby including liquids and other soft materials. This change in 1978 has led to explosive growth in condensed matter physics during the past 30 years, and the material for supplementing the available textbooks has risen exponentially. In addition, research in various areas has accelerated rapidly, fueled by grants and a need for fast development in computer memory and storage as well as other applications of nanoscience and nanotechnology. The subject, which has now become multidisciplinary, includes materials science, solid state chemistry, and electrical engineering.

Recently, I wrote a book called *Heavy-Fermion Systems*, which is a part of the book series "Handbook of Metal Physics," of which I am the series editor. A large number of distinguished physicists and chemists contributed to the book series and I have learned much while editing their work. These are advanced research–level books, but it became obvious that there is a need for a one-year (two-semester) graduate-level textbook in condensed matter physics that includes material on some of the new topics covered in this book series as well as in many other advanced research–level books and research reviews in prestigious journals. A graduate student should have the choice to select a topic for research after being taught in the classroom in order to acquire enough background on the topic. I have endeavored to do just that in this textbook, which has been limited to 18 chapters and 3 appendices. The project has taken several years, much longer than I had originally planned. I have learned a lot during this period, including the fact that the boundaries between the various disciplines in physics, chemistry, electrical engineering, and materials science are getting blurred.

The book has three objectives:

1. To present a coherent, clear, and intelligible picture of simple models of crystalline solids in the first few chapters. The properties of real solids, which are more complicated, are dealt with in later chapters. The more advanced topics are dealt with in the later part of each chapter (after the first few introductory chapters). Each chapter includes a collection of problems in order to enable students to have a grasp of the topics taught in the chapter. The problems at the end of each chapter are designed to make the students derive some of the formulas of analytical development with no intrinsic interest. The objective is to keep the book within a reasonable length, but more importantly, with the belief that the mathematical steps are better understood if they are derived by the students with the aid of hints and suggestions. In the second part of the book (Research Applications), some of the problems at the end of the chapter are extensions of the advanced topics covered in the chapter. In this part, some other problems are designed to make the applications of the topics more clear. It is up to the instructor to choose and assign the problems, and some instructors have their own list of problems. However, students should at least read all the problems even if they do not have any motivation or intention to solve them.
2. To present a comprehensive account of the modern topics in condensed matter physics by including introductory accounts of the areas where intense research is going on at present. To be able to do so, I have included chapters on Spintronics (Chapter 11), Heavy Fermions (Chapter 15), Metallic Nanoclusters (Chapter 16), and Novel Materials (Chapter 18). In addition, I have included sections on ZnO (Section 9.9), graphene (Section 10.7), graphene-based electronics (Section 10.8), quantum hall effect (Section 12.5), fractional quantum hall effect (Section 12.6), high-temperature superconductivity (Section 14.9), liquid ^{3}He (Section 17.3), and quasicrystals (Section 17.5). Most of these topics are normally not included in standard textbooks in condensed matter physics. In fact, condensed matter physics is rapidly growing as an interdisciplinary subject because of its application in nanoscience and other areas of fast-growing science and technology. The objective of this book is to present the fundamental concepts as well as the methods for advanced research in this area.
3. To keep the size of the book within a reasonable length so that it can be taught as a two-semester course, I have avoided too many diagrams as well as excluded material not usually taught but included in most standard textbooks. I have also avoided including too many tables that list the properties of solids because these can be easily found in books specifically designed to provide such information. In addition, I have made a comprehensive review of many important topics such as band-structure calculations (Chapter 5), but left the details for students to learn if they are interested in doing research involving such topics.

I have consulted a large number of research papers and books while writing this textbook. It is not possible to acknowledge all these books and research papers at the appropriate places as is usually done in advanced research–level books. I have acknowledged whenever I have reprinted a figure with the permission of the author/publisher from a research paper published in a research journal or a book. I have also acknowledged at appropriate places whenever I have used any material published in research journals. There is a list of references at the end of each chapter where I have acknowledged the books and research papers I have used as primary sources of reference while writing this textbook.

ACKNOWLEDGMENTS

I learned the skills to do research in theoretical solid state physics from Professor Laura M. Roth who was my Ph.D. advisor at Tufts University. I have improved those skills by working as a postdoctoral research associate with Professor Leonard Kleinman of the University of Texas at Austin. I learned a lot of basic techniques as well as gained physical insight to solve a variety of research problems during my 10 years of collaboration with late Professor Joseph Callaway of Louisiana State University. I am also thankful to Professor S. D. Mahanti of Michigan State University with whom I have collaborated and published several important research papers on applications of many-body theory. I am thankful to the distinguished physicists and chemists who have contributed to the book series "Hand Book of Metal Physics," of which I had the privilege to be the Series Editor. I am thankful to the large number of colleagues and friends with whom I have consulted while writing this book, especially on their opinion as to what subjects should be included in a two-semester graduate-level textbook. I am also thankful to the graduate students who have worked on their Ph.D. theses under my supervision, for their patience as well as confidence in my ability to simultaneously work on a variety of research topics in condensed matter physics.

I am grateful to Ms. Patricia Osborn, Acquisitions Editor for Mathematics and Physics at Elsevier, who has made many helpful suggestions as well as helped me whenever I have requested her. She has been very gracious and prompt in her replies to my numerous questions. This book could not have been published without her help and support. I am grateful to Mr. Gavin Becker, Assistant Editor of Mathematics and Physics at Elsevier, for his help. I am also grateful to Ms. Sarah Binns, Associate Project Manager at Elsevier, for helping me at every step in successfully completing this book.

I am particularly grateful to Professor Larry Pinsky (Chair) and Professor Gemunu Gunaratne (Associate Chair) of the Department of Physics of the University of Houston for their hospitality, encouragement, and continuing help. I am also grateful to Professor C. S. Ting of the Centre of High T_c at the University of Houston for his friendship and generosity.

Finally, I express my gratitude to my wife and children who have loved me all these years even though I have spent most of my time in the physics department(s) learning physics, doing research, supervising graduate students and post-docs, publishing research papers, writing grant proposals and books, as well as editing a book series on Metal Physics. There is no way I can compensate for the lost time except to dedicate this textbook to them. I am thankful to my daughter-in-law Roopa, who has helped me significantly in my present endeavor. My fondest hope is that when my grandchildren Annika and Millan attend college in 2021 and Kishen and Nirvaan in 2024, at least one of them would be attracted to study condensed matter physics when they see this book in their college library.

CHAPTER

Basic Properties of Crystals

1

CHAPTER OUTLINE

1.1 Crystal Lattices 2
1.1.1 Primitive Cell 3
1.1.2 Unit Cell 3
1.1.3 Wigner–Seitz Cell 3
1.1.4 Lattice Point Group 3
1.2 Bravais Lattices in Two and Three Dimensions 4
1.2.1 Simple Cubic (sc) Lattice 4
1.2.2 Lattice Constants 5
1.2.3 Coordination Numbers 5
1.2.4 Body-Centered Cubic (bcc) Lattice 5
1.2.5 Face-Centered Cubic (fcc) Lattice 7
1.2.6 Other Bravais Lattices 9
1.3 Lattice Planes and Miller Indices 11
1.4 Bravais Lattices and Crystal Structures 13
1.4.1 Crystal Structure 13
1.4.2 Lattice with a Basis 13
1.4.3 Packing Fraction 14
1.5 Crystal Defects and Surface Effects 14
1.5.1 Crystal Defects 14
1.5.2 Surface Effects 14
1.6 Some Simple Crystal Structures 15
1.6.1 Sodium Chloride Structure 15
1.6.2 Cesium Chloride Structure 15
1.6.3 Diamond Structure 16
1.6.4 Zincblende Structure 17
1.6.5 Hexagonal Close-Packed (hcp) Structure 17
1.7 Bragg Diffraction 19
1.8 Laue Method 20
1.9 Reciprocal Lattice 21
1.9.1 Definition 21
1.9.2 Properties of the Reciprocal Lattice 22
1.9.3 Alternative Formulation of the Laue Condition 25

1.10 Brillouin Zones 27
1.10.1 Definition 27
1.10.2 One-Dimensional Lattice 28
1.10.3 Two-Dimensional Square Lattice 28
1.10.4 bcc Lattice 29
1.10.5 fcc Lattice 30
1.11 Diffraction by a Crystal Lattice with a Basis 31
1.11.1 Theory 31
1.11.2 Geometrical Structure Factor 32
1.11.3 Application to bcc Lattice 32
1.11.4 Application to fcc Lattice 33
1.11.5 The Atomic Scattering Factor or Form Factor 33
Problems 34
References 35

1.1 CRYSTAL LATTICES

A crystal lattice (Bravais lattice) is defined as an infinite array of discrete points that appear exactly the same from whichever of the points the array is viewed. If one starts from some point, all other points can be reached from it by the basic translations known as the lattice sites. For a three-dimensional lattice, these are defined by the set

$$\mathbf{R} = n_1\mathbf{a}_1 + n_2\mathbf{a}_2 + n_3\mathbf{a}_3, \tag{1.1}$$

where $\mathbf{R}$ is the lattice translation vector and $\mathbf{a}_1$, $\mathbf{a}_2$, and $\mathbf{a}_3$ are three fundamental translation vectors that are not in the same plane. They are also known as primitive vectors that generate the lattice. Here, n_1, n_2, and n_3 are integers that can be zero, positive, or negative. In fact, a lattice is a mathematical concept used to identify crystal structures. Theoretically, a lattice spans the entire space.

It may be noted from Figure 1.1 that there is an infinite number of nonequivalent choices of primitive vectors and consequently primitive cells for any Bravais lattice (in two dimensions).

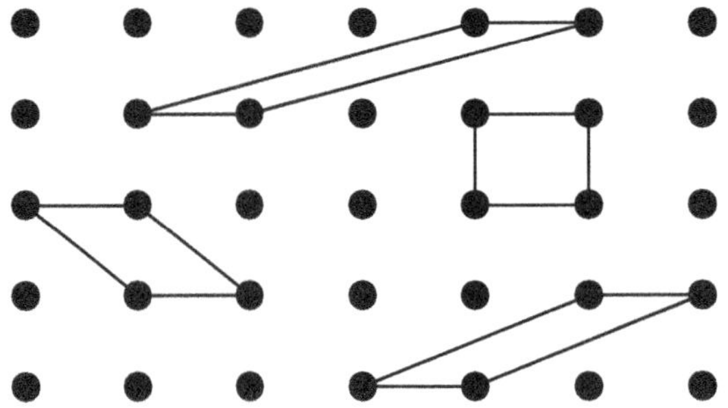

FIGURE 1.1

Possible choices of primitive cells for a two-dimensional square Bravais lattice.

1.1.1 Primitive Cell

The parallelepiped defined by the primitive axes $\mathbf{a}_1$, $\mathbf{a}_2$, and $\mathbf{a}_3$ is called a primitive lattice cell. The volume of a primitive cell is $\mathbf{a}_1 \cdot (\mathbf{a}_2 \times \mathbf{a}_3)$, and it has a density of one lattice point per unit cell. There are a variety of ways in which a primitive cell with the symmetry of the Bravais lattice can be chosen.

1.1.2 Unit Cell

A unit cell is defined as a cell that would define all space under the action of suitable crystal translation operators. Thus, a primitive cell is a minimum-volume unit cell. The difference between a primitive cell and a unit cell is shown in Figure 1.2. As we will see later, it is sometimes more convenient (especially in the case of cubic lattices) to define three-dimensional lattices in terms of unit cells rather than primitive cells.

1.1.3 Wigner–Seitz Cell

The Wigner–Seitz cell is obtained by drawing lines to connect a lattice point to all the neighboring lattice points and then by drawing new lines or planes at the midpoint and normal to these lines. The Wigner–Seitz cell is the smallest volume enclosed in this way. Figure 1.3 illustrates a Wigner–Seitz cell for a two-dimensional Bravais lattice. The Wigner–Seitz cell about a lattice point is the region of space that is closer to that point than to any other lattice point (except for points on the common surface of two or more Wigner–Seitz cells).

1.1.4 Lattice Point Group

A lattice point group is defined as the collection of the symmetry operations that leave the lattice invariant when applied about a lattice point. They include one-, two-, three-, four-, and six-fold rotations that correspond to rotations by 2π, π, $2\pi/3$, $\pi/2$, and $\pi/3$ radians as well as integral multiples of these rotations. These rotation axes are denoted as symbols 1, 2, 3, 4, and 6. A lattice point group also includes mirror reflections **m** about a lattice point. The inversion operation consists of a rotation of π and a reflection in a plane normal to the rotation axis such that $\mathbf{r}$ is replaced by $-\mathbf{r}$.

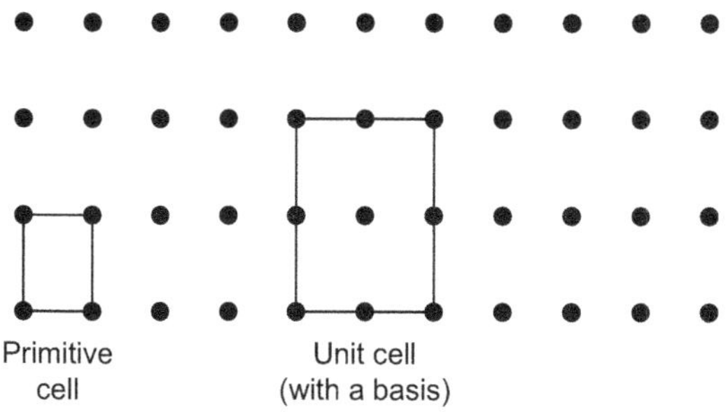

FIGURE 1.2

Difference between a primitive cell and a unit cell for a rectangular lattice.

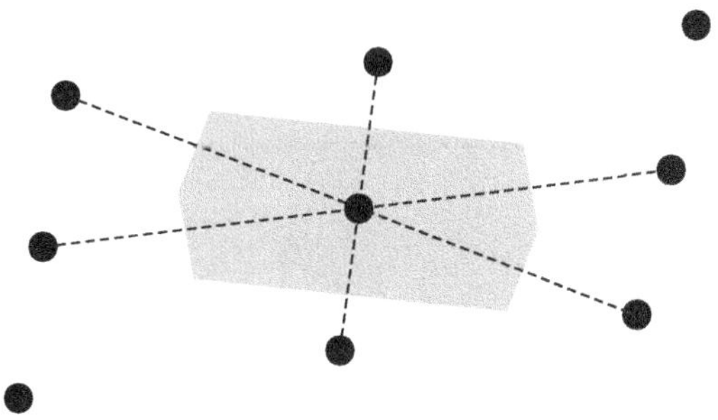

FIGURE 1.3

Wigner–Seitz primitive cell for a two-dimensional lattice.

1.2 BRAVAIS LATTICES IN TWO AND THREE DIMENSIONS

It can be shown that there are 5 distinct Bravais lattice types in two dimensions and 14 distinct Bravais lattices in three dimensions. These 14 Bravais lattices in three dimensions can be grouped into seven types of conventional unit cells. They are known as cubic (3), tetragonal (2), orthorhombic (4), monoclinic (2), triclinic (1), trigonal (1), and hexagonal (1). Later, we will discuss some of these Bravais lattices, an understanding of which is necessary in the study of commonly used solids. The three cubic Bravais lattices are simple cubic (sc), body-centered cubic (bcc), and face-centered cubic (fcc) cells. These are shown in Figure 1.4. They all have cubic point groups, but their space groups are not equivalent.

1.2.1 Simple Cubic (sc) Lattice

The simple cubic lattice (Figure 1.4a) is generated by the primitive lattice vectors $a\hat{\mathbf{x}}$, $a\hat{\mathbf{y}}$, and $a\hat{\mathbf{z}}$, where a is a side of the cube (known as the lattice constant) and $\hat{\mathbf{x}}$, $\hat{\mathbf{y}}$, and $\hat{\mathbf{z}}$ are the three orthonormal vectors. It may be easily seen that the entire cubic lattice can be obtained by using the lattice translation vectors $\mathbf{R}$, as defined in Eq. (1.1), to connect any lattice point to another lattice point. In fact, the simple cubic (sc) lattice is the simplest three-dimensional Bravais lattice.

It may be noted that although there are eight lattice points at the corners of each cubic primitive cell, each lattice point is shared by eight such primitive cells. Considering that eight lattice points are shared by eight primitive cells, on the average, each primitive cell has one lattice point. However, not one of the lattice points belongs uniquely to any simple cubic primitive cell.

The Wigner–Seitz cell of a simple cubic Bravais lattice is also a simple cubic cell. However, the primitive cell is closer to its own lattice point except for points on the nearest-neighbor surface of two or more Wigner–Seitz cells. Therefore, each lattice point has its own Wigner–Seitz cell. When a Wigner–Seitz cell is translated by all the lattice vectors, it will fill the lattice without overlapping. In that sense, the Wigner–Seitz cell is extremely convenient to describe a primitive cell around each lattice point.

Only one known element, the alpha phase of polonium, crystallizes in the simple cubic form.

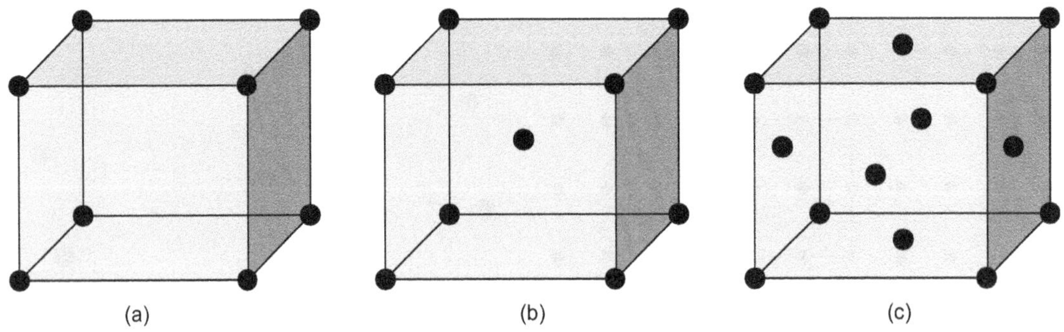

FIGURE 1.4

Unit cells of the three cubic lattices: (a) simple cubic lattice (sc), (b) body-centered cubic lattice (bcc), and (c) face-centered cubic lattice (fcc).

1.2.2 Lattice Constants

The lattice constants are the numbers that specify the size of a unit cell. For example, the lattice constant for cubic crystals is a, the side of the cube. The maximum number of lattice constants for a unit cell can be three. These are, in general, on the order of a few angstroms and are experimentally determined by crystallographers using X-rays. There are a variety of reference texts in which the lattice constants of different crystal structures have been noted.

1.2.3 Coordination Numbers

The number of nearest neighbors of each lattice point in a Bravais lattice is the same because of its periodic nature and is known as the coordination number. The coordination number is a property of the lattice. For example, the coordination number of a square or rectangular lattice (in two dimensions) is 4, whereas the coordination number of a simple cubic lattice is 6.

1.2.4 Body-Centered Cubic (bcc) Lattice

The body-centered cubic (bcc) lattice (Figure 1.4b) can be obtained by adding a second lattice point at the center of each cubic cell of a simple cubic lattice. Thus, the unit cell of each bcc lattice can be considered as two interpenetrating simple cubic primitive lattices. In fact, there are two alternate ways of considering a bcc lattice, either with a simple cubic lattice formed from the corner points with a lattice point at the cube center, or with the simple cubic lattice formed from the lattice points at the center and the corner points located at the center of the new cubic lattice. In either case, each one of the eight lattice points at the corner of a cubic cell is shared by eight adjacent cubic cells, while the lattice point at the center of the cubic cell exclusively belongs to that cell. Therefore, the bcc lattice can be considered as a unit cubic cell with two lattice points per cell. The number of nearest neighbors of each lattice point is 8. Alternately, one can state that the coordination number is 8.

However, the primitive cell of a bcc lattice can also be easily obtained. In fact, there are a variety of ways in which the primitive vectors of the bcc lattice can be described. The most symmetric set of primitive vectors is given as follows:

$$\mathbf{a}_1 = \frac{a}{2}(\hat{\mathbf{y}} + \hat{\mathbf{z}} - \hat{\mathbf{x}}), \ \mathbf{a}_2 = \frac{a}{2}(\hat{\mathbf{z}} + \hat{\mathbf{x}} - \hat{\mathbf{y}}), \ \mathbf{a}_3 = \frac{a}{2}(\hat{\mathbf{x}} + \mathbf{y} - \hat{\mathbf{z}}), \tag{1.2}$$

where a is the lattice constant (the side of the unit cubic cell), and $\hat{\mathbf{x}}$, $\hat{\mathbf{y}}$, and $\hat{\mathbf{z}}$ are orthonormal vectors. It is important to note that $\mathbf{a}_1$, $\mathbf{a}_2$, and $\mathbf{a}_3$ are not orthogonal vectors. The parallelepiped drawn with these three vectors (shown in Figure 1.5) is the primitive cell of the bcc lattice. The eight corners of this primitive cell have eight lattice points, each shared by eight primitive cells.

It can be shown that the volume of the primitive Bravais cell is (Problem 1.1)

$$V = \mathbf{a}_1 \cdot (\mathbf{a}_2 \times \mathbf{a}_3) = \frac{a^3}{2}. \tag{1.3}$$

Because the volume of the unit cubic cell is a^3, and each unit cell has two lattice points, the primitive cell of the bcc lattice is half of the volume of the unit cell. However, not one of the lattice points uniquely belongs to any primitive cell shown in Figure 1.5. The lattice constants of bcc lattices at low temperatures are shown in Table 1.1. To be able to specify the primitive cell around each lattice point, one has to draw the Wigner–Seitz cell of the bcc lattice.

It can be easily shown that the Wigner–Seitz cell of a body-centered cubic Bravais lattice is a truncated octahedron (see Figure 1.6). The octahedron has four square faces and four hexagonal faces. The square faces bisect the lines joining the central point of a cubic cell to the central points of the six neighboring cubic cells. The hexagonal faces bisect the lines joining the central point of a

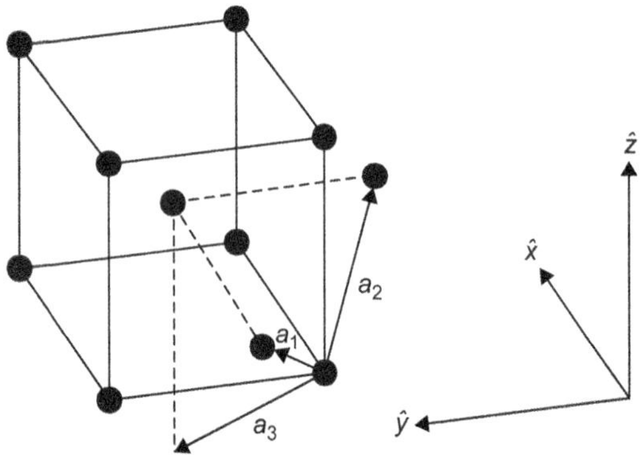

FIGURE 1.5

Symmetric set of primitive vectors for the bcc Bravais lattice.

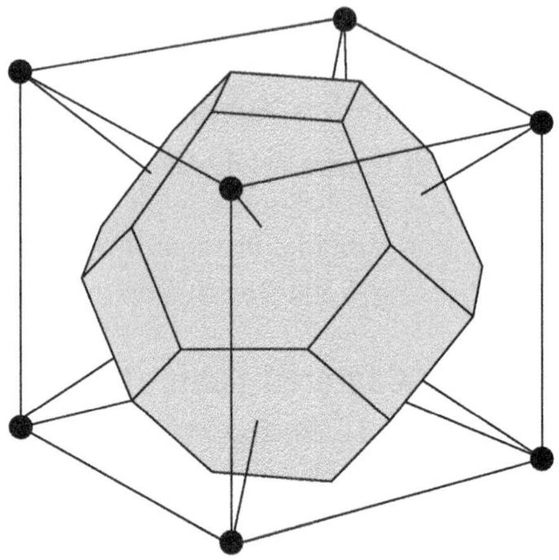

FIGURE 1.6

Wigner–Seitz cell of a bcc Bravais lattice.

Table 1.1 Lattice Constants of bcc Lattices at Low Temperatures

Element	Lattice Constant (A°)
Barium	5.02
Chromium	2.88 (Cr also has fcc and hcp phases)
Cesium	6.05
Europium	4.61
Iron	2.87 (Fe also has fcc phase)
Potassium	5.23
Lithium	3.50
Molybdenum	3.15 (Mo also has fcc phase)
Sodium	4.29
Niobidium	3.30
Rubidium	5.59
Tantalum	3.31
Thallium	3.88 (T1 also has fcc and hcp phases)
Uranium	3.47
Vanadium	3.02
Tungsten	3.16

Source: R. W. G. Wyckoff, *Crystal Structures*, vol. 1 (J. Wiley, 1963).

cubic cell to the eight corner points of the same cubic cell. The lattice point at the center of the bcc lattice is also at the center of this octahedron. Any point in the space within the octahedron (except for points on the common surface at two or more Wigner–Seitz cells) is closer to this central lattice point than any other central lattice point.

1.2.5 Face-Centered Cubic (fcc) Lattice

The face-centered cubic (fcc) Bravais lattice (Figure 1.4c) can be constructed from a simple cubic lattice (Figure 1.4a) by adding a lattice point in the center of each square face. Thus, there are eight lattice points, one each at the corner of the cubic unit cell, and six more lattice points, one each at the center of each square face. Each lattice point at the corner is shared by eight cubic cells, and each lattice point at the square face is shared by two cubic cells. Thus, the fcc lattice can be considered as a unit cubic cell with four lattice points per unit cubic cell. However, not one of the lattice points exclusively belongs to any unit cell. Each lattice point has 12 nearest neighbors. Therefore, the coordination number of a fcc lattice is 12.

Alternately, one can obtain the primitive Bravais cell for the fcc lattice. There are a variety of ways in which these primitive vectors can be obtained. The most symmetric set of primitive lattice vectors of a fcc lattice is

$$\mathbf{a}_1 = \frac{a}{2}(\hat{\mathbf{y}}+\hat{\mathbf{z}}),\ \mathbf{a}_2 = \frac{a}{2}(\hat{\mathbf{z}}+\hat{\mathbf{x}}),\ \mathbf{a}_3 = \frac{a}{2}(\hat{\mathbf{x}}+\hat{\mathbf{y}}). \tag{1.4}$$

Here, a is the side of the cubic unit cell (the lattice constant), $\hat{\mathbf{x}}$, $\hat{\mathbf{y}}$, and $\hat{\mathbf{z}}$ are orthonormal vectors, but $\mathbf{a}_1$, $\mathbf{a}_2$, and $\mathbf{a}_3$ are not orthogonal vectors. These vectors have been drawn in Figure 1.7. The lattice constants of fcc lattices at low temperatures are shown in Table 1.2.

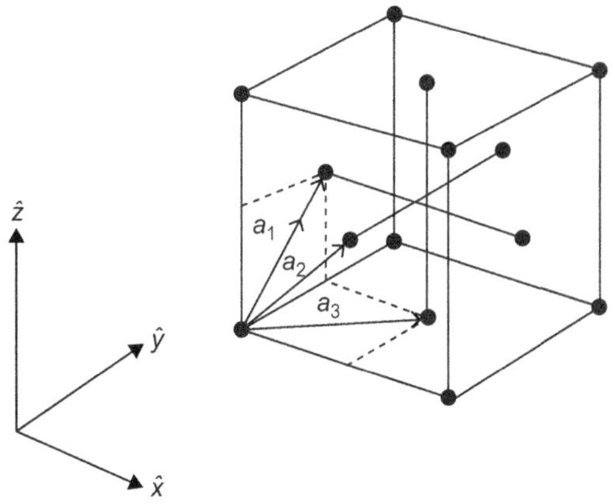

FIGURE 1.7

Symmetric set of primitive vectors for the fcc Bravais lattice.

The volume of the primitive Bravais cell is given by (Problem 1.2)

$$V = \mathbf{a}_1 \cdot (\mathbf{a}_2 \times \mathbf{a}_3) = \frac{a^3}{4}. \tag{1.5}$$

Because the volume of the unit cubic cell is a^3, and each unit cell has four lattice points, it is appropriate that the volume of the primitive cell of a fcc lattice is $a^3/4$. However, not one of the lattice points of the fcc Bravais lattice uniquely belongs to a primitive cell. The Wigner–Seitz cell for a fcc Bravais lattice is shown in Figure 1.8.

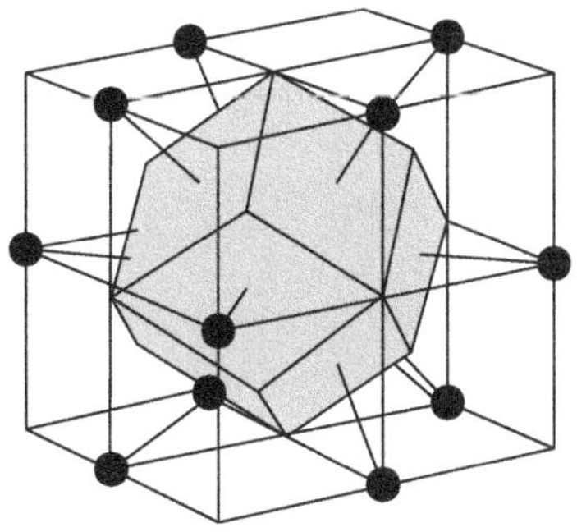

FIGURE 1.8

Wigner–Seitz cell for the fcc Bravais lattice.

As shown in Figure 1.8, the Wigner–Seitz cell of a fcc Bravais lattice is a "rhombic dodecahedron." The lattice points are at the center of the cube as well as at the center of the 12 edges. The Wigner–Seitz cell has 12 congruent faces, each of which is perpendicular to the line joining the center of an edge to the central point. Any point in the "rhombic dodecahedron" (except at the common surface of two or more Wigner–Seitz cells) is closer to this central point than any other central point of an adjacent Wigner–Seitz cell.

Table 1.2 Lattice Constants of fcc Lattices at Low Temperatures

Element	Lattice Constant (A°)
Actinium	5.31
Silver	4.09
Aluminum	4.05
Americium	4.89
Argon	5.26
Gold	4.08
Calcium	5.58
Cerium	5.16 (Ce also has two hcp structures)
Cobalt	3.55 (Co also has two hcp structures)
Chromium	3.68 (Cr also has a bcc and an hcp structure)
Copper	3.61
Iron	3.59 (Fe also has a bcc structure)
Iridium	3.84
Krypton	5.72
Lanthanum	5.30 (La also has two hcp structures)
Molybdenum	4.16 (Mo also has a bcc structure)
Neon	4.43
Nickel	3.52 (Ni also has two hcp structures)
Lead	4.95
Palladium	3.89
Praseodymium	5.16 (Pr also has two hcp structures)
Platinum	3.92
Rhodium	3.80
Scandium	4.54 (Sc also has two hcp structures)
Strontium	6.08
Thorium	5.08
Thallium	4.84 (Th also has two hcp structures and a bcc structure)
Xenon	6.20
Ytterbium	5.49

Source: R. W. G. Wyckoff, *Crystal Structures*, vol. 1 (J. Wiley, 1963).

1.2.6 Other Bravais Lattices

There are 7 crystal systems and 14 Bravais lattices in three dimensions. We have already discussed the 3 cubic Bravais lattices. We will discuss the primitive cells of the other 11 Bravais lattices. We note that sometimes it is more convenient as well as conventional to use a larger unit cell that involves atoms in the end, fcc, or bcc positions. In such cases, the advantage is that one can use orthogonal axes.

(a) Tetragonal Systems

There are two tetragonal systems. If a cube is stretched to make four of the sides into rectangles, an object with the symmetry of the tetragonal group is obtained. The solid is symmetric under reflections about planes that bisect it although the three-fold symmetry and the 90° rotation symmetry about two of the axes are lost. If a simple cubic lattice is stretched, a simple tetragonal lattice, shown in Figure 1.9a, is obtained. It has the sides $a = b \neq c$ and the angles $\alpha = \beta = \gamma$. If either a fcc lattice or a bcc lattice is stretched, a centered tetragonal lattice shown in Figure 1.9b is obtained. It also has the sides $a = b \neq c$ and the angles $\alpha = \beta = \gamma$ and, in addition to the above, a body-centered lattice point.

(b) Orthorhombic Systems

If the top and bottom squares of the tetragonal solid are deformed into rectangles, the 90° rotational symmetry is eliminated. One obtains a solid with orthorhombic symmetry. There are four orthorhombic systems. When a simple tetragonal lattice (Figure 1.9a) is deformed along one of its axes, a simple orthorhombic lattice, as shown in Figure 1.10a, is obtained. When one stretches a simple tetragonal lattice along face diagonals, the base-centered orthorhombic lattice shown in Figure 10b is obtained. When a centered tetragonal lattice (Figure 1.9b) is deformed, a body-centered orthogonal lattice (Figure 1.10c) is obtained. When one stretches the centered tetragonal lattice along the face diagonals, the face-centered orthorhombic lattice (Figure 1.10d) is obtained. All these Bravais lattices shown in Figure 1.10 have lattice constants $a \neq b \neq c$ and $\alpha = \beta = \gamma = 90°$.

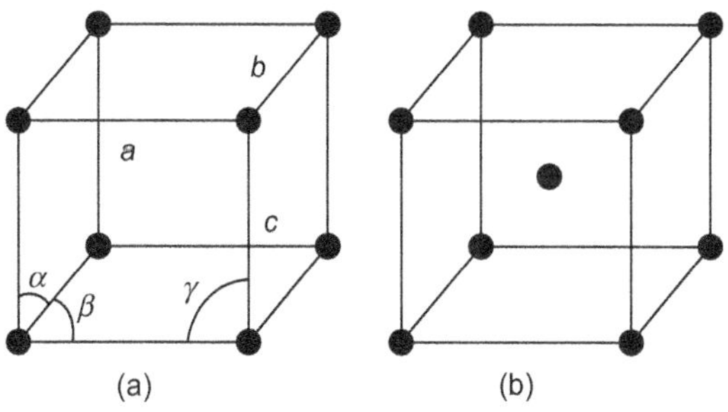

FIGURE 1.9

The two tetragonal Bravais lattices: (a) simple and (b) centered.

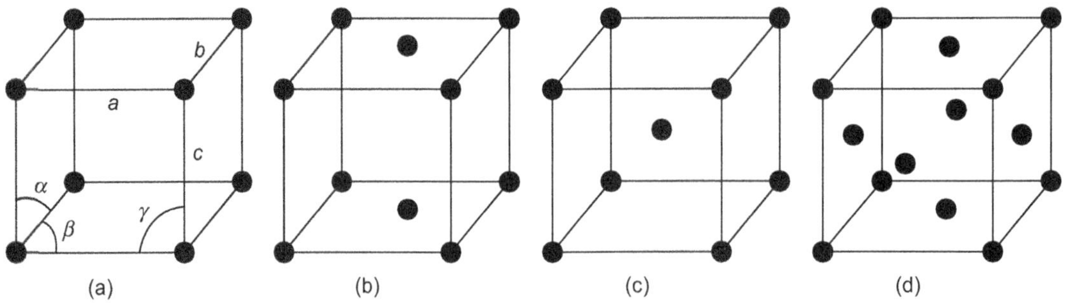

FIGURE 1.10

The Bravais lattices for the four orthorhombic systems.

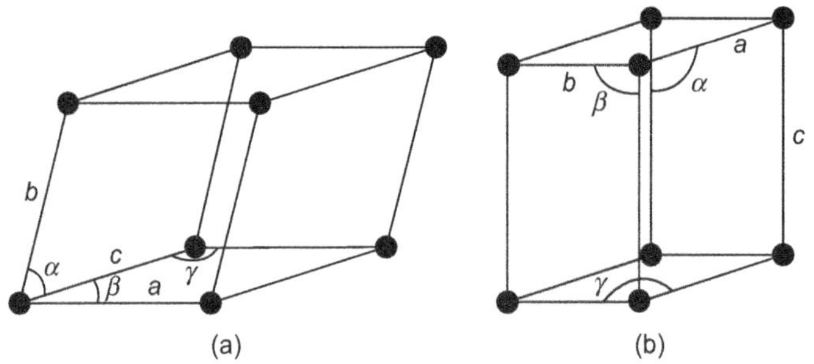

FIGURE 1.11

The Bravais lattices for the rhombohedral or trigonal systems and the hexagonal systems.

(c) Rhombohedral or Trigonal Systems

If a cube is stretched across a body diagonal, one obtains a solid with rhombohedral or trigonal symmetry. Stretching any of the three cubic Bravais lattices produces the same Bravais lattice known as the rhombohedral lattice or the trigonal lattice. The rhombohedral system is shown in Figure 1.11a. In the rhombohedral system, the lattice constants are $a = b = c$ and the angles are $\alpha = \beta = \gamma \neq 90°$.

(d) Hexagonal Systems

A solid can be formed with a hexagon at the base and perpendicular walls such that it has hexagonal symmetry. This Bravais lattice is called the hexagonal lattice. In the hexagonal system, shown in Figure 1.11b, the lattice constants are $a = b \neq c$ and the angles are $\alpha = \beta = 90°$, $\gamma = 120°$.

(e) Monoclinic and Triclinic Systems

A solid with monoclinic symmetry can be generated by squeezing a tetragonal solid across a diagonal in a manner such that the 90° angles on the top and bottom faces are eliminated. However,

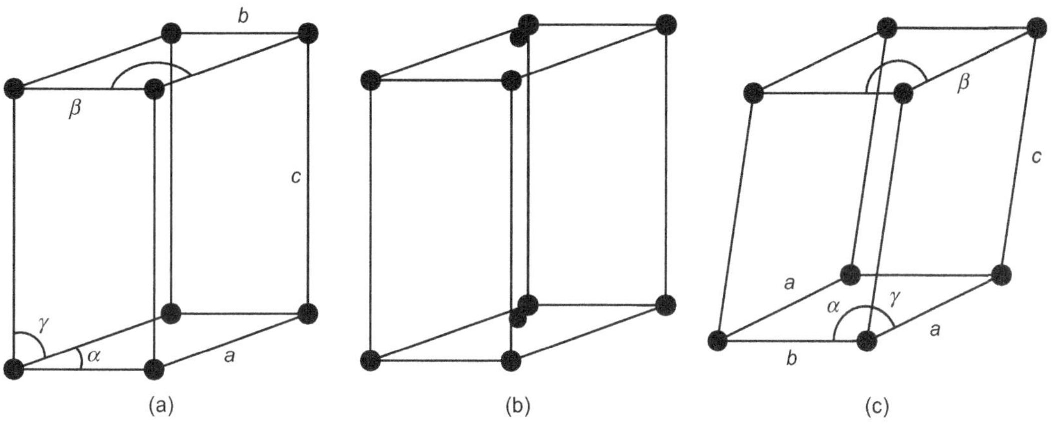

FIGURE 1.12

The Bravais lattices for the monoclinic and triclinic systems.

the sides are built out of rectangles. The simple monoclinic lattice shown in Figure 1.12a is obtained by distorting either the simple orthorhombic lattice or the base-centered orthorhombic lattice. The centered monoclinic lattice shown in Figure 1.12b is obtained from the distortion of either the face-centered orthorhombic lattice or the body-centered orthorhombic lattice. For the two monoclinic systems, the lattice constants are $a \neq b \neq c$ and the angles are $\alpha = \gamma = 90°$, $\beta \neq 90°$. The triclinic system, shown in Figure 1.12c, is obtained by pulling the top of a monoclinic solid sideways relative to the bottom. Thus, all the faces become diamonds, and the only remaining symmetry is inversion symmetry. For the triclinic system, the lattice constants are $a \neq b \neq c$ and the angles are $\alpha, \beta, \gamma \neq 90°$.

1.3 LATTICE PLANES AND MILLER INDICES

A lattice plane is defined to be a plane that has at least three noncollinear Bravais lattice points. In fact, such a plane would contain an infinite number of two-dimensional Bravais lattice points. A three-dimensional Bravais lattice is represented as a family of parallel equally spaced lattice planes. Such lattice planes can be constructed in a lattice in a variety of ways. Figures 1.13a and b represent two ways of representing the same simple cubic Bravais lattice as a family of lattice planes.

The Miller indices are used to label a crystal plane. They are obtained by using the following procedure. All the Bravais lattice points lie on a chosen assembly of equally spaced parallel lattice planes (whether cubic or otherwise). Certain planes of the assembly (sometimes all of them) will always intersect the coordinate axes at the lattice points. Therefore, every plane in this chosen set of parallel planes would intercept the coordinate axes that bear a definite rational ratio to one another. To define the Miller indices, one adopts the following procedure. One

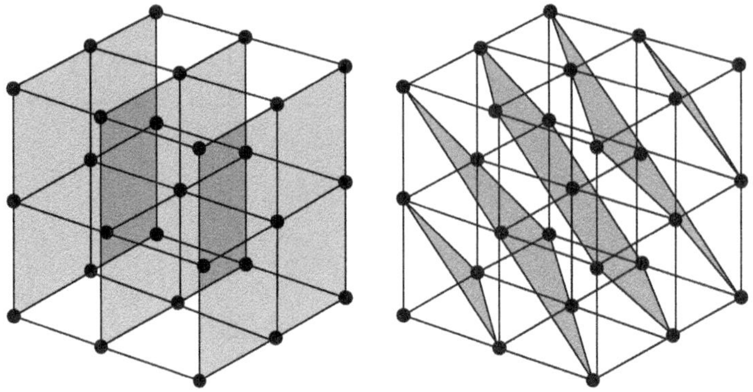

FIGURE 1.13

Lattice planes in a simple cubic lattice. The shaded regions in (a) and (b) are two of the infinite ways in which a family of lattice planes can be represented in the same cubic lattice.

of the lattice points is chosen as the origin, and the coordinate axes are chosen such that they are not coplanar. In some cases, such as the cubic lattices, the coordinate axes are chosen as orthogonal and parallel to the sides of the cubic unit cell. This property of the Bravais lattice was used to define the Miller indices by crystallographers and was later used for the proper definition of reciprocal lattice vectors. The Miller indices are obtained by using the following prescription:

1. Determine the intercepts on the axes $\mathbf{a}_1$, $\mathbf{a}_2$, $\mathbf{a}_3$ in the units of lattice constants. Here, $\mathbf{a}_1$, $\mathbf{a}_2$, $\mathbf{a}_3$ need not be primitive vectors. For example, determine $\frac{a_1}{a}, \frac{a_2}{b}, \frac{a_3}{c}$, where a, b, c are the lattice constants of the three-dimensional lattice (note that $a = b = c$ for a cubic lattice). These intercepts have rational ratios although they are not, in general, integers.
2. Take the reciprocal of each number, i.e., $\frac{a}{a_1}, \frac{b}{a_2}, \frac{c}{a_3}$.
3. Then reduce these numbers to the three smallest integers h, k, l that have the same ratio. The integers h, k, l are called the Miller indices.
4. The parentheses (hkl) denote a single crystal plane or a set of parallel planes.
5. If a plane cuts an axis on the negative side of the origin, the negative index is indicated by placing a minus sign above the index $(h\bar{k}l)$. If a plane is parallel to a particular axis, because the intercept with the axis occurs at infinity, the corresponding Miller index is zero (reciprocal of infinity). Thus, the cube faces of a cubic crystal are (100), (010), (001), $(\bar{1}00)$, $(0\bar{1}0)$, and $(00\bar{1})$.
6. Planes equivalent by symmetry are denoted by curly brackets $\{hkl\}$. Thus, the set of cube faces of a cubic crystal described previously can also be described as $\{100\}$.
7. A direction in a lattice is denoted by square brackets $[hkl]$. In cubic crystals, the direction $[hkl]$ is always perpendicular to a plane (hkl) having the same directions. However, this is not true for other crystals. Figure 1.14 shows the lattice planes and their Miller indices in a simple cubic lattice.

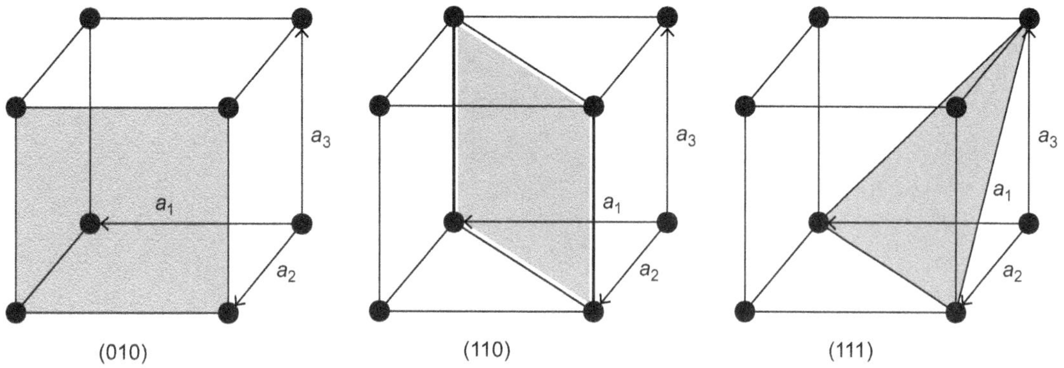

FIGURE 1.14

Lattice planes and their Miller indices in a simple cubic lattice.

1.4 BRAVAIS LATTICES AND CRYSTAL STRUCTURES

1.4.1 Crystal Structure

A crystal structure is obtained when an identical basis of atoms is attached to each lattice point. We will give a simple example of two atoms A and B (of different types) to illustrate a crystal structure.

Figure 1.15 shows an example of a two-dimensional crystal structure with two types of atoms A and B. Figure 1.15a shows a two-dimensional rectangular lattice. Figure 1.15b shows a basis of two different types of atoms, A and B. It has to be emphasized that the basis is the definition of the same physical unit of atoms or ions (ranging from 1 for some elements to nearly 100 for some complex proteins) that is located symmetrically at each point of the Bravais lattice. Thus, a basis can be translated through all the vectors of a Bravais lattice, and another identical basis with the same location around another lattice point would be reached. The crystal structure is defined (Figure 1.15c) as a lattice with a basis. In fact, this symmetry of the crystals makes it possible to study their physical properties. We note that the lattice points in the rectangular lattice can be symmetrically shifted in the crystal structure as long as the basis of atoms is grouped symmetrically around each lattice point.

1.4.2 Lattice with a Basis

We also note that there is a dichotomy in the usage of the words "lattice with a basis." In addition to describing a crystal structure, in which case the basis is the description of identical atoms or ions based symmetrically around each lattice point, this expression is also used to describe a Bravais lattice as a lattice with a basis. As examples of the most commonly used lattices, the bcc Bravais lattice can be described as a simple cubic lattice (the unit cell) with a two-point basis

$$\mathbf{0},\ \frac{a}{2}(\hat{\mathbf{x}}+\hat{\mathbf{y}}+\hat{\mathbf{z}}), \tag{1.6}$$

and the fcc lattice can be described as a four-point basis

$$\mathbf{0},\ \frac{a}{2}(\hat{\mathbf{x}}+\hat{\mathbf{y}}),\ \frac{a}{2}(\hat{\mathbf{y}}+\hat{\mathbf{z}}),\ \frac{a}{2}(\hat{\mathbf{z}}+\hat{\mathbf{x}}). \tag{1.7}$$

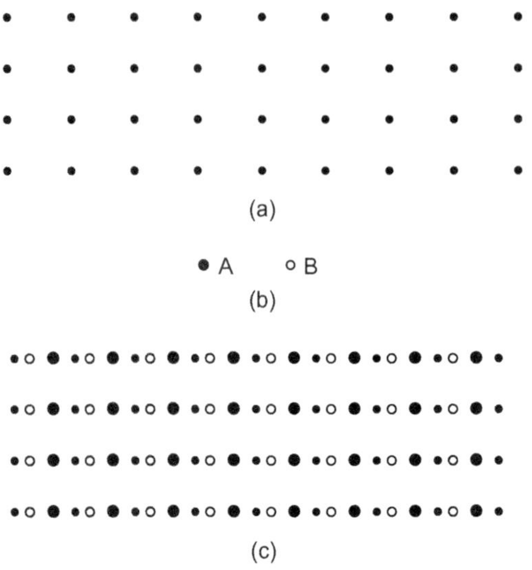

FIGURE 1.15

(a) A two-dimensional rectangular lattice; (b) basis of two atoms, A and B; (c) Crystal structure (two dimensions): lattice + basis.

1.4.3 Packing Fraction

The packing fraction of a crystal structure is defined as the fraction of space occupied by the atoms or ions considered as contacting hard spheres. For example, a bcc crystal has a packing fraction of 0.68, and a fcc crystal has a packing fraction of 0.74.

1.5 CRYSTAL DEFECTS AND SURFACE EFFECTS

1.5.1 Crystal Defects

It has to be noted at this point that no crystal, either grown naturally or in the laboratory, is perfect. There are a certain number of impurities, varying in range depending on the laboratory conditions during crystal growth, that are invariably present. The average number of impurities in reasonably pure crystal is 1 in 10,000. There are also lattice defects in the crystal in the sense that an atom is found not at the designated lattice point, but either it is missing or found at an interstitial. This is called a lattice defect, and such defects are also in the range of 1 in 10,000. The crystal impurities and lattice defects, together known as crystal defects, do play a significant role in certain physical properties of the crystal.

1.5.2 Surface Effects

There is another significant difference between a lattice and a crystal structure even if there are only single atoms or ions (called monatomic lattice) at each lattice point in the crystal structure. As defined earlier, a lattice is a mathematical concept that extends to infinity in all dimensions, whereas every crystal has finite dimensions. In general, a crystal has a huge number of atoms (around 10^{23}). This difference between the infinite lattice and the finite lattice structure is not significant inside the crystal, and as we will discuss later, one gets around the finite size of the crystal by using periodic boundary conditions. Therefore, the concept of a Bravais lattice can be extended to a crystal. However, this concept breaks down when one approaches the surface of the crystal because the periodicity of the lattice, which is the cornerstone of most of the basic theoretical concepts of solid state physics, is no more valid. Therefore, the physical properties at or near the surface of a crystal are very different. The study of these surface properties, constrained by the nonperiodicity in a third dimension perpendicular to the surface, requires application of different techniques and is by itself a vast and fascinating field.

1.6 SOME SIMPLE CRYSTAL STRUCTURES

1.6.1 Sodium Chloride Structure

The sodium chloride structure is shown in Figure 1.16. There are equal numbers of sodium and chlorine ions placed at alternate points of a simple cubic lattice. Each ion has six of the other kind of ions as its nearest neighbor. Thus, the coordination number is 6. The crystal structure can be described as a Bravais fcc lattice with a basis. The basis consists of one sodium ion at $\mathbf{0}$ and one chlorine ion at the center of the cubic unit cell; i.e., at $(a/2)(\hat{\mathbf{x}}+\hat{\mathbf{y}}+\hat{\mathbf{z}})$.

The compounds that have the sodium chloride structure include AgBr, AgCl, AgF, BaO, BaS, BaSe, BaTe, CaO, CaS, CaSe, CaTe, CsF, KBr, KCl, KF, KI, LiBr, LiCl, LiF, LiH, LiI, MgO, MgS, MgSe, MnO, NaBr, NaCl, NaF, NaI, RbBr, PbS, RbCl, RbF, RbI, SrO, SrS, SrSe, SrTe, and UO.

1.6.2 Cesium Chloride Structure

The cesium chloride structure is shown in Figure 1.17. There is one molecule per unit cubic cell with the ions in the body-centered positions; i.e., Cs^+: 000 and Cl^-: $\frac{1}{2}\frac{1}{2}\frac{1}{2}$. Thus, the cesium chloride

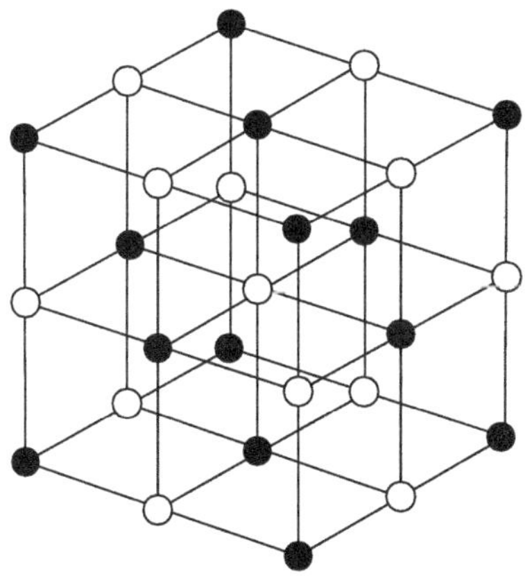

FIGURE 1.16

The sodium chloride structure.

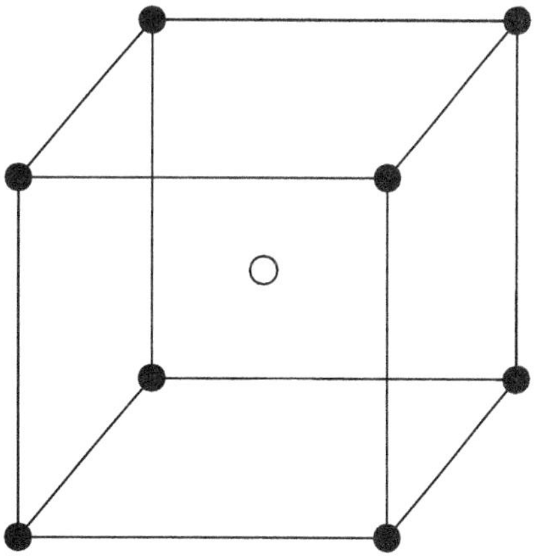

FIGURE 1.17

CsCl structure.

structure can be described as a simple cubic lattice with a basis. The cesium ion is at the origin **0**, and the chlorine ion is at $(a/2)(\hat{\mathbf{x}}+\hat{\mathbf{y}}+\hat{\mathbf{z}})$. Because each ion is at the center of a cube of ions of the other type of ions, the coordination number is 8.

The compounds that have the cesium chloride structure include AlNi, BeCu, CsBr, CsCl, CsI, LiHg, NH_4Cl, RbCl, TlBr, TlCl, and TlI.

1.6.3 Diamond Structure

The diamond structure is shown in Figure 1.18. The diamond cubic structure (Ref. 2a) consists of two interpenetrating fcc Bravais lattices displaced from each other by one-quarter of a body diagonal. Alternately, it can be considered as a fcc lattice with basis at **0** and at $(a/4)(\hat{\mathbf{x}}+\hat{\mathbf{y}}+\hat{\mathbf{z}})$. The diamond structure is a result of covalent bonding. The covalent bond between two atoms is a very strong bond between two electrons, one from each atom, with directional properties. The spins of the two electrons are antiparallel, and the electrons forming the bond tend to be localized in the region between the two atoms. Because the C, Ge, and Si atoms each lack four electrons to form filled shells, these elements can have attractive interaction due to charge overlap, a type of interaction not found between atoms with filled shells because of the Pauli exclusion principle and the consequent repulsive interaction. There are eight atoms in a unit cube, and each atom has 4 nearest neighbors and 12 next nearest neighbors in the diamond lattice. Hence, the diamond lattice, which is not a Bravais lattice, is relatively empty. The 4 nearest neighbors of each atom form the vertices of a regular tetrahedron, and the covalent bonding between the neighboring atoms in the diamond structure is also known as tetrahedral bonding. The maximum proportion of the volume that is available to be filled by hard spheres is 0.34. The elements that crystallize in the diamond structure are C (diamond), Si, Ge, and α-Sn (grey). The coordination number of the diamond lattice is 4, and the packing fraction is 0.34.

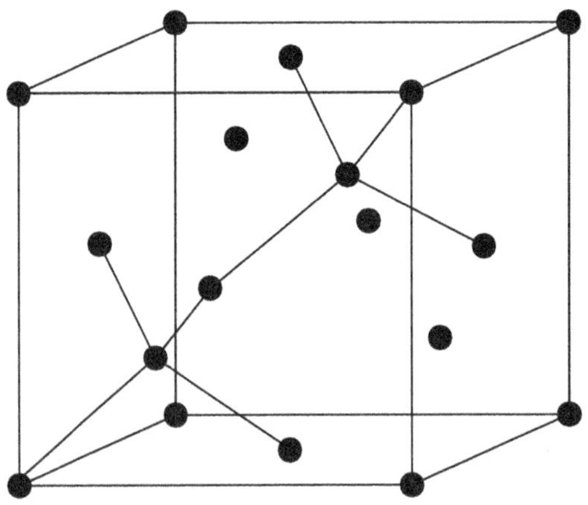

FIGURE 1.18

The diamond cubic structure.

1.6.4 Zincblende Structure

The zincblende (or cubic zinc sulfide) structure, shown in Figure 1.19, is obtained from the diamond structure (Figure 1.18) when Zn ions are placed on one fcc lattice and S ions are placed on the other fcc lattice. Thus, each ion has four of the opposite type as nearest neighbors.

Some crystals with zincblende structure are AgI, AlAs, AlP, AlSb, BeS, BeSe, BeTe, CdS, CdTe, CuBr, CuCl, CuF, CuI, GaAs, GaP, GaSb, HgS, HgSe, HgTe, InAs, InP, InSb, MnS(red), MnSe, SiC, ZnS, ZnSe, and ZnTe.

1.6.5 Hexagonal Close-Packed (hcp) Structure

The hexagonal close-packed (hcp) structure is not a Bravais lattice, but a large number of elements crystallize in this form. If we assume atoms to be hard spheres, close-packed planes with hexagonal symmetry (see Figure 1.20) can be formed.

If one starts with the atom in position A, there are two other kinds of spaces between the atoms, B or C. If the second plane is placed on the B positions, a third nesting layer can be placed either over site A or site C. The stacking ABABAB… yields the hcp structure while the stacking ABCABC… gives the fcc structure (see Problems 1.8 and 1.11). Both types of stacking have the same density of packing, and the packing fraction is 0.74.

The unit cell of the hcp structure is the hexagonal primitive cell, and the basis contains two elements of the same type. The ideal c/a ratio for hcp structures is $\sqrt{\frac{8}{3}} = 1.633$. Figures 1.21 and 1.22 represent two alternate ways of representing the hcp structure.

In Figure 1.21, the unit cell of the hcp structure is shown. The unit cell of the hcp structure is the hexagonal primitive cell, and the basis contains two atoms. One atom of the basis is at the origin (000), and the other atom is at $(\frac{2}{3}\frac{1}{3}\frac{1}{2})$, which means at $\mathbf{r} = 2/3\mathbf{a} + 1/3\mathbf{b} + 1/2\mathbf{c}$. The c/a ratio of hexagonal

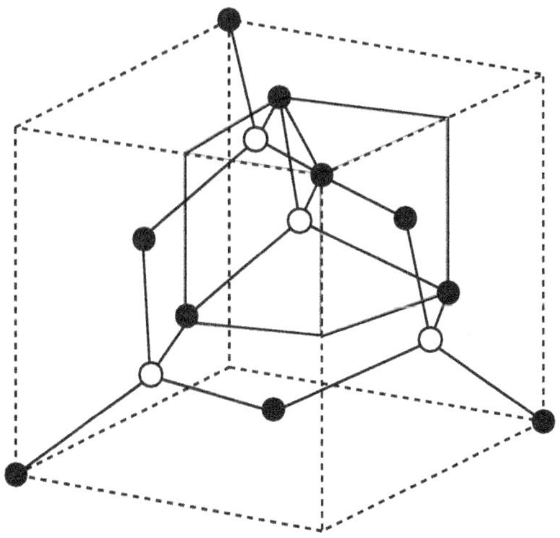

FIGURE 1.19

Zincblende structure.

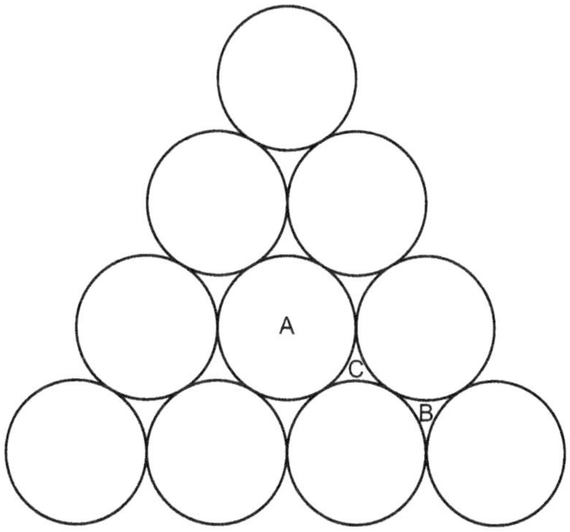

FIGURE 1.20

The three sites A, B, and C in a close-packed plane of atomic spheres (the [111] plane of a fcc structure or the basal plane of a hcp structure).

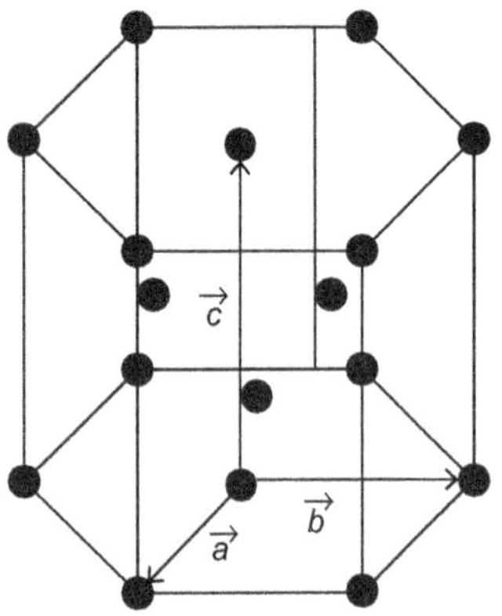

FIGURE 1.21

The hexagonal close-packed (hcp) structure.

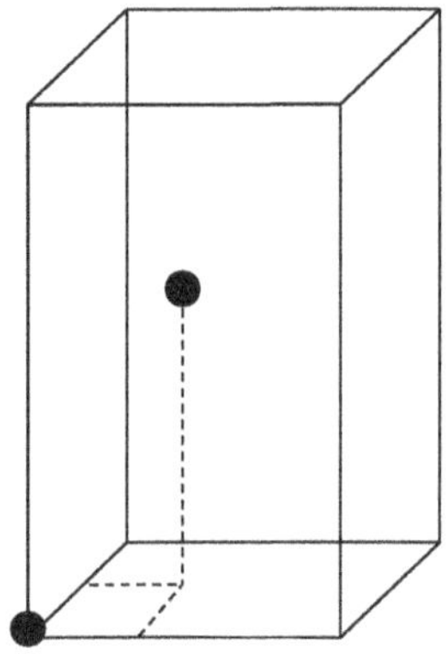

FIGURE 1.22

The basic unit of hcp structure as a trigonal cell with two identical atoms as the basis at (000) and $(\frac{1}{3}\frac{1}{3}\frac{1}{2})$. (Note the differences in origin between Figures 1.21 and 1.22.)

close-packing of spheres is 1.633. However, several crystals, of which the *c*/*a* ratio is quite different, such as zinc with a *c*/*a* ratio of 1.85, are also called hcp structures. The lattice constants of elements with hcp structure are shown in Table 1.3.

The coordination number is the same for both fcc and hcp structures. Therefore, at appropriate temperatures, many metals transform easily between fcc and hcp structures. This is called martinestic transformation. However, the alternate way to describe the basic unit of hcp structure is a trigonal cell containing two atoms, which is shown in Figure 1.22.

The basic unit of the unit cell of the hcp structure described in Figure 1.22 is the trigonal cell containing two identical atoms at (000) and $(\frac{1}{3}\frac{1}{3}\frac{1}{2})$. We note that the origins and the lattice vectors are chosen differently in Figures 1.21 and 1.22. The choice of any one of the two alternate methods of representing the same hcp crystal structure depends on convenience.

Table 1.3 Lattice Constants of Elements with hcp Structure

Element	a(A°)	c(A°)
Beryllium	2.29	3.58
Cadmium	2.98	5.62
Cerium	3.65	5.96
Chromium	2.72	4.43
Cobalt	2.51	4.07
Dysprosium	3.59	5.65
Erbium	3.55	5.59
Gadolinium	3.56	5.80
Hafnium	3.20	5.06
Helium (2 K and 26 atm)	3.57	5.83

(Continued)

Table 1.3 Lattice Constants of Elements with hcp Structure—cont'd

Element	a(A°)	c(A°)
Holmium	3.58	5.62
Hydrogen (molecule)	3.75	6.49
Lanthanum	3.75	6.07
Lutetium	3.50	5.55
Magnesium	3.21	5.21
Neodymium	3.66	5.90
Nickel	2.65	4.33
Osmium	2.74	4.32
Praseodymium	3.67	5.92
Rhenium	2.76	4.46
Ruthenium	2.70	4.28
Scandium	3.31	5.27
Terbium	3.60	5.69
Titanium	2.95	4.69
Thallium	3.46	5.53
Thulium	3.54	5.55
Yttrium	3.65	5.73
Zinc	2.66	4.95
Zirconium	3.23	5.15

Source: R. W. G. Wyckoff, *Crystal Structures,* vol. 1 (J. Wiley, 1963).

1.7 BRAGG DIFFRACTION

W. L. Bragg considered a crystal as made up of a set of parallel lattice planes of ions spaced equal distances d apart, as shown in Figure 1.23.

If the incident waves are reflected specularly, the reflected rays (also known as diffracted rays) would interfere constructively if the path difference is

$$2d \sin \theta = n\lambda \,, \tag{1.8}$$

where θ is the angle of incidence and n is an integer, also known as the order of the corresponding reflection. This is the famous Bragg law of X-ray diffraction, (Ref. 2a) although the same law is valid for other types of waves (such as electron waves) as long as there is specular reflection from the ions in the crystal lattice. The Bragg law considers the periodicity of the lattice but does not include the basis of ions or atoms at each lattice point. In fact, it is the composition of the basis that determines the intensity of diffraction for various orders of n. In addition, it is pertinent to note that the crystal planes can be arranged in an infinite number of ways. Even for the same incident ray, both the direction and intensity of the reflected (note that we are using *diffraction* and *reflection* interchangeably) rays would depend on the orientation of the crystal planes.

One immediate consequence of the Bragg law (Eq. 1.8) is that Bragg reflection can occur only for wavelengths $\lambda \leq 2d$. Originally, Bragg diffraction was used by crystallographers to study the

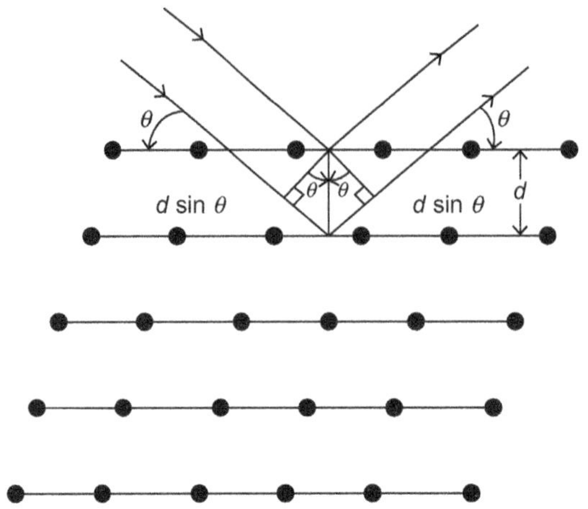

FIGURE 1.23

Bragg reflection from a family of lattice planes.

crystal structure of solids. Later, after the discovery of diffraction of electrons from a periodic crystal by Davisson and Germer, Bragg diffraction became important in understanding the theory of crystalline solids.

1.8 LAUE METHOD

Von Laue considered the X-ray diffraction from a crystal by considering it as composed of identical atoms or ions (basis) placed at the lattice sites **R** of a Bravais lattice. Each ion or atom at these sites would reradiate the incident radiation in all directions (Laue did not make any assumptions about specular reflections). However, sharp peaks would be observed only at appropriate wavelengths and directions when the scattered radiations from the ions or atoms (basis) placed at sites **R** would interfere constructively.

We consider two ions, separated by the vector **d** (see Figure 1.24).

If $\hat{\mathbf{n}}$ is the direction of the incident radiation of wavelength λ, the incident wave vector $\mathbf{k} = 2\pi\, \hat{\mathbf{n}}/\lambda$. If the path difference between the radiation scattered by each of the two ions is $m\lambda$, where m is an integer, the scattered radiation (assuming elastic scattering) would be observed in the direction $\hat{\mathbf{n}}'$ with the same wavelength λ and wave vector $\mathbf{k}' = 2\pi\hat{\mathbf{n}}'/\lambda$. This leads to constructive interference, the condition for which is (Figure 1.24)

$$\mathbf{d} \cdot (\hat{\mathbf{n}} - \hat{\mathbf{n}}') = m\lambda. \quad (1.9)$$

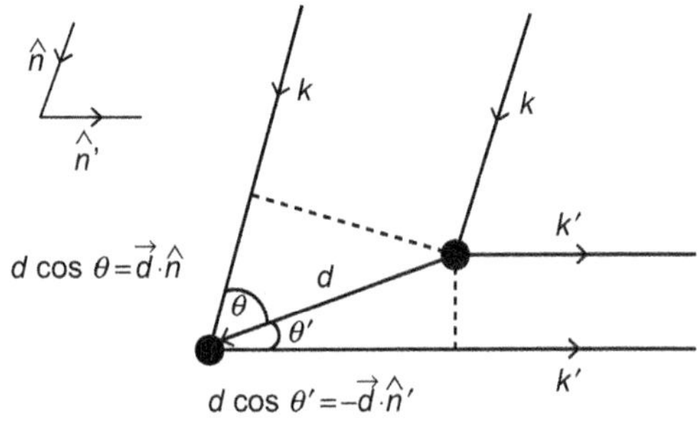

FIGURE 1.24

Path difference of X-rays scattered by two points separated by **d**.

Multiplying Eq. (1.9) by $2\pi/\lambda$ and using the definitions of $\mathbf{k}$ and $\mathbf{k}'$, we can write Eq. (1.9) in the alternate form

$$\mathbf{d} \cdot (\mathbf{k} - \mathbf{k}') = 2\pi m. \tag{1.10}$$

If we consider all the ions at each site of the Bravais lattice, the condition of constructive interference of all the scattered radiations is obtained from a generalization of Eq. (1.10), which can be written in the form

$$\mathbf{R} \cdot (\mathbf{k} - \mathbf{k}') = 2\pi m, \tag{1.11}$$

where $\mathbf{R}$ is a direct lattice vector of the Bravais lattice. If we write

$$\mathbf{K} = \mathbf{k} - \mathbf{k}' = \mathbf{k}' - \mathbf{k}, \tag{1.12}$$

where $\mathbf{K}$ is the set of wave vectors that satisfy the condition of constructive interference of all the scattered radiations (Eq. 1.11), i.e.,

$$\mathbf{K} \cdot \mathbf{R} = 2\pi m. \tag{1.13}$$

In Section 1.9, we will study the properties of the reciprocal lattice constructed from the set of wave vectors $\mathbf{K}$ by first constructing them using the basic properties of the direct lattice vectors $\mathbf{R}$.

1.9 RECIPROCAL LATTICE

1.9.1 Definition

If $\mathbf{R}$ is a set of points that constitute a Bravais lattice as defined in Eq. (1.1),

$$\mathbf{R} = n_1\mathbf{a}_1 + n_2\mathbf{a}_2 + n_3\mathbf{a}_3,$$

there would exist a particular set of wave vectors $\mathbf{K}$ that would yield plane waves with the periodicity of the lattice. The analytic definition of the set of vectors $\mathbf{K}$ would be

$$e^{i\mathbf{K}\cdot(\mathbf{r}+\mathbf{R})} = e^{i\mathbf{K}\cdot\mathbf{r}}, \tag{1.14}$$

for any vector $\mathbf{r}$ and all direct lattice vectors $\mathbf{R}$ in the Bravais lattice. Eq. (1.14) can be written in the alternate form

$$e^{i\mathbf{K}\cdot\mathbf{R}} = 1. \tag{1.15}$$

The set of wave vectors $\mathbf{K}$ that satisfy Eq. (1.14) for any direct lattice vector $\mathbf{R}$ that generates a Bravais lattice is defined as the reciprocal lattice of that Bravais lattice. It is obvious that each Bravais lattice has its own reciprocal lattice.

It can be shown that the reciprocal lattice can be generated from the three primitive vectors $\mathbf{b}_1$, $\mathbf{b}_2$, and $\mathbf{b}_3$, where $\mathbf{b}_1$, $\mathbf{b}_2$, $\mathbf{b}_3$ are defined by

$$\begin{aligned} \mathbf{b}_1 &= 2\pi \frac{\mathbf{a}_2 \times \mathbf{a}_3}{\mathbf{a}_1 \cdot (\mathbf{a}_2 \times \mathbf{a}_3)}, \\ \mathbf{b}_2 &= 2\pi \frac{\mathbf{a}_3 \times \mathbf{a}_1}{\mathbf{a}_1 \cdot (\mathbf{a}_2 \times \mathbf{a}_3)}, \\ \mathbf{b}_3 &= 2\pi \frac{\mathbf{a}_1 \times \mathbf{a}_2}{\mathbf{a}_1 \cdot (\mathbf{a}_2 \times \mathbf{a}_3)}. \end{aligned} \tag{1.16}$$

It is easy to show from Eq. (1.16) that (Problem 1.13)

$$\mathbf{a}_i \cdot \mathbf{b}_j = 2\pi\delta_{ij}, \tag{1.17}$$

where δ_{ij} is the Kronecker delta function, defined by $\delta_{ij} = 0$ for $i \neq j$ and $\delta_{ij} = 1$ for $i = j\cdot$

We also note from Eqs. (1.16) and (1.17) that the vectors $\mathbf{b}_1, \mathbf{b}_2$, and $\mathbf{b}_3$, are in the reciprocal space and are not in the same plane because the primitive lattice vectors $\mathbf{a}_1, \mathbf{a}_2$, and $\mathbf{a}_3$ are not coplanar. Therefore, any vector $\mathbf{k}$ in the reciprocal space can be written as

$$\mathbf{k} = k_1\mathbf{b}_1 + k_2\mathbf{b}_2 + k_3\mathbf{b}_3. \tag{1.18}$$

From Eqs. (1.1), (1.16), and (1.18), we obtain

$$\mathbf{k} \cdot \mathbf{R} = 2\pi(k_1 n_1 + k_2 n_2 + k_3 n_3). \tag{1.19}$$

For Eq. (1.15) to hold true for any $\mathbf{K}$,

$$\mathbf{K} \cdot \mathbf{R} = 2\pi m, \tag{1.20}$$

where m is an integer. Because n_1, n_2, and n_3 are integers, and we obtain from Eqs. (1.19) and (1.20) that

$$m = k_1 n_1 + k_2 n_2 + k_3 n_3, \tag{1.21}$$

whenever $\mathbf{k} = \mathbf{K}$. Eq. (1.21) must hold good for any choice of the integers n_i. It follows that k_1, k_2, and k_3 are also integers whenever $\mathbf{k} = \mathbf{K}$. Thus, we can write m in the alternate form

$$m = m_1 + m_2 + m_3, \tag{1.22}$$

where each m_i is an integer. If we write $m_1 = hn_1, m_2 = kn_2$, and $m_3 = \ln_3$, where h, k, l are integers, the reciprocal lattice vectors can be expressed as

$$\mathbf{K} = h\mathbf{b}_1 + k\mathbf{b}_2 + l\mathbf{b}_3, \tag{1.23}$$

Later, we will identify h, k, l as the Miller indices introduced earlier. As mentioned earlier, any Bravais lattice has its own reciprocal lattice. It has sometimes been compared to each person having his or her own shadow.

1.9.2 Properties of the Reciprocal Lattice

We can enumerate some of the properties of the reciprocal lattice from the previous definitions.

a. Each vector of the reciprocal lattice is normal to a set of lattice planes of the direct lattice.

Proof:

We have obtained from Eq. (1.20), $\mathbf{K} \cdot \mathbf{R} = 2\pi m$, where m is an integer. Therefore, the projection of vector $\mathbf{R}$ on the direction of $\mathbf{K}$ has the length d,

$$d = \frac{2\pi m}{|\mathbf{K}|}. \tag{1.24}$$

However, there is an infinite number of points in the direct lattice with the same property. To show this, let us consider a lattice point $\mathbf{R}'$ represented by the integers

$$n_1' = n_1 - pl;\ n_2' = n_2 - pl;\ n_3' = n_1 + p(h + k), \tag{1.25}$$

where p is an arbitrary integer; n_1, n_2, and n_3 are the set of integers that represent a lattice point **R**, and h, k, and l are the integers that define a reciprocal lattice vector. It can be easily shown that

$$\mathbf{K}\cdot\mathbf{R}' = \mathbf{K}\cdot\mathbf{R} = 2\pi m. \tag{1.26}$$

Therefore, **R**′ has the same projection on **K** as **R** and is on the plane normal to **K**, at a distance d from the origin. Thus, there is an infinite number of lattice points on this plane if it has at least one lattice point.

b. $|\mathbf{K}|$ is inversely proportional to the spacing of the lattice planes normal to **K**, if the components of **K** have no common factor.

Proof:
We have shown in Eq. (1.24) that $d = \frac{2\pi m}{|\mathbf{K}|}$. If (h, k, l), the components of **K** have no common factor, then a lattice vector **R**″ with components (n_1'', n_2'', n_3'') can always be found such that

$$\mathbf{K}\cdot\mathbf{R}'' = 2\pi(m+1). \tag{1.27}$$

Here,

$$m+1 = hn_1'' + kn_2'' + ln_3''. \tag{1.28}$$

Thus, the lattice plane containing **R**″ is at a distance

$$d'' = \frac{2\pi(m+1)}{|\mathbf{K}|} \tag{1.29}$$

from the origin. Comparing Eqs. (1.24) and (1.29), we note that the lattice plane containing **R**″ is spaced $2\pi/|\mathbf{K}|$ from the lattice plane containing **R**. The simplest way of identifying the planes of a direct lattice is by their normals, which are the vectors of the reciprocal lattice. The planes that are most densely populated with lattice sites are usually the most prominent planes in a direct lattice. These are also the most widely separated because the density of direct lattice sites is constant. Therefore, the most prominent planes are those with the smallest reciprocal lattice vectors.

c. The Miller indices h, k, l, which identify the direct lattice planes, are also the integers that identify the reciprocal lattice vectors normal to those planes.

Proof:
Consider a lattice plane with normal **K** such that $\mathbf{K}\cdot\mathbf{R} = 2\pi m$ (Eq. 1.20) is satisfied by all lattice points identified by **R** in that plane. If a lattice point has $n_2 = n_3 = 0$, from Eq. (1.21), we obtain

$$n_1 = m/h. \tag{1.30}$$

The intercept of this plane along the $\mathbf{a}_1$ axis has the length

$$d_1 = n_1 a_1 = \left(\frac{m}{h}\right) a_1. \tag{1.31}$$

One can similarly obtain

$$d_2 = n_2 a_2 = \left(\frac{m}{k}\right) a_2, \tag{1.32}$$

and

$$d_3 = n_3 a_3 = \left(\frac{m}{l}\right) a_3. \tag{1.33}$$

Thus, the intercepts of this plane among the axes, measured in the units of the corresponding basis vectors, are inversely proportional to h, k, l. These integers are precisely the definition of the Miller indices (after removal of common factors) of the plane.

d. The volume of a unit cell of the reciprocal lattice is inversely proportional to the volume of the unit cell of a direct lattice.

Proof:

The primitive vectors of the reciprocal lattice are $\mathbf{b}_1$, $\mathbf{b}_2$, and $\mathbf{b}_3$. The volume of a unit cell of the reciprocal lattice is

$$\mathbf{b}_1 \cdot (\mathbf{b}_2 \times \mathbf{b}_3) = \frac{2\pi(\mathbf{a}_2 \times \mathbf{a}_3) \cdot (\mathbf{b}_2 \times \mathbf{b}_3)}{\mathbf{a}_1 \cdot (\mathbf{a}_2 \times \mathbf{a}_3)}, \tag{1.34}$$

which can be simplified as

$$\mathbf{b}_1 \cdot (\mathbf{b}_2 \times \mathbf{b}_3) = \frac{-2\pi \mathbf{b}_2 \cdot (\mathbf{a}_2 \times \mathbf{a}_3) \times \mathbf{b}_3}{\mathbf{a}_1 \cdot (\mathbf{a}_2 \times \mathbf{a}_3)}. \tag{1.35}$$

By using the vector identity $(\mathbf{B} \times \mathbf{C}) \times \mathbf{A} = -\mathbf{B}(\mathbf{A} \cdot \mathbf{C}) + \mathbf{C}(\mathbf{A} \cdot \mathbf{B})$, we obtain

$$(\mathbf{a}_2 \times \mathbf{a}_3) \times \mathbf{b}_3 = -\mathbf{a}_2(\mathbf{b}_3 \cdot \mathbf{a}_3) + \mathbf{a}_3(\mathbf{b}_3 \cdot \mathbf{a}_2). \tag{1.36}$$

From Eqs. (1.17) and (1.36), we obtain

$$(\mathbf{a}_2 \times \mathbf{a}_3) \times \mathbf{b}_3 = -2\pi \mathbf{a}_2. \tag{1.37}$$

From Eqs. (1.17), (1.35), and (1.37), we obtain

$$\mathbf{b}_1 \cdot (\mathbf{b}_2 \times \mathbf{b}_3) = \frac{(2\pi)^3}{\mathbf{a}_1 \cdot (\mathbf{a}_2 \times \mathbf{a}_3)}. \tag{1.38}$$

Because the volume of the primitive cell in the direct lattice is

$$v = \mathbf{a}_1 \cdot (\mathbf{a}_2 \times \mathbf{a}_3), \tag{1.39}$$

we obtain from Eqs. (1.32) and (1.33),

$$\mathbf{b}_1 \cdot (\mathbf{b}_2 \times \mathbf{b}_3) = \frac{(2\pi)^3}{v}. \tag{1.40}$$

e. The reciprocal lattice of a simple cubic Bravais lattice with side a is a simple cubic lattice with cubic primitive cell of side $2\pi/a$.
Proof:
For a simple cubic (sc) lattice, the primitive lattice vectors are

$$\mathbf{a}_1 = a\hat{\mathbf{x}}, \quad \mathbf{a}_2 = a\hat{\mathbf{y}}, \quad \mathbf{a}_3 = a\hat{\mathbf{z}}. \tag{1.41}$$

From Eqs. (1.11) and (1.35), we obtain the reciprocal lattice vectors of the sc lattice,

$$\mathbf{b}_1 = \frac{2\pi}{a}\hat{\mathbf{x}}, \quad \mathbf{b}_2 = \frac{2\pi}{a}\hat{\mathbf{y}}, \quad \mathbf{b}_3 = \frac{2\pi}{a}\hat{\mathbf{z}}. \tag{1.42}$$

f. The direct lattice is the reciprocal of its own reciprocal lattice.
Proof:
This can be easily shown by inspection of Eq. (1.40) or (1.42).

g. The reciprocal lattice of a bcc Bravais lattice with conventional unit cell of side a is a fcc lattice with conventional unit cell of side $4\pi/a$.
Proof:
We have seen in Eq. (1.2) that for a bcc lattice, the symmetric set of primitive vectors is

$$\mathbf{a}_1 = \frac{a}{2}(\hat{\mathbf{y}} + \hat{\mathbf{z}} - \hat{\mathbf{x}}); \quad \mathbf{a}_2 = \frac{a}{2}(\hat{\mathbf{z}} + \hat{\mathbf{x}} - \hat{\mathbf{y}}); \quad \mathbf{a}_3 = \frac{a}{2}(\hat{\mathbf{x}} + \hat{\mathbf{y}} - \hat{\mathbf{z}}).$$

From Eqs. (1.11), the reciprocal lattice vectors are

$$\mathbf{b}_1 = \frac{4\pi}{a}\frac{1}{2}(\hat{\mathbf{y}} + \hat{\mathbf{z}}); \quad \mathbf{b}_2 = \frac{4\pi}{a}\frac{1}{2}(\hat{\mathbf{z}} + \hat{\mathbf{x}}); \quad \mathbf{b}_3 = \frac{4\pi}{a}\frac{1}{2}(\hat{\mathbf{x}} + \hat{\mathbf{y}}). \tag{1.43}$$

This has the form of the fcc primitive vectors (Eq. 1.3), provided the side of the cubic cell is taken to be $4\pi/a$.

h. One can similarly show (Problem 1.14) that the reciprocal lattice of the fcc Bravais lattice with conventional unit cell of side a is a bcc lattice with conventional unit cell of side $4\pi/a$.

i. The unit cell of the reciprocal lattice need not be a parallelepiped.

1.9.3 Alternative Formulation of the Laue Condition

We have derived the Laue condition (Eq. 1.12) for constructive interference of the incident radiation reradiated by the ions or atoms in all directions as $\mathbf{K} = \mathbf{k}' - \mathbf{k}$, where $\mathbf{K}$ is a reciprocal lattice vector. Because the incident and scattered radiations have the same wavelengths $\lambda = \lambda'$ for elastic scattering, it follows that

$$|\mathbf{k}| = |\mathbf{k}'| = k. \tag{1.44}$$

From Eqs. (1.12) and (1.44), we obtain

$$k' = |\mathbf{k} - \mathbf{K}| = k. \tag{1.45}$$

By squaring Eq. (1.45), we obtain

$$k^2 = k^2 + K^2 - 2\mathbf{k} \cdot \mathbf{K}, \tag{1.46}$$

which can be written in the alternate form

$$\mathbf{k} \cdot \hat{\mathbf{K}} = 1/2K. \tag{1.47}$$

Eq. (1.47) implies that the incident wave vector **k** would satisfy the Laue condition only if the tip of **k** is on a plane that is a perpendicular bisector of the line joining the origin to **K**. This is shown in Figure 1.25, and such planes in k space are known as Bragg planes.

In Figure 1.26, the Laue condition is shown in an alternate way such that its equivalence to Bragg refection can be demonstrated.

In fact, it can be easily shown that the Bragg and Laue formulations of X-ray diffraction from a crystal are equivalent. We write $\mathbf{K} = n\mathbf{K}_0$, where n is an integer and $\mathbf{K}_0$ is the shortest reciprocal lattice vector parallel to **K**. Further, Eq. (1.24) can be written in the alternate form

$$K = \frac{2\pi n}{d}. \tag{1.48}$$

We can easily show from Figure 1.26,

$$K = 2k \sin\theta. \tag{1.49}$$

From Eqs. (1.48) and (1.49), we obtain

$$k \sin\theta = \frac{n\pi}{d}. \tag{1.50}$$

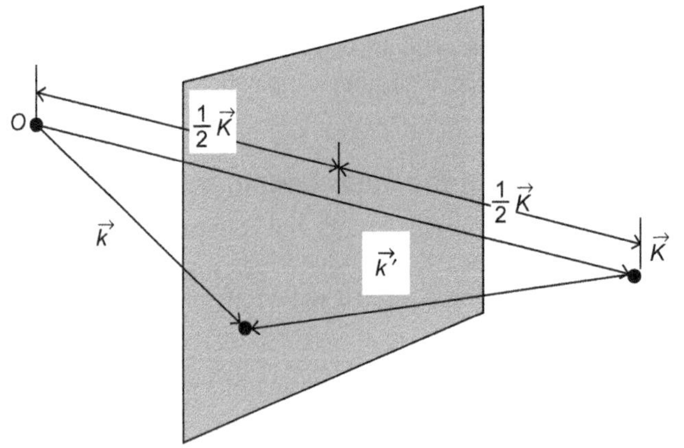

FIGURE 1.25

The Laue condition. A typical Bragg plane is shown in the diagram.

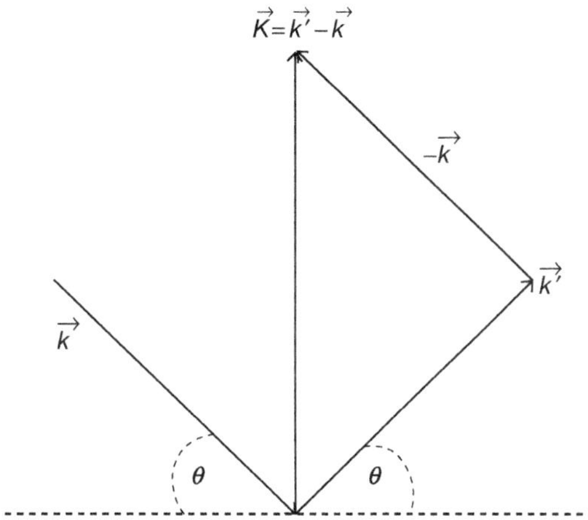

FIGURE 1.26

Because $\mathbf{K} = \mathbf{k}' - \mathbf{k}$ and $|\mathbf{k}| = |\mathbf{k}'|$, $\mathbf{k}$ and $\mathbf{k}'$ have the same angle θ with the plane perpendicular to $\mathbf{K}$.

Using the expression $k = 2\pi/\lambda$, we can write Eq. (1.50) as

$$2d \sin\theta = n\lambda, \tag{1.51}$$

which is precisely Eq. (1.8), the Bragg condition. We have also shown that the order n of the Bragg reflection is $n = |\mathbf{K}|/|\mathbf{K}_0|$.

1.10 BRILLOUIN ZONES

1.10.1 Definition

The Brillouin zone is a very important concept in solid state physics; it plays a major role in the theoretical understanding of the elementary ideas of electronic energy bands. The first Brillouin zone is defined as the Wigner–Seitz primitive cell of the reciprocal lattice. Thus, it is the set of points in the reciprocal space that is closer to $K = 0$ than to any other reciprocal lattice point. We have shown in Figure 1.21 that the Bragg planes bisect the lines joining 0 (the origin) to the reciprocal lattice points. Thus, we can also define the first Brillouin zone as the set of points that can be reached from 0 without crossing any Bragg planes. Here, the points common to the surface of two or more zones have not been considered.

The second Brillouin zone is the set of points that can be reached from the first Brillouin zone by crossing only one plane. Similarly, the nth Brillouin zone can be defined as the set of points that can be reached by crossing $n - 1$ Bragg planes. We will first describe Brillouin zones of one- and two-dimensional (square) lattices to explain the fundamental methods of obtaining the Brillouin zones before describing the Brillouin zones of some important lattices.

1.10.2 One-Dimensional Lattice

Consider a one-dimensional (linear) lattice of lattice constant a, which is taken in the $\hat{\mathbf{x}}$ direction. By definition, the direct lattice vector is $\mathbf{R} = n_1 a\hat{\mathbf{x}}$, where n_1 is an integer. The reciprocal lattice vectors for this linear lattice are

$$\mathbf{K} = (2\pi/a)h\hat{\mathbf{x}}, \tag{1.52}$$

where h is an integer. Thus, the reciprocal lattice is also a linear lattice of side $b = 2\pi/a$. As described earlier, the Bragg planes (points in one dimension also known as zone boundaries) bisect the lines joining a reciprocal lattice point with its neighbors. The first three Brillouin zones of the linear lattice are shown in Figure 1.27.

As we can see from Eq. (1.52), the shortest reciprocal lattice vector $|\mathbf{K}_0| = 2\pi/a$. If the origin $\mathbf{0}$ is chosen at the center, the first Bragg plane (zone boundary) is at $-\pi/a$ and π/a. Similarly, the second Bragg plane is at $-2\pi/a$ and $2\pi/a$, and the third Bragg plane is at $-3\pi/a$ and $3\pi/a$. Thus, the first Brilloun zone (shown by horizontal lines) extends from $-\pi/a$ to π/a, the second Brillouin zone (shown by lines \\\\\\\\) is between $-2\pi/a$ and $-\pi/a$ as well as between π/a and $2\pi/a$. Similarly, the third Brillouin zone (shown by lines ////////) is between $-3\pi/a$ and $-2\pi/a$ as well as between $2\pi/a$ and $3\pi/a$. These are consistent with the definition of the Brillouin zones; i.e., a point in the first zone does not cross any Bragg plane (point in one dimension), a point in the second zone crosses only one Bragg plane (zone boundary), and a point in the third Brillouin zone crosses two Bragg planes (zone boundaries).

1.10.3 Two-Dimensional Square Lattice

The direct lattice vectors of a two-dimensional square Bravais lattice are

$$\mathbf{R} = n_1 a\hat{\mathbf{x}} + n_2 a\hat{\mathbf{y}}, \tag{1.53}$$

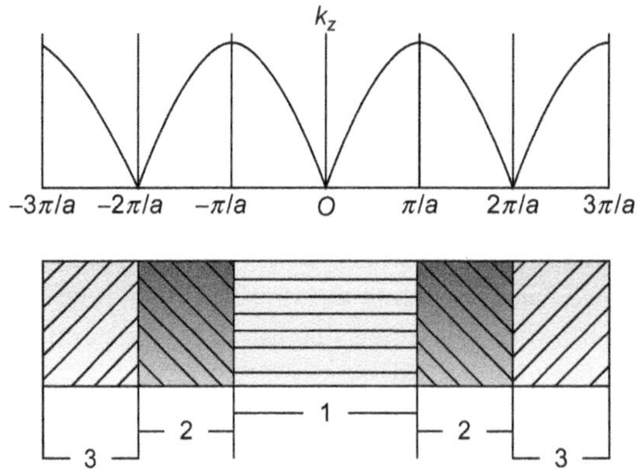

FIGURE 1.27

The Brillouin zones of the linear lattice.

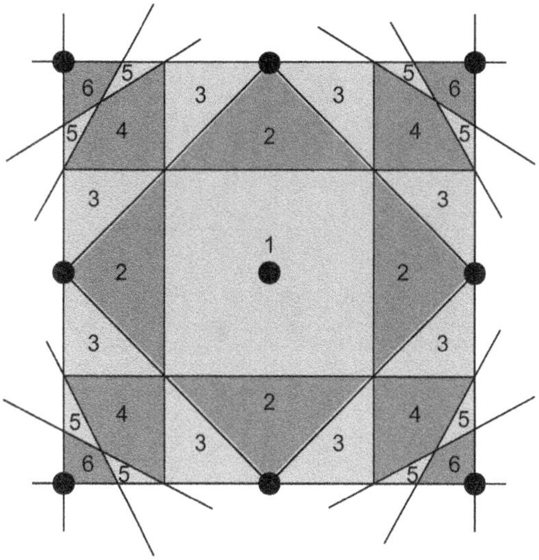

FIGURE 1.28

The Brillouin zones for a two-dimensional square lattice of side $4\pi/a$ [[$4\pi/\alpha$]].

where n_1 and n_2 are integers and a is the lattice constant. The reciprocal lattice vectors are

$$\mathbf{K} = \frac{2\pi}{a}(h\hat{\mathbf{x}} + k\hat{\mathbf{y}}), \tag{1.54}$$

where h and k are integers. Thus, the reciprocal lattice is also a square lattice of side $b = 2\pi/a$. The Brillouin zones are constructed according to the method outlined earlier. In Figure 1.28, all the Bragg lines (in two dimensions) that are in a square of side $4\pi/a$ centered on the origin are shown. The Bragg lines divide the square into regions belonging to different zones.

1.10.4 bcc Lattice

We have shown in Eq. (1.43) that the primitive translational vectors of the reciprocal lattice of a bcc lattice are given by

$$\mathbf{b}_1 = (2\pi/a)(\hat{\mathbf{y}} + \hat{\mathbf{z}});\ \mathbf{b}_2 = (2\pi/a)(\hat{\mathbf{z}} + \hat{\mathbf{x}});\ \mathbf{b}_3 = (2\pi/a)(\hat{\mathbf{x}} + \hat{\mathbf{y}}).$$

The reciprocal lattice vectors are

$$\begin{aligned}\mathbf{K} &= h\mathbf{b}_1 + k\mathbf{b}_2 + l\mathbf{b}_3 \\ &= (2\pi/a)[(k+l)\hat{\mathbf{x}} + (h+l)\hat{\mathbf{y}} + (h+k)\hat{\mathbf{z}}]\cdot\end{aligned} \tag{1.55}$$

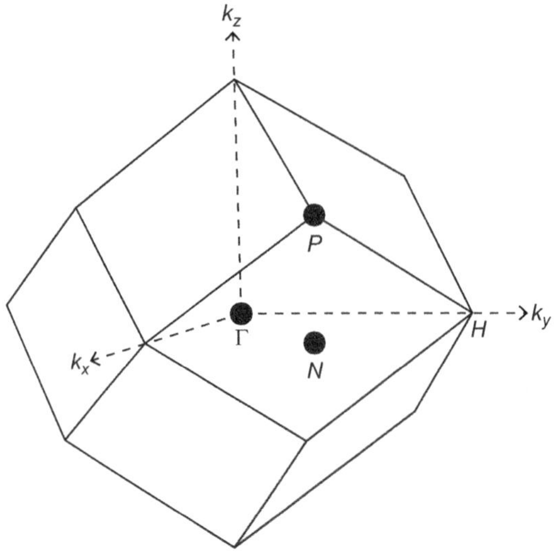

FIGURE 1.29

The first Brillouin zone for the bcc lattice.

From Eq. (1.55), we note that the shortest nonzero **K**'s are the 12 vectors

$$(2\pi/a)(\pm\hat{\mathbf{x}}\pm\hat{\mathbf{y}});\quad (2\pi/a)(\pm\hat{\mathbf{y}}\pm\hat{\mathbf{z}});\quad (2\pi/a)(\pm\hat{\mathbf{z}}\pm\hat{\mathbf{x}}). \tag{1.56}$$

The first Brillouin zone is the primitive cell formed from the planes normal to the 12 vectors (the Bragg planes) of Eq. (1.56). Thus, the 12 vectors from the origin to the center of each face of the first Brillouin zone are

$$(\pi/a)(\pm\hat{\mathbf{x}}\pm\hat{\mathbf{y}});\quad (\pi/a)(\pm\hat{\mathbf{y}}\pm\hat{\mathbf{z}});\quad (\pi/a)(\pm\hat{\mathbf{z}}\pm\hat{\mathbf{x}}). \tag{1.57}$$

The first Brillouin zone for the bcc lattice is shown in Figure 1.29. This regular 12-faced solid is a rhombic dodecahedron. The important symmetry points are marked by conventional symbols. By convention, Γ is denoted as the center of a zone.

1.10.5 fcc Lattice

The primitive translation vectors $\mathbf{b}_1, \mathbf{b}_2, \mathbf{b}_3$ of the reciprocal lattice of the fcc lattice are (Problem 1.14)

$$\begin{aligned}\mathbf{b}_1 &= (2\pi/a)(\hat{\mathbf{y}}+\hat{\mathbf{z}}-\hat{\mathbf{x}}),\\ \mathbf{b}_2 &= (2\pi/a)(\hat{\mathbf{z}}+\hat{\mathbf{x}}-\hat{\mathbf{y}}),\\ \mathbf{b}_3 &= (2\pi/a)(\hat{\mathbf{x}}+\hat{\mathbf{y}}-\hat{\mathbf{z}})\cdot\end{aligned} \tag{1.58}$$

Thus, the reciprocal lattice vectors of the fcc lattice are

$$\begin{aligned}\mathbf{K} &= h\mathbf{b}_1 + k\mathbf{b}_2 + l\mathbf{b}_3\\ &= (2\pi/a)[(-h+k+l)\hat{\mathbf{x}} + (h-k+l)\hat{\mathbf{y}} + (h+k-l)\hat{\mathbf{z}}]\cdot\end{aligned} \tag{1.59}$$

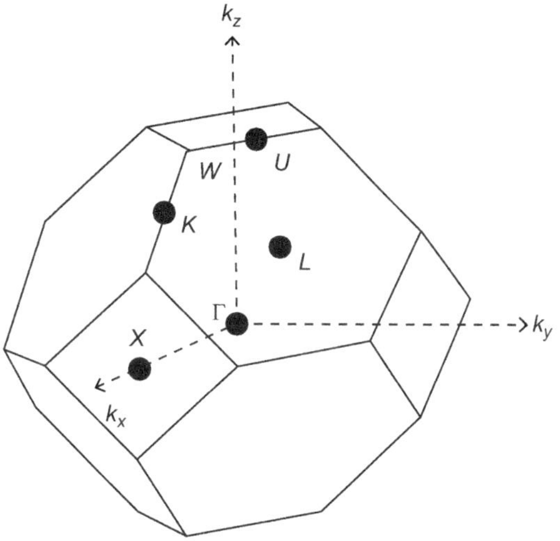

FIGURE 1.30

The first Brillouin zone for the fcc lattice.

While the shortest nonzero **K**'s are the eight vectors

$$(2\pi/a)(\pm\hat{\mathbf{x}} \pm \hat{\mathbf{y}} \pm \hat{\mathbf{z}}), \tag{1.60}$$

the corners of the octahedron formed by the eight planes normal to these vectors at their midpoints are truncated by the planes that are the perpendicular bisectors of the six reciprocal lattice vectors

$$\pm(4\pi/a)\hat{\mathbf{x}}; \quad \pm(4\pi/a)\hat{\mathbf{y}}; \quad \pm(4\pi/a)\hat{\mathbf{z}}. \tag{1.61}$$

The first Brillouin zone of the fcc lattice is the truncated octahedron shown in Figure 1.30. The conventional symbols of the important symmetry points are also shown in the figure.

1.11 DIFFRACTION BY A CRYSTAL LATTICE WITH A BASIS

The analysis of diffraction of the incident radiation by a crystal lattice with a basis and, consequently, the intensity of radiation in a given Bragg peak is more complicated than a monoatomic crystal because we have to consider the scattering of the radiation from each ion or atom of the basis at each lattice point. A crystal lattice with a basis can have two or more ions or atoms, either of the same type or of different types. In general, each lattice point in a primitive cell would have a basis of atoms or ions associated with it, and the crystal structure consists of a repetitive unit of this basis.

1.11.1 Theory

We consider n scatterers at positions $\mathbf{r}_1, \mathbf{r}_2, ..., \mathbf{r}_n$, in each primitive cell with the lattice point considered as the origin of $\mathbf{r}_j$. It is important to note that in an ideal crystal, each lattice point has an identical basis around it so that the scatterers are located at the same relative position in each cell. We have already noted that a Bragg peak is associated with a change in the wave vector $\mathbf{K} = \mathbf{k}' - \mathbf{k}$ and the phase difference between two rays scattered from $\mathbf{r}_i$ and $\mathbf{r}_j$ would be $\mathbf{K} \cdot (\mathbf{r}_i - \mathbf{r}_j)$. The amplitudes of the rays scattered at $\mathbf{r}_1, ..., \mathbf{r}_n$ would be in the ratios $e^{i\mathbf{K}\cdot\mathbf{r}_1}, ..., e^{i\mathbf{K}\cdot\mathbf{r}_n}$. Thus, the amplitude of the scattered ray from the primitive cell, which is the sum of the scattered rays from the n identical scatterers of the primitive cell, will be equal to

$$S = \sum_{\mathbf{R}} \sum_{j} f_j e^{-i(\mathbf{R}+\mathbf{r}_j)\cdot\Delta\mathbf{k}}. \tag{1.62}$$

The amplitude of the scattered rays (Eq. 1.62) can be expressed as

$$S = \sum_{\mathbf{R}} e^{-i\mathbf{R}\cdot\Delta\mathbf{k}} S_K, \tag{1.63}$$

where S_K is the structure factor obtained by the Bragg condition $\Delta\mathbf{k} = \mathbf{K}$,

$$S_{\mathbf{K}} = \sum_{j=1}^{n} f_j e^{i\mathbf{K}\cdot\mathbf{r}_j}. \tag{1.64}$$

Here, f_j is the atomic scattering factor or form factor that is a measure of the scattering power of the jth atom or ion in the basis.

1.11.2 Geometrical Structure Factor

We will first consider the basis as consisting of n identical atoms or ions around each lattice point. This is equivalent to stating that each atomic form factor f_j has the same value $f_j \equiv f$. Then we can write Eq. (1.64) in the form

$$S_{\mathbf{K}} = f \sum_{j=1}^{n} e^{i\mathbf{K}\cdot\mathbf{r}_j} \equiv f\mathbb{S}_{\mathbf{K}}. \tag{1.65}$$

$\mathbb{S}_{\mathbf{K}}$ is known as the geometrical structure factor. Because $\mathbb{S}_{\mathbf{K}}$ is proportional to the amplitude, $|\mathbb{S}_{\mathbf{K}}|^2$ is proportional to the intensity of the Bragg peak and indicates the extent to which the waves scattered from the ions or atoms in the basis within the primitive cell interfere to reduce it.

Next, we will discuss the structure factor of a few important lattices.

1.11.3 Application to bcc Lattice

If we consider the bcc lattice as a simple cubic lattice of side a with identical atoms or ions at $\mathbf{r}_1 = \mathbf{0}$ and at $\mathbf{r}_2 = (a/2)(\hat{\mathbf{x}} + \hat{\mathbf{y}} + \hat{\mathbf{z}})$, the structure factor (Eq. 1.65) can be expressed as

$$\mathbb{S}_{\mathbf{K}} = (1 + e^{i\mathbf{K}\cdot[1/2a(\hat{\mathbf{x}}+\hat{\mathbf{y}}+\hat{\mathbf{z}})]}). \tag{1.66}$$

Further, for a simple cubic lattice,

$$\mathbf{K} = \frac{2\pi}{a}(h\hat{\mathbf{x}} + k\hat{\mathbf{y}} + l\hat{\mathbf{z}}). \tag{1.67}$$

From Eqs. (1.66) and (1.67), we obtain

$$\begin{aligned}\mathbb{S}_{\mathbf{K}} &= \left(1+e^{i\pi(h+k+l)}\right)\\ &= 1+(-1)^{\mathrm{h+k+l}}\end{aligned} \tag{1.68}$$

$$= \begin{Bmatrix} 0, & h+k+l & \text{odd} \\ 2, & h+k+l & \text{even} \end{Bmatrix}. \tag{1.69}$$

From Eq. (1.69), we note that whenever $h+k+l$ is odd, there is no Bragg diffraction (reflection). Because in this derivation, the bcc lattice has been considered as a simple cubic lattice with a basis, this implies that the odd reciprocal lattice vectors $h+k+l$ of the reciprocal simple cubic lattice do not exist. Therefore, the actual reciprocal lattice is a fcc lattice, which could have been obtained directly if we had considered the bcc lattice as the primitive lattice.

1.11.4 Application to fcc Lattice

If we consider a fcc lattice as a simple cubic lattice of side a and basis at

$$\mathbf{r}_1 = \mathbf{0},\ \mathbf{r}_2 = \frac{a}{2}(\hat{\mathbf{y}}+\hat{\mathbf{z}}),\ \mathbf{r}_3 = \frac{a}{2}(\hat{\mathbf{z}}+\hat{\mathbf{x}}),\ \mathbf{r}_4 = \frac{a}{2}(\hat{\mathbf{x}}+\hat{\mathbf{y}}),$$

the structure factor (Eq. 1.65) can be expressed as

$$\mathbb{S}_{\mathbf{K}} = 1+e^{i\mathbf{K}\cdot a/2(\hat{\mathbf{y}}+\hat{\mathbf{z}})}+e^{i\mathbf{K}\cdot a/2(\hat{\mathbf{z}}+\hat{\mathbf{x}})}+e^{i\mathbf{K}\cdot a/2(\hat{\mathbf{x}}+\hat{\mathbf{y}})}. \tag{1.70}$$

From Eqs. (1.67) and (1.70), we obtain

$$\mathbb{S}_{\mathbf{K}} = 1+e^{i\pi(k+l)}+e^{i\pi(h+l)}+e^{i\pi(h+k)}. \tag{1.71}$$

Eq. (1.71) can be written in the alternate form

$$\mathbb{S}_{\mathbf{K}} = 1+(-1)^{k+l}+(-1)^{h+l}+(-1)^{h+k}. \tag{1.72}$$

From Eq. (1.72), we note that $\mathbb{S}_{\mathbf{K}} = 4$, if each value of the integers h, k, l is either even or odd. However, $\mathbb{S}_{\mathbf{K}} = 0$, if either only one of the three integers h, k, l is even and the other two are odd, or one of the three integers is odd and the other two are even. Thus, there can be no Bragg reflection if the indices h, k, l are partly odd and partly even. In contrast, there is Bragg reflection if each of the indices h, k, l is either even or odd. This is true for a reciprocal bcc lattice. Hence, the reciprocal lattice of a fcc lattice is a bcc lattice, which we stated earlier, and has been assigned as a problem.

1.11.5 The Atomic Scattering Factor or Form Factor

In Eq. (1.64), we defined f_j as the atomic scattering factor or form factor, which was a component of the geometrical structure factor. However, in subsequent discussions, we considered that all the atoms or ions in a basis are identical.

Thus, each $f_j = f$ and f are factored out of the summation. We wrote the geometrical structure factor as $S_{\mathbf{K}} = f\ \mathbb{S}_{\mathbf{K}}$, where $\mathbb{S}_{\mathbf{K}}$ is the structure factor.

However, if the atoms or the ions in a basis are not identical, the atomic form factor f_j at the site $\mathbf{d}_j$ in the basis is

$$f_j(\mathrm{K}) = -\frac{1}{e}\int d\mathbf{r}\, e^{i\mathbf{K}\cdot\mathbf{r}}\rho_j(\mathbf{r}). \tag{1.73}$$

Here, $\rho_j(\mathbf{r})$ is the electronic charge density of the ion of type j placed at $\mathbf{r} = \mathbf{0}$. It is evident from Eqs. (1.64) and (1.73) that the geometrical structure factor would no longer vanish.

PROBLEMS

1.1. Show that the volume of the primitive Bravais cell of a bcc lattice is $a^3/2$, where a is the side of the unit cube.

1.2. Show that the volume of the primitive Bravais cell of a fcc lattice is $a^3/4$, where a is the side of the unit cube.

1.3. A plane in a lattice with primitive vectors $\mathbf{a}_1, \mathbf{a}_2$, and $\mathbf{a}_3$ has intercepts at $3\mathbf{a}_1$, $2\mathbf{a}_2$, and $-2\mathbf{a}_3$. Calculate the Miller indices of the plane. Label the direction perpendicular to this plane.

1.4. Draw a sketch of the $(10\bar{2})$ plane in a simple cubic lattice.

1.5. Prove that in a cubic crystal, a direction $[hkl]$ is perpendicular to the plane (hkl) having the same indices.

1.6. Show that the actual volume occupied by the spheres in the simple cubic structure (assuming that they are contacting hard spheres) is 52.4% of the total volume.

1.7. Show that the actual volume occupied by the spheres in the bcc structure (packing fraction: assuming that they are contacting hard spheres) is 0.68.

1.8. Show that the actual volume occupied by the spheres in the fcc structure (packing fraction: assuming that they are contacting hard spheres) is 0.74.

1.9. Show that the angle between any two of the lines (bonds) joining a site of the diamond lattice to its four nearest neighbors is $\cos^{-1}(-1/3)$.

1.10. Show that the ideal c/a ratio of the hexagonal close-packed structure is $\sqrt{8/3}$.

1.11. Show that the packing fraction of the hexagonal close-packed structure is 0.74.

1.12. Sodium transforms from bcc to hcp at about 23 K, which is also known as the "martenistic" transformation. The lattice constant in the cubic phase is $a = 4.23$ A°. Determine the lattice constant a of the hexagonal phase. Assume that the c/a ratio in the hexagonal phase is indistinguishable.

1.13. Show that if $\mathbf{a}_i$ are the three direct lattice primitive vectors and $\mathbf{b}_j$ are the three primitive vectors of the reciprocal lattice (as defined in Eq. 1.16), then

$$\mathbf{a}_i \cdot \mathbf{b}_j = 2\pi\delta_{ij},$$

where δ_{ij} is the Kronecker delta function.

1.14. Show that the primitive translation vectors $\mathbf{b}_1, \mathbf{b}_2, \mathbf{b}_3$ of the reciprocal lattice of the fcc lattice are $\mathbf{b}_1 = (2\pi/a)(\hat{\mathbf{y}} + \hat{\mathbf{z}} - \hat{\mathbf{x}}), \mathbf{b}_2 = (2\pi/a)(\hat{\mathbf{z}} + \hat{\mathbf{x}} - \hat{\mathbf{y}}), \mathbf{b}_3 = (2\pi/a)(\hat{\mathbf{x}} + \hat{\mathbf{y}} - \hat{\mathbf{z}})$, and prove that the reciprocal lattice of a fcc lattice of side a is a bcc lattice of side $4\pi/a$.

1.15. Show that for a monatomic diamond lattice, the structure factor is

$$\mathbb{S}_{\mathbf{K}} = 1 + e^{i[2\pi(h+k+l)]}$$

$$= \left\{ \begin{array}{lll} 2, & h+k+l & \text{is twice an even number,} \\ 1 \pm i, & h+k+l & \text{is odd,} \\ 0, & h+k+l & \text{is twice an odd number.} \end{array} \right\}.$$

Interpret these conditions geometrically.

1.16. a. Show that the reciprocal lattice of the sodium chloride structure is bcc and a reciprocal lattice vector can be written as

$$\mathbf{K} = \frac{4\pi}{a}(n_1 \hat{\mathbf{x}} + n_2 \hat{\mathbf{y}} + n_3 \hat{\mathbf{z}}),$$

where a is the side of the cube and all the coefficients n_i of a of a reciprocal lattice vector are integers or integer+1/2.

b. If the atomic form factors of the two types of ions are f_1 and f_2, show that the geometrical structure factor $\mathrm{S}_{\mathbf{K}} = f_1 + f_2$ if each n_i in a set is an integer and $\mathrm{S}_{\mathbf{K}} = f_1 - f_2$ if each n_i in a set is an integer+1/2.

References

1. Aschroft NW, Mermin ND. *Solid state physics.* New York: Brooks/Cole; 1976.

2a. Bragg WL. Structure of some crystals as indicated by their diffraction of x-rays. *Proc Roy Soc.* London; 1913; **A89**.

2b. Bragg WH and Bragg WL. Structure of the diamond. *Proc Roy Soc.* London; 1913;**A89**:277.

3. Buerger MJ. *Crystal structure analysis.* New York: John Wiley & Sons; 1960.

4. Kittel C. *Introduction to solid state physics.* New York: John Wiley & Sons; 1976.

5. Marder MC. *Condensed matter physics.* New York: John Wiley & Sons; 2000.

6. Myers HP. *Introduction to solid state physics.* London: Taylor & Francis; 1990.

7. Wyckoff RWG. *Crystal structures, vol. 1.* New York: John Wiley & Sons; 1963.

CHAPTER

Phonons and Lattice Vibrations

2

CHAPTER OUTLINE

2.1 Lattice Dynamics ... 37
2.1.1 Theory ... 37
2.1.2 Normal Modes of a One-Dimensional Monoatomic Lattice ... 41
2.1.3 Normal Modes of a One-Dimensional Chain with a Basis ... 44
2.2 Lattice Specific Heat ... 48
2.2.1 Theory ... 48
2.2.2 The Debye Model of Specific Heat ... 49
2.2.3 The Einstein Model of Specific Heat ... 52
2.3 Second Quantization ... 53
2.3.1 Occupation Number Representation ... 53
2.3.2 Creation and Annihilation Operators ... 54
2.3.3 Field Operators and the Hamiltonian ... 58
2.4 Quantization of Lattice Waves ... 61
2.4.1 Formulation ... 61
2.4.2 Quantization of Lattice Waves ... 65
Problems ... 66
References ... 68

2.1 LATTICE DYNAMICS

2.1.1 Theory

In a crystalline solid, at finite temperatures, the ions or the atoms are not stationary but vibrate around an equilibrium position. In 1907, Einstein proposed a theory of the heat capacity of a solid based on Planck's quantum hypothesis. He assumed that each atom of the solid vibrates around its equilibrium position with a frequency ν_E, known as the Einstein frequency. Each atom vibrates like a simple harmonic oscillator that is in the potential well of the force field of its neighbors. The atoms have the same frequency ν_E and vibrate independent of the other atoms. Thus, a mole of solid with N atoms is assumed to have $3N$ independent harmonic oscillators. The excitation spectrum of the crystalline solid is composed of levels that are spaced at a distance $\hbar\nu_E$ from each other.

The basic assumption of the Einstein model was that the atoms vibrate independent of each other, which could be justified only if the temperature is very high. However, at normal temperatures this assumption breaks down, because if two or more atoms move in unison, the restoring forces between them, which tend to restore each of them to their equilibrium position, would be reduced. In such a scenario, the required energy to excite a quantum would be reduced. In fact, the correlation between the motion of the adjacent atoms would play a significant role in solving the problem of lattice vibrations.

We consider the ground state of the lattice with a basis as the state at zero temperature, where each ion or atom of mass M_n is located at the equilibrium position $\mathbf{d}_n$, which is a vector connecting the local origin of the cell with a basis to the atom or ion of mass M_n. We assume that if the lattice has only one atom per unit cell, the atom is located at the lattice point and $\mathbf{d}_n = 0$. At a finite temperature, the displacement of the nth atom or the ion in the ith unit cell (the local origin of which is located at the direct lattice vector $\mathbf{R}_i$, and we have assumed that the origin of the lattice is one of the lattice points) from its equilibrium position at $\mathbf{d}_n$ is the vector $\mathbf{u}_{ni}$. The definition of the vector $\mathbf{u}_{ni}$ is schematically shown in Figure 2.1. We note that $\mathbf{u}_{ni}$ can also be written as

$$\mathbf{u}_{ni} = \mathbf{r} - \mathbf{R}_i - \mathbf{d}_n, \tag{2.1}$$

where $\mathbf{r}$ is the instantaneous position of the atom or ion of mass M_n located in the unit cell of which the local origin is at $\mathbf{R}_i$.

The kinetic energy of the crystalline lattice can be written as

$$T = \sum_{ni} \frac{1}{2} M_n \left| \dot{\mathbf{u}}_{ni} \right|^2. \tag{2.2}$$

The potential energy of the crystal depends on the structure of the cell as well as the interatomic forces. However, we assume that at any given instant, the function $V(\mathbf{u}_{ni})$ describes the potential energy of the crystal in terms of the instantaneous positions of all atoms, i.e., in terms of their actual displacements from the equilibrium positions. In addition, it is assumed that the perfect lattice is a configuration of stable equilibrium. The study of phonons is based on the assumption that the deviation $\mathbf{u}_{ni}$ of the ions or atoms from their equilibrium position $\mathbf{R}_i + \mathbf{d}_n$ is so small that one can make a Taylor expansion in powers of the variables $\mathbf{u}_{ni}$. Thus, we can write

$$V = V_0 + \sum_{ni\alpha} \left[\frac{\partial V}{\partial u^{\alpha}_{ni}} \right]_0 u^{\alpha}_{ni} + \frac{1}{2} \sum_{nn'} \sum_{ii'} \sum_{\alpha\beta} \left[\frac{\partial^2 V}{\partial u^{\alpha}_{ni} \partial u^{\beta}_{n'i'}} \right]_0 u^{\alpha}_{ni} u^{\beta}_{n'i'} + \ldots, \tag{2.3}$$

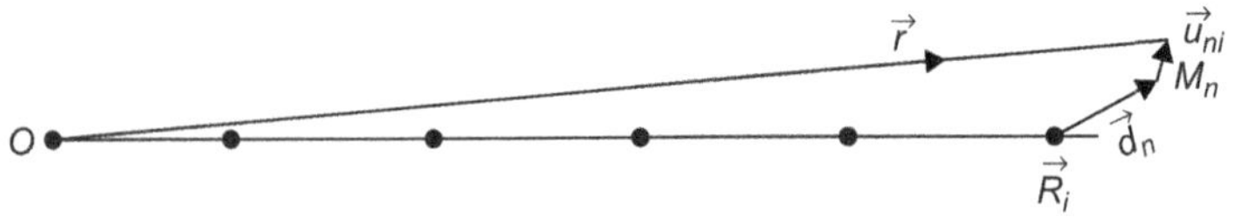

FIGURE 2.1

Definition of $\mathbf{u}_{ni}$ for a lattice with a basis.

where u^{α}_{ni} are the Cartesian components of $\mathbf{u}_{ni}$ and α and β are Cartesian indices running over the basis vectors of three-dimensional space.

The first term V_0 is essentially the cohesive energy of the crystal, but for the present purpose it is a constant term that can be neglected without affecting the calculation. Because the lowest energy state is a minimum as a function of the locations of the ions or atoms, the linear term in Eq. (2.3) must vanish near equilibrium. The first important term in Eq. (2.3) is the quadratic term in Eq. (2.3). This is known as the harmonic approximation. If one considers the higher-order terms for study of certain properties of the crystal, those are known anharmonic terms. However, in the present discussion, we shall restrict ourselves to the harmonic approximation.

We follow the Lagrangian procedure in classical mechanics to solve Eqs. (2.2) and (2.3). If we define the Lagrangian function (Symon, 1971, p. 366)

$$L = T - V, \tag{2.4}$$

the Lagrangian equations are

$$\frac{d}{dt}\left(\frac{\partial L}{\partial \dot{q}_k}\right) - \frac{\partial L}{\partial q_k} = 0, \quad k = 1, \ldots, 3N. \tag{2.5}$$

From Eqs. (2.2), (2.3), and (2.5), we obtain (Problem 2.1)

$$M_n \ddot{u}^{\alpha}_{ni} = -\sum_{n'i'\beta} \left[\frac{\partial^2 V}{\partial u^{\alpha}_{ni} \partial u^{\beta}_{n'i'}}\right]_0 \mathbf{u}^{\beta}_{n'i'}. \tag{2.6}$$

We define a Cartesian tensor Φ such that its components are obtained by the relation

$$\Phi^{\alpha\beta}_{ni;n'i'} \equiv \left[\frac{\partial^2 V}{\partial u^{\alpha}_{ni} \partial u^{\beta}_{n'i'}}\right]_0. \tag{2.7}$$

From Eqs. (2.6) and (2.7), we obtain the vector equation

$$M_n \ddot{u}_{ni} = -\sum_{n'i'} \Phi_{ni;n'i'} \cdot \mathbf{u}_{n'i'}. \tag{2.8}$$

Eq. (2.8) can be interpreted as the force acting on the nth atom in the ith cell (the cell of which the local origin is at a distance $\mathbf{R}_i$ from the origin of the lattice, where $\mathbf{R}_i$ is a direct lattice vector) due to the displacement $\mathbf{u}_{n'i'}$ of the atom on the n'th site of the i'th cell. In fact, much of the theory of phonons can be developed without considering how to calculate $\Phi_{ni;n'i'}$. However, $\Phi_{ni;n'i'}$ cannot depend on the absolute position of $\mathbf{R}_i$ and $\mathbf{R}_{i'}$ in the crystalline lattice. Thus, the tensor Φ has to be a function of their relative position $\mathbf{R}_i$ and $\mathbf{R}_{i'}$. If we write

$$\mathbf{R}_{i'} - \mathbf{R}_i = \mathbf{R}_l, \tag{2.9}$$

the Cartesian tensor can be expressed in the alternate form

$$\Phi_{ni,n'i'} = \Phi_{nn'}(\mathbf{R}_l). \tag{2.10}$$

From Eqs. (2.8) and (2.10), we obtain

$$M_n \ddot{\mathbf{u}}_{ni} = -\sum_{n'l} \Phi_{nn'}(\mathbf{R}_l) \cdot \mathbf{u}_{n',i+l}. \tag{2.11}$$

Here, $\mathbf{u}_{n,i+l}$ denotes the instantaneous displacement of the nth atom in the $i+l$th cell. We note that because the summation over the direct lattice vector $\mathbf{R}_l$ spans the entire lattice, Eq. (2.11) is translationally invariant. For example, if we change the label $\mathbf{R}_l$ to $\mathbf{R}_{l'}$, we obtain

$$M_n \ddot{\mathbf{u}}_{ni} = -\sum_{n'l'} \Phi_{nn'}(\mathbf{R}_{l'}) \cdot \mathbf{u}_{n',i+l'}, \tag{2.12}$$

which yields the same set of equations as Eq. (2.11). We also note another property of the tensor Φ is that the energy of the crystal cannot change if all the ions (or atoms) are displaced by a single vector, i.e.,

$$\sum_{l'} \Phi_{nn'}(\mathbf{R}_{l'}) = 0. \tag{2.13}$$

Because Eq. (2.11) must satisfy Bloch's theorem, if the set of functions of the time that describes the value of $\mathbf{u}_{ni}(t)$ for each value of $\mathbf{R}_i$ has been found, then according to the Bloch condition, there would be a wave vector $\mathbf{q}$ such that

$$\mathbf{u}_{ni}(t) = e^{i\mathbf{q}\cdot\mathbf{R}_i}\mathbf{u}_{n,0}(t), \tag{2.14}$$

where $\mathbf{u}_{n,0}(t)$ is the displacement of the nth atom in the cell that has been chosen as the origin for the direct lattice vectors $\mathbf{R}_i$. It is obvious from Eq. (2.14) that the atom or ion located at every site $\mathbf{d}_n$ (measured from the local origin of the unit cell, as shown in Figure 2.1) moves with the same amplitude and direction. However, the phase would vary for each cell. From Eqs. (2.11) and (2.14), we obtain

$$M_n \ddot{\mathbf{u}}_{n,0} e^{i\mathbf{q}\cdot\mathbf{R}_i} = -\sum_{n'l} \Phi_{nn'}(\mathbf{R}_l) \cdot \mathbf{u}_{n',0} e^{i\mathbf{q}\cdot\mathbf{R}_l} e^{i\mathbf{q}\cdot\mathbf{R}_i}. \tag{2.15}$$

In Eq. (2.15), we cancel the $e^{i\mathbf{q}\cdot\mathbf{R}_i}$ from both sides, and because the origin 0 is arbitrary, we consider a solution with a definite value of $\mathbf{q}$ by writing

$$\mathbf{u}_{n,0} = \mathbf{U}_{n,\mathbf{q}}, \tag{2.16}$$

and obtain

$$M_n \ddot{\mathbf{U}}_{n,\mathbf{q}} = -\sum_{n'} \left[\sum_{l} \Phi_{nn'}(\mathbf{R}_l) e^{i\mathbf{q}\cdot\mathbf{R}_l}\right] \cdot \ddot{\mathbf{U}}_{n',\mathbf{q}}. \tag{2.17}$$

We define the Fourier transform of the force tensor Φ as

$$\Phi_{nn'}(\mathbf{q}) \equiv \sum_{l} \Phi_{nn'}(\mathbf{R}_l) e^{i\mathbf{q}\cdot\mathbf{R}_l}, \tag{2.18}$$

and rewrite Eq. (2.17) as

$$M_n \ddot{\mathbf{U}}_{n,\mathbf{q}} = -\sum_{n'} \Phi_{nn'}(\mathbf{q}) \cdot \mathbf{U}_{n',\mathbf{q}}, \tag{2.19}$$

Eq. (2.19) is a set of $3m$ equations (assuming that there are m atoms per unit cell and there are three component equations—due to the three Cartesian components of the vectors—for each of the m values of n) in contrast to Eq. (2.8), which had a set of $3mN$ equations. This enormous simplification was possible because of the translational invariance.

Using the classical theory of vibrations,

$$U^{\alpha}_{n,\mathbf{q}}(t) = U^{\alpha}_{n,\mathbf{q}} e^{i\omega t}, \tag{2.20}$$

we obtain from Eqs. (2.19) and (2.20),

$$\sum_{n'\beta} [\Phi^{\alpha\beta}_{nn'}(\mathbf{q}) - \omega^2 M_n \delta_{nn'} \delta_{\alpha\beta}] U^{\alpha}_{n',\mathbf{q}} = 0. \tag{2.21}$$

Eq. (2.21) is an eigenvalue equation with $3m$ solutions that are solved by finding the roots of the equation in ω^2 when the determinant of the matrix [ ] is equal to zero. In a sense, we are solving $3m$ normal modes of vibration of m atoms in a unit cell that are assumed to interact via the force tensor $\Phi_{nn'}(\mathbf{q})$. This force tensor, $\Phi_{nn'}(\mathbf{q})$, which is different for each value of $\mathbf{q}$, is a sum of interactions of all n-type atoms (those atoms located at site n in each unit cell) with all the atoms on site n', and includes the effect of their relative phases.

2.1.2 Normal Modes of a One-Dimensional Monoatomic Lattice

We consider a set of ions of mass M (we have been using ions and atoms interchangeably for the lattice vibrations) located at the lattice points separated by a distance a. The one-dimensional Bravais lattice vectors are $\mathbf{R}_i = n_i a$. In this one-dimensional chain of lattice, one atom (or ion) of the same type is located at the lattice point. If the lattice constant is a, we define $L = Na$ such that periodic boundary conditions are applied to the linear chain. This boundary condition requires that the atoms located at s and at $s+N$ vibrate with the same amplitude and phase. The linear chain is shown in Figure 2.2.

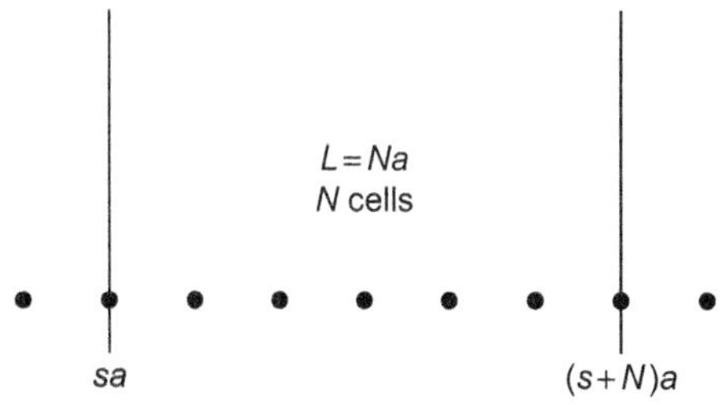

FIGURE 2.2

A linear chain of lattice points with periodic boundary conditions.

The linear boundary condition is best illustrated if we construct an endless circular chain of lattice points. This endless circular chain is illustrated in Figure 2.3. We note that although this illustration is possible for a one-dimensional lattice, such periodic boundary conditions have to be imagined for a three-dimensional lattice.

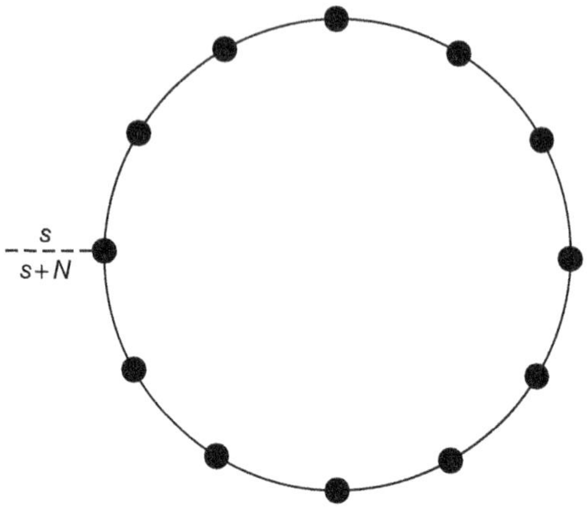

FIGURE 2.3

An endless circular chain of lattice points.

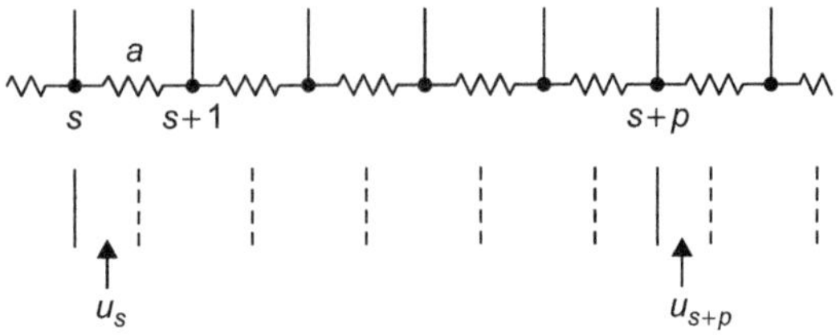

FIGURE 2.4

A linear chain of identical atoms, each located at a lattice point.

Figure 2.4 shows a linear chain of atoms of identical point mass m in a linear lattice of lattice constant a. The atoms are located at the lattice points. The atom (or ion) of which the equilibrium position is sa is displaced from equilibrium by an amount u_s, as shown in Figure 2.4.

For simplicity, we consider only nearest-neighbor interactions between the atoms (ions). The potential energy V in Eq. (2.3) is of the form

$$V^{harm} = \frac{1}{2}K\sum_s [u(sa) - u([s+1]a)]^2, \tag{2.22}$$

where K is the interaction energy of two ions (popularly known as the spring constant) $-\Phi_{i,i+1}$. Φ_{ii} is determined by the condition set in Eq. (2.13). The equations of motion are obtained from Eq. (2.22) as

$$\mathrm{M}\ddot{u}(sa) = -\frac{\partial V^{harm}}{\partial u(sa)} = -K[(u(sa) - 2u([s-1]a) - u([(s+1]a)]. \tag{2.23}$$

The solution of Eq. (2.23) is of the type

$$u(sa, t) = Ae^{i(qsa-\omega t)}. \tag{2.24}$$

Using the periodic boundary condition (Figure 2.3),

$$e^{iqNa} = 1, \tag{2.25}$$

we obtain the expression for q,

$$q = \frac{s}{N}\frac{2\pi}{a}, \tag{2.26}$$

where s is an integer and a is the lattice constant of the one-dimensional lattice. The q values lie between $-\frac{\pi}{a}$ and $\frac{\pi}{a}$. Substituting Eq. (2.24) in Eq. (2.23), we obtain

$$M\omega^2 = 2K(1-\cos qa) = 4K\ \sin^2(qa/2), \tag{2.27}$$

which leads to the solution

$$\omega(q) = 2\sqrt{\frac{K}{M}}\ \sin\left(\frac{qa}{2}\right), \tag{2.28}$$

where we have taken only the positive root of Eq. (2.27) because ω is an even function of q. We plotted ω as a function of q in Figure 2.5. We note that there is a maximum vibrational frequency $\omega = 2\sqrt{\frac{K}{M}}$ and the behavior is periodic with period $2\pi/a$. In fact, all possible vibrations are given by values of q in the range

$$-\frac{\pi}{a} < q \leq \frac{\pi}{a}, \tag{2.29}$$

which is the Brillouin zone (we will discuss the Brillouin zone in detail in Chapter 4 by using the nearly free electron model in a periodic lattice potential) for a linear lattice.

One can also count the number of modes in the following way. From Eq. (2.26), we obtain an expression for the density of modes in one-dimensional $\mathbf{q}$ space as

$$\frac{Na}{2\pi} = \frac{L}{2\pi}, \tag{2.30}$$

where L is the length of the sample. We also note that when $qa \ll 1$,

$$\frac{d\omega}{dq} = \frac{\omega}{q} = \sqrt{\frac{K}{M}}\ a = \text{constant}. \tag{2.31}$$

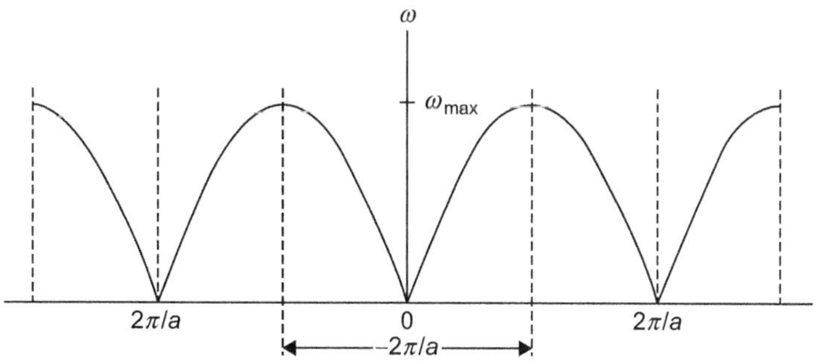

FIGURE 2.5

The vibrational frequencies of a linear chain of identical point masses.

Thus, the group velocity is equal to the phase velocity, and this proportionality of frequency to the wave number is also the property of the ordinary elastic waves in a continuum. This is also the velocity of sound waves and is known as the acoustic mode. However, at large values of q, the velocity of the wave is not constant. In fact, when $q = \pi/a = 2\pi/\lambda$, i.e., when the wavelength $\lambda = 2a$, ω_q bends over to a horizontal tangent (see Figure 2.5). This shows the property of dispersion.

2.1.3 Normal Modes of a One-Dimensional Chain with a Basis

We consider a one-dimensional Bravais lattice of lattice constant a with two ions or atoms of masses M_1 and M_2 per unit cell. This is shown in Figure 2.6. The basic assumption is that each ion interacts only with the nearest neighbors, which are at a distance $a/2$ from each other, and $M_1 > M_2$. Thus, the lattice constant of the linear chain is a.

If K is the force constant, from Eq. (2.21) we obtain

$$V^{harm} = \frac{K}{2}\sum_s [u_1(sa) - u_2(sa)]^2 + \frac{K}{2}\sum_s [u_2(sa) - u_1[s+1]a]^2, \tag{2.32}$$

where $u_1(sa)$ is the displacement of the ion that oscillates about the site sa and $u_2(sa)$ is the displacement of the ion that oscillates around $sa + d$. The equations of motion are

$$\begin{aligned} M_1\,\ddot{u}_1(sa) &= -\frac{\partial V^{harm}}{\partial u_1(sa)} = -K[2u_1(sa) - u_2(sa) - u_2([s-1]a)], \\ M_2\,\ddot{u}_2(sa) &= -\frac{\partial V^{harm}}{\partial u_2(sa)} = -K[2u_2(sa) - u_1(sa) - u_1([s+1]a)]. \end{aligned} \tag{2.33}$$

The solutions of Eq. (2.33) are of the type

$$u_1(sa, t) = \epsilon_1 e^{i(qsa - \omega t)}$$

and

$$u_2(sa, t) = \epsilon_2 e^{i(qsa - \omega t)}. \tag{2.34}$$

Substituting Eq. (2.34) in Eq. (2.33), we obtain

$$-\omega^2 M_1 \epsilon_1 e^{i(qsa-\omega t)} = K(\epsilon_2 - 2\epsilon_1 + \epsilon_2 e^{-iqa})\, e^{i(qsa-\omega t)}$$

and

$$-\omega^2 M_2 \epsilon_2\, e^{i(qsa-\omega t)} = K(\epsilon_1 e^{iqa} - 2\epsilon_2 + \epsilon_1)\, e^{i(qsa-\omega t)}. \tag{2.35}$$

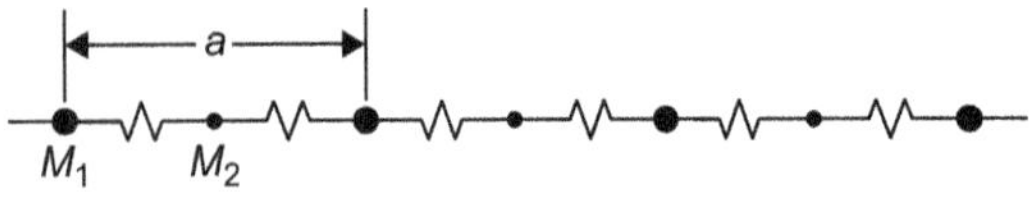

FIGURE 2.6

Diatomic linear chain of masses M_1 and M_2.

We cancel the $e^{i(qsa-\omega t)}$ term from both sides and solve the determinantal equation

$$\begin{vmatrix} 2K - M_1\omega^2 & -K(1+e^{-iqa}) \\ -K(1+e^{iqa}) & 2K - M_2\omega^2 \end{vmatrix} = 0. \tag{2.36}$$

There are two roots of the solution of Eq. (2.36), which yields (Problem 2.4)

$$\omega_{\pm}^2 = K\left(\frac{1}{M_1}+\frac{1}{M_2}\right) \pm K\sqrt{\left[\left(\frac{1}{M_1}+\frac{1}{M_2}\right)^2 - \frac{4\sin^2(qa/2)}{M_1M_2}\right]}. \tag{2.37}$$

The two solutions of Eq. (2.37) are the two branches of the phonon dispersion relation. For small q, the two roots of Eq. (2.37) are

$$\omega_- = \sqrt{\frac{K}{2(M_1+M_2)}}qa \tag{2.38}$$

and

$$\omega_+ = \sqrt{\frac{2K(M_1+M_2)}{M_1M_2}}. \tag{2.39}$$

We also note from Eq. (2.37) that if $qa = \pm\pi$ (the Brillouin zone boundary), the expressions for $\omega_{\pm}$ reduce to

$$\omega_{\pm}^2 = \left(K\left[\frac{1}{M_1}+\frac{1}{M_2}\right] \pm K\left[\frac{1}{M_2}-\frac{1}{M_1}\right]\right). \tag{2.40}$$

Thus, we obtain

$$\omega_+ = \sqrt{\frac{2K}{M_2}} \tag{2.41}$$

and

$$\omega_- = \sqrt{\frac{2K}{M_1}}. \tag{2.42}$$

At the Brillouin zone boundary, $q = \pm\frac{\pi}{a}$. We also note that because $M_1 > M_2, \omega_+ > \omega_-$. Another interesting point to note is that from Eq. (2.34), u_1 and u_2 are periodic with $q = \pm 2\pi/a$. Therefore, the dispersion relation repeats itself for each Brillouin zone. From the previous discussions, we obtain the following results.

The vibrational frequency of a diatomic linear chain of mass M_1 and M_2 is shown in Figure 2.7.

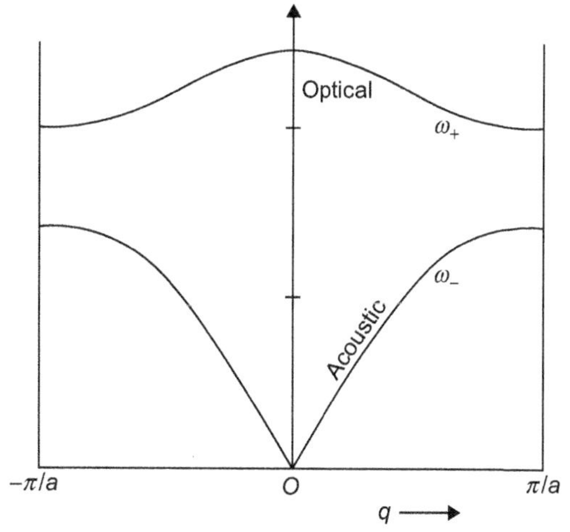

FIGURE 2.7

Optical and acoustic phonon branches of a diatomic linear chain.

The first branch, ω_-, which tends to become zero at $q = 0$, is known as the acoustic mode.

It is in fact the analogue of a long-wavelength vibration of the linear chain, conceived as an elastic continuum. The normal value of the velocity of sound is reproduced in this case just as in the case of a monoatomic lattice. In the acoustic mode, the atoms vibrate in unison, whereas in the optical mode, the atoms in the unit cell vibrate out of phase. Essentially, the two sublattices of the two types of atoms move rigidly in opposition to one another. This can also be stated in the alternate form that the diatomic molecules vibrate as if they are independent of the neighbors. If these crystals are ionic crystals, the two types of atoms are of opposite electric charge. This yields an oscillating dipole moment that is optically active. We also note that if the masses of the two atoms in the linear chain were equal ($M_1 = M_2$), there would be no gap at the zone boundaries.

The configuration of atoms in the acoustic and optical modes in the diatomic linear chain is shown in Figure 2.8. In the acoustic mode, atoms within a unit cell move in concert, whereas in the optical mode they vibrate against each other.

In the previous example, we discussed only a linear chain of atoms. Thus, we obtained dispersion relation for longitudinal acoustic (LA) and longitudinal optical (LO) phonons. However, if we consider two atoms per primitive cell in a three-dimensional lattice, as in the NaCl or diamond structure, for each polarization mode in a direction of propagation, there are two branches in the dispersion relation ω versus q. In addition to LA and LO phonons, of which the dispersion relations are essentially the same as shown in Figure 2.7, one obtains transverse acoustical (TA) and transverse optical (TO) phonons. Figure 2.9 shows the TA and TO modes for a diatomic linear lattice at the same wavelength.

The optical modes of oscillation of individual ions in both transverse and longitudinal modes are shown in Figure 2.10.

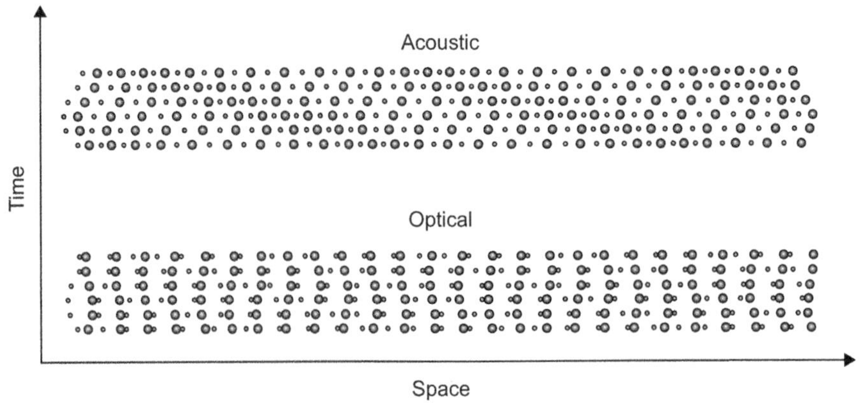

FIGURE 2.8

Configuration of atoms in acoustic and optical modes in the diatomic linear chain.

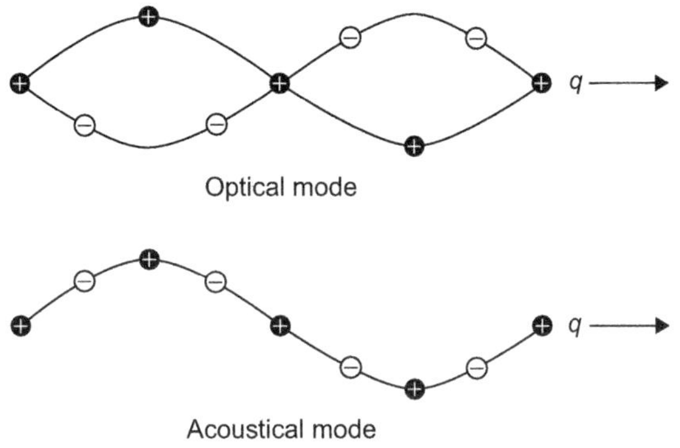

FIGURE 2.9

Transverse optical and acoustical waves for the same wavelength in a diatomic linear lattice.

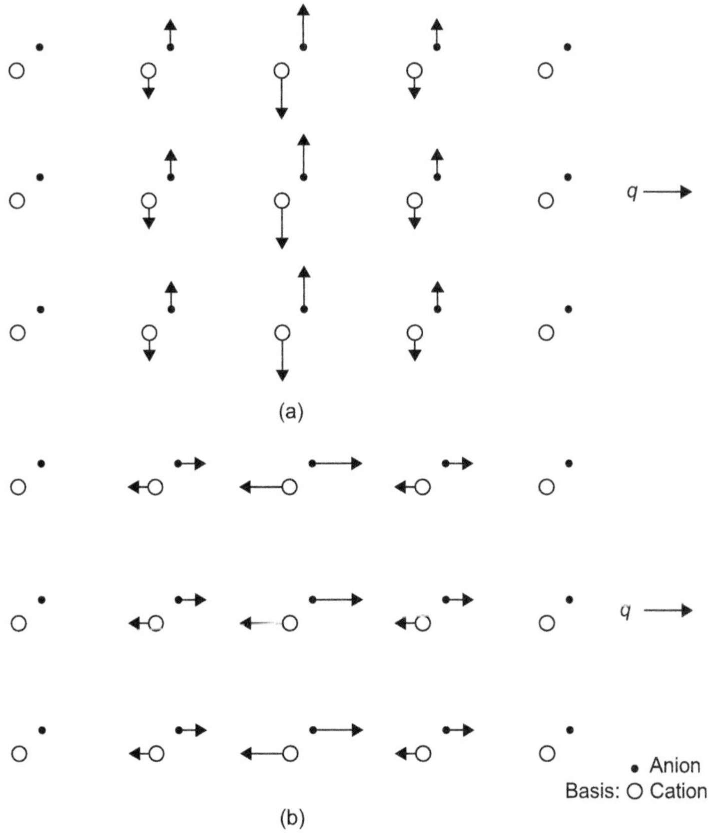

FIGURE 2.10

The direction of displacement of individual ions in (a) transverse optical (TO) and (b) longitudinal optical (LO) modes of oscillation.

2.2 LATTICE SPECIFIC HEAT

2.2.1 Theory

The thermal excitation of the lattice contributes to the specific heat of solids. To calculate the lattice specific heat, one has to use quantum mechanical operators instead of the classical coordinates $\mathbf{u}_{ni}$ and their momenta. In the earlier formulation of the theory of lattice vibrations, we obtained a set of equations of motion that are those of an assembly of independent simple harmonic oscillators. However, in a quantum mechanical treatment, the excitations are of the Bose–Einstein type. We introduced the term *phonons* to describe the normal modes of the crystal in the same manner the term *photons* was introduced to describe the quantum theory of electromagnetic field. In that theory, a normal mode of the radiation field in a cavity is given by $(n+\frac{1}{2})\hbar\omega$, where ω is the angular frequency of the mode. The *photons* are the quanta of radiation field that describes classical light. Similarly, the *phonons* are the quanta of the ions' displacement field that describes classical sound. The nomenclature of normal modes and phonons are equivalent although the latter is much more convenient.

If we want to specify the energy levels of an N-ion harmonic crystal that can be regarded as $3N$ independent oscillators, the contribution to the energy $\varepsilon_{\mathbf{q}s}$ of a particular normal mode with angular frequency $\omega_s(\mathbf{q})$ can have the discrete set of values

$$\varepsilon_{\mathbf{q}s} = \left(n_{\mathbf{q}s} + \frac{1}{2}\right)\hbar\omega_s(\mathbf{q}), \tag{2.43}$$

where there are $n_{\mathbf{q}s}$ phonons of type s with wave vector $\mathbf{q}$ present in the crystal. We note that an equivalent classical description would be that the normal mode of branch s with wave vector $\mathbf{q}$ is in its $n_{\mathbf{q}s}$th excited state. We note that the total energy E is the sum of the energies of the individual normal modes,

$$E = \sum_{\mathbf{q}s}\left(n_{\mathbf{q}s} + \frac{1}{2}\right)\hbar\omega_s(\mathbf{q}). \tag{2.44}$$

When we consider an assembly of independent simple harmonic oscillators, the excitations must be bosons. From the theory of Bose–Einstein statistics, $\bar{n}_{\mathbf{q}s}$, the mean number of bosons with energy $\hbar\omega_s(\mathbf{q})$ in thermal equilibrium at temperature T, when the chemical potential μ is taken to be zero ($\mu = 0$ because the total number of phonons in thermal equilibrium is determined by the temperature and hence is not an independent variable), is given by

$$\bar{n}_{\mathbf{q}s} = \frac{1}{e^{\beta\hbar\omega_s(\mathbf{q})} - 1}, \tag{2.45}$$

where

$$\beta = 1/k_B T. \tag{2.46}$$

The quanta in the $\mathbf{q}$th mode will contribute an energy

$$\bar{\varepsilon}_{\mathbf{q}s} = \left(\bar{n}_{\mathbf{q}s} + \frac{1}{2}\right)\hbar\omega_s(\mathbf{q}), \tag{2.47}$$

which includes the zero-point energy. From Eqs. (2.45) and (2.47), the average total energy of the system (neglecting the zero-point energy) is

$$\bar{\varepsilon} = \sum_{\mathbf{q}s} \frac{\hbar\omega_s(\mathbf{q})}{e^{\beta\hbar\omega_s(\mathbf{q})}-1}. \tag{2.48}$$

Here, the summation is over all polarizations s and different modes $\mathbf{q}$. Thus, the specific heat is given by

$$C_V = \frac{1}{V}\frac{\partial\bar{\varepsilon}}{\partial T} = \frac{1}{V}\sum_{\mathbf{q}s}\frac{\partial}{\partial T}\frac{\hbar\omega_s(\mathbf{q})}{e^{\beta\hbar\omega_s(\mathbf{q})}-1}, \tag{2.49}$$

where V is the volume of the crystal. We will now evaluate the specific heat in the high-temperature and intermediate-temperature. We will evaluate low-temperature limits in Problem 2.5.

1. In the high-temperature limit, $k_BT/\hbar$ is large compared with all the phonon frequencies. Thus, if we write $x = \frac{\hbar\omega}{k_BT} \ll 1$,

$$\frac{1}{e^x-1} \approx \frac{1}{x}\left[1-\frac{x}{2}+\frac{x^2}{12}+\cdots\right]. \tag{2.50}$$

   From Eqs. (2.49) and (2.50), we obtain

$$C_V \simeq \frac{3N}{V}, \tag{2.51}$$

   which is the classical law of Dulong and Petit. The higher-order terms in Eq. (2.50) yield the quantum corrections to the classical Dulong and Petit law.
2. In intermediate-temperature limits, both the Debye and Einstein models of specific heat are obtained by making different approximations. We note that in a large crystal, we can replace the sum in Eq. (2.49) by an integration over the wave vectors $\mathbf{q}$ that satisfy the Born–von Karman boundary conditions. Thus, Eq. (2.49) can be expressed as

$$C_V = \frac{\partial}{\partial T}\sum_s \int \frac{d\mathbf{q}}{(2\pi)^3}\frac{\hbar\omega_s(\mathbf{q})}{e^{\beta\hbar\omega_s(\mathbf{q})}-1}, \tag{2.52}$$

   where the integral is over the first Brillouin zone. We can retrieve both the Debye and the Einstein models of specific heat of solids from Eq. (2.52), although they were originally derived under very different assumptions.

2.2.2 The Debye Model of Specific Heat

The Debye model of specific heat can be obtained by making two basic assumptions. First, all branches of the vibration spectrum are replaced with three branches of the acoustic mode such that the linear dispersion

$$\omega(q) = sq, \tag{2.53}$$

where s is the velocity of sound. Second, the integral over the first Brillouin zone is replaced by an integral over a sphere of the same volume and radius q_D in reciprocal space. It is assumed that the sphere contains exactly N allowed wave vectors just as in the case of the first Brillouin zone. Thus, the radius of the Debye sphere, q_D, is such that

$$N = \frac{V}{8\pi^3}\frac{4}{3}\pi q_D^3, \tag{2.54}$$

and hence

$$q_D = \left(\frac{6\pi^2}{v_c}\right)^{1/3}, \tag{2.55}$$

where v_c is the volume of the Wigner–Seitz cell, which can be written in the alternate form

$$N = \frac{q_D{}^3}{6\pi^2}, \tag{2.56}$$

where n is the density of ions. From Eqs. (2.53) through (2.55), we obtain

$$C_V = \frac{\partial}{\partial T}\frac{3\hbar c}{2\pi^2}\int_0^{q_D}\frac{q^2 dq}{e^{\beta\hbar sq}-1}. \tag{2.57}$$

We define a Debye frequency

$$\omega_D = q_D s \tag{2.58}$$

and a Debye temperature

$$\Theta_D = \hbar\omega_D/k_B. \tag{2.59}$$

If we write

$$x = \hbar sq/k_B T, \tag{2.60}$$

Eq. (2.57) can be written as

$$C_V = 9Nk_B\left(\frac{T}{\Theta_D}\right)^3\int_0^{\Theta_D/T}\frac{x^4 e^x dx}{(x-1)^2}. \tag{2.61}$$

where N is the number of cells in the crystal. At low temperatures, the upper limit of the integral Θ_D/T tends to infinity, in which case the integral

$$\int_0^{\infty}\frac{x^4 e^x dx}{(x-1)^2} = \frac{4\pi^4}{5}. \tag{2.62}$$

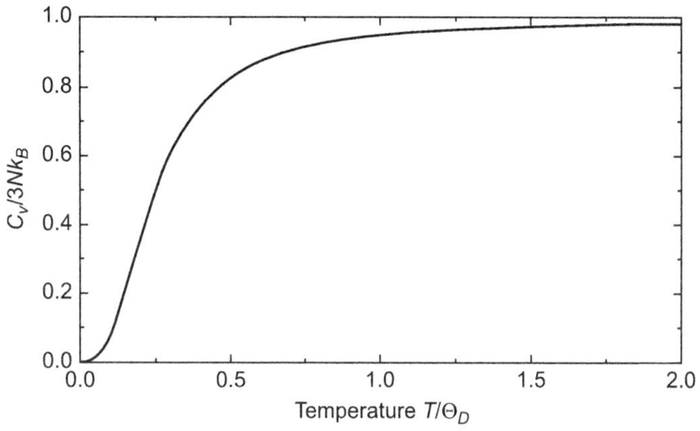

FIGURE 2.11

Specific heat in the Debye approximation.

Thus, we obtain from Eqs. (2.61) and (2.62),

$$C_V \approx \frac{12\pi^4}{5} N k_B \left(\frac{T}{\Theta_D}\right)^3, \tag{2.63}$$

which is the Debye's T^3—*law of lattice specific heats.* The Debye law of specific heat is shown schematically in Figure 2.11.

The Debye's theory of specific heat works very well for solids. However, the Debye temperature, Θ_D, is often interpolated from the observed specific heat, thereby allowing it to depend on the temperature T. In fact, from Eqs. (2.58) and (2.59), Θ_D is related to the velocity of sound, s, by the relation,

$$\Theta_D = \frac{\hbar q_D s}{k_B}, \tag{2.64}$$

and therefore should be calculated directly for a solid. In any case, Θ_D for most solids is listed in tables.

It is interesting to note that at temperatures well above Θ_D, the integrand form in Eq. (2.61) can be replaced by its form for small x and one obtains Dulong and Petit's law. Therefore, the Debye temperature, Θ_D, is a measure of the temperature separating the low-temperature region where quantum statistics is used from the high-temperature region where classical mechanics can be used.

However, Debye's theory of specific heat has its limitations. In most cases, the sharp cutoff frequency $\omega_D = \frac{k_B \Theta_D}{\hbar}$ is not justified. In general, there is a spread with several peaks that correspond to the modes of different polarization, which have different velocities. In addition, there is a peak at high frequencies due to the strong dispersion near the zone boundary. The contrast between the Debye spectrum and an actual lattice spectrum is shown in Figure 2.12.

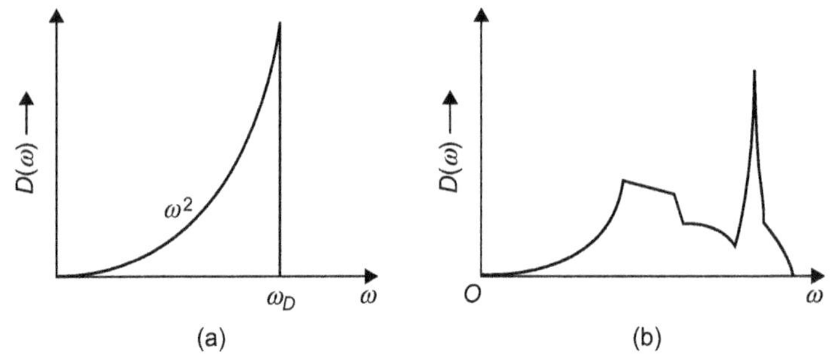

FIGURE 2.12

(a) Debye spectrum; (b) actual lattice spectrum.

2.2.3 The Einstein Model of Specific Heat

As we noted earlier, in a lattice with a basis, there are both acoustic modes and optical modes. The three acoustic modes can be easily treated by the Debye model. The optical modes can be treated by the Einstein model in which each mode has the same frequency, ω_E, which is independent of $\mathbf{q}$. Therefore, from Eq. (2.45), we obtain that each optical branch will contribute to the thermal energy density in the Einstein approximation,

$$\varepsilon = \frac{N\hbar\omega_E}{e^{\beta\hbar\omega_E} - 1}, \tag{2.65}$$

and if there are p such branches,

$$C_V^{optical} = \frac{1}{V}\frac{\partial E}{\partial T} = pNk_B \frac{(\beta\hbar\omega_E)^2 e^{\beta\hbar\omega_E}}{(e^{\beta\hbar\omega_E} - 1)^2}. \tag{2.66}$$

We can also define an *Einstein temperature* Θ_E by

$$\Theta_E = \frac{\hbar\omega_E}{k_B}. \tag{2.67}$$

It should be noted that Einstein was the first to derive the theory of specific heat of solids by using quantum statistics instead of classical statistics. From Eq. (2.66), we note that when $T \gg \Theta_E$, $\beta\hbar\omega_E \equiv \frac{\Theta_E}{T}$ is very small and we can write

$$C_V^{optical} = pNk_B, \tag{2.68}$$

so that each optical mode contributes k_B/V, as required by Dulong and Petit's law. However, when $T \ll \Theta_E$, $\beta\hbar\omega_E$ is very large so that the

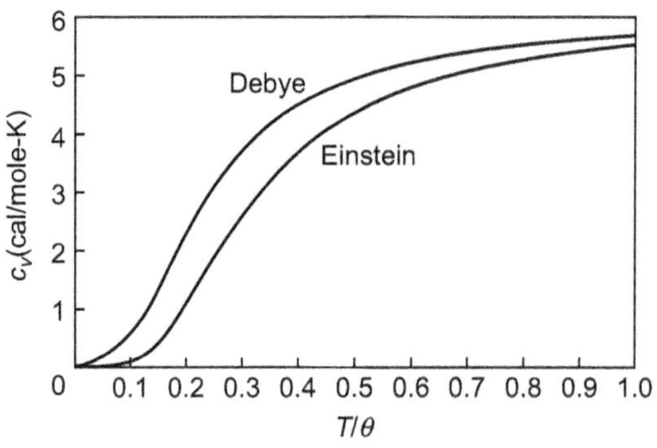

FIGURE 2.13

Comparison of Debye's and Einstein's specific heat as a function of $\frac{T}{\Theta}$.

contribution to C_V drops exponentially. Thus, it is very difficult to excite the optical modes at low temperatures.

A comparison of Debye's and Einstein's expressions for specific heat as a function of $\frac{T}{\Theta}$ is shown in Figure 2.13.

2.3 SECOND QUANTIZATION

2.3.1 Occupation Number Representation

The behavior of systems with a very large number of particles is studied in statistical physics. Suppose we have a system of N noninteracting particles that are found in states with wave functions $\varphi_1, \varphi_2, \ldots, \varphi_N$ which form a complete orthonormal set. For noninteracting particles, the wave function φ_i corresponds to one of the plane wave states

$$\varphi_{\mathbf{k}} = e^{i\mathbf{k}\cdot\mathbf{r}}. \tag{2.69}$$

They obey the orthonormality condition,

$$\int d\mathbf{r}\, \varphi_i^*(\mathbf{r})\varphi_j(\mathbf{r}) = \delta_{i,j}, \tag{2.70}$$

as well as

$$\sum_i \varphi_i^*(\mathbf{r})\varphi_j(\mathbf{r}') = \delta(\mathbf{r}-\mathbf{r}'). \tag{2.71}$$

The system can evidently be described by specifying the number of particles in states $\varphi_1, \varphi_2, \ldots, \varphi_N$. The complete wave function can be specified by using a new representation. We represent the basis function in the many-body wave function as

$$|n_1, n_2, n_3, \ldots, n_k \ldots >, \tag{2.72}$$

with n_k as the number of particles in state φ_k. The total number of particles

$$\sum n_i = N. \tag{2.73}$$

For Fermions (electrons, protons, neutrons, and He^3 atoms), which obey Fermi–Dirac statistics, the restriction is that

$$n_i = 0, \text{or } 1. \tag{2.74}$$

For bosons (photons, phonons, and He_4 atoms), which obey Bose–Einstein statistics, there is *no restriction* for n_i. For photons and phonons,

$$\sum_i n_i = N(T), \tag{2.75}$$

where $N(T)$ depends on the temperature T. Further,

$$N = 0 \text{ at } T = 0$$

and

$$N \neq 0 \text{ at } T \neq 0. \quad (2.76)$$

2.3.2 Creation and Annihilation Operators

(a) Bosons

We will first consider the *creation and annihilation operators* for bosons. We define the operators a_k and $a_k^\dagger$ by

$$a_k \,|n_1, n_2, \ldots, n_k, \ldots > = \sqrt{n_k}\,|n_1, n_2, \ldots, n_k - 1, \ldots > \quad (2.77)$$

and

$$a_k^\dagger \,|n_1 \cdot n_2, \ldots, n_k, \ldots > = \sqrt{n_k + 1}\,|n_1, n_2, \ldots, n_k + 1, \ldots >. \quad (2.78)$$

Here, a_k is the *annihilation operator* because it decreases

$$n_k \rightarrow n_k - 1, \quad (2.79)$$

and $a_k^\dagger$ is the *creation operator* because it increases

$$n_k \rightarrow n_k + 1. \quad (2.80)$$

The vacuum state is defined as

$$|0, 0, \ldots, 0, \ldots >, \quad (2.81)$$

whereas the other occupation numbers remain unchanged. From Eqs. (2.77) and (2.78), we obtain

$$\begin{aligned} a_k^\dagger a_k \,|n_1, n_2, \ldots, n_k, \ldots > &= \sqrt{n_k}\; a_k^\dagger \,|n_1, n_2, \ldots, n_k - 1, \ldots > \\ &= n_k \,|n_1, n_2, \ldots, n_k, \ldots >. \end{aligned} \quad (2.82)$$

From Eq. (2.82), we obtain that $a_k^\dagger a_k \Rightarrow n_k$ is the number operator in state **k**. Similarly, we obtain

$$\begin{aligned} a_k a_k^\dagger \,|n_1, n_2, \ldots, n_k, \ldots > &= \sqrt{n_k + 1}\, a_k \,|n_1, n_2, \ldots, n_k + 1, \ldots > \\ &= (n_k + 1)\,|n_1, n_2, \ldots, n_k, \ldots >. \end{aligned} \quad (2.83)$$

Subtracting Eq. (2.82) from Eq. (2.83), we obtain

$$(a_k a_k^\dagger - a_k^\dagger a_k)\ |n_1, n_2, ..., n_k, ... > = \ |n_1, n_2, ..., n_k, ... >. \tag{2.84}$$

Eq. (2.84) implies that

$$a_k a_k^\dagger - a_k^\dagger a_k = 1, \tag{2.85}$$

which can be written in the alternate form

$$[a_k, a_k^\dagger]_- = 1, \tag{2.86}$$

where $[A, B]_-$ is the commutator

$$[A, B]_- = AB - BA. \tag{2.87}$$

If $k \neq k'$,

$$\begin{aligned} a_{k'}^\dagger a_k\,|n_1, n_2, ..., n_{k'}, ..., n_k, ... > &= \sqrt{n_k}\ a_{k'}^\dagger\ |n_1, n_2, ..., n_{k'}, ..., n_{k-1}, ... > \\ &= \sqrt{n_k}\ \sqrt{n_{k'}+1}\,|n_1, n_2, ..., n_{k'+1}, ...,\ n_{k-1} ... >. \end{aligned} \tag{2.88}$$

$$\begin{aligned} a_k a_{k'}^\dagger\ |n_1, n_2, ..., n_{k'}, ..., n_k, ... > &= a_k\sqrt{n_{k'}+1}\,|n_1, n_2, ..., n_{k'+1}, ..., n_k ... > \\ &= \sqrt{n_k}\sqrt{n_{k'}+1}\,|n_1 \cdot n_2, ..., n_{k'+1}, ..., n_k ... >. \end{aligned} \tag{2.89}$$

Subtracting Eq. (2.89) from Eq. (2.88), we obtain

$$(a_{k'}^\dagger a_k - a_k a_{k'}^\dagger)|n_1, n_2, ..., n_{k'+1}, ..., n_k, ... > = 0. \tag{2.90}$$

From Eqs. (2.86) and (2.90), we obtain the commutation relation between the creation and annihilation operators of bosons,

$$[a_k, a_{k'}^\dagger]_- \ = \ \delta_{k,k'}, \tag{2.91}$$

where $\delta_{k,k'}$ is the Kronecker delta function, i.e.,

$$\delta_{k,k'} = 1 \text{ if } k = k', \text{ and } \delta_{k,k'} = 0 \text{ if } k \neq k'. \tag{2.92}$$

Similarly, we can obtain the other commutation relations for bosons,

$$\begin{aligned} &[a_k, a_{k'}]_- = 0, \\ \text{and} \quad &[a_k^\dagger, a_{k'}^\dagger]_- = 0. \end{aligned} \tag{2.93}$$

Eqs. (2.91) and (2.93) are the three commutation relations for bosons. The total number of bosons like He^4 atoms,

$$\sum_k <n_k> = N, \tag{2.94}$$

where $<n_k>$ is the thermally average occupation number representation in state **k** at temperature T or, equivalently, the number of particles in state **k** at temperature T. According to the Bose–Einstein distribution function,

$$<n_k> = \frac{1}{e^{\beta(\varepsilon_k - \mu)} - 1}, \tag{2.95}$$

where

$$\varepsilon_k = \hbar^2 k^2 / 2m, \tag{2.96}$$

and μ = chemical potential. For He^4 atoms, $\mu \neq 0$ at high temperatures. However, for photons and phonons,

$$\varepsilon_k = sk \text{ and } \mu \equiv 0. \tag{2.97}$$

In Eq. (2.97),

$$s = c = \text{speed of light for photons,}$$

and

$$s = \text{speed of sound for phonons.}$$

(b) Fermions

The creation and destruction operators for Fermions obey very different commutation relations because the total number of particles in a level n_i can be 0 or 1.

The basis function in the many-body representation is

$$|n_1, n_2, \ldots, n_i, \ldots>, \tag{2.98}$$

with the constraint, $n_i = 0$, or $n_i = 1$. The Fermion operators a_i and $a_i^\dagger$ are defined as

$$a_i\,|n_1, n_2, \ldots, n_i, \ldots> = \sqrt{n_i}\,(-1)^m |n_1, n_2, \ldots, n_i - 1, \ldots>, \tag{2.99}$$

where

$$m = \sum_{j<i} n_j \tag{2.100}$$

and

$$a_i^\dagger\,|n_1, n_2, \ldots, n_i, \ldots> = \sqrt{1 - n_i}\,(-1)^m |n_1, n_2, \ldots, n_i + 1, \ldots>. \tag{2.101}$$

Here, the function $(-1)^m$ comes from the requirement that the wave function is antisymmetric for Fermions. From Eqs. (2.99) and (2.101), we obtain

$$\begin{aligned} a_i^\dagger a_i \,|n_1, n_2, \ldots, n_i, \ldots > &= \sqrt{n_i}(-1)^m a_i^\dagger \,|n_1, n_2, \ldots, n_i - 1, \ldots > \\ &= \sqrt{n_i}\,\sqrt{2-n_i}\,(-1)^{2m} |n_1, n_2, \ldots, n_i \ldots >. \end{aligned} \tag{2.102}$$

Similarly,

$$\begin{aligned} a_i a_i^\dagger \,|n_1, n_2, \ldots, n_i \ldots > &= \sqrt{1-n_i}\,(-1)^m a_i |n_1, n_2, \ldots, n_i + 1, \ldots > \\ &= \sqrt{1-n_i}\,\sqrt{1+n_i}\,(-1)^{2m} |n_1, n_2, \ldots, n_i, \ldots >. \end{aligned} \tag{2.103}$$

If $n_i = 0$, we obtain from Eqs. (2.102) and (2.103),

$$(a_i^\dagger a_i + a_i a_i^\dagger)\,|n_1, n_2, \ldots, n_i, \ldots > = |n_1, n_2, \ldots, n_i, \ldots >. \tag{2.104}$$

Similarly, if $n_i = 1$, we obtain from Eqs. (2.102) and (2.103),

$$(a_i^\dagger a_i + a_i a_i^\dagger)\,|n_1, n_2, \ldots, n_i, \ldots > = |n_1, n_2, \ldots, n_i, \ldots >. \tag{2.105}$$

From Eqs. (2.104) and (2.105), we obtain

$$a_i^\dagger a_i + a_i a_i^\dagger = 1, \tag{2.106}$$

which can be written in the alternate form

$$[a_i, a_i^\dagger]_+ = 1. \tag{2.107}$$

In general, we can show that

$$[a_i, a_j^j]_+ = 0, \quad i \neq j,$$

and

$$[a_i, a_j]_+ = [a_i^\dagger, a_j^\dagger]_+ = 0. \tag{2.108}$$

From Eqs. (2.107) and (2.108), the commutation relations for Fermions can be written as

$$[a_i, a_j^\dagger]_+ = \delta_{ij},$$

and

$$[a_i, a_j]_+ = [a_i^\dagger, a_j^\dagger]_+ = 0. \tag{2.109}$$

We also obtain from Eq. (2.102),

$$a_i^\dagger a_i = \sqrt{n_i}\,\sqrt{2-n_i} = \bar{n}_i, \tag{2.110}$$

because if

$$n_i = 0, \quad \bar{n}_i = 0$$

and if
$$n_i = 1, \quad \bar{n}_i = 1. \tag{2.111}$$

Thus, $a_i^\dagger a_i = \bar{n}_i = n_i$ is the number operator for state i. From the Fermi–Dirac distribution function, we obtain the expression for the average number of particles in state k at temperature T,

$$<a_k^\dagger a_k> = <n_k> = \frac{1}{e^{\beta(\varepsilon_k - \mu)} + 1}. \tag{2.112}$$

2.3.3 Field Operators and the Hamiltonian

Here, we introduce the operators of a *field of particles*:

$$\psi(\xi) = \sum_i \varphi_i(\xi)\, a_i \tag{2.113}$$

and

$$\psi^\dagger(\xi) = \sum_i \varphi_i^*(\xi) a_i^\dagger. \tag{2.114}$$

Here, a_i and $a_i^\dagger$ are the second quantization operators, and $\varphi_i(\xi)$ is the particle in the state i. If a_i and $a_i^\dagger$ are operators for bosons,

$$\psi(\xi)\psi^\dagger(\xi') - \psi^\dagger(\xi')\psi(\xi) = \sum_{i,j} \varphi_i(\xi)\varphi_j^*(\xi')[a_i a_j^\dagger - a_j^\dagger a_i]. \tag{2.115}$$

From Eqs. (2.91) and (2.115), we obtain

$$\psi(\xi)\psi^\dagger(\xi') - \psi^\dagger(\xi')\psi(\xi) = \sum_i \varphi_i(\xi)\varphi_i^*(\xi). \tag{2.116}$$

We note that because $\varphi_i(\xi) \Rightarrow \varphi_1(\xi), \varphi_2(\xi), \varphi_3(\xi), \ldots,$ forms a complete orthonormal set, it satisfies

$$\sum_i \varphi_i(\xi)\varphi_i^*(\xi) = \delta(\xi - \xi'). \tag{2.117}$$

From Eqs. (2.116) and (2.117), we obtain

$$[\psi(\xi), \psi^\dagger(\xi')]_- = \delta(\xi - \xi') \tag{2.118}$$

By using a similar procedure, we can easily derive

$$[\psi(\xi), \psi(\xi')]_- = [\psi^\dagger(\xi), \psi^\dagger(\xi')]_- = 0. \tag{2.119}$$

If a_i and $a_i^\dagger$ are Fermion operators, we can derive, by using a similar procedure,

$$[\psi(\xi), \psi^\dagger(\xi')]_+ = \delta(\xi - \xi'), \tag{2.120}$$

and

$$[\psi(\xi), \psi(\xi')]_+ = [\psi^\dagger(\xi), \psi^\dagger(\xi')]_+ = 0. \tag{2.121}$$

We will now consider the Hamiltonian that describes a system of interacting particles (for example, the electrons in a solid):

$$H = \sum_i \frac{1}{2m} p_i{}^2 + \sum_i V(\mathbf{r}_i) + \frac{1}{2} \sum_{i \neq j} V^{(2)}(\mathbf{r}_i - \mathbf{r}_j). \tag{2.122}$$

Here, $p_j = -i\hbar \nabla_j = -i\hbar \frac{\partial}{\partial \mathbf{r}_j}$ is the momentum operator for the jth particle, $V(\mathbf{r}_i)$ is the periodic potential, and $V^{(2)}(\mathbf{r}_i - \mathbf{r}_j)$ is the interaction energy between the ith particle and the jth particle. We will denote $p_i{}^2/2m$ and $V(\mathbf{r}_i)$ as single-particle operators and $V^{(2)}(\mathbf{r}_i - \mathbf{r}_j)$ as two-particle operators.

The Hamiltonian in Eq. (2.122) can be expressed in terms of the second quantization operators

$$\begin{aligned} H = \sum_\alpha \int \left[-\frac{\hbar^2}{2m} \psi_\alpha^\dagger(\mathbf{r}) \nabla^2 \psi_\alpha(\mathbf{r}) + V(\mathbf{r}) \psi_\alpha^\dagger(\mathbf{r}) \psi_\alpha(\mathbf{r}) \right] d\mathbf{r} \\ + \frac{1}{2} \sum_{\alpha\beta} \iint \psi_\alpha^\dagger(\mathbf{r}) \psi_\beta^\dagger(\mathbf{r}') V^{(2)}(\mathbf{r}-\mathbf{r}') \psi_\beta(\mathbf{r}') \psi_\alpha(\mathbf{r}) \, d\mathbf{r} \, d\mathbf{r}'. \end{aligned} \tag{2.123}$$

In Eq. (2.123), we have written

$$\psi(\xi) = \psi_\alpha(\mathbf{r}),$$

where α is the spin index and $\mathbf{r}$ is the coordinate.

The expression for the Hamiltonian given by Eq. (2.123) is equivalent to the corresponding expression in Eq. (2.122). We also note that the total number of particles

$$N = \langle N \rangle = \sum_k n_k = \sum_k \frac{1}{e^{\beta(\varepsilon_k - \mu)} + 1}. \tag{2.124}$$

The chemical potential μ is determined by the condition that when

$$T = 0, \quad \mu = \varepsilon_F = \text{the Fermi energy}. \tag{2.125}$$

We consider a system of free particles moving in space. The single-particle wave function can be written as

$$\varphi_{\mathbf{k}}(\mathbf{r}) = \frac{1}{\sqrt{V}} e^{i\mathbf{k} \cdot \mathbf{r}}. \tag{2.126}$$

This leads to the free-particle Schrodinger equation,

$$\frac{-\hbar^2}{2m} \nabla^2 \varphi_{\mathbf{k}}(\mathbf{r}) = \varepsilon_{\mathbf{k}} \varphi_{\mathbf{k}}(\mathbf{r}) = \frac{\hbar^2 k^2}{2m} \varphi_{\mathbf{k}}(\mathbf{r}). \tag{2.127}$$

Here, $\mathbf{k}$ is the momentum of the particle, and its kinetic energy is $\frac{\hbar^2 k^2}{2m}$. For different $\mathbf{k}$ values, we have different states. Thus, the field operators become

$$\psi(\mathbf{r}) = \frac{1}{\sqrt{V}} \sum_{\mathbf{k}} a_{\mathbf{k}}\, e^{i\mathbf{k}\cdot\mathbf{r}} \tag{2.128}$$

and

$$\psi^{\dagger}(\mathbf{r}) = \frac{1}{\sqrt{V}} \sum a_{\mathbf{k}}^{\dagger}\, e^{-i\mathbf{k}\cdot\mathbf{r}}. \tag{2.129}$$

In Eqs. (2.128) and (2.129), the spin indices are neglected for the present. Substituting Eqs. (2.128) and (2.129) in Eq. (2.123), and letting $V = 1$, we obtain

$$H = \sum_{\mathbf{k}} \frac{\hbar^2 k^2}{2m} a_{\mathbf{k}}^{\dagger} a_{\mathbf{k}} + \sum_{\mathbf{kq}} V(\mathbf{q})\, a_{\mathbf{k}+\mathbf{q}}^{\dagger} a_{\mathbf{k}} + \frac{1}{2} \sum_{\mathbf{kpq}} V^{(2)}(\mathbf{q})\, a_{\mathbf{k}-\mathbf{q}}^{\dagger}\, a_{\mathbf{p}+\mathbf{q}}^{\dagger} a_{\mathbf{p}}\, a_{\mathbf{q}}. \tag{2.130}$$

We note that if $V(\mathbf{q}) = V^{(2)}(\mathbf{q}) = 0$, the Hamiltonian is diagonalized and Eq. (2.130) reduces to

$$H = \sum_{\mathbf{k}} \varepsilon_{\mathbf{k}}\, a_{\mathbf{k}}^{\dagger}\, a_{\mathbf{k}} = \sum_{\mathbf{k}} \varepsilon_{\mathbf{k}} n_{\mathbf{k}}. \tag{2.131}$$

Eq. (2.131) reduces to the previous result

$$a_{\mathbf{k}}^{\dagger} a_{\mathbf{k}} = n_{\mathbf{k}}. \tag{2.132}$$

If we define $V(\mathbf{q})$ as the Fourier transform of $V(\mathbf{r})$,

$$V(\mathbf{q}) = \int V(\mathbf{r}) e^{i\mathbf{q}\cdot\mathbf{r}}\, d\mathbf{r}, \tag{2.133}$$

and the interaction $V(\mathbf{q})\, a_{\mathbf{k}+\mathbf{q}}^{\dagger}\, a_{\mathbf{k}}$ can be represented by Figure 2.14.

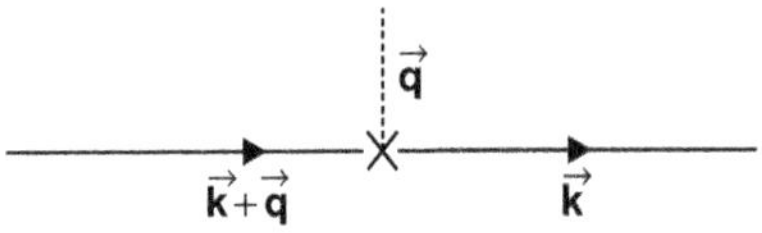

FIGURE 2.14

$V(\mathbf{q})\, a_{\mathbf{k}+\mathbf{q}}^{\dagger} a_{\mathbf{k}}$.

Similarly, we define the Fourier transform of $V^{(2)}(\mathbf{r})$ as

$$V^{(2)}(\mathbf{q}) = \int V^{(2)}(\mathbf{r})\, e^{i\mathbf{q}\cdot\mathbf{r}}\, d\mathbf{r}. \tag{2.134}$$

For example, if

$$V^{(2)}(\mathbf{r}) = \frac{e^2}{r}, \tag{2.135}$$

$$V^{(2)}(\mathbf{q}) = \frac{4\pi e^2}{q^2}. \tag{2.136}$$

The interaction $V^{(2)}(\mathbf{q})\, a_{\mathbf{k}+\mathbf{q}}^{\dagger}\, a_{\mathbf{p}-\mathbf{q}}^{\dagger}\, a_{\mathbf{p}}\, a_{\mathbf{k}}$ can be represented by Figure 2.15.

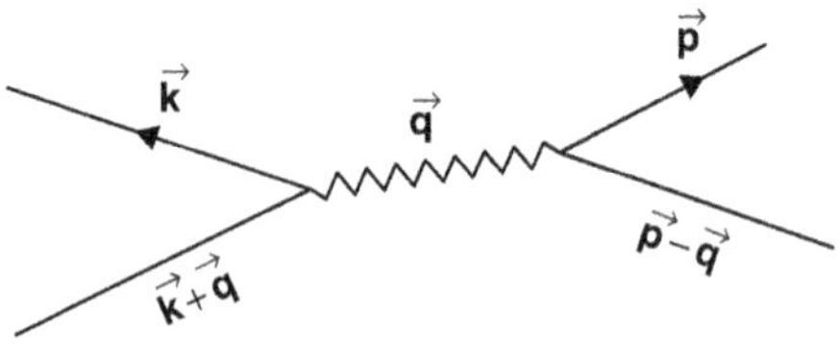

FIGURE 2.15

$V^{(2)}(\mathbf{q})\, a_{\mathbf{k}+\mathbf{q}}^{\dagger} a_{\mathbf{p}-\mathbf{q}}^{\dagger} a_{\mathbf{p}} a_{\mathbf{k}}$.

2.4 QUANTIZATION OF LATTICE WAVES

2.4.1 Formulation

We consider a cubic lattice of lattice constant a. We assume that at each lattice site, we have atoms with mass M vibrating around this site. The Hamiltonian that describes this system of atoms is

$$H = \sum_i \frac{p_i^2}{2M} + \sum_{i \neq j} \frac{1}{2} V(\mathbf{d}_i - \mathbf{d}_j), \tag{2.137}$$

where $V(\mathbf{d}_i - \mathbf{d}_j)$ is the interaction potential energy between atoms at the sites $\mathbf{d}_i$ and $\mathbf{d}_j$. Here, $\mathbf{d}_i = \mathbf{R}_i + \mathbf{u}_i$ and $\mathbf{d}_j = \mathbf{R}_j + \mathbf{u}_j$, where $\mathbf{u}_i$ and $\mathbf{u}_j$ are the displacements of the atoms from the lattice sites located at $\mathbf{R}_i$ and $\mathbf{R}_j$. Figure 2.16 shows these displacements, for convenience, on the surface of a cubic lattice.

For small $\mathbf{u}_i$ and $\mathbf{u}_j$($\mathbf{u}_i$ and $\mathbf{u}_j \ll a$), we have

$$\sum_{i \neq j} V(\mathbf{d}_i - \mathbf{d}_j) \approx \sum_{i \neq j} V(\mathbf{R}_i - \mathbf{R}_j) + \sum_{i \neq j} \mathbf{u}_i \cdot \frac{\partial^2 V}{\partial \mathbf{R}_i \partial \mathbf{R}_j} \cdot \mathbf{u}_j. \tag{2.138}$$

The first term on the right side of Eq. (2.138) is a constant term and can be ignored. If we define the tensor

$$\overleftrightarrow{\mathbf{G}} \equiv \frac{\partial^2 V}{\partial \mathbf{R}_i \partial \mathbf{R}_j}, \tag{2.139}$$

Eq. (2.138) can be written in the alternate form

$$\sum_{i \neq j} V(\mathbf{d}_i - \mathbf{d}_j) = \sum_{i \neq j} \frac{1}{2} \mathbf{u}_i \cdot \overleftrightarrow{\mathbf{G}} \cdot \mathbf{u}_j. \tag{2.140}$$

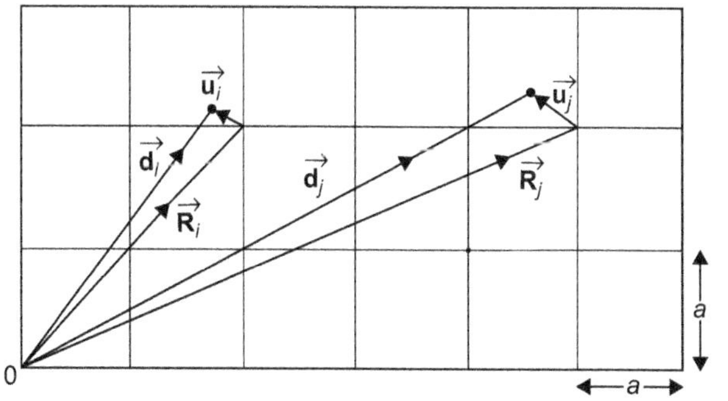

FIGURE 2.16

Definitions of $\mathbf{d}_i$ and $\mathbf{u}_i$ in a cubic lattice of lattice constant a.

Thus, the Hamiltonian can be written as

$$H = \sum_i \frac{P_i^2}{2M} + \sum_{i \neq j} \frac{1}{2} \mathbf{u}_i \cdot \overleftrightarrow{\mathbf{G}} \cdot \mathbf{u}_j = T + V. \tag{2.141}$$

In Eq. (2.141),

$$\mathbf{P}_i = M\dot{\mathbf{u}}_i\,. \tag{2.142}$$

From Lagrange's equation of motion,

$$L = T - V \tag{2.143}$$

and

$$\frac{d}{dt}\frac{\partial L}{\partial \dot{\mathbf{u}}_l} - \frac{\partial L}{\partial \mathbf{u}_l} = 0, \tag{2.144}$$

we obtain from Eq. (2.141),

$$\mathrm{M}\ddot{\mathbf{u}}_l = -\sum_{l'} \overleftrightarrow{\mathbf{G}}_{ll'} \cdot \dot{\mathbf{u}}_{l'}. \tag{2.145}$$

We make a Fourier transformation,

$$\mathbf{u}_l = \sum_q \mathbf{U}_q\, e^{i\mathbf{q}\cdot\mathbf{l}}, \tag{2.146}$$

and obtain from Eqs. (2.145) and (2.146),

$$M\sum_q \ddot{\mathbf{U}}_l\, e^{i\mathbf{q}\cdot\mathbf{l}} = -\sum_{l'} \overleftrightarrow{\mathbf{G}}_{ll'} \cdot \sum_q \mathbf{U}_q e^{i\mathbf{q}\cdot\mathbf{l}'}, \tag{2.147}$$

which can be written in the alternate form

$$\sum_q \left[M\,\ddot{\mathbf{U}}_q\, e^{i\mathbf{q}\cdot\mathbf{l}} + \sum_{l'} \overleftrightarrow{\mathbf{G}}_{ll'} \cdot \mathbf{U}_q e^{i\mathbf{q}\cdot\mathbf{l}'} \right] = 0. \tag{2.148}$$

Eq. (2.148) implies

$$M\,\ddot{\mathbf{U}}_q + \sum_{l'} \overleftrightarrow{\mathbf{G}}_{ll'} \cdot \mathbf{U}_q e^{i\mathbf{q}\cdot(\mathbf{l}'-\mathbf{l})} = 0. \tag{2.149}$$

If we write

$$\mathbf{U}_q(t) = \mathbf{U}_q e^{i\omega_q t}, \tag{2.150}$$

Eq. (2.149) can be written as

$$M\omega_q^2 \hat{\epsilon}_q = \sum_{l'} \overleftrightarrow{\mathbf{G}}_{ll'} \cdot \hat{\epsilon}_q\, e^{i\mathbf{q}\cdot(\mathbf{l}-\mathbf{l}')}. \tag{2.151}$$

Here,

$$\hat{\epsilon}_q = \frac{\mathbf{U}_q}{|\mathbf{U}_q|} = \text{unit vector along } \mathbf{U}_q = \text{polarization vector.} \tag{2.152}$$

$\hat{\epsilon}_q$ could be along $\lambda = x, y, z$ directions. Thus,

$$\hat{\epsilon}_{q\lambda} \Rightarrow \omega_{q\lambda}. \tag{2.153}$$

Thus, Eq. (2.151) can be written in the alternate form

$$M\omega_{q\lambda}^2 \hat{\epsilon}_{q\lambda} = \sum_{l'} \overleftrightarrow{\mathbf{G}}_{ll'} \cdot \hat{\epsilon}_{q\lambda} e^{i\mathbf{q}\cdot(\mathbf{l}-\mathbf{l}')} = \overleftrightarrow{\mathbf{G}}(q) \cdot \hat{\epsilon}_{q\lambda}. \tag{2.154}$$

From Eqs. (2.141) and (2.154), we can write the Hamiltonian in the alternate form

$$H = \sum_l \frac{P_l^2}{2m} + \frac{1}{2} \sum_{l \neq l'} \mathbf{u}_l \cdot \overleftrightarrow{\mathbf{G}}_{ll'} \cdot \mathbf{u}_{l'}. \tag{2.155}$$

We can express

$$\mathbf{P}_l = \sqrt{\frac{M}{N}} \sum_{k\lambda} \hat{\epsilon}_{k\lambda} P_{k\lambda} e^{-i\mathbf{k}\cdot\mathbf{l}}, \tag{2.156}$$

where

$$\hat{\epsilon}_{kx} \perp \hat{\epsilon}_{ky} \perp \hat{\epsilon}_{kz}.$$

We also write

$$\mathbf{u}_\mathbf{l} = \sum_k \mathbf{U}_k e^{i\mathbf{k}\cdot\mathbf{l}}. \tag{2.157}$$

Here,

$$\mathbf{U}_k = \frac{1}{\sqrt{MN}} \hat{\epsilon}_{k\lambda} Q_{k\lambda}, \tag{2.158}$$

where N is the number of lattice sites (or atoms). From Eqs. (2.157) and (2.158), we obtain

$$\mathbf{u}_l = \frac{1}{\sqrt{MN}} \sum_k \hat{\epsilon}_{k\lambda} Q_{k\lambda} e^{i\mathbf{k}\cdot\mathbf{l}}. \tag{2.159}$$

From Eq. (2.156), we obtain

$$\sum_l \frac{P_l^2}{2M} = \frac{1}{2N} \sum_{k\lambda, k'\lambda',} \hat{\epsilon}_{k'\lambda'} \cdot \hat{\epsilon}_{k\lambda} P_{k'\lambda'} P_{k\lambda} \sum_l e^{-(\mathbf{k}+\mathbf{k}')\cdot\mathbf{l}}. \tag{2.160}$$

Using the relation

$$\sum_l e^{-i(\mathbf{k}+\mathbf{k}')\cdot\mathbf{l}} = N\delta_{\mathbf{k},-\mathbf{k}'}, \tag{2.161}$$

we can write Eq. (2.160) in the alternate form

$$\sum_l \frac{P_l^2}{2M} = \frac{1}{2} \sum_{k,\lambda,\lambda'} \hat{\epsilon}_{-k\lambda'} \cdot \hat{\epsilon}_{k\lambda} P_{-k\lambda'} P_{k\lambda}. \tag{2.162}$$

Because

$$\omega_{k\lambda} = \omega_{-k\lambda}, \tag{2.163}$$

$$\hat{\epsilon}_{k\lambda'} = \hat{\epsilon}_{-k\lambda'}. \tag{2.164}$$

Further,

$$\hat{\epsilon}_{k\lambda'} \cdot \hat{\epsilon}_{k\lambda} = \delta_{\lambda\lambda'}, \tag{2.165}$$

and we can substitute

$$P_{-k\lambda} = P_{k\lambda}^{\dagger}.$$

Substituting Eq. (2.164) in Eq. (2.162), we obtain

$$\sum_l \frac{P_l^2}{2M} = \frac{1}{2} \sum_{k\lambda} P_{k\lambda}^{\dagger} P_{k\lambda}. \tag{2.166}$$

We now consider the term

$$\frac{1}{2} \sum_{ll'} \mathbf{u}_l \cdot \overleftrightarrow{\mathbf{G}}_{ll'} \cdot \mathbf{u}_{l'} = \frac{1}{2MN} \sum_{k\lambda,\kappa'\lambda',ll'} \hat{\epsilon}_{k\lambda} Q_{k\lambda} \cdot \overleftrightarrow{\mathbf{G}}_{ll'} \cdot \hat{\epsilon}_{k'\lambda'} Q_{k'\lambda'} e^{i\mathbf{k}\cdot\mathbf{l}+i\mathbf{k}'\cdot\mathbf{l}'}. \tag{2.167}$$

We can write

$$e^{i\mathbf{k}\cdot\mathbf{l}+i\mathbf{k}'\cdot\mathbf{l}'} = e^{i\mathbf{k}'\cdot(\mathbf{l}'-\mathbf{l})} e^{i(\mathbf{k}+\mathbf{k}')\cdot\mathbf{l}}. \tag{2.168}$$

Eq. (2.154) can be written in the alternate form (by changing q to k', and λ to λ')

$$\mathrm{M}\omega_{k'\lambda'}^2 \hat{\epsilon}_{k'\lambda'} = \sum_{l'} \mathbf{G}_{ll'} \cdot \hat{\epsilon}_{k'\lambda'} e^{i\mathbf{k}'\cdot(\mathbf{l}'-\mathbf{l})}. \tag{2.169}$$

Further,

$$\sum_l e^{i(\mathbf{k}+\mathbf{k}')\cdot\mathbf{l}} = N\delta_{\mathbf{k},-\mathbf{k}'}. \tag{2.170}$$

From Eqs. (2.167) through (2.170), we obtain

$$\frac{1}{2} \sum_{ll'} \mathbf{u}_l \cdot \overleftrightarrow{\mathbf{G}}_{ll'} \cdot \mathbf{u}_{l'} = \frac{1}{2} \sum_{k\lambda,k'\lambda'} \hat{\epsilon}_{k\lambda} \cdot \hat{\epsilon}_{k'\lambda'} Q_{k\lambda} Q_{k'\lambda'} \delta_{\mathbf{k},-\mathbf{k}'} \omega_{k'\lambda'}^2, \tag{2.171}$$

$$= \frac{1}{2} \sum_{k,\lambda,\lambda'} \hat{\epsilon}_{k\lambda} \cdot \hat{\epsilon}_{-k\lambda'} Q_{k\lambda} Q_{-k\lambda'} \omega_{-k\lambda'}^2. \tag{2.172}$$

We use the identities

$$\hat{\epsilon}_{-k\lambda'} = \hat{\epsilon}_{k\lambda'}$$
$$\hat{\epsilon}_{k\lambda} \cdot \hat{\epsilon}_{k\lambda'} = \delta_{\lambda\lambda'} \tag{2.173}$$
$$\omega_{-k\lambda} = \omega_{k\lambda},$$

in Eq. (2.172), and obtain

$$\frac{1}{2}\sum_{ll'} \mathbf{u}_l \cdot \overleftrightarrow{\mathbf{G}}_{ll'} \cdot \mathbf{u}_{l'} = \frac{1}{2}\sum_{k\lambda} \omega_{k\lambda}^2 Q_{k\lambda} Q_{-k\lambda} = \frac{1}{2}\sum_{k\lambda} \omega_{k\lambda}^2 Q_{k\lambda}^{\dagger} Q_{k\lambda}. \tag{2.174}$$

From Eqs. (2.155), (2.166), and (2.174), we obtain

$$H = \frac{1}{2}\sum_{k\lambda} P_{k\lambda}^{\dagger} P_{k\lambda} + \frac{1}{2}\sum_{k\lambda} \omega_{k\lambda}^2 Q_{k\lambda}^{\dagger} Q_{k\lambda}. \tag{2.175}$$

In Eq. (2.175), we have expressed the Hamiltonian in normal coordinates $P_{k\lambda}$ and $Q_{k\lambda}$.

2.4.2 Quantization of Lattice Waves

The quantization of lattice waves starts from the fundamental ideas of quantum mechanics. If we define

$$[A, B] = AB - BA, \tag{2.176}$$

$$\mathbf{P}_l = (P_l^x, P_l^y, P_l^z), \tag{2.177}$$

and

$$\mathbf{u}_l = (u_l^x, u_l^y, u_l^z), \tag{2.178}$$

we have the commutation relations

$$[P_l^\alpha, u_{l'}^\beta] = \frac{\hbar}{i}\delta_{ll'}\delta_{\alpha\beta}, \quad \alpha, \beta = x, y, z \tag{2.179}$$

$$[P_l^\alpha, P_{l'}^\beta] = [u_l^\alpha, u_{l'}^\beta] = 0. \tag{2.180}$$

Introducing the creation and annihilation operators, $a_{k\lambda}^{\dagger}$ and $a_{k\lambda}$, and substituting

$$P_{k\lambda} = \left(\frac{\hbar\omega_{k\lambda}}{2}\right)^{\frac{1}{2}} i(a_{k\lambda}^{\dagger} - a_{-k\lambda}) \tag{2.181}$$

and

$$Q_{k\lambda} = \left(\frac{\hbar}{2\omega_{k\lambda}}\right)^{\frac{1}{2}} (a_{k\lambda} + a_{-k\lambda}^{\dagger}), \tag{2.182}$$

and using the commutation relations from Problem 2.9, we obtain

$$\left[a_{k\lambda}, a^{\dagger}_{k'\lambda'}\right] = \delta_{kk'}\delta_{\lambda\lambda'}. \tag{2.183}$$

and

$$\left[a_{k\lambda}, a_{k'\lambda'}\right] = \left[a^{\dagger}_{k\lambda}, a^{\dagger}_{k'\lambda'}\right] = 0. \tag{2.184}$$

These operators satisfy the commutation relations for bosons. Substituting Eqs. (2.181) and (2.182) in Eq. (2.175), we obtain (Problem 2.10)

$$H = \sum_{k\lambda} \hbar\omega_{k\lambda}\left(a^{\dagger}_{k\lambda} a_{k\lambda} + \frac{1}{2}\right). \tag{2.185}$$

We denote $a^{\dagger}_{k\lambda}a_{k\lambda} = n_{k\lambda}$, as the occupation number operator for phonons with energy $\hbar\omega_{k\lambda}$. Here, $\omega_{k\lambda}$ is the vibrational frequency in the state k and polarization λ. The term $\frac{1}{2}\sum_{k\lambda}\hbar\omega_{k\lambda}$ in H is the zero-point vibrational energy. The other definitions are

$$\begin{aligned}
|n_k> &= \text{the state with } n_k \text{ phonons},\\
a^{\dagger}_k a_k\,|n_k> &= |n_k>,\\
a_k\,|n_k> &= \sqrt{n_k}\;|n_k - 1>,\\
a^{\dagger}_k\,|n_k> &= \sqrt{n_k + 1}\;|n_k + 1>,\\
a_k\,|0> &= 0, \text{ no phonons to destroy.}
\end{aligned} \tag{2.186}$$

PROBLEMS

2.1. Derive Eq. (2.6) from Eqs. (2.2), (2.3), and (2.5).

2.2. Derive Eq. (2.21) from Eq. (2.20).

2.3. Derive Eq. (2.28) from Eq. (2.27).

2.4. Derive Eq. (2.37) from Eq. (2.36).

2.5. The general expression for specific heat of solids is given in Eq. (2.52):

$$C_v = \frac{\partial}{\partial T}\sum_s \int \frac{d\mathbf{q}}{(2\pi)^3}\,\frac{\hbar\omega_s(\mathbf{q})}{e^{\beta\hbar\omega_s(\mathbf{q})} - 1}, \tag{1}$$

where the integration is over the first Brillouin zone. Show that at very low temperatures, where the optical modes can be neglected, the three acoustic branches can be written as $\omega = \omega_s(\mathbf{q}) = c_s(\hat{\mathbf{q}})q$ ($c_s(\hat{\mathbf{q}})$ are the long-wavelength phase velocities of the acoustic mode), and the k-space integration over the first Brillouin zone can be replaced by an integration over all space,

$$C_v = \frac{\partial}{\partial T}\sum_s \int \frac{d\mathbf{q}}{(2\pi)^3}\,\frac{\hbar c_s(\hat{\mathbf{q}})q}{e^{\beta\hbar c_s(\hat{\mathbf{q}})q} - 1}. \tag{2}$$

In spherical coordinates, $d\mathbf{q} = q^2 dq\, d\Omega$. By writing $z \equiv \beta\hbar c_s(\hat{\mathbf{q}})q$, show that Eq. (2) can be written as

$$C_v = \frac{\partial}{\partial T} \frac{(k_B T)^4}{(\hbar s)^3} \frac{3}{2\pi^2} \int_0^\infty \frac{z^3 dz}{e^z - 1}, \tag{3}$$

where $\frac{1}{s^3}$ is the average of the third power of the long-wavelength phase velocities,

$$\frac{1}{s^3} = \frac{1}{3} \sum_s \frac{d\Omega}{4\pi} \frac{1}{c_s(\hat{\mathbf{q}})^3}. \tag{4}$$

Using the result of the integral,

$$\int_0^\infty \frac{x^3 dx}{e^x - 1} = \frac{\pi^4}{15}, \tag{5}$$

show from Eqs. (3) and (5) that at very low temperatures,

$$C_v = \frac{2\pi^2}{5} k_B \left(\frac{1}{\beta\hbar c} \right)^3. \tag{6}$$

2.6. The lattice specific heat is of the form Eq. (2.52) or Eq. (1) in Problem 2.5,

$$C_v = \frac{\partial}{\partial T} \sum_s \int \frac{d\mathbf{q}}{(2\pi)^3} F(\omega_s(\mathbf{q})). \tag{1}$$

Show that Eq. (1) can be written in the alternate form

$$C_v = \frac{\partial}{\partial T} \int D(\omega) F(\omega) d\omega, \tag{2}$$

where $D(\omega)$ is the density of normal modes per unit volume, i.e., the phonon density of levels. Hence, $D(\omega)d\omega$ is the total number of modes between ω and $\omega + d\omega$ in the crystal divided by its volume. Show that

$$D(\omega) = \sum_s \int \frac{d\mathbf{q}}{(2\pi)^3} \delta(\omega - \omega_s(\mathbf{q})). \tag{3}$$

It can also be shown that

$$D(\omega) = \sum_s \int \frac{d\Omega}{(2\pi)^3} \frac{1}{|\nabla\omega_s(\mathbf{q})|}, \tag{4}$$

where the integral is over the surface of the first Brillouin zone on which $\omega_s(\mathbf{q}) \equiv \omega$.

2.7. In the Debye approximation, all the wave vectors of the normal modes lie within a sphere of radius q_D, and in all three branches of the spectrum the linear dispersion relation $\omega = cq$. Show by using Eq. (3) of Problem 2.6 that the density of normal modes, $D(\omega)$, is given by

$$D(\omega) = \left\{ \frac{3}{2\pi^2} \frac{\omega^2}{s^3} \right\}, \qquad \omega = \omega_D = q_D c_s(\mathbf{q}),$$
$$= 0, \qquad \omega > \omega_D.$$

2.8. Derive Eq. (2.145) from Eqs. (2.141), (2.143), and (2.144).

2.9. Using the expressions

$$\mathbf{P}_l = \sqrt{\frac{M}{N}} \sum_{k\lambda} \hat{\epsilon}_{k\lambda} P_{k\lambda} e^{-\mathbf{k} \cdot \mathbf{l}} \tag{1}$$

and

$$\mathbf{Q}_l = \frac{1}{\sqrt{MN}} \sum_{k\lambda} \hat{\epsilon}_{k\lambda} Q_{k\lambda} e^{i\mathbf{k} \cdot \mathbf{l}}, \tag{2}$$

show that

$$[P_{k\lambda}, Q_{k'\lambda'}] = \frac{\hbar}{i} \delta_{kk'} \delta_{\lambda\lambda'}, \tag{3}$$

and

$$[P_{k\lambda}, P_{k'\lambda'}] = [Q_{k\lambda}, Q_{k'\lambda'}] \equiv 0. \tag{4}$$

2.10. Substituting Eqs. (2.181) and (2.182) in Eq. (2.175), show that

$$H = \sum_{k\lambda} \hbar\omega_{k\lambda} \left(a_{k\lambda}^{\dagger} a_{k\lambda} + \frac{1}{2} \right). \tag{1}$$

References

1. Ashcroft NW, Mermin ND. *Solid state physics*. New York: Brooks/Cole; 1976.
2. Born M, Huang K. *The dynamical theory of crystal lattices*. London: Oxford University Press; 1954.
3. Callaway J. *Quantum theory of the solid state*. New York: Academic Press; 1976.
4. Carruthers P. Theory of thermal conductivity of solids at low temperatures. *Rev Mod Phys* 1961;**33**:92.
5. Kittel C. *Quantum theory of solids*. New York: John Wiley & Sons; 1963.
6. Kittel C. *Introduction to solid state physics*. New York: John Wiley & Sons; 1976.
7. Madelung O. *Introduction to solid state theory*. New York: Springer-Verlag; 1978.
8. Mahan GD. *Many-particle physics*. New York: Plenum Press; 1986.
9. Maradudin A. Lattice dynamical aspects of the resonance absorption of gamma rays by nuclei bound in a crystal. *Rev Mod Phys* 1964;**36**:417.

10. Maradudin A, Montroll EW, Weiss GH, Ipatova IP. *Theory of lattice dynamics in the harmonic approximation, solid state physics, suppl. 3*. New York: Academic Press; 1971.
11. Marder MP. *Condensed matter physics*. New York: John Wiley & Sons; 2000.
12. Messiah A. *Quantum mechanics, vol. 1*. Amsterdam: Elsevier; 1961.
13. Peierls RE. *Quantum theory of solids*. London: Oxford University Press; 1955.
14. Schewber S. *An introduction to relativistic quantum field theory*. New York: Harper; 1961.
15. Symon KR. *Mechanics*. Reading, MA: Addison-Wesley; 1971.
16. Wannier G. *Elements of solid state theory*. London: Cambridge University Press; 1959.
17. Ziman JM. *Principles of the theory of solids*. Cambridge: Cambridge University Press; 1972.

CHAPTER

Free Electron Model

3

CHAPTER OUTLINE

3.1 The Classical (Drude) Model of a Metal ... 71
3.2 Sommerfeld Model ... 73
3.2.1 Introduction ... 73
3.2.2 Fermi Distribution Function ... 74
3.2.3 Free Electron Fermi Gas ... 77
3.2.4 Ground-State Energy of the Electron Gas ... 79
3.2.5 Density of Electron States ... 81
3.3 Fermi Energy and the Chemical Potential ... 82
3.4 Specific Heat of the Electron Gas ... 84
3.5 DC Electrical Conductivity ... 86
3.6 The Hall Effect ... 87
3.7 Failures of the Free Electron Model ... 89
Problems ... 90
References ... 93

3.1 THE CLASSICAL (DRUDE) MODEL OF A METAL

Around 1900, Drude developed the theory of electrical and thermal conductivity of metals by considering a metal as a classical gas of electrons, the properties of which were governed by Maxwell–Boltzmann statistics. According to his model, the electrons were wandering around in a metal with a background of immobile but heavy positive charges. A sketch of his model is shown in Figure 3.1.

Drude had no idea as to the nature or origin of these positive charges, although he recognized that in order to be electrically neutral, a metal would have to have an equal number of positive and negative charges. Later, these heavy positive immobile charges were identified as the ions (constituted of the nucleus and the surrounding core electrons) of the neutrally charged atom stripped of its valence electron(s). The conduction electrons in a metal were the valence electrons stripped from their parent atoms because of the strong attractive interaction between the valence electron(s) of an atom and the positively charged ions surrounding it. In what follows, we will treat these positive charges as the ions described previously, although this concept was introduced by Bohr in his model of the atom long after Drude formulated a theory of metals. This concept of a sea of valence electrons moving around in a background of static positively charged ions and holding them together like a glue was in a sense an extension of the model of a hydrogen molecule in which the

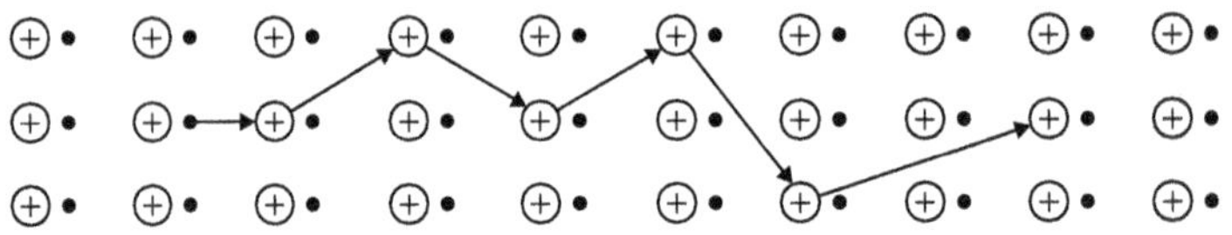

FIGURE 3.1

Drude model of scattering of electrons in a metal.

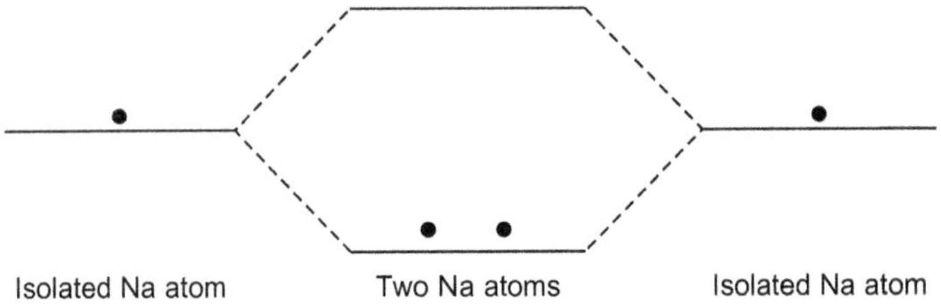

FIGURE 3.2a

A system of two sodium atoms.

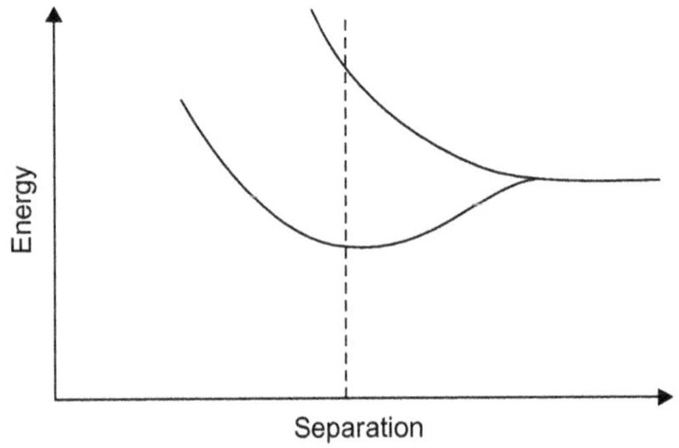

FIGURE 3.2b

The energy levels of two 3s electrons of sodium atoms as function of separation.

two valence electrons were stripped from their parent atoms and moving around the positively charged protons, thereby forming a strong bond.

An example of formation of bands in a sodium atom is shown Figure 3.2. Figure 3.2a shows a system of two sodium atoms. The two 3s electrons occupy a lower energy level than the isolated atoms.

Figure 3.2b shows the energy levels of the outer 3s electrons in a pair of sodium atoms as a function of the separation between the two atoms. This figure shows how the energy level widens with smaller distance.

When there are N sodium atoms, where N is a very large number, these energy levels form bands. In Figure 3.2c, we show the 1s, 2s, 2p, and 3s energy bands of sodium. However, the bands become wider for electrons that are less tightly bound to the parent nucleus. We note the significant difference between the 1s and the 3s bands.

Drude introduced the concept of electron density $n = N/V$, where V is the volume of the metal and N is the number of atoms per mole (the Avogadro's number). He also introduced the electron density r_s, which is defined as the radius of a sphere of which the volume is equal to the available volume for each conduction electron. According to his model,

$$\frac{1}{n} = \frac{4\pi r_s^3}{3} = \frac{V}{N}, \tag{3.1}$$

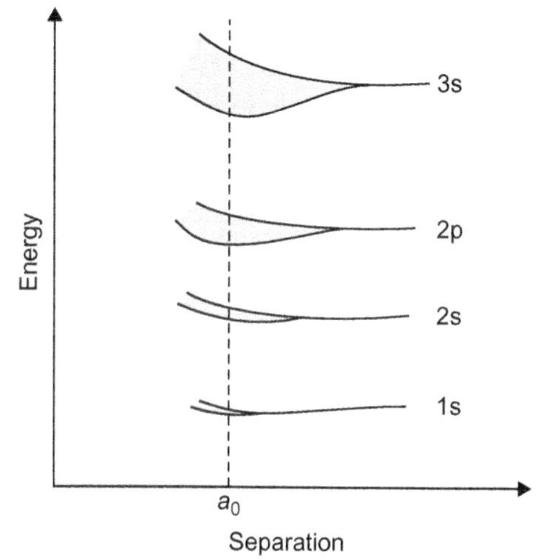

FIGURE 3.2c

The formation of 1s, 2s, 2p, and 3s energy bands in sodium. Note that 1s is in the lowest energy state, whereas 3s is in the highest energy state.

from which it follows that

$$r_s = \left(\frac{3}{4\pi n}\right)^{1/3}. \tag{3.2}$$

Drude applied the kinetic theory of gas to these conduction electrons even though their electron densities are approximately a thousand times greater than those of a classical gas at normal pressure and temperature. He introduced several interesting approximations (in his kinetic theory model) to develop theories of both electrical and thermal conductivity of metals. First, the electron gas was assumed to be free and independent in the sense that there was no electron–electron or electron–ion interactions. An electron moved in a straight line (as per Newton's law) in the absence of an external electric field. Second, when an external field is applied, the electron continues to move in a straight line between collisions with the ions (the collisions with other electrons were neglected in his theory). The velocity of an electron would change immediately after a collision, which was assumed to be an instantaneous event. Third, the time between two collisions was the same value τ (often known as the relaxation time) for each electron. Fourth, immediately after each collision, an electron's velocity was not related to the velocity with which it was traveling before the collision. The new speed of the electron depended on the temperature at the time of collision, and the direction of the new velocity was randomly directed. In a sense, the electron achieved thermal equilibrium with its surroundings.

The two major achievements of Drude's model were the derivation of electrical and thermal conductivity (see problems) in a very simplistic way. However, it had several drawbacks. Some of these were solved by Sommerfeld, who recognized the fact the free electrons in a metal were not classical but quantum gas, for which the application of Fermi–Dirac statistics (instead of Maxwell–Boltzmann statistics) was appropriate.

3.2 SOMMERFELD MODEL

3.2.1 Introduction

The major failure of Drude's theory was the prediction that the specific heat of a metal was $C_v = 3/2Nk_B$, where N is the number of electrons in a metal and k_B is the Boltzmann constant. In practice, the specific heat of a metal was more than 100 times smaller than this value at room temperature. This anomaly was explained by Sommerfeld only after the introduction of quantum mechanics. The Pauli exclusion principle requires that no two electrons can have identical quantum numbers, which implies that a quantum state can be occupied by at most one electron. Sommerfeld recognized the significance of the Pauli exclusion principle, and as a consequence, the Fermi–Dirac statistics (instead of the classical Maxwell–Boltzmann statistics) were applicable to a system of identical electron gas (or other fermions) in thermal equilibrium. Sommerfeld essentially adapted Drude's model of the metal as a free electron gas moving in a background of static positively charged ions but modified it by using the Fermi–Dirac distribution function for the electron velocity. In the process, he was able to explain the (apparently) anomalous low specific heat of the electron gas as well as other thermal properties.

3.2.2 Fermi Distribution Function

(a) Grand Canonical Ensemble

The Fermi distribution function is easily derived from the concept of a Grand Canonical Ensemble, which is an ensemble of $\mathbb{N}$ identical systems (labeled as $1, 2, \ldots, \mathbb{N}$) mutually sharing a total number of particles $\mathbb{N}\bar{N}$ and a total amount of energy $\mathbb{N}\bar{E}$. If $n_{r,s}$ denotes the number of systems that have, at any time t, the number $\mathbb{N}_r$ of particles and the amount of energy $E_s (r, s = 0, 1, 2, \ldots)$,

$$\sum_{r,s} n_{r,s} = \mathbb{N}, \tag{3.3}$$

$$\sum_{r,s} n_{r,s} N_r = \mathbb{N}\bar{N}, \tag{3.4}$$

$$\sum_{r,s} n_{r,s} E_s = \mathbb{N}\bar{E}. \tag{3.5}$$

We consider this ensemble of $\mathbb{N}$ identical systems that are characterized by a Hamiltonian operator $\hat{H}$. At any time t, the physical states of the various systems in the ensemble are characterized by the wave functions $\psi(\mathbf{r}, t)$. Let $\psi^k(\mathbf{r}, t)$ denote the normalized wave function characterizing the physical state in which the kth state of the system of the ensemble happens to be at the time. We can write

$$\hat{H}\psi^k(\mathbf{r}, t) = i\hbar\dot{\psi}^k(\mathbf{r}, t). \tag{3.6}$$

We introduce a complete set of orthonormal functions $\phi_n(\mathbf{r}) \equiv \phi_n(\mathbf{r}_1, \mathbf{r}_2, \ldots, \mathbf{r}_N)$ and express

$$\psi^k(\mathbf{r}, t) = \sum_n a_n^k(t)\phi_n(\mathbf{r}). \tag{3.7}$$

We can write

$$a_n^k(t) = \int \phi_n^*(\mathbf{r})\psi^k(\mathbf{r}, t)\, d\tau, \tag{3.8}$$

where $d\tau$ is the volume element of the coordinate space of the system. Thus, $a_n^k(t)$ are the probability amplitudes for the various systems of the ensemble to be in the states $\phi_n(\mathbf{r})$, and $|a_n^k(t)|^2$ represents the probability that a measurement of time t finds the kth system of the ensemble to be in the particular state $\phi_n(\mathbf{r})$. Thus, we obtain

$$\sum_n |a_n^k(t)|^2 = 1 \text{ for all } k. \tag{3.9}$$

Eq. (3.9) can also be derived from Eq. (3.7) in a straightforward manner. From Eq. (3.7), we obtain

$$\begin{aligned} \int |\psi^k(\mathbf{r}, t)|^2\, d\tau &= \sum_{n,m} a_n^{k*}(t)a_m^k(t) \int \phi_n^*(\mathbf{r})\phi_m(\mathbf{r})\, d\tau \\ &= \sum_{n,m} a_n^{k*}(t)a_m^k(t)\, \delta_{n,m} = \sum_n |a_n^k(t)|^2 = 1. \end{aligned} \tag{3.10}$$

(b) Density Operator

The density operator $\hat{\rho}(t)$ is defined by the matrix elements

$$\rho_{mn}(t) = \frac{1}{\mathbb{N}} \sum_{k=1}^{\mathbb{N}} \{a_m^k(t) a_n^{k*}(t)\}. \tag{3.11}$$

The diagonal element $\rho_{nn}(t)$, which is the ensemble average of the probability $|a_n^k(t)|^2$, represents the probability of a system, chosen at random from the ensemble at time t, and is found to be in the state ϕ_n. From Eqs. (3.10) and (3.11), we obtain

$$\sum_n \rho_{nn} = 1. \tag{3.12}$$

One can show that if the system is in a state of equilibrium,[8]

$$i\hbar \dot{\hat{\rho}} = [\hat{H}, \hat{\rho}] = 0, \tag{3.13}$$

$$<\hat{G}> = tr(\hat{\rho}\hat{G}), \tag{3.14}$$

and

$$tr(\hat{\rho}) = 1, \tag{3.15}$$

where $<\hat{G}>$ is the expectation value of a physical quantity G.

In the Grand Canonical Ensemble, the density operator $\hat{\rho}$ operates on a Hilbert space in an indefinite number of particles. Therefore, the density operator commutes not only with the Hamiltonian operator $\hat{H}$, but also with a number operator $\hat{n}$ of which the eigenvalues are 0, 1, 2. We can write[8]

$$\hat{\rho} = \frac{1}{Z(\mu, V, T)} e^{-\beta(\hat{H} - \mu\hat{n})}, \tag{3.16}$$

where the Grand partition function $Z(\mu, V, T)$ is defined as

$$Z(\mu, V, T) = \sum_{r,s} e^{-\beta(E_r - \mu N_s)} = tr\left\{e^{-\beta(\hat{H} - \mu\hat{n})}\right\}. \tag{3.17}$$

The chemical potential μ is defined as

$$\mu = \left(\frac{\partial F}{\partial N}\right)_{T,V,N}, \tag{3.18}$$

where F is the Helmholtz free energy defined by the relation

$$F = E - TS. \tag{3.19}$$

Here, E is the energy, S is the entropy, T is the absolute temperature, and V is the volume.

For a free electron gas,

$$\hat{H} = \sum_{\mathbf{k}} \varepsilon(\mathbf{k}) c_{\mathbf{k}\sigma}^{\dagger} c_{\mathbf{k}\sigma} \tag{3.20}$$

and

$$N_s = 0 \text{ or } 1. \tag{3.21}$$

Therefore,

$$\begin{aligned} E_r &= N_s\, \varepsilon(\mathbf{k}) \\ &= 0 \text{ or } \varepsilon(\mathbf{k}). \end{aligned} \tag{3.22}$$

From Eqs. (3.17) and (3.20) through (3.22), we obtain (omitting spin $[\sigma]$),

$$Z_k = 1 + e^{-\beta(\varepsilon(\mathbf{k})-\mu)} \tag{3.23}$$

and

$$\hat{\rho}_k = \frac{1}{Z_k} e^{-\beta(\varepsilon(\mathbf{k})-\mu)c_{\mathbf{k}}^{\dagger}c_{\mathbf{k}}}. \tag{3.24}$$

The average number in state **k** is

$$\begin{aligned} \bar{n}_{\mathbf{k}} &= tr\{n_{\mathbf{k}}\hat{\rho}_k\} = \sum_{n_{\mathbf{k}}} < n_{\mathbf{k}} | n_{\mathbf{k}}\hat{\rho}_k | n_k > = <1|\hat{\rho}_k|1> \\ &= \frac{e^{-\beta(\varepsilon(\mathbf{k})-\mu)}}{1+e^{-\beta(\varepsilon(\mathbf{k})-\mu)}} = \frac{1}{e^{\beta(\varepsilon(\mathbf{k})-\mu)}+1} = f(\varepsilon(\mathbf{k}), T). \end{aligned} \tag{3.25}$$

Here, $f(\varepsilon(\mathbf{k}), T)$ is the Fermi distribution, which is a function of energy and temperature. The Fermi distribution function $f(\varepsilon(\mathbf{k}), T)$ gives the probability that a state at energy ε_i is occupied in an ideal electron gas in thermal equilibrium. We also note that the Fermi distribution is an eigenvalue of the statistical operator $\hat{\rho}_k$.

In practice, because the electrons in an electron gas are indistinguishable and are constantly moving around, the probability that a state is occupied by an electron is the average number of electrons in the state and

$$\sum_i f(\varepsilon_i) = N, \tag{3.26}$$

where N is the total number of electrons in the free electron gas (note that we do not include the core electrons of the atoms that are bound to the parent nuclei and are not free to move around). In Eq. (3.26), ε_i are the energy levels. In semiconductor physics, μ is also called the Fermi level.

The Fermi energy ε_F is defined as the energy of the topmost filled level of the electron states at $T=0$. All the electron states having energy greater than the Fermi energy are empty (there are no electrons in these states) at $T=0$. From Eq. (3.26), we note that the alternate definition of the Fermi energy at $T=0$ is

$$\begin{aligned} f(\varepsilon_i) &= 1, \quad \varepsilon_i \leq \varepsilon_F, \\ &= 0, \quad \varepsilon_i > \varepsilon_F. \end{aligned} \tag{3.27}$$

We also note from Eq. (3.23) that

$$\begin{aligned} \lim_{T\to 0} f(\varepsilon_i) &= 1, \quad \varepsilon_i \leq \mu, \\ &= 0, \quad \varepsilon_i > \mu. \end{aligned} \tag{3.28}$$

Comparing Eqs. (3.27) and (3.28), we note that at $T=0$, $\varepsilon_F = \mu$. Thus, in the limit $T \to 0$, when $\varepsilon_i = \varepsilon_F = \mu$, the Fermi–Dirac distribution function changes abruptly from 1 to 0. At higher temperatures ($T>0$), $f(\varepsilon_i) = 1/2$ when $\varepsilon_i = \mu$. The distribution of the Fermi distribution function and its derivative at $T=0$ and at finite temperature is shown in Figure 3.3.

From Figure 3.3, we note that the Fermi distribution function at a large temperature T is significantly different from that at zero temperature. When the metal is heated from $T=0$, electrons are transferred from the region $\varepsilon/\mu<1$ to the region $\varepsilon/\mu>1$. The rest of the electrons deep inside the

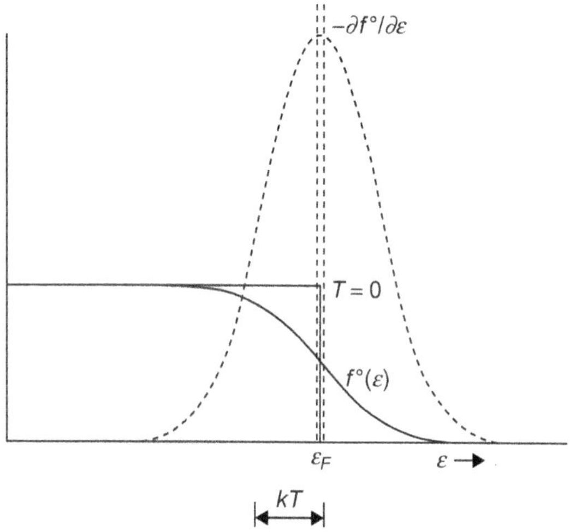

FIGURE 3.3

Fermi distribution function and its derivative at $T=0$ and at a finite temperature.

Fermi level are not affected when the electron gas is heated. However, as the temperature increases, there is a significant change in the Fermi distribution function. We will discuss this in Section 3.3. On the other hand, each electron in a classical electron gas would gain energy when the system is heated. This immediately explains the anomalous results for the specific heat of the electron gas. As we will see, this property of the fermions is very significant in the derivation of other physical quantities of metals.

3.2.3 Free Electron Fermi Gas

By definition, a free electron gas consists of N independent electrons confined in a volume V. There are positively charged static ions inside the metal to keep it electrically neutral. For simplicity, we will consider this volume to be a cube of side L. An electron is completely free to move inside this cube. There are no other interactions, either with other electrons or with the positively charged background (the lattice). Because there is no potential energy term, the Schrodinger equation for an electron in state **k** can be written as

$$-\frac{\hbar^2}{2m}\nabla^2\psi_{\mathbf{k}}(\mathbf{r}) = \varepsilon(\mathbf{k})\psi_{\mathbf{k}}(\mathbf{r}). \tag{3.29}$$

Here, $\psi_{\mathbf{k}}(\mathbf{r})$ is the wave function and $\varepsilon(\mathbf{k})$ is the energy of the electron in the state **k**. The normalized solution of Eq. (3.29) is easily obtained, by requiring $\psi_{\mathbf{k}}(\mathbf{r})$ to vanish at the surface of the cube (the particle in a box problem in quantum mechanics), as

$$\psi_{\mathbf{k}}(\mathbf{r}) = (8/V)^{1/2} \sin(n_1\pi x/L)\sin(n_2\pi y/L)\sin(n_3\pi z/L), \tag{3.30}$$

where L is the side of the cubic metal of volume $V(V=L^3)$ and n_1, n_2, and n_3 are positive integers. For simplicity, we have considered the metal as a cube, although we could have chosen a parallelepiped, which would yield the same results but the derivation would have been more complicated. However, we note that the solutions obtained in Eq. (3.30) are standing-wave solutions and the probability density $|\psi_{\mathbf{k}}(\mathbf{r})|^2$ of the electron in **k** state varies with its position in real space, thereby yielding an unrealistic picture for free electrons. To obtain more realistic solutions for free electron gas, we introduced the Born–von Karman (or periodic) boundary conditions. These conditions are

$$\begin{aligned}\psi(x+L,y,z) &= \psi(x,y,z),\\ \psi(x,y+L,z) &= \psi(x,y,z),\\ \psi(x,y,z+L) &= \psi(x,y,z).\end{aligned} \tag{3.31}$$

These periodic boundary conditions were imposed by imagining that each face of the cube is joined to the face opposite it. When an electron meets the surface of the metal, instead of being reflected by the surface, it emerges at an equivalent point on the opposite surface. This is obviously an improbable situation in a three-dimensional solid, but the periodic boundary conditions yield free electron wave functions for the free electron gas. It may be noted that these approximations hold good as long as one is not considering the physical properties of the electrons on or close to the surface, in which case they have to be modified. It may be further noted that it is not necessary to consider the metal as a cube of side L; the results would still hold good for a parallelepiped.

The general solutions of Eq. (3.29) are the well-known plane waves

$$\psi_{\mathbf{k}}(\mathbf{r}) = \frac{1}{\sqrt{V}} e^{i\mathbf{k}\cdot\mathbf{r}}. \tag{3.32}$$

In general, **k** is a continuous variable in the reciprocal space. However, if we invoke the periodic boundary conditions as stated in Eq. (3.31), we obtain only discrete values of **k**, i.e.,

$$e^{ik_xL} = e^{ik_yL} = e^{ik_zL} = 1. \tag{3.33}$$

Eq. (3.33) implies that $k_x, k_y,$ and k_z are discrete variables,

$$k_x = \frac{2\pi n_1}{L},\ k_y = \frac{2\pi n_2}{L},\ k_z = \frac{2\pi n_3}{L}. \tag{3.34}$$

Here, n_1, n_2, and n_3 are integers (zero, positive, or negative). Thus, the components of the wave vector **k** are discrete in the k-space and, along with the spin components m_s, are the quantum numbers of the problem.

We also obtain from Eqs. (3.29) and (3.32),

$$\varepsilon(\mathbf{k}) = \frac{\hbar^2 k^2}{2m}. \tag{3.35}$$

The plane wave is also an eigenfunction of the linear momentum **p**,

$$\mathbf{p}\psi_{\mathbf{k}}(\mathbf{r}) = -i\hbar\nabla \frac{1}{\sqrt{V}} e^{i\mathbf{k}\cdot\mathbf{r}} = \hbar\mathbf{k}\psi_{\mathbf{k}}(\mathbf{r}). \tag{3.36}$$

Thus, the electron velocity **v** in the state **k** is given by

$$\mathbf{v} = \hbar\mathbf{k}/\mathrm{m}. \tag{3.37}$$

We can now use Eq. (3.34) to build the k-space. The k-space is a three-dimensional space with an extremely large number (N) of allowed discrete k points. Figure 3.4 illustrates the points in a two-dimensional k-space that are separated by $\frac{2\pi}{L}$ in both k_x and k_y directions (as per Eq. 3.34).

As seen in Figure 3.4, the discrete k values are separated by $2\pi/L$ in each dimension. In a three-dimensional k-space, the volume per each **k**-point is $(2\pi/L)^3 = 8\pi^3/V$, where the volume of the metal $V = L^3$. Therefore, a k-space of volume Ω will contain $\Omega V/8\pi^3$ allowed values of **k**.

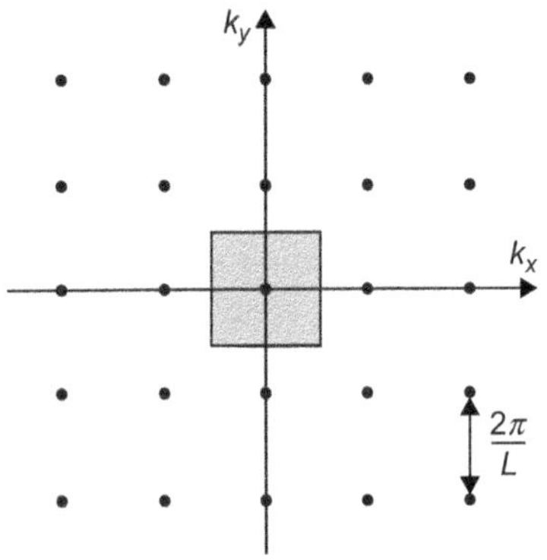

FIGURE 3.4

Points in a two-dimensional k-space.

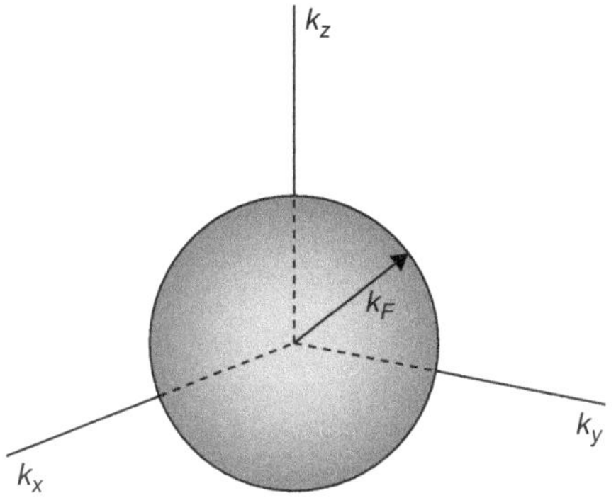

FIGURE 3.5

Schematic diagram of the Fermi sphere.

Thus, the density of states, ρ, in the k-space (number of allowed k-values per unit volume of k-space) is

$$\rho = V/8\pi^3. \tag{3.38}$$

We note that each **k** state can contain exactly two electrons (spin up or spin down) at zero temperature as per the Fermi–Dirac distribution function in Eq. (3.25). Thus, when we build the filled k-space at zero temperature in three dimensions, we are essentially filling up these discrete **k** states starting with the lowest energy, which is zero. Because the **k** states are densely packed and the total number of electrons is N, which is a huge number (~10^{23}), the constant energy levels of the electrons can be considered as the surface of a sphere (as per Eq. 3.33). Thus, all the N electrons will fill a sphere of volume $\Omega = 4\pi/3\, k_F^3$, which is known as the Fermi sphere. k_F, the radius of the sphere, is known as the Fermi wave vector. The surface of the Fermi sphere that separates the filled **k** states from the empty **k** states (at zero temperature) is known as the Fermi surface. The energy of an electron on the Fermi surface is called the Fermi energy $\varepsilon_F = \frac{\hbar^2 k_F^2}{2m}$, and the momentum $\mathbf{p}_F = \hbar \mathbf{k}_F$ is known as the Fermi momentum. A schematic diagram of the Fermi sphere is shown in Figure 3.5.

The volume of the Fermi sphere is $\Omega = 4\pi k_F^3/3$ and, as shown earlier, a k-space of volume Ω will contain $\Omega V/8\pi^3$ allowed values of **k**. Hence, the allowed values of **k** in the Fermi sphere are

$$\frac{\Omega V}{8\pi^3} = \left(\frac{V}{8\pi^3}\right)\left(\frac{4\pi k_F^3}{3}\right) = \frac{k_F^3 V}{6\pi^2}. \tag{3.39}$$

Because there are two electrons per each **k** state and the total number of electrons is N, we obtain from Eq. (3.39)

$$\frac{N}{2} = \frac{k_F^3 V}{6\pi^2}. \tag{3.40}$$

The electron density $n \equiv N/V$ is easily obtained from Eq. (3.40),

$$n = \frac{k_F^3}{3\pi^2}. \tag{3.41}$$

3.2.4 Ground-State Energy of the Electron Gas

To make a summation over allowed values of **k**, we use the following procedure. If $F(\mathbf{k})$ is a function of **k** states, we can write, using Eq. (3.38),

$$\sum_{\mathbf{k}} F(\mathbf{k}) = \frac{V}{8\pi^3} \sum_{\mathbf{k}} F(\mathbf{k}) \Delta \mathbf{k}. \tag{3.42}$$

In the limit $V \to \infty$, $\Delta\mathbf{k} \to 0$ and the summation over $\mathbf{k}$ on the right side, $\sum_{\mathbf{k}} F(\mathbf{k})\Delta\mathbf{k}$ approaches the integral $\int F(\mathbf{k})d\mathbf{k}$. Thus, we can write Eq. (3.42) in the alternate form

$$\lim_{V\to\infty} \sum_{\mathbf{k}} F(\mathbf{k}) = \frac{V}{8\pi^3}\int F(\mathbf{k})d\mathbf{k}. \tag{3.43}$$

In general, the approximation $V \to \infty$ is fairly good because the volume of a metal is infinitely large compared to the volume of a primitive cell.

The ground-state energy of N free electrons in a metal is calculated by adding the energies of the $\mathbf{k}$ states inside the Fermi sphere and multiplying by 2 (because each $\mathbf{k}$ state has two electrons of opposite spin),

$$E = 2\sum_{\mathbf{k}} \varepsilon(\mathbf{k}). \tag{3.44}$$

From Eqs. (3.35) and (3.44),

$$E = 2\sum_{k \le k_F} \frac{\hbar^2 k^2}{2m}. \tag{3.45}$$

From Eqs. (3.43) and (3.45),

$$E = \frac{V}{4\pi^3}\int_{k\le k_F} d\mathbf{k}\,\frac{\hbar^2 k^2}{2m}. \tag{3.46}$$

Because $d\mathbf{k} = 4\pi k^2 dk$, Eq. (3.46) can be easily integrated, and we obtain

$$E = \frac{V\hbar^2 k_F^5}{10\pi^2 m}. \tag{3.47}$$

Here, E is the total energy of the electron gas. To find the energy per electron, E/N, we obtain from Eqs. (3.40) and (3.47),

$$\frac{E}{N} = \frac{3\hbar^2 k_F^2}{10m}, \tag{3.48}$$

which can be written in the alternate form

$$\frac{E}{N} = \frac{3}{5}\varepsilon_F. \tag{3.49}$$

The average energy of an electron at zero temperature, derived in Eq. (3.49), differs significantly from the energy of a classical ideal gas, which is zero at $T = 0$. This significant difference is due to the fact that the electrons obey the Pauli exclusion principle, and therefore, the Fermi–Dirac distribution function has to be applied to the electron gas instead of the Maxwell–Boltzmann distribution function, which is applicable for a classical ideal gas.

We now define the Fermi temperature using the relation

$$\varepsilon_F = k_B T_F. \tag{3.50}$$

From Eqs. (3.49) and (3.50), we obtain

$$T_F = \frac{5E}{3Nk_B}. \tag{3.51}$$

The Fermi temperature T_F is of the order of 10^4 K for simple metals (see Problem 3.7), an incredibly large value for the temperature of the electrons on the Fermi surface. We note that for a classical ideal gas, the temperature $T = 0$ irrespective of the location of the gas molecule.

The wavelength of an electron at the Fermi surface is given by

$$\lambda_F = \frac{2\pi}{k_F}. \tag{3.52}$$

3.2.5 Density of Electron States

Because the **k** states are discrete variables, normally one makes a summation over all the **k** states. In Eq. (3.43), we approximated the summation of a function $F(\mathbf{k})$ over all the **k** states as an integration over the **k** states. An alternate way to consider the summation over all the **k** states is to consider the same summation as integration over the energy levels. Such integrations over the energy levels are done by using the concept of the density of electron states.

As an example, we consider the summation $\sum_{\mathbf{k}} F(\varepsilon(\mathbf{k}))$, where $F(\varepsilon(\mathbf{k}))$ is a function of the energy of the electrons in the filled **k** states. We can write Eq. (3.43) in the alternate form

$$\frac{1}{V}\sum_{\mathbf{k}} F(\varepsilon(\mathbf{k})) = \frac{1}{4\pi^3}\int d\mathbf{k} F(\varepsilon(\mathbf{k})). \tag{3.53}$$

We note that because each **k** state has two electrons with the same energy, we have multiplied the right side of Eq. (3.43) by a factor of 2. Because $\varepsilon(\mathbf{k}) = \hbar^2 k^2/2m$, we can write

$$d\mathbf{k} = 4\pi k^2 dk. \tag{3.54}$$

From Eqs. (3.53) and (3.54), we obtain

$$\begin{aligned}\frac{1}{4\pi^3}\int d\mathbf{k} F(\varepsilon(\mathbf{k})) &= \int_0^\infty \frac{k^2 dk}{\pi^2} F(\varepsilon(\mathbf{k})) \\ &= \int_\infty^{-\infty} d\varepsilon g(\varepsilon) F(\varepsilon).\end{aligned} \tag{3.55}$$

Here,

$$\begin{aligned} g(\varepsilon) &= \frac{m}{\pi^2\hbar^2}\sqrt{\frac{2m\varepsilon}{\hbar^2}}, \quad \varepsilon > 0 \\ &= 0, \qquad \varepsilon \le 0. \end{aligned} \tag{3.56}$$

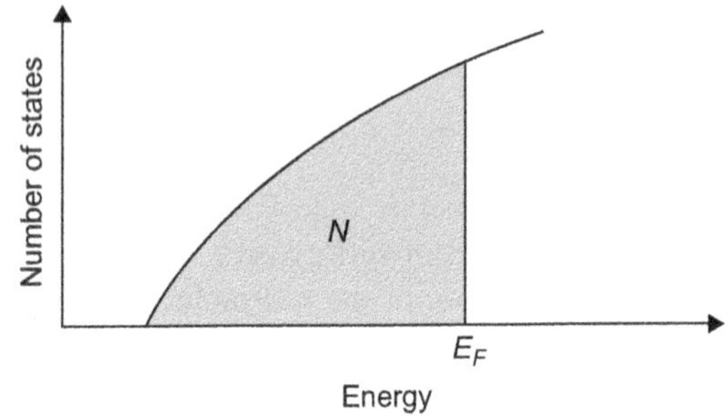

FIGURE 3.6

Density of states versus energy.

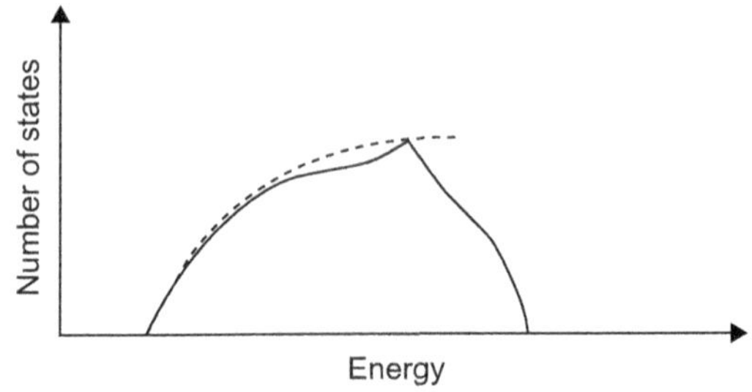

FIGURE 3.7

Comparison of density of states for a typical metal (solid line) with the predicted value (dashed curve).

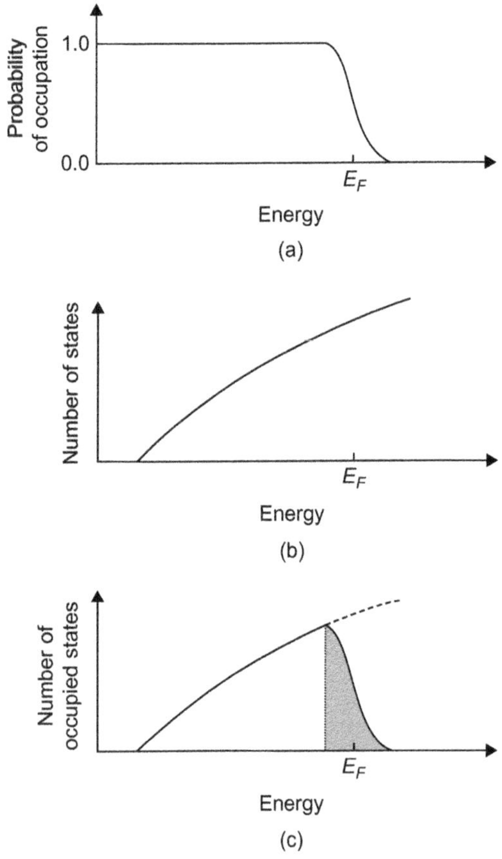

FIGURE 3.8

The density of occupied states (c) obtained from the Fermi–Dirac distribution (a) and the density of states (b).

Thus, the density of electron states $g(\varepsilon)d\varepsilon$ is defined as the number of one-electron levels per unit volume of the metal in the energy range ε and $\varepsilon + d\varepsilon$. From Eqs. (3.41) and (3.56), we can rewrite $g(\varepsilon)$ in the alternate form

$$g(\varepsilon) = \frac{3n}{2}\left(\frac{\varepsilon}{\varepsilon_F^3}\right)^{1/2}, \quad \varepsilon > 0 \qquad (3.57)$$
$$= 0, \qquad \varepsilon \leq 0.$$

The number of single electron states N is equal to the area under the density of states curve $g(\varepsilon)$ up to the Fermi energy. The plot of $g(\varepsilon)$ versus ε is shown in Figure 3.6.

However, the density of states for a typical metal, which is plotted in Figure 3.7, is different from the curve shown in Figure 3.6.

3.3 FERMI ENERGY AND THE CHEMICAL POTENTIAL

We will derive an expression between the chemical potential μ and the Fermi energy ε_F at a reasonably low temperature (we note that $\mu = \varepsilon_F$ at $T = 0$).

The electron density $n = N/V$ can be written in the form

$$n = \frac{1}{4\pi^3}\int d\mathbf{k} f(\varepsilon(\mathbf{k})). \qquad (3.58)$$

From Eqs. (3.55) and (3.58), we obtain

$$n = \int_{-\infty}^{\infty} g(\varepsilon) f(\varepsilon) d\varepsilon. \tag{3.59}$$

Using the Sommerfeld expansion (Problem 3.8) and retaining terms of the order T^2, we obtain

$$n = \int_{-\infty}^{\infty} g(\varepsilon) d\varepsilon + \frac{\pi^2}{6}(k_B T)^2 g'(\varepsilon_F). \tag{3.60}$$

From Eqs. (3.57) and (3.60), we obtain

$$n = \int_{0}^{\mu} g(\varepsilon) d\varepsilon + \frac{\pi^2}{6}(k_B T)^2 g'(\varepsilon_F). \tag{3.61}$$

Eq. (3.61) can be written in the alternate form

$$n = \int_{0}^{\varepsilon_F} g(\varepsilon) d\varepsilon + \int_{\varepsilon_F}^{\mu} g(\varepsilon) d\varepsilon + \frac{\pi^2}{6}(k_B T^2) g'(\varepsilon_F). \tag{3.62}$$

We note that at zero temperature, the electron density n can also be expressed as

$$n = \int_{0}^{\varepsilon_F} g(\varepsilon) d\varepsilon. \tag{3.63}$$

In addition, at low temperatures, $g(\varepsilon) \approx g(\varepsilon_F)$ in the small energy range of ε_F to μ and the integration of $g(\varepsilon)$ in this range can be approximated as

$$\int_{\varepsilon_F}^{\mu} g(\varepsilon) d\varepsilon \approx (\mu - \varepsilon_F) g(\varepsilon_F). \tag{3.64}$$

From Eqs. (3.62) through (3.64), we obtain

$$(\mu - \varepsilon_F) g(\varepsilon_F) + \frac{\pi^2}{6}(k_B T)^2 g'(\varepsilon_F) = 0. \tag{3.65}$$

From Eq. (3.65), the chemical potential μ can be expressed as

$$\mu = \varepsilon_F - \frac{\pi^2}{6}(k_B T)^2 \frac{g'(\varepsilon_F)}{g(\varepsilon_F)}. \tag{3.66}$$

From Eqs. (3.64) and (3.66), we obtain

$$\mu = \varepsilon_F \left[1 - \frac{\pi^2}{12} \left(\frac{k_B T}{\varepsilon_F} \right)^2 \right]. \tag{3.67}$$

Eq. (3.67) expresses the chemical potential μ in terms of the Fermi energy ε_F (up to the order of T^2).

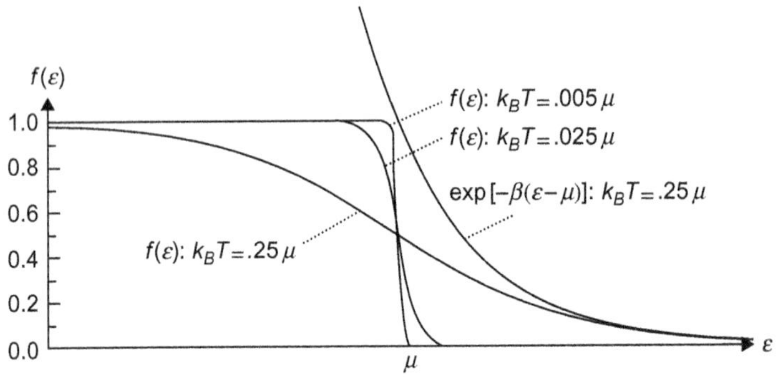

FIGURE 3.9

Variation of the Fermi–Dirac distribution function with temperature.

The variation of the Fermi–Dirac distribution function for various values of $k_B T$ is shown in Figure 3.9. The Fermi function differs slightly from a step function around the chemical potential μ. It differs significantly from 0 or 1 in the region in which the width is $k_B T$. For $\varepsilon \gg \mu$, it is hard to distinguish the Fermi function from the classical Boltzmann distribution function $e^{-\beta\varepsilon}$.

3.4 SPECIFIC HEAT OF THE ELECTRON GAS

At a finite temperature T, the **k** states will be filled by $2\,f(\varepsilon(\mathbf{k}))$ electrons, where $f(\varepsilon(\mathbf{k}))$ is the Fermi–Dirac distribution function, which determines the probability that a **k** state is occupied by an electron. The factor 2 is multiplied because each **k** state can have two electrons of opposite spin as prescribed by the Pauli exclusion principle. Thus, the total internal energy of the electrons at temperature T is given by

$$E_{el}(T) = 2\sum_{\mathbf{k}} \varepsilon(\mathbf{k}) f(\varepsilon(\mathbf{k})). \tag{3.68}$$

The electronic specific heat of the metal is given by

$$C_{el} = \frac{1}{V}\left(\frac{\partial E_{el}}{\partial T}\right)_N. \tag{3.69}$$

By converting the summation over **k** states to integration and following the procedure outlined in Eq. (3.53), we can rewrite Eq. (3.68) as

$$\frac{E_{el}}{V} = \frac{1}{4\pi^3}\int d\mathbf{k}\, \varepsilon(\mathbf{k}) f(\mathbf{k}). \tag{3.70}$$

From Eqs. (3.55) and (3.70), we obtain

$$\frac{E_{el}}{V} = \int_{-\infty}^{\infty} \varepsilon g(\varepsilon) d\varepsilon. \tag{3.71}$$

By using the Sommerfeld expansion (Problem 3.8), we obtain from Eq. (3.71)

$$\frac{E_{el}}{V} = \int_{-\infty}^{\mu} \varepsilon g(\varepsilon) d\varepsilon + \frac{\pi^2}{6}(k_B T)^2 \frac{d}{d\varepsilon}(\varepsilon g(\varepsilon)) \,|_{\varepsilon=\mu} + \text{higher order terms in } T. \tag{3.72}$$

We note from Eq. (3.57) that the density of one-electron levels $g(\varepsilon) = 0$ *for* $\varepsilon \leq 0$. Using this condition, we rewrite Eq. (3.72) as

$$\frac{E_{el}}{V} = \int_{0}^{\varepsilon_F} \varepsilon g(\varepsilon) d\varepsilon + \int_{\varepsilon_F}^{\mu} \varepsilon g(\varepsilon) d\varepsilon + \frac{\pi^2}{6}(k_B T)^2 [g(\mu) + \mu g'(\mu)] + \text{higher order terms in } T. \tag{3.73}$$

From Eqs. (3.67), which specifies the relation between μ and ε_F, retaining the second-order terms in T, we obtain

$$\int_{\varepsilon_F}^{\mu} \varepsilon g(\varepsilon) d\varepsilon \approx \varepsilon_F(\mu - \varepsilon_F) g(\varepsilon_F). \tag{3.74}$$

Substituting Eq. (3.74) in Eq. (3.73) and consistently retaining terms only up to the second order in T $g(\mu) \approx g(\varepsilon_F)$ and $\mu \approx \varepsilon_F$ in Eq. (3.73), we obtain

$$\frac{E_{el}}{V} = \int_{0}^{\varepsilon_F} \varepsilon g(\varepsilon) d\varepsilon + \frac{\pi^2}{6}(k_B T)^2 g(\varepsilon_F) + \varepsilon_F \left\{ (\mu - \varepsilon_F) g(\varepsilon_F) + \frac{\pi^2}{6}(k_B T)^2 g'(\varepsilon_F) \right\}. \tag{3.75}$$

From Eqs. (3.65) and (3.75),

$$\frac{E_{el}}{V} = \int_{0}^{\varepsilon_F} \varepsilon g(\varepsilon) d\varepsilon + \frac{\pi^2}{6}(k_B T)^2 g(\varepsilon_F). \tag{3.76}$$

From Eqs. (3.69) and (3.76), noting that the first term on the right side of Eq. (3.76) is independent of temperature for constant N, we obtain an expression for the electronic specific heat for a free electron gas in the metal,

$$C_{el} = \frac{\pi^2}{3} k_B{}^2 T g(\varepsilon_F). \tag{3.77}$$

From Eqs. (3.57) and (3.77),

$$C_{el} = \frac{n\pi^2 k_B^2 T}{2\varepsilon_F}. \tag{3.78}$$

We note the result for the electronic specific heat obtained by using Fermi–Dirac statistics is significantly different and approximately 100 times smaller than Drude's result for a classical electron gas ($C_v = 3/2\ nk_B$) obtained by using Maxwell–Boltzmann statistics. The major difference is that while each electron in a classical gas gains an energy 3/2 k_BT when heated to a temperature T, only the electrons in a range k_BT near the Fermi level are excited to the higher energy states, whereas the rest of the electrons inside the Fermi surface are unaffected.

3.5 DC ELECTRICAL CONDUCTIVITY

We have shown in Eq. (3.36) that

$$\mathbf{p} = \hbar\mathbf{k}. \tag{3.79}$$

Using Newton's second law, we can see the force **F** on the electron with momentum **p** is

$$\mathbf{F} = \frac{d\mathbf{p}}{dt} = \hbar\frac{d\mathbf{k}}{dt}. \tag{3.80}$$

In an external electric field **E**, the force on each electron of charge $-e$ is

$$\mathbf{F} = -e\mathbf{E}. \tag{3.81}$$

From Eqs. (3.80) and (3.81), we obtain

$$\hbar\frac{d\mathbf{k}}{dt} = -e\mathbf{E}. \tag{3.82}$$

If the electric field **E** is applied to the metal at time $t = 0$, after a time τ, the center of the Fermi sphere (within which all the electrons are located at $T = 0$), which was at the origin of the **k** space, would shift by

$$\Delta\mathbf{k} = -\frac{e\mathbf{E}\tau}{m}. \tag{3.83}$$

The displacement of the Fermi sphere due to an applied electric field is shown in Figure 3.10.

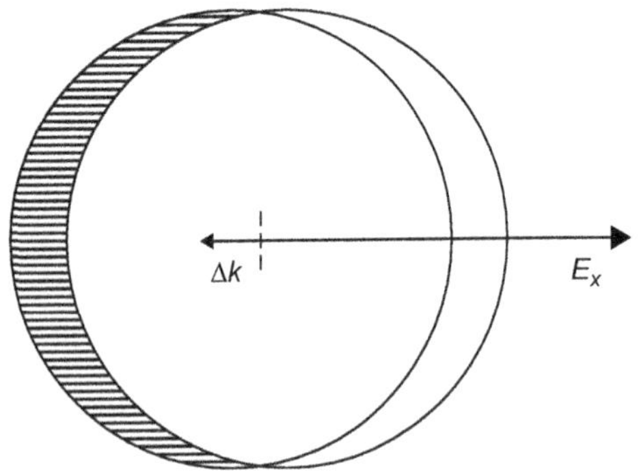

FIGURE 3.10

Displacement of the Fermi sphere in an applied electric field.

Because the change of momentum of each electron is

$$\Delta\mathbf{p} = \hbar\Delta\mathbf{k} = m\Delta\mathbf{v}, \tag{3.84}$$

and the electric current density of n electrons per unit volume is

$$\mathbf{j} = -ne\Delta\mathbf{v}, \tag{3.85}$$

we obtain from Eqs. (3.83) through (3.85)

$$\mathbf{j} = \frac{ne^2\tau}{m}\mathbf{E}. \tag{3.86}$$

The electrical conductivity σ is defined by (note that conductivity is a tensor, but we are treating it here as a scalar quantity)

$$\mathbf{j} = \sigma\mathbf{E}. \tag{3.87}$$

From Eqs. (3.86) and (3.87), we obtain

$$\sigma = \frac{ne^2\tau}{m}. \tag{3.88}$$

Eq. (3.88) for the electrical conductivity σ, derived by considering the displacement of the Fermi sphere in an electric field, is identical to the expression derived by using Drude's theory of metals (Problem 3.1), in which electrons were treated as a classical electron gas. However, one significant difference is that although the relaxation time defined by Drude was the average collision time between the electrons and the static positively charged ions while in the present model, the collisions of the electrons with impurities, lattice imperfections, and phonons can be included. The two results appear identical because of the approximation of a relaxation time τ in both models.

The following are the major drawbacks of the theory of DC electrical conductivity. First, it may be noted that in this derivation, the average relaxation time τ was not properly defined, although it was implicitly borrowed from the concept of the relaxation time defined by Drude. Second, because the origin of the collision time was left unanswered, the temperature dependence of the DC conductivity in metals could be explained only by introducing a temperature dependence in the relaxation time τ. Third, the current density **j** is not parallel to the electric field **E** in some metals, and the DC conductivity depends on the orientation of the metal with respect to **E**.

3.6 THE HALL EFFECT

Here, we consider a conductor in the shape of a rod that has a rectangular cross-section, which is placed under a magnetic field **B** in the z direction. There is a longitudinal electric field E_x. The electric and magnetic fields are so adjusted that the current cannot flow out of the rod in the y direction ($j_y = 0$). This configuration is shown in Figure 3.11.

The equation of motion of the displacement $\delta\mathbf{k}$, a Fermi sphere of particles acted on a force **F**, is given by

$$\mathbf{F} = \hbar\left(\frac{d}{dt} + \frac{1}{\tau}\right)\delta\mathbf{k}, \tag{3.89}$$

where τ is the relaxation time. The Lorentz force on carrier of charge q (which can be positive or negative) in an electric field **E** and a magnetic field **B** is

$$\mathbf{F} = q\left(\mathbf{E} + \frac{1}{c}\,\mathbf{V}\times\mathbf{B}\right). \tag{3.90}$$

Here, we note that $q = -e$ for electrons, but the Hall effect is equally applicable for determining the sign of other carriers in a solid.

If $\hbar\,\delta\mathbf{k} = m\delta\mathbf{V}$, Eq. (3.90) can be written as

$$m\left(\frac{d}{dt} + \frac{1}{\tau}\right)\delta\mathbf{V} = q\left(\mathbf{E} + \frac{1}{c}\,\delta\mathbf{V}\times\mathbf{B}\right). \tag{3.91}$$

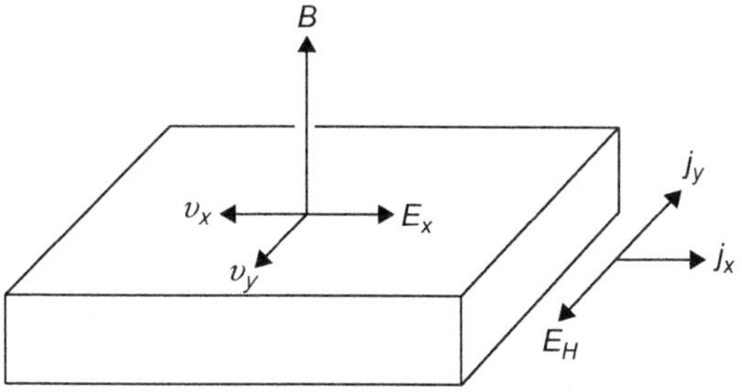

FIGURE 3.11

The geometry of the Hall effect.

Here, $\delta\mathbf{V}$ is the average of $\mathbf{V}$ over the Fermi sphere. In the steady state, the acceleration $\frac{d\delta\mathbf{V}}{dt} = 0$. We can write Eq. (3.91) in the steady state (writing $\mathbf{V}$ for $\delta\mathbf{V}$ for convenience),

$$\mathbf{V} = \frac{q\tau}{m}(\mathbf{E} + \mathbf{V} \times \mathbf{B}). \tag{3.92}$$

Because the static electric field $\mathbf{E}$ lies in the xy plane and $\mathbf{B}$ is along the z axis (Figure 3.7), we can write the components of Eq. (3.92) as

$$V_x = \frac{q\tau}{m} E_x - \omega_c \tau V_y, \tag{3.93}$$

and

$$V_y = \frac{q\tau}{m} E_y + \omega_c \tau V_x. \tag{3.94}$$

Here,

$$\omega_c \equiv \frac{-qB}{mc} \tag{3.95}$$

is the cyclotron frequency, usually defined for free electrons (in which case $q = -e$). By solving Eqs. (3.93) and (3.94), we obtain

$$V_x = \frac{q\tau/m}{1 + (\omega_c \tau)^2}(E_x - \omega_c \tau E_y), \tag{3.96}$$

and

$$V_y = \frac{q\tau/m}{1 + (\omega_c \tau)^2}(E_y + \omega_c \tau E_x). \tag{3.97}$$

The conductivity is

$$\sigma_0 = \frac{nq^2\tau}{m}, \tag{3.98}$$

and the components of charge current density from Eqs. (3.96) through (3.98),

$$j_x = nqV_x = \frac{\sigma_0}{1 + (\omega_c \tau)^2}(E_x - \omega_c \tau E_y), \tag{3.99}$$

$$j_y = nqV_y = \frac{\sigma_0}{1 + (\omega_c \tau)^2}(E_y + \omega_c \tau E_x), \tag{3.100}$$

and because the z-component of the current is not affected by the magnetic field,

$$j_z = nqV_z = \sigma_0 E_z. \tag{3.101}$$

In the Hall effect, the external fields are $E = E_x$ and $B_z = B$. There is a drift of the charges in the y direction due to the magnetic field B. However, no current can flow in this direction because of lack of circuit continuity. Instead, the charges pile up on the surface of the sample, thereby setting up an electric field E_y. This field, which nullifies the Lorentz force, is known as the "Hall field."

The Hall field is obtained by setting

$$j_y = nqv_y = 0. \tag{3.102}$$

From Eqs. (3.100) and (3.102), we obtain

$$\frac{\sigma_0}{1+\omega_c{}^2\tau^2}(\omega_c\tau E_x + E_y) = 0, \tag{3.103}$$

which yields

$$E_y = E_H = -\omega_c\tau E_x = \frac{qB}{m}\tau E_x. \tag{3.104}$$

From Eqs. (3.99) and (3.104), we obtain

$$j_x = \frac{\sigma_0}{1+\omega_c{}^2\tau^2}(E_x + \omega_c{}^2\tau^2 E_x) = \sigma_0 E_x. \tag{3.105}$$

The Hall coefficient R_H is defined as

$$R_H = \frac{E_H}{j_x B} = \frac{q\tau}{\sigma_0 m} = \frac{1}{nq}. \tag{3.106}$$

The Hall resistance is defined as

$$\mathfrak{R}_H = \frac{V_y}{I_x}. \tag{3.107}$$

The sign of the Hall coefficient is determined by the charge carrier q, and for metals, these are the electrons of which the charge is $-e$. Later, we will see the impact of the Hall effect in determining the sign of the charge carriers in semiconductors as well as study the importance of the quantum Hall effect.

3.7 FAILURES OF THE FREE ELECTRON MODEL

The failures of the free electron model can be briefly summarized as follows:

a. The temperature and the directional dependence of the DC electrical conductivity could not be adequately explained by the Sommerfeld theory.

b. The free electron model explains the contribution of the linear term in T to the specific heat of metals, but the magnitude of this contribution is inadequate for many metals in the sense that it is either too large or too small. In addition, there is a cubic term (T^3) that contributes to the specific heat of metals but cannot be explained by the free electron model.

c. The major failure of the free electron model is that it does not explain the significant difference between metals, insulators, and semiconductors. To be specific, it fails to explain why some elements crystallize as good conductors (metals) of which the conductivity decreases with the increase of temperature, some others crystallize as insulators that are very poor conductors, and the rest crystallize as semiconductors of which the conductivity is very poor at low temperatures but increases with increase of temperature unlike that of metals. In fact, it has been commented that the difference of electrical resistivity at room temperature between that of good conductors ($10^{-10}-10^{-6}$) ohm-cm and of good insulators ($10^{14}-10^{22}$) ohm-cm is one of the widest range of any physical property ever found in nature. In contrast, the electrical resistivity of semiconductors at room temperature is in the range ($10^{-2}-10^{9}$) ohm-cm but is strongly dependent on temperature.

To understand this major classification of solids, we have to include the effect of the positively charged stationary ions that are arranged in a periodic array. As we will show, the periodic array of ions leads to a periodic potential that results in the band theory of solids. The band theory of solids explains the difference between the electrical properties of metals, semiconductors, and insulators. In addition, it distorts the spherical Fermi surface to a more complex shape, and as we have noted, the Fermi surface determines many physical properties of the solid.

There are other significant deficiencies in the free electron model. First, it introduces the relaxation time of the electrons in an unrealistic manner in the sense that the role of the background positive ions is confined only to scatter an electron moving in an external field. Second, it ignores the electron–electron interactions, which has significant contribution to some physical properties. Third, the positively charged ions are not stationary but oscillate around their equilibrium position. Since these ions are positively charged, they affect the motion of the negatively charged electrons (the electron–phonon interaction). Fourth, the free electron model completely ignores the effects of lattice defects and lattice impurities. Fifth, the free electron model ignores the surface effects because of the periodic boundary condition invoked as a necessary condition of the model.

PROBLEMS

3.1. Show that in Drude's model, the average electron velocity in an external electric field $\mathbf{E}$ is $\mathbf{v}_{avg} = -\frac{e\mathbf{E}\tau}{m}$ and the current density $\mathbf{j} = -ne\mathbf{v} = \left(\frac{ne^2\tau}{m}\right)\mathbf{E}$. Here, n is the number of electrons per unit volume. Because $\mathbf{j} = \sigma\mathbf{E}$, where σ is the DC electrical conductivity of the metal, show that $\sigma = \frac{ne^2\tau}{m}$.

3.2. The basic assumption of the Drude model is that the thermal current in a metal is carried by the conduction electrons. If we define the thermal current density $\mathbf{j}^q = -\kappa\nabla T$, where κ is the thermal conductivity, it can be easily shown that $\kappa = \frac{1}{3}v^2\tau c_v$, where c_v is the electronic specific heat. By applying classical ideal gas laws to the electron gas, $c_v = \frac{3}{2}nk_B$, and $\frac{1}{2}mv^2 = \frac{3}{2}k_BT$ (where k_B is the Boltzmann constant), show that

$$\frac{\kappa}{\sigma} = \frac{3}{2}\left(\frac{k_B}{e}\right)^2 T.$$

This relation between the electrical and thermal conductivities of a metal is known as the Wiedemann–Franz law.

3.3. Show that the Fermi–Dirac distribution function reduces to the Maxwell–Bolzmann distribution function $f(\varepsilon) = e^{-(\varepsilon-\mu)/k_BT}$ for very high temperatures when $\varepsilon \ll \mu$. An alternate condition to make this classical approximation is $e^{-\mu/k_BT} \gg 1$.

3.4. Derive an expression for r_s (defined in Eq. 3.2) by using the classical approximation outlined in Problem 3.3.

3.5. Show that the wave function of a free electron confined in a cubic box of side L is

$$\psi_{\mathbf{k}}(\mathbf{r}) = \left(\frac{8}{L^3}\right)^{1/2} \sin\left(\frac{n_1 \pi x}{L}\right) \sin\left(\frac{n_2 \pi y}{L}\right) \sin\left(\frac{n_3 \pi z}{L}\right).$$

3.6. Show from Eqs. (3.2) and (3.40) that $k_F = \frac{(9\pi/4)^{1/3}}{r_s} = \frac{1.92}{r_s}$.

3.7. Show that the Fermi temperature T_F (defined in Eq. 3.50) can be written in the alternate form

$$T_F = \frac{58.2}{(r_s/a_0)^2} \times 10^4\, K,$$

where a_0 is the Bohr radius.

3.8. If a function $F(\varepsilon)$ does not vary rapidly in the energy range of the order of $k_B T$ about the chemical potential μ, by making a Taylor series expansion of the form,

$$F(\varepsilon) = \sum_{n=0}^{\infty} \frac{d^n}{d\varepsilon^n} F(\varepsilon)\,|_{\varepsilon=\mu} \frac{(\varepsilon-\mu)^n}{n!}. \tag{1}$$

Define a function

$$\phi(\varepsilon) = \int_{-\infty}^{\varepsilon} F(\varepsilon')d\varepsilon' \tag{2}$$

so that

$$F(\varepsilon) = \frac{d\phi(\varepsilon)}{d\varepsilon}. \tag{3}$$

Integrating by parts, show that

$$\int_{-\infty}^{\infty} F(\varepsilon) f(\varepsilon) d\varepsilon = \int_{-\infty}^{\infty} \phi(\varepsilon)\left(-\frac{\partial f}{\partial \varepsilon}\right) d\varepsilon, \tag{4}$$

where $f(\varepsilon)$ is the Fermi function. Expand $\phi(\varepsilon)$ in a Taylor series about $\varepsilon = \mu$ and show that

$$\phi(\varepsilon) = \phi(\mu) + \sum_{n=1}^{\infty} \left[\frac{(\varepsilon-\mu)^n}{n!}\right] \left[\frac{d^n \phi(\varepsilon)}{d\varepsilon^n}\right]_{\varepsilon=\mu}. \tag{5}$$

By using the identity

$$\int_{-\infty}^{\infty} (-\partial f/\partial \varepsilon) d\varepsilon = 1, \tag{6}$$

and the fact that $\partial f/\partial\varepsilon$ is an even function of $\varepsilon-\mu$ (only terms with even n in Eq. 5 contribute), show that

$$\int_{-\infty}^{\infty} d\varepsilon\, F(\varepsilon)f(\varepsilon) = \int_{-\infty}^{\mu} F(\varepsilon)d\varepsilon + \sum_{n=1}^{\infty} \int_{-\infty}^{\infty} \frac{(\varepsilon-\mu)^{2n}}{(2n)!}\left(-\frac{\partial f}{\partial\varepsilon}\right)d\varepsilon \frac{d^{2n-1}}{d\varepsilon^{2n-1}}F(\varepsilon)\,|_{\varepsilon=\mu}. \tag{7}$$

Show that the integration can also be expressed as

$$\int_{-\infty}^{\infty} F(\varepsilon)f(\varepsilon)d\varepsilon = \int_{-\infty}^{\mu} F(\varepsilon)d\varepsilon + \sum_{n=1}^{\infty} (k_BT)^{2n} a_n \frac{d^{2n-1}}{d\varepsilon^{2n-1}}F(\varepsilon)\,|_{\varepsilon=\mu}, \tag{8}$$

where

$$a_n = \int_{-\infty}^{\infty} \frac{x^{2n}}{(2n)!}\left(-\frac{d}{dx}\frac{1}{(e^x+1)}\right)dx. \tag{9}$$

This is known as the Sommerfeld expansion. Here, a_n are dimensionless constants of the order of unity and are related to the Riemann zeta function, $\xi(n)$, as

$$a_n = \left(2-\frac{1}{2^{2(n-1)}}\right)\xi(2n). \tag{10}$$

The Riemann zeta function, $\xi(n)$, is defined as

$$\xi(n) = 1+\frac{1}{2^n}+\frac{1}{3^n}+\frac{1}{4^n}+\ldots \tag{11}$$

3.9. From Eqs. (3.99) through (3.101), show that the current density can be written in the matrix form

$$\begin{pmatrix} j_x \\ j_y \\ j_z \end{pmatrix} = \frac{\sigma_0}{(1+\omega_c^2\tau^2)}\begin{pmatrix} 1 & -\omega_c\tau & 0 \\ \omega_c\tau & 1 & 0 \\ 0 & 0 & 1+\omega_c{}^2\tau^2 \end{pmatrix}\begin{pmatrix} E_x \\ E_y \\ E_z \end{pmatrix}. \tag{1}$$

The matrix on the right side of Eq. (1) gives the nine components of the magnetoconductivity tensor. Hence, show that

$$\sigma_{xx} = \sigma_{yy} = \frac{\sigma_0}{1+\omega_c{}^2\tau^2}, \tag{2}$$

and

$$\sigma_{xy} = -\sigma_{yx} = \frac{\sigma_0\omega_c\tau}{1+\omega_c{}^2\tau^2}. \tag{3}$$

Thus, the magnitude of the diagonal components σ_{xx} and σ_{yy} decrease monotonically as the magnetic field B is increased, whereas the off-diagonal components σ_{xy} and σ_{yx} at first increase and then decrease as the magnetic field is increased.

References

1. Ashcroft NW, Mermin ND. *Solid state physics*. New York: Brooks/Cole; 1976.
2. Harrison WA. *Solid state theory*. New York: McGraw-Hill; 1969.
3. Kittel C. *Introduction to solid state physics*. New York: John Wiley & Sons; 1967.
4. Landau LD, Lifshitz EM. *Statistical physics, part 1*. Oxford: Pergamon Press; 1980.
5. Madelung O. *Introduction to solid state theory*. New York: Springer-Verlag; 1978.
6. Marder MP. *Condensed matter physics*. New York: John Wiley & Sons; 2000.
7. Myers HP. *Introductory solid state physics*. London: Taylor & Francis; 1990.
8. Pathria RK. *Statistical mechanics*. Oxford: Butterworth-Heinemann; 2000.
9. Reif F. *Fundamentals of statistical and thermal physics*. New York: McGraw-Hill; 1965.
10. Seitz F. *The modern theory of solids*. New York: McGraw-Hill; 1940.
11. Ziman JM. *Electrons and phonons*. Oxford: Oxford University Press; 1960.
12. Ziman JM. *Principles of the theory of solids*. Cambridge: Cambridge University Press; 1972.

CHAPTER

Nearly Free Electron Model

4

CHAPTER OUTLINE

4.1 Electrons in a Weak Periodic Potential 96
4.1.1 Introduction 96
4.1.2 Plane Wave Solutions 97
4.2 Bloch Functions and Bloch Theorem 99
4.3 Reduced, Repeated, and Extended Zone Schemes 99
4.3.1 Reduced Zone Scheme 100
4.3.2 Repeated Zone Scheme 100
4.3.3 Extended Zone Scheme 101
4.4 Band Index 101
4.5 Effective Hamiltonian 102
4.6 Proof of Bloch's Theorem from Translational Symmetry 103
4.7 Approximate Solution Near a Zone Boundary 105
4.8 Different Zone Schemes 109
4.8.1 Reduced Zone Scheme 109
4.8.2 Extended Zone Scheme 110
4.8.3 Repeated Zone Scheme 111
4.9 Elementary Band Theory of Solids 111
4.9.1 Introduction 111
4.9.2 Energy Bands in One Dimension 112
4.9.3 Number of States in a Band 112
4.10 Metals, Insulators, and Semiconductors 112
4.11 Brillouin Zones 117
4.12 Fermi Surface 119
4.12.1 Fermi Surface (in Two Dimensions) 119
4.12.2 Fermi Surface (in Three Dimensions) 121
4.12.3 Harrison's Method of Construction of the Fermi Surface 121
Problems 124
References 130

4.1 ELECTRONS IN A WEAK PERIODIC POTENTIAL

4.1.1 Introduction

In the nearly free electron approximation, it is assumed that there are no electron–electron or electron–phonon interactions. This means that a valence electron, stripped from its parent atom due to the attractive interaction of the neighboring positively charged ions, does not interact either with other electrons or with the vibrating motion of the ions at a finite temperature. However, unlike the free electron approximation, the electron is subjected to a weak periodic potential due to the background of symmetric array of positively charged ions in the crystal lattice. We will first show that this potential is periodic with the periodicity of a lattice vector.

As an example, we consider a two-dimensional rectangular lattice, as shown in Figure 4.1.

If **OO′** = $\vec{R}_i$ and **OE** = $\vec{r}$, the potential energy at the electron E (charge $-e$) due to the positively charged ions of the crystal lattice (one ion of charge ze is assumed to be located at each lattice site) is given by

$$V(\vec{r}) = \sum_i \frac{-ze^2}{|\vec{r} - \vec{R}_i|}. \tag{4.1}$$

Here, we have considered the fundamental principle of electrostatics that in a spherical charge distribution, the potential at a point outside the sphere is the same as that of the potential due to the net charge considered to be located at the center of the sphere.

If $\vec{r}$ is translated by a lattice vector $\vec{R}_j$, Eq. (4.1) can be written in the alternate form

$$V(\vec{r} + \vec{R}_j) = \sum_i \frac{-ze^2}{|\vec{r} - \vec{R}_i + \vec{R}_j|}. \tag{4.2}$$

One can write $\vec{R}_i - \vec{R}_j = \vec{R}_l$, in which case Eq. (4.2) can be written in the alternate form

$$V(\vec{r} + \vec{R}_j) = \sum_l \frac{-ze^2}{|\vec{r} - \vec{R}_l|}. \tag{4.3}$$

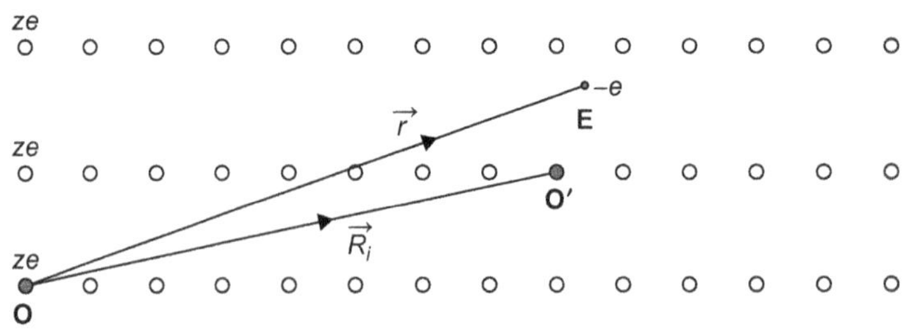

FIGURE 4.1

Two-dimensional rectangular lattice with **O** as the global origin and **O′** as the local origin of the unit cell within which the electron of charge $-e$ is located at **E**. The charge of each ion is ze.

Because the summation over both $\vec{R}_i$ and $\vec{R}_l$ spans the entire lattice vectors, we obtain from Eqs. (4.1) and (4.3)

$$V(\vec{r}) = V(\vec{r} + \vec{R}_j). \tag{4.4}$$

Eq. (4.4) clearly demonstrates that $V(\vec{r})$ is a periodic potential with the periodicity of a direct lattice vector. It may be noted that this proof was based on the simple assumption that an ion of charge ze is located at each lattice point. However, the proof can be generalized to an identical cluster of ions (a basis), located symmetrically around each lattice point.

4.1.2 Plane Wave Solutions

For simplicity, we consider a linear lattice of lattice constant a. Later, we will generalize our results to a three-dimensional lattice. From Eq. (4.4), the periodic potential in a one-dimensional lattice can be written as

$$V(x) = V(x + n\,a), \tag{4.5}$$

where n is an integer. If we express the periodic potential $V(x)$ as a Fourier series

$$V(x) = \sum_{q'} V(q')e^{iq'x}, \tag{4.6}$$

we obtain

$$V(x + na) = \sum_{q'} V(q')e^{iq'(x+na)}. \tag{4.7}$$

From Eqs. (4.5) through (4.7), it is easy to show that

$$\sum_{q'} V(q')e^{iq'x} = \sum_{q'} V(q')e^{iq'(x+na)}. \tag{4.8}$$

Eq. (4.8) has to be valid for each value of the integer n. This is possible only if

$$e^{iq'na} = 1, \tag{4.9}$$

for all values of n and q'. This condition is satisfied only when $q' = 2\pi m/a$, where m is any integer. This is precisely the definition of a reciprocal lattice vector K in one dimension and hence $q' = K$. The periodic potential $V(x)$ in Eq. (4.3) can therefore be expressed as

$$V(x) = \sum_{K} V(K)e^{iKx}. \tag{4.10}$$

The Schrodinger equation of the electron in a one-dimensional lattice is easily obtained,

$$\left(-\frac{\hbar^2}{2m}\frac{\partial^2}{\partial x^2} + \sum_{K} V(K)e^{iKx}\right)\psi(x) = E\psi(x), \tag{4.11}$$

where E is the energy eigenvalue and $\psi(x)$ is the wave function of the electron. The Born–von Karman boundary conditions imply that

$$\psi(x + Na) = \psi(x + L) = \psi(x). \tag{4.12}$$

$\psi(x)$ can also be expanded in terms of the plane waves, which are a complete set of functions,

$$\psi(x) = \sum_q a(q)e^{iqx}. \tag{4.13}$$

From Eqs. (4.11) and (4.13), we obtain

$$\sum_q \varepsilon_q^0 a(q)e^{iqx} + \sum_K \sum_q V(K)a(q)e^{i(q+K)x} = E\sum_q a(q)e^{iqx}, \tag{4.14}$$

where

$$\varepsilon_q^0 = \frac{\hbar^2 q^2}{2m}. \tag{4.15}$$

We assume that the one-dimensional crystal has a length L. Multiplying Eq. (4.14) by $e^{-iq'x}$ and integrating over x from 0 to L, we obtain

$$\sum_q \varepsilon_q^0 a(q)\int_0^L e^{i(q-q')x}dx + \sum_K \sum_q V(K)a(q)\int_0^L e^{i(q-q'+K)x}dx = E\sum_q a(q)\int_0^L e^{i(q-q')x}dx. \tag{4.16}$$

The Born–von Karman boundary conditions for a linear lattice lead to the conditions for the plane waves that $e^{iqx} = e^{iq(x+L)}$ and $e^{iq'x} = e^{iq'(x+L)}$. These conditions imply that both q and q' must satisfy

$$q = \frac{2\pi n}{L} \text{ and } q' = \frac{2\pi m}{L}, \tag{4.17}$$

where n and m are integers. The integration

$$I = \int_0^L e^{i(q-q')x}dx = \frac{e^{i(q-q')L} - 1}{i(q-q')} = \frac{[e^{i2\pi(n-m)} - 1]L}{2\pi i(n-m)} = L\delta_{n,m} = L\delta_{q,q'}, \tag{4.18}$$

where $\delta_{q,q'}$ is the Kronecker delta function. Similarly, one can show that

$$I' = \int_0^L e^{i(q-q'+K)x}dx = L\delta_{q,q'-K}. \tag{4.19}$$

Substituting the results of Eqs. (4.18) and (4.19) in (4.16), we obtain

$$\varepsilon_q'^0 a(q') + \sum_K V(K)a(q'-K) = Ea(q'), \tag{4.20}$$

which can be written in the alternate form, by substituting q for q',

$$(\varepsilon_q^0 - E)a(q) + \sum_K V(K)a(q-K) = 0. \tag{4.21}$$

Eq. (4.21) can be expressed as

$$a(q) = \sum_K \frac{V(K)a(q-K)}{E - \varepsilon_q^0}. \tag{4.22}$$

It is obvious from Eq. (4.22) that $a(q)$ is small unless $E \approx \varepsilon_q^0$.

We can also subtract an arbitrary reciprocal lattice vector K' from q in Eq. (4.21) and rewrite it as

$$(\varepsilon^0_{q-K} - E)a(q-K') + \sum_K V(K)a(q-K'-K) = 0. \tag{4.23}$$

Eq. (4.22) connects $a(q)$ with every Fourier coefficient $a(q-K)$, i.e., with the Fourier coefficients for which the wave vector differs from q by a reciprocal lattice vector K. This leads to a very important and useful result about the form of the eigenfunctions ψ. These wave functions of an electron in a periodic potential are called the Bloch functions.

The equivalent proof correlating $a(\mathbf{q})$ and $a(\mathbf{q}-\mathbf{K})$ for a three-dimensional crystal lattice is assigned as a homework problem (see Problem 4.4).

4.2 BLOCH FUNCTIONS AND BLOCH THEOREM

In Eq. (4.12), we considered an arbitrary wave vector that appears in the summation over q and denoted it as k. We note from Eq. (4.21) that instead of the continuous Fourier coefficients $a(q)$, only those of the form $a(k-K)$ enter into $\psi_k(x)$; i.e., the allowed K's in the wave function are of the form $k-K$. Thus, we can write

$$\psi_k(x) = \sum_K a(k-K)e^{i(k-K)x}. \tag{4.24}$$

Eq. (4.24) can be written in the alternate form

$$\psi_k(x) = \left(\sum_K a(k-K)e^{-iKx}\right)e^{ikx}. \tag{4.25}$$

If we introduce

$$u_k(x) \equiv \sum_K a(k-K)e^{-iKx}, \tag{4.26}$$

we obtain

$$\psi_k(x) = e^{ikx}u_k(x). \tag{4.27}$$

We note from Eq. (4.26) that if m is an integer,

$$u_k(x+ma) = \sum_K a(k-K)e^{-iK(x+ma)} = u_k(x) \tag{4.28}$$

because

$$e^{-iKma} = e^{-2im\pi} = 1. \tag{4.29}$$

$\psi_k(x)$ is referred to as a Bloch function and $u_k(x)$ is known as the periodic part of the Bloch function because it has the periodicity of the lattice.

4.3 REDUCED, REPEATED, AND EXTENDED ZONE SCHEMES

We will now discuss the three types of zone schemes (reduced, repeated, and extended) used to describe electrons in a crystal lattice. For simplicity, we will first discuss these schemes for free electrons in a one-dimensional lattice to introduce the concept of the band index before we extend our discussion to electrons in a three-dimensional lattice as well as to electrons in a periodic potential.

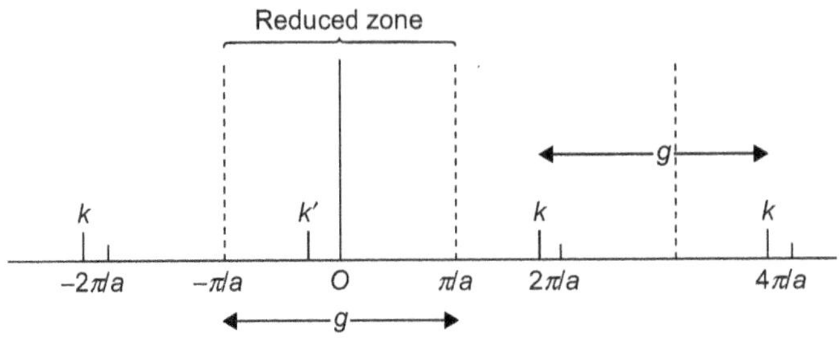

FIGURE 4.2

All k points reduce to k' in the one-dimensional reciprocal lattice.

4.3.1 Reduced Zone Scheme

If we consider a one-dimensional lattice, the reciprocal lattice vectors K can also be relabeled as g_n where n is an integer (positive or negative),

$$g_n = n\frac{2\pi}{a}. \tag{4.30}$$

If we restrict k to the first Brillouin zone, i.e., if we assign a state k, any wave number in the set

$$k = k' + \frac{2\pi}{a}n, \tag{4.31}$$

k is only defined modulo $(2\pi/a)$. Thus, all the k points in Figure 4.2 are equivalent.

One can therefore consider k' as the representative of all these k values, with $|k'|$ restricted to the first Brillouin zone. Thus, it is always possible to choose the value of k such that

$$-\frac{\pi}{a} < k \leq \frac{\pi}{a}. \tag{4.32}$$

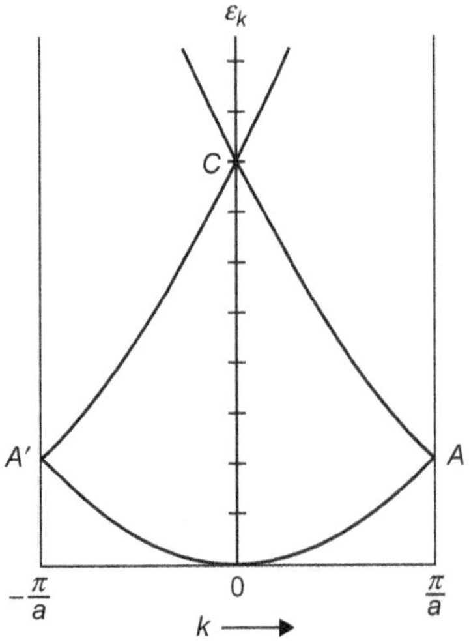

FIGURE 4.3

Reduced zone scheme for energy-wave number relations of free electrons in a one-dimensional lattice.

This way of restricting the wave numbers to the first Brillouin zone is known as the reduced zone scheme. In Figure 4.3, the reduced zone scheme is shown by drawing the energy-wave number relation for free electrons, $\varepsilon_0(k) = \hbar^2k^2/2m$.

We will now describe the energy-wave number relations $\varepsilon^0(k) = \hbar^2k^2/2m$ for the repeated and extended zone schemes.

4.3.2 Repeated Zone Scheme

It is often convenient to repeat the first Brillouin zone and the other zones reduced to the first Brillouin zone through all of k space. Thus, in the repeated zone scheme, $\varepsilon_k^0 = \varepsilon_{k+K}^0$ in one

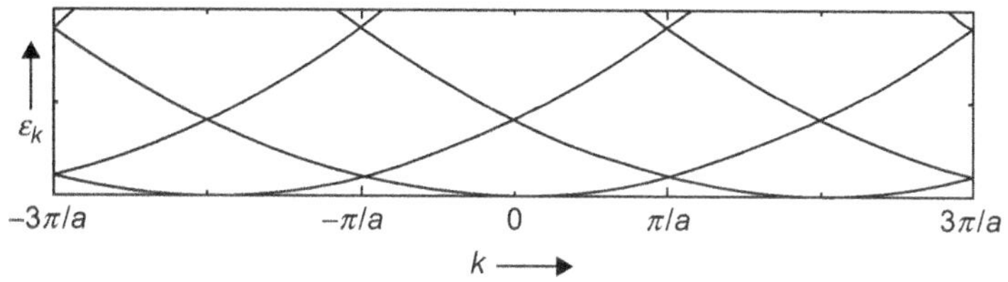

FIGURE 4.4

Energy-wave number relations for a one-dimensional lattice in the repeated zone scheme.

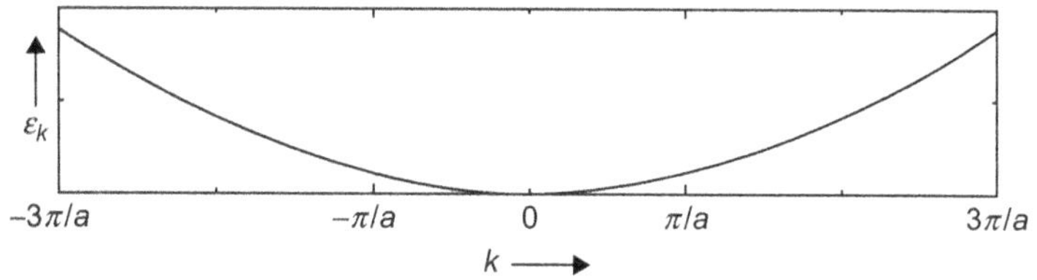

FIGURE 4.5

Energy-wave number relations in the extended zone scheme.

dimension and $\varepsilon^0_{\mathbf{k}} = \varepsilon^0_{\mathbf{k}+\mathbf{K}}$ in three dimensions. All values of ε^0_k, ε^0_{k+K}, ε^0_{k-K}, etc. are plotted against k, $k+K$, $k-K$, etc. in one dimension. The repeated zone scheme is useful in describing several physical properties of the solid, specifically the connectivity of electron orbits in a magnetic field. The repeated zone scheme for free electrons in a one-dimensional lattice is shown schematically in Figure 4.4.

4.3.3 Extended Zone Scheme

In the extended zone scheme, the k values extend throughout all reciprocal space, and the energy eigenvalues, $\varepsilon^0_k = \hbar^2 k^2/2m$, are plotted against the wave number k. Thus, for free electrons, one obtains a parabola. The extended zone scheme for free electrons in a one-dimensional lattice is shown schematically in Figure 4.5.

4.4 BAND INDEX

A large number of eigenfunctions and eigenvalues correspond to the same wave vector $\mathbf{k}$ (wave numbers k in one dimension) in the reduced zone scheme. To distinguish these eigenfunctions and eigenvalues in the reduced zone scheme, we introduce an additional index n (called band index). The band index has a much greater significance when one considers the eigenfunctions and eigenvalues by including the periodic potential in the Hamiltonian. The periodic potential opens up an energy gap at the zone boundaries, and the band index plays a much greater role in the classification of solids as metals, insulators, and semiconductors. The necessity of using a band index also follows as a natural consequence when we discuss the effective Hamiltonian.

4.5 EFFECTIVE HAMILTONIAN

The necessity of a band index n will be evident by constructing an effective Hamiltonian. The Schrodinger equation can be written as

$$H\psi_{\mathbf{k}}(\mathbf{r}) = \left(-\frac{\hbar^2}{2m}\nabla^2 + V(\mathbf{r})\right)e^{i\mathbf{k}\cdot\mathbf{r}}u_{\mathbf{k}}(\mathbf{r}) = E_{\mathbf{k}}e^{i\mathbf{k}\cdot\mathbf{r}}u_{\mathbf{k}}(\mathbf{r}). \tag{4.33}$$

We can rewrite Eq. (4.33) in the form

$$e^{i\mathbf{k}\cdot\mathbf{r}}\left\{\frac{\hbar^2}{2m}[-\nabla^2 - 2i\mathbf{k}\cdot\vec{\nabla} + k^2]u_{\mathbf{k}}(\mathbf{r}) + V(\mathbf{r})u_{\mathbf{k}}(\mathbf{r})\right\} = e^{i\mathbf{k}\cdot\mathbf{r}}E_{\mathbf{k}}(\mathbf{r})u_{\mathbf{k}}(\mathbf{r}). \tag{4.34}$$

Canceling $e^{i\mathbf{k}\cdot\mathbf{r}}$ from both sides, we obtain

$$\left\{\frac{\hbar^2}{2m}[-\nabla^2 - 2i\mathbf{k}\cdot\vec{\nabla} + k^2] + V(\mathbf{r})\right\}u_{\mathbf{k}}(\mathbf{r}) = E_{\mathbf{k}}(\mathbf{r})u_{\mathbf{k}}(\mathbf{r}). \tag{4.35}$$

Eq. (4.35) can be expressed as

$$H_{eff}u_{\mathbf{k}}(\mathbf{r}) = E_{\mathbf{k}}(\mathbf{r})u_{\mathbf{k}}(\mathbf{r}), \tag{4.36}$$

where the effective Hamiltonian

$$H_{eff} = \frac{\hbar^2}{2m}[-\nabla^2 - 2i\mathbf{k}\cdot\nabla + k^2] + V(\mathbf{r}). \tag{4.37}$$

The boundary conditions are that whenever $\mathbf{r}$ lies on one boundary of the unit cell and $\mathbf{r}+\mathbf{R}$ is another boundary of the unit cell, then

$$u_{\mathbf{k}}(\mathbf{r}+\mathbf{R}) = u_{\mathbf{k}}(\mathbf{r}) \tag{4.38}$$

and

$$\hat{n}(\mathbf{r})\cdot\vec{\nabla}u_{\mathbf{k}}(\mathbf{r}) = -\hat{n}(\mathbf{r}+\mathbf{R})\cdot\vec{\nabla}u_{\mathbf{k}}(\mathbf{r}+\mathbf{R}), \tag{4.39}$$

where $\hat{n}(\mathbf{r})$ is a unit normal to the cell boundary at $\mathbf{r}$.

Thus, Eq. (4.36) can be considered as a Hermitian eigenvalue problem that is restricted to a single primitive cell of the crystal. Because the eigenvalue problem is in a fixed finite volume, there will be an infinite family of solutions with discretely spaced eigenvalues. These are labeled with the band index n. The importance of the band index will become apparent when we consider the effect of the periodic potential by using perturbation theory.

If we include the band index n, Eq. (4.27) can now be rewritten as

$$\psi_{nk}(x) = e^{ikx}u_{nk}(x). \tag{4.40}$$

It is easy to generalize Eq. (4.27) to a three-dimensional crystal lattice (see Problem 4.4),

$$\psi_{\mathbf{k}}(\mathbf{r}) = e^{i\mathbf{k}\cdot\mathbf{r}}u_{\mathbf{k}}(\mathbf{r}), \tag{4.41}$$

where

$$u_{\mathbf{k}}(\mathbf{r}+\mathbf{R}) = u_{\mathbf{k}}(\mathbf{r}). \tag{4.42}$$

For a three-dimensional lattice, in analogy with Eq. (4.31), we can write

$$\mathbf{k} = \mathbf{k}' + \mathbf{K}, \tag{4.43}$$

where $\mathbf{k}'$ is restricted to the first Brillouin zone. From Eq. (4.41) and (4.43), we obtain

$$\begin{aligned}\psi_{\mathbf{k}}(\mathbf{r}+\mathbf{R}_l) &= e^{i(\mathbf{k}'+\mathbf{K})\cdot\mathbf{R}_l}\psi_{\mathbf{k}}(\mathbf{r}) \\ &= e^{i\mathbf{K}\cdot\mathbf{R}_l}e^{i\mathbf{k}'\cdot\mathbf{R}_l}\psi_{\mathbf{k}}(\mathbf{r}) \\ &= e^{i\mathbf{k}'\cdot\mathbf{R}_l}\psi_{\mathbf{k}}(\mathbf{r}).\end{aligned} \tag{4.44}$$

From Eq. (4.44), it is evident that $\psi_{\mathbf{k}}(\mathbf{r})$ satisfies the Bloch's theorem with the wave vector $\mathbf{k}'$. Thus, every state has a large number of possible wave vectors, differing from each other by the reciprocal lattice vectors $\mathbf{K}$. If we choose the value of $\mathbf{K}$ such that $\mathbf{k}'$ lies in the first Brillouin zone (which is the reduced zone scheme, in which we relabel $\mathbf{k}'$ as $\mathbf{k}$), there will be a large number of eigenfunctions and eigenvalues corresponding to the same wave vector $\mathbf{k}$.

If we introduce the band index n, which follows as a consequence of restricting the wave vector $\mathbf{k}$ to the first Brillouin zone, Eq. (4.44) can be rewritten as

$$\psi_{n\mathbf{k}}(\mathbf{r}) = e^{i\mathbf{k}\cdot\mathbf{r}}u_{n\mathbf{k}}(\mathbf{r}) \tag{4.45}$$

and

$$u_{n\mathbf{k}}(\mathbf{r}+\mathbf{R}) = u_{n\mathbf{k}}(\mathbf{r}). \tag{4.46}$$

Eq. (4.45) is known as the Bloch theorem; $\psi_{n\mathbf{k}}(\mathbf{r})$ is known as the Bloch function, and $u_{n\mathbf{k}}(\mathbf{r})$ is known as the periodic part of the Bloch function. The Bloch theorem can also be proved by using the translational symmetry of the crystal lattice.

4.6 PROOF OF BLOCH'S THEOREM FROM TRANSLATIONAL SYMMETRY

We will now prove Bloch's theorem by using the translational symmetry of the crystal lattice. Through use of a three-dimensional equivalence of Problem 4.1, it can be easily shown that the translation operator $\hat{T}(\mathbf{R}_i)$ is defined by

$$\hat{T}(\mathbf{R}_i)f(\mathbf{r}) = f(\mathbf{r}+\mathbf{R}_i). \tag{4.47}$$

The Hamiltonian of the electron in the periodic potential can be written as

$$\hat{H}(\mathbf{r}) = \frac{-\hbar^2}{2m}\nabla^2 + V(\mathbf{r}). \tag{4.48}$$

From Eqs. (4.4), (4.47), and (4.48), we obtain

$$\hat{T}(\mathbf{R}_i)\hat{H}(\mathbf{r})f(\mathbf{r}) = \hat{H}(\mathbf{r}+\mathbf{R}_i)f(\mathbf{r}+\mathbf{R}_i) = \hat{H}(\mathbf{r})f(\mathbf{r}+\mathbf{R}_i) = \hat{H}(\mathbf{r})\hat{T}(\mathbf{R}_i)f(\mathbf{r}). \tag{4.49}$$

Because $f(\mathbf{r})$ is any arbitrary function of $\mathbf{r}$, we obtain

$$\hat{T}(\mathbf{R}_i)\hat{H}(\mathbf{r}) = \hat{H}(\mathbf{r})\hat{T}(\mathbf{R}_i). \tag{4.50}$$

It is also easy to show that

$$\hat{T}(\mathbf{R}_i)\hat{T}(\mathbf{R}_j)f(\mathbf{r}) = f(\mathbf{r}+\mathbf{R}_i+\mathbf{R}_j) = \hat{T}(\mathbf{R}_j)\hat{T}(\mathbf{R}_i)f(\mathbf{r}), \tag{4.51}$$

from which we have

$$\hat{T}(\mathbf{R}_i)\hat{T}(\mathbf{R}_j) = \hat{T}(\mathbf{R}_j)\hat{T}(\mathbf{R}_i) = \hat{T}(\mathbf{R}_i+\mathbf{R}_j). \tag{4.52}$$

From Eqs. (4.49) and (4.52), we note that the Hamiltonian $\hat{H}$ and the translation operators $\hat{T}(\mathbf{R}_i)$ (corresponding to each Bravais lattice vector $\mathbf{R}_i$) form a mutually commuting set of operators. Therefore, according to an important theorem in quantum mechanics,[5] these operators will have a complete set of common eigenfunctions.

If $\psi(\mathbf{r})$ is one of the eigenfunctions of the Hamiltonian with eigenvalue ε,

$$\hat{H}\psi(\mathbf{r}) = \varepsilon\psi(\mathbf{r}), \tag{4.53}$$

it follows from the previous theorem that

$$\hat{T}(\mathbf{R}_i)\psi(\mathbf{r}) = C(\mathbf{R}_i)\psi(\mathbf{r}) = \psi(\mathbf{r}+\mathbf{R}_i). \tag{4.54}$$

Here, $C(\mathbf{R}_i)$ are the eigenvalues of the translation operators $\hat{T}(\mathbf{R}_i)$. It also follows from Eq. (4.52) and Eq. (4.53) that

$$C(\mathbf{R}_i)C(\mathbf{R}_j) = C(\mathbf{R}_j)C(\mathbf{R}_i) = C(\mathbf{R}_i+\mathbf{R}_j). \tag{4.55}$$

Because $\mathbf{R}_i$ and $\mathbf{R}_j$ are Bravais lattice vectors, they can be expressed as

$$\mathbf{R}_i = n_1\mathbf{a}_1 + n_2\mathbf{a}_2 + n_3\mathbf{a}_3$$

and

$$\mathbf{R}_j = m_1\mathbf{a}_1 + m_2\mathbf{a}_2 + m_3\mathbf{a}_3, \tag{4.56}$$

where $\mathbf{a}_1, \mathbf{a}_2$, and $\mathbf{a}_3$ are the three primitive vectors of the Bravais lattice and n_1, n_2, n_3, m_1, m_2, and m_3 are appropriate integers corresponding to the lattice vectors $\mathbf{R}_i$ and $\mathbf{R}_j$. From Eqs. (4.55) and (4.56), it is obvious that $C(\mathbf{a}_i)$ must be an exponential of the form

$$C(\mathbf{a}_i) = e^{p_i}, \tag{4.57}$$

where p_i, which could be a complex number, has to be determined. From Eqs. (4.56) and (4.57), we obtain

$$C(\mathbf{R}_i) = C(\mathbf{a}_1)^{n_1}C(\mathbf{a}_2)^{n_2}C(\mathbf{a}_3)^{n_3}. \tag{4.58}$$

From Eqs. (4.57) and (4.58), we obtain

$$C(\mathbf{R}_i) = e^{n_1p_1+n_2p_2+n_3p_3}. \tag{4.59}$$

We now restate the Born–von Karman boundary conditions (originally stated for a cubic crystal in Eq. 3.11) for the wave functions of the electrons in a more general form (instead of restricting these conditions to a cubic crystal),

$$\psi(\mathbf{r}+M_i\mathbf{a}_i) = e^{M_ip_i}\psi(\mathbf{r}) = \psi(\mathbf{r}), \quad i = 1,2,3, \tag{4.60}$$

where M_1, M_2, and M_3 are the number of primitive vectors in the directions $\mathbf{a}_1$, $\mathbf{a}_2$, and $\mathbf{a}_3$, respectively. Obviously, the total number of primitive cells in the crystal

$$N = M_1M_2M_3. \tag{4.61}$$

From Eq. (4.61), we obtain

$$e^{M_1p_1} = e^{M_2p_2} = e^{M_3p_3} = 1. \tag{4.62}$$

Eq. (4.62) yields the necessary condition that

$$p_i = \frac{2\pi i m_i}{M_i} \cdot i = 1,\ 2,\ 3, \tag{4.63}$$

where, m_1, m_2, and m_3 are a set of integers. If we define the Bloch wave vectors as

$$\mathbf{k} = \sum_{i=1}^{3} \frac{p_i}{2\pi i} \mathbf{b}_i, \quad 0 \leq m_i < M_i, \tag{4.64}$$

where $\mathbf{b}_i$, the primitive vectors of the reciprocal lattice, were defined in Eq. (1.16).

According to Eq. (4.64), the total number of $\mathbf{k}$ states in a Brillouin zone is $M_1M_2M_3$, which is also equal to the number of lattice points in the lattice. It is possible to have $\mathbf{k}$ outside the first Brillouin zone by allowing the integers to be greater than M_i. However, the resulting $\mathbf{k}$ would differ from the $\mathbf{k}$ within the first Brillouin zone by a reciprocal lattice vector $\mathbf{K}$. Because $e^{i\mathbf{k}\cdot\mathbf{R}_i} = e^{i(\mathbf{k}+\mathbf{K})\cdot\mathbf{R}_i}$, the resulting eigenfunction would be the same according to Eq. (4.45) as well as from Eq. (4.67) (which we will prove), and because $\mathbf{k}$ is a label for this eigenvalue, two such $\mathbf{k}$ values are physically identical. Thus, the number of physically distinct values of the Bloch wave vector $\mathbf{k}$ equals the number of lattice sites of the original Bravais lattice.

We obtain from Eqs. (4.60), (4.63), and (4.64),

$$e^{iM_i\mathbf{k}\cdot\mathbf{a}_i} = 1 \tag{4.65}$$

and

$$C(\mathbf{R}_i) = e^{i\mathbf{k}\cdot\mathbf{R}_i}. \tag{4.66}$$

We obtain from Eqs. (4.52) and (4.66), the Bloch theorem,

$$\psi(\mathbf{r}+\mathbf{R}_i) = C(\mathbf{R}_i)\psi(\mathbf{r}) = e^{i\mathbf{k}\cdot\mathbf{R}_i}\psi(\mathbf{r}). \tag{4.67}$$

If we identify the eigenfunctions $\psi(\mathbf{r})$ with a band index n and wave vector $\mathbf{k}$ (we will show the importance and the necessity of the band index), Eq. (4.67) can be written in the more general form

$$\psi_{n\mathbf{k}}(\mathbf{r}+\mathbf{R}_i) = e^{i\mathbf{k}\cdot\mathbf{R}_i}\psi_{n\mathbf{k}}(\mathbf{r}). \tag{4.68}$$

The Bloch theorem stated in Eq. (4.68) can also be rewritten in the alternate form

$$\psi_{n\mathbf{k}}(\mathbf{r}) = e^{i\mathbf{k}\cdot\mathbf{r}}u_{n\mathbf{k}}(\mathbf{r}), \tag{4.69}$$

where $u_{n\mathbf{k}}(\mathbf{r})$ is known as the periodic part of the Bloch function. In fact, from Eqs. (4.68) and (4.69), we obtain

$$u_{n\mathbf{k}}(\mathbf{r}+\mathbf{R}_i) = u_{n\mathbf{k}}(\mathbf{r}), \tag{4.70}$$

from which the term *periodic part* is self-evident.

4.7 APPROXIMATE SOLUTION NEAR A ZONE BOUNDARY

To understand the occurrence of band gaps, we consider the results of Problem 4.4 (Eq. 10),

$$(\varepsilon^0_{\mathbf{k}-\mathbf{K}} - \varepsilon)C_{\mathbf{k}-\mathbf{K}} + \sum_{\mathbf{K}'} V_{\mathbf{K}'-\mathbf{K}}C_{\mathbf{k}-\mathbf{K}'} = 0, \tag{4.71}$$

where

$$\varepsilon^0_{\mathbf{k}-\mathbf{K}} \equiv \frac{\hbar^2(\mathbf{k}-\mathbf{K})^2}{2m} \tag{4.72}$$

is the free electron energy eigenvalue for an electron of wave vector $\mathbf{k} - \mathbf{K}$. We will also denote $\varepsilon^0_{\mathbf{k}}$ as the free electron eigenvalue for wave vector $\mathbf{k}$.

We can rewrite Eq. (4.71) as

$$(\varepsilon - \varepsilon^0_{\mathbf{k-K}})C_{\mathbf{k-K}} = \sum_{\mathbf{K}' \neq \mathbf{K}} V_{\mathbf{K}'-\mathbf{K}} C_{\mathbf{k-K}'} \tag{4.73}$$

because we have assumed that $V_0 = 0$. Eq. (4.73) includes the terms $\mathbf{K} = 0$ and $\mathbf{K}' = 0$. If we use nondegenerate perturbation theory and assume that

$$\left| \varepsilon^0_{\mathbf{k-K}} - \varepsilon^0_{\mathbf{k-K}'} \right| \gg V_{\mathbf{K}'-\mathbf{K}}, \tag{4.74}$$

for all $\mathbf{K}' \neq \mathbf{K}$ and fixed $\mathbf{k}$, Eq. (4.73) can be rewritten as

$$C_{\mathbf{k-K}} = \sum_{\mathbf{K}' \neq \mathbf{K}} \frac{V_{\mathbf{K}'-\mathbf{K}} C_{\mathbf{k-K}'}}{\varepsilon - \varepsilon^0_{\mathbf{k-K}}}. \tag{4.75}$$

For another coefficient $C_{\mathbf{k-K}_1}$ corresponding to the reciprocal lattice vector $\mathbf{K}_1$ (where $\mathbf{K}_1$ satisfies the condition of Eq. 4.74),

$$C_{\mathbf{k-K}_1} = \frac{V_{\mathbf{K-K}_1} C_{\mathbf{k-K}}}{\varepsilon - \varepsilon^0_{\mathbf{k-K}_1}} + \sum_{\mathbf{K}' \neq \mathbf{K} \neq \mathbf{K}_1} \frac{V_{\mathbf{K}'-\mathbf{K}_1} C_{\mathbf{k-K}'}}{\varepsilon - \varepsilon^0_{\mathbf{k-K}'}}. \tag{4.76}$$

In deriving Eq. (4.76), we have made the basic assumption that the free electron eigenvalue $\varepsilon^0_{\mathbf{k-K}}$ is not nearly degenerate to any other $\varepsilon^0_{\mathbf{k-K}'}$ in the set. Otherwise, the expansion of the energy in Eq. (4.78) in second order and higher terms in V would not be valid.

From Eqs. (4.73) and (4.76), we obtain

$$(\varepsilon - \varepsilon^0_{\mathbf{k-K}})C_{\mathbf{k-K}} = \sum_{\mathbf{K}' \neq \mathbf{K}} \frac{V_{\mathbf{K}'-\mathbf{K}} V_{\mathbf{K-K}'}}{\varepsilon - \varepsilon^0_{\mathbf{k-K}'}} C_{\mathbf{k-K}} + \sum_{\mathbf{K}'' \neq \mathbf{K}' \neq \mathbf{K}} \frac{V_{\mathbf{K}'-\mathbf{K}} V_{\mathbf{K-K}'} V_{\mathbf{K-K}''}}{(\varepsilon - \varepsilon^0_{\mathbf{k-K}'})(\varepsilon - \varepsilon^0_{\mathbf{k-K}''})} C_{\mathbf{k-K}} + \text{higher-order terms in } V. \tag{4.77}$$

Because the perturbed energy ε differs from the free electron energy $\varepsilon^0_{\mathbf{k-K}}$ by $|V|^2$ or higher-order terms (in the specific case of energy values that are neither degenerate nor nearly degenerate), we retain the terms up to the second order in V, use the relation $V_{-\mathbf{K}} = V^*_{\mathbf{K}}$ in Eq. (4.77), substitute ε by $\varepsilon^0_{\mathbf{k-K}}$ in the denominator of the first term on the right, and obtain the expression for ε:

$$\varepsilon = \varepsilon^0_{\mathbf{k-K}} + \sum_{\mathbf{K}' \neq \mathbf{K}} \frac{|V_{\mathbf{K}'-\mathbf{K}}|^2}{\varepsilon^0_{\mathbf{k-K}} - \varepsilon^0_{\mathbf{k-K}'}} + O(V^3). \tag{4.78}$$

Eq. (4.78) is valid as long as nondegenerate perturbation theory can be applied to the problem, i.e., as long as $\varepsilon^0_{\mathbf{k-K}} \neq \varepsilon^0_{\mathbf{k-K}'}$ (or sufficiently close in values so that the perturbation theory breaks down). The simplest example is when $\mathbf{k}$ lies near a zone boundary, in which case the second-order perturbation theory breaks down.

If $\mathbf{k}$ lies near a zone boundary (for simplicity, we assume that it lies near the boundary bisecting the vector $\mathbf{K}$), the electron undergoes a Bragg reflection by the lattice, similar to the situation as if it would have been an external electron beam. In such a case, we will use degenerate perturbation theory, consider only $C_{\mathbf{k}}$ and $C_{\mathbf{k-K}}$, and neglect the other coefficients.

Eq. (4.73) can be rewritten as

$$\begin{aligned}(\varepsilon^0_{\mathbf{k}} - \varepsilon)C_{\mathbf{k}} + V_{\mathbf{K}}C_{\mathbf{k-K}} &= 0\\ V_{-\mathbf{K}}C_{\mathbf{k}} + (\varepsilon^0_{\mathbf{k-K}} - \varepsilon)C_{\mathbf{k-K}} &= 0.\end{aligned} \tag{4.79}$$

From Eqs. (4.77) and (4.79), we obtain

$$\begin{aligned}(\varepsilon^0_{\mathbf{k}} - \varepsilon)C_{\mathbf{k}} + V_{\mathbf{K}}C_{\mathbf{k-K}} &= 0\\ V^*_{\mathbf{K}}C_{\mathbf{k}} + (\varepsilon^0_{\mathbf{k-K}} - \varepsilon)C_{\mathbf{k-K}} &= 0.\end{aligned} \tag{4.80}$$

We have $\varepsilon^0_{\mathbf{k}} \approx \varepsilon^0_{\mathbf{k-K}}$ and $|\varepsilon^0_{\mathbf{k}} - \varepsilon^0_{\mathbf{k-K}}| \gg V$, when $\mathbf{K} \neq \mathbf{K}, \mathbf{0}$. This is possible only when $|\mathbf{k} - \mathbf{K}| = |\mathbf{k}|$. It is evident from Figure 4.6(a) that this is possible only when $\mathbf{k}$ lies on the Bragg plane that bisects the line joining the origin of $\mathbf{k}$ space to the reciprocal lattice point $\mathbf{K}$.

Eq. (4.80) can be solved from the determinant

$$\begin{vmatrix} \varepsilon^0_{\mathbf{k}} - \varepsilon & V_{\mathbf{K}} \\ V^*_{\mathbf{K}} & \varepsilon^0_{\mathbf{k-K}} - \varepsilon \end{vmatrix} = 0. \tag{4.81}$$

The solutions of the quadratic equation are

$$\varepsilon^{\pm}(\mathbf{k}) = \frac{1}{2}(\varepsilon^0_{\mathbf{k}} + \varepsilon^0_{\mathbf{k-K}}) \pm \left[\left(\frac{\varepsilon^0_{\mathbf{k}} - \varepsilon^0_{\mathbf{k-K}}}{2}\right)^2 + |V_{\mathbf{K}}|^2\right]^{\frac{1}{2}}. \tag{4.82}$$

Thus, the free electron states $e^{i\mathbf{k}\cdot\mathbf{r}}$ and $e^{i(\mathbf{k}-\mathbf{K})\cdot\mathbf{r}}$ with energy $\varepsilon^0_{\mathbf{k}}$ and $\varepsilon^0_{\mathbf{k-K}}$ are combined into two other states ψ^+ and $\psi-$ with energy $\varepsilon + (\mathbf{k})$ and $\varepsilon - (\mathbf{k})$.

It is easy to analyze Eq. (4.82) for points lying on the Bragg plane because $|\mathbf{k}| = |\mathbf{k} - \mathbf{K}|$ and $\varepsilon^0_{\mathbf{k}} = \varepsilon^0_{\mathbf{k-K}}$. This implies that $\mathbf{k}$ must lie on the Brillouin zone boundary (see Figure 4.6a). Further, at all points on the Bragg plane, one energy level is raised by $|V_{\mathbf{K}}|$, whereas the other energy level is lowered by $|V_{\mathbf{K}}|$. Thus, when $\mathbf{k}$ is on a single Bragg plane, we can write

$$\varepsilon^{\pm}(\mathbf{k}) = \varepsilon^0_{\mathbf{k}} \pm |V_{\mathbf{K}}|. \tag{4.83}$$

Hence, there is an energy gap of $2|V_{\mathbf{K}}|$ when $\mathbf{k} = \frac{1}{2}\mathbf{K}$. This is shown in Figure 4.7. This is known as the band gap because the energy levels are split into two bands. When $\mathbf{k}$ is closer to the origin (far away from the Bragg plane), the energy levels are practically the same as the free electron energy levels.

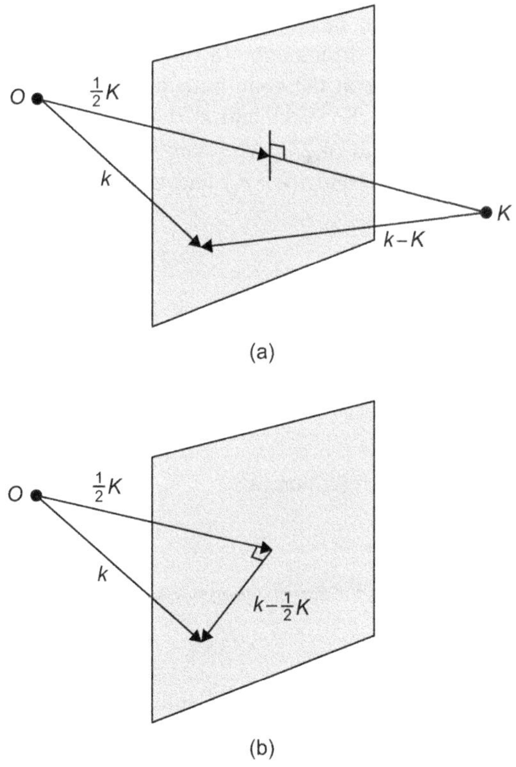

FIGURE 4.6

(a) $\mathbf{k}$ lies in the Bragg plane determined by $\mathbf{K}$ if $|\mathbf{k}| = |\mathbf{k} - \mathbf{K}|$; (b) $\mathbf{k} - \frac{1}{2}\mathbf{K}$ is parallel to the Bragg plane if $\mathbf{k}$ lies in the Bragg plane.

In addition, when $\varepsilon^0_{\mathbf{k}} = \varepsilon^0_{\mathbf{k-K}}$, we obtain from Eq. (4.82),

$$\frac{\partial \varepsilon}{\partial \mathbf{k}} = \frac{\hbar^2}{m}\left(\mathbf{k} - \frac{1}{2}\mathbf{K}\right). \tag{4.84}$$

Eq. (4.84) implies that when $\mathbf{k}$ is on the Bragg plane, the gradient of ε is parallel to the Bragg plane (see Figure 4.6b). Therefore, the constant-energy surfaces at the Bragg plane are perpendicular to the plane because the gradient is perpendicular to the surfaces on which a function is constant.

It is easy to plot the energy bands from Eq. (4.82) if $\mathbf{k}$ is parallel to $\mathbf{K}$ (see Figure 4.7). When $\mathbf{k} = ½\mathbf{K}$, the two bands are separated by a band gap $2|V_{\mathbf{K}}|$.

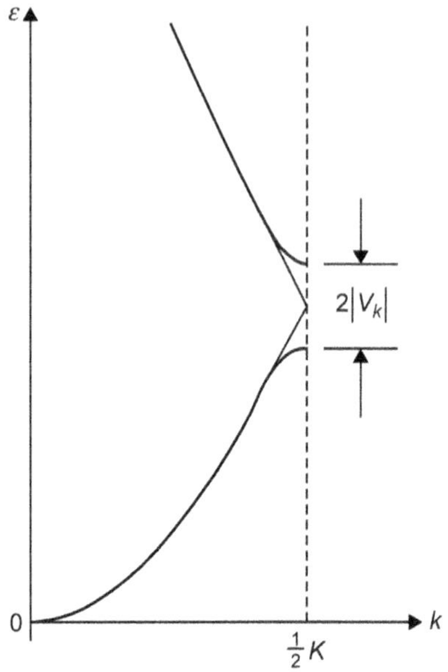

FIGURE 4.7

Energy bands when **k** is parallel to **K**.

It is much easier to consider the energy bands in one dimension. In one dimension, if we consider $\mathbf{k}$ at the zone boundary at $½\mathbf{K}$, we note that $(\mathbf{k} - \mathbf{K})^2 = \mathbf{k}^2$ and $\varepsilon^0_k = \varepsilon^0_{k-K}$. Thus, in Eqs. (4.20) and (4.23), we retain only the terms involving $a(k)$ and $a(k-K)$ and write $E = \varepsilon(k)$, and we obtain

$$(\varepsilon^0_k - \varepsilon(k))a(k) + V_K a(k-K) = 0, \tag{4.85}$$

$$V_{-K}a(k) + (\varepsilon^0_{k-K} - \varepsilon(k))a(k-K) = 0. \tag{4.86}$$

Because $V_K = V_{-K} = V_K *$, Eqs. (4.85) and (4.86) can be solved by the determinant equation

$$\begin{vmatrix} \varepsilon^0_k - \varepsilon(k) & V_K \\ V_K^* & \varepsilon^0_{k-K} - \varepsilon(k) \end{vmatrix} = 0, \tag{4.87}$$

which can be rewritten as

$$\varepsilon(k)^2 - \varepsilon(k)(\varepsilon^0_{k-K} + \varepsilon^0_k) + \varepsilon^0_{k-K}\varepsilon^0_k - |V_K|^2 = 0. \tag{4.88}$$

Thus, the two roots are

$$\varepsilon^{\pm}(k) = \frac{1}{2}(\varepsilon^0_{k-K} + \varepsilon^0_k) \pm \left[\frac{1}{4}(\varepsilon^0_{k-K} - \varepsilon^0_k)^2 + |V_K|^2\right]^{\frac{1}{2}}. \tag{4.89}$$

When $\mathbf{k} = \mathbf{K}/2$ (at the zone boundary),

$$\varepsilon^{\pm}(K/2) = \varepsilon^0_{K/2} \pm |V_K|. \tag{4.90}$$

Substituting Eq. (4.90) in Eqs. (4.85) and (4.86), we obtain

$$|\pm V_K|a(K/2) = -V_K a(-K/2), \tag{4.91}$$

for the two roots marked $\pm$.

The corresponding eigenstates $\psi^{\pm}(r)$ are obtained from Eqs. (4.24), (4.85), and (4.86). When $\mathbf{k} = \mathbf{K}/2$, assuming that V_K is negative, $a(K/2) = a(-K/2)$ for the negative root and $a(K/2) = -a(-K/2)$ for the positive root. Thus, we obtain

$$\psi^{-}(r) = a(K/2)[e^{i/2Kr} + e^{-i/2Kr}] \tag{4.92}$$

and

$$\psi^{+}(r) = a(K/2)[e^{i/2Kr} - e^{-i/2Kr}]. \tag{4.93}$$

Using the normalization conditions for the eigenstates, we obtain

$$\psi^{-}(r) = \frac{1}{\sqrt{2}}[e^{i/2Kr} + e^{-i/2Kr}] = \sqrt{2}\cos\frac{1}{2}Kr, \tag{4.94}$$

and

$$\psi^{+}(r) = \frac{1}{\sqrt{2}}[e^{i/2Kr} - e^{-i/2Kr}] = \sqrt{2}i\sin\frac{1}{2}Kr. \tag{4.95}$$

When k is near the zone boundary, we can define a wave vector δ, which measures the difference of k from the zone boundary by

$$\delta = K/2 - k. \tag{4.96}$$

From Eqs. (4.89) and (4.96), we obtain

$$\begin{aligned}\varepsilon^{\pm}(k) &= (\hbar^2/2m)\left(\frac{1}{4}K^2 + \delta^2\right) \pm [4\varepsilon^0_{K/2}(\hbar^2\delta^2/2m) + |V_K|^2]^{\frac{1}{2}} \\ &\cong (\hbar^2/2m)\left(\frac{1}{4}K^2 + \delta^2\right) \pm |V_K|[1 + 2(\varepsilon^0_{K/2}/|V_K|^2)(\hbar^2\delta^2/2m)]^{\frac{1}{2}}.\end{aligned} \tag{4.97}$$

From Eqs. (4.90) and (4.97), we obtain

$$\varepsilon^{\pm}(k) = \varepsilon^{\pm}(K/2) + (\hbar^2\delta^2/2m)[1 \pm 2(\varepsilon^0_{K/2}/|V_K|^2]. \tag{4.98}$$

When $k \sim 0$, or far from a zone boundary (in the extended zone scheme),

$$\varepsilon(k) \sim \varepsilon^0_k \sim \hbar^2k^2/2m, \tag{4.99}$$

which is a free electron parabola. We will represent these results in the reduced, extended, and repeated zone schemes described earlier for free electrons.

4.8 DIFFERENT ZONE SCHEMES

4.8.1 Reduced Zone Scheme

In the reduced zone scheme, the wave vector $\mathbf{k}$ always lies within the first Brillouin zone. If a wave vector $\mathbf{k}'$ lies outside the first Brillouin zone, one can always find a lattice vector $\mathbf{K}'$ such that $\mathbf{k} = \mathbf{k}' - \mathbf{K}'$ lies within the first Brillouin zone. We show in Figure 4.8 the energy bands of a linear lattice (with a periodic potential) in the reduced zone scheme.

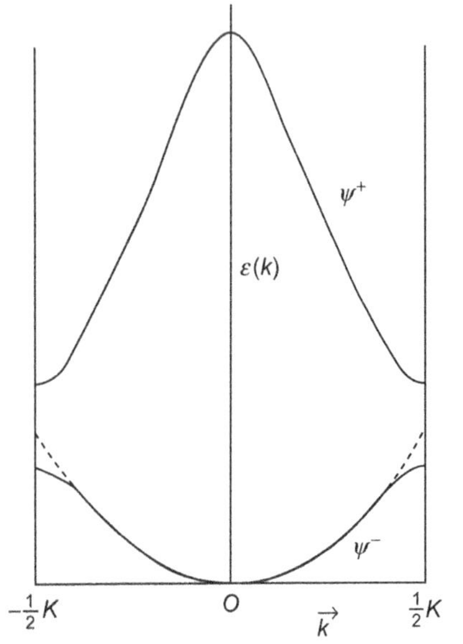

FIGURE 4.8

The energy bands of a linear lattice in the reduced zone scheme.

4.8.2 Extended Zone Scheme

In the extended zone scheme, the energy $\varepsilon(\mathbf{k})$ is plotted against the wave vector $\mathbf{k}$. We had seen that for free electrons, the curve is the free electron parabola because $\varepsilon(\mathbf{k}) = \frac{\hbar^2 k^2}{2m}$.

However, in the presence of a periodic potential, as we have seen, the parabola must meet the zone boundary normally, and an energy gap of $2|V_K|$ develops between the lower and the upper band. This gap increases as $\mathbf{K}$ increases. The energy bands in the extended zone scheme of a linear lattice with periodic potential are shown in Figure 4.9.

Figure 4.10 shows the energy bands for a linear lattice with periodic potential and succinctly demonstrates the development of the forbidden part of the zone (or more commonly referred to as band gap) that increases as the energy of the band increases.

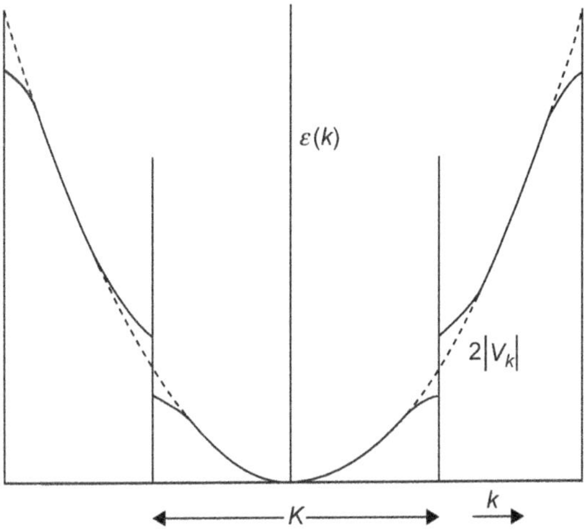

FIGURE 4.9

The energy bands of a linear lattice in the extended zone scheme.

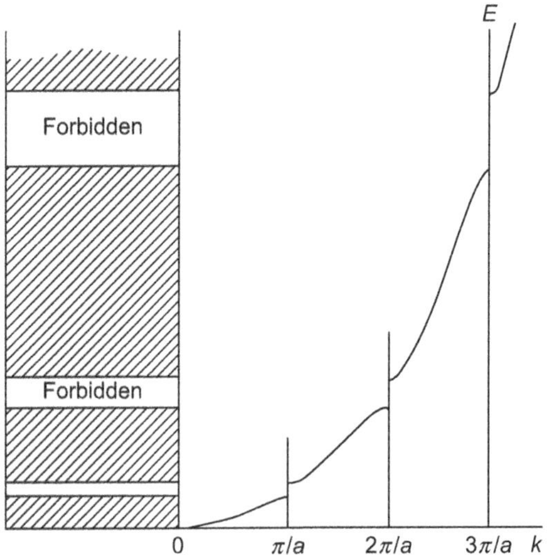

FIGURE 4.10

The energy gaps in a linear lattice. The width in energy increases as the energy of the band increases. In pure materials, there are no eigenstates for electrons with energy lying within these energy gaps.

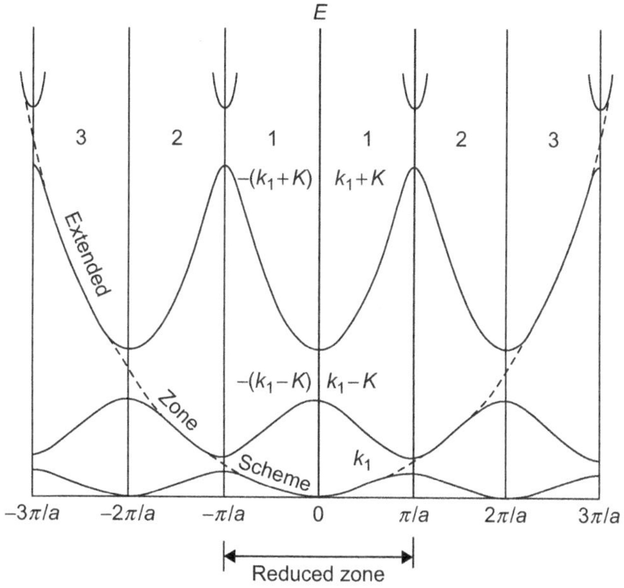

FIGURE 4.11

The repeated zone scheme for a linear lattice and its comparison with the reduced and extended zone schemes shown in Figures 4.9 and 4.10.

4.8.3 Repeated Zone Scheme

When the first Brillouin zone is periodically repeated through all **k** space, $\varepsilon(\mathbf{k}) = \varepsilon(\mathbf{k}+\mathbf{K})$. In fact, $\varepsilon(\mathbf{k}+\mathbf{K})$ is the same energy band as $\varepsilon(\mathbf{k})$. This type of construction of energy bands is known as the repeated zone scheme. This scheme is particularly useful in demonstrating the electron orbits in a magnetic field. The repeated zone scheme and its connection with the reduced and extended zone schemes is shown in Figure 4.11.

4.9 ELEMENTARY BAND THEORY OF SOLIDS

4.9.1 Introduction

We will now discuss the elementary band theory of solids using the one-dimensional lattice and analyze how crystalline solids, of which the basic components are negatively charged electrons and positively charged ions, have a wide diversity in physical properties. For example, some solids are metals that are good conductors, some others are metals but poor conductors, some crystallize as insulators, and the rest are crystals that are semiconductors. Each of these types of solids has widely divergent properties. For example, whereas some metals such as the alkali metals are very good conductors, some others such as the alkaline-earth metals are comparatively poor conductors. Another striking feature is that the ratio of the resistivity of metals (good conductors) and insulators is of the order of 10^{-20} at room temperature. As has been often remarked, this is one of the widest divergences in the physical properties occurring in nature. Another example is the difference between the temperature dependence of the resistivity of metals and semiconductors. The resistivity of metals, which is small at absolute zero, increases with increase of temperature. In contrast, the resistivity of pure semiconductors, which are insulators (if the material is pure) at absolute zero, decreases with increase of temperature. We will try to explain this wide variety of exotic properties by using a simple one-dimensional band theory. In subsequent chapters, we will discuss the various techniques used in the energy band theory of solids in three dimensions and discuss the properties of various types of solids in a more rigorous manner.

4.9.2 Energy Bands in One Dimension

We have discussed how a periodic potential breaks the free electron energy curve, which is a parabola, into discrete segments of interval π/a. Thus, there is a forbidden region for eigenstates of electrons in pure materials that is known as the energy gap. This energy gap increases with the increase in energy; i.e., the higher the energy of the band, the larger its width in energy. This significant fact, which is essentially the elementary band theory of solids, allows us to understand many of the characteristic features of solids. The basic idea of the formation of these energy bands was shown in Figure 4.10.

4.9.3 Number of States in a Band

Here, we consider a one-dimensional crystal constructed of primitive cells of lattice constant a. The length of the crystal is $L = Na$, where N is the number of primitive cells. As we have noted, in one dimension, the allowed values of the electron wave vector $\mathbf{k}$ in the first Brillouin zone are given by

$$k = 0; \pm\frac{2\pi}{L}; \pm\frac{4\pi}{L}; \ldots, \frac{N\pi}{L}. \tag{4.100}$$

We note that because $N\pi/L \equiv \pi/a$, the point, defined as $-N\pi/L \equiv -\pi/a$, is connected by a reciprocal lattice vector K with π/a, and hence cannot be counted as an independent point. The total number of points given in Eq. (4.100) is N. This result is also carried over to three dimensions; i.e., each primitive cell contributes one independent value of k to each energy band. If one considers the spin of the electron, each energy band will have $2N$ independent states.

4.10 METALS, INSULATORS, AND SEMICONDUCTORS

We can now discuss the reason crystalline solids have to be grouped into four extremely dissimilar varieties: metals (good conductors), semimetals (poor conductors), insulators, and semiconductors. First, by using the elementary band theory, we will discuss some of the general features that are responsible for distinguishing solids into these four categories. Later, we will discuss some specific examples in each category to illustrate the characteristic features.

If there is a single atom of valence one in each primitive cell, the first band (the bands are stacked above each other with [increasing] energy gaps between them, as shown in Figure 4.11) will be half-filled with electrons, and the solid will be a metal (good conductor) because there are enough empty states available for electrons to be excited whenever an electric field $\mathbf{E}$ is applied. In fact, when an electric field **E** is applied, the force on the electron of charge $-e$ is $\mathbf{F} = -e\mathbf{E}$. The force is also the rate of change of momentum,

$$\mathbf{F} = -e\mathbf{E} = \hbar \frac{d\mathbf{k}}{dt}. \tag{4.101}$$

Because the alkali metals and noble metals have one valence electron per primitive cell, they are good conductors. As an example, we consider sodium. Each Na atom has the atomic configuration $1s^2 2s^2 2p^6 3s^1$. Thus, there is one valence electron in the $3s$ state in each separated Na atom, while the $3s$ state could accommodate two valence electrons. When N such atoms are bound in a solid, the $3s$ energy band has N electrons, and therefore, it is only half-filled (a band can accommodate $2N$ electrons). Thus, sodium is a good conductor because there are a large number of energy levels available just above the filled ones and the valence electrons can be easily raised to a higher energy state by an electric field, as shown in Eq. (4.101). In fact, as a rule of thumb, all monovalent solids are good conductors.

According to the same rule of thumb, all divalent solids like the alkaline-earth metals that have two valence electrons per primitive cell should be insulators. However, this is not true if we consider a three-dimensional band picture. For example, we consider magnesium, of which the atomic configuration is $1s^2 2s^2 2p^6 3s^2$. The $3s$ energy band is full, and as per the one-dimensional band picture, magnesium should be an insulator. However, this is not true because in a three-dimensional band picture, there is an overlap between the $3s$ and $3p$ bands, which is shown in Figure 4.12. In fact, the same overlap was also there for sodium in a three-dimensional band picture, but we did not have to take that fact into consideration because the $3s$ band was only half full. Because of this overlap between $3s$ and $3p$ bands, magnesium, like all alkaline-earth metals, is a metal, but some

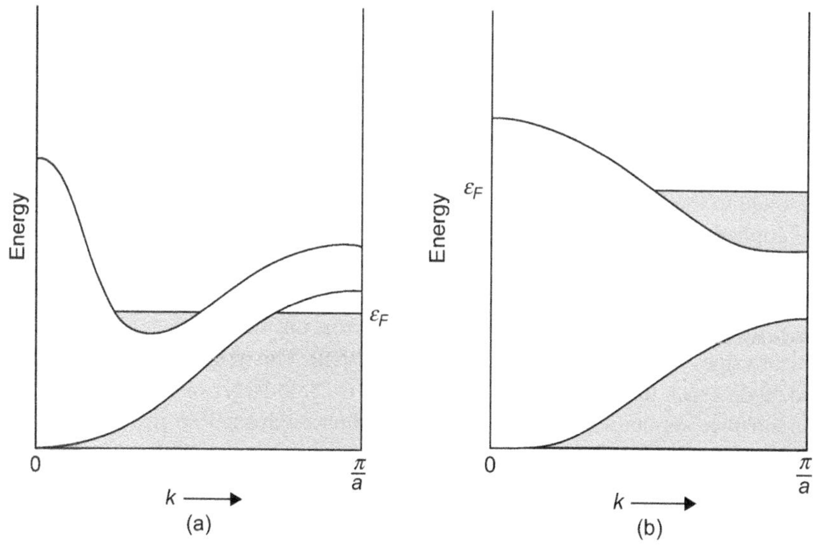

FIGURE 4.12

(a) The overlap of the $3s$ and $3p$ energy bands in three dimensions. (b) In Al, the $3s$ band is full, but the $3p$ band is not full. The upper energy level is the Fermi energy ε_F.

divalent solids such as Sr and Ba are poor conductors because the overlap is small and a relatively small number of electrons are excited when an electric field is applied.

We note in Figure 4.12a that the three components of the wave vector **k** are in different directions in the $3s$ and $3p$ bands. Thus, the lowest energy levels of the $3p$ band are lower in energy than the highest energy levels of the $3s$ band, a fact that would be impossible in a one-dimensional band picture. Because the electrons tend to occupy the lowest energy states, some electrons have spilled over to the $3p$ band. We also note that the highest energy level in both bands is the Fermi level ε_F, which is in conformity with the definition of the Fermi level.

All trivalent solids such as Al, Ga, In, and Tl are good metals because there are three valence electrons per primitive cell, and hence, they can fill one and a half bands. For example, the atomic configuration of aluminum is $1s^2 2s^2 2p^6 3s^2 3p^1$. The valence electrons are in the $3p$ band that is half empty and can be easily excited to higher energy states by an electric field. Therefore, aluminum is a very good conductor. This is schematically shown in Figure 4.12b.

A crystalline solid becomes an insulator if it has only completely filled bands, provided the energy gap between the last-filled band (known as the valence band) and the next allowed empty band (known as the conduction band) is very large. The condition that the energy gap between the valence and conduction bands must be very large is due to the fact that the electrons can be thermally excited to the conduction band at room temperature if the energy gap is small. In addition, when an electric field is applied, the electrons are excited due to the external force (Eq. 4.101). The energy gap has to be large enough to prevent the excitation of the electrons from the valence band to the conduction band at reasonably large electric fields. The ionic crystals are good examples of insulators. The energy bands of an ionic crystal such as NaCl are from the Na^+ ($1s^2 2s^2 2p^6$) and Cl^-($1s^2 2s^2 2p^6 3s^2 3p^6$) ions. Because both of these ions have a closed-shell structure, all the occupied bands of NaCl are full and the energy gap between the highest occupied band (valence band) and the next empty (conduction) band is very large. Another example of an insulator that is a tetravalent solid is diamond, but the energy gap is very large for it to become a semiconductor.

In an insulator, the energy gap is large enough to prevent the valence electrons to be excited to the conduction band. However, if the applied electric field is greater than a critical value (known as the critical field), such that the valence electrons gain energy that is equal to or greater than the energy gap, they can be excited to the conduction band. In such cases, the insulator behaves like a good conductor as long as the applied electric field is greater than the critical field. Thus, insulators, which have large energy gaps, are used as breaking devices in high-voltage transmission. When the external electric field is greater than the critical field required to cross the energy gap, there is good transmission of electric current because the insulator behaves as a good conductor, but when the applied electric field drops below the critical value, the insulator stops the flow of current. The transmission of electricity is restored by appropriate repairs such that the external electric field is again larger than the critical field.

In case of intrinsic semiconductors (pure semiconductors are called intrinsic in order to distinguish them from impurity [or doped] semiconductors), the energy gap between the valence band and the conduction band is sufficiently low (0.7 eV for Ge and 1.09 eV for Si). Although an intrinsic semiconductor behaves as an insulator at absolute zero temperature, some valence electrons are thermally excited to the conduction band, leaving behind an equal number of unoccupied states (holes) in the valence band. We will later show that these holes act like positive charges. In an applied electric field, both the (few) electrons in the conduction band and the holes in the valence band are excited and move in opposite directions, thereby conducting electricity. However, because the number of conducting electrons and holes is much smaller than that compared to metals, the resistivity of semiconductors is very large compared to that of metals. When the temperature is increased, more valence electrons are excited into the conduction band, leaving behind more (positively charged) holes in the valence band. Thus, the resistivity of semiconductors decreases with increase of temperature because of the increase in the number of carriers. In contrast, the resistivity of metals increases with the increase of temperature because the electrons are scattered by the lattice ions (phonons) and lattice impurities due to thermal vibrations.

We will now discuss the typical case of the most commonly used semiconductors such as Si ($1s^2 2s^2 2p^6 3s^2 3p^2$) and Ge ($1s^2 2s^2 2p^6 3s^2 3p^6 3d^{10} 4s^2 4p^2$). We note that Si has two $3s$ and two $3p$ electrons, and Ge has two $4s$ and two $4p$ electrons. Normally, we would expect Si and Ge to be conductors because each one of them has four unfilled p states. However, the $3s$ and $3p$ levels (for Si) and the $4s$ and $4p$ levels (for Ge) mix when they form covalent bonds. The energy of the electron levels corresponding to the four space-symmetric wave functions, one for the $2s$ levels and three for the $2p$ levels, is lowered. The energy of the other four levels, one $2s$ and three $2p$, is raised. Thus, the valence band has four levels per atom that are filled, whereas the conduction band is empty.

An interesting example is Sn, which is also a tetravalent solid. It has two phases: in one phase it is metallic, whereas in another phase it is a semiconductor. The shape of the Brillouin zone changes when the crystal structure is changed, and hence, it becomes possible to have large energy gaps to hold all the electrons. On the other hand, Pb, which is a tetravalent solid is a metal because of the band structure such that the electrons in the conduction band can be excited to higher energy states by an electric field. To summarize, the elements in Group IV of the periodic table have a wide range of properties. C in the form of a diamond is an insulator, Si and Ge are semiconductors, Sn can either be a metal or a semiconductor, whereas Pb is a metal.

The pentavalent solids such as As, Sb, and Bi have 5 electrons per atom. However, their crystal structure is such that there are 2 atoms per unit cell. Thus, there are 10 electrons per unit cell. These 10 electrons would normally fill 5 bands. However, due to the effect of the band structure, the fifth band is not quite full because there is a little overlap (schematically very similar to Figure 4.12a) with the sixth band. Therefore, even at zero temperature, a few electrons in both the fifth and the sixth bands are always available to be excited (to carry the current) when an external electric field is applied. These are poor conductors and are known as *semimetals.*

The iron group of the transition metals (Cr, Mn, Fe, Co, Ni) and the groups that are higher in the periodic table have incomplete d-shells. For example, only 6 out of the 10 states in the $3d$-shell of Fe atom are filled, while two more electrons fill the outer $4s$ state. The d-orbitals in a solid overlap to form a "d-band" that can be treated by a tight-binding or LCAO method (see Chapter 5). The two electrons that form an s-band (in some transition metals there are electrons in both s and p states, and the corresponding band is known as the *s-p-band*) hybridize with the narrow d-band that is capable of accommodating up to 10 electrons per atom. This hybridization between d-bands and s-bands is shown in Figure 4.13. These bands are called resonance bands, and the hybridization is important in understanding the magnetic phenomena. Because neither of the bands is full, these solids are metallic and the conduction is mainly metallic.

The four possible band structures for a solid are shown in Figure 4.14. Thus, it is possible to explain the occurrence of metals (good conductors), semimetals (poor conductors), insulators, and (intrinsic) semiconductors from a simple one-dimensional picture of band theory. We will later discuss in detail the characteristic properties of each of these solids.

We have summarized, by using the one-dimensional band theory of solids, the classification of solids into metals (good conductors), semimetals (poor conductors), insulators, and semiconductors. The rule of thumb is that each Brillouin zone has room for two electrons per primitive cell of a sample. If we consider a linear lattice that has one monovalent atom per primitive cell, the Brillouin zone is half filled. The electrons near the Fermi surface (the surface that separates the highest filled-energy states from the empty states) can be accelerated by an applied electric field, and because there are many empty states available, the metal is a good conductor. If there is one divalent atom per primitive cell, the first zone should be normally filled with electrons. However, in a three-dimensional band picture, there is usually overlap between the top of the electron states in the first zone and the bottom of the empty electron states in the second zone. The energy gap (in different k directions) disappears. This leads the electrons to spill over from the top of the first zone to the bottom of the second zone, and the Fermi surface is in both zones. Such metals are not very good metals because of the small number of electrons that are excited in an external electric field. If the atom in each lattice point is trivalent, the first Brillouin zone is completely filled, but the second zone is half full.

The Fermi surface is in the second zone, and because a large number of electron states above the Fermi surface are empty, the solid is a metal and a good conductor. If there is a quadrivalent atom per primitive cell, the solid is either an insulator or a semiconductor depending on the magnitude of the energy gap. If there are two quadrivalent atoms per primitive cell (examples

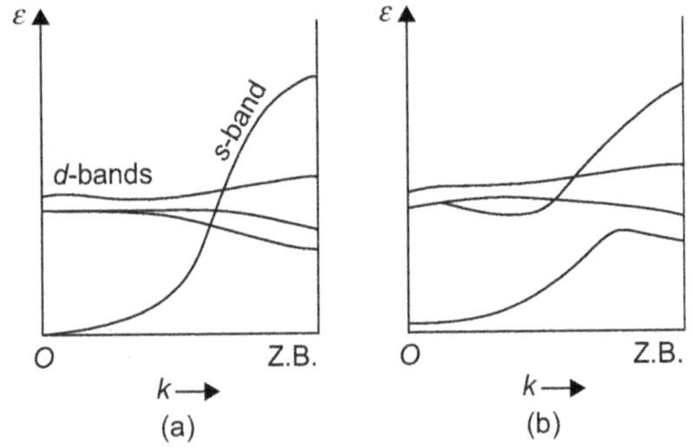

FIGURE 4.13

(a) *d*-bands crossing *s*-bands; (b) *s*-*d* hybridization.

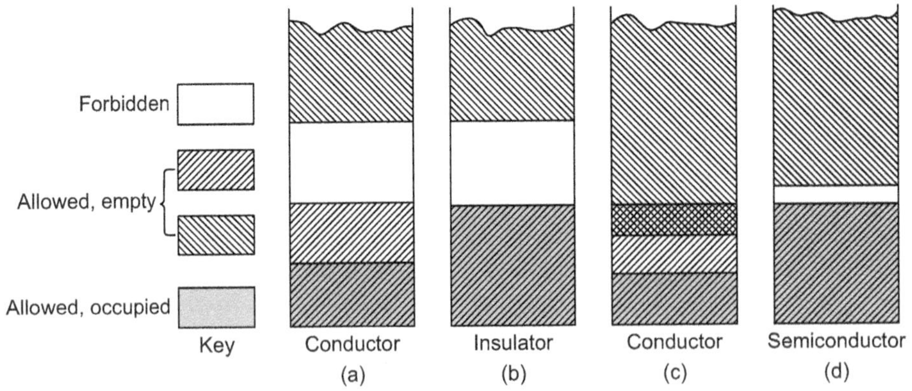

FIGURE 4.14

Four possible band structures for a solid: (a) conductor because the band is partially full, (b) insulator because of large energy gap between the filled and the empty bands, (c) semimetals because the allowed bands overlap, and (d) semiconductor because of the very small energy gap between the filled and the empty bands.

are diamond, silicon, and germanium), there are eight valence electrons per primitive cell. Because the bands do not overlap, diamond is an insulator (because of the large energy gap), and both silicon and germanium are intrinsic semiconductors because of the small energy gap. In both cases, there is no Fermi surface in the usual sense, but for semiconductors, the Fermi level is usually located at the center of the energy gap.

The electrons in metals in the highest occupied states have immediate access to the empty states, and the surface that separates these states is called the Fermi surface. However, the highest occupied electron states in insulators and semiconductors (at zero temperature) are separated from each other by energy gaps. Thus, the Fermi surface plays a vital role in determining the properties of metals. The Fermi surface of free electrons is a sphere in three dimensions. However, the Fermi surface is much more complex in a metal because of the periodic potential. To be able to understand the increasing complexity of the Fermi surface in such solids, we need to first understand the properties of two- and three-dimensional lattices and the Brillouin zones. In the nearly free electron approximation, a constant-energy surface is perpendicular to a Bragg plane when they intersect.

4.11 BRILLOUIN ZONES

We will first discuss the Brillouin zones for a two-dimensional square lattice, which we discussed in Chapter 1. The Bragg planes bisect the line joining the origin to points of the reciprocal lattice. The first Brillouin zone is defined as the set of points reached from the origin without crossing any Bragg plane (except that the points lying on the Bragg planes are common to two or more zones). The second Brillouin zone is the set of points that can be reached from the first zone by crossing only one Bragg plane. One can make a generalization of this definition and define the *n*th Brillouin zone as the set of points that can be reached from the origin by crossing no fewer than $n - 1$ Bragg planes. The first four zones of the two-dimensional square Bravais lattice are shown in Figure 4.15.

In general, a Brillouin zone can be constructed by using the rule that an incoming wave scatters strongly off a lattice with reciprocal lattice vector **K**, only when

$$\mathbf{k} \cdot \mathbf{K} = \tfrac{1}{2}K^2. \tag{4.102}$$

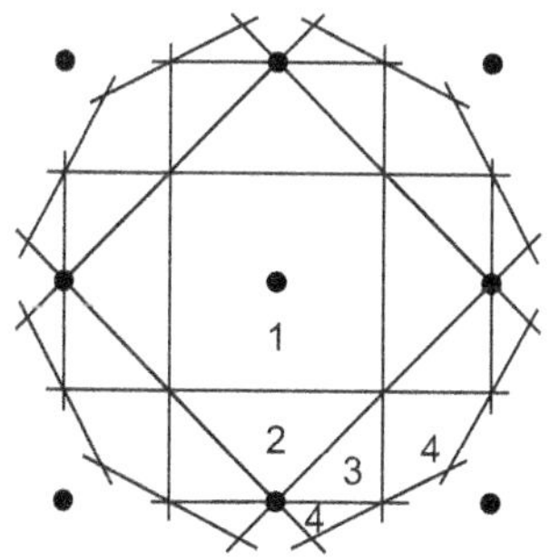

FIGURE 4.15

Brillouin zones for a two-dimensional square Bravais lattice. The first three zones are contained entirely in the square.

The set of points that satisfy Eq. (4.102) is a plane that is perpendicular to the vector connecting the origin to **K** and lying midway between 0 and **K**. When many such planes are constructed using all possible **K** values, the origin would be enclosed within a solid region. This is the first Brillouin zone because all points inside are closer to the origin than any reciprocal lattice vector. An example of this construction is the Brillouin zone of a two-dimensional centered rectangular lattice shown in Figure 4.16.

As explained earlier, the *n*th Brillouin zone is constituted of the set of points in reciprocal space that is closer to the $n - 1$ reciprocal points

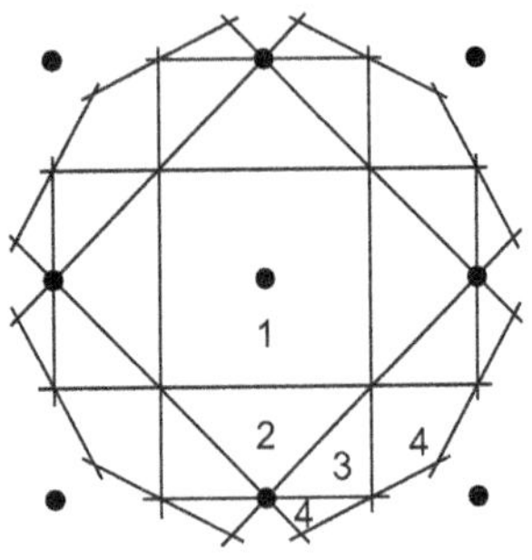

FIGURE 4.16

The zone boundaries for a two-dimensional centered rectangular lattice are obtained by drawing perpendicular bisectors between the origin and the nearby reciprocal points.

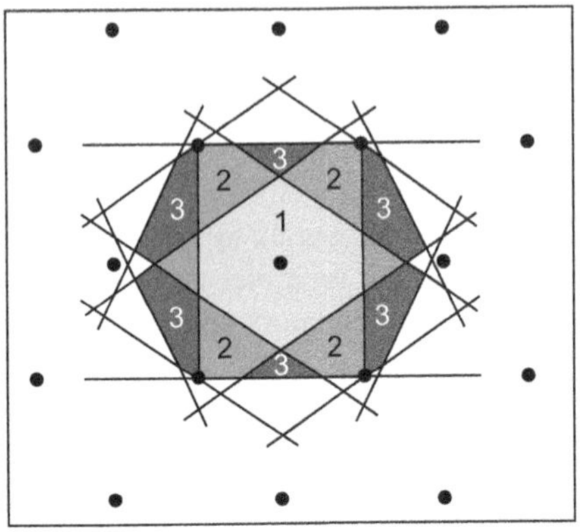

FIGURE 4.17

The first three Brillouin zones of a two-dimensional centered rectangular lattice.

than it is to the origin. The construction of the first three Brillouin zones for a rectangular centered lattice is obviously more complicated than a square lattice. Such construction for the first three Brillouin zones, shaded in different ways, is shown in Figure 4.17.

The first zone is the set of points closer to the origin than any other reciprocal lattice point. The second zone is constituted of the set of points that one reaches by crossing only one zone boundary. The third zone is the set of points that one reaches by crossing a minimum of two zone boundaries.

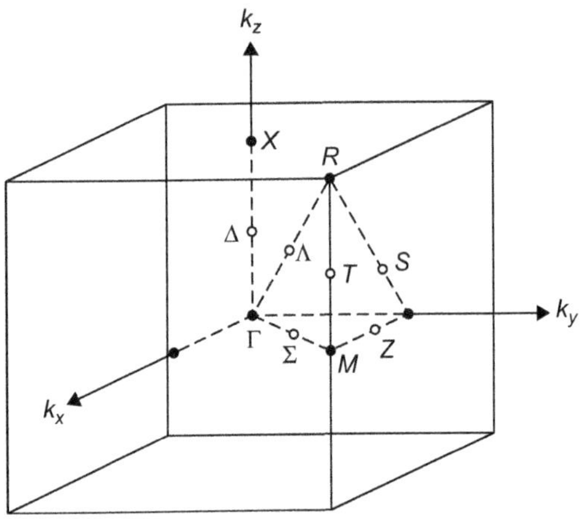

FIGURE 4.18

First Brillouin zone of the simple cubic lattice.

The construction of Brillouin zones for a three-dimensional lattice gets more complicated. For example, the first Brillouin zone of a simple cubic lattice is simple cubic, but the first Brillouin zones of a bcc and a fcc lattice are much more complicated. The first Brillouin zone of a simple cubic lattice with the symmetry points is shown in Figure 4.18. (The symmetry points are explained in Appendix A.)

In Figure 4.18, the point Γ is at the center of the zone. R is at the corner of the cube that is connected to the other corners so that all eight corners are a single point. Γ and R have the same representation, the cubic group. X is at the intersection of the k_z axis with the lower face of the cube. M is at the intersection of the k_xk_y plane with the vertical edges (there are three equivalent points to M). M and X have the same symmetry elements 4/mmm. T is equivalent to the three points on the other vertical edges. The points T and Δ have the same point group, 4mm. The point Λ has point group 3m. The points Σ and S are holomorphic to 2mm. The point Z has two mirror planes and a two-fold axis.

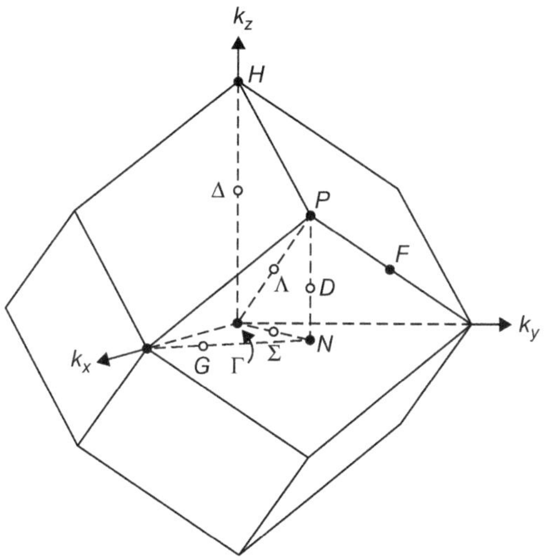

FIGURE 4.19

The first Brillouin zone of the bcc lattice.

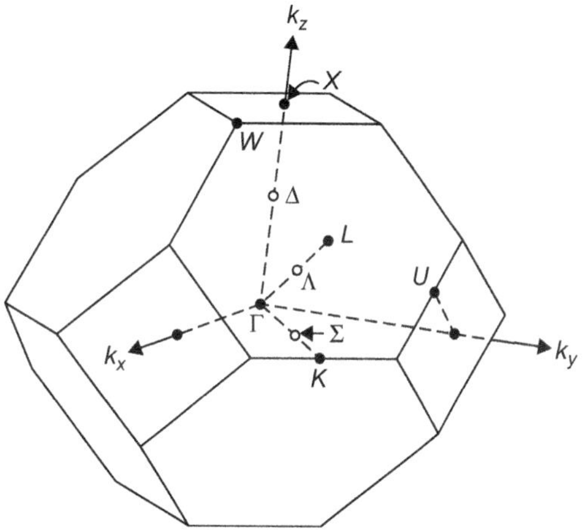

FIGURE 4.20

The Brillouin zone of the fcc lattice showing the symmetry points.

The first Brillouin zone of the body-centered cubic lattice, which is a rhombic dodecahedron, is shown in Figure 4.19 along with the symmetry points and the axes.

The symmetry operations of $\Gamma, \Delta, \Lambda, \Sigma$ are the same as the similar points in the simple cubic lattice shown in Figure 4.19. H has the full cubic symmetry like Γ.

The Brillouin zone of a face-centered cubic lattice is a truncated octahedron that is shown in Figure 4.20. Here, Γ is at the center of the zone, L is at the center of each hexagonal face, X is at the center of each square face, and W is at each corner formed from one square and two hexagons.

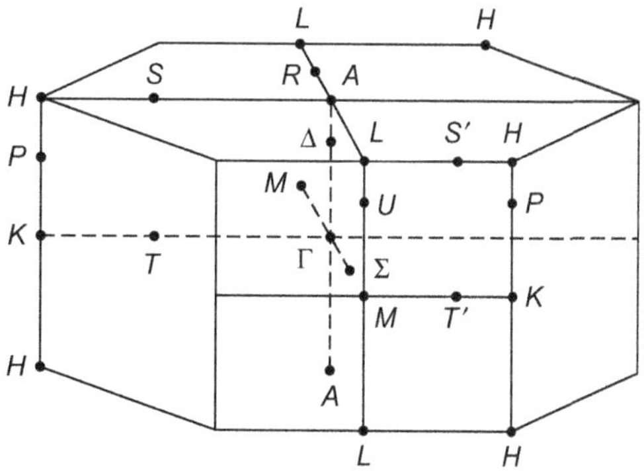

FIGURE 4.21

The Brillouin zone of the hcp structure.

The Brillouin zone of the hexagonal close-packed structure is shown with the symmetry points in Figure 4.21.

4.12 FERMI SURFACE

4.12.1 Fermi Surface (in Two Dimensions)

For free electrons, the Fermi surface is a circle in two dimensions and a sphere in three dimensions. The two-dimensional Fermi circles corresponding to one, two, and three electrons per atom for a square lattice are shown in Figure 4.22.

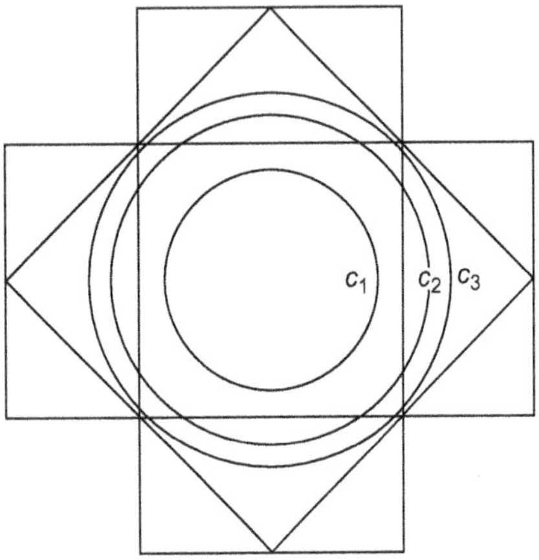

FIGURE 4.22

Two-dimensional Fermi circles corresponding to one, two, and three electrons per atom in the Brillouin zones of a square lattice (without distortion at the zone boundaries).

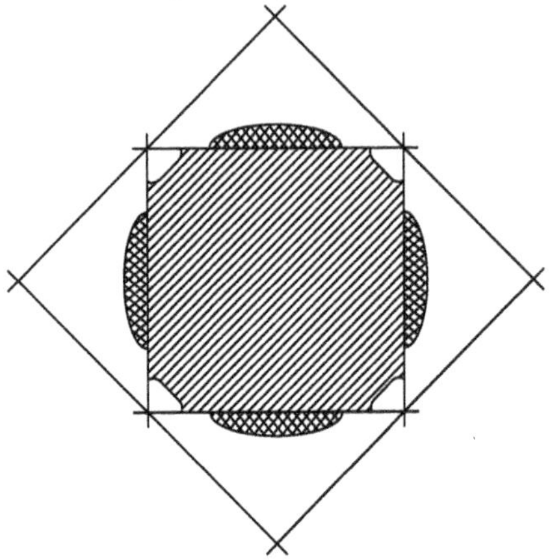

FIGURE 4.23

The distortion of the Fermi circle of a two-dimensional electron gas due to a periodic potential.

The weak periodic potential causes a distortion of the Fermi circle of a two-dimensional electron gas as it approaches the zone boundary. For example, in Figure 4.22, the free electron circle c_1 is entirely within the first Brillouin zone. However, the free electron circles c_2 and c_3 intersect the zone boundaries. Because there is an energy gap between the electron states in the zone boundaries, the Fermi circle is distorted as it approaches the zone boundaries. In addition, the energy curves must be normal to the zone boundaries and develop "necks." The distortion of the Fermi circle by the weak periodic potential is shown in Figure 4.23.

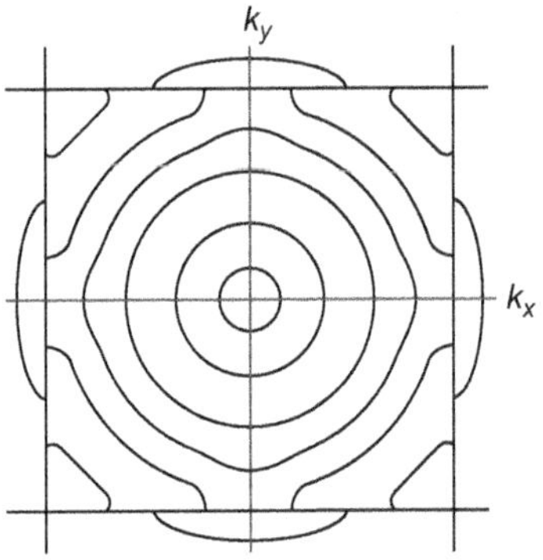

FIGURE 4.24

Distortion of the free electron Fermi circles as energy contours approach the zone boundaries.

However, the distortion of the Fermi circle does not suddenly develop only when the circle approaches the Brillouin zone boundary. The energy contours develop bumps that increase as they approach the zone boundary. The free electron Fermi circles are distorted much before they approach the zone boundary and develop "necks" at the zone boundary because of the energy gap. The distortion of these energy contours is shown in Figure 4.24.

We note from Figure 4.24 that there is a decrease in the constant energy contour when the Fermi circle has contact with the Brillouin zone.

4.12.2 Fermi Surface (in Three Dimensions)

The occupied states of the free electron gas lie within a sphere. The radius of this sphere is the Fermi radius, and the surface is the Fermi surface. In Figure 4.25a, it is shown that when $V_{\mathbf{K}} = 0$, the free electron Fermi sphere meets the zone boundary at a distance ½**K** from the origin O, but there is no distortion of the Fermi surface. In Figure 4.25b, $V_{\mathbf{K}} \neq 0$ and there is distortion of the Fermi sphere at the zone boundary. The Fermi surface intersects the plane in two circles.

It can be shown (Problem 4.6) that the radii r_1 and r_2 of these circles are related by the equation

$$(r_1^2 - r_2^2) = \frac{4m}{\hbar^2} |V_{\mathbf{K}}|. \quad (4.103)$$

In Figure 4.26, we show a free electron Fermi surface completely enclosing the first Brillouin zone of a two-dimensional centered rectangular lattice. We note that the shape of the Fermi surface is modified near the zone boundaries.

Figure 4.27 shows the portion of the Fermi surface in the second Brillouin zone that is mapped back into the first zone so that the energy surface is continuous. This is essentially achieved by using the reduced zone scheme. The portion of the Fermi surface is mapped back to the first Brillouin zone by appropriate translations through reciprocal lattice vectors so that the energy surface is contiguous, as shown in Figure 4.27. However, this method of mapping back the Fermi surface to the first Brillouin zone by any single reciprocal lattice vector becomes increasingly complicated even when there are electron states in the third Brillouin zone. In that case, it is not possible to map the contiguous portions of the third Brillouin zone into the first Brillouin zone by a single reciprocal lattice vector. In such cases, Harrison's method of construction of the Fermi surface becomes very useful.

4.12.3 Harrison's Method of Construction of the Fermi Surface

When the band structure of a solid gets more complicated and the number of valence electrons per atom is large, it becomes very difficult to draw the Fermi surface of a metal. As we have noted, the shape and contours of the Fermi surface are important in determining the physical properties of a metal. To make this task simpler, Harrison[2] proposed a method of constructing the Fermi surface of

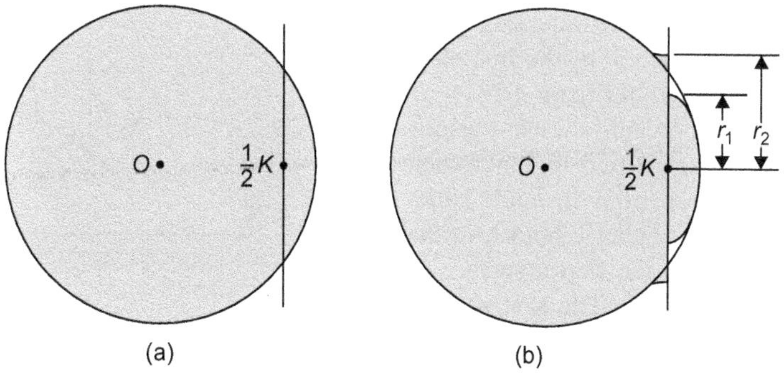

FIGURE 4.25

(a) Free electron sphere cutting Bragg plane when $V_{\mathbf{K}} = 0$; (b) free electron sphere cutting Bragg plane when $V_{\mathbf{K}} \neq 0$.

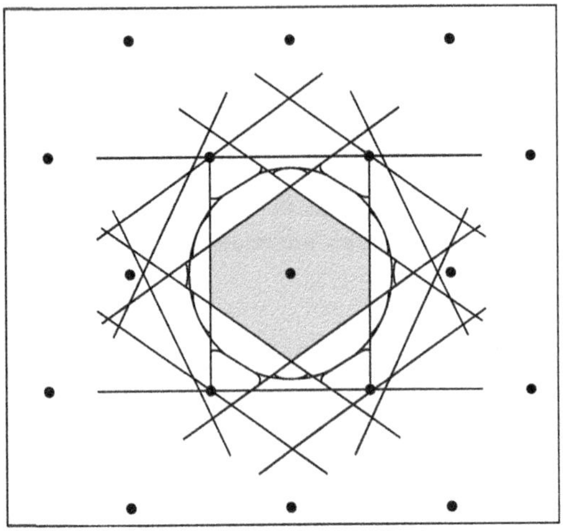

FIGURE 4.26

Free electron Fermi surface completely enclosing the Brillouin zone of a two-dimensional centered rectangular lattice.

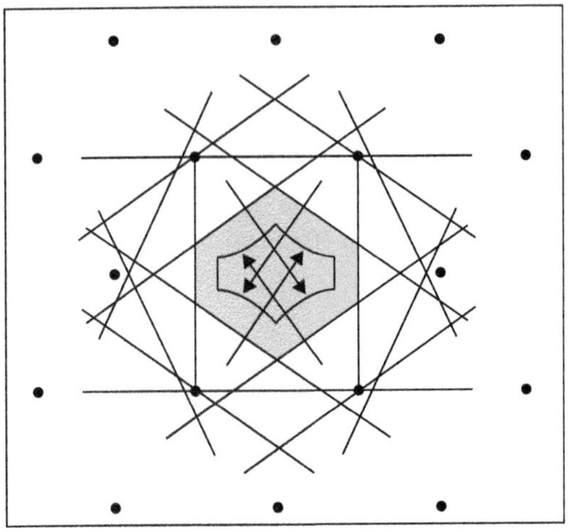

FIGURE 4.27

The portion of the Fermi surface in the second Brillouin zone mapped back to the first zone.

a metal of valence Z by using the periodic zone scheme. According to his method, if the perturbing potential is very small, the energy surfaces must be spheres. The radius of the sphere, which contains ½Z times the volume of a zone (there are two electrons per each **k** state because of spin), is drawn with the center at the origin. The same sphere is drawn about each point of the reciprocal lattice, and one obtains a pattern that has the periodicity of the repeated zone scheme. Harrison's construction of the free electron Fermi surface is shown in Figure 4.28.

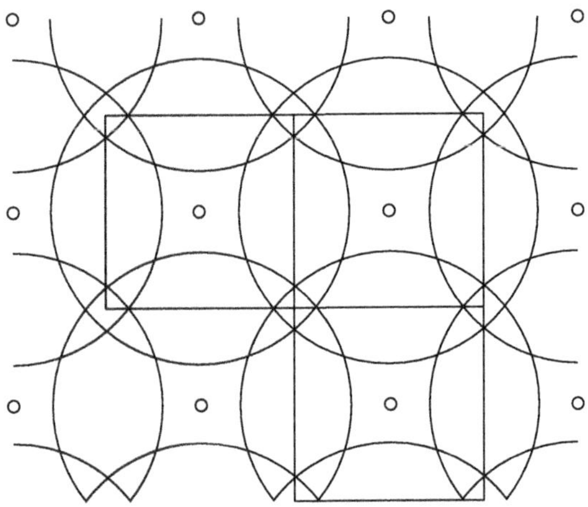

FIGURE 4.28

Free electron Fermi surface constructed by using Harrison's method.

From Figure 4.28, one can choose various parts that are continuously fitted together such that the surfaces are repeated in each zone. These different figures are either a branch of the Fermi surface or a part of the Fermi surface in the second and the third zone. The first zone is completely filled and therefore does not have a Fermi surface. The different parts of the Fermi surface for the second and the third zones are shown in Figures 4.29a and b.

We note from Figures 4.29a and b that these surfaces have cusps where the parts join because they are drawn for spherical Fermi surfaces (free electrons). However, in the nearly free electron model, the Fourier components of the potential would round off the corners, and one would obtain smooth

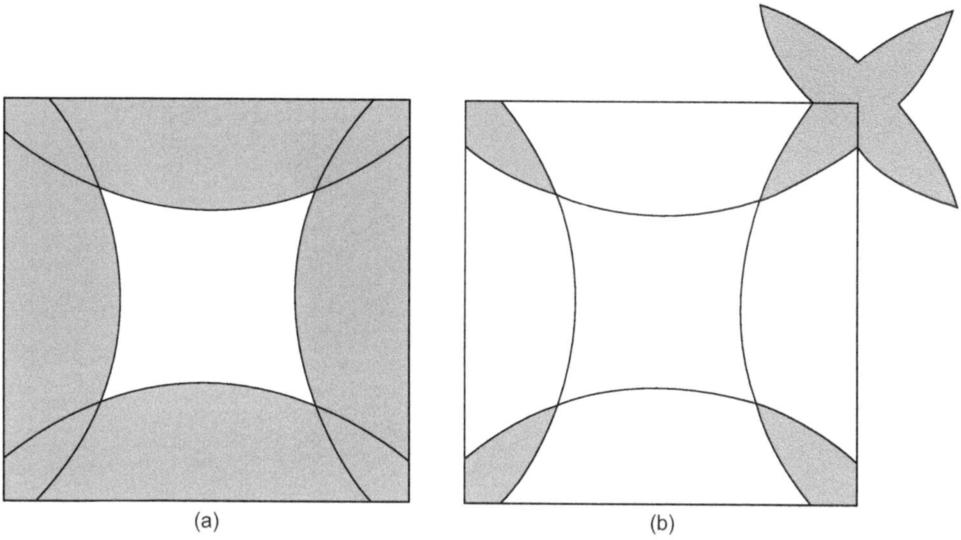

FIGURE 4.29

(a) Fermi surface of the second zone in the reduced zone obtained from Harrison's construction (Figure 4.28). The orbit is a hole orbit. (b) Fermi surfaces in the third Brillouin zone. The orbit in the top-right corner (rosettes) is in the reduced zone scheme.

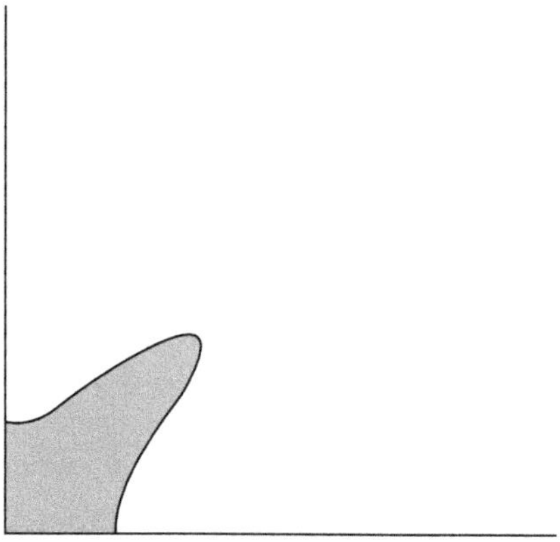

FIGURE 4.30

One corner of the third zone of Figure 4.29b due to the effect of the periodic potential.

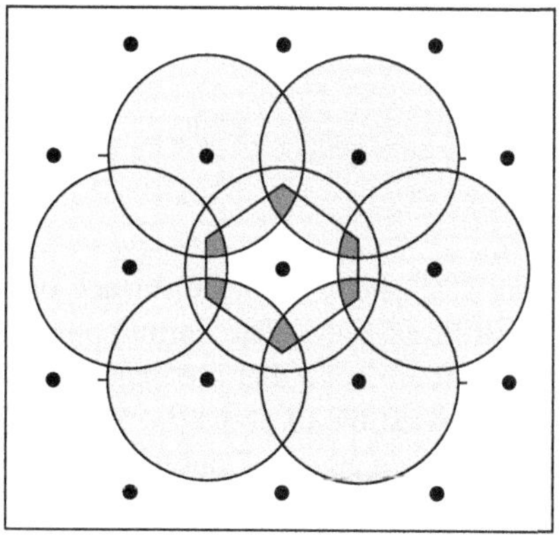

FIGURE 4.31

The Harrison construction of the Fermi surfaces for a two-dimensional centered rectangular lattice.

geometrical objects. This rounding off of the corners and the fact that the line of constant energy intersects the zone boundary at normal incidence are shown for a corner of the third zone in Figure 4.30.

A more visual construction of Harrison's method (see Figure 4.31) shows how the surface in the *n*th Brillouin zone looks when it is mapped into the first Brillouin zone (i.e., the reduced zone scheme).

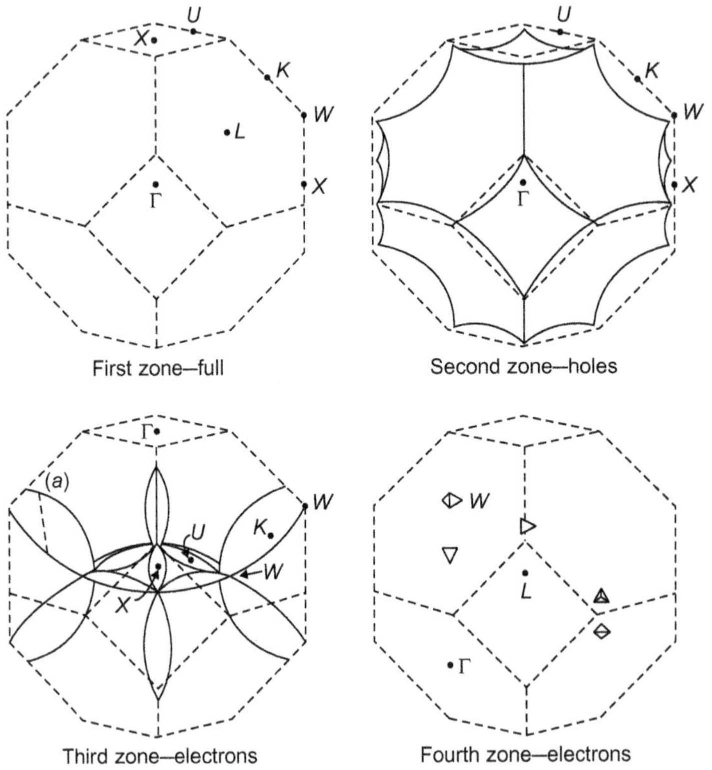

FIGURE 4.32

Free electron Fermi surface of aluminum in the reduced zone scheme obtained by Harrison.

Reproduced from Harrison[2] with the permission of the American Physical Society.

Figure 4.31 shows how the surface of the nth Brillouin zone looks when it is mapped back into the first Brillouin zone. The Fermi sphere in the second Brillouin zone is identified by all points in the first Brillouin zone that are inside two or more spheres. The Fermi sphere in the third Brillouin zone is identified by all points in the first Brillouin zone that are inside three or more spheres. One can extend this method of obtaining Fermi surfaces in three dimensions by using the operations of constructive solid geometry. The free electron Fermi surface of aluminum in the reduced zone scheme, as obtained by Harrison, is shown in Figure 4.32.

PROBLEMS

4.1. It can be shown in quantum mechanics that the momentum operator is the generator of infinitesimal translations ε (equivalent to an analogous relation in classical mechanics)

$$\hat{T}(\varepsilon) = 1 - \frac{i\varepsilon}{\hbar}\hat{p}. \quad (1)$$

Using $\varepsilon = a/N$ in Eq. (1), one obtains

$$\hat{T}(a/N) = 1 - \frac{ia}{\hbar N}\hat{p}. \tag{2}$$

By using the formula

$$e^{-ax} = \lim_{N\to\infty} \left(1 - \tfrac{ax}{N}\right)^N, \tag{3}$$

show that the operator $\hat{T}(a)$ corresponding to a finite translation a (in one dimension) can be obtained by

$$\hat{T}(a) = \lim_{N\to\infty} [\hat{T}(a/N)]^N = e^{-ia\hat{p}/\hbar}. \tag{4}$$

4.2. In general, any function $f(\mathbf{r})$ can be expanded in terms of the plane waves that form a complete set of functions. However, if a function $f(\mathbf{r}) = f(\mathbf{r}+\mathbf{R})$ for all $\mathbf{r}$ and all $\mathbf{k}$ in the Bravais lattice, then it is easy to show that

$$f(\mathbf{r}) = \sum_{\mathbf{K}} f_{\mathbf{K}} e^{i\mathbf{K}\cdot\mathbf{r}} \tag{1}$$

because only $e^{i\mathbf{K}\cdot\mathbf{r}}$ has the periodicity of the lattice. Show that the Fourier coefficients $f_{\mathbf{K}}$ are given by

$$f_{\mathbf{K}} = \frac{1}{v}\int_C d\mathbf{r}\, e^{-i\mathbf{K}\cdot\mathbf{r}} f(\mathbf{r}), \tag{2}$$

where v is the volume of the primitive cell C. To prove Eq. (2), first show that

$$\int_C d\mathbf{r} e^{i\mathbf{K}\cdot(\mathbf{r}+\mathbf{l})} = \int_{C'} d\mathbf{r} e^{i\mathbf{K}\cdot\mathbf{r}} = \int_C d\mathbf{r} e^{i\mathbf{K}\cdot\mathbf{r}}, \tag{3}$$

where C' is the translated cell when C is translated through a vector $\mathbf{l}$. From Eq. (3), one obtains

$$(e^{i\mathbf{K}\cdot\mathbf{l}} - 1)\int_C d\mathbf{r}\, e^{i\mathbf{K}\cdot\mathbf{r}} = 0, \tag{4}$$

from which it follows that (because $e^{i\mathbf{K}\cdot\mathbf{l}} \neq 1$)

$$\int_C d\mathbf{r}\, e^{i\mathbf{K}\cdot\mathbf{r}} = 0, \tag{5}$$

which is needed to prove Eq. (2) from Eq. (1).

4.3. By using the Born–von Karman boundary conditions, one obtains the periodicity for a crystal lattice (which can be considered as a very large Bravais lattice with the volume of the primitive cell V, the volume of the crystal),

$$f(\mathbf{r}) = f(\mathbf{r} + M_i\mathbf{a}_i),\ i = 1, 2, 3. \tag{1}$$

It has been shown that a vector of the reciprocal to this lattice has the form

$$\mathbf{k} = \sum_{i=1}^{3} \frac{m_i}{M_i}\mathbf{b}_i. \tag{2}$$

In a manner similar to Problem 4.2, show that if

$$f(\mathbf{r}) = \sum_{\mathbf{k}} f_{\mathbf{k}} e^{i\mathbf{k}\cdot\mathbf{r}}, \tag{3}$$

then

$$\int_V d\mathbf{r}\, e^{i\mathbf{k}\cdot\mathbf{r}} = 0, \tag{4}$$

and

$$f_{\mathbf{k}} = \frac{1}{V}\int d\mathbf{r} e^{-i\mathbf{k}\cdot\mathbf{r}} f(\mathbf{r}). \tag{5}$$

4.4. It was shown in Eq. (4.4) that the lattice potential $V(\mathbf{r})$ has the periodicity of the lattice,

$$V(\mathbf{r}) = V(\mathbf{r}+\mathbf{R}). \tag{1}$$

Therefore, using the results from Problem 4.2, one obtains

$$V(\mathbf{r}) = \sum_{\mathbf{K}} V(\mathbf{K}) e^{i\mathbf{K}\cdot\mathbf{r}}. \tag{2}$$

From Eq. (2) of Problem 4.2, one obtains

$$\mathrm{V}(\mathbf{K}) = \frac{1}{v}\int_C d\mathbf{r}\, e^{-i\mathbf{K}\cdot\mathbf{r}} V(\mathbf{r})\cdot \tag{3}$$

Assume that $V(0) = 0$. Show that because $V(\mathbf{r})$ is real and if the crystal has inversion symmetry,

$$V(\mathbf{K}) = V(-\mathbf{K}) = V(\mathbf{K})^*. \tag{4}$$

Because the wave function $\psi(\mathbf{r})$ can be expanded in the set of plane waves

$$\psi(\mathbf{r}) = \sum_{\mathbf{q}} C_{\mathbf{q}} e^{i\mathbf{q}\cdot\mathbf{r}}, \tag{5}$$

where the $\mathbf{q}$'s are wave vectors of the form $\mathbf{q} = \sum_{i=1}^{3} \frac{m_i}{M_i}\mathbf{b}_i$, show that the Schrodinger equation can be written as

$$\left(-\frac{\hbar^2}{2m}\nabla^2 + V(\mathbf{r}) - E\right)\psi(\mathbf{r}) = \sum_{\mathbf{q}}\left\{\left(\frac{\hbar^2}{2m}q^2 - E\right)C_{\mathbf{q}} + \sum_{\mathbf{K}'} V_{\mathbf{K}'} C_{\mathbf{q}-\mathbf{K}'}\right\} e^{i\mathbf{q}\cdot\mathbf{r}} = 0. \tag{6}$$

The coefficient of each term in Eq. (6) must vanish (because the plane waves are orthogonal),

$$\left(\frac{\hbar^2}{2m}q^2 - E\right)C_{\mathbf{q}} + \sum_{\mathbf{K}'} V_{\mathbf{K}'} C_{\mathbf{q}-\mathbf{K}'} = 0. \tag{7}$$

If $\mathbf{q} = \mathbf{k} - \mathbf{K}$, where $\mathbf{k}$ lies in the first Brillouin zone and changing the variables to $\mathbf{K}' \to \mathbf{K}' - \mathbf{K}$, show that Eq. (7) can be written as

$$\left(\frac{\hbar^2}{2m}(\mathbf{k}-\mathbf{K})^2 - E)C_{\mathbf{k}-\mathbf{K}}\right) + \sum_{\mathbf{K}'} V_{\mathbf{K}'-\mathbf{K}} C_{\mathbf{k}-\mathbf{K}'} = 0. \tag{8}$$

Eq. (8) shows that for a fixed $\mathbf{k}$ in the first Brillouin zone, only wave vectors that differ from $\mathbf{k}$ by a reciprocal lattice vector are coupled. Rewriting $\psi(\mathbf{r})$ as $\psi_{\mathbf{K}}(\mathbf{r})$ and E as ε, from Eqs. (5) and (8), show that

$$\psi_{\mathbf{k}}(\mathbf{r}) = e^{i\mathbf{k}\cdot\mathbf{r}}(\sum_{\mathbf{G}} C_{\mathbf{k}-\mathbf{K}} e^{-i\mathbf{K},\mathbf{r}}), \tag{9}$$

and

$$\left(\frac{\hbar^2}{2m}(\mathbf{k}-\mathbf{K})^2 - \varepsilon\right) C_{\mathbf{k}-\mathbf{K}} + \sum_{\mathbf{K}'} V_{\mathbf{K}'-\mathbf{K}} C_{\mathbf{k}-\mathbf{K}'} = 0. \tag{10}$$

Define

$$u_k(\mathbf{r}) = \sum_{\mathbf{K}} C_{\mathbf{k}-\mathbf{K}} e^{-i\mathbf{K}\cdot\mathbf{r}}, \tag{11}$$

and hence show from Eqs. (9) and (11) that

$$\psi_{\mathbf{k}}(\mathbf{r}) = e^{i\mathbf{k}\cdot\mathbf{r}} u_{\mathbf{k}}(\mathbf{r}). \tag{12}$$

From Eq. (11), show that

$$u_{\mathbf{k}}(\mathbf{r}+\mathbf{R}) = u_{\mathbf{k}}(\mathbf{r}). \tag{13}$$

Here, $\psi_{\mathbf{k}}(\mathbf{r})$ is the Bloch function, and $u_{\mathbf{k}}(\mathbf{r})$ is the periodic part of the Bloch function (in three dimensions).

4.5. In the Kronig–Penney model, an electron in a one-dimensional lattice is in the presence of a potential

$$V(x) = \sum_n V_0 \phi(x-nb)\varphi(nb+c-x), \tag{1}$$

where $b > c$, n is zero or a positive integer, $b + c = L$ and φ is the Heaviside unit-step function

$$\varphi(x) = \begin{Bmatrix} 0, & x<0. \\ 1, & x>0. \end{Bmatrix}. \tag{2}$$

The one-dimensional periodic potential can be represented as

$$V(x+L) = V(x), \tag{3}$$

as shown in Figurc P4.1. The potential energy as a function of distance is given in Figure P4.1 Because there is symmetry under the displacement by L, it can be shown that the eigenfunction is

$$\psi(x) = e^{ikx}u(x), \tag{4}$$

where $u(x+L) = u(x)$ and k is arbitrary, and as we will show, it is the propagation constant. In addition, we consider the one-dimensional periodic potential visualized as a ring of circumference NL, such that

$$\psi(x+NL) = \psi(x). \tag{5}$$

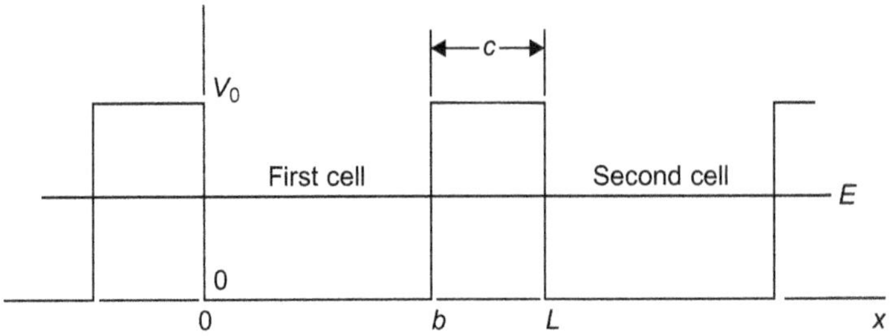

FIGURE P4.1

The Kronig–Penny model of the square potential lattice.

From Eqs. (4) and (5), we obtain

$$e^{ik(x+NL)} = e^{ikx}, \tag{6}$$

from which, we obtain

$$k = \frac{2\pi n}{NL}, \quad \text{where } n = 0, \pm 1, \pm 2, \ldots. \tag{7}$$

Here, k is called the propagation constant of the state. We represent the wave function as

$$\begin{aligned} \psi(x+L) &= e^{ikL}\psi(x), \\ \psi(x) &= e^{ikx}u(x), \end{aligned} \tag{8}$$

where

$$u(x+L) = u(x). \tag{9}$$

The advantage of Eq. (8) is that if we know $\psi(x)$ for any one cell of the periodic lattice, it can be calculated for any other cell.

To solve this problem, one can use the one-dimensional equivalence of Eq. (4.35) (except that k is now called the propagation constant of the state $k = \frac{2\pi n}{NL}$),

$$\frac{d^2u}{dx^2} + 2ik\frac{du}{dx} + \frac{2m}{\hbar^2}\left[E - V(x) - \frac{\hbar^2 k^2}{2m}\right]u(x) = 0. \tag{10}$$

From Figure P4.1 and Eq. (9), we have the periodicity condition

$$u(x+L) = u(x) \quad \text{and} \quad du(x+L)/dx = du(x)/dx. \tag{11}$$

Introducing the notations

$$k_1 = \left(\frac{2mE}{\hbar^2}\right)^{\frac{1}{2}} \quad \text{and} \quad k_2 = \left[\frac{2m}{\hbar^2}(V_0 - E)\right]^{\frac{1}{2}}, \tag{12}$$

the solution of Eq. (10) for the square lattice can be written as

$$u_1(x) = A\,e^{i(k_1-k)x} + Be^{-i(k_1+k)x}, \tag{13}$$

for the region of the well (from $x = 0$ to $x = b$), and

$$u_2(x) = Ce^{(k_2-ik)x} + De^{-(k_2+ik)x}, \tag{14}$$

for the region of the hill ($x = b$ to $x = L$). Using the periodicity conditions mentioned previously, show that

$$A + B = e^{-ikL}\left[Ce^{k_2L} + De^{-k_2L}\right], \tag{15}$$

and

$$ik_1(A - B) = k_2 e^{-ikL}\left[Ce^{k_2L} - De^{-k_2L}\right]. \tag{16}$$

Show from the continuity conditions that

$$Ae^{ik_1b} + Be^{-ik_1b} = Ce^{k_2b} + De^{-k_2b}, \tag{17}$$

and

$$ik_1\left[Ae^{ik_1b} - Be^{-ik_1b}\right] = k_2\left[Ce^{k_2b} - De^{-k_2b}\right]. \tag{18}$$

Eqs. (15) through (18) have nontrivial solutions only if the determinant of the matrix of the coefficients vanishes. Show that

$$\cos k_1b \cosh k_2c - \frac{k_1^2 - k_2^2}{2k_1k_2} \sin k_1b \sinh k_2c = \cos kL, \quad E < V_0. \tag{19}$$

It can also be shown that

$$\cos k_1b \cos k_2c - \frac{k_1^2 + k_2^2}{2k_1k_2} \sin k_1b \sin k_2c = \cos kL, \quad E > V_0. \tag{20}$$

Eqs. (19) and (20) can be solved numerically, and the results are shown in Figure P4.2. The remarkable feature is that the right sides of the eigenvalue equations are bound between −1 and +1. Thus, only those values of E that make the left side of these equations also lie in the same interval and all other values are excluded. This is the origin of the band structure in solids.

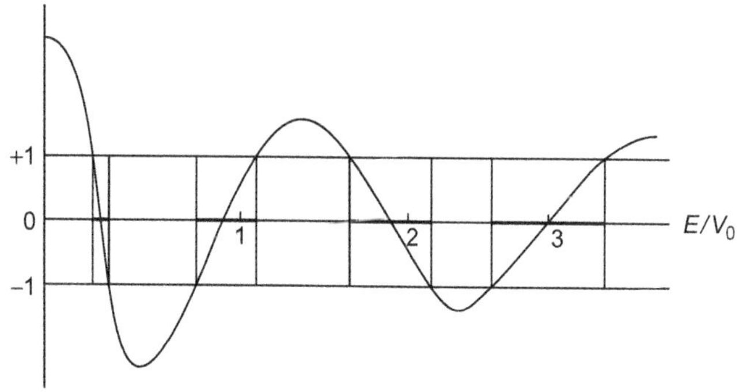

FIGURE P4.2

The left sides of Eqs. (19) and (20) are plotted as functions of E that join smoothly at $E = V_0$. The heavy lines display the allowed range of energy values.

4.6. If we write

$$\mathbf{k} = \tfrac{1}{2}\mathbf{K} + \mathbf{q},$$

Eq. (4.82) can be rewritten as

$$\varepsilon(\mathbf{k}) = \varepsilon^0_{\mathbf{K}/2} + \frac{\hbar^2 q^2}{2m} \pm \left(4\varepsilon^0_{\mathbf{K}/2}\frac{\hbar^2 q_{\parallel}^2}{2m} + |V_{\mathbf{K}}|^2\right)^{\frac{1}{2}}, \tag{1}$$

where $q_{\parallel}$ is the parallel component of $\mathbf{q}$. We can also write

$$\varepsilon_F = \varepsilon^0_{\mathbf{K}/2} - |V_{\mathbf{K}}| + \delta. \tag{2}$$

Show that:

a. When $\delta < 0$, no Fermi surface intersects the Bragg plane.

b. When $0 < \delta < 2|V_{\mathbf{K}}|$, the Fermi surface intersects the Bragg plane in a circle of radius,

$$r = \sqrt{\frac{2m\delta}{\hbar^2}}. \tag{3}$$

c. When $\delta > |2V_{\mathbf{K}}|$, the Fermi surface cuts the Bragg plane in two circles (because it lies in both bands) of radii r_1 and r_2 and

$$r_1^2 - r_2^2 = \frac{4m|V_{\mathbf{K}}|}{\hbar^2}. \tag{4}$$

References

1. Ashcroft NW, Mermin ND. *Solid state physics*. Toronto: Thomson Learning; 1976.
2. Harrison WA. Band structure of aluminum. *Phys Rev* 1960;**118**:1182.
3. Harrison WA. *Solid state theory*. New York: McGraw-Hill; 1969.
4. Kittel C. *Introduction to solid state physics*. New York: John Wiley & Sons; 1976.
5. Liboff RI. *Introductory quantum mechanics*. Reading, MA: Addison-Wesley; 1980.
6. Madelung O. *Introduction to solid state theory*. New York: Springer-Verlag; 1978.
7. Marder MP. *Condensed matter physics*. New York: John Wiley & Sons; 2000.
8. Myers HP. *Introductory solid state physics*. London: Taylor & Francis; 1990.
9. Seitz F. *The modern theory of solids*. New York: McGraw-Hill; 1940.
10. Wannier GH. *Elements of solid state theory*. Cambridge: Cambridge University Press; 1954.
11. Wilson AH. *The theory of metals*. Cambridge: Cambridge University Press; 1958.
12. Ziman JM. *Principles of the theory of solids*. Cambridge: Cambridge University Press; 1972.

CHAPTER

Band-Structure Calculations

5

CHAPTER OUTLINE

5.1 Introduction 131
5.2 Tight-Binding Approximation 131
5.3 LCAO Method 135
5.4 Wannier Functions 140
5.5 Cellular Method 142
5.6 Orthogonalized Plane-Wave (OPW) Method 145
5.7 Pseudopotentials 147
5.8 Muffin-Tin Potential 149
5.9 Augmented Plane-Wave (APW) Method 150
5.10 Green's Function (KKR) Method 152
5.11 Model Pseudopotentials 156
5.12 Empirical Pseudopotentials 157
5.13 First-Principles Pseudopotentials 158
Problems 160
References 163

5.1 INTRODUCTION

In the preceding chapter, the nearly free electron approximation was described in detail. In this approximation, the valence electrons are considered to be nearly free electrons moving in a background of positive charges that are arranged in a regular array such that the potential is periodic. The positive charges are actually the cores of the atoms in which the valence electrons have been stripped from the parent atoms. These cores either are located at the lattice points or a group of cores is located symmetrically around each lattice point. The cores are constituted from atomic orbitals that are essentially localized. In general, these atomic orbitals do not overlap. However, in some cases, the localized atomic orbitals overlap to an extent such that they form Bloch functions in the crystalline solid.

5.2 TIGHT-BINDING APPROXIMATION

It is useful to formulate a linear combination of these atomic orbitals such that they would be Bloch functions as required for wave functions in a crystalline solid. Thus, the tight-binding approximation, formulated for the core electrons, complements the nearly free electron approximation formulated for the valence electrons. The tight-binding method is very useful in calculating the energy bands of partially filled d-shells of transition metal atoms as well as for describing the electronic structure of insulators.

In its simplest form, the tight-binding method can be expressed as follows. We assume that $\phi_a(\mathbf{r})$ is an atomic orbital for a free atom located at a lattice point at the origin; i.e., it is the ground state of an electron moving in the potential $v_a(\mathbf{r})$ of an isolated atom such that the energy is ε_a. As a first approximation, we consider a monatomic lattice with one atom located at each lattice point. We assume that the free atom is in an s state since the bands obtained from the other $(p, d, \ldots)$ states are much more complicated because the atomic levels are degenerate. We further assume that the influence of one atom on another is small. If the bound levels of the atomic Hamiltonian (H_a) are localized, the Schrodinger equation can be written as

$$H_a \phi_a(\mathbf{r}) = \varepsilon_a \phi_a(\mathbf{r}), \tag{5.1}$$

where

$$H_a = -\frac{\hbar^2}{2m}\nabla^2 + v_a. \tag{5.2}$$

The range of $\phi_a(\mathbf{r})$ is very small when $\mathbf{r}$ exceeds the distance of a lattice constant. However, in the crystalline solid, the Hamiltonian H would differ from the atomic Hamiltonian because of the corrections to the atomic potential in the crystal lattice. If we write $\Delta V(\mathbf{r})$ as the (small) correction to the atomic potential in the lattice (the difference between the periodic potential in the crystal lattice and the potential of an isolated atom), the crystal Hamiltonian H can be written as

$$H = H_a + \Delta V(\mathbf{r}). \tag{5.3}$$

The periodic potential $V(\mathbf{r})$ is the sum of the atomic potential $v_a(\mathbf{r})$ and the correction term $\Delta V(\mathbf{r})$,

$$V(\mathbf{r}) = v_a(\mathbf{r}) + \Delta V(\mathbf{r}). \tag{5.4}$$

In Figure 5.1, $r\phi_a(\mathbf{r})$ and $\Delta V(\mathbf{r})$ are drawn along a linear chain of the atomic sites that are located at the lattice points.

The Schrodinger equation can be written as

$$\left(\frac{-\hbar^2}{2m}\nabla^2 + V(\mathbf{r})\right)\psi_{\mathbf{k}}(\mathbf{r}) = E_{\mathbf{k}}\psi_{\mathbf{k}}(\mathbf{r}). \tag{5.5}$$

We note that the eigenfunctions $\psi_{\mathbf{k}}(\mathbf{r})$ have to be Bloch functions. As a zero-order approximation, we assume that $\Delta V(\mathbf{r}) = 0$ in the region where $\phi_a(\mathbf{r}) \neq 0$. In that case, we can construct the Bloch functions as a linear combination of the atomic orbitals

$$\psi_{\mathbf{k}}(\mathbf{r}) = \sum_i e^{i\mathbf{k}\cdot\mathbf{R}_i}\phi_a(\mathbf{r} - \mathbf{R}_i), \tag{5.6}$$

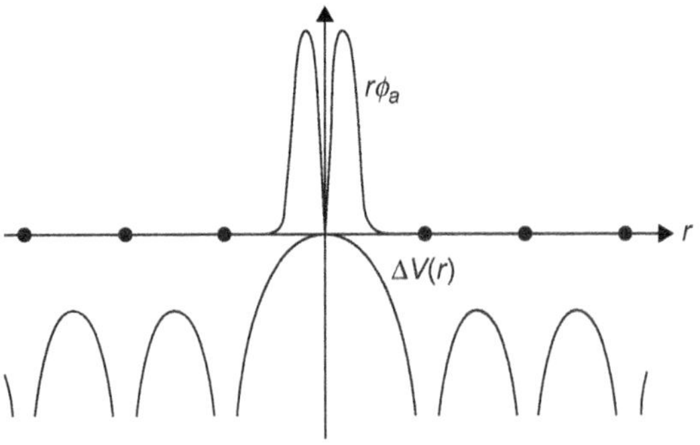

FIGURE 5.1

Here, $r\phi_a(\mathbf{r})$ and $\Delta V(\mathbf{r})$ are plotted against $\mathbf{r}$ in a linear chain of atomic sites.

where the $\mathbf{R}_i$ are the complete set of Bravais lattice vectors in the crystal lattice. We will now prove that $\psi_{\mathbf{k}}(\mathbf{r})$ are Bloch functions. From Eq. (5.6),

$$\begin{aligned}\psi_{\mathbf{k}}(\mathbf{r}+\mathbf{R}_j) &= \sum_i e^{i\mathbf{k}\cdot\mathbf{R}_i}\phi_a(\mathbf{r}+\mathbf{R}_j-\mathbf{R}_i)\\ &= e^{i\mathbf{k}\cdot\mathbf{R}_j}\sum_i e^{i\mathbf{k}\cdot(\mathbf{R}_i-\mathbf{R}_j)}\phi_a(\mathbf{r}+\mathbf{R}_j-\mathbf{R}_i)\\ &= e^{i\mathbf{k}\cdot\mathbf{R}_j}\sum_l e^{i\mathbf{k}\cdot\mathbf{R}_l}\phi_a(\mathbf{r}-\mathbf{R}_l)\\ &= e^{i\mathbf{k}\cdot\mathbf{R}_j}\psi_{\mathbf{k}}(\mathbf{r}),\end{aligned} \tag{5.7}$$

which are the criteria for the Bloch functions.

The expectation value of the energy can be written as

$$\varepsilon(\mathbf{k}) = \frac{\displaystyle\int \psi_{\mathbf{k}}^*\left[-\frac{\hbar^2}{2m}\nabla^2+V(\mathbf{r})\right]\psi_{\mathbf{k}}(\mathbf{r})d\mathbf{r}}{\displaystyle\int \psi_{\mathbf{k}}^*(\mathbf{r})\psi_{\mathbf{k}}(\mathbf{r})d\mathbf{r}}. \tag{5.8}$$

Assuming that the overlap between the neighboring cells is small,

$$\int \psi_{\mathbf{k}}^*(\mathbf{r})\psi_{\mathbf{k}}(\mathbf{r})d\mathbf{r}\approx 1. \tag{5.9}$$

We obtain from Eqs. (5.4) and (5.8),

$$\varepsilon(\mathbf{k})\approx\sum_{i,j} e^{i\mathbf{k}\cdot(\mathbf{R}_i-\mathbf{R}_j)}\int \phi_a^*(\mathbf{r}-\mathbf{R}_j)\left[-\frac{\hbar^2}{2m}\nabla^2+v_a(\mathbf{r})+\Delta V(\mathbf{r})\right]\phi_a(\mathbf{r}-\mathbf{R}_i)d\mathbf{r}. \tag{5.10}$$

From Eqs. (5.1) and (5.10), we can write $\varepsilon(\mathbf{k})$ in the alternate form

$$\varepsilon(\mathbf{k}) = \varepsilon_a + N\sum_l e^{-i\mathbf{k}\cdot\mathbf{R}_l}\int \phi_a(\mathbf{r}-\mathbf{R}_l)\Delta V(\mathbf{r})\phi_a(\mathbf{r})d\mathbf{r}. \tag{5.11}$$

Here, N is the number of atoms in the monatomic crystal lattice. Neglecting all integrals except those at the atomic sites and between the nearest-neighbor atomic sites and defining $\mathbf{R}_n$ as the nearest-neighbor lattice vectors of a primitive lattice in which the origin is located, we can rewrite Eq. (5.11) in the alternate form

$$\varepsilon(\mathbf{k})\approx\varepsilon_a-\beta-\gamma\sum_l e^{-i\mathbf{k}\cdot\mathbf{R}_l}, \tag{5.12}$$

where

$$\beta = -N\int \phi_a^*(\mathbf{r})\Delta V(\mathbf{r})\phi_a(\mathbf{r})d\mathbf{r}, \tag{5.13}$$

and

$$\gamma = -N\int \phi_a^*(\mathbf{r}-\mathbf{R}_n)\Delta V(\mathbf{r})\phi_a(\mathbf{r})d\mathbf{r}. \tag{5.14}$$

As an example, we consider a simple cubic lattice. The nearest-neighbor positions of a simple cubic lattice of lattice constant a are $\mathbf{R}_n = (\pm a, 0, 0)$, and $(0, \pm a, 0)$, $(0, 0, \pm a)$. Thus, Eq. (5.12) becomes

$$\varepsilon(\mathbf{k}) = \varepsilon_a - \beta - 2\gamma(\cos k_x a + \cos k_y a + \cos k_z a). \tag{5.15}$$

The energies are located in a band of width 12 γ. If $ka \ll 1$, Eq. (5.15) yields

$$\varepsilon(\mathbf{k}) \cong \varepsilon_a - \beta - 6\gamma + \gamma k^2 a^2. \tag{5.16}$$

In this derivation, we have considered only one state (the s state) of the free atom and obtained one band. We note that usually for s states, $\beta=0$. We also note that Eq. (5.15) is periodic in $\mathbf{k}$ and therefore only those values of $\mathbf{k}$ lying in the first Brillouin zone will define independent wave functions. The number of states in the first Brillouin zone that corresponds to a nondegenerate atomic level is equal to $2N$. In Figure 5.2, we schematically show $\varepsilon(\mathbf{k})$ plotted as a function of $\mathbf{k}$ along the cube axis by using the tight-binding method. Here, $\varepsilon(\mathbf{k})$ has a minimum at $\mathbf{k} = 0$ along the cube direction and a maximum at $\mathbf{k} = (\pi/a, 0, 0)$, which is the zone boundary.

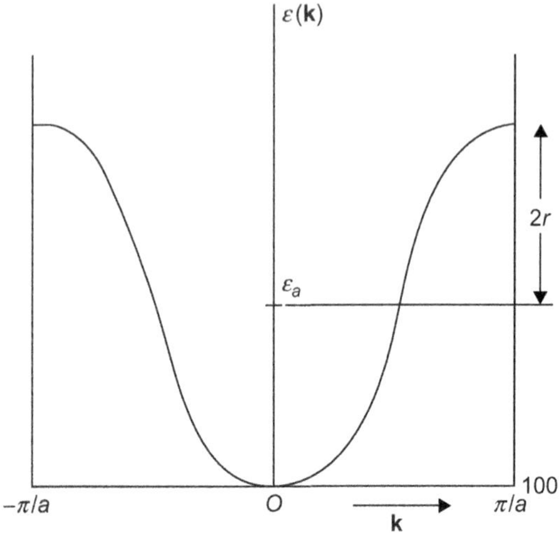

FIGURE 5.2

Here, $\varepsilon(\mathbf{k})$ is plotted as a function of $\mathbf{k}$ along the cube axis.

Thus, for every state of an electron in the free atom, there will be a band of energies in the crystal. The complexities of the problem increase for higher atomic levels. One can generalize by stating that when N identical atoms are kept far apart, and each atom has several different atomic levels (orbitals), there will be N-fold degenerate states for a single electron. When these atoms are brought closer, the atomic orbitals overlap, and a band that has N states is formed. This is shown in Figure 5.3.

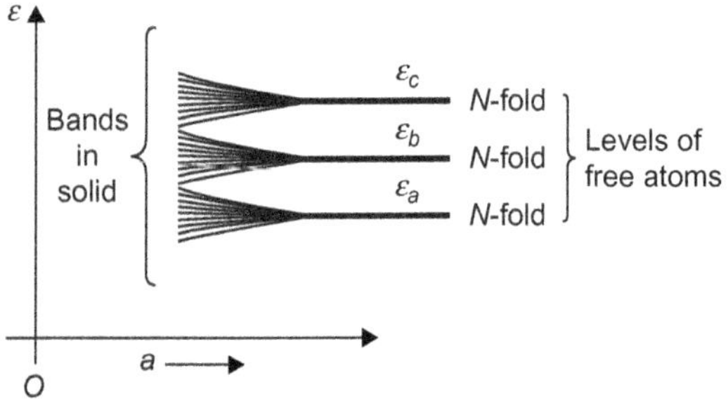

FIGURE 5.3

Formation of bands from atomic levels as the atoms come closer.

Thus, one can refer to the 3s-band, 3p-band, 3d-band, and so on. It is too simplistic to consider these higher atomic states in the simple tight-binding method discussed previously. However, in Figure 5.3, we have schematically shown the formation of bands in solids by using the tight-binding method. Figure 5.4 shows the broadening of the bands that ultimately overlap.

It is important to note at this stage that when these bands, formed from different atomic orbitals, broaden and overlap, the tight-binding method has to be modified and one uses the linear combination of atomic orbitals (LCAO method). We will discuss specific cases of such overlaps when we consider the p- and d-bands as well as the s-d "resonance" bands.

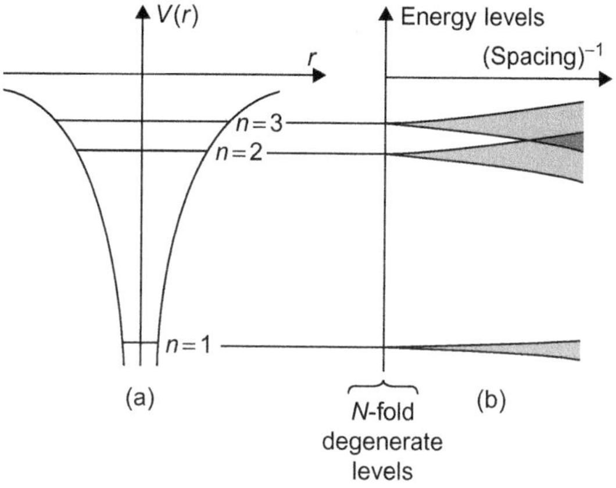

FIGURE 5.4

(a) Nondegenerate atomic levels; (b) energy levels of N such atoms in a periodic array plotted as a function of mean interatomic spacing.

5.3 LCAO METHOD

In the linear combination of atomic orbitals (LCAO) method (Ref. 18), the Bloch wave function of the crystalline solid is expanded in terms that are linear combinations of atomic orbitals. Thus, Eq. (5.1) is rewritten as

$$\psi_{\mathbf{k}}(\mathbf{r}) = \sum_i e^{i\mathbf{k}\cdot\mathbf{R}_i}\phi_c(\mathbf{r}-\mathbf{R}_i), \tag{5.17}$$

where

$$\phi_c(\mathbf{r}) = \sum_j \beta_j\phi_j(\mathbf{r}), \tag{5.18}$$

β_j are arbitrary constants to be determined, and $\phi_j(\mathbf{r})$ is one of the set of localized atomic orbitals on the atom located at the origin. In general, $\psi_{\mathbf{k}}(\mathbf{r})$ is used as a trial wave function and β_j are obtained by minimizing the ground-state energy.

The crystal Schrodinger equation can be rewritten as

$$H\psi_{\mathbf{k}}(\mathbf{r}) = \varepsilon(\mathbf{k})\psi_{\mathbf{k}}(\mathbf{r}), \tag{5.19}$$

where

$$H = H_{at} + \Delta V(\mathbf{r}). \tag{5.20}$$

Multiplying Eq. (5.19) by one of the atomic wave functions, $\phi_a{}^*(\mathbf{r})$, and integrating over all $\mathbf{r}$, we obtain

$$\int \phi_a{}^*(\mathbf{r})[H_{at} + \Delta V(\mathbf{r})]\psi_{\mathbf{k}}(\mathbf{r})d\mathbf{r} = \varepsilon(\mathbf{k})\int \phi_a{}^*(\mathbf{r})\psi_{\mathbf{k}}(\mathbf{r})d\mathbf{r}, \tag{5.21}$$

which can be rewritten in the alternate form

$$(\varepsilon(\mathbf{k})-\varepsilon_a)\int \phi_a^*(\mathbf{r})\psi_{\mathbf{k}}(\mathbf{r})d\mathbf{r}=\int \phi_a^*(\mathbf{r})\Delta V(\mathbf{r})\psi_{\mathbf{k}}(\mathbf{r})d\mathbf{r}, \tag{5.22}$$

where we have used Eq. (5.1) to obtain ε_a. Because the atomic orbitals are orthonormal, i.e.,

$$\int \phi_a^*(\mathbf{r})\phi_{a'}(\mathbf{r})d\mathbf{r}=\delta_{aa'}, \tag{5.23}$$

we obtain from Eqs. (5.17), (5.18), and (5.23),

$$\int \phi_a^*(\mathbf{r})\psi_{\mathbf{k}}(\mathbf{r})d\mathbf{r}=\beta_a+\sum_j \beta_j\left(\sum_{i\neq 0}\int \phi_a^*(\mathbf{r})\phi_j(\mathbf{r}-\mathbf{R}_i)e^{i\mathbf{k}\cdot\mathbf{R}_i}d\mathbf{r}\right). \tag{5.24}$$

From Eqs. (5.21), (5.23), and (5.24), we obtain

$$\begin{aligned}\left(\varepsilon(\mathbf{k})-\varepsilon_a\right)\beta_a &= (\varepsilon_a-\varepsilon(\mathbf{k}))\sum_j \beta_j\left(\sum_{i\neq 0}\int \phi_a^*(\mathbf{r})\phi_j(\mathbf{r}-\mathbf{R}_i)e^{i\mathbf{k}\cdot\mathbf{R}_i}d\mathbf{r}\right)\\ &+\sum_j \beta_j\left(\int \phi_a^*(\mathbf{r})\Delta V(\mathbf{r})\phi_j(\mathbf{r})d\mathbf{r}\right)\\ &+\sum_j \beta_j\left(\sum_{i\neq 0}\int \phi_a^*(\mathbf{r})\Delta V(\mathbf{r})\phi_j(\mathbf{r}-\mathbf{R}_i)e^{i\mathbf{k}\cdot\mathbf{R}_i}d\mathbf{r}\right).\end{aligned} \tag{5.25}$$

We note that each of the three terms on the right side of the equation is small (for different reasons) unless the atomic levels are degenerate. The first term contains the overlap integrals $\int \phi_a^*(\mathbf{r})\phi_j(\mathbf{r}-\mathbf{R}_i)d\mathbf{r}$ where $\mathbf{R}_i\neq 0$. The atomic wave functions are centered on different lattice sites, and therefore, the overlap is very small compared to unity because the atomic orbitals are well localized. The second term on the right side of the equation is small because at large distances, where $\Delta V(\mathbf{r})$ (which is the difference between the periodic and atomic potentials) is significant, the atomic wave functions are small. The third term on the right side is small for the same reason as the first term, because they also contain atomic wave functions centered at different sites.

If the atomic levels are nondegenerate, i.e., for an s-level, Eq. (5.25) is essentially the same as Eq. (5.13). For bands arising from atomic p-levels, which are triple-degenerate, Eq. (5.25) would give a set of three homogeneous equations. One has to solve a 3×3 secular problem. The eigenvalues would give $\varepsilon(\mathbf{k})$ for the three bands, and $\beta(\mathbf{k})$, the appropriate linear combination of the atomic orbitals at the various $\mathbf{k}$ values in the Brillouin zone, would be obtained from the solutions. Similarly, the d-levels are five-fold degenerate and yield five homogeneous equations. One has to solve a 5×5 problem to obtain the eigenvalues as well as the values of $\beta(\mathbf{k})$. In transition metals, the s and d electrons overlap in energy. In addition, because the s electrons have a large degree of plane wave character, they have essentially a uniform spatial distribution. Therefore, there is a spatial overlap between the s and the d orbitals. There is an intra-atomic s-d resonance that hybridizes the s- and d-bands due to this overlap in both real space and energy. Therefore, for transition metals, one has to solve a 6×6 secular problem that includes both d- and s-levels.

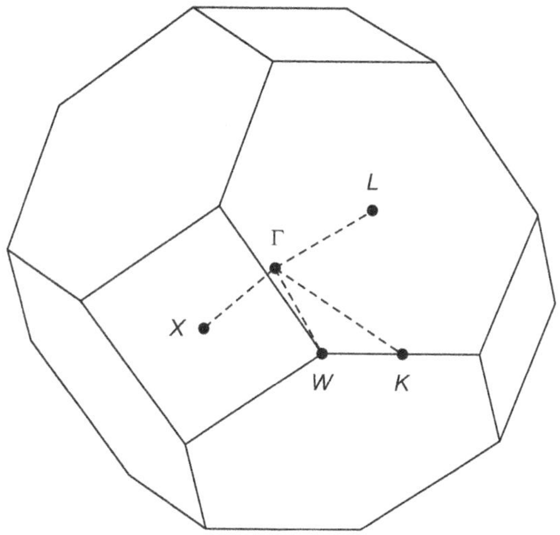

FIGURE 5.5

ΓX direction of the first Brillouin zone of the fcc structure.

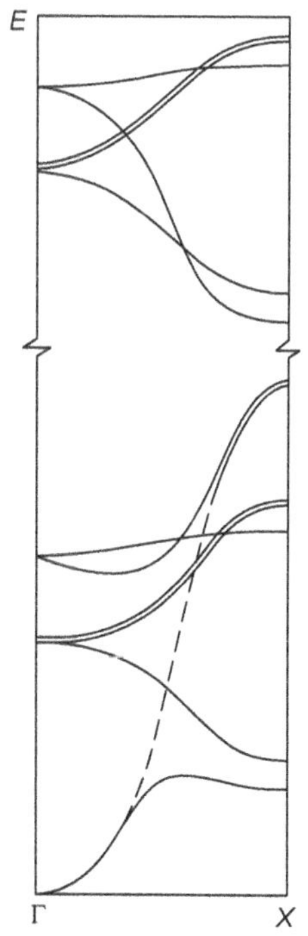

FIGURE 5.6

The five *d* sub-bands of a transition metal atom in the ΓX direction of a fcc structure.

The five *d* sub-bands of a transition metal in the ΓX direction in the fcc structure that is shown in Figure 5.5 are shown in Figure 5.6.

The progression of the 3*d*-band through the Fermi energy for elements in and near the long series of transition metals is shown in Figure 5.7. The bands that are primarily of *sp* character are shown by the heavier lines.

The tight-binding approximation is more complicated in crystalline solids in which the Bravais lattice is not monatomic. For example, the hexagonal close-packed metals are simple hexagonal with a two-point basis. There are two ways of solving this problem. One procedure is to consider the two-point basis as a molecule, of which the wave functions are known, and treat the problem in the same manner as we have done for a monatomic lattice, except that in this case, molecular wave functions are used instead of atomic wave functions. For *s*-levels, for which the nearest-neighbor overlaps are small, the overlap would be small in the molecule. Thus, an atomic *s*-level will give rise to two nearly degenerate molecular levels and hence will yield two tight-binding bands. The alternate

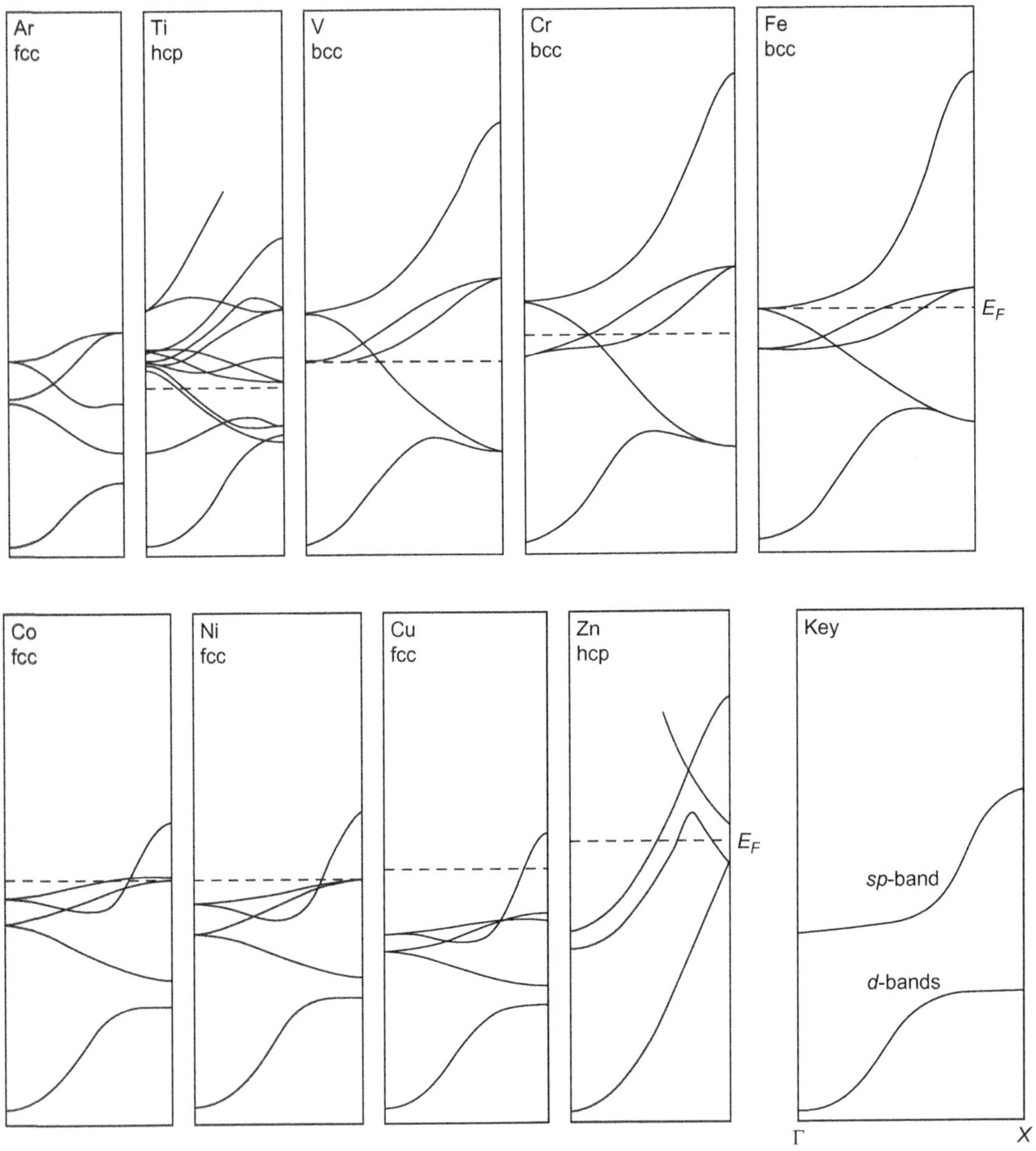

FIGURE 5.7

The progression of the 3*d*-band through the Fermi energy for elements in and near the first long series of transition metals.

method is to make linear combinations of atomic levels centered at the Bravais lattice points and at the basis points. The LCAO wave function would be of the form

$$\psi_{\mathbf{k}}(\mathbf{r}) = \sum_i \left(a_1\phi(\mathbf{r}-\mathbf{R}_i) + a_2\phi(\mathbf{r}-\mathbf{R}_i+\mathbf{l})\right)e^{i\mathbf{k}\cdot\mathbf{R}_i}, \tag{5.26}$$

where $\mathbf{l}$ is the vector connecting the two basis atoms in the Bravais lattice.

In these methods of tight-binding calculations, we have considered only the spin-independent linear combinations of atomic orbitals, thereby implicitly neglecting spin-orbit coupling, which is important in calculating the atomic levels in the heavier elements. We can include spin-orbit coupling by considering the interaction of the electron spin on the orbital of its parent atom located at the origin as well as by including the interaction between the spin of that electron and the electric field of all the other ions in $\Delta V(\mathbf{r})$. In this method, we must use linear combinations of the spin-dependent atomic wave functions.

The LCAO method is not very satisfactory for the quantitative calculation of Bloch functions in solids. It is very difficult to calculate the three-center integrals and nonorthogonal basis functions. In addition, representing the valence electron states in a metal or a semiconductor by an expression such as Eq. (5.17) is basically wrong. We consider the interstitial region, the creation of which is illustrated in Figures 5.8 and 5.9. Figure 5.8 shows the bound atomic orbitals of two free atoms.

In Figure 5.9, these two bound atomic orbitals are superposed, thereby creating an interstitial region with a small constant potential.

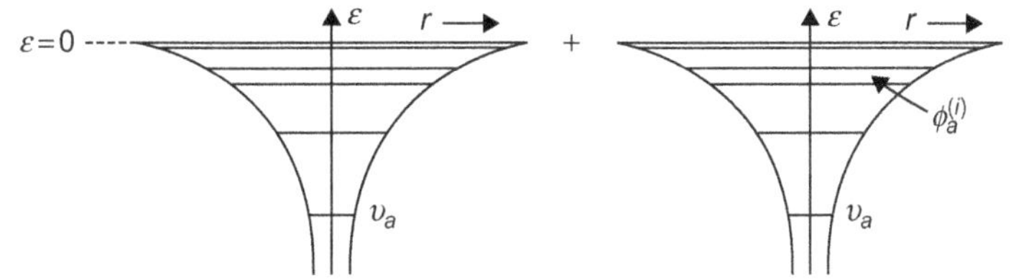

FIGURE 5.8

Potential wells centered on the nuclei of two free atoms.

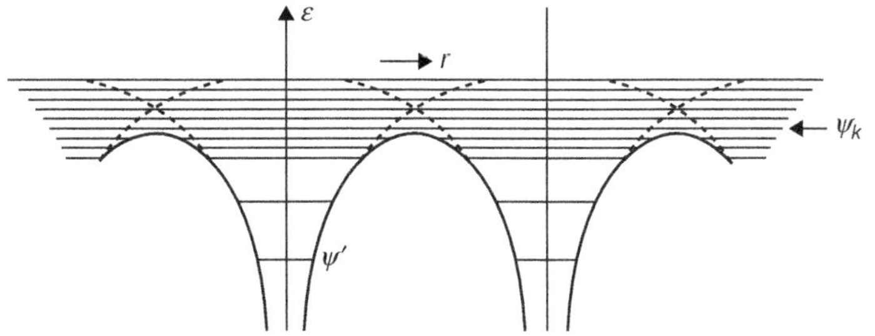

FIGURE 5.9

The Bloch functions in a crystalline lattice.

The individual atomic orbitals, used in Eq. (5.17), do not exist in the interstitial regions between the main potential wells centered on the atomic nuclei because they would lie above the energy of the barriers between the atomic spheres. It is also inappropriate to represent a Bloch function, which is a consequence of a periodic potential $V(\mathbf{r})$ in the crystalline solid, by a linear combination of the atomic orbitals $\phi_a(\mathbf{r})$, which arise due to an atomic potential $v_a(\mathbf{r})$ and rapidly tend to zero as $\mathbf{r}\to\infty$.

Finally, the set of functions used in the LCAO method is incomplete because it does not include the scattered-wave eigenstates in the Schrodinger equation in the continuum, above the energy zero of $v_a(\mathbf{r})$. They are useful in the construction of the Bloch functions inside the atomic core but inappropriate in the interstitial region, where the use of a linear combination of plane waves is more appropriate.

The main disadvantage of the tight-binding as well as the LCAO methods is that the independent electron approximation is one of the bases of these approximations. This approximation works very well for both insulators and the low-lying bands in metals. In such cases, the tight-binding bands are very low in energy and completely filled. However, the independent electron approximation in the formulation of the tight-binding method fails when there are partially filled bands that are obtained from localized atomic orbitals with small overlap integrals. In particular, when we consider the properties of the narrow tight-binding bands that are obtained from partially filled atomic shells, the failure of the independent electron approximation becomes more apparent and leads to anomalous results. When one uses the tight-binding method for the *d*- and *f*-shells in metals, the inclusion of the electron–electron interaction becomes very important. In particular, when there is a magnetic structure, this failure is even more apparent.

One of the important reasons for going beyond the tight-binding approximation is that when there is a second electron at a given atomic site, the strong repulsion between the first electron and the second electron at the same site cannot be treated by the independent electron approximation.

An interesting consequence of the tight-binding approximation is the hypothetical question as to what would happen if the distance between the atoms is continuously increased so that there is a slow but steady transition from the metallic to the atomic state. The overlap integrals would become smaller with the increase of the lattice constant, and eventually all bands, including the partially filled conduction band(s), become very narrow tight-binding bands. The decrease of electrical conductivity of a metal would continuously drop with the decrease of the overlap integrals as the lattice constant increases and eventually would become zero. Thus, the metal would become an insulator. This is known as a Mott transition. In practice, if electron–electron interactions are included, the conductivity would abruptly become zero at a Mott transition.

The Mott transition has indeed been observed in certain transition metal oxides. They are normally insulators but suddenly become good conductors above a certain temperature.

5.4 WANNIER FUNCTIONS

One of the main problems of the tight-binding formulation is caused by the difficulties of orthogonalization since a linear combination of the atomic orbitals is used. This orthogonalization problem can be circumvented by defining a set of orthonormal wave functions that can be constructed from Bloch functions and are localized on the atomic sites. These functions are known as Wannier functions. The Wannier function (Ref. 19) centered at the lattice site $\mathbf{R}_i$ is defined as

$$w_n(\mathbf{r}-\mathbf{R}_i) = \frac{1}{\sqrt{N}}\sum_{\mathbf{k}} e^{-i\mathbf{k}\cdot\mathbf{R}_i}\psi_{n\mathbf{k}}(\mathbf{r}), \tag{5.27}$$

where N is the number of lattice sites as well as the number of $\mathbf{k}$ states in the first Brillouin zone. We will now show that the Wannier functions form an orthonormal set:

$$\int d\mathbf{r}\, w_m^*(\mathbf{r}-\mathbf{R}_j)w_n(\mathbf{r}-\mathbf{R}_i) = \int d\mathbf{r} \sum_{\mathbf{k}} \sum_{\mathbf{k}'} \frac{1}{N} e^{-i\mathbf{k}\cdot\mathbf{R}_i + i\mathbf{k}'\cdot\mathbf{R}_j} \psi_{m\mathbf{k}'}^*(\mathbf{r})\psi_{n\mathbf{k}}(\mathbf{r}). \tag{5.28}$$

Because the Bloch functions are orthonormal (Problem 5.3), i.e.,

$$\int d\mathbf{r}\, \psi_{m\mathbf{k}'}^*(\mathbf{r})\psi_{n\mathbf{k}}(\mathbf{r})d\mathbf{r} = \delta_{m,n}\delta_{\mathbf{k},\mathbf{k}'}, \tag{5.29}$$

Eq. (5.28) can be rewritten in the alternate form

$$\int d\mathbf{r}\, w_m^*(\mathbf{r}-\mathbf{R}_j)w_n(\mathbf{r}-\mathbf{R}_i) = \frac{1}{N}\sum_{\mathbf{k}}\sum_{\mathbf{k}'} e^{-i\mathbf{k}\cdot\mathbf{R}_i + i\mathbf{k}'\cdot\mathbf{R}_j}\delta_{m,n}\delta_{\mathbf{k},\mathbf{k}'}. \tag{5.30}$$

Because it can be easily shown that (Problem 5.4)

$$\sum_{\mathbf{k}} e^{-i\mathbf{k}\cdot(\mathbf{R}_i-\mathbf{R}_j)} = N\delta_{i,j}, \tag{5.31}$$

we obtain from Eqs. (5.30) and (5.31) the orthonormal conditions of the Wannier functions,

$$\int d\mathbf{r}\, w_m^*(\mathbf{r}-\mathbf{R}_j)w_n(\mathbf{r}-\mathbf{R}_i) = \delta_{m,n}\delta_{i,j}. \tag{5.32}$$

If the Wannier functions are known, one can obtain the Bloch functions from Eq. (5.27) by multiplying both sides by $e^{i\mathbf{k}'\cdot\mathbf{R}_i}$ and summing over all direct lattice vectors $\mathbf{R}_i$,

$$\sum_{\mathbf{R}_i} e^{i\mathbf{k}'\cdot\mathbf{R}_i} w_n(\mathbf{r}-\mathbf{R}_i) = \frac{1}{\sqrt{N}}\sum_{\mathbf{R}_i}\sum_{\mathbf{k}} e^{-i(\mathbf{k}-\mathbf{k}')\cdot\mathbf{R}_i}\psi_{n\mathbf{k}}(\mathbf{r}). \tag{5.33}$$

It can be easily shown that (Problem 5.5)

$$\sum_{\mathbf{R}_i} e^{-(\mathbf{k}-\mathbf{k}')\cdot\mathbf{R}_i} = N\delta_{\mathbf{k},\mathbf{k}'}. \tag{5.34}$$

From Eqs. (5.33) and (5.34), we obtain

$$\psi_{n\mathbf{k}}(\mathbf{r}) = \frac{1}{\sqrt{N}}\sum_{\mathbf{R}_i} e^{i\mathbf{k}\cdot\mathbf{R}_i} w_n(\mathbf{r}-\mathbf{R}_i). \tag{5.35}$$

It may be noted that any Bloch function is determined only within an overall phase factor. So the Wannier functions defined in Eq. (5.27) can also be defined as

$$w_n(\mathbf{r}-\mathbf{R}_i) = \frac{1}{\sqrt{N}}\sum_{\mathbf{k}} e^{-i\mathbf{k}\cdot\mathbf{R}_i + i\varphi(\mathbf{k})}\psi_{n\mathbf{k}}(\mathbf{r}), \tag{5.36}$$

where $\varphi(\mathbf{k})$ is an arbitrary real function. The Wannier functions can be optimized by making them drop off as fast as possible when $\mathbf{r}-\mathbf{R}_i$ starts increasing. This can be achieved by manipulating the arbitrary function $\varphi(\mathbf{k})$. The Wannier functions on adjacent sites can be schematically shown as in Figure 5.10.

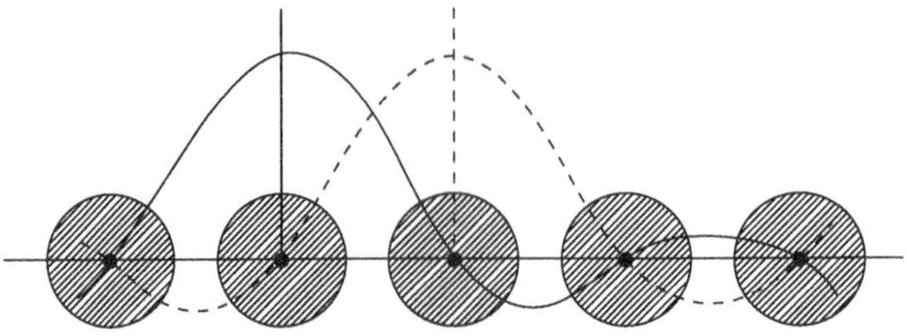

FIGURE 5.10

The Wannier functions on adjacent sites.

5.5 CELLULAR METHOD

The cellular method was originally proposed by Wigner and Seitz (Ref. 20) to calculate the band structure of sodium. Later, with the availability of powerful computers that could solve differential equations, the cellular method was considerably improved and generalized to calculate the band structure of metals. The essential principle behind the cellular method is that because of the Bloch condition

$$\psi_{\mathbf{k}}(\mathbf{r}+\mathbf{R}_i) = e^{i\mathbf{k}\cdot\mathbf{R}_i}\psi_{\mathbf{k}}(\mathbf{r}), \tag{5.37}$$

it is sufficient to solve the Schrodinger equation in Eq. (5.5) within a single primitive cell C. The wave function in other primitive cells can be obtained from its values in C by using Eq. (5.37). However, both $\psi_{\mathbf{k}}(\mathbf{r})$ and $\nabla\psi_{\mathbf{k}}(\mathbf{r})$ must be continuous as $\mathbf{r}$ crosses the cell boundary.

In the original Wigner–Seitz formulation, the Wigner–Seitz cell of a fcc or a bcc lattice is a polyhedron (see Figure 5.11) that could be approximated as a sphere of equal volume of which the radius is r_s. We will denote this sphere as the Wigner–Seitz sphere.

In addition, they considered the lowest state energy of sodium metal so that $\mathbf{k}=0$. Thus, the boundary conditions reduce to

$$\psi_{\mathbf{k}}(\mathbf{r}+\mathbf{R}_i) = \psi_{\mathbf{k}}(\mathbf{r}), \tag{5.38}$$

and because $\psi_{\mathbf{k}}(\mathbf{r})$ is periodic from cell to cell, it should have a horizontal tangent at the surface of the Wigner–Seitz sphere

$$\left[\frac{\partial \psi_{\mathbf{k}}}{\partial \mathbf{r}}\right]_{r=r_s} = 0. \tag{5.39}$$

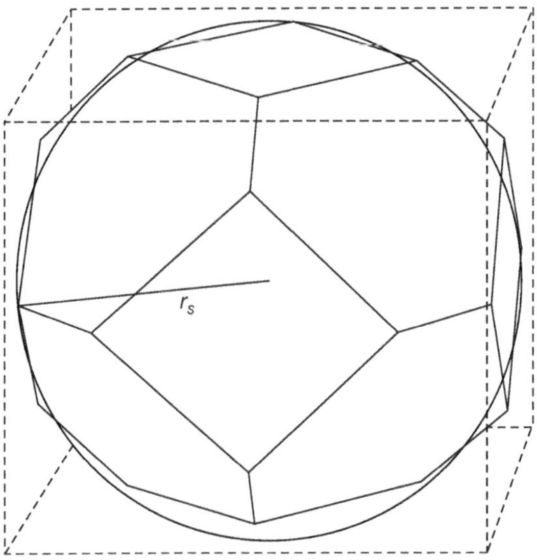

FIGURE 5.11

Wigner–Seitz cell of sodium metal (bcc structure) approximated as a sphere.

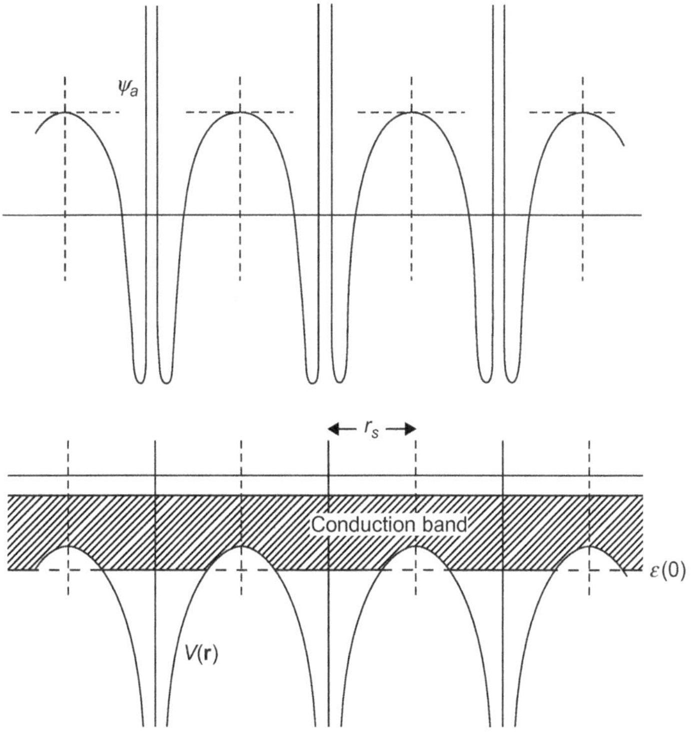

FIGURE 5.12

The ground-state energy $\varepsilon(0)$ of sodium metal in the Wigner–Seitz method.

For a potential that is symmetric inside the sphere, the energy $\varepsilon(0)$ can be defined as shown in Figure 5.12.

One can also write, by using the Bloch condition,

$$\psi_{\mathbf{k}}(\mathbf{r}) = e^{i\mathbf{k}\cdot\mathbf{r}} u_{\mathbf{k}}(\mathbf{r}), \tag{5.40}$$

where $u_{\mathbf{k}}(\mathbf{r})$ is the periodic part of the Bloch function. From Eqs. (5.5) and (5.40),

$$-\frac{\hbar^2}{2m}(\nabla^2 + 2i\mathbf{k}\cdot\nabla - k^2)u_{\mathbf{k}}(\mathbf{r}) + V(\mathbf{r})u_{\mathbf{k}}(\mathbf{r}) = \varepsilon(\mathbf{k})u_{\mathbf{k}}(\mathbf{r}), \tag{5.41}$$

where $u_{\mathbf{k}}(\mathbf{r})$ has a horizontal tangent at $r = r_s$. The shape of the band can be approximately obtained by using this method. However, the structure of the solid is neglected in this procedure.

There has been considerable improvement in the techniques of the cellular method for calculation of the band structure since Wigner and Seitz's original papers in 1933 and 1934 (Ref. 21). The starting point is the boundary conditions that $\psi_{\mathbf{k}}(\mathbf{r})$ *and* $\nabla\psi_{\mathbf{k}}(\mathbf{r})$ must be continuous as $\mathbf{r}$ crosses the boundary of the primitive cell C (usually located at the origin, i.e., $\mathbf{R} = 0$) to the neighboring cell located at the lattice point **R**. These boundary conditions, which introduce the wave vector **k** in the solution as well as retain the discrete set of energies $\varepsilon_n(\mathbf{k}) = \varepsilon$, can be restated as

$$\psi_{\mathbf{k}}(\mathbf{r}) = e^{-i\mathbf{k}\cdot\mathbf{R}}\psi_{\mathbf{k}}(\mathbf{r}+\mathbf{R}) \tag{5.42}$$

and

$$\nabla\psi_{\mathbf{k}}(\mathbf{r}) = e^{-i\mathbf{k}\cdot\mathbf{R}}\nabla\psi_{\mathbf{k}}(\mathbf{r}+\mathbf{R}) \tag{5.43}$$

for pairs of points on the surface separated by **R**. If the point **r** is located on the surface of the primitive cell C, the normals $\hat{\mathbf{n}}$ to surface of the cell at **r** and **r** + **R** are oppositely directed. Therefore, the continuity condition (Eq. 5.43) can be rewritten as (Problem 5.7)

$$\hat{\mathbf{n}}(\mathbf{r})\cdot\nabla\psi_{\mathbf{k}}(\mathbf{r}) = -e^{-i\mathbf{k}\cdot\mathbf{R}}\hat{\mathbf{n}}(\mathbf{r}+\mathbf{R})\cdot\nabla\psi_{\mathbf{k}}(\mathbf{r}+\mathbf{R}). \tag{5.44}$$

In the cellular method, the periodic potential $V(\mathbf{r})$ in the Wigner–Seitz cell C (see Figure 5.13) is replaced by a potential $V'(\mathbf{r})$ that has spherical symmetry (see Figure 5.14) about the origin.

Because we have made the approximation that the potential $V'(\mathbf{r})$ is spherically symmetric inside the cell C, a complete set of solutions to the Schrodinger equation (Eq. 5.5) is of the form (see Goswami[7] or any book on quantum mechanics)

$$\psi_{\varepsilon lm}(\mathbf{r}) = R_{\varepsilon l}(r)Y_{lm}(\theta,\phi), \tag{5.45}$$

where $Y_{lm}(\theta,\phi)$ are the spherical harmonics and $R_{\varepsilon l}(r)$ satisfies the differential equation

$$\left[\frac{1}{r^2}\frac{d}{dr}\left(r^2\frac{d}{dr}\right) + \frac{2m}{\hbar^2}\left(\varepsilon - V'(r) - \frac{\hbar^2 l(l+1)}{2mr^2}\right)\right]R_{\varepsilon l}(r) = 0. \tag{5.46}$$

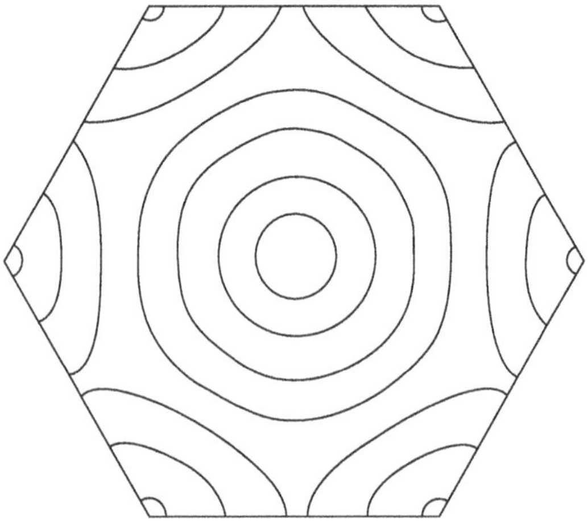

FIGURE 5.13

The actual potential $V(\mathbf{r})$ inside a primitive cell.

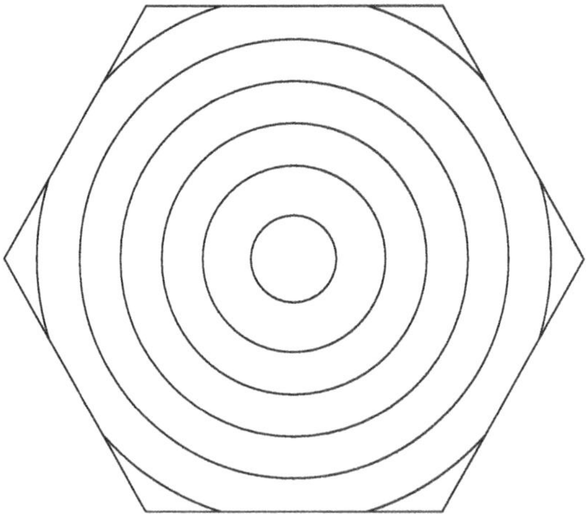

FIGURE 5.14

The approximate potential $V'(\mathbf{r})$ with spherical symmetry.

Here, we have written ε instead of n (which is appropriate for an atom) for the quantum number related to energy. Any linear combination of Eq. (5.45) would also be a solution of Eq. (5.5). Thus, we obtain

$$\psi(\mathbf{r}, \varepsilon) = \sum_{lm} B_{lm} R_{\varepsilon l}(r) Y_{lm}(\theta, \phi), \tag{5.47}$$

where the arbitrary coefficients B_{lm} are determined from the boundary conditions in Eqs. (5.42) and (5.44). In actual computations, a finite number of B_{lm} are chosen, and an equal number of points on the surface of the cell are chosen where the boundary conditions in Eqs. (5.42) and (5.44) are imposed. One obtains a set of **k**-dependent linear homogeneous equations for B_{lm}. The energies $\varepsilon_n(\mathbf{k})$ are those values of ε for which the determinant of the homogeneous equations for B_{lm} vanishes.

5.6 ORTHOGONALIZED PLANE-WAVE (OPW) METHOD

The nearly free electron method does not account for the rapid oscillatory behavior in the core region of the atom. On the other hand, the tight-binding method, which matches reasonably well to $\psi_{\mathbf{k}}(\mathbf{r})$ inside the atomic core, does not account for the scattered-wave eigenstates of the Schrodinger equation in the continuum above the energy zero of the atomic potential $v_a(\mathbf{r})$. The Bloch states in the interstitial region must behave like a combination of free electron plane wave states. This anomaly of finding a complete set of wave functions that can account for the entire region in the crystal was resolved by the orthogonalized plane-wave (OPW) method first introduced by Herring. This method is based on the basic fact that the eigenfunctions of the crystalline Hamiltonian must be orthogonal at every point in the crystal.

In the OPW method, one starts with the construction of a complete set of Bloch functions for the core states. Here, the core states are defined as the atomic orbitals of the ions. A tight-binding combination of the core states, which are Bloch functions, as constructed in Eq. (5.17), can be written as

$$\phi_{c\mathbf{k}}(\mathbf{r}) = \sum_{\mathbf{R}_i} e^{i\mathbf{k}\cdot\mathbf{R}_i} \phi_c(\mathbf{r} - \mathbf{R}_i), \tag{5.48}$$

where $\phi_c(\mathbf{r})$ is one of the core orbital c. Here, the core wave functions are localized around the atomic sites and are assumed to be known. In fact, $\phi_c(\mathbf{k})$ is a linear combination of degenerate one-electron states. Thus, Eq. (5.48) is one of the solutions of the Schrodinger equation of the crystal with energy $\varepsilon_c(\mathbf{k})$. As we have noted in the discussion of the tight-binding method, this would form a fully occupied narrow band.

If we denote $\chi_{\mathbf{k}}(\mathbf{r})$ as a wave function for one of the higher states, it should be orthogonal to the core states defined in Eq. (5.48). In addition, the higher states $\chi_{\mathbf{k}}(\mathbf{r})$ must have the properties of a free electron wave function (plane wave)$e^{i\mathbf{k}\cdot\mathbf{r}}$ in the interstitial region. Thus, we can define the OPW as

$$\chi_{\mathbf{k}}(\mathbf{r}) = e^{i\mathbf{k}\cdot\mathbf{r}} - \sum_c b_c \phi_{c\mathbf{k}}(\mathbf{r}), \tag{5.49}$$

where the sum is over all the core levels. The coefficients b_c are determined from the orthogonalization condition

$$\int d\mathbf{r}\,\phi^*_{c\mathbf{k}}(\mathbf{r})\chi_{\mathbf{k}}(\mathbf{r}) = 0. \tag{5.50}$$

From Eqs. (5.49) and (5.50), we obtain

$$b_c = \int d\mathbf{r}\,\phi^*_{c\mathbf{k}}(\mathbf{r})e^{i\mathbf{k}\cdot\mathbf{r}}. \tag{5.51}$$

In deriving Eq. (5.51), we have made the basic assumption that the core states are orthogonal, i.e.,

$$\int d\mathbf{r}\,\phi^*_{c\mathbf{k}}(\mathbf{r})\phi_{c'\mathbf{k}}(\mathbf{r}) = \delta_{c,c'}. \tag{5.52}$$

The OPW state $\chi_{\mathbf{k}}(\mathbf{r})$ defined in Eq. (5.49) is appropriate for the interstitial region because the core states are localized, and consequently, each $\phi_{c\mathbf{k}}$ is small. Thus, $\chi_{\mathbf{k}}(\mathbf{r})$ behaves like the plane wave $e^{i\mathbf{k}\cdot\mathbf{r}}$ in the interstitial region. However, because $\chi_{\mathbf{k}}(\mathbf{r})$ is orthogonal to the core state within the core, it is appropriately a higher atomic orbital than the occupied atomic states. For example, if the core states are $2s$ and $2p$ orbitals, $\chi_{\mathbf{k}}(\mathbf{r})$ will behave like a $3s$ or $3p$ orbital with an extra nodal surface.

A schematic representation of an OPW is shown in Figure 5.15.

The eigenfunctions of the Schrodinger equation in the crystalline lattice can be expressed as a linear combination of OPWs that would form the basis states. The wave function can be written as

$$\psi_{\mathbf{k}}(\mathbf{r}) = \sum_{\mathbf{K}} C_{\mathbf{k}-\mathbf{K}}\chi_{\mathbf{k}-\mathbf{K}}(\mathbf{r}). \tag{5.53}$$

The coefficients $C_{\mathbf{k}-\mathbf{K}}$ are obtained by using the variational principle, to minimize the expectation value of the energy. The variational principle is briefly explained here.

If we define the energy functional

$$\varepsilon[\phi] = \frac{\int\left(\frac{\hbar^2}{2m}\,|\nabla\phi(\mathbf{r})\,|^2 + V(\mathbf{r})\,|\phi(\mathbf{r})\,|^2\right)d\mathbf{r}}{\int|\phi(\mathbf{r})|^2 d\mathbf{r}}, \tag{5.54}$$

it follows from Eq. (11) of Problem 5.8 that

$$\varepsilon[\psi_{\mathbf{k}}] = \varepsilon_{\mathbf{k}}, \tag{5.55}$$

which is the variational principle.

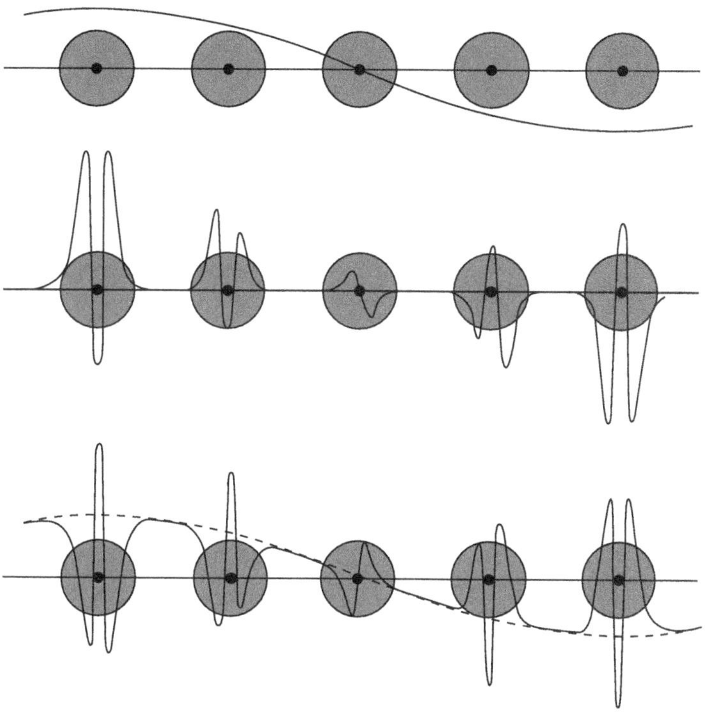

FIGURE 5.15

Plane waves, core functions, and orthogonalized plane waves.

5.7 PSEUDOPOTENTIALS

In 1959, Phillips and Kleinman[14] developed the pseudopotential method that is extensively used to calculate the band structure of metals. This method has been refined through the past 50 years. We will essentially discuss the original theory proposed by Phillips and Kleinman. They showed that advantage can be taken of the crystal symmetry to construct wave functions $\phi_{\mathbf{k}}(\mathbf{r})$, which are smooth parts of the symmetrized Bloch functions. The wave equation satisfied by $\phi_{\mathbf{k}}(\mathbf{r})$ contains an additional term of simple character that corresponds to the usual OPW terms and has a simple physical interpretation as an effective repulsive potential. The sum of the crystal potential and this repulsive potential is known as the pseudopotential. The cancellation between the attractive periodic potential and the repulsive potential in the core region is responsible for the rapid convergence of the OPW calculations for s states.

The pseudopotential method starts with the assumption that $\psi_{\mathbf{k}}(\mathbf{r})$, as expressed in Eq. (5.53) and with appropriate coefficients $C_{\mathbf{k}-\mathbf{K}}$ obtained by the use of the variational method, is the exact crystal wave function. We construct a linear combination of the plane waves,

$$\varphi_{\mathbf{k}}(\mathbf{r}) = \sum_{\mathbf{K}} C_{\mathbf{k}-\mathbf{K}} e^{i(\mathbf{k}-\mathbf{K})\cdot\mathbf{r}}, \tag{5.56}$$

which have the same coefficients $C_{\mathbf{k}-\mathbf{K}}$. Because $\psi_{\mathbf{k}}(\mathbf{r})$ must be orthogonal to the core states $\phi_{c\mathbf{k}}(\mathbf{r})$ described in Eq. (5.48), we obtain

$$\psi_{\mathbf{k}}(\mathbf{r}) = \varphi_{\mathbf{k}}(\mathbf{r}) - \sum_{c} < \phi_{c\mathbf{k}}(\mathbf{r}) \,|\, \varphi_{\mathbf{k}}(\mathbf{r}) > \phi_{c\mathbf{k}}(\mathbf{r}). \tag{5.57}$$

The Schrodinger equation in the crystal is

$$H\psi_{\mathbf{k}}(\mathbf{r}) = \varepsilon_{\mathbf{k}}\psi_{\mathbf{k}}(\mathbf{r}). \tag{5.58}$$

From Eqs. (5.57) and (5.58), we obtain

$$H\varphi_{\mathbf{k}}(\mathbf{r}) - \sum_{c} < \phi_{c\mathbf{k}}(\mathbf{r}) \,|\, \varphi_{\mathbf{k}}(\mathbf{r}) > H\phi_{c\mathbf{k}}(\mathbf{r}) = \varepsilon_{\mathbf{k}}\varphi_{\mathbf{k}}(\mathbf{r}) - \varepsilon_{\mathbf{k}}\sum_{c} < \phi_{c\mathbf{k}}(\mathbf{r}) \,|\, \varphi_{\mathbf{k}}(\mathbf{r}) > \phi_{c\mathbf{k}}(\mathbf{r}). \tag{5.59}$$

Using the relation

$$H\phi_{c\mathbf{k}}(\mathbf{r}) = \varepsilon_{c\mathbf{k}}(\mathbf{r})\phi_{c\mathbf{k}}(\mathbf{r}), \tag{5.60}$$

in Eq. (5.59) and rearranging the terms, we obtain

$$H\varphi_{\mathbf{k}}(\mathbf{r}) + \sum_{c}(\varepsilon_{\mathbf{k}} - \varepsilon_{c\mathbf{k}}) < \phi_{c\mathbf{k}}(\mathbf{r}) \,|\, \varphi_{\mathbf{k}}(\mathbf{r}) > \phi_{c\mathbf{k}}(\mathbf{r}) = \varepsilon_{\mathbf{k}}\varphi_{\mathbf{k}}(\mathbf{r}). \tag{5.61}$$

Eq. (5.61) can be expressed as

$$(H + V_R)\varphi_{\mathbf{k}}(\mathbf{r}) = \varepsilon_{\mathbf{k}}\varphi_{\mathbf{k}}(\mathbf{r}), \tag{5.62}$$

where V_R is the repulsive part of the potential,

$$V_R\varphi_{\mathbf{k}}(\mathbf{r}) \equiv \sum_{c}(\varepsilon_{\mathbf{k}} - \varepsilon_{c\mathbf{k}}) < \phi_{c\mathbf{k}}(\mathbf{r}) \,|\, \varphi_{\mathbf{k}}(\mathbf{r}) > \phi_{c\mathbf{k}}(\mathbf{r}). \tag{5.63}$$

Thus, Eq. (5.62) is an effective Schrodinger equation satisfied by $\varphi_{\mathbf{k}}(\mathbf{r})$, which is the smooth part of the Bloch function. Because

$$H + V_R = -\frac{\hbar^2}{2m}\nabla^2 + V(\mathbf{r}) + V_R, \tag{5.64}$$

the pseudopotential is defined as

$$V_{ps} = V(\mathbf{r}) + V_R. \tag{5.65}$$

From Eqs. (5.62), (5.64), and (5.65), we obtain

$$H_{ps}\varphi_{\mathbf{k}}(\mathbf{r}) = \left(-\frac{\hbar^2}{2m}\nabla^2 + V_{ps}\right)\varphi_{\mathbf{k}}(\mathbf{r}) = \varepsilon_{\mathbf{k}}\varphi_{\mathbf{k}}(\mathbf{r}). \tag{5.66}$$

The "smoothed" wave function $\varphi_{\mathbf{k}}(\mathbf{r})$ is often referred to as the pseudo wave function that satisfies the effective Schrodinger equation, which has a pseudo-Hamiltonian H_{ps} in which the potential is relatively small because the pseudopotential V_{ps} is the sum of the lattice potential and the repulsive potential V_R. However, we note from Eq. (5.63) that V_R is a nonlocal operator. Thus, we define

$$V_R(\mathbf{r},\mathbf{r}') = \sum_c (\varepsilon_{\mathbf{k}} - \varepsilon_{c\mathbf{k}})\phi_{c\mathbf{k}}(\mathbf{r})\phi^*_{c\mathbf{k}}(\mathbf{r}') \tag{5.67}$$

such that

$$V_R\varphi_{\mathbf{k}}(\mathbf{r}) = \int V_R(\mathbf{r},\mathbf{r}')\varphi_{\mathbf{k}}(\mathbf{r}')d\mathbf{r}'. \tag{5.68}$$

Thus, V_R is different when it operates on different functions with different angular momentum. This implies that the effect of the pseudopotential V_{ps} on a wave function is not just to multiply it by a function of **r**. Further, because V_{ps} depends on the energy $\varepsilon_{\mathbf{k}}$ that is supposed to be the quantity being calculated, one cannot apply some of the basic theorems of quantum mechanics. For example, one cannot apply the theorem that the eigenfunctions belonging to different eigenvalues of H_{ps} are orthogonal.

The pseudopotential formulation has another problem in the sense that there is no unique method to construct the pseudopotential. One can easily show (Problem 5.9) that the valence eigenvalues of the Hamiltonian $H + V_R$ are the same for any operator of the form

$$V_R\varphi_{\mathbf{k}}(\mathbf{r}) = \sum_c < \theta_{c\mathbf{k}}(\mathbf{r}) \,|\, \varphi_{\mathbf{k}}(\mathbf{r}) > \phi_{c\mathbf{k}}(\mathbf{r}), \tag{5.69}$$

where $\theta_{c\mathbf{k}}(\mathbf{r})$ are arbitrary functions. This is also an advantage because one can appropriately choose the functions $\theta_{c\mathbf{k}}(\mathbf{r})$ such that there is good cancellation between $V(\mathbf{r})$ and V_R such that the pseudopotential is small.

As an example, consider

$$\theta_{c\mathbf{k}}(\mathbf{r}) = -V(\mathbf{r})\phi_{c\mathbf{k}}(\mathbf{r}). \tag{5.70}$$

From Eqs. (5.65), (5.69), and (5.70), we obtain

$$V_{ps}\varphi_{\mathbf{k}}(\mathbf{r}) = V(\mathbf{r})\varphi_{\mathbf{k}}(\mathbf{r}) - \sum_c < \phi_{c\mathbf{k}} \,|\, V(\mathbf{r}) \,|\, \varphi_{\mathbf{k}}(\mathbf{r}) > \phi_{c\mathbf{k}}(\mathbf{r}). \tag{5.71}$$

Thus, one can subtract from $V(\mathbf{r})$ any sum of the core functions. This cancellation is at the core of the pseudopotential theory because it explains why the valence electrons in metals and semiconductors appear not to interact strongly with the ions of the crystal lattice. The success of the nearly free electron model is a consequence of the cancellation principle implicitly included in the pseudopotential concept. There has been significant improvement in the application of the pseudopotential theory in the band calculations in metals, but here we have essentially discussed the original theory developed by Phillips and Kleinman.

5.8 MUFFIN-TIN POTENTIAL

The muffin-tin potential is a very convenient description of the metal in the sense that it represents an isolated ion within a sphere of radius r_i about each lattice point and constant (zero) everywhere else. Therefore, the periodic potential in the metals considered as a muffin-tin potential is shown in Figure 5.16.

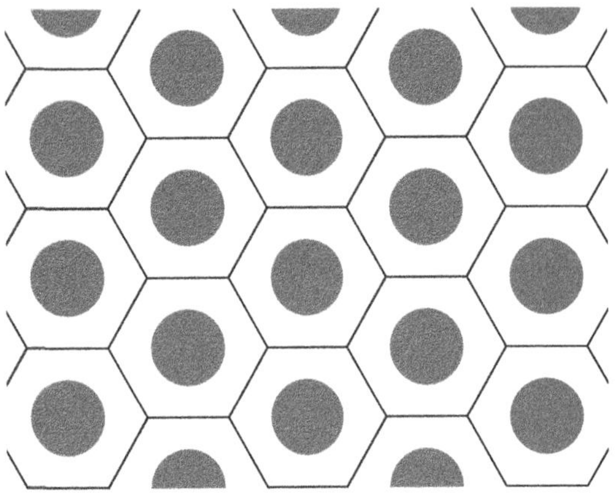

FIGURE 5.16

Muffin-tin potential.

The muffin-tin potential $U(\mathbf{r})$ is considered to be a constant (zero) in the interstitial region and spherically symmetrical within radius r_i about each ion (the core or the atomic region). We assume that there is one ion located at the center of each Wigner–Seitz cell and the spheres do not overlap; i.e., the radius r_i of the sphere is smaller than the Wigner–Seitz radius r_s. Thus, the periodic potential of the lattice can be defined as

$$V(\mathbf{r}) = \sum_i U(|\mathbf{r} - \mathbf{R}_i|). \tag{5.72}$$

The Schrodinger equation can be solved exactly within each sphere because of the spherical symmetry as well as in the interstitial region where the potential is zero. These solutions are matched on the surface of each sphere, and the Schrodinger equation for the crystal is obtained. The basis states $\phi_{\mathbf{k}\varepsilon}(\mathbf{r})$ are defined as follows:

1. $\phi_{\mathbf{k}\varepsilon} = e^{i\mathbf{k}.\mathbf{r}}$ in the interstitial region (outside the muffin hole). We note that because $\phi_{\mathbf{k}\varepsilon}$ is obtained by matching the solutions at the boundaries of the sphere, there is no precondition that $\varepsilon = \frac{\hbar^2 k^2}{2m}$.
2. In the spherical region about $\mathbf{R}_i$, $\phi_{\mathbf{k}\varepsilon}$ satisfies the atomic Schrodinger equation

$$-\frac{\hbar^2}{2m}\nabla^2\phi_{\mathbf{k}\varepsilon}(\mathbf{r}) + U(|\mathbf{r} - \mathbf{R}_i|)\phi_{\mathbf{k}\varepsilon}(\mathbf{r}) = \varepsilon\phi_{\mathbf{k}\varepsilon}(\mathbf{r}),\ |\mathbf{r} - \mathbf{R}_i| < r_i. \tag{5.73}$$

3. $\phi_{\mathbf{k}\varepsilon}$ is continuous at the boundary of the sphere. There are two elegant methods to solve the eigenvalue problem of this type of Schrodinger equation: the augmented plane-wave (APW) method and the Green's function (KKR) method proposed independently by Korringa, and Kohn and Rostoker (Ref. 11). We will first describe the APW method.

5.9 AUGMENTED PLANE-WAVE (APW) METHOD

The augmented plane-wave (APW) method was proposed by Slater[15]. The potential $U(|\mathbf{r} - R_i|)$ is a spherically symmetric function within the sphere (muffin-tin hole), and therefore, we write it as $U(\mathbf{r})$. We express the solutions $\psi_\varepsilon(\mathbf{r}) = R_{\varepsilon l}(r)Y_{lm}(\hat{r})$ of an electron in a spherically symmetric potential (note that the energy quantum number is written as ε instead of n and the spherical harmonics $Y_{l,m}(\theta,\phi)$ are expressed as $Y_{l\cdot m}(\hat{r})$). From atomic physics, we know that the radial part of the Schrodinger equation for an atom satisfies

$$-\frac{\hbar^2}{2mr^2}\frac{\partial}{\partial r}r^2\frac{\partial}{\partial r}R_{\varepsilon l}(r) + \left[U(r) + \frac{l(l+1)\hbar^2}{2mr^2}\right]R_{\varepsilon l}(r) = \varepsilon R_{\varepsilon l}(r). \tag{5.74}$$

Ignoring the solution of Eq. (5.74) that diverges at the origin, one can rewrite Eq. (5.73) as

$$\phi_{\mathbf{k}\varepsilon} = \sum_{l=0}^{\infty} \sum_{m=-l}^{l} B_{lm}(k) Y_{lm}(\hat{r}) R_{\varepsilon l}(r). \tag{5.75}$$

The coefficients $B_{lm}(k)$ are arbitrary, which are obtained from the boundary condition that the wave function is continuous across the boundary of each sphere (muffin hole). The wave function in the interstitial is a plane wave $e^{i\mathbf{k}\cdot\mathbf{r}}$. It is well known that the plane wave can be expanded in terms of the spherical harmonics,

$$e^{i\mathbf{k}\cdot\mathbf{r}} = 4\pi \sum_{l=0}^{\infty} \sum_{m=-l}^{l} i^l j_l(kr) Y^*_{lm}(\hat{k}) Y_{lm}(\hat{r}), \tag{5.76}$$

where $j_l(kr)$ is the spherical Bessel function. At the boundary of each sphere, $r = R_i$, where R_i is the radius of each sphere. Matching Eqs. (5.75) and (5.76) at the surface of each sphere where $r = R_i$, we obtain

$$B_{lm}(k) = 4\pi \frac{i^l j_l(kR_i) Y^*_{lm}(\hat{k})}{R_{\varepsilon l}(R_i)}. \tag{5.77}$$

From Eqs. (5.75) and (5.77), we obtain

$$\phi_{\mathbf{k}\varepsilon}(\mathbf{r}) = 4\pi \sum_{l=0}^{\infty} \sum_{m=-l}^{l} \frac{i^l j_l(kR_i) Y^*_{lm}(\hat{k})}{R_{\varepsilon l}(R_i)} Y_{lm}(\hat{r}) R_{\varepsilon l}(r). \tag{5.78}$$

We have a function $\phi_{\mathbf{k}\varepsilon}(\mathbf{r})$ for each value of $\mathbf{k}$ and ε. These APW functions have a discontinuity in slope at the boundary of the muffin tin. The boundary conditions at the edge of the primitive cell can be matched (Problem 5.7) by noting that the APW functions are plane waves at the edge of the primitive cell. The plane waves obey the Bloch condition. Therefore, one can write the solution of the Schrodinger equation of the crystal by making a linear combination of the APWs, all of the same energy ε:

$$\psi_{\mathbf{k}}(\mathbf{r}) = \sum_{\mathbf{K}} b_{\mathbf{k}+\mathbf{K}} \phi_{\mathbf{k}+\mathbf{K},\varepsilon(\mathbf{k})}(\mathbf{r}), \tag{5.79}$$

where the coefficients $b_{\mathbf{k}+\mathbf{K}}$ are to be determined. The best way is to use the variational principle (Problem 5.8), which is briefly described here.

We define an energy functional,

$$E[\psi] = \frac{\int \left(\frac{\hbar^2}{2m} |\nabla\psi(\mathbf{r})|^2 + V(\mathbf{r}) |\psi(\mathbf{r})|^2 \right) d\mathbf{r}}{\int |\psi(\mathbf{r})|^2 d\mathbf{r}}. \tag{5.80}$$

In the variational technique (Problem 5.8), $E[\psi] = E[\psi_{\mathbf{k}}] = \varepsilon(\mathbf{k})$ when Eq. (5.80) is stationary with respect to a differentiable function $\psi(\mathbf{r})$ that satisfies the Bloch condition with wave vector $\mathbf{k}$. From Eqs. (5.79) and (5.80), one can show that the condition that $E[\psi_{\mathbf{k}}]$ is stationary leads to

$\partial E/\partial b_{\mathbf{k}+\mathbf{K}} = 0$, which yields a set of homogeneous equations in $b_{\mathbf{k}+\mathbf{K}}$. When the determinant of the coefficients of these equations is equal to zero, one obtains an equation of which the roots determine $\varepsilon(\mathbf{k})$.

5.10 GREEN'S FUNCTION (KKR) METHOD

Korringa (1947) and Kohn and Rostoker[11] independently proposed a Green's function method to calculate the band structure of metals. The KKR method essentially uses a Green's function method to solve the Schrodinger equation of a crystalline solid with a periodic potential.

The KKR method starts with the objective to find the "propagating" solutions of the Schrodinger equation in the lattice

$$\left[\frac{-\hbar^2}{2m} + V(\mathbf{r}) - E\right]\psi_{\mathbf{k}}(\mathbf{r}) = 0. \tag{5.81}$$

Here, $V(\mathbf{r})$ is the periodic potential, and the boundary conditions in the central polyhedron (Wigner–Seitz cell) that surrounds the origin are

$$\psi_{\mathbf{k}}(\mathbf{r}_1) = e^{i\mathbf{k}\cdot\mathbf{R}}\psi_{\mathbf{k}}(\mathbf{r}) \tag{5.82}$$

and

$$\partial\psi_{\mathbf{k}}(\mathbf{r}_1)/\partial n_1 = -e^{-i\mathbf{k}\cdot\mathbf{R}}\partial\psi_{\mathbf{k}}(\mathbf{r})/\partial n. \tag{5.83}$$

The conjugate points $\mathbf{r}$ and $\mathbf{r}_1$ are defined (see Figure 5.17) as the points on the surface of the polyhedron separated by the lattice translation vector $\mathbf{R}$.

The Green's function is defined by

$$\left(\frac{\hbar^2}{2m}\nabla^2 + \varepsilon\right)G_\varepsilon(\mathbf{r}-\mathbf{r}') = \delta(\mathbf{r}-\mathbf{r}'). \tag{5.84}$$

For conjugate boundary points $\mathbf{r}$ and $\mathbf{r}_1$,

$$G_\varepsilon(\mathbf{r}_1-\mathbf{r}') = e^{i\mathbf{k}\cdot\mathbf{R}}G_\varepsilon(\mathbf{r}-\mathbf{r}') \tag{5.85}$$

and

$$\partial G_\varepsilon(\mathbf{r}_1-\mathbf{r}')/\partial n_1 = -e^{i\mathbf{k}\cdot\mathbf{R}}\partial G_\varepsilon(\mathbf{r}-\mathbf{r}')/\partial n. \tag{5.86}$$

It can be shown that $G_\varepsilon(\mathbf{r}-\mathbf{r}')$ can be rewritten in the alternate form

$$G_\varepsilon(\mathbf{r}-\mathbf{r}') = -\frac{2m}{\hbar^2}\frac{e^{i\kappa|\mathbf{r}-\mathbf{r}'|}}{4\pi|\mathbf{r}-\mathbf{r}'|}, \tag{5.87}$$

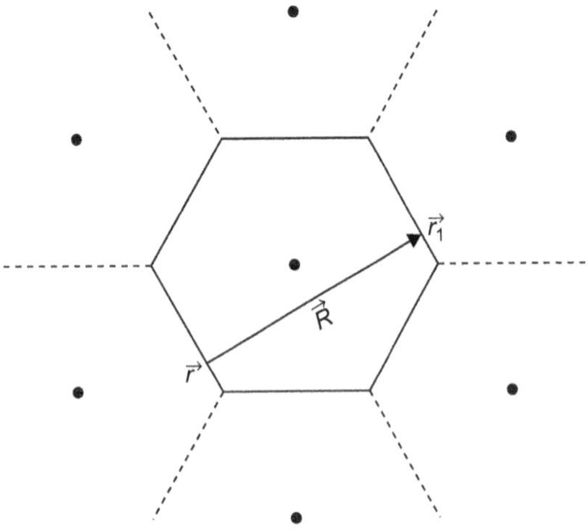

FIGURE 5.17

Conjugate boundary points $\mathbf{r}$ *and* $\mathbf{r}_1$.

where

$$\kappa = \sqrt{\frac{2m\varepsilon}{\hbar^2}},\ \varepsilon > 0, \tag{5.88}$$

and

$$= i\sqrt{\frac{2m(-\varepsilon)}{\hbar^2}},\ \varepsilon < 0. \tag{5.89}$$

An alternate way of expressing G is obtained by multiplying Eq. (5.81) by $G^*(\mathbf{r}-\mathbf{r}')$ and the complex conjugate of Eq. (5.84) by $\psi_{\mathbf{k}}(\mathbf{r})$, subtracting and integrating over $\mathbf{r}$ in the interior of the polyhedron (Problem 5.10) to obtain

$$\psi_{\mathbf{k}}(\mathbf{r}) = \int d\mathbf{r}' G_{\varepsilon(\mathbf{k})}(\mathbf{r}-\mathbf{r}')V(\mathbf{r}')\psi_{\mathbf{k}}(\mathbf{r}'). \tag{5.90}$$

Consider the muffin-tin potential

$$\begin{aligned} \hat{V}(\mathbf{r}) &= V(r) - V_0,\ r \leq r_i \\ &= 0, r > r_i, \end{aligned} \tag{5.91}$$

Here, r_i is the radius of the inscribed sphere, and V_0 is the average value of the constant potential $V(r)$ in the space between the inscribed sphere and the boundary of the polyhedron. Thus, we can write the periodic potential as

$$V(r) = \sum_i \hat{V}(\mathbf{r}-\mathbf{R}_i). \tag{5.92}$$

From Eqs. (5.87), (5.90), and (5.92), we obtain

$$\psi_{\mathbf{k}}(\mathbf{r}) = \sum_i \int G_{\varepsilon(\mathbf{k})}(\mathbf{r}-\mathbf{r}')\hat{V}(\mathbf{r}'-\mathbf{R}_i)\psi_{\mathbf{k}}(\mathbf{r}')d\mathbf{r}'. \tag{5.93}$$

We move the origin such that $\mathbf{r}'' = \mathbf{r}' - \mathbf{R}_i$, and Eq. (5.93) can be rewritten as

$$\psi_{\mathbf{k}}(\mathbf{r}) = \sum_i \int G_{\varepsilon(\mathbf{k})}(\mathbf{r}-\mathbf{r}''-\mathbf{R}_i)\hat{V}(\mathbf{r}'')\psi_{\mathbf{k}}(\mathbf{r}''+\mathbf{R}_i). \tag{5.94}$$

Using the Bloch theorem $\psi_{\mathbf{k}}(\mathbf{r}''+\mathbf{R}) = e^{i\mathbf{k}\cdot\mathbf{R}}\psi_{\mathbf{k}}(\mathbf{r}'')$ and replacing $\mathbf{r}''$ with $\mathbf{r}'$ in Eq. (5.94), we obtain

$$\psi_{\mathbf{k}}(\mathbf{r}) = \int_{r'<r_i-\varepsilon} d\mathbf{r}'\, G^0_{\mathbf{k},\varepsilon(\mathbf{k})}(\mathbf{r}-\mathbf{r}')\hat{V}(\mathbf{r}')\psi_{\mathbf{k}}(\mathbf{r}'), \tag{5.95}$$

where the integration is done over a single cell, and the structural Green function or Greenian is defined as

$$G^0_{\mathbf{k},\varepsilon(\mathbf{k})}(\mathbf{r}-\mathbf{r}') = \sum_i G_{\varepsilon(\mathbf{k})}(\mathbf{r}-\mathbf{r}'-\mathbf{R}_i)e^{i\mathbf{k}\cdot\mathbf{R}_i}. \tag{5.96}$$

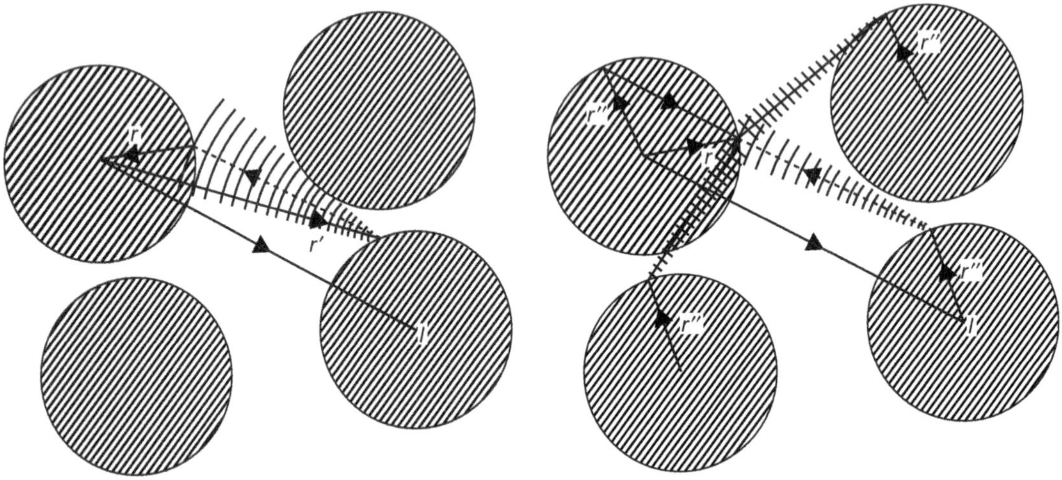

(a) $G(\mathbf{r}-\mathbf{r}')$ propagating from $\mathbf{r}'$ to $\mathbf{r}$

(b) $G^0_{\mathbf{k},\varepsilon(\mathbf{k})}(\mathbf{r}-\mathbf{r}')$ combines waves from all points in the lattice.

FIGURE 5.18

(a) The ordinary Green function; (b) the structural Green function.

From Eqs. (5.87) and (5.96), $G^0_{\mathbf{k},\varepsilon(\mathbf{k})}(\mathbf{r}-\mathbf{r}')$ can also be rewritten in the alternate form

$$G^0_{\mathbf{k},\varepsilon(\mathbf{k})}(\mathbf{r}-\mathbf{r}') = -\frac{2m}{\hbar^2}\frac{1}{4\pi}\sum_i \frac{e^{i\kappa|\mathbf{r}-\mathbf{r}'-\mathbf{R}_i|}}{|\mathbf{r}-\mathbf{r}'-\mathbf{R}_i|}e^{i\mathbf{k}\cdot\mathbf{R}_i}. \tag{5.97}$$

The ordinary and structural Green functions are shown in Figure 5.18.

From Eq. (5.96), it can be easily shown that

$$\left(\frac{\hbar^2}{2m}\nabla'^2 + \varepsilon\right)G^0_{\mathbf{k},\varepsilon(\mathbf{k})}(\mathbf{r}-\mathbf{r}') = \delta(\mathbf{r}-\mathbf{r}'), \quad r,\ r' < r_i, \tag{5.98}$$

where r_i is the radius of the spherical region in muffin-tin potential. We can also write by using the mathematical identity

$$G^0_{\mathbf{k},\varepsilon(\mathbf{k})}\nabla'^2\psi_{\mathbf{k}}(\mathbf{r}') = \nabla'\cdot(G^0_{\mathbf{k},\varepsilon(\mathbf{k})}\nabla'\psi_{\mathbf{k}}(\mathbf{r}') - \nabla'\psi_{\mathbf{k}}(\mathbf{r}')G^0_{\mathbf{k},\varepsilon(\mathbf{k})}) + \psi_{\mathbf{k}}(\mathbf{r}')\nabla'^2 G_{\mathbf{k},\varepsilon(\mathbf{k})}. \tag{5.99}$$

From Eqs. (5.95), (5.98), and (5.99), we can show that

$$\int\limits_{r'-\varepsilon} d\mathbf{r}'\ \nabla'\cdot[G^0_{\mathbf{k},\varepsilon(\mathbf{k})}(\mathbf{r}-\mathbf{r}')\nabla'\psi_{\mathbf{k}}(\mathbf{r}') - \psi_{\mathbf{k}}(\mathbf{r}')\nabla' G^0_{\mathbf{k},\varepsilon(\mathbf{k})}(\mathbf{r}-\mathbf{r}')] = 0. \tag{5.100}$$

Using the Gauss theorem to transform the volume integral to a surface integral over the sphere of radius $r_i - \varepsilon$, we obtain

$$\int_{r'-\varepsilon} ds' \left[G^0_{\mathbf{k},\varepsilon(\mathbf{k})}(r_i\theta\phi, r_i\theta'\phi') \frac{\partial}{\partial r'} \psi_{\mathbf{k}}(r'\theta'\phi') \,|_{r'=r_i} - \psi_{\mathbf{k}}(r_i\theta'\phi') \frac{\partial}{\partial r'} G_{\mathbf{k},\varepsilon(\mathbf{k})}(r_i\theta\phi, r'\theta'\phi') \,|_{r'=r_i} \right] = 0. \tag{5.101}$$

It can be easily shown[11] that for

$$r < r' < r_i,$$

$$\psi(\mathbf{r}) = \sum_{l=0}^{\infty} \sum_{m=-l}^{m=l} C_{lm} R_l(r) Y_{lm}(\theta, \phi), \tag{5.102}$$

and

$$G^o_{\mathbf{k},\varepsilon(\mathbf{k})}(\mathbf{r}-\mathbf{r}') = \sum_{l,m} \sum_{l',m'} [A_{lm,l'm'} j_l(\kappa r) j_{l'}(\kappa r') + \kappa \delta_{ll'} \delta_{mm'} j_l(\kappa r) n_l(\kappa r')] Y_{lm}(\theta, \phi) Y^*_{l'm'}(\theta', \phi'). \tag{5.103}$$

Here,

$$\begin{aligned} j_l(x) &= (\pi/2x)^{1/2} J_{l+1/2}(x) \\ n_l(x) &= (\pi/2x)^{1/2} J_{-l-1/2}(x), \end{aligned} \tag{5.104}$$

$J_\nu(x)$ are the Bessel functions, and $A_{lm;l'm'}$ are functions of $\mathbf{k}$ and ε, which are characteristics for the lattice under consideration. Substituting Eqs. (5.102) and (5.103) in Eq. (5.101), multiplying by $Y^*_{lm}(\theta, \phi)$, integrating over the sphere $r = r_i - \in$, using the normalization condition $R_l(r_i) = 1$, and finally letting $\in \to 0$, we obtain

$$\sum_{l',m'} j_l [A_{lm;l'm'}(j_{l'} L_{l'} - j'_{l'}) + \kappa \delta_{ll'} \delta_{mm'} (n_{l'} L_{l'} - n'_{l'})] C_{l'm'} = 0, \tag{5.105}$$

where

$$\begin{aligned} L_l &= \frac{dR_l(r)}{dr} / R_l(r) \,|_{r=r_i} \\ j'_l &= \frac{dj_l(\kappa r)}{dr} \,|_{r=r_i} \\ n'_l &= \frac{dn_l(\kappa r)}{dr} \,|_{r=r_i}. \end{aligned} \tag{5.106}$$

Before equating the determinant of Eq. (5.105) to zero, we divide each row by j_l and each column by $(j_{l'} L_{l'} - j'_{l'})$ and obtain the secular equation

$$Det \left| A_{lm;l'm'} + \kappa \delta_{ll'} \delta_{mm'} \frac{(n_l L_l - n'_l)}{(j_l L_l - j'_l)} \right| = 0. \tag{5.107}$$

Eq. (5.107) is used by first tabulating the structure constants $A_{lm,l'm'}$ as functions of ε and **k** for each type of lattice. The logarithmic derivatives L_l for the first few l are obtained as functions of energy. The convenient way of solving Eq. (5.107) is to fix ε (and hence κ) to find those **k**'s (**k** enters through the A's) that make Eq. (5.107) vanish. It may be noted that the same expression (Eq. 5.107) can be also obtained by using a variational method.

One may note a good deal of similarity between the Green function (KKR) and the APW method. In the APW method, the expansion is in terms of spherical harmonics. Then a secular determinant is solved for contributions from different lattice vectors. In contrast, in the KKR method, the summation is over lattice vectors (in practice, the summation is over reciprocal lattice vectors by first making a Fourier transformation of $G^0_{\mathbf{k},\varepsilon(\mathbf{k})}(\mathbf{r}-\mathbf{r}')$) and then have a secular determinant in the contributions from different spherical harmonics. Anderson (Ref. 1) has proposed an approximate first-principle method for calculating the band structure of closely-packed structures. In his atomic sphere model (ASM), the atomic polyhedra are replaced by spheres of the same volume and the potential is spherically symmetric within each sphere. The ASM model has no interstitial region.

5.11 MODEL PSEUDOPOTENTIALS

It is obvious from the preceding discussions that the APW and KKR methods do not depend on the atomic potentials, but only on the gradient of R_l at the surface of the atomic sphere. It can be shown from the partial wave theory of scattering that for a spherical potential, the radial solution $R_l(r, \varepsilon)$ can be matched to a free electron wave of the same energy $\varepsilon = \hbar^2\kappa^2/2m$ and angular momentum l, through a phase shift $\eta_l(\varepsilon)$ defined by

$$L_l \equiv \frac{R_l'(R_i, \varepsilon)}{R_l(R_i, \varepsilon)} = \frac{j_l'(\kappa r) - \tan\eta_l(\varepsilon)\cdot n_l'(\kappa r)}{j_l(\kappa r) - \tan\eta_l(\varepsilon)\cdot n_l'(\kappa r)}\Big|_{r=R_i}, \tag{5.108}$$

where R_i for a muffin-tin potential is defined in Eq. (5.73) and the spherical Bessel functions and their derivatives were defined earlier. In the model pseudopotential approach, each atomic potential v_a is replaced by a weak potential w_a, which has the same scattering amplitude as conduction electrons. The energy $\varepsilon(\mathbf{k})$ of the crystal would be identical to this hypothetical material. Figure 5.19 shows how the model potential w, with wave function ϕ inside the atom, replaces the effect of the true potential V with the true wave function ψ.

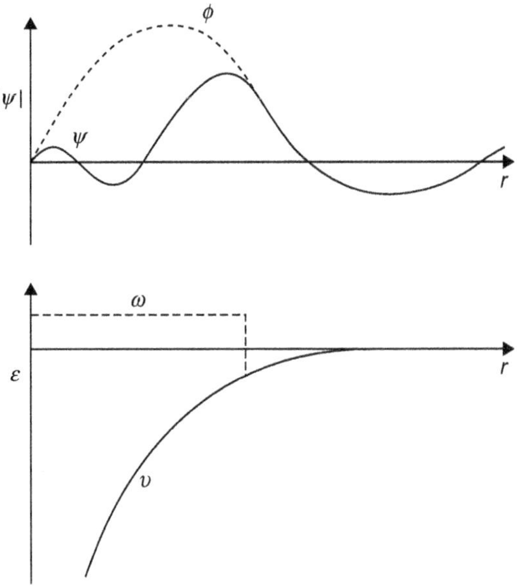

FIGURE 5.19

The true potential V with wave function ψ is replaced by model potential w with wave function ϕ inside the atom.

The analytical pseudopotentials of Phillips and Kleinman are nonlocal, energy dependent, and arbitrary. Thus, one can obtain an infinitely

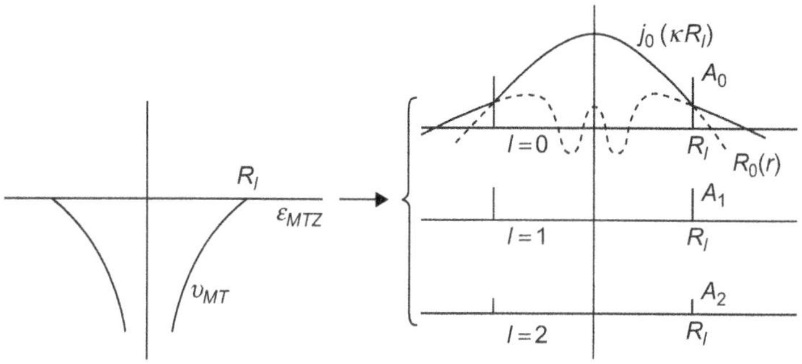

FIGURE 5.20

Model pseudopotential for KKR matrix elements.

large number of localized potentials, $w_l(r, \varepsilon)$, that can be used to obtain the true radial functions R_l outside the atomic core. However, the same model potential $w_l(r, \varepsilon)$ would not work for a different ε or l. Thus, the various angular momentum components in a wave function have to be separated by introduction of appropriate operators. However, it is more convenient to introduce the functional form of each model pseudopotential $w_l(\mathbf{r}, \varepsilon)$ for computational convenience. The appropriate choice for a muffin-tin lattice is a delta-function singularity for each value of l, to reproduce the scattering phase shift $\eta_l(\varepsilon)$ at the surface of the sphere.

The model pseudopotential for the KKR matrix elements is shown in Figure 5.20.

It can be shown, by the use of the analytical properties of spherical Bessel functions and plane waves, that the pseudopotential matrix elements used in the KKR method can be written as

$$V_{ps}^{KKR}(\mathbf{K}, \mathbf{K}') = -\frac{4\pi N}{\kappa} \sum_l (2l+1) \tan \eta_l' \frac{j_l(|\mathbf{k}-\mathbf{K}|R_i) j_l(|\mathbf{k}-\mathbf{K}'|R_i)}{[j_l(\kappa R_i)]^2} P_l(\cos \theta_{\mathbf{KK}'}), \tag{5.109}$$

where

$$\cot \eta_l' \equiv \cot \eta_l - n_l(\kappa R_i)/j_l(\kappa R_i). \tag{5.110}$$

5.12 EMPIRICAL PSEUDOPOTENTIALS

The pseudopotential theory of Phillips and Kleinman established the existence of weak potentials due to cancellations. The empirical pseudopotential concept is based on the assumption that one can choose weak potentials to match the important features of experimental results. An example of the empirical pseudopotential is the empty-core potential. The Ashcroft pseudopotential $U(r)$ is obtained from the approximation that

$$U(r) = 0, \quad r < R_c, \tag{5.111}$$

and

$$U(r) = -U_0 e^{-r/d}/r, \quad r > R_c, \tag{5.112}$$

where R_c is a cutoff radius, and d is the exponential decay length. The Ashcroft (Ref. 2) empty-core pseudopotential is shown in Figure 5.21.

The three parameters in the empty-core pseudopotential are $U_0, R_c,$ and d. These parameters are adjusted to fit the measurement of magnetic or optical experimental results. In general, these types of potentials yield satisfactory results for alkali metals or aluminum.

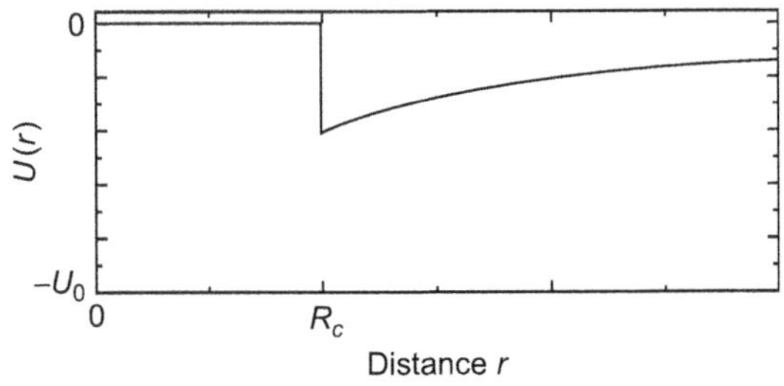

FIGURE 5.21

The Ashcroft empty-core pseudopotential.

5.13 FIRST-PRINCIPLES PSEUDOPOTENTIALS

The starting point for a convenient method of obtaining first-principles pseudopotentials (probably a misnomer) is to start with the Kohn–Sham equations[12] obtained by using the local density approximation (LDA),

$$-\frac{\hbar^2}{2m}\nabla^2\psi_i(\mathbf{r}) + \left[V(\mathbf{r}) + \int d\mathbf{r}'\frac{e^2 n(\mathbf{r}')}{|\mathbf{r}-\mathbf{r}'|} + \frac{\delta\varepsilon_{xc}(n)}{\delta n}\right]\psi_i(\mathbf{r}) = \varepsilon_l\psi_i(\mathbf{r}), \tag{5.113}$$

where $\psi_i(\mathbf{r})$ is one of N single-electron wave functions, $n(\mathbf{r})$ is the density defined by

$$n(\mathbf{r}) = \sum_{i=1}^{N} |\psi_i(\mathbf{r})|^2, \tag{5.114}$$

and ε_{xc} is the exchange-correlation energy of the uniform electron gas, which means that it can be chosen freely to ensure that properties of the uniform electron gas are obtained correctly. In the first-principle pseudopotential method, one chooses an atom and makes the approximation that $n(\mathbf{r})$ is spherically symmetric about the nucleus. Thus, the atomic potential is $V(\mathbf{r}) = -\frac{Ze^2}{|r|}$, and all the solutions are of the form $\psi(\mathbf{r}) = R_{nl}(r)Y_{lm}(\theta,\varphi)$, where $R_{nl}(r)$ is the radial wave function and $Y_{lm}(\theta,\varphi)$ is a spherical harmonic. With these approximations, Eq. (5.113) can be rewritten as

$$-\frac{\hbar^2}{2m}\left[\frac{1}{r}\frac{\partial^2}{\partial r^2}r - \frac{l(l+1)}{r^2}\right]R_{nl}(r) + \left[\int\frac{e^2 n(\mathbf{r})}{|\mathbf{r}-\mathbf{r}'|}d\mathbf{r}' - \frac{Ze^2}{r} + \frac{\delta\varepsilon_{xc}}{\delta n} - \varepsilon_{ni}\right]R_{nl}(r) = 0. \tag{5.115}$$

Eq. (5.115) is solved for all the electrons of the atom, and the energies of these states are ε_{ni}.

The outermost states that are in partially filled shells are singled out for special treatment because they contribute to the bonding between atoms and solids. The radial wave functions and the pseudo radial wave functions of such a solid are schematically shown in Figure 5.22.

The radial wave functions for a solid with partially filled shells can be obtained from Eq. (5.115) by using the appropriate values for $n(\mathbf{r})$, Z, ε_{xc}, and ε_{ni}. The corresponding pseudo radial function $R_{nl}^{ps}(r)$ is obtained from each radial function $R_{nl}(r)$ by picking a point beyond the rightmost node and drawing a smooth curve into the origin. This pseudo radial function $R_{nl}^{ps}(r)$

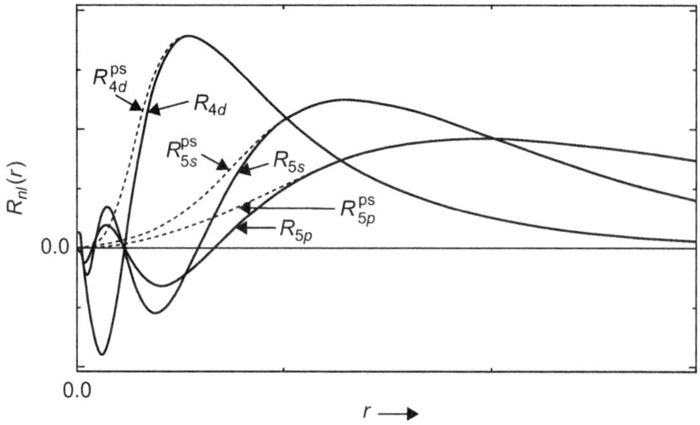

FIGURE 5.22

Schematic diagram of the radial wave functions $R_{nl}(r)$ and pseudo radial wave functions $R^{ps}_{nl}(r)$ for $4d$, $5s$, and $5p$ states of a solid with partially filled shells.

should join on to the actual radial function $R_{nl}(r)$ with two continuous derivatives, should be without any nodes, but vanish as r^l at the origin. In addition, the pseudo wave functions built from these pseudo radial functions have to be normalized.

We can now replace the Coulomb potential $-\frac{ze^2}{r}$ in Eq. (5.115) with the pseudopotential

$$V_l^{ps}(r) = \frac{\hbar^2}{2m}\left[\frac{1}{r'R_{nl}^{ps}}\frac{\partial^2 r'R_{nl}^{ps}}{\partial r^2} - \frac{l(l+1)}{r^2}\right] - \left[\int \frac{e^2 n^{ps}(\mathbf{r}')}{|\mathbf{r}-\mathbf{r}'|}d\mathbf{r}' + \frac{\delta\varepsilon_{xc}}{\delta n^{ps}} - \varepsilon_{nl}\right], \tag{5.116}$$

such that when one solves the Kohn–Sham equations, the radial functions R^{ps} are obtained instead of R. However, there is a different $V_l^{ps}(r)$ for each angular momentum state. To be able to resolve this problem that the pseudopotential is nonlocal, we write an arbitrary wave function as a sum of its angular momentum components,

$$\psi(\mathbf{r}) = \sum_{lm} \psi_{lm}(r) Y_{lm}(\theta,\phi). \tag{5.117}$$

We multiply Eq. (5.117) by $Y^*_{l'm'}(\theta,\phi)$, and integrating over θ and ϕ, we obtain

$$\int d\theta d\phi\, Y^*_{l'm'}(\theta,\phi)\psi(\mathbf{r}) = \sum_{lm}\psi_{lm}(r)\int d\theta d\phi\, Y^*_{l'm'}(\theta,\phi)Y_{lm}(\theta,\phi). \tag{5.118}$$

Using the orthonormality condition of the spherical harmonics,

$$\int d\theta d\phi Y^*_{l'm'}(\theta,\phi)Y_{l,m}(\theta,\phi) = \delta_{l,l'}\delta_{m\cdot m'}, \tag{5.119}$$

we can rewrite Eq. (5.118) as

$$\psi_{lm}(r) = \int d\theta d\phi \, Y^*_{lm}(\theta,\phi)\psi(\mathbf{r}). \tag{5.120}$$

The pseudopotential $V_l^{ps}(r)$ defined in Eq. (5.116) is multiplied by $\psi_{lm}(r)$ defined in Eq. (5.120) in forming the Hamiltonian.

PROBLEMS

5.1. The 12 nearest neighbors of the origin of a fcc crystal are

$$\mathbf{R} = \frac{a}{2}(\pm 1, \pm 1, 0), \ \frac{a}{2}(\pm 1, 0, \pm 1), \ \frac{a}{2}(0, \pm 1, \pm 1). \tag{1}$$

Show that for an s-level, of which the wave function depends only on the magnitude $\mathbf{r}$,

$$\varepsilon(\mathbf{k}) = \varepsilon_s - \beta - 4\gamma\left(\cos\frac{1}{2}k_x a \cos\frac{1}{2}k_y a + \cos\frac{1}{2}k_y a \cos\frac{1}{2}k_z a + \cos\frac{1}{2}k_z a \cos\frac{1}{2}k_x a\right), \tag{2}$$

where

$$\gamma = -\int d\mathbf{r}\, \phi^*(x,y,z)\Delta U(x,y,z)\phi(x - \frac{1}{2}a, y - \frac{1}{2}a, z). \tag{3}$$

Show that in the limit of small ka, Eq. (2) can be rewritten as

$$\varepsilon(\mathbf{k}) = \varepsilon_s - \beta - 12\gamma + \gamma k^2 a^2. \tag{4}$$

5.2. Show that the tight-binding expression for energies of an s-band in a fcc crystal can be written as

$$\varepsilon = \varepsilon_s - \beta - 4\gamma(1 + 2\cos\mu\pi) \quad \text{along } \Gamma X,$$

where $0 \le \mu \le 1$.

5.3. Show that the Bloch functions of different bands and different $\mathbf{k}$ are orthonormal, i.e.,

$$\int \psi^*_{m\mathbf{k}'}(\mathbf{r})\psi_{n\mathbf{k}}(\mathbf{r})d\mathbf{r} = \delta_{n,m}\delta_{\mathbf{k},\mathbf{k}'}.$$

5.4. Show that $\sum_{\mathbf{k}} e^{-i\mathbf{k}\cdot(\mathbf{R}_i - \mathbf{R}_j)} = N\delta_{i,j}$, where the summation is over all values of $\mathbf{k}$ in the first Brillouin zone and $\mathbf{R}_i$ and $\mathbf{R}_j$ are direct lattice vectors.

5.5. Show that $\sum_{\mathbf{R}_i} e^{-i(\mathbf{k}-\mathbf{k}')\cdot\mathbf{R}_i} = N\delta_{\mathbf{k},\mathbf{k}'}$, where the summation is over all the direct lattice vectors $\mathbf{R}_i$.

5.6. Show that for a simple cubic lattice of side **a**, the Wannier function at the origin is

$$w_n(\mathbf{r}) = \sqrt{N}u_{n\mathbf{k}}(\mathbf{r})\frac{\sin(\pi x/a)\sin(\pi y/a)\sin(\pi z/a)}{(\pi x/a)(\pi y/a)(\pi z/a)}, \tag{1}$$

where $u_{n\mathbf{k}}(\mathbf{r})$ is the periodic part of the Bloch function.

5.7. The periodic part of the Bloch function has the property

$$u_{\mathbf{k}}(\mathbf{r}) = u_{\mathbf{k}}(\mathbf{r}+\mathbf{R}), \tag{1}$$

where $\mathbf{R}$ is a Bravais lattice vector. If $\mathbf{r}$ lies on the boundary of the unit cell and $\mathbf{r}+\mathbf{R}$ is another boundary point of the cell, then

$$\hat{n}(\mathbf{r})\cdot\nabla u_{\mathbf{k}}(\mathbf{r}) = -\hat{n}(\mathbf{r}+\mathbf{R})\cdot\nabla u_{\mathbf{k}}(\mathbf{r}+\mathbf{R}), \tag{2}$$

where $\hat{n}(\mathbf{r})$ is normal to the cell boundary. Because

$$\psi_{\mathbf{k}}(\mathbf{r}) = e^{i\mathbf{k}\cdot\mathbf{r}}u_{\mathbf{k}}(\mathbf{r}), \tag{3}$$

show that

$$e^{i\mathbf{k}\cdot\mathbf{R}}\hat{n}(\mathbf{r})\cdot\nabla\psi_{\mathbf{k}}(\mathbf{r}) = -\hat{n}(\mathbf{r}+\mathbf{R})\cdot\nabla\psi_{\mathbf{k}}(\mathbf{r}+\mathbf{R}). \tag{4}$$

5.8. Define the functional

$$\varepsilon[\phi] = \frac{\displaystyle\int\left(\frac{\hbar^2}{2m}|\nabla\phi(\mathbf{r})|^2 + V(\mathbf{r})|\phi(\mathbf{r})|^2\right)d\mathbf{r}}{\displaystyle\int|\phi(\mathbf{r})|^2 d\mathbf{r}}, \tag{1}$$

where

$$\phi(\mathbf{r}) = \psi_{\mathbf{k}}(\mathbf{r}) + \delta\phi(\mathbf{r}). \tag{2}$$

Here, $\psi_{\mathbf{k}}(\mathbf{r})$ is the wave function in the Schrodinger equation,

$$-\frac{\hbar^2}{2m}\nabla^2\psi_{\mathbf{k}}(\mathbf{r}) + V(\mathbf{r})\psi_{\mathbf{k}}(\mathbf{r}) = \varepsilon_{\mathbf{k}}\psi_{\mathbf{k}}(\mathbf{r}), \tag{3}$$

$V(\mathbf{r})$ is the periodic potential in the lattice, and both $\phi(\mathbf{r})$ and $\psi_{\mathbf{k}}(\mathbf{r})$ satisfy the Bloch condition. From Eqs. (1) and (2), show that

$$\varepsilon[\phi] = \varepsilon[\psi_{\mathbf{k}}] + O(\delta\phi)^2. \tag{4}$$

Define the functional

$$f[\theta,\varphi] = \int d\mathbf{r}\left(\frac{\hbar^2}{2m}\nabla\theta^*\cdot\nabla\varphi + V(\mathbf{r})\theta^*\varphi\right). \tag{5}$$

Using the Dirac notation, show that

$$\varepsilon[\phi] = \frac{f[\phi,\phi]}{<\phi\,|\,\phi>}. \tag{6}$$

Using the integration-by-parts formulas for any functions $\varphi_1(\mathbf{r})$ and $\varphi_2(\mathbf{r})$ that have the periodicity of the Bravais lattice,

$$\int_c d\mathbf{r}\,\varphi_1\nabla\varphi_2 = -\int_c d\mathbf{r}\,\varphi_2\nabla\varphi_2 \tag{7}$$

and

$$\int_c d\mathbf{r}\,\varphi_1\nabla^2\varphi_2 = \int_c d\mathbf{r}\,\varphi_2\nabla^2\varphi_1, \tag{8}$$

show that

$$f[\varphi,\psi_{\mathbf{k}}] = \varepsilon_{\mathbf{k}}<\varphi\,|\,\psi_{\mathbf{k}}> \tag{9}$$

and

$$f[\psi_{\mathbf{k}},\varphi] = \varepsilon_{\mathbf{k}}<\psi_{\mathbf{k}}\,|\,\varphi>, \tag{10}$$

where φ also satisfies the Bloch condition. Hence, show that

$$f[\phi,\phi] = \varepsilon_{\mathbf{k}}\{<\psi_{\mathbf{k}}\,|\,\psi_{\mathbf{k}}> + <\psi_{\mathbf{k}}\,|\,\delta\phi> + <\delta\phi\,|\,\psi_{\mathbf{k}}>\} + o(\delta\phi)^2, \tag{11}$$

$$<\phi\,|\,\phi> = <\psi_{\mathbf{k}}\,|\,\psi_{\mathbf{k}}> + <\psi_{\mathbf{k}}\,|\,\delta\phi> + <\delta\phi\,|\,\psi_{\mathbf{k}}> + o(\delta\phi)^2, \tag{12}$$

and

$$\varepsilon[\phi] = \frac{f[\phi,\phi]}{<\phi\,|\,\phi>} = \varepsilon_{\mathbf{k}} + O(\delta\phi)^2. \tag{13}$$

5.9. Show that the valence eigenvalues of the Hamiltonian $H + V_R$ are the same for any operator of the form

$$V_R\varphi_{\mathbf{k}}(\mathbf{r}) = \sum_c <\theta_{c\mathbf{k}}(\mathbf{r})\,|\,\varphi_{\mathbf{k}}(\mathbf{r})>\phi_{c\mathbf{k}}(\mathbf{r}), \tag{1}$$

where $\theta_{c\mathbf{k}}$ are completely arbitrary functions.

5.10. The Schrodinger equation for a crystalline solid with a periodic potential $V(\mathbf{r})$ is

$$\left(\frac{-\hbar^2}{2m}\nabla^2 + V(\mathbf{r}) - \varepsilon\right)\psi(\mathbf{r}) = 0, \tag{1}$$

and the Green's function is defined as

$$\left(-\frac{\hbar^2}{2m}\nabla^2-\varepsilon\right)G(\mathbf{r}-\mathbf{r}')=-\delta(\mathbf{r}-\mathbf{r}'). \tag{2}$$

Multiply Eq. (1) by $G^*(\mathbf{r}-\mathbf{r}')$ and the complex conjugate of Eq. (2) by $\psi(\mathbf{r})$, subtract and integrate over **r**, use the Hermitian property of the Green's function

$$G_\varepsilon(\mathbf{r}-\mathbf{r}')=G_\varepsilon^*(\mathbf{r}'-\mathbf{r}), \tag{3}$$

and show that

$$\psi_\mathbf{k}(\mathbf{r})=\int d\mathbf{r}'G_{\varepsilon(\mathbf{k})}(\mathbf{r}-\mathbf{r}')V(\mathbf{r}')\psi_\mathbf{k}(\mathbf{r}'). \tag{4}$$

References

1. Anderson OK. Simple approach to the band structure problem. *Solid State Commun* 1973;**13**:133.
2. Ashcroft NW. Electron-ion pseudopotentials in metals. *Phys Lett* 1966;**23**:48.
3. Ashcroft NW, Mermin ND. *Solid state physics*. New York: Brooks/Cole; 1976.
4. Bullett DW. *Solid State Phys: Adv Res Appl* 1982;**35**:129.
5. Callaway JC. *Quantum theory of the solid state*. New York: Academic Press; 1976.
6. Car R, Parrinello M. United approach for molecular dynamics and density functional theory. *Phys Rev Lett* 1985;**55**:2471.
7. Goswami A. *Quantum mechanics*. Dubuque: Wm. C. Brown; 1997.
8. Harrison WA. *Pseudopotentials in the theory of metals*. New York: Benjamin; 1966.
9. Heine V. *Solid State Phys: Adv Res Appl* 1970;**24**:1.
10. Heine V, Wearie D. *Solid State Phys: Adv Res Appl* 1970;**24**:250.
11. Kohn W, Rostoker N. *Phys Rev* 1954;**94**:1111.
12. Kohn W, Sham LJ. Self-consistent equations including exchange and correlation effects. *Phys Rev* 1965;**140**:A1133.
13. Kohn W, Rostoker N. Solution of the Schrodinger equation in periodic lattice with an application to metallic lithium. *Phys Rev* 1954;**94**:1111.
14. Korringa J. On the calculation of the energy of a Bloch wave in a metal. *Physica* 1947;**13**:392.
15. Marder MP. *Condensed Matter Physics*. New York: Brooks/Cole; 2000.
16. Phillips JC, Kleinman L. New Method for calculating wave functions in crystals and molecules. *Phys Rev* 1959;**116**:287.
17. Slater JC. Wave funcions in a periodic potential. *Phys Rev* 1937;**51**:846.
18. Slater JC, Koster GF. Wave functions in a periodic potential. *Phys Rev* 1956;**94**:1498.
19. Wannier GH. *Elements of solid state theory*. Cambridge: Cambridge University Press; 1959.
20. Wigner E, Seitz F. On the constitution of metallic sodium. *Phys Rev* 1933;**43**:804.
21. Wigner E, Seitz F. *Solid State Phys: Adv Res Appl* 1955;**1**:97.
22. Ziman JM. *Principles of the theory of solids*. Cambridge: Cambridge University Press; 1972.

CHAPTER

Static and Transport Properties of Solids

6

CHAPTER OUTLINE

6.1 Band Picture ... 166
6.2 Bond Picture ... 167
6.3 Diamond Structure ... 168
6.4 Si and Ge ... 168
6.5 Zinc-Blende Semiconductors ... 170
6.6 Ionic Solids ... 172
6.7 Molecular Crystals ... 174
6.7.1 Molecular Solids ... 174
6.7.2 Hydrogen-Bonded Structures ... 174
6.8 Cohesion of Solids ... 174
6.8.1 Molecular Crystals: Noble Gases ... 174
6.8.2 Ionic Crystals ... 176
6.8.3 Covalent Crystals ... 177
6.8.4 Cohesion in Metals ... 178
6.9 The Semiclassical Model ... 179
6.10 Liouville's Theorem ... 182
6.11 Boltzmann Equation ... 183
6.12 Relaxation Time Approximation ... 184
6.13 Electrical Conductivity ... 186
6.14 Thermal Conductivity ... 187
6.15 Weak Scattering Theory of Conductivity ... 188
6.15.1 Relaxation Time and Scattering Probability ... 188
6.15.2 The Collision Term ... 188
6.15.3 Impurity Scattering ... 189
6.16 Resistivity Due to Scattering by Phonons ... 192
Problems ... 194
References ... 196

6.1 BAND PICTURE

In Chapter 4, we discussed how the different types of solids can be classified as metals, semimetals, insulators, and semiconductors according to the band picture. In Chapter 5, we considered some of the methods used for calculation of the energy bands of solids. To summarize the band picture of solids, one constructs Brillouin zones, which are the Wigner–Seitz cells of the reciprocal lattice. Each Brillouin zone can have N **k** states (where N is the number of primitive cells in the crystal) and hence can accommodate $2N$ electrons (because each **k** state can contain two electrons of opposite spin). The electrons in the lattice move in a periodic potential in a background of the positively charged ions (because the valence electrons have been stripped off from their parent atoms) that opens up an energy gap $2|V_{\mathbf{K}}|$ at each Brillouin zone boundary. In pure materials, there are no eigenstates for electrons with energies lying within these energy gaps. This can be restated as "the electron states are forbidden in this energy gap." Normally, these energy gaps appear in the free electron parabola at the Brillouin zone boundaries because of the periodic potential. This is called the extended zone scheme because it is extended over the entire reciprocal space. However, any **k** state in the extended zone scheme can be written as

$$\mathbf{k} = \mathbf{k}_1 \pm \mathbf{K}, \tag{6.1}$$

where $\mathbf{K}$ is a reciprocal lattice vector, and $\mathbf{k}_1$ is a vector in the first Brillouin zone. Therefore, we can map the different segments of the extended zone into the first zone and use a band index n to identify them. These are called bands because each segment is an independent region of energy levels. For example, the state $\mathbf{k}_i^n$ in the nth zone can be written as

$$\mathbf{k}_{in} = \mathbf{k}_i^n - \mathbf{K}, \tag{6.2}$$

where $\mathbf{k}_{in}$ is the corresponding state in the reduced zone. In the reduced zone scheme, these electron energy bands are stacked above each other in the first Brillouin zone. The energy gap between the bands increases as one proceeds from the lowest zone to the higher zones. This is shown in Figure 6.1.

It is easy to classify the solids in the band picture (as we did in Chapter 4) by counting the number of electrons per unit cell. Thus, all monovalent and trivalent solids are good metals because they are either half-filled or one-and-a-half-filled bands. A divalent solid, which normally would be an insulator, is actually a poor conductor because of the overlap of the bands in three dimensions. Solids with five electrons per atom such as As, Sb, and Bi have two atoms per unit cell that would fill five bands, but due to a small overlap of the fifth and sixth band, there are a few electrons in both bands that can be excited to higher energy states, and hence these are poor conductors. These are known as semimetals. Solids with four electrons per atom range from insulators (diamond) to semiconductors (Ge, Si). Sn is either a metal or a semiconductor, depending on the phase in which it crystallizes, whereas Pb is a pure metal.

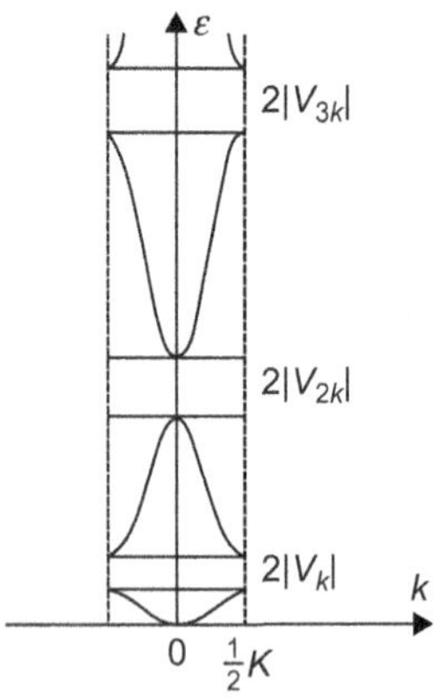

FIGURE 6.1

The bands stacked above each other in the reduced zone scheme.

6.2 BOND PICTURE

To be able to discuss the properties of diamond as well as that of semiconductors such as Ge and Si, which crystallize in the "diamond structure," we need to discuss the bond picture of solids. The simplest bond picture is that of a covalent bond shown in Figure 6.2, which is formed by a hydrogen molecule.

There is a repulsive force between the two nuclei, but the attractive force between the two electrons and the nuclei lowers the energy of the system. This is the simplest example of a covalent bond. The simplest example of a tetrahedral bond is that of a methane molecule, shown in Figure 6.3.

Diamond, Si, and Ge crystallize in the diamond structure, which is essentially a fcc lattice with a basis of two atoms at (0, 0, 0) and (¼, ¼, ¼) per unit cell. It can be visualized as having an atom at each lattice point of a fcc lattice and another atom at each point of a second fcc lattice of which the corner is at one-fourth of the distance along the main diagonal of the cube. The valence electrons in the ground state of the free atoms have the configuration $2s^2 2p^2$, $3s^2 3p^2$, and $4s^2 4p^2$ for diamond, Si, and Ge, respectively. In the crystal, the ground state is formed from the configuration $nsnp^3$, where $n = 2, 3,$ and 4 for diamond, Si, and Ge, respectively. The valence electrons form directed sp^3 tetrahedral bonding orbitals (from the orthonormal s and p orbitals) of the form (assuming no overlap)

$$\begin{aligned}
\phi_1 &= \frac{1}{2}(s+p_x+p_y+p_z), \\
\phi_2 &= \frac{1}{2}(s-p_x+p_y+p_z), \\
\phi_3 &= \frac{1}{2}(s+p_x-p_y-p_z), \\
\phi_4 &= \frac{1}{2}(s-p_x-p_y-p_z).
\end{aligned} \tag{6.3}$$

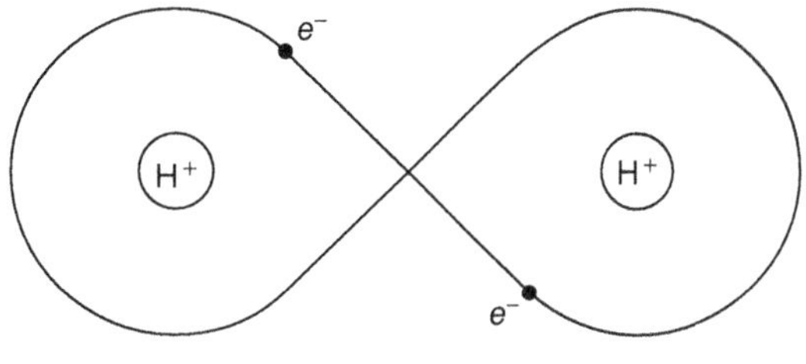

FIGURE 6.2

Covalent bond of a hydrogen molecule.

There is a tetrahedral bond between the atom that forms the center of the tetrahedron and the neighboring four atoms. In fact, there is only one way in which these tetrahedra can be arranged so that each atom forms bonds with four others. In the diamond structure, each atom is at the center of a tetrahedron with the nearest-neighbor atoms at the vertices. The diamond structure is shown in Figure 6.4.

The positions of the atoms in the unit cell of the diamond structure projected on a cube face are shown in Figure 6.5.

The points 0 and ½ are on the fcc lattice, and those at ¼ and ¾ are on a similar lattice displaced among the body diagonal by one-fourth of its length.

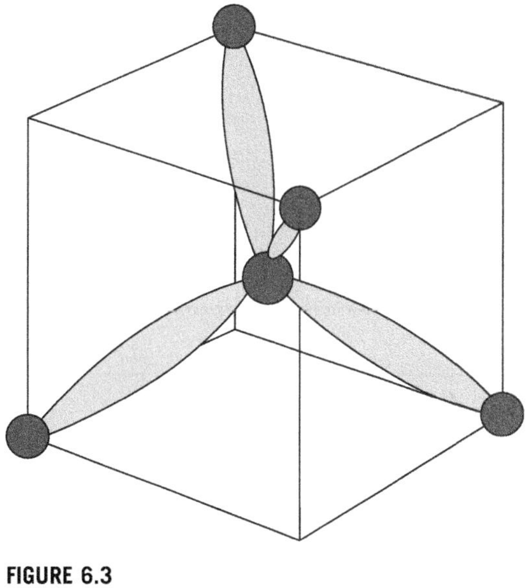

FIGURE 6.3

A simple example of tetrahedral bonding (methane molecule).

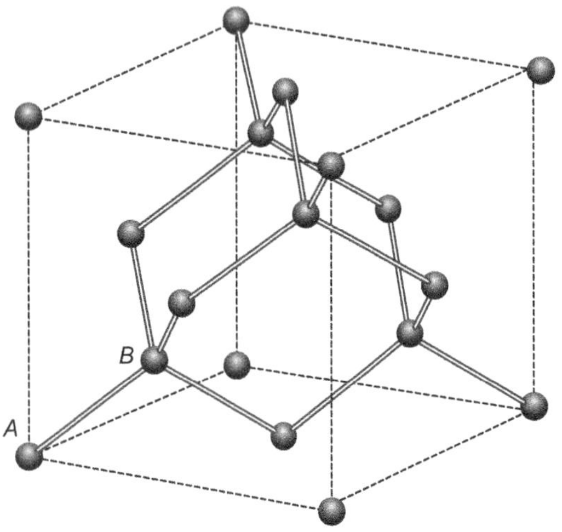

FIGURE 6.4

The diamond structure. Each atom forms the center of a tetrahedron.

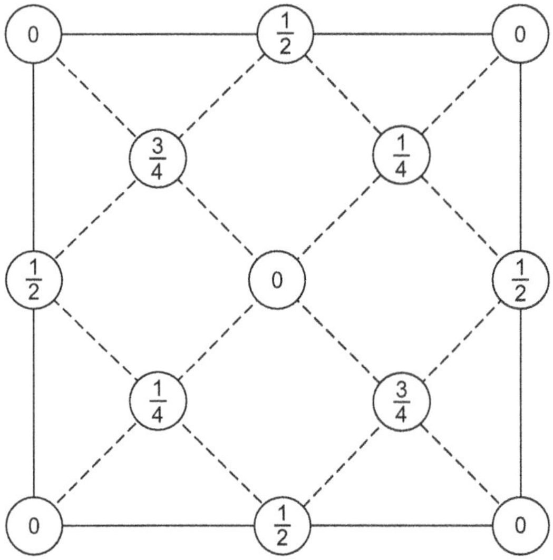

FIGURE 6.5

Positions of the atoms in the unit cell of the diamond structure projected on a cube face.

6.3 DIAMOND STRUCTURE

Diamond obviously crystallizes in the diamond structure. The orbital electrons of carbon atom have the structure $(2s^2 2p^2)$. When they form the diamond crystal, the valence electrons are in the state $(2s2p^3)$. They form the directional sp^3 orbitals as shown in Eq. (6.3). The tetrahedral bonding in diamond is due to the bonding between a carbon atom and its four nearest carbon atoms. However, diamond is an insulator because of the large energy gap between the filled valence band and the empty conduction band. At room temperature, the valence band is completely filled with electrons, whereas the conduction band is empty because the energy gap is too large for electrons to be excited to the conduction band. When an external electric field is applied, the electrons cannot be excited to the higher energy states that lie in the energy gap.

6.4 Si AND Ge

Si and Ge crystallize in the diamond structure. The Bravais lattice of the diamond structure has a basis of two atoms, each of which has eight *sp* electron states, but only four of them are occupied by electrons. Thus, the Brillouin zones have to accommodate 16 electron states, but only eight electrons (from the two atoms) fill them. Consequently, the band structure has eight sub-bands. Four of these are completely filled, and the other four are completely empty at 0° K.

Si and Ge are typical examples of semiconductors because the energy gap between the valence band and the conduction band is sufficiently small. At absolute zero, these are insulators, but at room temperature, a few electrons are excited from the valence band to the conduction band. The conductivity is small but increases with increase of temperature. The property of these intrinsic semiconductors can be easily understood from the bond picture that will help us understand the property of impurity or doped semiconductors as well as that of the *p-n* junctions and their importance in electronics.

A simple example can illustrate the formation of the energy gap in these solids by considering the fact that Si and Ge have covalent bonding. The *sp* states in a free atom form an eight-fold-degenerate level. In an Si dimer, the electron states would interact and form bonding and antibonding levels, as shown in Figure 6.6.

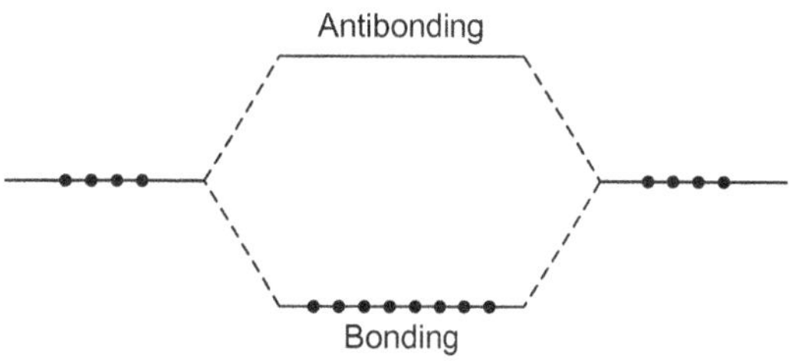

FIGURE 6.6

Bonding and antibonding levels in a silicon dimer.

The tetrahedral bonding of Si is shown in Figure 6.7. An Si atom ($3s^2 3p^2$) has four valence electrons. In the crystalline form, a Silicon atom has the valence configuration ($3s3p^3$), which forms four sp^3 hybrids and results in the tetrahedral bonding with the four nearest Si atoms. Therefore, each bond has two electrons shared by two Si atoms. In principle, the wave functions of these electrons would extend throughout the crystalline solid. However, the electron cloud, which is essentially a pair of electrons of opposite spin, is mostly located in the bond between the neighboring Si ions, as shown schematically in Figure 6.7. This is also confirmed by band calculations.

Ge also crystallizes in the diamond structure. In the Ge atom, the valence electrons have the ($4s^2 4p^2$) configuration, whereas in the crystalline form, the valence electrons have the ($4s4p^3$) configuration. These form the sp^3 hybrids, as explained in Eq. (6.3). Each of these four orbital wave functions is directed toward the vertex of a tetrahedron, and hence, the electronic structure of Ge is almost identical to that of Si. The band structures of Si and Ge, calculated by Chelikowsky and Cohen[3] using the empirical nonlocal pseudopotential method, are shown in Figures 6.8 and 6.9.

We note that the energy gaps in both Ge and Si are indirect. The highest occupied state lies at Γ. However, the lowest lying unoccupied state is near the *X* point, and thus the gap is "indirect."

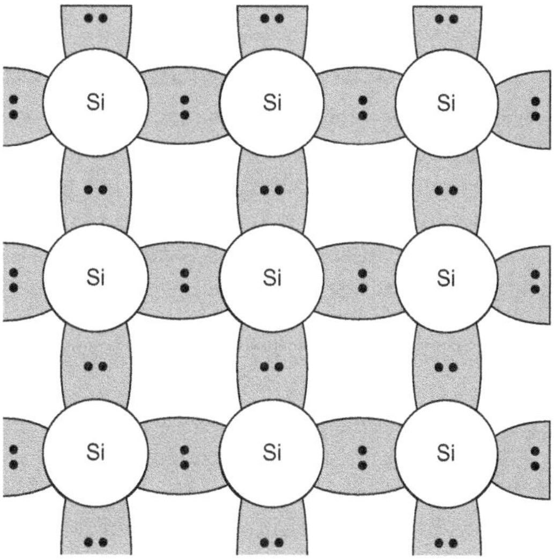

FIGURE 6.7

Electronic structure (schematic) of Si.

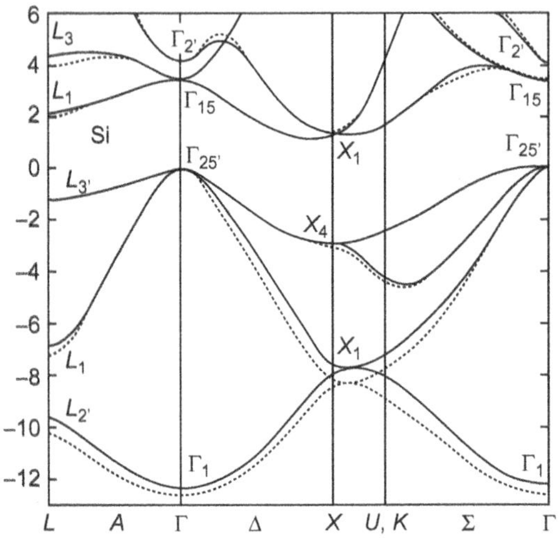

FIGURE 6.8

Band structure of Si calculated by Chelikowsky and Cohen by using nonlocal pseudopotential. The dashed lines are results obtained by using local pseudopotential.

Reproduced from Chelikowsky and Cohen[3] with the permission of the American Physical Society.

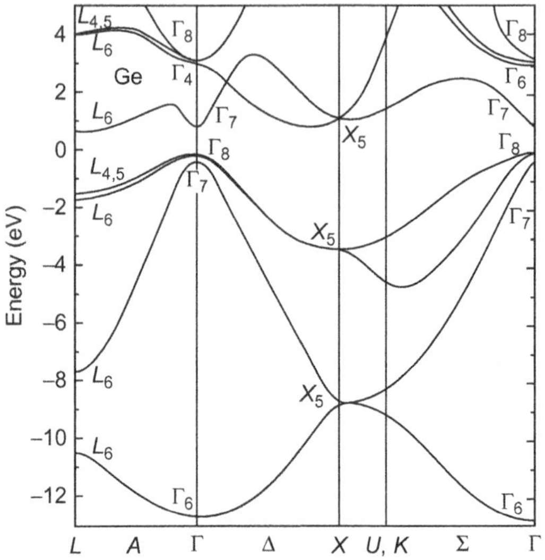

FIGURE 6.9

Band structure of Ge calculated by Chelikowsky and Cohen by using nonlocal pseudopotential.

Reproduced from Chelikowsky and Cohen[3] with the permission of the American Physical Society.

6.5 ZINC-BLENDE SEMICONDUCTORS

The III–V zinc-blende semiconductors such as GaAs and InSb have the same basis core of closed shells as Si and Ge. However, Ga and In have three outer electrons, whereas As and Sb have five outer electrons. In addition, Ga or In occupies all the A sites in the diamond structure, whereas As or Sb occupies all the B sites (see Figure 6.4). The bond picture of GaAs is shown in Figure 6.10.

The bond structure of InSb is similar to that of GaAs. We note that the bond structures of GaAs and InSb are different from that of Si or Ge in the sense that because Ga (or In) has three outer electrons and As (or Sb) has five outer electrons, there is a shrinking of the charge clouds toward As (or Sb), and away from Ga (or In). These compounds are also semiconductors like Si and Ge, but with a smaller energy gap. Their band structures, as calculated by Chelikowsky and Cohen using a nonlocal pseudopotential method, are shown in Figures 6.11 and 6.12.

ZnS, which is a II–VI compound, has the zinc-blende structure (in fact, the name is derived from it). It is also a semiconductor like the III–V compounds. In ZnS, Zn has two electrons, and S has six electrons in the outermost shells. However, the S^{6+} ions attract the eight electrons surrounding it and try to pull them further in through a process known as electron affinity. The S^{6+} ions try to form a closed shell, and the Zn^{2+} ions barely hold on to the two outermost electrons. In the process, the two electrons in each bond are much closer to S than to Zn because the bond electrons are greatly polarized by the residual charge on the ions. The zinc sulphide bond is shown schematically in Figure 6.13.

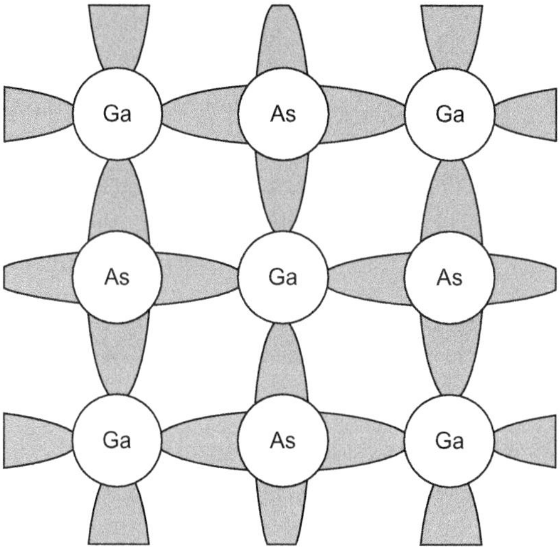

FIGURE 6.10

Bond structure of GaAs.

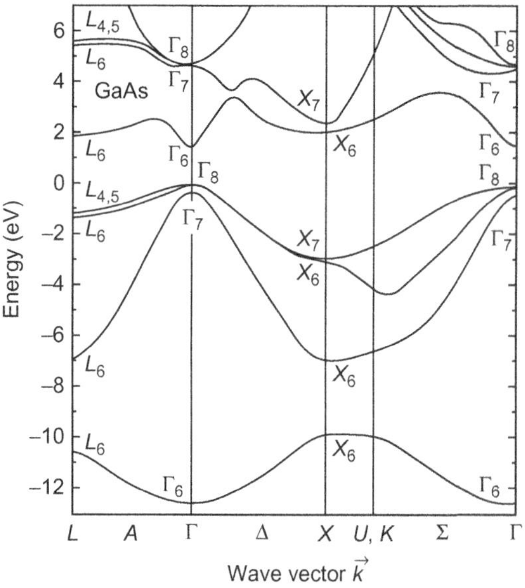

FIGURE 6.11

Band structure of GaAs calculated by Chelikowsky and Cohen.

Reproduced from Chelikowsky and Cohen[3] with the permission of the American Physical Society.

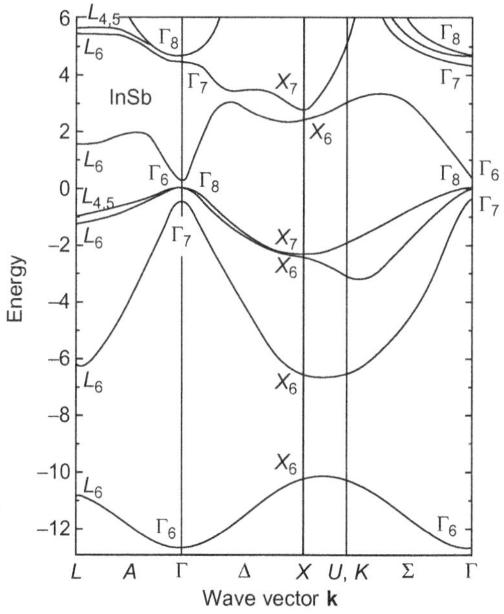

FIGURE 6.12

Band structure of InSb calculated by Chelikowsky and Cohen.

Reproduced from Chelikowsky and Cohen[3] with the permission of the American Physical Society.

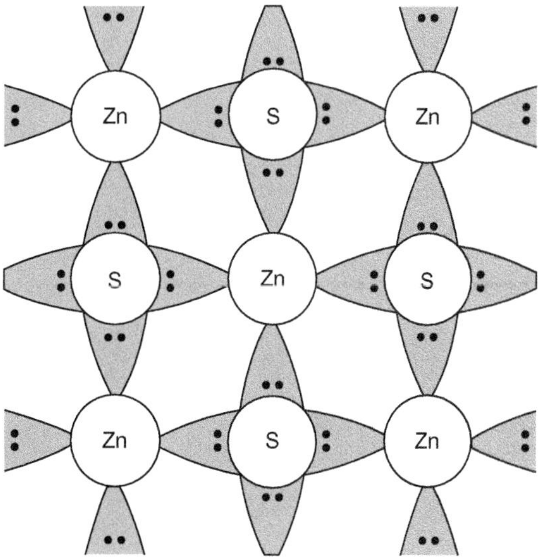

FIGURE 6.13

Schematic picture of zinc sulphide bond.

6.6 IONIC SOLIDS

In CuCl, the copper atom has one electron in its outermost shell, whereas the chlorine atom has seven electrons. In the crystalline form, the Cu atom loses its electron, which is captured by the chlorine atom. CuCl crystallizes in an alternating lattice. The attraction is between the Cu^+ and Cl^- ions. In fact, there are two alternate crystal structures in which the ionic crystals crystallize: the NaCl structure and the CsCl structure. The important factor in the crystal structure in an ionic crystal is to have as many negative ions as possible around a positive ion. A schematic diagram of sodium chloride is shown in Figure 6.14.

The sodium chloride crystal structures are shown in Figure 6.15. In the sodium chloride structure, the space lattice is *fcc*, and the basis has one Na^+ ion at (000) and one Cl^- ion at (½½½).

The cesium chloride crystal structure is shown in Figure 6.16. The space lattice is simple cubic, and the basis has one Cs^+ ion at (000) and one Cl^- ion at (½½½).

We have discussed the progression of the band structure from perfectly covalent silicon in which the four electrons are distributed around the Si^{4+} ion cores in tetrahedral bonding, to covalent gallium arsenide, where there is more charge of the electron cloud surrounding the As^{5+} core than Ga^{3+}. Consequently, GaAs has a small ionic character. The zinc sulphide crystal is weakly covalent because Zn^{2+} has very few valence electrons while S^{6+} is almost full; therefore, the electron cloud is mostly around S^{6+}. ZnS is also partly ionic in character. Sodium chloride is perfectly ionic because the Na^+ ion has lost its valence electron to the chlorine atom that becomes the Cl^- ion. The attraction between the ions of opposite sign results in the ionic bond.

A schematic diagram of the progression from perfectly covalent to perfectly ionic crystals is shown in Figure 6.17.

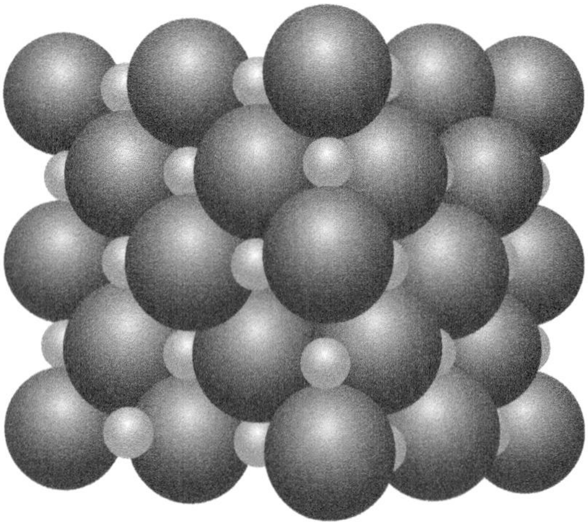

FIGURE 6.14

A schematic diagram of sodium chloride.

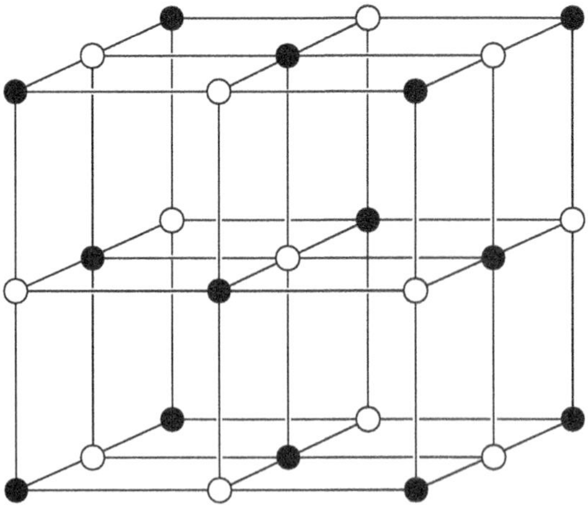

FIGURE 6.15

Sodium chloride crystal structure.

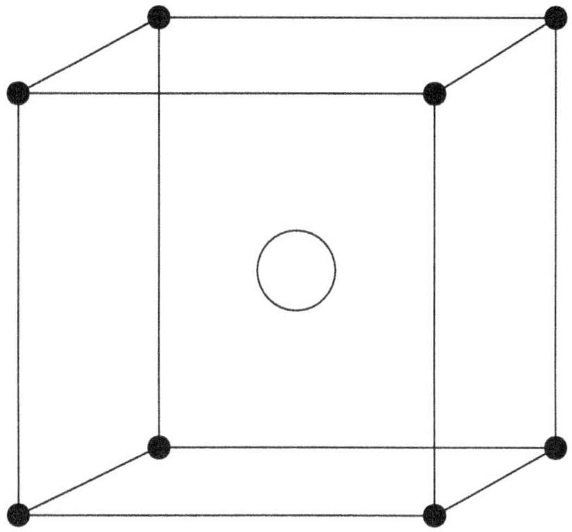

FIGURE 6.16

Cesium chloride structure.

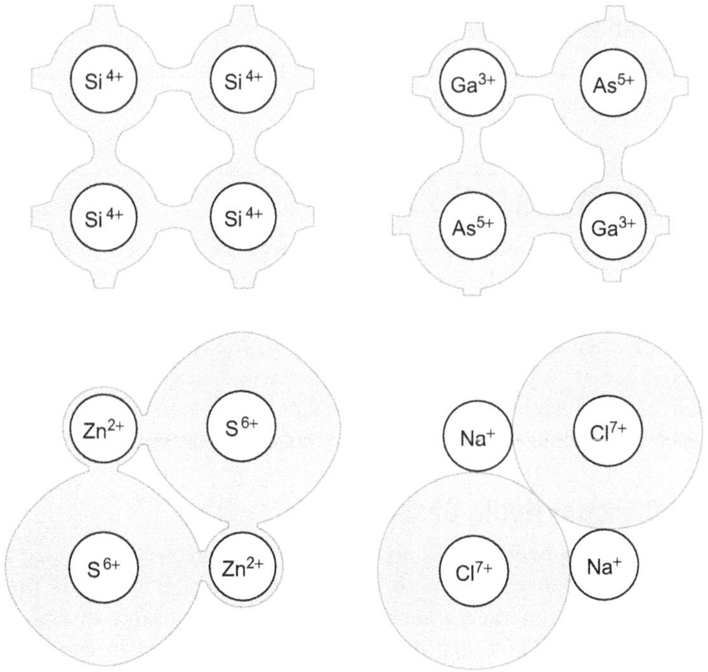

FIGURE 6.17

A schematic diagram of progression from perfectly covalent to perfectly ionic crystals.

6.7 MOLECULAR CRYSTALS

6.7.1 Molecular Solids

The column VIII elements are good examples of molecular solids. The solid noble gases (except helium) crystallize in monatomic fcc Bravais lattices. Each atom has a stable closed-shell type, which gets deformed in the solid. However, the solid is held together by the van der Waals or fluctuating dipole forces. These forces are very weak and can be qualitatively calculated by considering two atoms (A and B) at a distance r. At any instant, if atom A has an instantaneous dipole moment $\mathbf{p}_A$ (of which the time-averaged value vanishes), then the electric field at atom B due to atom A is proportional to p_A/r^3. The dipole moment at atom B due to this electric field is given by

$$p_B \sim \frac{\alpha p_A}{r^3}, \tag{6.4}$$

where α is the polarizability of the atom. The interaction energy between two dipoles of moment p_A and p_B is given by

$$V_i = -\beta \frac{p_A p_B}{r^3} \approx -\alpha\beta \frac{p_A^2}{r^6} = -\frac{A}{r^6}, \tag{6.5}$$

where β is the polarizability of atom B and $A = \alpha\beta p_A^2$. Thus, the lowering of energy between two atoms due to the van der Waals force is very small.

6.7.2 Hydrogen-Bonded Structures

In the hydrogen-bonded structures, the forces between the molecular groups are through a shared proton. The chemical properties of the molecules dominate the properties of these structures.

6.8 COHESION OF SOLIDS

The cohesive energy is the difference between the energy of a solid and the energy of a gas of widely separated atoms from which the solid is eventually constituted. Essentially, the determination of the cohesive energy answers two separate questions. First, it explains how a large group of atoms form a crystalline solid when they come together. Second, it explains which crystal structure leads to the lowest energy.

As we have seen, crystals can be divided into five categories: *molecular, ionic, covalent, metallic,* and *hydrogen bonded.* We will discuss the cohesive energy of each type of solids.

6.8.1 Molecular Crystals: Noble Gases

The van der Waals interaction between two molecules, which are well separated at a distance $\mathbf{r}$, is derived (using a classical theory) in Eq. (6.5). However, the term that is proportional to r^{-6} (Eq. 6.5) represents the interaction energy between the molecules at large distances that do not have a permanent dipole moment. We note that the induced dipole moment in van der Waals interaction is such that it is always attractive and hence lowers the energy of the system. It may be noted that quantum or thermal fluctuations always induce small dipole moments in atoms that normally do not have any dipole moments. However, when the atoms come close together such that their separation is of the same order as that of the atomic radii, there is an additional repulsive force between them. This additional force is proportional to r^{-12}. The resulting potential, which is the sum of the attractive force arising out of the van der Waals term (Eq. 6.5) and the repulsive force, can be written as

$$\phi(r) = -\frac{A}{r^6} + \frac{B}{r^{12}}, \tag{6.6}$$

where A and B are positive constants. For convenience, we introduce the parameters $\sigma = \left(\frac{B}{A}\right)^{1/6}$ and $\in = \frac{A^2}{4B}$, so that Eq. (6.6) can be written as

$$\phi(r) = 4\in\left[\left(\frac{\sigma}{r}\right)^{12} - \left(\frac{\sigma}{r}\right)^{6}\right]. \tag{6.7}$$

This is known as the Lennard–Jones 6-12 potential. The Lennard–Jones parameters $\in$ and σ are appropriately chosen to reproduce the thermodynamic properties of gaseous neon, argon, krypton, and xenon at low temperatures. The Leonard–Jones potential $\phi/4\in$ has been plotted against r/a in Figure 6.18.

The total potential energy U of a crystal of N atoms separated by $\mathbf{R}$ (using periodic boundary conditions so that each atom has identical surroundings) is given by (from Eq. 6.7)

$$U = \frac{1}{2}N\sum_{\mathbf{R}\neq 0}\phi(\mathbf{R}) = 2N\in\sum_{\mathbf{R}\neq 0}\left[\left(\frac{\sigma}{R}\right)^{12} - \left(\frac{\sigma}{R}\right)^{6}\right], \tag{6.8}$$

where the sum is over all nonzero vectors for the bcc lattice (He) or the fcc lattice (all other noble gases). Here, the factor ½ is multiplied because the interaction energy of each atom has been counted twice. The potential energy per atom is given by

$$u = U/N = 2\in\sum_{\mathbf{R}\neq 0}\left[\left(\frac{\sigma}{R}\right)^{12} - \left(\frac{\sigma}{R}\right)^{6}\right]. \tag{6.9}$$

If the nearest-neighbor distance is r, we can rewrite Eq. (6.9) as

$$u = 2\in\left[A_{12}\left(\frac{\sigma}{r}\right)^{12} - A_6\left(\frac{\sigma}{r}\right)^{6}\right], \tag{6.10}$$

where

$$A_n = \sum_{\mathbf{R}\neq 0}\left(\frac{r}{R}\right)^{n}. \tag{6.11}$$

It can be shown that $A_6 = 14.45$ for a fcc structure and 12.25 for a bcc structure while $A_{12} = 12.13$ for a fcc structure and 9.11 for a bcc structure.

The nearest-neighbor equilibrium spacing r_0 is easily obtained from Eq. (6.10) from the expression,

$$\partial u/dr = 0\,|_{r=r_0}. \tag{6.12}$$

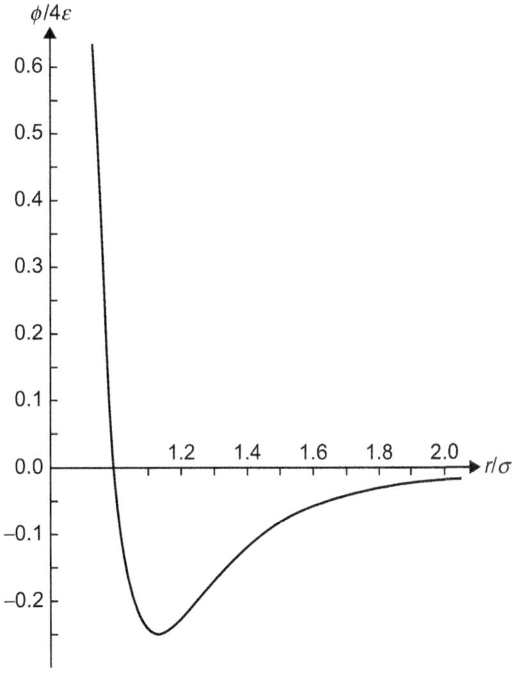

FIGURE 6.18

The Leonard–Jones 6-12 potential.

From Eqs. (6.10) and (6.12), we obtain,

$$r_0 = \sigma\left(\frac{2A_{12}}{A_6}\right)^{1/6}. \tag{6.13}$$

The cohesive energy per ion pair at equilibrium, $u_0 \equiv \varepsilon/N$, is obtained from Eqs. (6.10) and (6.13),

$$u_0 = -\in\frac{A_6{}^2}{2A_{12}} \approx -8.6\in, \tag{6.14}$$

for a fcc lattice. The Bulk modulus B is given by

$$B = V\frac{\partial^2 \varepsilon}{\partial V^2}. \tag{6.15}$$

For a fcc lattice,

$$\frac{V}{N} = \frac{r^3}{\sqrt{2}} = v, \quad \frac{\partial}{\partial v} = \frac{\sqrt{2}}{3r^3}\frac{\partial}{\partial r}. \tag{6.16}$$

Therefore, at equilibrium,

$$B_0 = \frac{\sqrt{2}}{9r_0}\frac{\partial^2 u}{\partial r^2}\Big|_{r=r_0} = \frac{4\in}{\sigma^3}A_{12}\left(\frac{A_6}{A_{12}}\right)^{5/2}. \tag{6.17}$$

6.8.2 Ionic Crystals

The ionic crystals crystallize in four common lattice structures: the sodium chloride, the cesium chloride, the zinc-blende, and the wurtzite structures. The largest term in the interaction energy is the Coulomb interaction because the particles in the ionic crystals are electrically charged ions. The dipole interaction, considered for molecular crystals, which is inversely proportional to the sixth power of the interionic distance, is easily neglected. However, the short-range core–core repulsion due to the Pauli principle prevents the crystal from collapsing. We assume that in the first unit cell, the negative ion is at the origin, and the positive ion is at a distance **d** (**d** is a translation vector through $a/2$ of a cubic side). The total cohesive energy per ion pair is

$$u(d) = u^{coulomb}(d) + u^{core}(d). \tag{6.18}$$

As an example, we consider the sodium chloride structure that is a fcc Bravais of negative ions at site $\mathbf{R}_i$ and a second Bravais lattice of positive cations displaced by **d** from the first lattice. The total potential energy of a single anion or cation is given by

$$u = -\frac{e^2}{d}\left\{1 + \sum_{i\neq 0}\left(\frac{d}{|\mathbf{R}_i + \mathbf{d}|} - \frac{d}{R_i}\right)\right\}. \tag{6.19}$$

The total potential energy for $N/2$ ions pair (or N ions) is given by

$$U = -\frac{Ne^2}{2d}\left\{1 + \sum_{i\neq 0}\left(\frac{d}{|\mathbf{R}_i + \mathbf{d}|} - \frac{d}{R_i}\right)\right\}. \tag{6.20}$$

The energy per ion pair is obtained by dividing Eq. (6.20) by $N/2$ ions, and we obtain

$$u^{coul}(d) = -\frac{e^2}{d}\left\{1 + \sum_{i \neq 0}\left(\frac{d}{|\mathbf{R}_i + \mathbf{d}|} - \frac{d}{R_i}\right)\right\}. \tag{6.21}$$

The evaluation of the summation is complicated and depends on the lattice structure in addition to containing diverging terms. It can be shown, by performing a Ewald summation (Problem 6.4), that

$$u^{coul}(d) = -\frac{\alpha e^2}{d}, \tag{6.22}$$

where α is known as the Madelung constant, which depends on the crystal structure.

The repulsive term $u_{core}(d)$, which is essential in preventing a collapse, can only be estimated approximately by a phenomenological term,

$$u_{core}(d) = \frac{A}{12d}. \tag{6.23}$$

The total energy per pair of ions is obtained from Eqs. (6.22) and (6.23),

$$u(d) = -\alpha\frac{e^2}{d} + \frac{A}{d^{12}}. \tag{6.24}$$

The equilibrium distance d_0 is obtained from the relation

$$\frac{\partial u(d)}{\partial d} = 0. \tag{6.25}$$

From Eqs. (6.24) and (6.25), we obtain

$$\alpha\frac{e^2}{d_0^2} = 12\frac{A}{d_0^3}, \tag{6.26}$$

which leads to

$$d_0 = \left(\frac{12A}{\alpha e^2}\right)^{1/11}. \tag{6.27}$$

The phenomenological constant A is obtained by matching the minimum of the potential to the experimental results.

6.8.3 Covalent Crystals

The theory of cohesion in the covalent crystals involves much more complex calculation than that of ionic or molecular crystals. The main reason is that in the crystalline form, the valence electrons in covalent crystals are distributed in a very different way than they are in a group of isolated atoms or ions. Previously, for noble gases and ionic crystals, we obtained reasonable results by calculating the potential energy of a large number of deformed ions or atoms that were arranged in the relevant crystal structure. To calculate the cohesive energy of covalent crystals, we have to calculate their band structure by following techniques discussed in Chapter 5. In fact, the theory of chemical bonding in molecules is used to describe the cohesion in covalent insulators.

We can make a qualitative discussion by considering a specific case. For example, we place a set of carbon atoms in the sites of a diamond structure, where the lattice constant is large enough so that there is no interaction between the ions, and consequently, the cohesive energy is zero. As the lattice constant is gradually reduced, the atomic orbitals would start overlapping, and the interaction energy between the atoms placed at the designated lattice sites would correspondingly decrease. If the outermost shell of the isolated atom were filled, this overlap would lead to core–core repulsion due to the Pauli exclusion principle, and the energy would increase. However, because the outermost shells of the carbon atoms are partially filled, the electrons are flexible enough to rearrange themselves whenever there is overlap between neighboring wave functions. The primary reason for this flexibility is due to the fact that in the same atomic shell, there are empty energy levels. Thus, when the lattice constant of the (hypothetical) diamond structure is reduced to an appropriate value (the actual lattice constant of diamond), the cohesion energy will be sufficient to form the crystal. The overlap of the outermost shells leads to lowering of the energy. The new levels formed by the electrons are no longer localized around a single atom or ion. The bond picture of covalent solids in general, and the tetrahedral bonding in diamond structure in particular, partially explains the cohesion in such crystals. However, as noted earlier, the theory of cohesion in covalent crystals requires more complex calculations.

6.8.4 Cohesion in Metals

In general, the theory of the cohesion of metals also requires band theory calculations. However, the explanation of cohesion for the alkali metals is comparatively easier because one can consider an alkali metal to consist of a sea of electrons moving around in a background of localized positively charged atomic cores that are located at the lattice sites. There are three types of contributions to the cohesive energy of alkali metals: the interaction between the sea of electrons and the background of positive ions, the average kinetic energy of the electrons, and the exchange energy.

The interaction between the ion cores and the sea of electrons can be expressed as

$$\varepsilon_{el} = -\int d\mathbf{r}\, n(\mathbf{r}) \sum_i \frac{e^2}{|\mathbf{r}-\mathbf{R}_i|} + \frac{e^2}{2}\sum_{i\neq j}\frac{1}{|\mathbf{R}_i-\mathbf{R}_j|} + \frac{1}{2}\int d\mathbf{r}_1 d\mathbf{r}_2 \frac{e^2 n(\mathbf{r}_1)n(\mathbf{r}_2)}{|\mathbf{r}_1-\mathbf{r}_2|}, \tag{6.28}$$

where $n(\mathbf{r})$ is the electron density. We note that the first term in Eq. (6.28) is the sum of the attractive Coulomb interaction between the electrons and the ions; the second term is the sum of the repulsive interaction between the positively charged ions of charge e at the lattice sites $\mathbf{R}_i$ and $\mathbf{R}_j$ (the factor 1/2 prevents double counting); and the third term is the direct electron–electron interaction term that is a function of the electron densities $n(\mathbf{r}_1)$ and $n(\mathbf{r}_2)$. If $n(\mathbf{r})$ is taken as a constant, $n = N/V$. It can be easily shown by adapting a technique used similar to that for ionic crystals in Problem 6.4,

$$\frac{\varepsilon_{el}}{N} = -\frac{1}{2}\alpha\frac{e^2}{r_s}, \tag{6.29}$$

where α is the Madelung constant and r_s (defined in Chapter 3) is the free space around each electron in an electron gas,

$$r_s = \left(\frac{3V}{4\pi N}\right)^{1/3}. \tag{6.30}$$

We also showed in Chapter 3 that the average kinetic energy of the electrons is given by

$$\frac{\varepsilon_k}{N} = \frac{3}{5}\varepsilon_F = \frac{3\hbar^2 k_F{}^2}{10m} = \frac{3\hbar^2}{10m}\left(\frac{9\pi}{4}\right)^{2/3}\frac{1}{r_s{}^2}. \tag{6.31}$$

The exchange term arises due to the fact that electrons at $\mathbf{r}_1$ and $\mathbf{r}_2$ flip places while interacting with each other. Because the wave function is antisymmetric, a negative sign is introduced due to this interaction. The exchange integral can be written as

$$\varepsilon_{ex} = -\int \frac{e^2 d\mathbf{r}_1 d\mathbf{r}_2}{|\mathbf{r}_1 - \mathbf{r}_2|} \sum_{n<m} [\phi_n^*(\mathbf{r}_1)\phi_m^*(\mathbf{r}_2)\phi_n(\mathbf{r}_2)\phi_m(\mathbf{r}_1)\,\delta_{\chi_i\chi_j}], \tag{6.32}$$

where $\phi's$ are the one-particle wave functions and $\chi's$ are the spin index. In the jellium model, the exchange energy can be shown to be

$$\frac{\varepsilon_{ex}}{N} = -\frac{3e^2}{4\pi}\left(\frac{9\pi}{4}\right)^{1/3}\frac{1}{r_s}. \tag{6.33}$$

Adding Eqs. (6.29), (6.31), and (6.33); using the value $\alpha = 1.792$ for fcc, bcc, and hcp lattices; expressing the results in terms of Bohr radius $a_0 = 0.529\,A°$; and measuring the energy in units of electron volts per atom, we obtain

$$\frac{\varepsilon}{N} = \left[\frac{-24.35}{(r_s/a_0)} + \frac{30.1}{(r_s/a_0)^2} - \frac{12.5}{(r_s/a_0)}\right] eV/atom. \tag{6.34}$$

The minimum of $\frac{\varepsilon}{N}$ occurs when $\frac{r_s}{a_0} \approx 1.6$. However, for alkali metals, the observed values of $\frac{r_s}{a_0}$ are between 2 and 6. The significant difference between the theoretical value and the experimental results is due to the neglect of correlation energy in this derivation. The treatment of correlation energy is beyond the scope of the present discussions because they are treated by using quantum field theory. Qualitatively, one can summarize the effect of correlation by stating that the correlations tend to keep the electrons apart, thereby further decreasing the energy of the system derived in Eq. (6.34).

We will now formulate the semiclassical model of electron dynamics and then derive the Boltzmann equations, which are essential for understanding the study of solids.

6.9 THE SEMICLASSICAL MODEL

The semiclassical model predicts the position $\mathbf{r}$ and the wave vector $\mathbf{k}$ of each electron evolved in the presence of external electric field $\mathbf{E}$ and magnetic field $\mathbf{B}$ in the absence of collisions. The basic assumptions of the semiclassical model are as follows:

1. The band index n is a constant of motion, and there are no "interband transitions."
2. The position of an electron in a crystal with inversion symmetry evolves according to

$$\dot{\mathbf{r}} = \mathbf{v}_n(\mathbf{k}) = \frac{1}{\hbar}\frac{\partial \varepsilon_{n\mathbf{k}}}{\partial \mathbf{k}}. \tag{6.35}$$

3. The electron wave vector obeys the equation of motion

$$\hbar\dot{\mathbf{k}} = -e\mathbf{E}(\mathbf{r},t) - \frac{e}{c}\dot{\mathbf{r}}\times\mathbf{B}(\mathbf{r},t), \tag{6.36}$$

where $\mathbf{E}(\mathbf{r},t)$ and $\mathbf{B}(\mathbf{r},t)$ are electric and magnetic fields that may vary spatially. Eqs. (6.35) and (6.36) can be derived from the Lagrangian formulation in classical mechanics as follows (see Symon[10]).

If the forces acting on a dynamical system depend on the velocities, we can define a function $U(q_1, q_2, ..., q_f;\ \dot{q}_1, \dot{q}_2, ..., \dot{q}_f\ t)$, called the velocity-dependent potential (which includes the ordinary potential energy V), such that the generalized force Q_k associated with the coordinate q_k is given by

$$Q_k = \frac{d}{dt}\frac{\partial U}{\partial \dot{q}_k} - \frac{\partial U}{\partial q_k}, \quad k = 1, ..., f. \tag{6.37}$$

The Lagrangian function is defined as

$$L = T - U, \tag{6.38}$$

and the equations of motion can be written as

$$\frac{d}{dt}\frac{\partial L}{\partial \dot{q}_k} - \frac{\partial L}{\partial q_k} = 0. \tag{6.39}$$

If we consider a system described in terms of a fixed system of coordinates, the kinetic energy T is a homogeneous quadratic function of the generalized velocities $\dot{q}_1, \dot{q}_2, ..., \dot{q}_n$, we obtain from Euler's theorem,

$$\sum_{k=1}^{f} \dot{q}_k \frac{\partial T}{\partial \dot{q}_k} = 2T. \tag{6.40}$$

If V, the potential energy, is a function of the coordinates $q_1, q_2, ..., q_f$, the total energy from Eqs. (6.40) and (6.41) is

$$E = T + V. \tag{6.41}$$

The Hamiltonian function is defined as

$$H = \sum_{k=1}^{f} p_k \dot{q}_k - L. \tag{6.42}$$

We can also write

$$\begin{aligned} dL &= \sum_{k=1}^{f}\left(\frac{\partial L}{\partial \dot{q}_k}d\dot{q}_k + \frac{\partial L}{\partial q_k}dq_k\right) + \frac{\partial L}{\partial t}dt \\ &= \sum_{k=1}^{f}(p_k d\dot{q}_k + \dot{p}_k dq_k) + \frac{\partial L}{\partial t}dt. \end{aligned} \tag{6.43}$$

From Eqs. (6.42) and (6.43), we obtain

$$dH = \sum_{k=1}^{f}(\dot{q}_k dp_k - \dot{p}_k dq_k) - \frac{\partial L}{\partial t}dt. \tag{6.44}$$

Thus, we obtain

$$\dot{q}_k = \frac{\partial H}{\partial p_k} \text{ and } \dot{p}_k = -\frac{\partial H}{\partial q_k}, \quad k = 1, \ldots, f, \tag{6.45}$$

Eq. (6.42) can be written in the generalized form that whenever there is a Lagrangian $L(\mathbf{Q}, \dot{\mathbf{Q}})$, it is also possible to derive a Hamiltonian, using the formulae

$$H = \sum_l \dot{Q}_l P_l - L \tag{6.46}$$

and

$$P_l = \frac{\partial L}{\partial \dot{Q}_l}. \tag{6.47}$$

Here, the Q's are the three components of $\mathbf{r}_c$ and the three components of $\mathbf{k}_c$ (the subscript c is for classical values that we will drop in the future).

In an electromagnetic field, the electromagnetic force on an electron of charge $-e$ can be written as

$$\mathbf{F} = -e\mathbf{E} - \frac{e}{c}\mathbf{V} \times \mathbf{B}. \tag{6.48}$$

The Hamiltonian can also be written by replacing the Hamiltonian in a pure perfect crystal lattice by an equivalent Hamiltonian operator $\varepsilon_n(-i\nabla)$, where n is the band index. Thus, the electron is treated as free because the effect of the lattice potential is included in $\varepsilon_n(-i\nabla)$, the modified kinetic energy. If the potential of the external field is $U(\mathbf{r})$, the equivalent Hamiltonian operator can be written in the alternate form,

$$H(\mathbf{r}, \mathbf{p}) = \varepsilon_n(-i\nabla) + U(\mathbf{r}). \tag{6.49}$$

According to the quantum mechanical formulation, the Hamiltonian operator is obtained by replacing the classical momentum variable $\mathbf{p}$ by $-i\hbar\nabla$ in the classical Hamiltonian function. Thus, by reversing the steps, we obtain the equivalent classical Hamiltonian from Eq. (6.49) by replacing $-i\nabla$ by $\mathbf{p}/\hbar$,

$$H(\mathbf{r}, \mathbf{p}) = \varepsilon_n(\mathbf{p}/\hbar) + U(\mathbf{r}). \tag{6.50}$$

Here, the basic assumption is that the energy functions $\varepsilon_n(\mathbf{k})$ are known from band structure calculations, and thus, there is no need to include the explicit form of the periodic potential of the ions.

The equations of motion can be written in the canonical Hamiltonian form (Eq. 6.45),

$$\dot{\mathbf{r}} = \frac{\partial H}{\partial \mathbf{p}}, \tag{6.51}$$

and

$$\dot{\mathbf{p}} = -\frac{\partial H}{\partial \mathbf{r}}. \tag{6.52}$$

From Eqs. (6.50) and (6.51), we obtain

$$\dot{\mathbf{r}} \equiv \mathbf{v}_n(\mathbf{k}) = \frac{\partial H}{\partial \mathbf{p}} = \frac{\partial}{\partial \mathbf{p}}\{\varepsilon_n(\mathbf{p}/\hbar) + U(\mathbf{r})\} = \frac{\partial \varepsilon_n(\mathbf{p}/\hbar)}{\partial \mathbf{p}} = \frac{1}{\hbar}\frac{\partial \varepsilon_n(\mathbf{k})}{\partial \mathbf{k}}. \tag{6.53}$$

From Eqs. (6.48), (6.50), and (6.52), we obtain

$$\dot{\mathbf{p}} = -\frac{\partial H}{\partial \mathbf{r}} = -\frac{\partial U}{\partial \mathbf{r}} = \mathbf{F} = -e\left[\mathbf{E}(\mathbf{r},t) + \frac{1}{c}\mathbf{v}_n(\mathbf{k}) \times \mathbf{B}(\mathbf{r},t)\right]. \tag{6.54}$$

An alternate way to approach the problem is by defining the vector and scalar potentials $\mathbf{A}$ and ϕ as

$$\mathbf{B} = \nabla \times \mathbf{A} \tag{6.55}$$

and

$$\mathbf{E} = -\nabla\phi - \frac{1}{c}\frac{\partial \mathbf{A}}{\partial t}. \tag{6.56}$$

From Eqs. (6.48), (6.55), and (6.56), we can write $\mathbf{F}$ as

$$\mathbf{F} = e\nabla\phi + \frac{e}{c}\frac{\partial \mathbf{A}}{\partial t} - \frac{e}{c}\nabla(\mathbf{v}\cdot\mathbf{A}). \tag{6.57}$$

One can show from Eqs. (6.41) and (6.42) that the Hamiltonian for the electrons in the nth band can be written as

$$H(\mathbf{r},\mathbf{p}) = \varepsilon_n\left(\frac{1}{\hbar}\left[\mathbf{p} + \frac{e}{c}\mathbf{A}(\mathbf{r},t)\right]\right) - e\phi(\mathbf{r},t) \tag{6.58}$$

and

$$\hbar\dot{\mathbf{k}} = -e\mathbf{E} - \frac{e\mathbf{v}}{c} \times \mathbf{B}, \tag{6.59}$$

where the variables $\hbar\mathbf{k}$ are defined as

$$\hbar\mathbf{k} = \mathbf{p} + \frac{e}{c}\mathbf{A}(\mathbf{r},t). \tag{6.60}$$

The canonical crystal momentum $\mathbf{p}$ (which is the canonical momentum in the Hamiltonian formulation) can be written as

$$\mathrm{p} = \hbar\mathbf{k} - \frac{e}{c}\mathbf{A}(\mathbf{r},t). \tag{6.61}$$

6.10 LIOUVILLE'S THEOREM

We will now briefly discuss Liouville's theorem before proceeding further. The configuration and the motion of a system is specified by the coordinates and momenta, $q_1, q_2, \ldots, q_f : p_1, p_2, \ldots, p_f$. The $2f$-dimensional space is called the phase space of the mechanical system. The velocity of the phase points is given by the Hamilton equations in Eq. (6.45). The possible state of the mechanical system is represented by each phase point that is occupied by a "particle" that moves according to the equation of motion (Eq. 6.45). These particles trace out paths that represent the history of the mechanical system. The solutions of the equations of motion are uniquely determined because the positions and momenta are given, and hence, there is only one possible path through each phase point. Liouville's theorem states that the phase "particles" move as an incompressible fluid. The phase volume occupied by a set of "particles" is a constant. The proof of Liouville's theorem can be found in any standard book of classical mechanics (see Symon[10]).

6.11 BOLTZMANN EQUATION

The Boltzmann equation can be obtained by first considering the continuity equation. Here, we consider the motion of particles illustrated in Figure 6.19. We consider a large number of a conserved collection of identical particles at position x, moving with a velocity $v(x)$ and density $g(x)$. In a time Δt, the total number of particles moving into a small region dx minus the number of particles that are moving out is equal to

$$\Delta g(x) = v(x)g(x)\Delta t - v(x+dx)g(x+dx)\Delta t. \tag{6.62}$$

In the limit $\Delta t \to 0$,

$$\frac{\partial g}{\partial t} = -\frac{\partial}{\partial x}v(x)g(x,t). \tag{6.63}$$

Eq. (6.63) can be generalized for many variables as

$$\frac{\partial g}{\partial t} = -\sum_i \frac{\partial}{\partial x_i} v_i(\vec{x})g(\vec{x},t). \tag{6.64}$$

We define $g_{\mathbf{rk}}(t)$ as the occupation number of electrons at time t with position $\mathbf{r}$ and wave vector $\mathbf{k}$. Thus, $g_{\mathbf{rk}}(t)$ is the probability that a state is occupied at time t. From Eq. (6.64), we obtain

$$\frac{\partial g_{\mathbf{rk}}}{\partial t} = -\frac{\partial}{\partial \mathbf{r}}\cdot\dot{\mathbf{r}}g_{\mathbf{rk}} - \frac{\partial}{\partial \mathbf{k}}\cdot\dot{\mathbf{k}}g_{\mathbf{rk}}. \tag{6.65}$$

In an electromagnetic field, we have derived

$$\begin{aligned}\dot{\mathbf{r}} &= \frac{1}{\hbar}\frac{\partial \varepsilon}{\partial \mathbf{k}} = \mathbf{v},\\ \hbar\dot{\mathbf{k}} &= -e\mathbf{E} - \frac{e}{c}\mathbf{v}\times\mathbf{B} = \mathbf{F}(\mathbf{r},\mathbf{k}).\end{aligned} \tag{6.66}$$

$\mathbf{E}$ does not depend on $\mathbf{k}$, and $\mathbf{v}$ does not depend on $\mathbf{r}$. Further,

$$\frac{\partial}{\partial \mathbf{k}}\cdot\mathbf{v}\times\mathbf{B} = \mathbf{B}\cdot\frac{\partial}{\partial \mathbf{k}}\times\mathbf{v} = 0 \text{ because } \mathbf{v} = \frac{1}{\hbar}\frac{\partial\varepsilon}{\partial \mathbf{k}}. \tag{6.67}$$

From Eqs. (6.65) through (6.67), we obtain

$$\frac{\partial g}{\partial t} = -\dot{\mathbf{r}}\cdot\frac{\partial}{\partial \mathbf{r}}g - \dot{\mathbf{k}}\cdot\frac{\partial}{\partial \mathbf{k}}g. \tag{6.68}$$

From Eqs. (6.66) and (6.68), we obtain

$$\frac{\partial g}{\partial t} + \mathbf{v}\cdot\frac{\partial}{\partial \mathbf{r}}g + \frac{1}{\hbar}\mathbf{F}\cdot\frac{\partial}{\partial \mathbf{k}}g = 0. \tag{6.69}$$

However, there are momentum transfers caused by thermal fluctuations and impurities, which are not included in Eq. (6.69). To include these,

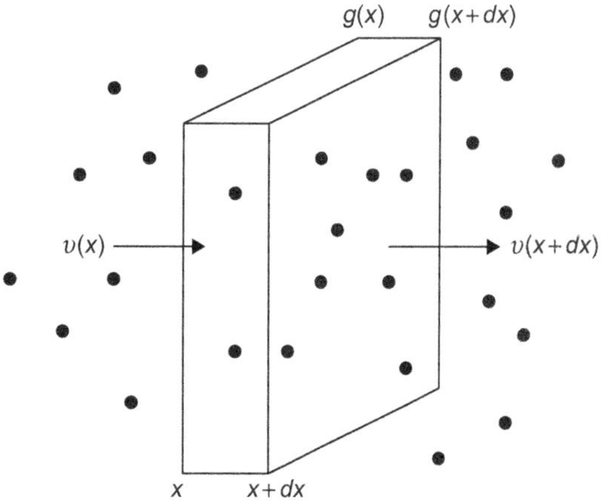

FIGURE 6.19

Density of particles at x and $x+dx$.

Boltzmann added a collision term to Eq. (6.69), and the resulting equation, known as the Boltzmann equation, can be written as

$$\frac{\partial g}{\partial t}+\mathbf{v}\cdot\frac{\partial}{\partial \mathbf{r}}g+\frac{1}{\hbar}\mathbf{F}\cdot\frac{\partial}{\partial \mathbf{k}}g=\left(\frac{\partial g}{\partial t}\right)_{coll}. \tag{6.70}$$

The terms on the left side of the equation are known as the drift terms, and the term on the right side is known as the collision term.

6.12 RELAXATION TIME APPROXIMATION

In thermal equilibrium, the distribution function reduces to the Fermi function

$$f_{\mathbf{rk}}=\frac{1}{e^{\beta_{\mathbf{r}}(\varepsilon_{\mathbf{k}}-\mu_{\mathbf{r}})}+1}, \tag{6.71}$$

where $T(\mathbf{r})$ is the temperature at $\mathbf{r}$ and $\mu_{\mathbf{r}}$ is the constant that produces the correct density $n(\mathbf{r})$. The collision term $\left(\frac{\partial g}{\partial t}\right)_{coll}$ in Eq. (6.70) must be such that the distribution g must relax toward thermal equilibrium. The simplest approximation for this to happen is the relaxation time approximation,

$$\left(\frac{\partial g}{\partial t}\right)_{coll}=-\frac{1}{\tau_\varepsilon}(g_{r\mathbf{k}}-f_{r\mathbf{k}}), \tag{6.72}$$

where τ_ε is defined as the relaxation time. In the simplest solutions, τ_ε is considered as a constant τ.

If we approach the same problem in an alternate method, we can argue that because $g=g(\mathbf{r},\mathbf{k},t)$, we can write

$$\frac{dg}{dt}=\frac{\partial g}{\partial t}+\dot{\mathbf{r}}\cdot\frac{\partial g}{\partial \mathbf{r}}+\dot{\mathbf{k}}\cdot\frac{\partial g}{\partial \mathbf{k}}. \tag{6.73}$$

From Eqs. (6.70), (6.72), and (6.73), we obtain

$$\frac{dg}{dt}=-\frac{g-f}{\tau_\varepsilon}. \tag{6.74}$$

Integrating Eq. (6.74), we obtain

$$g(\mathbf{r},\mathbf{k},t)=\int_{-\infty}^{t}dt' f(\mathbf{r}(t'),\mathbf{k}(t'))\frac{e^{-(t-t')/\tau_\varepsilon}}{\tau_\varepsilon}. \tag{6.75}$$

Here, $\mathbf{r}(t')$ and $\mathbf{k}(t')$, which are solutions of the semiclassical equations, evolve in time such that $\mathbf{r}(t)=\mathbf{r}$ and $\mathbf{k}(t)=\mathbf{k}$. Integrating Eq. (6.75) by parts, we obtain

$$g(\mathbf{r},\mathbf{k},t)=f(\mathbf{r},\mathbf{k})-\int_{-\infty}^{t}dt'\, e^{-(t-t')/\tau_\varepsilon}\frac{d}{dt'}f(\mathbf{r}(t'),\mathbf{k}(t')). \tag{6.76}$$

Because f is the Fermi function, which is evaluated at local temperature and chemical potential,

$$\frac{d}{dt'}f(\mathbf{r}(t'),\mathbf{k}(t'))=\left(\dot{\mathbf{r}}(t')\cdot\frac{\partial}{\partial \mathbf{r}}+\dot{\mathbf{k}}(t')\cdot\frac{\partial}{\partial \mathbf{k}}\right)f(\mathbf{r}(t'),\mathbf{k}(t')). \tag{6.77}$$

From Eq. (6.71), we obtain

$$\frac{\partial f}{\partial \mathbf{r}} = \frac{\partial f}{\partial \varepsilon}\left[-\frac{\partial \mu}{\partial \mathbf{r}} - \frac{(\varepsilon-\mu)}{T}\frac{\partial T}{\partial \mathbf{r}}\right], \tag{6.78}$$

$$\frac{\partial f}{\partial \mathbf{k}} = \frac{\partial f}{\partial \varepsilon}\frac{\partial \varepsilon}{\partial \mathbf{k}} = \hbar \mathbf{v_k}\frac{\partial f}{\partial \varepsilon}, \tag{6.79}$$

and

$$\frac{\partial f}{\partial \varepsilon} = -\frac{\partial f}{\partial \mu}. \tag{6.80}$$

In addition, from Eq. (6.66), we have

$$\dot{\mathbf{k}} \cdot \frac{\partial f}{\partial \mathbf{k}} = -e\mathbf{v_k} \cdot \mathbf{E}. \tag{6.81}$$

Substituting Eqs. (6.77) through (6.81) in Eq. (6.76), we obtain

$$g(\mathbf{r},\mathbf{k},t) = f(\mathbf{r},\mathbf{k}) - \int_{-\infty}^{t} dt'\, e^{-(t-t')/\tau_\varepsilon}\mathbf{v_k}\cdot\left[e\mathbf{E} + \frac{\partial \mu}{\partial \mathbf{r}} + \frac{\varepsilon_{\mathbf{k}}-\mu}{T}\frac{\partial T}{\partial \mathbf{r}}\right]\frac{\partial f(t')}{\partial \mu}. \tag{6.82}$$

Neglecting the time dependence of all terms except the exponential factor, the integral can be easily done, and we obtain

$$g(\mathbf{r},\mathbf{k}) = f(\mathbf{r},\mathbf{k}) - \tau_\varepsilon \mathbf{v_k}\cdot\left[e\mathbf{E} + \vec{\nabla}\mu + \frac{\varepsilon_{\mathbf{k}}-\mu}{T}\vec{\nabla}T\right]\frac{\partial f}{\partial \mu}. \tag{6.83}$$

The current density can be written as

$$\mathbf{J} = -2e\int e\mathbf{v_k}g(\mathbf{k})d\mathbf{k}, \tag{6.84}$$

which can be rewritten in the alternate form

$$\mathbf{J} = -2\int e\mathbf{V}_k f(\mathbf{k})d\mathbf{k} - 2\int e^2\tau_e \mathbf{V_k}(\mathbf{V_k}\cdot\mathbf{E})\frac{\partial f}{\partial \varepsilon} - \frac{e}{T}(\mathbf{K}_1\cdot(-\nabla T)), \tag{6.85}$$

where

$$\mathbf{K}_n = \frac{1}{4\pi^2}\frac{\tau_e}{\hbar}\iint \mathbf{V_k}\mathbf{V_k}(\varepsilon-\mu)^n\left(-\frac{\partial f}{\partial \varepsilon_n}\right)\frac{dS}{v_{\mathbf{k}}}d\varepsilon\cdot(-\nabla T), \tag{6.86}$$

and we have used $\iint d\mathbf{k} = \frac{1}{8\pi^2}\iint \frac{dS}{\hbar v_{\mathbf{k}}}d\varepsilon$ for integration over a constant energy surface (Problem 6.7). Eq. (6.86) indicates that even if there were no electric field, the thermal gradient, ∇T, would give rise to an electric current. This is known as the thermo-electric effect, which we will discuss later. We note that Eq. (6.85) can be rewritten in the alternate form

$$\mathbf{J} = -2\int e\mathbf{V}_k f(\mathbf{k})d\mathbf{k} + e^2\mathbf{K}_0\cdot\mathbf{E} - \frac{e}{T}\mathbf{K}_1\cdot(-\nabla T). \tag{6.87}$$

6.13 ELECTRICAL CONDUCTIVITY

In an electric field $\mathbf{E}$, which is independent of time, there is no variation of g, f, μ, and T with position. We can rewrite Eq. (6.83) in the simpler form

$$g(\mathbf{k}) = f(\mathbf{k}) - \tau_\varepsilon \mathbf{v}_\mathbf{k} \cdot [e\mathbf{E}] \frac{\partial f}{\partial \mu}. \tag{6.88}$$

To calculate the electrical conductivity, we need to first calculate the current density,

$$\mathbf{J} = -2e \int e\mathbf{v}_\mathbf{k} g(\mathbf{k}) d\mathbf{k}. \tag{6.89}$$

From Eqs. (6.88) and (6.89), we obtain

$$\mathbf{J} = -2 \int e\mathbf{v}_\mathbf{k} f(\mathbf{k}) d\mathbf{k} - 2 \int e^2 \tau_\varepsilon \mathbf{v}_\mathbf{k} \cdot (\mathbf{v}_\mathbf{k} \cdot \mathbf{E}) \frac{\partial f}{\partial \varepsilon}, \tag{6.90}$$

where we have used the identity,

$$\frac{\partial f}{\partial \mu} = -\frac{\partial f}{\partial \varepsilon}. \tag{6.91}$$

In Eq. (6.90), because $f_\mathbf{k}$ is the equilibrium Fermi distribution function,

$$-\int e\mathbf{v}_\mathbf{k} f_\mathbf{k} d\mathbf{k} \equiv 0. \tag{6.92}$$

From Eqs. (6.90) and (6.92), we obtain

$$\mathbf{J} = \frac{1}{4\pi^3} \iint e^2 \tau_\varepsilon \mathbf{v}_\mathbf{k} (\mathbf{v}_\mathbf{k} \cdot \mathbf{E}) \left(-\frac{\partial f}{\partial \varepsilon}\right) \frac{dS}{\hbar v_k} d\varepsilon, \tag{6.93}$$

where we have changed an integral over a volume of $\mathbf{k}$ space into integrations over surfaces of constant energy (Problem 6.7).

In a metal,

$$-\left(\frac{\partial f}{\partial \varepsilon}\right) = \delta(\varepsilon - \varepsilon_F). \tag{6.94}$$

We obtain from Eqs. (6.93) and (6.94),

$$\mathbf{J} = \frac{1}{4\pi^3} \frac{e^2 \tau_\varepsilon}{\hbar} \int \frac{\mathbf{v}_\mathbf{k} \mathbf{v}_\mathbf{k} dS_F}{v_\mathbf{k}} \cdot \mathbf{E}. \tag{6.95}$$

The equation for electric conductivity tensor can be expressed as

$$\mathbf{J} = \sigma \cdot \mathbf{E}. \tag{6.96}$$

From Eqs. (6.95) and (6.96), we obtain

$$\sigma = \frac{1}{4\pi^3} \frac{e^2 \tau_\varepsilon}{\hbar} \int \frac{\mathbf{v}_\mathbf{k} \mathbf{v}_\mathbf{k} dS_F}{v_\mathbf{k}}. \tag{6.97}$$

In crystals with cubic symmetry, the conductivity tensor is a scalar and if **E** and **J** are both in the same x direction,

$$(\mathbf{v_k v_k} \cdot \mathbf{E})_x = E v_x{}^2 = \frac{1}{3} v^2 E. \tag{6.98}$$

From Eqs. (6.97) and (6.98), we obtain

$$\sigma = \frac{e^2}{12\hbar\pi^3} \int \Lambda \, dS_F, \tag{6.99}$$

where Λ is the mean free path

$$\Lambda = \tau v. \tag{6.100}$$

6.14 THERMAL CONDUCTIVITY

The thermal conductivity experiment is usually done by keeping the specimen in an open circuit. This implies that the specimen has an electric field **E**, but there is no electric current; i.e., $\mathbf{J} = 0$. From Eqs. (6.87) and (6.92), we have an electric field in the wire

$$\mathbf{E} = -\frac{1}{e} \mathbf{K}_0^{-1} \mathbf{K}_1 \frac{1}{T} \cdot \nabla T. \tag{6.101}$$

The thermal conductivity can be calculated from the total flux of heat (per unit volume)

$$\mathbf{U} = 2 \int f_{\mathbf{k}} \{\varepsilon(\mathbf{k}) - \mu\} \mathbf{v_k} d\mathbf{k}. \tag{6.102}$$

From Eqs. (6.83), (6.86), and (6.102), we obtain

$$\mathbf{U} = -e\mathbf{K}_1 \cdot \mathbf{E} + \frac{1}{T} \mathbf{K}_2 \cdot (-\nabla T). \tag{6.103}$$

From Eqs. (6.101) and (6.103), we obtain

$$\mathbf{U} = \frac{1}{T} (\mathbf{K}_1 \mathbf{K}_0^{-1} \mathbf{K}_1 \cdot \nabla T - \mathbf{K}_2 \cdot \nabla T), \tag{6.104}$$

which can be rewritten in the alternate form

$$\mathbf{U} = \frac{1}{T} (\mathbf{K}_2 - \mathbf{K}_1 \mathbf{K}_0^{-1} \mathbf{K}_1) \cdot (-\nabla T). \tag{6.105}$$

The thermal conductivity κ is defined as

$$\mathbf{U} = \kappa \cdot (-\nabla T). \tag{6.106}$$

If we neglect the second term of Eq. (6.105) for metals, we obtain from Eqs. (6.105) and (6.106)

$$\kappa \approx \frac{1}{T} \mathbf{K}_2. \tag{6.107}$$

Using the expansion (Problem 6.8)

$$\int F(\varepsilon) d\varepsilon \left(\frac{-\partial f_0}{\partial \varepsilon} \right) d\varepsilon = F(\mu) + \frac{1}{6} \pi^2 (k_B T)^2 \left[\frac{\partial^2 F(\varepsilon)}{\partial \varepsilon^2} \right]_{\varepsilon = \mu} + \ldots, \tag{6.108}$$

and $\sigma = e^2 \mathbf{K}_0$ [from Eq. (6.87) under isothermal conditions], we can write

$$\kappa = \frac{\pi^2}{3} \frac{k^2}{e^2} T\sigma. \tag{6.109}$$

This is also known as the Wiedemann–Franz law.

6.15 WEAK SCATTERING THEORY OF CONDUCTIVITY

6.15.1 Relaxation Time and Scattering Probability

We will now consider a more realistic description of the collisions by assuming that there is a scattering probability

$$\frac{W_{\mathbf{k},\mathbf{k}'} d\mathbf{k}' dt}{(2\pi)^3}, \tag{6.110}$$

which is the probability in an infinitesimal time interval dt that an electron with wave vector $\mathbf{k}$ is scattered into any one of the group of levels with the same spin in an infinitesimal k-space volume element $d\mathbf{k}'$ about $\mathbf{k}'$. These levels are not forbidden by the exclusion principle in the sense that they are not occupied. Because, by definition, $g(\mathbf{k}')$ is the fraction of levels in the volume element $d\mathbf{k}'$ about $\mathbf{k}'$ that are occupied, the fraction that is unoccupied is $1-g(\mathbf{k}')$, and hence, the total probability per unit time for a collision (which is precisely the definition of $\frac{1}{\tau(\mathbf{k})}$), is given by (from Eq. 6.110)

$$\frac{1}{\tau(\mathbf{k})} = \int \frac{d\mathbf{k}'}{(2\pi)^3} W_{\mathbf{k},\mathbf{k}'} [1 - g(\mathbf{k}')]. \tag{6.111}$$

We note that $\tau(\mathbf{k})$ is not the usual relaxation time defined in Eq. (6.72) but explicitly depends on the distribution function $g(\mathbf{k})$.

6.15.2 The Collision Term

The number of electrons scattered out of the volume element $d\mathbf{k}$ about $\mathbf{k}$ in time interval dt is given by

$$dn_{out} = -\left(\frac{dg(\mathbf{k})}{dt}\right)_{out} \frac{d\mathbf{k}}{(2\pi)^3} dt. \tag{6.112}$$

The total number of electrons per unit volume in $d\mathbf{k}$ about $\mathbf{k}$ that suffer a collision in time interval dt is also obtained from Eq. (6.111) as

$$dn_{out} = \frac{dt}{\tau(\mathbf{k})} g(\mathbf{k}) \frac{d\mathbf{k}}{(2\pi)^3}. \tag{6.113}$$

From Eqs. (6.111) through (6.113), we obtain

$$\left(\frac{dg(\mathbf{k})}{dt}\right)_{out} = -\frac{g(\mathbf{k})}{\tau(\mathbf{k})} = -g(\mathbf{k}) \int \frac{d\mathbf{k}'}{(2\pi)^3} W_{\mathbf{k},\mathbf{k}'} [1-g(\mathbf{k}')]. \tag{6.114}$$

The electrons not only scatter out of the level **k**, but are also scattered into it from other levels. By making a similar argument as before (see Problem 6.8), we obtain

$$\left(\frac{dg(\mathbf{k})}{dt}\right)_{in} = \frac{g^0(\mathbf{k})}{\tau(\mathbf{k})} = \left(1 - g(\mathbf{k})\right) \int \frac{d\mathbf{k}'}{(2\pi)^3} W_{\mathbf{k}',\mathbf{k}} g(\mathbf{k}'). \tag{6.115}$$

If we define

$$\left(\frac{dg(\mathbf{k})}{dt}\right)_{coll} = \left(\frac{dg(\mathbf{k})}{dt}\right)_{in} + \left(\frac{dg(\mathbf{k})}{dt}\right)_{out}, \tag{6.116}$$

we obtain from Eqs. (6.114) through (6.116)

$$\left(\frac{dg(\mathbf{k})}{dt}\right)_{coll} = \int \frac{d\mathbf{k}'}{(2\pi)^3} \{g(\mathbf{k}')[1 - g(\mathbf{k})] W_{\mathbf{k}',\mathbf{k}} - g(\mathbf{k})[1 - g(\mathbf{k}')] W_{\mathbf{k},\mathbf{k}'}\}. \tag{6.117}$$

In the relaxation-time approximation, Eq. (6.104) can be rewritten as

$$\left(\frac{dg(\mathbf{k})}{dt}\right)_{coll} = \left[\frac{g^0(\mathbf{k}) - g(\mathbf{k})}{\tau(\mathbf{k})}\right]. \tag{6.118}$$

6.15.3 Impurity Scattering

In any real specimen, at very low temperatures, the main source of collisions would be impurities because the scattering of electrons by the thermal vibrations of the ions and the electron–electron scattering become increasingly weak as the temperature is lowered. This impurity-electron scattering will be elastic as long as the energy gap between the impurity ground state and the excited state is large compared to $k_B T$. We consider the case of elastic scattering by fixed substitutional impurities that are located at random lattice sites **R** in the crystal. The basic assumptions made in this calculation are as follows:

1. There is a relatively small number of impurities in the crystal, such that one can consider the electrons to be interacting with one impurity at a time.
2. There are very few excited impurity ions that lose energy to electrons in collisions.
3. There are very few empty electronic levels that are low enough in energy to receive an electron after it has lost enough energy to excite an impurity ion.
4. The scattering potential U, which describes the interaction between an electron and a single impurity, is spherically symmetrical and weak.
5. The energies of the occupied electronic states $\varepsilon_{\mathbf{k}}$ are isotropic and depend only on the magnitude of **k**.
6. The probability of scattering between two levels **k** and **k**$'$ vanishes unless $k = k'$ and depends on the common value of their energies and on the angle between **k** and **k**$'$.

The rate at which an electron at **k** makes a transition to **k**$'$ is given by Fermi's golden rule (see, for example, Goswami[4]),

$$W_{\mathbf{k},\mathbf{k}'} = \frac{2\pi}{\hbar} \delta(\varepsilon(\mathbf{k}) - \varepsilon(\mathbf{k}')) |<\mathbf{k}|U_t(\mathbf{r})|\mathbf{k}'>|^2, \tag{6.119}$$

where

$$U_t(\mathbf{r}) = \sum_{\mathbf{R}} U(\mathbf{r}-\mathbf{R}). \tag{6.120}$$

Here, $U(\mathbf{r}-\mathbf{R})$ are spherically symmetric potentials centered at sites $\mathbf{R}$. In simple cases, one can write

$$U_t(\mathbf{r}) = n_i U(\mathbf{r}), \tag{6.121}$$

where n_i is the number of impurities per unit volume. In more rigorous derivations, one can consider $U_t(\mathbf{r})$ as the sum of pseudopotentials of impurity atoms added to a perfect crystal or of atoms dislocated from their regular positions in the crystal. Here,

$$<\mathbf{k}|U|\mathbf{k}'> = \int d\mathbf{r}\, \psi^*_{n\mathbf{k}}(\mathbf{r}) U(\mathbf{r}) \psi_{n\mathbf{k}}(\mathbf{r}), \tag{6.122}$$

and the Bloch functions $\psi_{n\mathbf{k}}(\mathbf{r})$ are normalized,

$$\int_{cell} d\mathbf{r} |\psi_{n\mathbf{k}}(\mathbf{r})|^2 = v_{cell}. \tag{6.123}$$

We note from Eq. (6.119) that $W_{\mathbf{k},\mathbf{k}'} = 0$ unless $\varepsilon(\mathbf{k}) = \varepsilon(\mathbf{k}')$. In addition, $W_{\mathbf{k},\mathbf{k}'}$ is independent of the electron distribution function in the independent electron approximation. Further, because $U(\mathbf{r})$ is spherically symmetric, $W_{\mathbf{k},\mathbf{k}'}$ depends only on the angle between $\mathbf{k}$ and $\mathbf{k}'$, and therefore, we obtain

$$W_{\mathbf{k},\mathbf{k}'} = W_{\mathbf{k}',\mathbf{k}}. \tag{6.124}$$

This symmetry is also known as "detailed balancing." From Eqs. (6.117) and (6.124), we obtain

$$\left(\frac{\partial g(\mathbf{k})}{\partial t}\right)_{coll} = -\int \frac{d\mathbf{k}'}{(2\pi)^3} W_{\mathbf{k},\mathbf{k}'} [g(\mathbf{k}) - g(\mathbf{k}')]. \tag{6.125}$$

We further assume that the solution of the Boltzmann equation g has the form

$$g_{\mathbf{k}} = f_{\mathbf{k}} + \mathbf{a}(\varepsilon) \cdot \mathbf{k}, \tag{6.126}$$

where $\mathbf{a}$ is a vector that depends on $\mathbf{k}$ only through its magnitude $\varepsilon(\mathbf{k})$ and $f_{\mathbf{k}}$ is the local equilibrium distribution function (Fermi function). We define the relaxation time through

$$\frac{1}{\tau_\varepsilon}[g(\mathbf{k}) - f(\mathbf{k})] = \int \frac{d\mathbf{k}'}{(2\pi)^3} W_{\mathbf{k},\mathbf{k}'} [g(\mathbf{k}) - g(\mathbf{k}')]. \tag{6.127}$$

Because $W_{\mathbf{k},\mathbf{k}'}$ vanishes unless $\varepsilon(\mathbf{k}) = \varepsilon(\mathbf{k}'), \mathbf{a}(\varepsilon) = \mathbf{a}(\varepsilon')$ and we obtain from Eqs. (6.126) and (6.127),

$$\frac{1}{\tau_\varepsilon}\mathbf{a}(\varepsilon) \cdot \mathbf{k} = \mathbf{a}(\varepsilon) \cdot \int \frac{d\mathbf{k}'}{(2\pi)^3} W_{\mathbf{k},\mathbf{k}'} (\mathbf{k} - \mathbf{k}'). \tag{6.128}$$

From Eq. (6.128) and Eq. (3) of Problem 6.8, we obtain

$$\frac{1}{\tau_\varepsilon} = \int \frac{d\mathbf{k}'}{(2\pi)^3} W_{\mathbf{k},\mathbf{k}'} (1 - \hat{\mathbf{k}} \cdot \hat{\mathbf{k}}'). \tag{6.129}$$

When **k** and **k**′ are on the Fermi surface, $k = k' = k_F$. If we define

$$\mathbf{q} = \mathbf{k} - \mathbf{k}', \tag{6.130}$$

we obtain

$$q^2 = 2k_F^2(1 - \hat{\mathbf{k}} \cdot \hat{\mathbf{k}}') \tag{6.131}$$

or

$$(1 - \hat{\mathbf{k}} \cdot \hat{\mathbf{k}}') = 2\left(\frac{q}{2k_F}\right)^2. \tag{6.132}$$

We can now write from Eqs. (6.119) and (6.130),

$$W_{\mathbf{k},\mathbf{k}'} = \frac{2\pi}{\hbar}\left|\int \frac{d\mathbf{r}}{V} e^{i\mathbf{q}\cdot\mathbf{r}} \sum_{\mathbf{R}} U(\mathbf{r} - \mathbf{R})\right|^2 \delta(\varepsilon_F - \varepsilon(\mathbf{k}')). \tag{6.133}$$

Eq. (6.133) can be rewritten in the alternate form

$$W_{\mathbf{k},\mathbf{k}'} = \frac{2\pi}{\hbar}\frac{1}{V^2}\left|\sum_{\mathbf{R}} e^{i\mathbf{q}\cdot\mathbf{R}} \int d\mathbf{r} e^{i\mathbf{q}\cdot\mathbf{r}}\, U(\mathbf{r})\right|^2 \delta(\varepsilon_F - \varepsilon(\mathbf{k}')). \tag{6.134}$$

We define the static structure factor

$$S(\mathbf{q}) = \frac{1}{N_s}\left|\sum_{\mathbf{R}} e^{i\mathbf{q}\cdot\mathbf{R}}\right|^2, \tag{6.135}$$

and the Fourier transform of the scattering potential $U(\mathbf{r})$

$$U(\mathbf{q}) = \int d\mathbf{r}\, e^{i\mathbf{q}\cdot\mathbf{r}}\, U(\mathbf{r}), \tag{6.136}$$

where N_s is the number of scatterers. From Eqs. (6.130) and (6.134) through (6.136), we obtain

$$W_{\mathbf{k},\mathbf{k}'} = \frac{2\pi N_s}{\hbar V^2} S(\mathbf{q})|U(\mathbf{q})|^2\, \delta(\varepsilon_F - \varepsilon(|\mathbf{k} - \mathbf{q}|)). \tag{6.137}$$

Substituting Eqs. (6.132) and (6.137) in Eq. (6.129), we obtain

$$\frac{1}{\tau_\varepsilon} = \frac{2\pi N_s}{\hbar V}\int \frac{d\mathbf{k}'}{(2\pi)^3} S(\mathbf{q})|U(\mathbf{q})|^2\, \delta(\varepsilon_F - \varepsilon(|\mathbf{k} - \mathbf{q})|))2\left(\frac{q}{2k_F}\right)^2. \tag{6.138}$$

Eq. (6.138) can be rewritten in the alternate form,

$$\frac{1}{\tau_\varepsilon} = \frac{N_s}{\hbar V}\frac{1}{4\pi}\int \frac{q^4}{k_F^2} dq\, S(\mathbf{q})|U(\mathbf{q})|^2 \int_{-1}^{1} d(\cos\theta)\, \delta\left(\varepsilon_F - \varepsilon\left(\sqrt{k_F^2 + q^2 - 2k_F q\cos\theta}\right)\right). \tag{6.139}$$

We also use the results of the angular integration,

$$\int_{-1}^{1} d(\cos\theta)\delta\left(\varepsilon_F - \varepsilon\left(\sqrt{k_F^2 + q^2 - 2k_F q\cos\theta}\right)\right) = \frac{\theta(2k_F - q)}{q\,(\partial\varepsilon/\partial k_F)}, \tag{6.140}$$

where $\theta(x-a)$ is the step function,

$$\theta(x-a) = 1, \quad \text{for } x > a \\ = 0, \quad \text{for } x < a. \tag{6.141}$$

The step function is related to the Dirac delta function,

$$\delta(x-a) = d\theta(x-a)/dx. \tag{6.142}$$

From Eqs. (6.139) through (6.142), we obtain (Problem 6.10)

$$\frac{1}{\tau_\varepsilon} = \frac{1}{4\pi\hbar^2 k_F^2 v_F} \frac{N_s}{V} \int_0^{2k_F} dq\, q^3 S(\mathbf{q}) |U(\mathbf{q})|^2. \tag{6.143}$$

The resistivity (in the relaxation time approximation) is given by

$$\rho = \frac{m}{ne^2 \tau_\varepsilon} = \frac{3\pi N_s}{4e^2 \hbar v_F^2 k_f^4 V} \int_0^{2k_F} dq\, q^3\, S(\mathbf{q}) |U(\mathbf{q})|^2. \tag{6.144}$$

6.16 RESISTIVITY DUE TO SCATTERING BY PHONONS

We consider only one atom per unit cell that is located at the lattice site (its equilibrium position) at zero temperature. At a finite temperature, the deviation of the atom or ion from equilibrium position is (from Chapter 2)

$$\mathbf{u}_l = \mathbf{r} - \mathbf{R}_l - \mathbf{d}. \tag{6.145}$$

The static structure factor is defined in Eq. (1.64). We obtain

$$\sum_l e^{i\mathbf{q}\cdot(\mathbf{R}_l + \mathbf{u}_l)} = \sum_l e^{i\mathbf{q}\cdot\mathbf{R}_l} [1 + i\mathbf{q}\cdot\mathbf{u}_l + \ldots]. \tag{6.146}$$

The structure factor of the unperturbed lattice, $\sum_l e^{i\mathbf{q}\cdot\mathbf{R}_l}$, has no contribution to the resistivity. From Eqs. (2.158) and (2.182), we have the Fourier transform of

$$\hat{\mathbf{u}}_l = \frac{1}{\sqrt{MN}} \sum_{\mathbf{k}} \hat{\epsilon}_{\vec{k}\lambda} \hat{Q}_{\vec{k}\lambda} e^{i\mathbf{k}\cdot\mathbf{R}_l}, \tag{6.147}$$

where

$$\hat{Q}_{\vec{k}\lambda} = \left[\frac{\hbar}{2\omega_{\vec{k}\lambda}}\right]^{\frac{1}{2}} \left(\hat{a}_{\vec{k}\lambda} + \hat{a}^\dagger_{-\vec{k}\lambda}\right). \tag{6.148}$$

We can rewrite from Eqs. (6.147) and (6.148),

$$\mathbf{u}_l = \frac{1}{\sqrt{N}} \sum_{\mathbf{k}\lambda} [\hat{u}_{\mathbf{k}\lambda} e^{i\mathbf{k}\cdot\mathbf{R}_l} + \hat{u}^\dagger_{\mathbf{k}\lambda} e^{-i\mathbf{k}\cdot\mathbf{R}_l}], \tag{6.149}$$

where

$$\hat{u}_{\mathbf{k}\lambda} = \sqrt{\frac{\hbar}{2M\omega_{\overrightarrow{k\lambda}}}}\overrightarrow{\in}_{\mathbf{k}\lambda}\hat{a}_{\mathbf{k}\lambda}. \tag{6.150}$$

From Eqs. (6.146) and (6.149), we obtain

$$\sum_l e^{i\mathbf{q}\cdot(\mathbf{R}_l+\mathbf{u}_l)} = \frac{1}{\sqrt{N}}\sum_{l\mathbf{k}} i[\hat{u}_{\mathbf{k}}\cdot\mathbf{q}e^{i(\mathbf{k}+\mathbf{q})\cdot\mathbf{R}_l} + \hat{u}_{\mathbf{k}}^{\dagger}\cdot\mathbf{q}e^{i(\mathbf{q}-\mathbf{k})\cdot\mathbf{R}_l}], \tag{6.151}$$

where only the longitudinal mode ($\mathbf{q}\|\mathbf{k}$) is retained. Because

$$\sum_l e^{i(\mathbf{q}+\mathbf{k})\cdot\mathbf{R}_l)} = N\delta_{\mathbf{K},\mathbf{q}+\mathbf{k}}, \tag{6.152}$$

Eq. (6.151) can be rewritten in the alternate form

$$\sum_l e^{i\mathbf{q}\cdot(\mathbf{R}_l+\mathbf{u}_l)} = \sqrt{N}\sum_{\mathbf{Kk}} i[\hat{u}_{\mathbf{k}}\cdot\mathbf{q}\delta_{\mathbf{K},\mathbf{q}+\mathbf{k}} + \hat{u}_{\mathbf{k}}^{\dagger}\cdot\mathbf{q}\delta_{\mathbf{K},\mathbf{q}-\mathbf{k}}]. \tag{6.153}$$

In the N-process (normal scattering), $\mathbf{K}=0$ and $\mathbf{q}=\mathbf{k}$; i.e., the wave vectors of electrons and phonons are equal. In the Umklapp process (U-process), $\mathbf{K}\neq 0$. In the following discussion, we will consider the N-process.

From Eqs. (6.135) and (6.153), we obtain

$$\begin{aligned} S(\mathbf{q}) &= \frac{1}{N}\sum_l \quad <|e^{i\mathbf{q}\cdot(\mathbf{R}_l+\mathbf{u}_l)}|^2> \\ &\approx <|\hat{u}_{\mathbf{q}}^{\dagger}\cdot\mathbf{q}|^2>, \end{aligned} \tag{6.154}$$

because $\delta_{0,\mathbf{q}-\mathbf{k}}\neq 0$ can be nonzero only when both **k** and **q** are in the first Brillouin zone.

From Eqs. (6.137) and (6.141), we obtain

$$S(\mathbf{q}) = \frac{\hbar}{2M\omega_{\mathbf{q}}}|\overrightarrow{\in}.\overrightarrow{q}|^2<\hat{a}_{\mathbf{q}}\hat{a}_{\mathbf{q}}^{\dagger} + \hat{a}_{\mathbf{q}}^{\dagger}\hat{a}_{\mathbf{q}}>, \tag{6.155}$$

where $\overrightarrow{\in}$ is the polarization vector of the longitudinal mode. $S(\mathbf{q})$ can be rewritten in the alternate form,

$$S(\mathbf{q}) = \frac{\hbar q^2}{2M\omega_{\mathbf{q}}}(2n_{\mathbf{q}}+1). \tag{6.156}$$

From Eq. (6.144) and (6.156), we obtain

$$\rho = \frac{3\pi}{e^2\hbar v_F^2}\left(\frac{N_s}{V}\right)\frac{1}{4k_F^4}\int_0^{2k_F} dq\, q^3\frac{\hbar q^2}{2M\omega_{\mathbf{q}}}(2n_{\mathbf{q}}+1)|U(q)|^2. \tag{6.157}$$

It can be shown after some algebra that ρ can be written as

$$\rho = \frac{3\pi}{e^2\hbar v_F^2}\left(\frac{N_s}{V}\right)\frac{1}{4k_F^4}\frac{\hbar}{2Mc}\left(\frac{k_BT}{\hbar c}\right)^5\int_0^{2\Theta/T} dz\, z^4\frac{e^z+1}{e^z-1}\left|U\left(\frac{zk_FT}{\Theta}\right)\right|^2. \tag{6.158}$$

In Eq. (6.158), $\Theta = \frac{\hbar k_F c}{k_B}$; $z = \frac{q\Theta}{k_F T}$, Θ is the Debye temperature, c is the velocity of the longitudinal wave, and M is the mass of the ion. At large temperatures, ρ is linear in T, while at very low temperatures, $\rho \sim T^5$.

PROBLEMS

6.1. Derive Eq. (6.10) from Eq. (6.9).

6.2. Two noble gas atoms of charge Ze, each surrounded by Z electrons, are at **0** and **R**, where **R** is so large that there is no interaction between the electronic charges around the two nuclei. The Hamiltonian of the system can be written as

$$H = H_1 + H_2 + H_I, \tag{1}$$

where the interaction Hamiltonian H_I is given by

$$H_I = e^2\left[\frac{Z^2}{R} - \sum_{j=1}^{z}\left(\frac{Z}{|\mathbf{r}_j^{(1)} - \mathbf{R}|} + \frac{Z}{r_j^{(2)}}\right) + \sum_{i,j=1}^{z}\frac{1}{|\mathbf{r}_i^{(1)} - \mathbf{r}_j^{(2)}|}\right]. \tag{2}$$

a. Show that the effect of the first-order perturbation is exponentially small because the ground state of the two atoms is spherically symmetric and the charge distributions hardly overlap because they are far apart.

b. Show that in the second-order perturbation theory, all contributions to the relevant integrals are small unless $|\mathbf{r}_j^{(1)}| \ll R$ and $|\mathbf{r}_j^{(2)} - \mathbf{R}| \ll R$. Thus, one can expand Eq. (2) as

$$H_I \approx -\frac{e^2}{R^3}\sum_{i,j}\left[\frac{3(\mathbf{r}_j^{(1)}\cdot\mathbf{R})(\mathbf{r}_i^{(2)} - \mathbf{R})}{R^2} - \mathbf{r}_j^{(1)}\cdot(\mathbf{r}_i^{(2)} - \mathbf{R})\right]. \tag{3}$$

c. Hence, show that the leading term in the second-order perturbation theory, $\Delta E = <0|H_I|0> + \sum_k \frac{|<0|H_I|k>|^2}{E_0 - E_k}$, varies as $-\frac{1}{R^6}$.

This is essentially the quantum theory of the van der Waals force.

6.3. Derive Eq. (6.17) for the Bulk modulus at equilibrium.

6.4. We have derived in Eq. (6.21)

$$u^{coul}(d) = -\frac{e^2}{d}\left\{1 + \sum_{i\neq 0}\left(\frac{d}{|\mathbf{R}_i + \mathbf{d}|} - \frac{d}{R_i}\right)\right\}, \tag{1}$$

Rewrite the expression as

$$u^{coul}(d) = \frac{e^2}{d}\{dS(0) - dS(\mathbf{d}) - 1\}. \tag{2}$$

$S(\mathbf{d})$ can be written as (because $\mathbf{R}_i$ is a vector)

$$S(\mathbf{d}) = \sum_{i\neq 0}\frac{1}{|\mathbf{d} - \mathbf{R}_i|}, \tag{3}$$

which can be expressed in the integral form

$$S(\mathbf{d}) = \int_0^\infty \frac{2\,dr}{\sqrt{\pi}} \sum_{i\neq 0} e^{-r^2|\mathbf{d}-\mathbf{R}_i|^2}. \tag{4}$$

Eq. (4) can also be rewritten as

$$S(\mathbf{d}) = \int_0^\infty \frac{2\,dr}{\sqrt{\pi}} \int \frac{d\mathbf{k}}{r^3\sqrt{\pi^3}} \sum_{\mathbf{R}_i\neq 0} e^{-k^2/r^2 + 2i\mathbf{k}\cdot(\mathbf{d}-\mathbf{R}_i)}. \tag{5}$$

Using the relation,

$$\sum_i e^{i\mathbf{q}\cdot\mathbf{R}_i} = (2\pi)^3 \frac{1}{\Omega} \sum_{\mathbf{K}} \delta(\mathbf{q}-\mathbf{K}), \tag{6}$$

where $\delta(\mathbf{q}-\mathbf{K})$ is the Dirac delta function and Ω is the volume of the unit cell, show that Eq. (5) can be rewritten as

$$S(\mathbf{d}) = \int_0^\infty \frac{2\,dr}{\sqrt{\pi}} \int \frac{d\mathbf{k}}{r^3\sqrt{\pi^3}} \left[\left\{\sum_{\mathbf{K}} \frac{(2\pi)^3}{\Omega} \delta(2\mathbf{k}-\mathbf{K})\right\} - 1\right] e^{-k^2/r^2 + 2i\mathbf{k}\cdot\mathbf{d}}. \tag{7}$$

Show that Eq. (7) can be simplified as

$$S(\mathbf{d}) = \int_0^\infty \frac{2\,dr}{\sqrt{\pi}} \left[\frac{\pi^3}{r^3\sqrt{\pi^3}} \sum_{\mathbf{K}} \frac{1}{\Omega} e^{-K^2/4r^2 + i\mathbf{K}\cdot\mathbf{d}} - e^{d^2r^2}\right], \tag{8}$$

$$= \sum_{\mathbf{K}} \frac{4\pi}{\Omega K^2} e^{i\mathbf{K}\cdot\mathbf{d}} - \frac{1}{d}. \tag{9}$$

In deriving Eqs. (4) and (9), we have used the identity

$$\frac{1}{d} = \int_0^\infty \frac{2\,dr}{\sqrt{\pi}} e^{-d^2r^2}. \tag{10}$$

Eq. (9) diverges for the $\mathbf{K}=0$ term. $S(0)$ also diverges. However, $S(\mathbf{d}) - S(0)$ does not diverge, which can be shown by rewriting $S(\mathbf{d})$ from Eqs. (4) and (8) as

$$S(\mathbf{d}) = \int_0^G \frac{2\,dr}{\sqrt{\pi}} \left[\frac{\pi^3}{r^3\sqrt{\pi^3}} \sum_{\mathbf{K}\neq 0} \frac{1}{\Omega} e^{-K^2/4r^2 + i\mathbf{K}\cdot\mathbf{d}} - e^{d^2r^2}\right] + \int_G^\infty \frac{2\,dr}{\sqrt{\pi}} \sum_{\mathbf{R}_i\neq 0} e^{-r^2(\mathbf{d}-\mathbf{R}_i)^2}, \tag{11}$$

where G is of the order of a reciprocal lattice vector. Eq. (11) can be rewritten by using the procedure followed earlier as

$$S(\mathbf{d}) = \sum_{\mathbf{K}\neq 0} \frac{4\pi}{\Omega K^2} e^{-K^2/4G^2 + i\mathbf{K}\cdot\mathbf{d}} - \int_0^G \frac{2\,dr}{\sqrt{\pi}} e^{-d^2r^2} + \int_G^\infty \frac{2\,dr}{\sqrt{\pi}} \sum_{\mathbf{R}_i\neq 0} e^{-r^2(\mathbf{d}-\mathbf{R}_i)^2}. \tag{12}$$

Each of the terms in Eq. (12) converges. Similarly, one can show that $S(0)$ also converges. Thus, $dS(\mathbf{d})-dS(0)$ converges, although the individual summations in Eq. (9) do not converge. Hence, we define

$$dS(\mathbf{d}) - dS(0) + 1 = \alpha, \tag{13}$$

where α is the Madelung constant.

6.5. Derive Eq. (6.35) from Eq. (6.34).

6.6. Derive Eq. (6.82) from Eqs. (6.72) through (6.81).

6.7. Derive Eq. (6.93) from Eq. (6.92).

6.8. According to the Sommerfeld expansion rule,

$$\int_{-\infty}^{\infty} d\varepsilon\, H(\varepsilon) f(\varepsilon) = \int_{\infty}^{\mu} H(\varepsilon) d\varepsilon + \sum_{n=1}^{\infty} a_n (k_b T)^{2n} \frac{d^{2n-1}}{d\xi^{2n-1}} H(\varepsilon)\,|_{\varepsilon=\mu}. \tag{1}$$

Here, $\xi(2n) = 2^{2n-1} \frac{\pi^{2n}}{(2n)!} B_n$, $f(\varepsilon)$ is the Fermi distribution function, ξ is the Riemann zeta function, and B_n are the Bernoulli numbers. If

$$H(\varepsilon) = \frac{dF(\varepsilon)}{d\varepsilon}, \tag{2}$$

show (by using the values $\xi(2) = \frac{\pi^2}{6}$ and $B_1 = \frac{1}{6}$) that

$$\int_{-\infty}^{\infty} d\varepsilon\, F(\varepsilon) \left(\frac{-\partial f}{\partial \varepsilon} \right) = F(\mu) + \frac{\pi^2}{6} (k_B T)^2 \left[\frac{\partial^2 F(\varepsilon)}{\partial \varepsilon^2} \right]_{\varepsilon=\mu}. \tag{3}$$

6.9. Derive Eq. (6.115) by making the same arguments made for Eq. (6.114).

6.10. Write the vector $\mathbf{k}'$ in terms of its components

$$\mathbf{k}' = \mathbf{k}'_{\parallel} + \mathbf{k}'_{\perp} = (\hat{\mathbf{k}} \cdot \mathbf{k}')\hat{\mathbf{k}} + \mathbf{k}'_{\perp}. \tag{1}$$

Show that because the collision probability $W_{\mathbf{k},\mathbf{k}'}$ depends only on the angle between $\mathbf{k}$ and $\mathbf{k}'$ (for elastic scattering),

$$\int d\mathbf{k}'\, W_{\mathbf{k},\mathbf{k}'} \mathbf{k}'_{\perp} = 0. \tag{2}$$

Hence, show that because for $k \neq k'$, $W_{\mathbf{k},\mathbf{k}'} = 0$,

$$\int d\mathbf{k}'\, W_{\mathbf{k},\mathbf{k}'} = \mathbf{k} \int d\mathbf{k}'\, W_{\mathbf{k},\mathbf{k}'} (\hat{\mathbf{k}} \cdot \hat{\mathbf{k}}'). \tag{3}$$

6.11. Derive Eq. (6.143) from (6.139) through (6.142).

References

1. Aschroft NW, Mermin ND. *Solid state physics*. New York: Brooks/Cole; 1976.
2. Bernandes N. Theory of Solid Ne, A, Kr and Xe at 0 K. *Phys Rev* 1958;**112**:1534.

3. Chelikowsky JR, Cohen ML. Nonlocal pseudopotential calculation for the electronic structure of eleven diamond and zinc-blende structures. *Phys Rev B* 1976;**14**:556.
4. Goswami A. *Quantum mechnics.* Boston: Wm. C. Brown; 1997.
5. Hirshfelder JO, Curtiss CF, Bird BB. *Molecular theory of gases and liquids.* New York: John Wiley & Sons; 1954.
6. Kittel C. *Introduction to solid state physics.* New York: John Wiley & Sons; 1967.
7. Lennard Jones JE. *Proc R Soc (London)* 1924;**A106**:463.
8. Madelung O. *Introduction to solid state theory.* New York: Springer-Verlag; 1978.
9. Marder MP. *Condensed matter physics.* New York: John Wiley & Sons; 2000.
10. Symon KR. *Mechanics.* Reading, MA: Addison-Wesley; 1971.
11. Ziman JM. *Principles of theory of solids.* Cambridge: Cambridge University Press; 1972.

CHAPTER

Electron–Electron Interaction 7

CHAPTER OUTLINE

7.1 Introduction 199
7.2 Hartree Approximation 200
7.3 Hartree–Fock Approximation 203
7.3.1 General Formulation 203
7.3.2 Hartree–Fock Theory for Jellium 204
7.4 Effect of Screening 207
7.4.1 General Formulation 207
7.4.2 Thomas–Fermi Approximation 208
7.4.3 Lindhard Theory of Screening 209
7.5 Friedel Sum Rule and Oscillations 214
7.6 Frequency and Wave-Number-Dependent Dielectric Constant 217
7.7 Mott Transition 222
7.8 Density Functional Theory 223
7.8.1 General Formulation 223
7.8.2 Local Density Approximation 224
7.9 Fermi Liquid Theory 225
7.9.1 Quasiparticles 225
7.9.2 Energy Functional 227
7.9.3 Fermi Liquid Parameters 230
7.10 Green's Function Method 232
7.10.1 General Formulation 232
7.10.2 Finite-Temperature Green's Function Formalism for Interacting Bloch Electrons 233
7.10.3 Exchange Self-Energy in the Band Model 234
Problems 235
References 241

7.1 INTRODUCTION

In the nearly free electron model, we assumed that the motion of the electrons is independent of each other. The independent electron was assumed to obey the one-electron Schrodinger equation, and the lattice potential was considered to be due to the motion of the electron in a background of static positive charges in the lattice. In this approximation, we neglect the electron–electron interaction as well as the motion of the ions at a finite temperature.

However, in a rigorous solution of the underlying quantum mechanical problem, the Hamiltonian of the solid should include the motion of the interacting N electrons as well as the motion of the nuclei of which the mass is much heavier than that of the electrons. Because the nuclei are much more massive than the electrons and the interaction between the conduction electrons and the lattice waves is a complex problem, as a first step in solving the Hamiltonian, we will apply the Born–Oppenheimer approximation. According to this approximation, the electronic problem is first solved by assuming that the nuclei are static, classical potentials. The motion of the nuclei gradually increases around their equilibrium positions in the lattice when the temperature increases. As we saw in Chapter 2, the quanta of these lattice vibrations are called phonons. In the Born–Oppenheimer approximation, it is assumed that the cloud of negative charge of the electrons follow the nuclei in their motion. Because this motion is followed by charge redistribution, the energies involved in moving the nuclei also depend on the electron energy. This interaction between the conduction electrons and lattice waves is a dynamical problem that can be treated by perturbation theory. In the adiabatic approximation of Born and Oppenheimer, the eigenfunction of the Hamiltonian can be written as a product of the N-electron wave function $\Psi(\mathbf{r}_1 s_1, \mathbf{r}_2 s_2, \ldots, \mathbf{r}_N s_N)$ and $\Phi(\mathbf{u})$, where $\Phi(\mathbf{u})$ satisfies a Schrodinger equation for the wave functions of the ions and $\Psi(\mathbf{r}_1 s_1, \mathbf{r}_2 s_2, \ldots, \mathbf{r}_N s_N)$ is the wave function of the N electrons in a static lattice, frozen with the lth ion at the point $\mathbf{R}_l$ of the Bravais lattice. Here, the instantaneous positions of the N electrons are $\mathbf{r}_i (i = 1, 2, \ldots, N)$ and the corresponding spins are s_i. We will consider the ionic motion in a later section, where we will show that we would need to add an additional term $\varepsilon_e(\mathbf{u})$, which is the total energy of the electrons as a function of the ions, in the Schrodinger equation for the ions.

If we include the electron–electron interactions, the Schrodinger equation for the N-electron wave function can be written as

$$H\Psi = \sum_{i=1}^{N}\left(-\frac{\hbar^2}{2m}\nabla_i^2\Psi - ze^2\sum_l \frac{1}{|\mathbf{r}_i - \mathbf{R}_l|}\Psi\right) + \frac{1}{2}\sum_{i \neq j}\frac{e^2}{|\mathbf{r}_i - \mathbf{r}_j|}\Psi. \tag{7.1}$$

It is impossible to solve Eq. (7.1), even with the fastest computer available to physicists, because the total number of electrons is of the order of 10^{23}. The only possible way to solve Eq. (7.1) is by making drastic approximations. We will start with the simplest approximation, known as the Hartree approximation.

7.2 HARTREE APPROXIMATION

There are two alternate methods to derive the Hartree approximation. We will first consider the simplest method. We proceed with the basic idea of "deriving" from Eq. (7.1) the one-electron Schrodinger equation for the electrons in a lattice potential of the form

$$-\frac{\hbar^2}{2m}\nabla^2\psi(\mathbf{r}) + U(\mathbf{r})\psi(\mathbf{r}) = \varepsilon\psi(\mathbf{r}), \tag{7.2}$$

by obviously making suitable (and drastic) approximations in Eq. (7.1). The potential due to the static ions in the lattice is

$$U_{ion}(\mathbf{r}) = -ze^2\sum_l \frac{1}{|\mathbf{r} - \mathbf{R}_l|}. \tag{7.3}$$

The effective electron–electron potential $U_{ee}(\mathbf{r})$ can be approximated as an electron moving in a field produced by the sum of all other electrons. If the electrons are considered as a smooth distribution of negative charge of charge density $-en(\mathbf{r})$, where $n(\mathbf{r})$ is the number density of electrons,

$$n(\mathbf{r}) = \sum_j |\psi_j(\mathbf{r})|^2, \tag{7.4}$$

the potential energy of the electron in this field would be

$$U_{ee}(\mathbf{r}) = \int d\mathbf{r}' \frac{e^2 n(\mathbf{r}')}{|\mathbf{r}-\mathbf{r}'|}. \tag{7.5}$$

The one-electron Schrodinger equation is obtained by adding $U = U_{ion} + U_{ee}$,

$$-\frac{\hbar^2}{2m}\nabla^2\psi_i(\mathbf{r}) - Ze^2\sum_l \frac{1}{|\mathbf{r}-\mathbf{R}_l|}\psi_i(\mathbf{r}) + \left[e^2\sum_j \int d\mathbf{r}' \, |\psi_j(\mathbf{r}')|^2 \frac{1}{|\mathbf{r}-\mathbf{r}'|}\right]\psi_i(\mathbf{r}) = \varepsilon_i\psi_i(\mathbf{r}). \tag{7.6}$$

Eq. (7.6) is known as the Hartree equation. The Hartree equation can also be derived from Eq. (7.1) by following a variational method as follows. We write the N-electron wave function as

$$\Psi(\mathbf{r}_1 s_1, \mathbf{r}_2 s_2, \ldots, r_N s_N) = \psi_1(\mathbf{r}_1 s_1)\,\psi_2(\mathbf{r}_2 s_2)\ldots\psi_N(\mathbf{r}_N s_N), \tag{7.7}$$

where $\psi_i(\mathbf{r}_i s_i)$ are a set of orthonormal one-electron wave functions. We have to consider Ψ as a trial wave function for a variational calculation and find the equation that is satisfied by the single-particle functions $\psi_i(\mathbf{r}_i, s_i)$ so that the trial function (Eq. 7.7) minimizes

$$<H> = \sum_{s_1}\sum_{s_2}\ldots\sum_{s_N}\int d\mathbf{r}_1 d\mathbf{r}_2 \ldots d\mathbf{r}_N \Psi^* H \Psi. \tag{7.8}$$

Here, Ψ is normalized because the set $\psi_i(\mathbf{r}_i s_i)$ is orthonormal. If we rewrite the Hamiltonian in Eq. (7.1) as

$$H = \sum_i H_i + \frac{1}{2}\sum_{i\neq j} V_{ij}, \tag{7.9}$$

we note that H_i operates only on the coordinate of the ith electron and V_{ij} operates on the two-body coordinates of both i and j. From Eqs. (7.7) through (7.9), we obtain

$$<H> = \sum_i \int \psi_i^* H_i \psi_i d\mathbf{r} + \frac{1}{2}\sum_{i\neq j}\int\int d\mathbf{r}_i\, d\mathbf{r}_j\, \psi_i^*\psi_j^* V_{ij}\psi_i\psi_j. \tag{7.10}$$

If we minimize $<H>$ with respect to variation of ψ_i^*, we obtain

$$\delta <H> = \sum_i \int \delta\psi_i^* \left[H_i + \sum_{j\neq i}\int \psi_j^* V_{ij}\,\psi_j\, d\mathbf{r}_j\right]\psi_i\, d\mathbf{r}_i = 0, \tag{7.11}$$

where the variations $\delta\psi_i^*$ satisfy the equation

$$\int \delta\psi_i^*\, \psi_i\, d\mathbf{r}_i = 0, \tag{7.12}$$

because of their normalization conditions. Eq. (7.12) acts as a constraint on the variation of $<H>$. We follow the method of Lagrangian multipliers by multiplying Eq. (7.12) by a multiplier $\in_i$ and subtract the sum from Eq. (7.11). We obtain

$$\sum_i \int \delta\psi_i^* \left[H_i + \sum_{j \neq i} \int \psi_j^* V_{ij} \psi_j \, d\mathbf{r}_j - \in_i \right] \psi_i d\mathbf{r}_i = 0. \tag{7.13}$$

Because the variations $\delta\psi_i^*$ are independent, the coefficient of each $\delta\psi_i^*$ must vanish; i.e., we must have

$$\left(H_i + \sum_{j \neq i} \int \psi_j^* V_{ij} \psi_j \, d\mathbf{r}_j \right) \psi_i = \in_i \psi_i, \tag{7.14}$$

which, along with Eqs. (7.1) and (7.9), leads to the Hartree equation,

$$-\frac{\hbar^2}{2m} \nabla^2 \psi_i(\mathbf{r}_i) - Ze^2 \sum_l \frac{1}{|\mathbf{r}_i - \mathbf{R}_l|} \psi_i(\mathbf{r}_i) + \left[e^2 \sum_{j \neq i} \int d\mathbf{r}_j \, |\psi_j(\mathbf{r}_j)|^2 \frac{1}{|\mathbf{r}_i - \mathbf{r}_j|} \right] \psi_i(\mathbf{r}_i) = \varepsilon_i \psi_i(\mathbf{r}_i). \tag{7.15}$$

Comparing Eqs. (7.6) and (7.15), we note that the result of the formal derivation of the Hartree equation (Eq. 7.15) is almost identical to that derived earlier (Eq. 7.6) by using a qualitative argument except that each electron interacts not with the full charge density of the system (Eq. 7.6), but with the charge density minus the density of the electron itself. However, in general, the set of equations in Eq. (7.6) is known as the Hartree equations and is easier to solve. In what follows, we will assume that Eq. (7.6) is the Hartree equation even though it fails to represent the way in which a particular configuration of the $N-1$ electrons affects the electron under consideration.

The nonlinear Hartree equations (Eq. 7.6) for one-electron wave functions and energies are usually solved by first guessing a form for U_{ee} (the term in the square bracket in Eq. 7.6) and the one-particle Schrodinger equation is solved. The value of U_{ee} is again computed from the new wave functions $\psi_i(\mathbf{r})$, and the Schrodinger equation is again solved using these values. This iteration procedure is continued until there is no significant change in the potential. In fact, for this reason, the Hartree approximation is also known as the "self-consistent field approximation."

The major failure of the Hartree approximation is that in the formal derivation (by using the variational principle), we started with Eq. (7.7) in which the basic approximation is that the full N-electron wave function Ψ is a product of the one-electron levels. However, this simple assumption is not compatible with the Pauli principle according to which the N-electron wave function is antisymmetric; that is, the sign of Ψ changes when two of its arguments are interchanged, i.e.,

$$\Psi(\mathbf{r}_1 s_1, \ldots, \mathbf{r}_i s_i, \ldots, \mathbf{r}_l s_l, \ldots, \mathbf{r}_N s_N) = -\Psi(\mathbf{r}_1 s_1, \ldots, \mathbf{r}_l s_l, \ldots, \mathbf{r}_i s_i, \ldots, \mathbf{r}_N s_N). \tag{7.16}$$

Eq. (7.7) does not satisfy the constraints imposed by Eq. (7.16) unless $\Psi = 0$ identically.

In addition to the noncompliance of the Pauli principle, the other failure of the Hartree approximation is that it does not include the well-known effects such as "exchange," "correlation," and "screening." In what follows, we will discuss "exchange" by deriving the Hartree–Fock equations and "screening" by discussing "Thomas–Fermi theory" and "Lindhard theory," but we will ignore "correlation" because it is much harder to discuss and is in the realm of "many-body problems," which requires the use of "field theory," a subject beyond the scope of this book.

7.3 HARTREE–FOCK APPROXIMATION

7.3.1 General Formulation

Fock and Slater showed that the simplest way to ensure that the N-electron wave function obeys the Pauli principle is to construct a Slater determinant of orthonormal one-electron wave functions that is antisymmetric:

$$\Psi(\mathbf{r}_1 s_1, r_2 s_2, \ldots, \mathbf{r}_N s_N) = \frac{1}{\sqrt{N!}} \begin{vmatrix} \psi_1(\mathbf{r}_1 s_1) & \psi_1(\mathbf{r}_2 s_2) & \cdots & \psi_1(\mathbf{r}_N s_N) \\ \vdots & \vdots & & \vdots \\ \psi_N(\mathbf{r}_1 s_1) & \psi_N(\mathbf{r}_2 s_2) & \cdots & \psi_N(\mathbf{r}_N s_N) \end{vmatrix}. \tag{7.17}$$

The Slater determinant can be rewritten in the alternate form,

$$\Psi(\mathbf{r}_1 s_1, \mathbf{r}_2 s_2, \ldots, \mathbf{r}_N s_N) = \frac{1}{\sqrt{N!}} \sum_n (-1)^n \psi_{n_1}(\mathbf{r}_1 s_1)\, \psi_{n_2}(\mathbf{r}_2 s_2) \ldots \psi_{n_N}(\mathbf{r}_N s_N), \tag{7.18}$$

where the sum is over all permutations n of $1 \ldots N$. Eq. (7.18) can be rewritten in the alternate form

$$\Psi(\mathbf{r}_1 s_1 \ldots \mathbf{r}_N s_\mathbf{N}) = \frac{1}{\sqrt{N!}} \sum_n (-1)^n \prod_j \psi_{n_j}(\mathbf{r}_j s_j). \tag{7.19}$$

Here, if the Hamiltonian does not involve the spin explicitly, $\psi_i(\mathbf{r}_j s_j)$ has the simple form

$$\psi_i(\mathbf{r}_j s_j) = \phi_i(\mathbf{r}_j)\chi_i(s_j), \tag{7.20}$$

where $\chi_i(s_j) = \delta_{s_j,1}$ for "spin-up" functions and $\chi_i(s_j) = \delta_{s_j,-1}$ for "spin-down" functions.
It can be shown, by using the Slater determinant (Eq. 7.19) for the wave function $\Psi(\mathbf{r}_1 s_1, \mathbf{r}_2 s_2, \ldots, \mathbf{r}_N s_N)$ in Eq. (7.1) and a variational technique, very similar to that used for the Hartree approximation but involving much more tedious algebra, that the Hartree–Fock equation can be rewritten as (Problem 7.1)

$$\begin{aligned} &\frac{-\hbar^2}{2m}\nabla^2\phi_i(\mathbf{r}) - ze^2\sum_l \frac{1}{|\mathbf{r}-\mathbf{R}_l|}\phi_i(\mathbf{r}) + \left[e^2\sum_j \int d\mathbf{r}'\, |\phi_j(\mathbf{r}')|^2 \frac{1}{|\mathbf{r}-\mathbf{r}'|}\right]\phi_i(\mathbf{r}) \\ &-\sum_j \int d\mathbf{r}' \frac{e^2}{|\mathbf{r}-\mathbf{r}'|}\phi_j^*(\mathbf{r}')\phi_i(\mathbf{r}')\phi_j(\mathbf{r})\delta_{\chi_i,\chi_j} = \varepsilon_i\phi_i(\mathbf{r}). \end{aligned} \tag{7.21}$$

Eq. (7.21) can also be expressed as

$$-\frac{\hbar^2}{2m}\nabla^2\phi_i(\mathbf{r}) + U^{ion}(\mathbf{r})\phi_i(\mathbf{r}) + U^{el}(\mathbf{r})\phi_i(\mathbf{r}) + U^{ex}(\mathbf{r})\phi_i(\mathbf{r}) = \varepsilon_i\phi_i(\mathbf{r}). \tag{7.22}$$

The Hartree–Fock equation derived in Eq. (7.21) differs from the Hartree equation derived in Eq. (7.15) in the sense that there is an additional term (the last term) on the left side of the equation known as the exchange term. It can be interpreted by stating that particles 1 and 2 flip places during the course of interaction and the negative sign in the exchange integral is due to the antisymmetry of the wave function. The exchange term is in fact an integral operator of the type $U^{ex}(\mathbf{r})\phi(\mathbf{r}) = \int U(\mathbf{r},\mathbf{r}')\phi(\mathbf{r}')d\mathbf{r}'$. Thus, the Hartree–Fock equations are a complicated set of nonlinear equations that can only be solved numerically.

7.3.2 Hartree–Fock Theory for Jellium

The only case in which the Hartree–Fock equations can be solved exactly is the jellium model where the electrons in a metal are considered to be a set of free electrons of which the solutions are the familiar plane waves. In the jellium model, the positively charged ions are represented by a uniform distribution of positive charge with the same density as the electronic charge.

If there are N electrons in a volume V, the wave function of the free electrons is the plane waves

$$\phi_i(\mathbf{r}) = \frac{e^{i\mathbf{k}_i\cdot\mathbf{r}}}{\sqrt{V}}. \tag{7.23}$$

The kinetic energy term is

$$-\frac{\hbar^2}{2m}\nabla^2\left(\frac{1}{\sqrt{V}}e^{i\mathbf{k}_i\cdot\mathbf{r}}\right) = \frac{1}{\sqrt{V}}\frac{\hbar^2 k_i{}^2}{2m}e^{i\mathbf{k}_i\cdot\mathbf{r}} = \frac{\hbar^2 k_i{}^2}{2m}\phi_i(\mathbf{r}). \tag{7.24}$$

The potential due to the interaction of the ions with the electrons in Eq. (7.22) in the jellium model can be expressed from Eq. (7.21) as

$$U^{ion}(\mathbf{r}) = -\frac{N}{V}\int d\mathbf{r}'\frac{e^2}{|\mathbf{r}-\mathbf{r}'|}. \tag{7.25}$$

The potential due to the Coulomb interaction in Eq. (7.22) in the jellium model can be expressed from Eq. (7.21) as

$$U^{el}(\mathbf{r}) = \int d\mathbf{r}'\sum_{j=1}^{N}\frac{e^2\,|\phi_j(\mathbf{r})|^2}{|\mathbf{r}-\mathbf{r}'|}. \tag{7.26}$$

For plane waves,

$$\sum_{j=1}^{N}|\phi_j(\mathbf{r})|^2 = \frac{N}{V}. \tag{7.27}$$

From Eqs. (7.26) and (7.27),

$$U^{el}(\mathbf{r}) = \frac{N}{V}\int d\mathbf{r}'\frac{e^2}{|\mathbf{r}-\mathbf{r}'|}. \tag{7.28}$$

From Eqs. (7.25) and (7.28), we obtain

$$U^{ion} + U^{el} = 0. \tag{7.29}$$

We note that this type of cancellation is true only for the jellium model. We will now discuss the exchange term that arises due to the antisymmetry of the wave function and distinguishes the Hartree approximation from the Hartree–Fock approximation. The exchange term can be written as

$$U^{ex}\phi_i(\mathbf{r}) = -\sum_{j=1}^{N}\left[\int d\mathbf{r}' \frac{e^2}{|\mathbf{r}-\mathbf{r}'|}\phi_j^*(\mathbf{r}')\phi_i(\mathbf{r}')\right]\phi_j(\mathbf{r})\delta_{\chi_i,\chi_j}. \tag{7.30}$$

From Eqs. (7.23) and (7.30), we obtain

$$U^{ex}\phi_i(\mathbf{r}) = -\frac{e^2}{V^{3/2}}\sum_{j=1}^{N}\left[\int d\mathbf{r}' \frac{e^{i(\mathbf{k}_i-\mathbf{k}_j)\cdot\mathbf{r}'}e^{i\mathbf{k}_j\cdot\mathbf{r}}}{|\mathbf{r}-\mathbf{r}'|}\right]\delta_{\chi_i,\chi_j}. \tag{7.31}$$

Changing the variable of integration $\mathbf{r}'' = \mathbf{r}' - \mathbf{r}$, we can rewrite Eq. (7.31) as

$$U^{ex}\phi_i(\mathbf{r}) = -\frac{e^2}{V^{3/2}}\sum_{j=1}^{N}\left[\int d\mathbf{r}'' \frac{e^{i(\mathbf{k}_i-\mathbf{k}_j)\cdot\mathbf{r}''}}{r''}\right]e^{i\mathbf{k}_i\cdot\mathbf{r}}\,\delta_{\chi_i,\chi_j}. \tag{7.32}$$

Using the relation

$$\phi_i(\mathbf{r}) = \frac{1}{\sqrt{V}}e^{i\mathbf{k}\cdot\mathbf{r}_i}, \tag{7.33}$$

and the Fourier transformation

$$\int d\mathbf{r}'' \frac{e^{i(\mathbf{k}_i-\mathbf{k}_j)\cdot\mathbf{r}''}}{r''} = \frac{4\pi}{|\mathbf{k}_i-\mathbf{k}_j|^2}, \tag{7.34}$$

Eq. (7.32) can be rewritten as

$$U^{ex}\phi_i(\mathbf{r}) = -e^2\phi_i(\mathbf{r})\sum_{j=1}^{N}\frac{1}{V}\frac{4\pi}{|\mathbf{k}_i-\mathbf{k}_j|^2}\delta_{\chi_i,\chi_j}. \tag{7.35}$$

Before we proceed further, it is appropriate to mention that there is a divergence when $\mathbf{k}_i \to \mathbf{k}_j$. This divergence arises due to the unphysical assumption in the Hartree–Fock approximation, which treats only two electrons at a time, that the Coulomb interaction $\frac{e^2}{|\mathbf{r}-\mathbf{r}'|}$ decays very slowly at long distances. In practice, the effective interaction between two electrons falls off much more rapidly due to the effect of screening, i.e., where the rest of the electrons play a role in reducing this interaction. We will treat the effect of screening in a subsequent section.

The sum over $\mathbf{k}_j$ can be converted to an integration over $\mathbf{k}$ space as outlined in Chapter 3:

$$\sum_{\mathbf{k}} F(\mathbf{k}) = \frac{V}{8\pi^3}\int d\mathbf{k}\, F(\mathbf{k}). \tag{7.36}$$

The density of states is usually multiplied by a factor of 2 because each **k** state can have two electrons of opposite spin. However, the Kronecker delta function δ_{χ_i,χ_j} reduces the density of states by a factor of $\frac{1}{2}$. Thus, Eq. (7.35) can be rewritten as

$$U^{ex}\phi_i(\mathbf{r}) = -e^2\phi_i(\mathbf{r})\int^{k_F} \frac{d\mathbf{k}}{8\pi^3}\frac{4\pi}{(k^2+k_i^2-2kk_i\cos\theta)}. \tag{7.37}$$

Because $d\mathbf{k} = k^2dk\,\sin\theta\,d\theta\,d\phi$, integrating Eq. (7.37) over $d\phi$ yields a factor of 2π. Then integrating over $d(\cos\theta)d\theta$ and subsequently over $d\mathbf{k}$ (Problem 7.2), we can easily show that

$$U^{ex}\phi_i(\mathbf{r}) = -e^2\phi_i(\mathbf{r})\frac{1}{2\pi k_i}\left[(k_F^2-k_i^2)\ln\left(\frac{k_F+k_i}{k_F-k_i}\right)+2k_ik_F\right]. \tag{7.38}$$

We define a Lindhard dielectric function as

$$F(x) = \frac{1}{2}+\frac{(1-x^2)}{4x}\ln\left|\frac{1+x}{1-x}\right| \tag{7.39}$$

and express Eq. (7.38) as

$$U^{ex}\phi_i(\mathbf{r}) = -\frac{2e^2}{\pi}k_F F\left(\frac{k_i}{k_F}\right)\phi_i(\mathbf{r}). \tag{7.40}$$

From Eqs. (7.22), (7.24), (7.29), and (7.40), we obtain

$$\varepsilon_i(\mathbf{k}) = \frac{\hbar^2k_i^2}{2m}-\frac{2e^2}{\pi}k_F F\left(\frac{k_i}{k_F}\right). \tag{7.41}$$

We note that the group velocity of the electrons at the Fermi surface, $\frac{1}{\hbar}\left(\frac{\partial\varepsilon}{\partial k_i}\right)\Big|_{k_i=k_F}$, is infinite, an unphysical result due to the fact that screening has been neglected in deriving the Hartree–Fock results.

The total energy of the N-electron system is obtained by summing over $\varepsilon_i(\mathbf{k})$, multiplying the first term in Eq. (7.41) by 2 (because each **k** has two spin levels), and then dividing the second term by 2 (to avoid counting each electron pair twice while summing the interaction energy of an electron with all electrons). We obtain

$$E = 2\sum_{k\le k_F}\frac{\hbar^2k^2}{2m}-\frac{e^2k_F}{\pi}\sum_{k\le k_F}\left[1+\frac{k_F^2-k^2}{2kk_F}\ln\left|\frac{k_F+k}{k_F-k}\right|\right]. \tag{7.42}$$

The first term was evaluated in Chapter 1, and the second term is evaluated in Problem 7.3. We obtain

$$E = N\left[\frac{3}{5}\varepsilon_F-\frac{3}{4}\frac{e^2k_F}{\pi}\right]. \tag{7.43}$$

Eq. (7.43) can also be rewritten as (Problem 7.4)

$$\begin{aligned}\frac{E}{N} &= \frac{e^2}{2a_0}\left[\frac{3}{5}(k_F a_0)^2 - \frac{3}{2\pi}(k_F a_0)\right] \\ &= \left[\frac{2.21}{(r_s/a_0)^2} - \frac{0.916}{(r_s/a_0)}\right] \text{Ry},\end{aligned} \tag{7.44}$$

where a_0 is the Bohr radius and r_s is the radius of the average free space (sphere) for each electron (Chapter 1). It may be noted that the second term in Eq. (7.44) is comparable to the first term for simple metals, and hence, the electron–electron interaction term cannot be ignored in the theory of metals.

However, as we indicated earlier, there is a logarithmic singularity in the expression for energy when $k = k_F$. The singularity arises due to the fact that we have ignored the presence of other electrons when considering the effect of electron–electron interaction in the Hartree–Fock approximation. One has to consider the effect of screening to eliminate this divergence. There are two theories to include the effect of screening. We will first discuss the general effect of screening and then discuss both the Thomas–Fermi and Lindhard theories of screening.

7.4 EFFECT OF SCREENING

7.4.1 General Formulation

It is shown in Problem 7.5 that the Fourier transform of the external potential due to an external positive charge (for example, substituting a Zn ion for a Cu ion), $\phi^e(\mathbf{q})$, is related to the Fourier transform of the total potential (external plus induced charge), $\phi(\mathbf{q})$, by the relation

$$\phi(\mathbf{q}) = \frac{\phi^e(\mathbf{q})}{\in(\mathbf{q})}, \tag{7.45}$$

where $\in(\mathbf{q})$ is the wave-vector-dependent dielectric constant. This type of relation is normally used in dielectric materials where the wave vector dependence is not considered because the fields are uniform.

The induced charge density $\rho^i(\mathbf{r})$ is linearly related to the total potential $\phi(\mathbf{r}) = \phi^e(\mathbf{r}) + \phi^i(\mathbf{r})$, provided $\phi(\mathbf{r})$ is sufficiently weak. Here, $\phi^e(\mathbf{r})$ is the potential due to the extra positively charged particle introduced in the metal, and $\phi^i(\mathbf{r})$ is the potential due to the cloud of screening electrons induced by it. In such a case, one can write Fourier transforms as

$$\rho^i(\mathbf{q}) = \chi(\mathbf{q})\phi(\mathbf{q}). \tag{7.46}$$

We will now derive a relation between $\chi(\mathbf{q})$ and $\in(\mathbf{q})$, the dielectric constant. The Poisson's equations for the particle's charge density $\rho^e(\mathbf{r})$ can be written as

$$-\nabla^2\phi^e(\mathbf{r}) = 4\pi\rho^e(\mathbf{r}), \tag{7.47}$$

the Fourier transform of which is

$$q^2\phi^e(\mathbf{q}) = 4\pi\rho^e(\mathbf{q}). \tag{7.48}$$

If the total charge density $\rho(\mathbf{r}) = \rho^e(\mathbf{r}) + \rho^i(\mathbf{r})$, the Poisson's equation is

$$-\nabla^2\phi(\mathbf{r}) = 4\pi\rho(\mathbf{r}), \tag{7.49}$$

the Fourier transform of which is

$$q^2\phi(q) = 4\pi\rho(\mathbf{q}). \tag{7.50}$$

From Eqs. (7.46), (7.48), and (7.50), we obtain

$$\phi(\mathbf{q}) = \frac{\phi^e(\mathbf{q})}{\left(1 - \frac{4\pi}{q^2}\chi(\mathbf{q})\right)}. \tag{7.51}$$

Comparing Eqs. (7.45) and (7.51), we derive the relation between $\in(\mathbf{q})$ and $\chi(\mathbf{q})$,

$$\in(\mathbf{q}) = 1 - \frac{4\pi}{q^2}\chi(\mathbf{q}). \tag{7.52}$$

7.4.2 Thomas–Fermi Approximation

Thomas and Fermi argued that if the total local potential $\phi(\mathbf{r})$ is slowly varying, the energy of an electron, $\varepsilon(\mathbf{k})$, will be modified from the free electron value. Treating this modification in a classical approximation, we obtain

$$\varepsilon(\mathbf{k}) = \frac{\hbar^2k^2}{2m} - e\phi(\mathbf{r}). \tag{7.53}$$

The induced charge density is

$$\rho^i(\mathbf{r}) = -e[n(\mathbf{r}) - n_0], \tag{7.54}$$

where $-en(\mathbf{r})$ is the charge density when there is a local potential $\phi(\mathbf{r})$, and en_0 is the charge density of the positive background ($\phi(\mathbf{r}) = 0$). We can rewrite Eq. (7.54) in the alternate form

$$\rho^i(\mathbf{r}) = -e[n_0(\mu + e\phi(\mathbf{r})) - n_0(\mu)], \tag{7.55}$$

where

$$n_0(\mu) = \frac{1}{4\pi^3}\int d\mathbf{k}\,\frac{1}{e^{[\beta((\hbar^2k^2/2m)-\mu)]} + 1}. \tag{7.56}$$

Because $\phi(\mathbf{r})$ is small, we make a Taylor expansion of the first term in Eq. (7.55). The leading terms cancel, and Eq. (7.55) can be rewritten in the alternate form

$$\rho^i(\mathbf{r}) \approx -e^2\frac{\partial n_0}{\partial\mu}\phi(\mathbf{r}). \tag{7.57}$$

We make a Fourier transformation of $\rho^i(\mathbf{r})$ and $\phi(\mathbf{r})$, and from Eqs. (7.46) and (7.57), we obtain

$$\chi_{TF}(\mathbf{q}) = -e^2 \frac{\partial n_0}{\partial \mu}. \tag{7.58}$$

From Eqs. (7.52) and (7.58), we obtain

$$\in_{TF}(\mathbf{q}) = 1 + \frac{4\pi e^2}{q^2} \frac{\partial n_0}{\partial \mu}, \tag{7.59}$$

where $\in_{TF}(\mathbf{q})$ is the dielectric constant in the Thomas–Fermi approximation. If we define a screening factor

$$\lambda_{TF}^2 = 4\pi e^2 \frac{\partial n_0}{\partial \mu}, \tag{7.60}$$

Eq. (7.60) can be rewritten as

$$\in_{TF}(\mathbf{q}) = 1 + \frac{\lambda_{TF}^2}{q^2}. \tag{7.61}$$

As an example, if there is a point charge Q at $\mathbf{r}$, it can be shown that (Problem 7.6)

$$\phi(\mathbf{r}) = \phi_e(\mathbf{r}) e^{-\lambda_{TF} r}. \tag{7.62}$$

This is known as the screened Coulomb potential because there is an exponential damping factor λ_{TF}.

7.4.3 Lindhard Theory of Screening

In a quantum mechanical treatment of Eq. (7.53), the one-electron Schrodinger equation can be written as

$$-\frac{\hbar^2}{2m} \nabla^2 \psi_{\mathbf{k}}(\mathbf{r}) - e\phi(\mathbf{r})\psi_{\mathbf{k}}(\mathbf{r}) = \varepsilon_{\mathbf{k}} \psi_{\mathbf{k}}(\mathbf{r}), \tag{7.63}$$

where $-e\phi(\mathbf{r})$ is the potential energy. In the absence of the potential, the one-electron wave function is a plane wave that we denote by $|\mathbf{k}>$. Using first-order perturbation theory, we obtain

$$\psi_{\mathbf{k}}(\mathbf{r}) = |\mathbf{k}> + \sum_{\mathbf{q}} b_{\mathbf{k}+\mathbf{q}} |\mathbf{k}+\mathbf{q}>, \tag{7.64}$$

where

$$b_{\mathbf{k}+\mathbf{q}} = \frac{<\mathbf{k}+\mathbf{q}| - e\phi(\mathbf{r}) |\mathbf{k}>}{\varepsilon^0(\mathbf{k}) - \varepsilon^0(\mathbf{k}+\mathbf{q})} \tag{7.65}$$

and

$$|\mathbf{k}> = e^{i\mathbf{k}\cdot\mathbf{r}}. \tag{7.66}$$

The charge density is

$$\rho(\mathbf{r}) = -e\sum_{\mathbf{k}} f^0(\mathbf{k})\,|\psi_{\mathbf{k}}(\mathbf{r})|^2 = \rho^0(\mathbf{r}) + \rho^i(\mathbf{r}), \tag{7.67}$$

where $f^0(\mathbf{k})$ is the equilibrium Fermi distribution. We can write the charge density as

$$\rho(\mathbf{r}) = -e\sum_{\mathbf{k}} [f^0(\mathbf{k})\,|\psi_{\mathbf{k}}(\mathbf{r})|^2] \tag{7.68}$$

$$= -e\sum_{\mathbf{k}} f^0(\mathbf{k})\left[\left\{e^{-i\mathbf{k}\cdot\mathbf{r}} + \sum_{\mathbf{q}} b^*_{\mathbf{k}+\mathbf{q}} e^{-i(\mathbf{k}+\mathbf{q})\cdot\mathbf{r}}\right\}\left\{e^{i\mathbf{k}\cdot\mathbf{r}} + \sum_{\mathbf{q}} b_{\mathbf{k}+\mathbf{q}} e^{i(\mathbf{k}+\mathbf{q})\cdot\mathbf{r}}\right\}\right]. \tag{7.69}$$

Eq. (7.69) can be approximated as

$$\rho(\mathbf{r}) \approx -e\sum_{\mathbf{k}} f_0(\mathbf{k})\left[1 + \sum_{\mathbf{q}} b_{\mathbf{k}+\mathbf{q}} e^{i\mathbf{q}\cdot\mathbf{r}} + \sum_{\mathbf{q}} b^*_{\mathbf{k}+\mathbf{q}} e^{-i\mathbf{q}\cdot\mathbf{r}}\right] \tag{7.70}$$

or

$$\rho(\mathbf{r}) \approx \rho^0(\mathbf{r}) + \sum_{\mathbf{k}}\sum_{\mathbf{q}}\left\{\frac{e^2\phi(\mathbf{q})[f^0(\mathbf{k}) - f^0(\mathbf{k}+\mathbf{q})]e^{i\mathbf{q}\cdot\mathbf{r}}}{\varepsilon(\mathbf{k}) - \varepsilon(\mathbf{k}+\mathbf{q})}\right\}, \tag{7.71}$$

where we have written $\mathbf{k}$ for $\mathbf{k}-\mathbf{q}$, $\mathbf{k}+\mathbf{q}$ for $\mathbf{k}$, and $e^{i\mathbf{q}\cdot\mathbf{r}}$ for $e^{-i\mathbf{q}\cdot\mathbf{r}}$ as the labels for the second term.

Because

$$\rho(\mathbf{r}) = \rho^0(\mathbf{r}) + \rho^i(\mathbf{r}), \tag{7.72}$$

$$\rho^i(\mathbf{r}) = \sum_{\mathbf{q}} \rho^i(\mathbf{q})e^{i\mathbf{q}\cdot\mathbf{r}}, \tag{7.73}$$

from Eqs. (7.71) through (7.73), we obtain

$$\rho^i(\mathbf{q}) = e^2\sum_{\mathbf{k}}\frac{f^0(\mathbf{k}) - f^0(\mathbf{k}+\mathbf{q})}{\varepsilon(\mathbf{k}) - \varepsilon(\mathbf{k}+\mathbf{q})}\phi(\mathbf{q}). \tag{7.74}$$

From Eqs. (7.46) and (7.74), we obtain

$$\chi_L(\mathbf{q}) = e^2\sum_{\mathbf{k}}\frac{f^0(\mathbf{k}) - f^0(\mathbf{k}+\mathbf{q})}{\varepsilon(\mathbf{k}) - \varepsilon(\mathbf{k}+\mathbf{q})}, \tag{7.75}$$

where $\chi_L(\mathbf{q})$ is the expression derived from Lindhard theory. From Eqs. (7.52) and (7.75), we obtain the expression for the dielectric constant for static screening in the Lindhard theory:

$$\epsilon_L(\mathbf{q}) = 1 - \frac{4\pi e^2}{q^2}\sum_{\mathbf{k}}\frac{f^0(\mathbf{k}) - f^0(\mathbf{k}+\mathbf{q})}{\varepsilon(\mathbf{k}) - \varepsilon(\mathbf{k}+\mathbf{q})}. \tag{7.76}$$

We first consider the dielectric constant when $\mathbf{q} \to 0$. We can write

$$\varepsilon(\mathbf{k}) - \varepsilon(\mathbf{k}+\mathbf{q}) \approx -\mathbf{q} \cdot \nabla_{\mathbf{k}} \varepsilon(\mathbf{k}) \tag{7.77}$$

and

$$f^0(\mathbf{k}) - f^0(\mathbf{k}+\mathbf{q}) \approx -\mathbf{q} \cdot \frac{\partial f^0(\mathbf{k})}{\partial \varepsilon} \nabla_{\mathbf{k}} \varepsilon(\mathbf{k}). \tag{7.78}$$

From Eqs. (7.76) through (7.78), we obtain

$$\in_L(\mathbf{q}) \approx 1 - \frac{4\pi e^2}{q^2} \sum_{\mathbf{k}} \frac{\partial f^0(\mathbf{k})}{\partial \varepsilon}, \tag{7.79}$$

which can be rewritten in the alternate form,

$$\in_L(\mathbf{q}) \approx 1 + \frac{4\pi e^2}{q^2} \int \left(-\frac{\partial f^0(\mathbf{k})}{\partial \varepsilon} \right) g(\varepsilon) d\varepsilon, \tag{7.80}$$

where $g(\varepsilon)$ is the density of states. Because, at very low temperatures,

$$\left(-\frac{\partial f^0(\mathbf{k})}{\partial \varepsilon} \right) \approx \delta(\varepsilon - \varepsilon_F), \tag{7.81}$$

Eq. (7.80) can be rewritten in the alternate form

$$\in_L(\mathbf{q}) \approx 1 + \frac{4\pi^2}{q^2} g(\varepsilon_F). \tag{7.82}$$

From Eq. (7.82), we obtain $\in \to \infty$ as $q \to 0$. Thus, an external field of long wavelength is almost entirely screened due to the flow of electrons. We can show that the same results are obtained through a simpler method by using the Thomas–Fermi approximation.

We can also write

$$\varepsilon(\mathbf{k}) - \varepsilon(\mathbf{k}+\mathbf{q}) \approx -\hbar^2 (\mathbf{k} \cdot \mathbf{q}/m) \tag{7.83}$$

and

$$\sum_{\mathbf{k}} F(\mathbf{k}) = \int \frac{d\mathbf{k}}{4\pi^3} F(\mathbf{k}). \tag{7.84}$$

We note that Eq. (7.84) is different from Eq. (7.34) because we have considered $V = 1$ and multiplied the density of states by a factor of 2 to be able to accommodate two electrons of opposite spin in each $\mathbf{k}$ state.

We obtain from Eqs. (7.71) through (7.73) and (7.83) through (7.84),

$$\rho^i(\mathbf{q}) = -e^2 \int \frac{d\mathbf{k}}{4\pi^3} \frac{f_0(\mathbf{k}) - f_0(\mathbf{k}+\mathbf{q})}{\hbar^2 (\mathbf{k} \cdot \mathbf{q}/m)} \phi(\mathbf{q}). \tag{7.85}$$

Comparing with the definition of $\rho^i(\mathbf{q}) = \chi(\mathbf{q})\phi(\mathbf{q})$, in Eq. (7.46), we obtain

$$\chi_L(\mathbf{q}) = -e^2 \int \frac{d\mathbf{k}}{4\pi^3} \frac{f_0(\mathbf{k}) - f_0(\mathbf{k}+\mathbf{q})}{\hbar^2(\mathbf{k}\cdot\mathbf{q})/m}. \tag{7.86}$$

We note that the equilibrium Fermi function for free electrons is

$$f_0(\mathbf{k}) = \frac{1}{e^{\beta(\hbar^2 k^2/2m - \mu)} + 1}, \tag{7.87}$$

$$f_0(\mathbf{k}) - f_0(\mathbf{k}+\mathbf{q}) \approx \frac{\hbar^2}{m}\mathbf{k}\cdot\mathbf{q}\frac{\partial}{\partial\mu} f_0(\mathbf{k}) + O(q^2). \tag{7.88}$$

From Eqs. (7.86) and (7.88), we obtain

$$\chi_L(\mathbf{q}) \approx -e^2 \frac{\partial n_0}{\partial \mu} = \chi_{TF}(\mathbf{q}), \tag{7.89}$$

which is identical to the Thomas–Fermi result derived in Eq. (7.58).

However, the general result for $\chi_L(\mathbf{q})$ in Lindhard theory can be obtained by integrating Eq. (7.86) at $T = 0$. It can be shown that (Problem 7.7)

$$\chi_L(\mathbf{q}) = -e^2 \left(\frac{mk_F}{\hbar^2\pi^2}\right)\left[\frac{1}{2} + \frac{1-x^2}{4x} \ln\left|\frac{1+x}{1-x}\right|\right], \tag{7.90}$$

where $x = \frac{q}{2k_F}$. We note that when $T << T_F$,

$$\frac{\partial n_0}{\partial \mu} = g(\varepsilon_F) = \frac{mk_F}{\hbar^2\pi^2}. \tag{7.91}$$

From Eqs. (7.90) through (7.91), we obtain

$$\chi_L(\mathbf{q}) = \chi_{TF}(\mathbf{q}) F(x). \tag{7.92}$$

We have shown that the function in the square bracket, which is 1 at $x = 0$ and is equal to the function $F(x)$ appearing in the Hartree–Fock energy, is the Lindhard correction to the Thomas–Fermi result. Thus, the dielectric constant $\epsilon_L = 1 - 4\pi\chi_L/q^2$ is not analytic at $q = 2k_F$. In fact, the static dielectric constant in the Lindhard theory can be written as (Problem 7.8)

$$\epsilon_L(\mathbf{q}) = 1 + \frac{4\pi e^2}{q^2}\frac{mk_F}{\hbar^2\pi^2}\left[\frac{1}{2} + \frac{4k_F^2 - q^2}{8k_F q} \ln\left|\frac{2k_F + q}{2k_F - q}\right|\right]. \tag{7.93}$$

It can also be shown that at large distances and at $T = 0$, the screened potential $\phi(\mathbf{r})$ of a point charge is of the form

$$\phi(\mathbf{r}) \sim \frac{1}{r^3} \cos 2k_F r. \tag{7.94}$$

If we write

$$\in_L(\mathbf{q}) = 1 + \frac{\lambda_L^2}{q^2}, \tag{7.95}$$

where λ_L is the screening parameter, we obtain

$$\lambda_L^2 = 4\pi e^2 \frac{mk_F}{\hbar^2\pi^2}\left[\frac{1}{2} + \frac{4k_F^2 - q^2}{8k_F q}\ln\left|\frac{2k_F + q}{2k_F - q}\right|\right]. \tag{7.96}$$

The variation of λ_L^2 as a function of q is shown in Figure 7.1.

We note from Eqs. (7.60) and (7.96) that as $q \to 0$, $\lambda_L \to \lambda_{TF}$. The effective screening length $1/\lambda_L$ increases as q increases, and it becomes increasingly difficult to screen out the potentials of short wavelength.

When $q = q_c = 2k_F$, there is a logarithmic singularity in the expression for λ_L^2. This singularity arises due to the term $f^0(\mathbf{k}) - f^0(\mathbf{k}+\mathbf{q})$ occurring in the summation over $\mathbf{k}$ in the expression for $\in_L(\mathbf{q})$ in Eq. (7.76). If we consider the values of $\mathbf{k}$, where either $|\mathbf{k}>$ is occupied and $|\mathbf{k}+\mathbf{q}>$ is empty or vice versa, these lie in two regions covering the surface of the Fermi sphere. Figure 7.2 shows these regions of the Fermi sphere.

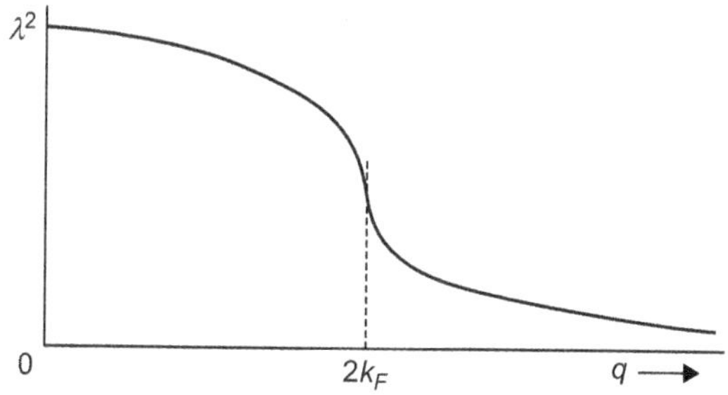

FIGURE 7.1

Variation of Lindhard screening parameter with q.

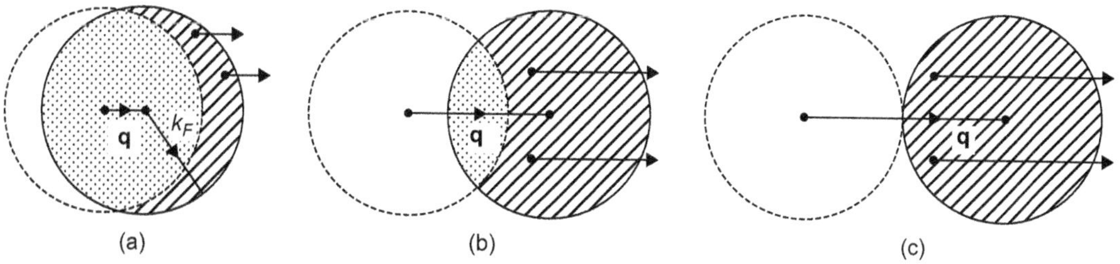

FIGURE 7.2

Contribution to the dielectric constant from different regions of the Fermi sphere.

These regions gradually expand as q increases. Thus, the sum over $\mathbf{k}$ increases until a value of $q = q_c = 2k_F$ is reached and the whole Fermi sphere contributes to the sum. Because there are no additional terms in the sum over $\mathbf{k}$, the functional form of the sum changes. However, each term remains a continuous function of $\mathbf{q}$. The singularity is not serious because the contribution from the last few points on the sphere before reaching $q = q_c = 2k_F$ is small. However, at $q = 2k_F$, $\partial \epsilon_L / \partial q \to \infty$.

If we consider the effect of an extra positive point charge in the jellium model, the "external field" is

$$V(\mathbf{r}) = e^2/r, \tag{7.97}$$

and the Fourier transform in a three-dimensional box is

$$V(\mathbf{q}) = 4\pi e^2/q^2. \tag{7.98}$$

The "screened" potential is

$$U(\mathbf{q}) = V(\mathbf{q})/\epsilon_L(\mathbf{q}) \tag{7.99}$$

and

$$U(\mathbf{r}) = \int U(\mathbf{q}) e^{-i\mathbf{q}\cdot\mathbf{r}} d\mathbf{q}. \tag{7.100}$$

From Eqs. (7.93) and (7.98) through (7.100), we obtain

$$U(\mathbf{r}) = 4\pi e^2 \int \left\{ q^2 + \frac{4me^2 k_F}{\hbar^2 \pi} \left[\frac{1}{2} + \frac{4k_F^2 - q^2}{8k_F q} \ln \left| \frac{2k_F + q}{2k_F - q} \right| \right] \right\}^{-1} e^{-i\mathbf{q}\cdot\mathbf{r}} d\mathbf{q}. \tag{7.101}$$

Due to the singularity at $q = 2k_F$, there will be a special contribution to $U(\mathbf{r})$ that will contain oscillations of wave number $2k_F$. These oscillations are known as Friedel oscillations or Ruderman–Kittel oscillations depending on the context. We will now derive the Friedel oscillations using a much simpler approach.

7.5 FRIEDEL SUM RULE AND OSCILLATIONS

Friedel derived an equivalent formula for the oscillations by considering the effect of a spherically symmetric potential $U(\mathbf{r})$ (impurity) placed in the electron gas. First, he derived a sum rule for the valence difference between the impurity and the solvent metal. Then he derived an expression for the oscillating charge density associated with the singularity in Eq. (7.101). It is well known in quantum mechanics[14] that when an incident plane wave (for convenience, we consider it moving in the z direction) is scattered by a spherically symmetric potential (see Figure 7.3), the wave function at a distance far from the scattering target ($r \to \infty$) can be expressed as

$$\psi_k(r, \theta) = e^{ikz} + \frac{f(\theta) e^{ikr}}{r}. \tag{7.102}$$

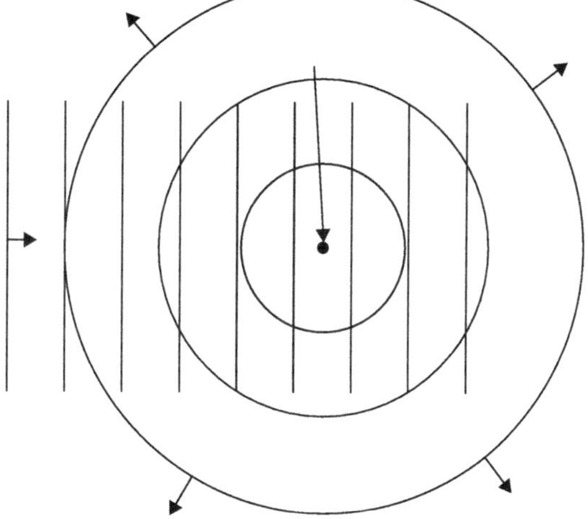

FIGURE 7.3

Plane wave is incident in the *z* direction on spherically symmetric potential at the center. The scattered outgoing wave is spherical.

It has been shown in Liboff—Eqs. (14.11) through (14.13)—that at large distances,

$$\psi_k(r,\theta) = \sum_{l=0}^{\infty} i^l (2l+1) e^{i\delta_l} \frac{1}{kr} \sin(kr - \frac{l\pi}{2} + \delta_l)\, P_l(\cos\theta). \tag{7.103}$$

Here, $P_l(\cos\theta)$ is a Legendre polynomial, and δ_l is a phase shift for the lth partial wave

$$\psi_{k,l}(r,\theta) = A_{k,l} \frac{1}{r} \sin(kr - \frac{l\pi}{2} + \delta_l)\, P_l(\cos\theta). \tag{7.104}$$

Friedel used the boundary condition that $\psi_k = 0$ at $r = \mathfrak{R}$, as if the spherically symmetric potential was placed at the center of a sphere of radius $\mathfrak{R}$. Thus, the values of k that are appropriate are

$$k\mathfrak{R} - \frac{l\pi}{2} + \delta_l = m\pi, \tag{7.105}$$

where m is an integer. We note that the phase shift δ_l would have been zero if the sphere had been empty. In that case, the "allowed" values of k would be

$$k_m = \frac{\left(m + \frac{l}{2}\right)\pi}{\mathfrak{R}}. \tag{7.106}$$

There are many values of l for each value of m, and there are $(2l+1)$ values of ψ for each l. However, the total number of states with $\varepsilon \le \varepsilon_F$, i.e., states of which the wave vector is less than k_F,

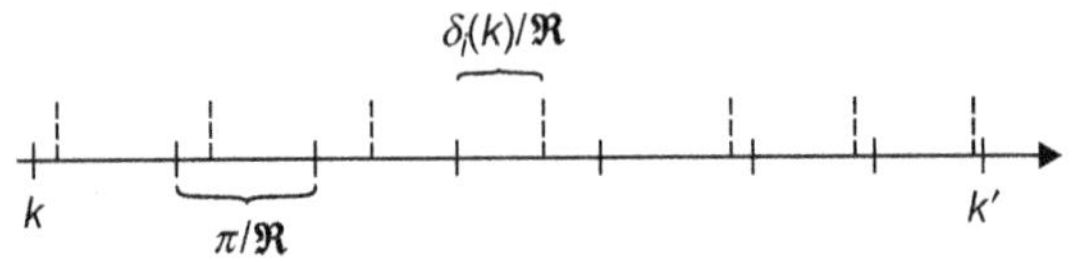

FIGURE 7.4

New "allowed" values between k and k'.

would be the same as that for the cubic lattice considered in Chapter 1. When the impurity with the spherically symmetric potential is at the center of the sphere, $\delta_l \neq 0$ and the "allowed" values of k do not become equal to k_m. They will be shifted by $\delta_l/\Re$. In addition, δ_l varies with k. This is shown schematically in Figure 7.4.

Thus, between k and k', the set will have $[\delta_l(k) - \delta_l(k')]/\pi$ new "allowed" values. Because $\delta_l(k=0) = 0$, counting a factor of 2 for spin states of an electron and the fact that for each l there are $(2l+1)$ states, the total number of extra electrons needed to fill up to the Fermi wave vector k_F is

$$\xi = \frac{2}{\pi}\sum_l (2l+1)\delta_l(k_F). \tag{7.107}$$

Here, ξ is the difference between the valence of the spherically symmetric impurity and the metal, which is the number of electrons necessary to neutralize the charge of the impurity. Eq. (7.107) is the Friedel sum rule, which was derived with two basic assumptions that the charge of a spherically symmetric impurity must be neutralized by extra electrons within a finite distance, and at large distances, k_F is the same as in the crystal without any impurity.

At large distances from the spherical potential at the center of the sphere $\Re$, the change in the electron charge density associated with the phase-shifted waves in Eq. (7.104) is given by $\delta\rho = -e\delta n$, where δn is the change in electron density,

$$\delta\rho = -e\sum_l (2l+1)\int_0^{k_F} \left[\left|\psi_{k,l}(\delta_l)\right|^2 - \left|\psi_{k,l}(0)\right|^2\right]\frac{2\Re}{\pi}dk. \tag{7.108}$$

Here, we have used the expression from Eq. (7.106),

$$k_m - k_{m-1} = \frac{\pi}{\Re}, \tag{7.109}$$

and there are two spin states for each k.

The wave function in Eq. (7.104) has been normalized in a sphere of radius $\Re$ in which the terms going to zero more rapidly at large values of r have been neglected,

$$\int |\psi_{k,l}(r, \delta_l)|^2 r^2\, dr \sin\theta\, d\theta\, d\phi = A_l^2 \int_0^{\Re}\int_0^{\pi}\int_0^{2\pi} dr \sin^2(kr - \tfrac{l\pi}{2} + \delta_l)P_l^2(\cos\theta)\sin\theta\, d\theta\, d\phi. \tag{7.110}$$

Because $\int_{-1}^{1} P_l^2(x)dx = \frac{2}{2l+1}$ and $\sin^2\varphi = \frac{1}{2}(1 - \cos 2\varphi)$, neglecting the oscillatory terms, we obtain

$$A_l \approx \frac{(2l+1)^{1/2}}{(2\pi\mathfrak{R})^{1/2}}. \tag{7.111}$$

From Eqs. (7.104) and (7.111), we obtain

$$\psi_{k,l}(r, \delta_l) \approx \frac{(2l+1)^{1/2}}{(2\pi\mathfrak{R})^{1/2}} \frac{1}{r} \sin\left(kr - \frac{l\pi}{2} + \delta_l\right). \tag{7.112}$$

From Eqs. (7.108) and (7.112), we obtain

$$\delta\rho = -\frac{e}{\pi^2}\sum_l (2l+1) \int_0^{k_F} \left[\sin^2\left(kr - \frac{l\pi}{2} + \delta_l\right) - \sin^2\left(kr - \frac{l\pi}{2}\right)\right] \frac{1}{r^2} dk. \tag{7.113}$$

Using the formula

$$\sin^2(\alpha) - \sin^2(\beta) = \sin(\alpha+\beta)\sin(\alpha-\beta), \tag{7.114}$$

we obtain from Eq. (7.113)

$$\delta\rho = -\frac{e}{2\pi^2}\sum_l (2l+1) \int_0^{k_f} \sin(2kr - l\pi + \delta_l)\sin\delta_l \frac{1}{r^3} d(2kr). \tag{7.115}$$

By integrating Eq. (7.115), we obtain

$$\delta\rho \approx \frac{-e}{2\pi^2}\sum_l (2l+1)(-1)^l \sin\delta_l \frac{\cos(2k_F r + \delta_l)}{r^3}. \tag{7.116}$$

Here, the terms going to zero more rapidly at r (the limit of the integral at $k=0$) have been dropped. Eq. (7.116) gives the oscillating charge density that arises due to the singularity obtained earlier in Eq. (7.101). The oscillating charge is not a negligible effect because it varies as $1/r^3$. Further, the electrons are driven away by the spherically symmetric impurity in some regions because $\delta\rho$ becomes negative.

7.6 FREQUENCY AND WAVE-NUMBER-DEPENDENT DIELECTRIC CONSTANT

We consider a homogeneous electron gas that has a number density $n = N/v$, where v is the volume of the crystal. When an external perturbation $\phi_e(\mathbf{r}, t)$ is applied, fluctuations in the electron gas are induced such that

$$n = n_0 + \delta n. \tag{7.117}$$

The fluctuations give rise to an internal potential $\phi_i(\mathbf{r}, t)$, which essentially describes the screening effect of the electron gas on an electron at $\mathbf{r}$. These fluctuations are obtained by the Poisson equation

$$\nabla^2\phi_i(\mathbf{r}, t) = -4\pi e^2 \delta n(\mathbf{r}, t). \tag{7.118}$$

The total potential acting on the electron is

$$\phi(\mathbf{r}, t) = \phi_i(\mathbf{r}, t) + \phi_e(\mathbf{r}, t). \tag{7.119}$$

The statistical mean values of any physical quantity G are determined through the statistical operator $<\hat{G}> = tr(\hat{\rho}\hat{G})$ (Eq. 3.14) where $\hat{\rho}_k$ is defined in Eq. (3.24) as

$$\hat{\rho}_k = Z_k^{-1} e^{-\beta(\varepsilon(\mathbf{k})-\mu)}, \tag{7.120}$$

where Z_k is the grand partition function defined in Eq. (3.23) as

$$\hat{Z}_k = 1 + e^{-\beta(\varepsilon(\mathbf{k})-\mu)}, \tag{7.121}$$

where the spin has been omitted in both Eqs. (7.120) and (7.121).

It can be easily shown that the equilibrium state of the electron gas (index 0) is given by

$$\hat{H}_0 \,|\mathbf{k}> = \varepsilon(\mathbf{k})\,|\mathbf{k}>, \tag{7.122}$$

and

$$\hat{\rho}_0 \,|\mathbf{k}> = f_0(\mathbf{k})\,|\mathbf{k}>, \tag{7.123}$$

where $f_0(\mathbf{k})$ is the Fermi distribution function. In the presence of the time-dependent external perturbation,

$$\hat{H} = \hat{H}_0 + \hat{\phi}(\mathbf{r}, t) \tag{7.124}$$

and

$$\hat{\rho} = \hat{\rho}_0 + \delta\hat{\rho}(\mathbf{r}, t). \tag{7.125}$$

We have

$$i\hbar\delta\dot{\hat{\rho}}\,|\mathbf{k}> = \left[\hat{H}, \hat{\rho}\right]|\mathbf{k}> = \left[\hat{H}_0, \delta\hat{\rho}\right]|\mathbf{k}> + \left[\hat{\phi}, \hat{\rho}_0\right]|\mathbf{k}>. \tag{7.126}$$

Here, we have neglected the term $[\hat{\phi}, \delta\hat{\rho}]\,|\mathbf{k}>$ in Eq. (7.126). We note that $\hat{H}_0$ commutes with $\hat{\rho}_0$. We can rewrite Eq. (7.126) as

$$i\hbar <\mathbf{k}'|\delta\dot{\hat{\rho}}|\mathbf{k}> = [\varepsilon(\mathbf{k}') - \varepsilon(\mathbf{k})] <\mathbf{k}'|\delta\hat{\rho}|\mathbf{k}> - [f_0(\mathbf{k}') - f_0(\mathbf{k})] <\mathbf{k}'|\hat{\phi}|\mathbf{k}>. \tag{7.127}$$

The $\mathbf{q}$th Fourier component of ϕ is given by

$$\phi_{\mathbf{q}}(t) = \frac{1}{v}\int e^{-i\mathbf{q}\cdot\mathbf{r}}\,\phi(\mathbf{r}, t)\,d\mathbf{r}, \tag{7.128}$$

where

$$\mathbf{q} = \mathbf{k}' - \mathbf{k}. \tag{7.129}$$

We can express

$$\phi(\mathbf{r}, t) = \iint \phi(\mathbf{q}, \omega) e^{i\mathbf{q}\cdot\mathbf{r}} e^{i\omega t} e^{\alpha t}\, d\mathbf{q}\, d\omega, \tag{7.130}$$

where the term $e^{\alpha t}$ has been multiplied so that the oscillation of frequency ω and wave vector $\mathbf{q} \equiv \mathbf{k}' - \mathbf{k}$ grows slowly with a time constant α. In the final result, $\alpha \to 0$. Because $\delta\rho$ has the same time dependence as $\phi(\mathbf{r}, t)$, from Eqs. (7.127) and (7.128), we obtain for each Fourier component (Problem 7.10)

$$[\varepsilon(\mathbf{k}+\mathbf{q}) - \varepsilon(\mathbf{k}) - i\hbar(i\omega + \alpha)] < \mathbf{k}+\mathbf{q} \,|\, \delta\hat{\rho} \,|\, \mathbf{k} > = [f_0(\mathbf{k}+\mathbf{q}) - f_0(\mathbf{k})]\phi_{\mathbf{q}}(t), \tag{7.131}$$

where

$$\phi_{\mathbf{q}}(t) = \phi_{e\mathbf{q}}(t) + \phi_{i\mathbf{q}}(t) = [\phi_e(\mathbf{q}, \omega) + \phi_i(\mathbf{q}, \omega)] e^{(i\omega t + \alpha t)}. \tag{7.132}$$

From Eqs. (7.118), (7.128), and (7.132), we obtain

$$q^2 \phi_{i\mathbf{q}}(t) = 4\pi e^2\, \delta n_{\mathbf{q}}. \tag{7.133}$$

It can be easily shown that the particle concentration $\delta n(\mathbf{r}_0, t)$ is given by

$$\delta n(\mathbf{r}_0, t) = \sum_{\mathbf{q}} e^{i\mathbf{q}\cdot\mathbf{r}_0}\, \delta n_{\mathbf{q}}. \tag{7.134}$$

We can also express from Eqs. (7.121) and (7.125)

$$\begin{aligned} \delta n(\mathbf{r}_0, t) = tr[\delta(\mathbf{r} - \mathbf{r}_0)\delta\rho] &= \sum_{\mathbf{k}\mathbf{k}'} < \mathbf{k} \,|\, \delta(\mathbf{r} - \mathbf{r}_0) > |\, \mathbf{k}' > < \mathbf{k}' \,|\, \delta\rho \,|\, \mathbf{k} > \\ &= \frac{1}{v} \sum_{\mathbf{k}\mathbf{k}'} e^{i(\mathbf{k}-\mathbf{k}')\cdot\mathbf{r}_0} < \mathbf{k}' \,|\, \delta\rho \,|\, \mathbf{k} > \\ &= \frac{1}{v} \sum_{\mathbf{q}} e^{i\mathbf{q}\cdot\mathbf{r}_0} \sum_{\mathbf{k}} < \mathbf{k}+\mathbf{q} \,|\, \delta\rho \,|\, \mathbf{k} >. \end{aligned} \tag{7.135}$$

From Eqs. (7.133) through (7.135), we obtain (taking $v = 1$)

$$\phi_{i\mathbf{q}}(t) = \frac{4\pi e^2}{q^2} \sum_{\mathbf{k}} < \mathbf{k}+\mathbf{q} \,|\, \delta\rho \,|\, \mathbf{k} >. \tag{7.136}$$

From Eqs. (7.131), (7.132), and (7.136), we obtain

$$\begin{aligned} \phi_i(\mathbf{q}, \omega) &= \phi(\mathbf{q}, \omega) - \phi_e(\mathbf{q}, \omega) \\ &= \frac{4\pi e^2}{q^2} \sum_{\mathbf{k}} \frac{f_0(\mathbf{k}+\mathbf{q}) - f_0(\mathbf{k})}{\varepsilon(\mathbf{k}+\mathbf{q}) - \varepsilon(\mathbf{k}) + \hbar\omega - i\hbar\alpha} \phi(\mathbf{q}, \omega). \end{aligned} \tag{7.137}$$

We obtain from Eq. (7.137),

$$\phi_e(\mathbf{q},\omega) = \phi(\mathbf{q},\omega)\left[1 - \frac{4\pi e^2}{q^2}\sum_{\mathbf{k}}\frac{f_0(\mathbf{k}+\mathbf{q}) - f_0(\mathbf{k})}{\varepsilon(\mathbf{k}+\mathbf{q}) - \varepsilon(\mathbf{k}) + \hbar\omega - i\hbar\alpha}\right]. \tag{7.138}$$

If we define the dielectric constant as

$$\in(\mathbf{q},\omega) = \frac{\phi_e(\mathbf{q},\omega)}{\phi(\mathbf{q},\omega)}, \tag{7.139}$$

we obtain from Eqs. (7.138) and (7.139)

$$\in(\mathbf{q},\omega) = 1 - \frac{4\pi e^2}{q^2}\sum_{\mathbf{k}}\frac{f^0(\mathbf{k}+\mathbf{q}) - f^0(\mathbf{k})}{\varepsilon(\mathbf{k}+\mathbf{q}) - \varepsilon(\mathbf{k}) + \hbar\omega - i\hbar\alpha}. \tag{7.140}$$

Eq. (7.140) is the Lindhard equation for the dielectric constant of the electron gas. We note that for the case of static screening ($\omega = 0, \alpha = 0$), Eq. (7.140) becomes identical to Eq. (7.76) derived by using time-independent perturbation theory.

We can write

$$Lim\ \frac{1}{z - i\alpha} = P\left(\frac{1}{z}\right) + i\pi\delta(z). \tag{7.141}$$

Thus, we obtain

$$\in(\mathbf{q},\omega) = \in_1(\mathbf{q},\omega) + i\in_2(\mathbf{q},\omega), \tag{7.142}$$

where

$$\in_1(\mathbf{q},\omega) = 1 - \frac{4\pi e^2}{q^2}P\left[\sum_{\mathbf{k}}\frac{f^0(\mathbf{k}+\mathbf{q}) - f^0(\mathbf{k})}{\varepsilon(\mathbf{k}+\mathbf{q}) - \varepsilon(\mathbf{k}) + \hbar\omega}\right] \tag{7.143}$$

and

$$\in_2(\mathbf{q},\omega) = \frac{4\pi^2 e^2}{q^2}\sum_{\mathbf{k}}[f^0(\mathbf{k}+\mathbf{q}) - f^0(\mathbf{k})]\delta(\varepsilon(\mathbf{k}+\mathbf{q}) - \varepsilon(\mathbf{k}) - \hbar\omega). \tag{7.144}$$

The imaginary part $\in_2(\mathbf{q},\omega)$ is related to the absorption constant of the electron gas and pair excitations are involved in this absorption. The conservation of energy indicated by the $\delta-$ function indicates that in Eq. (7.143), the denominator always has

$$\hbar\omega > [\varepsilon(\mathbf{k}+\mathbf{q}) - \varepsilon(\mathbf{k})] \tag{7.145}$$

for each term. We divide the summation over $\mathbf{k}$ into two parts and introduce new summation indices $\mathbf{k}+\mathbf{q}$ and $-\mathbf{k}$ into the first and second part, respectively. We bring the two parts together and neglect all energy differences between states $\mathbf{k} + \mathbf{q}$ and $\mathbf{k}$.

We obtain

$$\epsilon_1(\mathbf{q}, \omega) = 1 - \left\{ \frac{4\pi e^2}{q^2} \sum_{\mathbf{k}} \frac{f^0(\mathbf{k})}{\varepsilon(\mathbf{k}) - \varepsilon(\mathbf{k}-\mathbf{q}) + \hbar\omega} - \sum_{-\mathbf{k}} \frac{f^0(-\mathbf{k})}{\varepsilon(-\mathbf{k}) - \varepsilon(-\mathbf{k}+\mathbf{q}) - \hbar\omega} \right\}, \tag{7.146}$$

$$= 1 + \frac{4\pi e^2}{q^2} \sum_{\mathbf{k}} \frac{2f^0(\mathbf{k})\{\varepsilon(\mathbf{k}) - \varepsilon(\mathbf{k}-\mathbf{q})\}}{\hbar^2\omega^2}. \tag{7.147}$$

Here, we have rearranged the sum over **k** and neglected the energy differences in the denominator because $\hbar\omega \gg [\varepsilon(\mathbf{k}) - \varepsilon(\mathbf{k}-\mathbf{q})]$ in the plasma oscillation region.

Expanding $\varepsilon(\mathbf{k}) - \varepsilon(\mathbf{k}-\mathbf{q})$ in powers of **q**, the first term vanishes. We obtain

$$\varepsilon(\mathbf{q}) - \varepsilon(\mathbf{k}-\mathbf{q}) \approx -\frac{\hbar^2 q^2}{2m} \tag{7.148}$$

and

$$\sum_{\mathbf{k}} f^0(\mathbf{k}) = n. \tag{7.149}$$

Substituting Eqs. (7.147) and (7.148) in Eq. (7.149), we obtain

$$\epsilon_1(\omega) = 1 - \frac{4\pi n e^2}{m\omega^2} = 1 - \frac{{\omega_p}^2}{\omega^2}, \tag{7.150}$$

where ω_p is the plasma frequency. Eq. (7.150) shows that when $\omega \to \omega_p, \epsilon_1 \to 0$. However, from Eq. (7.138),

$$\phi(\mathbf{q}, \omega) = \frac{\phi_e(\mathbf{q}, \omega)}{\epsilon(\mathbf{q}, \omega)} \to \infty. \tag{7.151}$$

An infinitesimal external field $\phi_e(\mathbf{q}, \omega)$ gives rise to an extremely large effective field. Thus, the system is self-exciting. ω_p is a natural mode of oscillation of the electron gas. It is called a plasma mode.

We note that Eq. (7.140) is also known as Lindhard dielectric susceptibility.

Thus, if $\phi_e(\mathbf{r}, t)$ is the actual external potential, the Fourier transformation is given by

$$\phi_e(\mathbf{r}, t) = \iint \phi_e(\mathbf{q}, \omega) e^{i\mathbf{q}\cdot\mathbf{r}} e^{i\omega t} d\mathbf{q}\, d\omega. \tag{7.152}$$

The effective potential $U(\mathbf{r}, t)$ seen by the electrons is given by

$$U(\mathbf{r}, t) = \iint \frac{\phi_e(\mathbf{q}, \omega)}{\epsilon_L(\mathbf{q}, \omega)} e^{i\mathbf{q}\cdot\mathbf{r}} e^{i\omega t} d\mathbf{q}\, d\omega. \tag{7.153}$$

7.7 MOTT TRANSITION

The Mott transition is an example of a metal-insulator transition that has been actually observed in transition metal oxides. This involves discussion of the Hubbard model, which we will consider later. However, we will consider the basic argument for metal-insulator transition first proposed by Mott. He argued that if one has a collection of hydrogen or other monovalent atoms, it is unlikely that they will form a conductor even if they are far apart from each other. Individual hydrogen atoms are insulators because the electron is bound to the parent atom. If, through some process, an electron is taken out from one of the array of hydrogen (or any other group of monovalent) atoms, it will leave behind a positively charged ion. However, the electron will still be attracted to the positively charged ion and might even form a bound state. Thus, it would not be able to carry current. The solid would still be an insulator. There has to be a certain criterion when such types of insulators would become conductors (metals).

When many electrons are excited, the electron gas would screen the electron–ion interaction. The potential is of the form (because of screening of the ion)

$$V(r) = -\frac{e^2}{r}e^{-\lambda r}. \tag{7.154}$$

If the density of ionized electrons is n, we have derived in Eq. (7.96) an expression for the screening parameter λ_L, which can be written for small q values as

$$\lambda^2 \approx \frac{4e^2 m k_F}{\hbar^2 \pi}. \tag{7.155}$$

Because $k_F = (3\pi^2 n)^{1/3}$,

$$\lambda^2 \approx \frac{4me^2 n^{\frac{1}{3}}}{\hbar^2}. \tag{7.156}$$

The radius of the ground state of a hydrogen atom is given by

$$a_o = \frac{\hbar^2}{me^2}. \tag{7.157}$$

There cannot be any bound state if $1/\lambda < a_0$. The criterion that the ionized atom, of which the potential is expressed in Eq. (7.154), is capable of recapturing the electron that has been removed from it is

$$\lambda < a_0^{-1}. \tag{7.158}$$

From Eqs. (7.156) through (7.158), this condition can be expressed as

$$n^{-1/3} > 4a_0. \tag{7.159}$$

Thus, if the average spacing of the atoms is greater than 4 atomic units, the system would be an insulator. As the atoms come closer, the transition from insulator to metal would be quite sharp. This is known as a Mott transition.

7.8 DENSITY FUNCTIONAL THEORY

7.8.1 General Formulation

Hohenberg and Kohn hypothesized that the electron density $n(\mathbf{r})$ of a many-electron system at point $\mathbf{r}$ has all the information about the many-electron wave function. In this case, $n(\mathbf{r})$ is defined as

$$n(\mathbf{r}) = <\Psi \,|\, \sum_{l=1}^{N} \delta(\mathbf{r}-\mathbf{r}_l) \,|\, \Psi>. \tag{7.160}$$

Here, $\Psi(\mathbf{r}_1, \mathbf{r}_2, ..., \mathbf{r}_N)$ is the many-electron wave function defined earlier both in the context of the Hartree and Hartree–Fock approximations. Eq. (7.160) can be rewritten in the alternate form

$$n(\mathbf{r}) = N \int d\mathbf{r}_1 d\mathbf{r}_2 \ldots d\mathbf{r}_N \, \Psi^*(\mathbf{r}_1, \mathbf{r}_2, ..., \mathbf{r}_N) \, \delta(\mathbf{r}-\mathbf{r}_1) \, \Psi(\mathbf{r}_1, \mathbf{r}_2, ..., \mathbf{r}_N). \tag{7.161}$$

According to Hohenberg and Kohn, the electronic density $n(\mathbf{r})$ determines the external potentials $U(\mathbf{r})$ and the number of electrons that are inside these potentials. If the kinetic energy of the electrons is T, the ground-state energy is ε, the electrons that obey Schrodinger's equation interact via the Coulomb potentials U_{ee}, and the potential due to the ions is U, one can write an expression for the functional for the ground-state energy $\varepsilon[n]$,

$$\varepsilon[n] = T[n] + U[n] + U_{ee}[n]. \tag{7.162}$$

If the functional $\varepsilon[n]$ is found, the ground-state energy density $n(\mathbf{r})$ minimizes it. However, there is a constraint that the total number of electrons is

$$N = \int d\mathbf{r} \, n(\mathbf{r}). \tag{7.163}$$

Because $U[n]$ depends only on the density, one can also write

$$U[n] = \int d\mathbf{r} \, n(\mathbf{r}) U(\mathbf{r}). \tag{7.164}$$

If we write

$$F_{HK}[n] = T[n] + U_{ee}[n], \tag{7.165}$$

it is a universal function for all systems of N particles. We obtain from Eqs. (7.162) and (7.165),

$$\varepsilon[n] = F_{HK}[n] + U[n]. \tag{7.166}$$

We can also define a functional $F[n]$, which is the minimum over all wave functions producing density $n(\mathbf{r})$ of F:

$$F[n] \equiv \min_{\Psi \to n} <\Psi \,|\, T + U_{ee} \,|\, \Psi>. \tag{7.167}$$

In theory, after the universal function $F[n]$ is found, any set of nuclei can be added to the many-particle system through the potential $U(\mathbf{r})$. Then one has to find the function $n(\mathbf{r})$, which minimizes it to solve the Schrodinger equation. In practice, $F[n]$ can be obtained by making appropriate approximations such that the results would agree with the experimental results, and hence, the form of $F[n]$ varies with the nature of the experiment.

7.8.2 Local Density Approximation

Kohn and Sham introduced the local density approximation (LDA), in which they proposed to express the electron density as a function of a set of N single-electron wave functions $\psi_i(\mathbf{r})$ instead of a function of all material properties,

$$n(\mathbf{r}) = \sum_{i=1}^{N} |\psi_i(\mathbf{r})|^2. \tag{7.168}$$

The kinetic energy term of the energy functional is written as

$$T[n] = \sum_i \frac{\hbar^2}{2m} |\nabla\psi_i|^2. \tag{7.169}$$

The rest of the quantities (including exchange) are obtained from the results of a homogeneous electron gas. We derived (the second term in Eq. 7.43) that

$$\varepsilon_{ex} = -\frac{3}{4}N\frac{e^2k_F}{\pi}, \tag{7.170}$$

and we derived in Chapter 1 that

$$n = \frac{k_F^3}{3\pi^2}, \tag{7.171}$$

from which we obtain

$$\varepsilon_{ex}[n] = -\frac{3e^2}{4}\left(\frac{3}{\pi}\right)^{1/3} \int n^{4/3}(\mathbf{r})d\mathbf{r}. \tag{7.172}$$

From Eqs. (7.169) and (7.172) and using standard expressions for $U_{ee}[n]$ and $U[n]$, we obtain

$$\varepsilon[n] = \sum_{i=1}^{N} \frac{\hbar^2}{2m} |\nabla\psi_i|^2 + \int d\mathbf{r} n(\mathbf{r})U(\mathbf{r}) + \frac{1}{2}\int d\mathbf{r}\, d\mathbf{r}' \frac{e^2 n(\mathbf{r})n(\mathbf{r}')}{|\mathbf{r}-\mathbf{r}'|} - \frac{3e^2}{4}\left(\frac{3}{\pi}\right)^{1/3} \int d\mathbf{r} n^{4/3}(\mathbf{r}). \tag{7.173}$$

If we vary Eq. (7.173) with respect to ψ_i^* (because we know the density as a functional of the wave function from Eq. 7.168), we obtain

$$\frac{-\hbar^2}{2m}\nabla^2\psi_i(\mathbf{r}) + \left\{U(\mathbf{r}) + \int d\mathbf{r}' \frac{e^2 n(\mathbf{r}')}{|\mathbf{r}-\mathbf{r}'|} - e^2\left(\frac{3}{\pi}n(\mathbf{r})\right)^{1/3}\right\}\psi_i(\mathbf{r}) = \varepsilon_i\psi_i(\mathbf{r}). \tag{7.174}$$

In general, if one wants to add the correlation terms (any corrections to the exchange terms because of the many-body nature of the N-electron system), one can rewrite Eq. (7.174) in the alternate form

$$\frac{-\hbar^2}{2m}\nabla^2\psi_i(\mathbf{r})+\left\{U(\mathbf{r})+\int d\mathbf{r}'\frac{e^2 n(\mathbf{r}')}{|\mathbf{r}-\mathbf{r}'|}+\frac{\partial\varepsilon_{xc}(n)}{\partial n}\right\}\psi_i(\mathbf{r})=\varepsilon_i\psi_i(\mathbf{r}). \tag{7.175}$$

Here, $\varepsilon_{xc}(n)$ is the exchange-correlation energy of the electron gas.

The approximation of the form (Eq. 7.175) is known as the local density approximation (LDA). A variety of band calculations are based on the density functional theory in general and the LDA in particular. The success of these band calculations primarily depends on the use of appropriate exchange-correlation functions. In fact, the correlation functions are a consequence of many-body theory and can be calculated only by using field-theoretical methods. In band calculations, one can only make intelligent guesses of these functions depending on the available experimental results. However, a variety of first principles calculations of the exchange functionals are available to calculate the total energies of atoms. It has been shown that the recent calculations using these functionals yield more accurate results for the energy of atoms than the results obtained by using the Hartree–Fock approximation. In fact, Kohn was awarded the Nobel prize in chemistry for his pioneering work on density functional theory.

7.9 FERMI LIQUID THEORY

7.9.1 Quasiparticles

The Fermi liquid theory was originally developed by Landau to explain the properties of liquid ^{3}He. However, it has also been applied to the theory of electron–electron interactions in metals. In particular, the Fermi liquid theory explains the success of the independent electron approximation even though the electron–electron interactions are significant. The Fermi liquid theory also indicates how the effect of electron–electron interactions can be qualitatively taken into account in the study of the properties of metals in general and transport properties in particular.

In Landau's argument, if one starts with a set of noninteracting electrons and gradually turns on the interactions between electrons, there would be two types of effects. The energy of the one-electron level would be modified, and it can be treated by the Hartree–Fock and other types of approximations described earlier in this chapter. However, in the Hartree–Fock approximation, the one-electron levels are stationary in spite of the interacting system. In contrast, when the interactions between the electrons are turned on, the electrons would be scattered in and out of the single-electron levels, which are no longer stationary. If the scattering is sufficiently low, one can introduce a relaxation time and try to solve the problem by using a relaxation-time mechanism used to treat transport theories. However, the electron–electron relaxation time is usually much larger than those used in other transport theories. The basic idea of the Landau Fermi Liquid theory is to consider the excitations of the strongly interacting system instead of concentrating on the nature of the ground state. The scattering rate of the fermions is considerably reduced due to the Pauli exclusion principle. In addition, Landau termed the elementary excitations, which act like particles, as quasiparticles. These quasiparticles interact with each other, but not as strongly as the particles from which they are constructed.

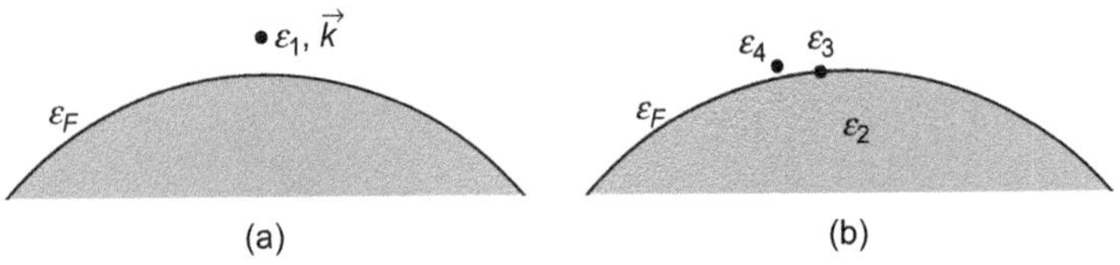

FIGURE 7.5

(a) An excited state $\varepsilon_1 > \varepsilon_F$ of wave vector **k**. (b) $\varepsilon_2 < \varepsilon_F, \varepsilon_3 > \varepsilon_F, \varepsilon_4 > \varepsilon_F$.

To highlight Landau's theory, we consider an N-electron state at $T=0$ filling a Fermi sphere of radius k_F and energy ε_F. In addition, there is a single excited electron (quasiparticle) in a level $\varepsilon_1 > \varepsilon_F$. To be scattered, this electron can interact with an electron of energy ε_2, but because only electron levels of energy less than ε_F are occupied, $\varepsilon_2 < \varepsilon_F$. Both of these electrons can scatter only into energy states ε_3 and ε_4, which are unoccupied; i.e., these states are above the Fermi sphere. Thus, $\varepsilon_3 > \varepsilon_F$ and $\varepsilon_4 > \varepsilon_F$. These are schematically shown in Figure 7.5.

The law of conservation of energy requires that

$$\varepsilon_1 + \varepsilon_2 = \varepsilon_3 + \varepsilon_4. \tag{7.176}$$

Because $\varepsilon_1 > \varepsilon_F$ at $T=0$, some phase space is available for scattering, and $\varepsilon_2, \varepsilon_3$, and ε_4 can vary within a shell of which the thickness is on the order of $|\varepsilon_1 - \varepsilon_F|$ near the Fermi surface. After ε_2 and ε_3 are fixed within the shell $|\varepsilon_1 - \varepsilon_F|$, there cannot be any new choice for the value of ε_4 because of Eq. (7.176). Thus, at $T \to 0$, the scattering rate is given by

$$\frac{1}{\tau}\Big|_{T=0} \approx a_1(\varepsilon_1 - \varepsilon_F)^2. \tag{7.177}$$

If we consider the scattering at a temperature T, there would be partially occupied levels of width $k_B T$ around ε_F. Thus, there will be an additional scattering rate proportional to $(k_B T)^2$. The scattering rate near the Fermi surface at temperature T is given by

$$\frac{1}{\tau} \approx a_1(\varepsilon_1 - \varepsilon_F)^2 + a_2(k_B T)^2. \tag{7.178}$$

These derivations are the same as that for an independent N-electron system. However, we will discuss Landau's formulation of quasiparticles and his arguments that lead to a one-to-one correspondence with the fermions in the N-electron system. We will first derive an approximate formula for the temperature-dependent part of the relaxation time in Eq. (7.178). If $\varepsilon_1 \sim \varepsilon_F$, the first term in Eq. (7.178) can be neglected compared to the second term; τ will essentially be proportional to $1/T^2$. We further assume that the dependence of τ on the electron–electron interaction is through the Fourier transform of the interaction potential, which can be estimated by the Thomas–Fermi screened potential defined in Eq. (7.61). Because $\frac{\partial n_0}{\partial \mu} > 1$, according to Eq. (7.61), the Thomas–Fermi screened potential is everywhere less than $\left(\frac{4\pi e^2}{\lambda_{TF}^2}\right)$. Thus, neglecting the first term in Eq. (7.178), we assume that

$$\frac{1}{\tau} \propto \left(\frac{4\pi e^2}{\lambda_{TF}^2}\right)(k_B T)^2. \tag{7.179}$$

From Eqs. (7.60), (7.91), and (7.179), we obtain

$$\frac{1}{\tau} \propto \left(\frac{\hbar^2\pi^2}{mk_F}\right)^2 (k_BT)^2. \tag{7.180}$$

The proportionality constant 'a' in Eq. (7.180) can be written as a dimensionless number by multiplying Eq. (7.180) by $m^3/\hbar^7$ such that $\frac{1}{\tau}$ has the dimensions of inverse time,

$$\frac{1}{\tau} = \frac{a(k_BT)^2}{\hbar\varepsilon_F}. \tag{7.181}$$

It has been estimated that the range of the dimensionless number 'a' is between 1 and 100.

The empirical formula for scattering in the presence of electron–electron interaction, using the Thomas–Fermi theory for screening, was derived essentially by turning on electron–electron interaction on an N-particle independent electron system. Landau validated these arguments by using the concept of quasiparticles, according to which the low-lying states of the strongly interacting N-electron system evolve in a continuous way such that there is a one-to-one correspondence with the noninteracting electron system. In a noninteracting electron system, if n electrons of wave vectors $\mathbf{k}_1, \ldots, \mathbf{k}_n$ above k_F have been excited from the states $\mathbf{k}'_1, \ldots, \mathbf{k}'_n$ below k_F, the energy of the excited state is

$$E = E_g + \varepsilon(\mathbf{k}_1) + \cdots + \varepsilon(\mathbf{k}_n) - \varepsilon(\mathbf{k}'_1) - \cdots - \varepsilon(\mathbf{k}'_n), \tag{7.182}$$

where E_g is the ground-state energy, and for free electrons, $\varepsilon(\mathbf{k}_i) = \hbar^2 k_i^2/2m$. The quasiparticles are the states of the interacting system where n quasiparticles have been excited out of levels with wave vectors $\mathbf{k}'_1, \ldots, \mathbf{k}'_n$ below k_F and the same number of excited quasiparticles with wave vectors $\mathbf{k}_1, \ldots, \mathbf{k}_n$ are present above k_F. In this case, we have the relation

$$E = E_g + \varepsilon(\mathbf{k}_1) + \cdots + \varepsilon(\mathbf{k}_n) - \varepsilon(\mathbf{k}'_1) - \cdots - \varepsilon(\mathbf{k}'_n), \tag{7.183}$$

but $\varepsilon(\mathbf{k}_i)$ are very different from the free electron energies. In fact, the quasiparticle relation $\varepsilon(\mathbf{k}_i)$ versus $\mathbf{k}_i$ is very difficult to determine because particles with the same spin have an interaction that differs from the interaction between particles with a different spin. However, there is indeed a one-to-one correspondence between the free electrons and quasiparticles. Landau showed, by using Green's function methods, that for all orders of perturbation theory, every interacting Fermi system is normal in the sense that the quasiparticle representation is valid. Landau's argument is very complex, and we will not discuss his theory any further here.

The summary of Landau's model of quasiparticles is that there exists a set of wave functions that are in one-to-one correspondence with low-lying excited states of the noninteracting Fermi gas. They behave under translations like noninteracting particles with index $\mathbf{k}$. Similarly, states that correspond to low-energy holes of the noninteracting electron gas can be constructed. Although these states are not true eigenfunctions, they decay very slowly near the Fermi surface.

7.9.2 Energy Functional

Landau proposed a phenomenological description of a quantum state with a large number of quasiparticles. Let $n_{\mathbf{k}\sigma}$ describe the occupation number of state $\mathbf{k}$. In the ground state, $n_{\mathbf{k}\sigma} = 1$ below the

Fermi surface and $n_{\mathbf{k}\sigma} = 0$ above it. The difference between the occupation of the state $\mathbf{k}\sigma$ and its occupation in the ground state is given by $\delta n_{\mathbf{k}\sigma}$. Suppose that the energy of the quantum state can be expanded in terms of the occupation number $\delta n_{\mathbf{k}\sigma}$ by

$$\varepsilon[\delta n] = \varepsilon_0 + \sum_{\mathbf{k}\sigma} \varepsilon_{\mathbf{k}}^{(0)} \delta n_{\mathbf{k}\sigma} + \frac{1}{2} \sum_{\mathbf{k}\sigma,\mathbf{k}'\sigma'} \delta n_{\mathbf{k}\sigma} f(\mathbf{k}\sigma,\mathbf{k}'\sigma') \delta n_{\mathbf{k}'\sigma'} + \cdots, \tag{7.184}$$

where $\delta n_{\mathbf{k}\sigma}$ is the change in the occupation number of the quasiparticle mode of wave vector $\mathbf{k}$ and spin σ in the ground state, and $\delta n_{\mathbf{k}'\sigma'}$ corresponds to the higher energy level. The function $\varepsilon_{\mathbf{k}}^{(0)}$ is the energy of the noninteracting particles. For a metal, it is the energy of the Bloch electrons. The f-function, which depends on the mutual spins of the two quasiparticles, describes interactions between the quasiparticles. The exchange force favors parallel alignment of spin because the Pauli exclusion principle keeps two electrons with the same spin away from the same point in space. Thus, the correlation effect is enhanced while the mutual electrostatic potential energy is reduced. However, in general, it is difficult to compute the f-function.

At zero temperature, the energy needed to add one quasiparticle $\delta n_{\mathbf{k}\sigma}$ above the Fermi surface is

$$\varepsilon_{\mathbf{k}\sigma} \equiv \varepsilon_{\mathbf{k}\sigma}^{(0)} + \sum_{\mathbf{k}'\sigma'} f(\mathbf{k}\sigma, \mathbf{k}'\sigma') \delta n_{\mathbf{k}'\sigma'}. \tag{7.185}$$

Because the chemical potential μ increases when a quasiparticle is added, the system will be filled with quasiparticles until the cost of adding them rises above μ. Similarly, the quasiparticle states will empty out until $\mu = \varepsilon_{\mathbf{k}\sigma}$. At $T = 0$, $\mu = \varepsilon_F$ and the occupation numbers $n_{\mathbf{k}\sigma}$ are

$$n_{\mathbf{k}\sigma}^{(0)} \equiv \theta(\varepsilon_F - \varepsilon_{\mathbf{k}\sigma}). \tag{7.186}$$

The Fermi wave vector is defined by $\varepsilon_{\mathbf{k}_F} = \varepsilon_F$ for an isotropic system. Because there is a one-to-one correspondence between the quasiparticles and the free particles,

$$n = \frac{N}{V} = \frac{1}{V} \sum_{\mathbf{k}\sigma} f_{\mathbf{k}\sigma} = \frac{k_F^3}{3\pi^2}. \tag{7.187}$$

It can also be shown after some algebra that

$$\delta n_{\mathbf{k}\sigma} = n_{\mathbf{k}\sigma} - n_{\mathbf{k}\sigma}^{(0)} = \frac{1}{e^{\beta(\varepsilon_{\mathbf{k}\sigma} - \mu)} + 1} - n_{\mathbf{k}\sigma}^{(0)}. \tag{7.188}$$

Eq. (7.188) indicates the probability that the quasiparticle states are occupied is given by an implicit expression for the Fermi function with the value of energy given in Eq. (7.185).

The effective mass of the quasiparticles can be obtained from the relation

$$v_F \equiv \left| \frac{1}{\hbar} \frac{\partial \varepsilon_{\mathbf{k}}}{\partial \mathbf{k}} \Big|_{k_F} \right| \equiv \frac{\hbar k_F}{m^*}, \tag{7.189}$$

where $\varepsilon_{\mathbf{k}}$ was defined in Eq. (7.185), from which it becomes obvious that the effective mass m^* is different from the free electron mass.

The effective mass can be computed by calculating the particle current of quasiparticles in two different ways and then by equating them.

Because the quasiparticle states are eigenfunctions of the momentum, the total flow of particles $\mathbf{J}_N$ is given by

$$\mathbf{J}_N = \sum_{\mathbf{k}\sigma} \frac{\hbar \mathbf{k}_\sigma}{m} n_{\mathbf{k}\sigma}, \tag{7.190}$$

which can be rewritten in the alternate form

$$\mathbf{J}_N = \sum_{\mathbf{k}\sigma} \frac{\hbar \mathbf{k}_\sigma}{m} \delta n_{\mathbf{k}\sigma} \tag{7.191}$$

because $n_{\mathbf{k}\sigma}^{(0)}$ is spherically symmetric. The particle current can also be calculated by calculating the change in energy when their momenta change by a small amount. It can be shown that

$$\mathbf{J}_N = \sum_{\mathbf{k}\sigma} \mathbf{v}_{\mathbf{k}\sigma} n_{\mathbf{k}\sigma}, \tag{7.192}$$

where

$$\mathbf{v}_{\mathbf{k}\sigma} = \frac{1}{\hbar} \frac{\partial \varepsilon_{\mathbf{k}\sigma}}{\partial \mathbf{k}}. \tag{7.193}$$

From Eqs. (7.185), (7.192), and (7.193), we obtain

$$\mathbf{J}_N = \frac{1}{\hbar} \sum_{\mathbf{k}\sigma} \frac{\partial \varepsilon_{\mathbf{k}\sigma}^{(0)}}{\partial \mathbf{k}} n_{\mathbf{k}\sigma} + \frac{1}{\hbar} \sum_{\mathbf{k}\mathbf{k}'\sigma\sigma'} n_{\mathbf{k}\sigma} \frac{\partial}{\partial \mathbf{k}} \left[f(\mathbf{k}\sigma, \mathbf{k}'\sigma') \delta n_{\mathbf{k}'\sigma'} \right]. \tag{7.194}$$

We can write Eq. (7.194) by using Eq. (7.188) in the alternate form

$$\mathbf{J}_N = \frac{1}{\hbar} \sum_{\mathbf{k}\sigma} \frac{\partial \varepsilon_{\mathbf{k}\sigma}^{(0)}}{\partial \mathbf{k}} \delta n_{\mathbf{k}\sigma} + \frac{1}{\hbar} \sum_{\mathbf{k}\mathbf{k}'\sigma\sigma'} \left[\delta n_{\mathbf{k}\sigma} + n_{\mathbf{k}\sigma}^{0} \right] \frac{\partial}{\partial \mathbf{k}} \left[f(\mathbf{k}\sigma, \mathbf{k}'\sigma') \delta n_{\mathbf{k}'\sigma'} \right]. \tag{7.195}$$

Eq. (7.195) can be rewritten with the help of Eq. (7.185),

$$\mathbf{J}_N = \sum_{\mathbf{k}\sigma} \mathbf{v}_{\mathbf{k}\sigma} \delta n_{\mathbf{k}\sigma} + \frac{1}{\hbar} \sum_{\mathbf{k}\mathbf{k}'\sigma\sigma'} n_{\mathbf{k}\sigma}^{0} \frac{\partial}{\partial \mathbf{k}} \left[f(\mathbf{k}\sigma, \mathbf{k}'\sigma') \delta n_{\mathbf{k}'\sigma'} \right]. \tag{7.196}$$

Integrating the second term by parts, we obtain

$$\mathbf{J}_N = \sum_{\mathbf{k}\sigma} \mathbf{v}_{\mathbf{k}\sigma} \delta n_{\mathbf{k}\sigma} - \frac{1}{\hbar} \sum_{\mathbf{k}\mathbf{k}'\sigma\sigma'} \frac{\partial n_{\mathbf{k}\sigma}^{(0)}}{\partial \mathbf{k}} f(\mathbf{k}\sigma, \mathbf{k}'\sigma') \delta n_{\mathbf{k}'\sigma'}, \tag{7.197}$$

which can be rewritten in the alternate form by interchanging $\mathbf{k}\sigma$ and $\mathbf{k}'\sigma'$,

$$\mathbf{J}_N = \sum_{\mathbf{k}\sigma} \mathbf{v}_{\mathbf{k}\sigma} \delta n_{\mathbf{k}\sigma} + \sum_{\mathbf{k}\mathbf{k}'\sigma\sigma'} f(\mathbf{k}\sigma, \mathbf{k}'\sigma') \mathbf{v}_{\mathbf{k}'\sigma'} \delta(\varepsilon_{\mathbf{k}'\sigma'}^{(0)} - \varepsilon_F) \delta n_{\mathbf{k}\sigma}. \tag{7.198}$$

Comparing Eqs. (7.190) and (7.198) for the case when one $\delta n_{\mathbf{k}\sigma}$ is nonzero, we obtain

$$\frac{\hbar\mathbf{k}}{m} = \mathbf{v}_{\mathbf{k}} + \sum_{\mathbf{k}'\sigma'} f(\mathbf{k}\sigma, \mathbf{k}'\sigma')\mathbf{v}_{\mathbf{k}'\sigma'}\delta(\varepsilon^{(0)}_{\mathbf{k}'\sigma'} - \varepsilon_F). \tag{7.199}$$

Using the definition of the effective mass from Eq. (7.189), we obtain

$$\frac{\hbar\mathbf{k}}{m} = \frac{\hbar\mathbf{k}}{m^*} + \sum_{\mathbf{k}'\sigma'} f(\mathbf{k}\sigma, \mathbf{k}'\sigma')\frac{\hbar\mathbf{k}'}{m^*}\delta(\varepsilon^{(0)}_{\mathbf{k}'\sigma'} - \varepsilon_F). \tag{7.200}$$

We take dot products on both sides with $\mathbf{k}$ and assume that the Fermi surface is spherical. In addition, because the Fermi liquid theory is valid near the Fermi surface, the magnitudes of both $\mathbf{k}$ and $\mathbf{k}'$ are approximately k_F. Thus, we can rewrite Eq. (7.200) in the alternate form

$$\frac{m^*}{m} \approx 1 + \sum_{\mathbf{k}'\sigma'} f(\mathbf{k}\sigma, \mathbf{k}'\sigma')\frac{\mathbf{k}\cdot\mathbf{k}'}{k_F^2}\delta(\varepsilon^{(0)}_{\mathbf{k}'\sigma'} - \varepsilon_F). \tag{7.201}$$

The summation over $\mathbf{k}'$ can be converted as an integration, and we can write

$$\frac{m}{m^*} = 1 + V\int dk'k'^2 g(\mathbf{k}')\delta(\varepsilon^{(0)}_{\mathbf{k}'\sigma'} - \varepsilon_F)\int_0^{\theta}\int_0^{2\pi} d\theta \sin\theta\, d\phi\, f(\mathbf{k}\sigma, \mathbf{k}'\sigma')\cos\theta, \tag{7.202}$$

where θ is the angle between $\hat{\mathbf{k}}$ and $\hat{\mathbf{k}}'$. Because the density of states $g(\varepsilon)$ normally includes the angular integral, dividing $g(\varepsilon)$ by 4π, we obtain

$$\int g(\mathbf{k}')k'^2\delta(\varepsilon^0_{\mathbf{k}'} - \varepsilon_F)dk' = \frac{1}{4\pi}\int g(\varepsilon^{(0)}_{\mathbf{k}'})\delta(\varepsilon^{(0)}_{\mathbf{k}'} - \varepsilon_F)d\varepsilon = \frac{g(\varepsilon_F)}{4\pi}. \tag{7.203}$$

Further, because $f(\mathbf{k}\sigma, \mathbf{k}'\sigma')$ depends on the angle between $\mathbf{k}$ and $\mathbf{k}'$ near the Fermi surface, we obtain from Eqs. (7.202) and (7.203)

$$\frac{m^*}{m} = 1 + \frac{1}{2}Vg(\varepsilon_F)\int_{-1}^{1} d(\cos\theta)\, f(\mathbf{k}\sigma,\mathbf{k}'\sigma')\cos\theta\cdot \tag{7.204}$$

We note that the effective mass depends on the weighted average of the interactions over the Fermi surface, but the latter is not known for strongly correlated systems. We will now derive an expression for the effective mass from Eq. (7.204) and the Fermi liquid parameters.

7.9.3 Fermi Liquid Parameters

If we assume that the particles with the same spin have the possibility of a different interaction compared to the particles with different spin, we define

$$f(\mathbf{k}\uparrow, \mathbf{k}'\uparrow) = f(\mathbf{k}\downarrow, \mathbf{k}'\downarrow) = f^s_{\mathbf{k}\mathbf{k}'} + f^a_{\mathbf{k}\mathbf{k}'}, \tag{7.205}$$

$$f(\mathbf{k}\uparrow, \mathbf{k}'\downarrow) = f(\mathbf{k}\downarrow, \mathbf{k}'\uparrow) = f^s_{\mathbf{k}\mathbf{k}'} - f^a_{\mathbf{k}\mathbf{k}'}, \tag{7.206}$$

where the symmetric and antisymmetric functions are represented by s and a. These formulae may be inverted,

$$f^s_{\mathbf{kk}'} = \sum_{l=0}^{\infty} f^s_l P_l(\cos\theta) \tag{7.207}$$

and

$$f^a_{\mathbf{kk}'} = \sum_{l=0}^{\infty} f^a_l P_l(\cos\theta), \tag{7.208}$$

where $P_l(\cos\theta)$ is the Legendre polynomial. We can invert Eqs. (7.207) and (7.208) to write

$$f^s_l = \frac{2l+1}{2} \int_{-1}^{1} d(\cos\theta) P_l(\cos\theta) f^s_{\mathbf{kk}'} \tag{7.209}$$

and

$$f^a_l = \frac{2l+1}{2} \int_{-1}^{1} d(\cos\theta) P_l(\cos\theta) f^a_{\mathbf{kk}'}. \tag{7.210}$$

We define the dimensionless Fermi liquid parameters

$$F^s_l \equiv Vg(\varepsilon_F) f^s_l \tag{7.211}$$

and

$$F^a_l \equiv Vg(\varepsilon_F) f^a_l. \tag{7.212}$$

Here, $Vg(\varepsilon_F)$ is the density of energy states at the Fermi surface. F^s_l and F^a_l are known as Fermi liquid parameters. For example, it can be shown that the effective mass defined in Eq. (7.204) and the Fermi liquid parameters defined in Eqs. (7.205) and (7.206) have the form

$$\frac{m^*}{m} = 1 + \frac{1}{2} Vg(\varepsilon_F) \int_{-1}^{1} d(\cos\theta)\, P_1(\cos\theta) \left[\frac{f(\mathbf{k}\uparrow, \mathbf{k}'\uparrow) + f(\mathbf{k}\uparrow, \mathbf{k}'\downarrow)}{2} \right]. \tag{7.213}$$

From Eqs. (7.209), (7.211), and (7.212), we obtain

$$\frac{m^*}{m} = 1 + \frac{1}{3} F^s_1. \tag{7.214}$$

The effective mass equation is widely used in Fermi liquid theory, especially in highly correlated systems.

7.10 GREEN'S FUNCTION METHOD

7.10.1 General Formulation

The Green's function method has been widely used in solving many-body problems that go beyond the electron–electron interactions. It starts with the idea that amplitude for finding a particle at site $|\mathbf{R}>$ at time t, when it was at site $|0>$ at time 0, is given by

$$<\mathbf{R}|\hat{G}(t)|0>=<\mathbf{R}|e^{-i\hat{H}t/\hbar}|0>. \tag{7.215}$$

The Fourier transformation of $\hat{G}(t)$ is given by

$$\hat{G}(\xi) = \frac{1}{i\hbar}\int_0^\infty dt\, e^{i\xi t/\hbar}\hat{G}(t). \tag{7.216}$$

Eq. (7.216) converges if ξ has a positive imaginary part, and hence, ξ varies in a complex plane. In fact, the physical significance of the Green's functions depends on the complex part of the energy. From Eqs. (7.125) and (7.126), we obtain

$$\hat{G}(\xi) = (\xi - \hat{H})^{-1}. \tag{7.217}$$

If the Hamiltonian is perturbed, we can write

$$\hat{H} = \hat{H}_0 + \hat{H}_1. \tag{7.218}$$

If we define

$$\hat{G}_0 = (\xi - \hat{H}_0)^{-1}, \tag{7.219}$$

we obtain from Eqs. (7.217) and (7.219)

$$\hat{G} = (\xi - \hat{H}_0 - \hat{H}_1)^{-1} \tag{7.220}$$

$$= ((\varepsilon - \hat{H}_0)(1 - (\xi - \hat{H}_0)^{-1}\hat{H}_1))^{-1} \tag{7.221}$$

$$= (1 - \hat{G}_0\hat{H}_1)^{-1}\hat{G}_0 \tag{7.222}$$

$$= \hat{G}_0 + \hat{G}_0\hat{H}_1\hat{G}_0 + \hat{G}_0\hat{H}_1\hat{G}_0\hat{H}_1\hat{G}_0 + \cdots \tag{7.223}$$

$$= \hat{G}_0 + \hat{G}_0\hat{H}_1\hat{G} = \hat{G}\hat{H}_1\hat{G}_0 \equiv \hat{G}_0 + \hat{G}_0\hat{T}\hat{G}_0\cdot \tag{7.224}$$

Here, the $\hat{T}$ matrix is the operator that satisfies the equality in Eq. (7.224).

7.10.2 Finite-Temperature Green's Function Formalism for Interacting Bloch Electrons

In the finite-temperature Green's function formalism for an interacting system of Bloch electrons in the presence of a periodic potential $V(\mathbf{r})$, the one-particle propagator G satisfies the equation

$$(\xi_l - H)G(\mathbf{r},\mathbf{r}',\xi_l) + \int d\mathbf{r}''\Sigma(\mathbf{r},\mathbf{r}',\xi_l)\,G(\mathbf{r}'',\mathbf{r}',\xi_l) = \delta(\mathbf{r}-\mathbf{r}'), \tag{7.225}$$

where Σ is the exact proper self-energy operator, ξ_l is the complex energy,

$$\xi_l = (2l+1)i\pi/\beta + \mu, \tag{7.226}$$

μ is the chemical potential, $l = 0, \pm 1, \pm 2, \ldots$, and $\hat{H}$ is the one-particle Hamiltonian,

$$\hat{H} = \frac{p^2}{2m} + V(\mathbf{r}). \tag{7.227}$$

Both G and Σ have the symmetry

$$G(\mathbf{r}+\mathbf{R},\mathbf{r}'+\mathbf{R},\xi_l) = G(\mathbf{r},\mathbf{r}',\xi_l) \tag{7.228}$$

and

$$\Sigma(\mathbf{r}+\mathbf{R},\mathbf{r}'+\mathbf{R},\xi_l) = \Sigma(\mathbf{r},\mathbf{r}',\xi_l). \tag{7.229}$$

We can write the equation of motion in the Bloch representation, i.e., in terms of the basis functions,

$$\phi_{n\mathbf{k}\rho}(\mathbf{r}) = e^{i\mathbf{k}\cdot\mathbf{r}}U_{n\mathbf{k}\rho}(\mathbf{r}), \tag{7.230}$$

where $U_{n\mathbf{k}\rho}(\mathbf{r})$ is a periodic two-component function, n is the band index, $\mathbf{k}$ is the reduced wave vector, and ρ is the spin index. Using the Bloch representation, we can rewrite Eq. (7.225) as

$$\begin{aligned}
&\sum_{n'',\rho'',\mathbf{k}',\mathbf{k}''} \int d\mathbf{r}\,d\mathbf{r}'d\mathbf{r}''e^{-i\mathbf{k}\cdot\mathbf{r}}U^*_{n\mathbf{k}\rho}(\mathbf{r})\left(\xi_l - \frac{p^2}{2m} - V(\mathbf{r})\right)e^{i\mathbf{k}''\cdot(\mathbf{r}-\mathbf{r}'')}U_{n''\mathbf{k}''\rho''}(\mathbf{r})U^*_{n''\mathbf{k}''\rho''}(\mathbf{r}'') \\
&\qquad \times G(\mathbf{r}'',\,\mathbf{r}',\,\xi_l)U_{n'\mathbf{k}'\rho'}(\mathbf{r}')e^{i\mathbf{k}'\cdot\mathbf{r}'} \\
&+ \sum_{n'',\rho'',\mathbf{k}',\mathbf{k}''} \int d\mathbf{r}d\mathbf{r}'d\mathbf{r}''d\mathbf{r}'''e^{-i\mathbf{k}\cdot\mathbf{r}}U^*_{n\mathbf{k}\rho}(\mathbf{r})\Sigma(\mathbf{r},\mathbf{r}'',\xi_l)e^{i\mathbf{k}''\cdot(\mathbf{r}''-\mathbf{r}''')}U_{n''\mathbf{k}''\rho''}(\mathbf{r}'')U^*_{n''\mathbf{k}''\rho''}(\mathbf{r}''') \\
&\qquad \times G(\mathbf{r}''',\mathbf{r}',\xi_l)U_{n'\mathbf{k}'\rho'}(\mathbf{r}')e^{i\mathbf{k}'\cdot\mathbf{r}'} = \delta_{nn'}\delta_{\rho\rho'}.
\end{aligned} \tag{7.231}$$

Eq. (7.231) can be rewritten in the alternate form (Problem 7.11)

$$\sum_{n'',\rho''} [\xi_l - H(\mathbf{k}',\xi_l)]_{n\mathbf{k}\rho,n''\mathbf{k}\rho''}G_{n''\mathbf{k}\rho'',n'\mathbf{k}\rho'}(\mathbf{k}',\xi_l)\,|_{\mathbf{k}'=\mathbf{k}} = \delta_{nn'}\delta_{\rho\rho'}, \tag{7.232}$$

where

$$H(\mathbf{k}', \xi_l) = \frac{1}{2m}(\mathbf{p} + \hbar\mathbf{k}')^2 + V(\mathbf{r}) + \Sigma(\mathbf{k}', \xi_l), \tag{7.233}$$

$$\Sigma_{n\mathbf{k}\rho,n''\mathbf{k}\rho''}(\mathbf{k}', \xi_l) = \int d\mathbf{r} d\mathbf{r}'\, U^*_{n\mathbf{k}\rho}(\mathbf{r}) e^{-i\mathbf{k}'\cdot(\mathbf{r}-\mathbf{r}')} \Sigma(\mathbf{r}, \mathbf{r}', \xi_l)\, U_{n''\mathbf{k}\rho''}(\mathbf{r}'), \tag{7.234}$$

and

$$G_{n''\mathbf{k}\rho'',n'\mathbf{k}\rho'}(\mathbf{k}', \xi_l) = \int d\mathbf{r} d\mathbf{r}'\, U^*_{n''\mathbf{k}\rho''}(\mathbf{r}) G(\mathbf{r}, \mathbf{r}', \xi_l) e^{-i\mathbf{k}'\cdot(\mathbf{r}-\mathbf{r}')} U_{n'\mathbf{k}\rho'}(\mathbf{r}'). \tag{7.235}$$

Because the $U_{n\mathbf{k}\rho}$'s form a complete set of periodic functions, Eq. (7.232) can be rewritten in the alternate form

$$[\xi_l - H(\mathbf{k}, \xi_l)]\, G(\mathbf{k}, \xi_l) = I. \tag{7.236}$$

Eq. (2.236) can also be rewritten as

$$G(\mathbf{k}, \xi_l) = \frac{1}{\xi_l - H(\mathbf{k}, \xi_l)}. \tag{7.237}$$

7.10.3 Exchange Self-Energy in the Band Model

The exchange contribution to the self-energy is nonlocal in **r** space,

$$\Sigma(\mathbf{r}, \mathbf{r}', \xi_l) = -\frac{1}{\beta} \sum_{\xi_l} v_{eff}(\mathbf{r}, \mathbf{r}') G(\mathbf{r}, \mathbf{r}', \xi_l - \xi_{l'}), \tag{7.238}$$

where a simple static screening approximation is made in obtaining $v_{eff}(\mathbf{r}, \mathbf{r}')$ from $v(\mathbf{r}, \mathbf{r}')$. In this approximation, the self-energy is independent of ξ_l, and one can write

$$\Sigma(\mathbf{r}, \mathbf{r}') = -\frac{1}{\beta} \sum_{\xi_l} v_{eff}(\mathbf{r}, \mathbf{r}') G(\mathbf{r}, \mathbf{r}', \xi_l). \tag{7.239}$$

Σ and G can be expanded in terms of Bloch states as follows:

$$\Sigma(\mathbf{r}, \mathbf{r}') = \sum_{n,m,\mathbf{k},\rho,\rho'} \Sigma_{n\rho,m\rho'}(\mathbf{k}) \psi_{n\mathbf{k}\rho}(\mathbf{r}) \psi^*_{m\mathbf{k}\rho'}(\mathbf{r}') \tag{7.240}$$

and

$$G(\mathbf{r}, \mathbf{r}') = \sum_{n,m,\mathbf{k},\rho,\rho'} G_{n\rho,m\rho'}(\mathbf{k}) \psi_{n\mathbf{k}\rho}(\mathbf{r}) \psi^*_{m\mathbf{k}\rho'}(\mathbf{r}'). \tag{7.241}$$

Substituting Eqs. (7.240) and (7.241) in Eq. (7.239), we obtain

$$\sum_{n,m,\rho,\rho'} \Sigma_{n\rho,m\rho'}(\mathbf{k})\,\psi_{n\mathbf{k}\rho}(\mathbf{r})\,\psi^*_{m\mathbf{k}\rho'}(\mathbf{r}') = -\frac{1}{\beta}\sum_{\xi_l}\sum_{p,q,\mathbf{k}',\overline{\rho},\overline{\rho}'} v_{eff}(\mathbf{r},\mathbf{r}')G_{p\overline{\rho},q\overline{\rho}'}(\mathbf{k}')\psi_{p\mathbf{k}'\overline{\rho}}(\mathbf{r})\psi_{q\mathbf{k}'\overline{\rho}'}(\mathbf{r}'). \tag{7.242}$$

If the effective electron–electron interaction is spin independent, then $\rho = \overline{\rho}$, $\rho' = \overline{\rho}'$, and we have

$$\Sigma_{n\rho,m\rho'}(\mathbf{k}) = -\frac{1}{\beta}\sum_{\mathbf{k}',\xi_l,p,q} <nm\,|\,v_{eff}(\mathbf{k},\mathbf{k}')\,|\,pq>_{\rho\rho'} G_{p\rho,q\rho'}(\mathbf{k},\xi_l), \tag{7.243}$$

where

$$<nm\,|\,v_{eff}(\mathbf{k},\mathbf{k}')\,|\,pq>_{\rho\rho'} = \int \psi^*_{n\mathbf{k}\rho}(\mathbf{r})\,\psi_{m\mathbf{k}\rho'}(\mathbf{r}')\,v_{eff}(\mathbf{r},\mathbf{r}')\psi_{p\mathbf{k}'\rho}(\mathbf{r})\psi^*_{q\mathbf{k}'\rho'}(\mathbf{r}')d\mathbf{r}\,d\mathbf{r}'. \tag{7.244}$$

Equation (7.243) is the expression for exchange self-energy in the band model. The problem of exchange self-energy in the band model, which includes the effect of a magnetic field where the self-energy has been expanded in different orders in the magnetic field, is discussed in more detail in Chapter 12 (Section 12.4.5).

PROBLEMS

7.1. In Eq. (7.1), we obtained

$$H\Psi = \sum_{i=1}^{N}\left(-\frac{\hbar^2}{2m_e}\nabla_i^2\Psi - ze^2\sum_l \frac{1}{|\mathbf{r}_i - \mathbf{R}_l|}\Psi\right) + \frac{1}{2}\sum_{i\neq j}\frac{e^2}{|\mathbf{r}_i - \mathbf{r}_j|}\Psi. \tag{1}$$

The Hamiltonian can be written as

$$H = \sum_i H_i + \frac{1}{2}\sum_{i\neq j} V_{ij}, \tag{2}$$

where H_i operates on the coordinate of the ith electron, and V_{ij} operates on the two-body coordinates of both i and j.

In Eq. (7.19), we showed that

$$\Psi(\mathbf{r}_1 s_1 \ldots \mathbf{r}_N) = \frac{1}{\sqrt{N!}}\sum_n (-1)^n \prod_k \psi_{n_k}(\mathbf{r}_k s_k). \tag{3}$$

Show that the expectation value of the one-electron terms in the Hamiltonian $<H_i>$ is

$$\sum_{s_1\ldots s_N}\int d^N\mathbf{r}\frac{1}{N!}\sum_{nn'}(-1)^{n+n'}\left[\prod_k \psi^*_{n_k}(\mathbf{r}_k s_k)\right]\sum_i\left(\frac{-\hbar^2\nabla_i^2}{2m} - \sum_l \frac{ze^2}{|\mathbf{r}_i - \mathbf{R}_l|}\right)\left[\prod_{k'}\psi_{n'_{k'}}(\mathbf{r}_{k'}s_{k'})\right]. \tag{4}$$

Because the ψ's are orthonormal, only $n = n'$ terms survive the summation and integration $\sum_{s_1\ldots s_N}\int d^N\mathbf{r}$. The sum over n results in a factor $(N-1)!$ for all indices other than i.

Hence, show that Eq. (4) can be rewritten as

$$\sum_i \sum_{s_i} \int d\mathbf{r}_i \frac{1}{N!} \sum_n \psi^*_{n_i}(\mathbf{r}_i s_i) \left(\frac{-\hbar^2 \nabla_i^2}{2m} - \sum_l \frac{ze^2}{|\mathbf{r}_i - \mathbf{R}_l|} \right) \psi_{n_i}(\mathbf{r}_i s_i). \tag{5}$$

The sum over n results in a factor $(N-1)!$ for all indices other than i. However, s_i ranges over all values that can be written as a sum over i', and one can drop the dummy index i when integrating over $\mathbf{r}_i s_i$. Show that Eq. (5) can be rewritten as

$$\sum_i \sum_s \int d\mathbf{r} \frac{1}{N} \sum_{i'} \psi^*_{i'}(\mathbf{r}s) \left(\frac{-\hbar^2 \nabla^2}{2m} - \sum_l \frac{ze^2}{|\mathbf{r} - \mathbf{R}_l|} \right) \psi_{i'}(\mathbf{r}s). \tag{6}$$

Because $\psi_i(\mathbf{r}_i s_i) = \phi_i(\mathbf{r}_i)\chi_i(s_i)$, the sum over the spin index can be eliminated. The sum over i yields a factor of N, and the sum over i' can be rewritten as a sum over i. Show that Eq. (6) can be expressed as

$$\sum_{i=1}^{N} \int d\mathbf{r}\, \phi^*_i(\mathbf{r}) \left(\frac{-\hbar^2 \nabla^2}{2m} - \sum_l \frac{-ze^2}{|\mathbf{r} - \mathbf{R}_l|} \right) \phi_i(\mathbf{r}). \tag{7}$$

The expectation value of the Coulomb interaction term in the Hamiltonian is obtained from Eqs. (2) and (3) as

$$\sum_{s_1 \ldots s_N} \int d^N\mathbf{r} \sum_{n,n'} \frac{1}{N!} \frac{1}{2} \sum_{i \neq j} \frac{e^2 (-1)^{n+n'}}{|\mathbf{r}_i - \mathbf{r}_j|} \prod_{k,k'} \psi^*_{n_k}(\mathbf{r}_k s_k)\, \psi_{n'_{k'}}(\mathbf{r}_{k'} s_{k'}). \tag{8}$$

Show that Eq. (8) can be rewritten in the alternate form

$$\sum_{s_1 \ldots s_N} \int d^N\mathbf{r} \sum_{n,n'} \frac{1}{N!} \frac{1}{2} \sum_{i \neq j} \frac{e^2 (-1)^{n+n'}}{|\mathbf{r}_i - \mathbf{r}_j|} \left[\psi^*_{n_i}(\mathbf{r}_i)\, \psi^*_{n_j}(\mathbf{r}_j)\, \psi_{n'_i}(\mathbf{r}_i)\, \psi_{n'_j}(\mathbf{r}_j) \prod_{k,k' \neq i,j} \psi^*_{n_k}(\mathbf{r}_k)\, \psi_{n'_{k'}}(\mathbf{r}_{k'}) \right], \tag{9}$$

where $\psi_{n_i}(\mathbf{r}_i) \equiv \psi_{n_i}(\mathbf{r}_i, s_i)$ etc. for brevity. Integrating over all the terms except $\mathbf{r}_i$ and $\mathbf{r}_j$ (which leaves two permutations n' for given n), show that Eq. (9) can be rewritten as

$$\frac{1}{2} \sum_{i \neq j} \sum_{s_i s_j} \int d\mathbf{r}_i\, d\mathbf{r}_j \sum_n \frac{1}{N!} \frac{e^2}{|\mathbf{r}_i - \mathbf{r}_j|} \left[|\psi_{n_i}(\mathbf{r}_i)|^2 |\psi_{n_j}(\mathbf{r}_j)|^2 - \psi^*_{n_i}(\mathbf{r}_i)\, \psi^*_{n_j}(\mathbf{r}_j)\, \psi_{n_i}(\mathbf{r}_j)\, \psi_{nj}(\mathbf{r}_i) \right]. \tag{10}$$

Show that because the integrations can be performed over the dummy variables 1 and 2, the sum over $i \neq j$ yields a factor of $N(N-1)$. Hence, show that Eq. (10) can be rewritten as

$$\sum_{s_1 s_2} \int \frac{d\mathbf{r}_1 d\mathbf{r}_2}{2(N-2)!} \frac{e^2}{|\mathbf{r}_1 - \mathbf{r}_2|} \sum_n \left[|\psi_{n_1}(\mathbf{r}_1)|^2 |\psi_{n_2}(\mathbf{r}_2)|^2 - \psi^*_{n_1}(\mathbf{r}_1)\, \psi^*_{n_2}(\mathbf{r}_2)\, \psi_{n_1}(\mathbf{r}_2)\, \psi_{n_2}(\mathbf{r}_1) \right]. \tag{11}$$

Summing over the permutations over n, show that Eq. (11) can be rewritten as

$$\sum_{s_1 s_2} \frac{1}{2} \int \frac{e^2 d\mathbf{r}_1 d\mathbf{r}_2}{|\mathbf{r}_1 - \mathbf{r}_2|} \sum_{i \neq j} \left[|\psi_i(\mathbf{r}_1)|^2 \, |\psi_j(\mathbf{r}_2)|^2 - \psi_i^*(\mathbf{r}_1)\psi_j^*(\mathbf{r}_2)\psi_i(\mathbf{r}_2)\psi_j(\mathbf{r}_1) \right]. \tag{12}$$

Using Eq. (7.20), show that Eq. (12) can be rewritten as

$$\frac{1}{2} \int \frac{e^2 d\mathbf{r}_1 d\mathbf{r}_2}{|\mathbf{r}_1 - \mathbf{r}_2|} \sum_{i \neq j} \left[|\phi_i(\mathbf{r}_1)|^{2|} \, |\phi_j(\mathbf{r}_2)|^2 - \phi_i^*(\mathbf{r}_1)\phi_j^*(\mathbf{r}_2)\phi_i(\mathbf{r}_2)\phi_j(\mathbf{r}_1)\delta_{\chi_i,\chi_j} \right]. \tag{13}$$

From Eqs. (7) and (13), we obtain

$$\begin{aligned}
\langle H \rangle &= \sum_{i=1}^{N} \int d\mathbf{r}\phi_i^*(\mathbf{r}) \left(\frac{-\hbar^2\nabla^2}{2m} - \sum_l \frac{ze^2}{|\mathbf{r} - \mathbf{R}_l|} \right) \phi_i(\mathbf{r}) \\
&\quad + \frac{1}{2} \iint \frac{e^2 d\mathbf{r}_1 d\mathbf{r}_2}{|\mathbf{r}_1 - \mathbf{r}_2|} \sum_{i \neq j} \left[|\phi_i(\mathbf{r}_1)|^2 \, |\phi_j(\mathbf{r}_2)|^2 - \phi_i^*(\mathbf{r}_1)\phi_j^*(\mathbf{r}_2)\phi_i(\mathbf{r}_2)\phi_j(\mathbf{r}_1)\delta_{\chi_i,\chi_j} \right] \\
&= \sum_i \int \phi_i^* H_1 \phi_i d\mathbf{r} + \frac{1}{2}\sum_{i \neq j} \iint d\mathbf{r}_1 d\mathbf{r}_2 \phi_i^*(\mathbf{r}_1)\phi_j^*(\mathbf{r}_2) H_{12} \phi_i(\mathbf{r}_1)\phi_j(\mathbf{r}_2) \\
&\qquad - \frac{1}{2}\sum_{i \neq j} \iint d\mathbf{r}_1 d\mathbf{r}_2 \phi_i^*(\mathbf{r}_1)\phi_j^*(\mathbf{r}_2) H_{12} \phi_i(\mathbf{r}_2)\phi_j(\mathbf{r}_1)\delta_{\chi_i,\chi_j},
\end{aligned}$$

where

$$H_1 \equiv \left(\frac{-\hbar^2\nabla^2}{2m} - \sum_l \frac{ze^2}{|\mathbf{r} - \mathbf{R}_l|} \right)$$

and

$$H_{12} \equiv \frac{e^2}{|\mathbf{r}_1 - \mathbf{r}_2|}.$$

Following a variational technique similar to that adopted in Eqs. (7.11) through (7.15), derive the Hartree–Fock equation in Eq. (7.21).

7.2. We derived

$$\begin{aligned}
U^{ex}\phi_i(\mathbf{r}) &= -e^2\phi_i(\mathbf{r}) \int^{k_F} \frac{d\mathbf{k}}{8\pi^3} \frac{4\pi}{(k^2 + k_i^2 - 2kk_i \cos\theta)} \\
&= -e^2\phi_i(\mathbf{r}) \int_0^{k_F}\int_0^{\pi} \frac{k^2 dk \sin\theta d\theta}{\pi(k_i^2 + k^2 - 2kk_i\cos\theta)}.
\end{aligned} \tag{1}$$

First, integrate over $d\theta$ and then integrate over dk to obtain

$$U^{ex}\phi_i(\mathbf{r}) = -e^2\phi_i(\mathbf{r}) \frac{1}{2\pi k_i} \left[(k_F^2 - k_i^2) \ln\left(\frac{k_F + k_i}{k_F - k_i} \right) + 2k_i k_F \right]. \tag{2}$$

7.3. Show, by integration, that the total energy of N electrons

$$E = 2\sum_{k \leq k_F} \frac{\hbar^2 k^2}{2m} - \frac{e^2 k_F}{\pi} \sum_{k \leq k_F} \left[1 + \frac{k_F^2 - k^2}{2kk_F} \ln \left| \frac{k_F + k}{k_F - k} \right| \right] \tag{1}$$

$$= N \left[\frac{3}{5}\varepsilon_F - \frac{3}{4}\frac{e^2 k_F}{\pi} \right]. \tag{2}$$

The first term was evaluated in Chapter 1.

7.4. We derived in Eq. (7.43),

$$E = N \left[\frac{3}{5}\varepsilon_F - \frac{3}{4}\frac{e^2 k_F}{\pi} \right]. \tag{1}$$

Show that Eq. (1) can also be rewritten as

$$\begin{aligned} \frac{E}{N} &= \frac{e^2}{2a_0}\left[\frac{3}{5}(k_F a_0)^2 - \frac{3}{2\pi}(k_F a_0) \right] \\ &= \left[\frac{2.21}{(r_s/a_0)^2} - \frac{0.916}{(r_s/a_0)} \right] \text{Ry}. \end{aligned} \tag{2}$$

$$(1\,\text{Ry} = e^2/2a_0 = 13.6\,\text{eV}).$$

7.5. Assume that a positively charged particle is placed at position $\mathbf{r}$ (an example is the substitution of a Zn ion for a copper ion in a metal lattice), which creates a surplus of negative charge, thereby screening its field. The electrostatic potential can be written as

$$\phi(\mathbf{r}) = \phi^e(\mathbf{r}) + \phi^i(\mathbf{r}). \tag{1}$$

Here, $\phi^e(\mathbf{r})$ is potential due to the positively charged particle, and $\phi^i(\mathbf{r})$ arises due to the induced charge density because of the presence of the positively charged particle. The Poisson's equation can be written as

$$-\nabla^2\phi(\mathbf{r}) = 4\pi\rho(\mathbf{r}), \tag{2}$$

where the total charge density $\rho(\mathbf{r})$ can be expressed as

$$\rho(\mathbf{r}) = \rho^e(\mathbf{r}) + \rho^i(\mathbf{r}). \tag{3}$$

Here, $\rho^e(\mathbf{r})$ and $\rho^i(\mathbf{r})$ are the external and induced charge densities, respectively.

Show that in a spatially uniform electron gas, $\phi^e(\mathbf{r})$ and $\phi(\mathbf{r})$ are linearly related through the difference between their position, in an equation of the form

$$\phi^e(\mathbf{r}) = \int d\mathbf{r}' \in(\mathbf{r} - \mathbf{r}')\phi(\mathbf{r}'). \tag{4}$$

By making appropriate Fourier transformations,

$$\in(\mathbf{q}) = \int d\mathbf{r}\, e^{-i\mathbf{q}\cdot\mathbf{r}} \in(\mathbf{r}), \tag{5}$$

$$\phi^e(\mathbf{q}) = \int d\mathbf{r}\, e^{-i\mathbf{q}\cdot\mathbf{r}} \phi^e(\mathbf{r}), \tag{6}$$

and

$$\phi(\mathbf{r}') = \int \frac{d\mathbf{q}}{(2\pi)^3} e^{i\mathbf{q}\cdot\mathbf{r}'} \phi(\mathbf{q}), \tag{7}$$

show that

$$\phi^e(\mathbf{q}) = \phi(\mathbf{q}) \in(\mathbf{q}). \tag{8}$$

Eq. (8) can be rewritten in the alternate form

$$\phi(\mathbf{q}) = \frac{\phi^e(\mathbf{q})}{\in(\mathbf{q})}, \tag{9}$$

where $\in(\mathbf{q})$ is the dielectric constant of the metal.

7.6. If the external potential in a metal (in the jellium model) is due to a positive point charge Q located at $\mathbf{r}$, show that

$$\phi^e(\mathbf{q}) = \frac{4\pi Q}{q^2} \tag{1}$$

Hence, show from Eq. (7.61) that

$$\phi(\mathbf{q}) = \frac{\phi^e(\mathbf{q})}{\in(\mathbf{q})} = \frac{4\pi Q}{q^2 + \lambda^2}. \tag{2}$$

Using Fourier transformation, show that

$$\phi(\mathbf{r}) = \int \frac{d\mathbf{q}}{(2\pi)^3} \phi(\mathbf{q}) e^{i\mathbf{q}\cdot\mathbf{r}} = \frac{Q}{r} e^{-\lambda r}. \tag{3}$$

7.7. Show that at $T = 0$, the integral

$$\chi(\mathbf{q}) = -e^2 \int \frac{d\mathbf{k}}{4\pi^3} \frac{f_0(\mathbf{k}) - f_0(\mathbf{k}+\mathbf{q})}{\hbar^2(\mathbf{k}\cdot\mathbf{q})/m} \tag{1}$$

can be performed explicitly to give

$$\chi(\mathbf{q}) = -e^2 \left(\frac{mk_F}{\hbar^2\pi^2}\right) \left[\frac{1}{2} + \frac{1-x^2}{4x} \ln\left|\frac{1+x}{1-x}\right|\right], \tag{2}$$

where

$$x = \frac{q}{2k_F}.$$

7.8. The expression for the dielectric constant for static screening in the Lindhard theory, derived in Eq. (7.76), is given by

$$\epsilon_L(\mathbf{q}) = 1 - \frac{4\pi e^2}{q^2} \sum_{\mathbf{k}} \frac{f^0(\mathbf{k}) - f^0(\mathbf{k}+\mathbf{q})}{\varepsilon(\mathbf{k}) - \varepsilon(\mathbf{k}+\mathbf{q})}. \quad (1)$$

By adjusting the summation over $f^0(\mathbf{k}+\mathbf{q})$ to $f^0(\mathbf{k})$ and converting the summation over $\mathbf{k}$ to an integration by using polar coordinates, show that (Eq. 7.93)

$$\epsilon_L(\mathbf{q}) = 1 + \frac{4\pi e^2}{q^2} \frac{mk_F}{\hbar^2 \pi^2} \left[\frac{1}{2} + \frac{4k_F^2 - q^2}{8k_F q} \ln \left| \frac{2k_F + q}{2k_F - q} \right| \right]. \quad (2)$$

7.9. If we define

$$f^s_{\mathbf{kk}'} = \sum_{l=0}^{\infty} f^s_l P_l(\cos\theta) \quad (1)$$

and

$$f^a_{\mathbf{kk}'} = \sum_{l=0}^{\infty} f^a_l P_l(\cos\theta), \quad (2)$$

by using the properties of the Legendre polynomial, show that

$$f^s_l = \frac{2l+1}{2} \int_{-1}^{1} d(\cos\theta) P_l(\cos\theta) f^s_{\mathbf{kk}'} \quad (3)$$

and

$$f^a_l = \frac{2l+1}{2} \int_{-1}^{1} d(\cos\theta) P_l(\cos\theta) f^a_{\mathbf{kk}'}. \quad (4)$$

7.10. Derive from Eq. (7.127)

$$i\hbar <\mathbf{k}'|\delta\dot{\rho}|\mathbf{k}> = [\varepsilon(\mathbf{k}') - \varepsilon(\mathbf{k})] <\mathbf{k}'|\delta\rho|\mathbf{k}> - [f_0(\mathbf{k}') - f_0(\mathbf{k})] <\mathbf{k}'|\phi|\mathbf{k}>, \quad (1)$$

and the fact that $\delta\rho$ has the same time dependence as $\phi(\mathbf{r}, t)$ in Eq. (7.130), show that

$$[\varepsilon(\mathbf{k}+\mathbf{q}) - \varepsilon(\mathbf{k}) - i\hbar(i\omega + \alpha)] <\mathbf{k}+\mathbf{q}|\delta\rho|\mathbf{k}> = [f_0(\mathbf{k}+\mathbf{q}) - f_0(\mathbf{k})]\phi_{\mathbf{q}}(t), \quad (2)$$

where

$$\phi_{\mathbf{q}}(t) = \phi_{e\mathbf{q}}(t) + \phi_{i\mathbf{q}}(t) = [\phi_e(\mathbf{q}, \omega) + \phi_i(\mathbf{q}, \omega)]e^{(i\omega t + \alpha t)}. \quad (3)$$

7.11. We derived in Eq. (7.231)

$$\sum_{n'',\rho'',\mathbf{k}',\mathbf{k}''} \int d\mathbf{r}d\mathbf{r}'d\mathbf{r}'' e^{-i\mathbf{k}\cdot\mathbf{r}} U^*_{n\mathbf{k}\rho}(\mathbf{r})\left(\xi_l - \frac{p^2}{2m} - V(\mathbf{r})\right)e^{i\mathbf{k}''\cdot(\mathbf{r}-\mathbf{r}'')} U_{n''\mathbf{k}''\rho''}(\mathbf{r}) U^*_{n''\mathbf{k}''\rho''}(\mathbf{r}'')$$
$$\times G(\mathbf{r}'', \mathbf{r}', \xi_l) U_{n'\mathbf{k}'\rho'}(\mathbf{r}')e^{i\mathbf{k}'\cdot\mathbf{r}'}$$
$$+ \sum_{n'',\rho'',\mathbf{k}',\mathbf{k}''} \int d\mathbf{r}d\mathbf{r}'d\mathbf{r}''d\mathbf{r}''' e^{-i\mathbf{k}\cdot\mathbf{r}} U^*_{n\mathbf{k}\rho}(\mathbf{r})\Sigma(\mathbf{r}, \mathbf{r}'', \xi_l)e^{i\mathbf{k}''\cdot(\mathbf{r}''-\mathbf{r}''')} U_{n''\mathbf{k}''\rho''}(\mathbf{r}'') U^*_{n''\mathbf{k}''\rho''}(\mathbf{r}''') \quad (1)$$
$$\times G(\mathbf{r}''', \mathbf{r}', \xi_l) U_{n'\mathbf{k}'\rho'}(\mathbf{r}')e^{i\mathbf{k}'\cdot\mathbf{r}'} = \delta_{nn'}\delta_{\rho\rho'}.$$

Show that Eq. (1) can be rewritten in the alternate form

$$\sum_{n'',\rho''} [\xi_l - H(\mathbf{k}', \xi_l)]_{n\mathbf{k}\rho, n''\mathbf{k}\rho''} G_{n''\mathbf{k}\rho'', n'\mathbf{k}\rho'}(\mathbf{k}', \xi_l)\,|_{\mathbf{k}'=\mathbf{k}} = \delta_{nn'}\delta_{\rho\rho'}, \quad (2)$$

where

$$H(\mathbf{k}', \xi_l) = \frac{1}{2m}(\mathbf{p} + \hbar\mathbf{k}')^2 + V(\mathbf{r}) + \Sigma(\mathbf{k}', \xi_l), \quad (3)$$

$$\Sigma_{n\mathbf{k}\rho, n''\mathbf{k}\rho''}(\mathbf{k}', \xi_l) = \int d\mathbf{r}d\mathbf{r}'\, U^*_{n\mathbf{k}\rho}(\mathbf{r})e^{-i\mathbf{k}'\cdot(\mathbf{r}-\mathbf{r}')}\Sigma(\mathbf{r}, \mathbf{r}', \xi_l)\, U_{n''\mathbf{k}\rho''}(\mathbf{r}'), \quad (4)$$

and

$$G_{n''\mathbf{k}\rho'', n'\mathbf{k}\rho'}(\mathbf{k}', \xi_l) = \int d\mathbf{r}d\mathbf{r}'\, U^*_{n''\mathbf{k}\rho''}(\mathbf{r})G(\mathbf{r}, \mathbf{r}', \xi_l)e^{-i\mathbf{k}'\cdot(\mathbf{r}-\mathbf{r}')} U_{n'\mathbf{k}\rho'}(\mathbf{r}'). \quad (5)$$

Because the $U_{n\mathbf{k}\rho}$'s form a complete set of periodic functions, show that Eq. (5) can be rewritten in the alternate form

$$[\xi_l - H(\mathbf{k}, \xi_l)]\, G(\mathbf{k}, \xi_l) = I. \quad (6)$$

References

1. Abrikosov AA. *Introduction to the theory of normal metals.* New York: Academic Press; 1972.
2. Ashcroft NA, Mermin ND. *Solid state physics.* New York: Brooks/Cole; 1976.
3. Born M, Oppenheimer JR. On the quantum theory of metals. *Ann Phys* 1927;**84**:457.
4. Callaway J. *Quantum theory of the solid state.* New York: Academic Press; 1976.
5. Dirac PAM. Quantum mechanics of many-electron systems. *Proc R Soc (Lond)* 1929;**A123**:714.

6. Ferrell RA. Rigorous validity criterion for testing approximations to the electron gas correlation energy. *Phys Rev Lett* 1958;**1**:444.
7. Fock V. A method for the solution of many-body problem in quantum mechanics. *Z Physik* 1930;**61**:126.
8. Hartree DR. The wave mechanics of an atom with a non-Coulomb central field. *Proc Camb Phil Soc* 1928;**24**:89.
9. Hohenberg PC, Kohn W. Inhomogeneous electron gas. *Phys Rev* 1964;**80**:864.
10. Kittel C. *Quantum theory of solids*. New York: John Wiley & Sons; 1987.
11. Kohn W, Sham LJ. Self-consistent equations including exchange and correlation effects. *Phys Rev* 1965;**140**:A1113.
12. Landau LP, Lifshitz EM. *Quantum mechanics, non-relativistic theory*. Reading, MA: Addison-Wesley; 1965.
13. Landau LD. The theory of a Fermi liquid. *Sov Phys JETP* 1957;**3**:920; Oscillations in a Fermi liquid. 1957;**5**:101.
14. Liboff RL. *Quantum mechanics*. Reading, MA: Addison-Wesley; 1980.
15. Lieb EH. Thomas-Fermi and related theories of atoms and molecules. *Rev Mod Phys* 1981;**53**:603.
16. Lindhard J. *Kgl Danske Videnskab. Selskab Mat.-Fys. Medd.* 1954;**28(8)**.
17. Marder MP. *Condensed matter physics*. New York: Brooks/Cole; 2000.
18. Noziers P, Blandin A. Kondo effect in real metals. *J Phys (Paris)* 1980;**41**:193.
19. Slater JC. A simplification of the Hartree-Fock method. *Phys Rev* 1951;**81**:385.
20. Varma CM, Nussinov Z, Sarloos W. Singular or non-Fermi liquids. *Phys Rep* 2002;**361**:267.
21. Wigner E. On the interaction of electrons in metals. *Phys Rev* 1934;**46**:1002.
22. Wilson AH. *The theory of metals*. Cambridge: Cambridge University Press; 1960.
23. Ziman JM. *Principles of the theory of solids*. Cambridge: Cambridge University Press; 1972.

CHAPTER

Dynamics of Bloch Electrons

8

CHAPTER OUTLINE

8.1 Semiclassical Model 243
8.2 Velocity Operator 244
8.3 k · p Perturbation Theory 245
8.4 Quasiclassical Dynamics 246
8.5 Effective Mass 247
8.6 Bloch Electrons in External Fields 248
8.6.1 Time Evolution of Bloch Electrons in an Electric Field 250
8.6.2 Alternate Derivation for Bloch Functions in an External Electric and Magnetic Field 252
8.6.3 Motion in an Applied DC Field 253
8.7 Bloch Oscillations 254
8.8 Holes 255
8.9 Zener Breakdown (Approximate Method) 258
8.10 Rigorous Calculation of Zener Tunneling 261
8.11 Electron–Phonon Interaction 264
Problems 271
References 274

8.1 SEMICLASSICAL MODEL

We determined in Chapter 4 (Eq. 4.45) that the eigenfunction of the Hamiltonian of an electron moving in a crystalline solid with a periodic potential $V(\mathbf{r})$ is a Bloch function $\psi_{n\mathbf{k}}(\mathbf{r})$ that can be expressed as

$$\psi_{n\mathbf{k}}(\mathbf{r}) = e^{i\mathbf{k}\cdot\mathbf{r}}u_{n\mathbf{k}}(\mathbf{r}), \tag{8.1}$$

where $u_{n\mathbf{k}}(\mathbf{r})$ is the periodic part of the Bloch function. Here, n is a band index, $\mathbf{k}$ is a vector in the first Brillouin zone in the restricted zone scheme but extends to infinity in the periodic zone scheme, and $u_{n\mathbf{k}}(\mathbf{r})$ has the unique property that it remains unchanged when translated by any direct lattice vector $\mathbf{R}_i$ (Eq. 4.46), i.e.,

$$u_{n\mathbf{k}}(\mathbf{r}+\mathbf{R}_i) = u_{n\mathbf{k}}(\mathbf{r}). \tag{8.2}$$

Thus, the Schrodinger equation

$$H\psi_{n\mathbf{k}}(\mathbf{r}) = \varepsilon_n(\mathbf{k})\psi_{n\mathbf{k}}(\mathbf{r}) \tag{8.3}$$

can be rewritten in the alternate form (Eq. 4.36)

$$H_{\mathbf{k}} u_{n\mathbf{k}}(\mathbf{r}) = \varepsilon_n(\mathbf{k}) u_{n\mathbf{k}}(\mathbf{r}), \tag{8.4}$$

where

$$H_{\mathbf{k}} = \left[\frac{\hbar^2}{2m}(-i\nabla + \mathbf{k})^2 + V(\mathbf{r})\right]. \tag{8.5}$$

In this chapter, we have written H_{eff} as $H_{\mathbf{k}}$ for convenience. Here, the boundary conditions for $u_{n\mathbf{k}}(\mathbf{r})$ are specified in Eq. (8.2). In the periodic zone scheme, which will be used in the derivations that follow, the eigenfunctions and eigenvalues are periodic functions of **k** in the reciprocal lattice, i.e.,

$$\psi_{n\mathbf{k}}(\mathbf{r}) = \psi_{n,\mathbf{k}+\mathbf{K}}(\mathbf{r}) \tag{8.6}$$

and

$$\varepsilon_n(\mathbf{k}) = \varepsilon_n(\mathbf{k} + \mathbf{K}). \tag{8.7}$$

The set of electronic levels $\varepsilon_n(\mathbf{k})$, for each n, is known as the energy band because the band index n is a constant of motion. In addition, the energy levels $\varepsilon_n(\mathbf{k})$ vary continuously as **k** varies because **k** is essentially a parameter in Eq. (8.5). Further, the size of the crystal is not a factor in Eq. (8.4), and as we noted earlier, the Born–von Karman boundary condition (Eq. 3.31),

$$\begin{aligned} \psi_{\mathbf{k}}(x+L, y, z) &= \psi_{\mathbf{k}}(x, y, z) \\ \psi_{\mathbf{k}}(x, y+L, z) &= \psi_{\mathbf{k}}(x, y, z) \\ \psi_{\mathbf{k}}(x, y, z+L) &= \psi_{\mathbf{k}}(x, y, z), \end{aligned} \tag{8.8}$$

specifies that **k** is continuous when $L \to \infty$ (Eq. 3.14). We will use the property that **k** is a continuous parameter in what follows. These same boundary conditions are also valid for the Bloch functions $\psi_{n\mathbf{k}}(\mathbf{r})$.

8.2 VELOCITY OPERATOR

For a crystal with inversion symmetry, the velocity operator

$$\mathbf{v} = \dot{\mathbf{r}} = -\frac{i}{\hbar}[\mathbf{r}, \mathbf{H}] \tag{8.9}$$

can be rewritten in the alternate form

$$\mathbf{v} = -\frac{i\hbar}{m}\nabla. \tag{8.10}$$

If we define the velocity operator using the effective Hamiltonian in Eq. (8.5),

$$\mathbf{v}_{\mathbf{k}} = -\frac{i}{\hbar}[\mathbf{r}, H_{\mathbf{k}}], \tag{8.11}$$

we obtain

$$\mathbf{v_k} = \frac{\hbar}{m}(-i\nabla + \mathbf{k}). \tag{8.12}$$

We will use this relation for the velocity operator to obtain an expression for the velocity of the Bloch electrons.

8.3 k · p PERTURBATION THEORY

We assume that **k** is increased by an infinitesimal amount $\Delta\mathbf{k}$. When $\mathbf{k} \to \mathbf{k}+\Delta\mathbf{k}$, Eq. (8.5) can be rewritten in the form

$$H_{\mathbf{k}+\Delta\mathbf{k}} = \left[\frac{\hbar^2}{2m}(-i\nabla + \mathbf{k} + \Delta\mathbf{k})^2 + V(\mathbf{r})\right], \tag{8.13}$$

which can be rewritten in the alternate form

$$H_{\mathbf{k}+\Delta\mathbf{k}} = H_{\mathbf{k}} + \frac{\hbar^2}{m}(-i\nabla + \mathbf{k})\cdot\Delta\mathbf{k} + \frac{\hbar^2}{2m}(\Delta\mathbf{k})^2. \tag{8.14}$$

We use nondegenerate second-order perturbation theory and write

$$H_{\mathbf{k}+\Delta\mathbf{k}} = H_{\mathbf{k}} + H'_{\mathbf{k}}. \tag{8.15}$$

The perturbation term $H'_{\mathbf{k}}$ is obtained from Eqs. (8.14) and (8.15),

$$H'_{\mathbf{k}} = \frac{\hbar^2}{m}(-i\nabla + \mathbf{k})\cdot\Delta\mathbf{k} + \frac{\hbar^2}{2m}(\Delta\mathbf{k})^2. \tag{8.16}$$

If we express the energy eigenvalue in different orders of ε as

$$\varepsilon_n(\mathbf{k}+\Delta\mathbf{k}) = \varepsilon_n(\mathbf{k}) + \varepsilon_n^{(1)}(\mathbf{k}) + \varepsilon_n^{(2)}(\mathbf{k}) + \ldots, \tag{8.17}$$

we obtain from second-order perturbation theory,

$$\varepsilon_n(\mathbf{k}) = \langle u_{n\mathbf{k}}|H_{\mathbf{k}}|u_{n\mathbf{k}}\rangle = \langle \psi_{n\mathbf{k}}|H|\psi_{n\mathbf{k}}\rangle \tag{8.18}$$

and

$$\varepsilon_n^{(1)}(\mathbf{k}) = \langle u_{n\mathbf{k}}|H'_{\mathbf{k}}|u_{n\mathbf{k}}\rangle = \langle u_{n\mathbf{k}}|\frac{\hbar^2}{m}(-i\nabla + \mathbf{k})\cdot\Delta\mathbf{k}|u_{n\mathbf{k}}\rangle, \tag{8.19}$$

where the term $\frac{\hbar^2}{2m}(\nabla\mathbf{k})^2$, which is second order in $\nabla\mathbf{k}$, has been included in $\varepsilon_n^{(2)}(\mathbf{k})$,

$$\varepsilon_n^{(2)}(\mathbf{k}) = \frac{\hbar^2}{2m}(\nabla\mathbf{k})^2 + \sum_{n'\neq n}\frac{|\langle u_{n\mathbf{k}}|\frac{\hbar^2}{m}\Delta\mathbf{k}\cdot(-i\nabla + \mathbf{k})|u_{n'\mathbf{k}}\rangle|^2}{\varepsilon_{n\mathbf{k}} - \varepsilon_{n'\mathbf{k}}}. \tag{8.20}$$

If we also expand $\varepsilon_n(\mathbf{k}+\Delta\mathbf{k})$ as

$$\varepsilon_n(\mathbf{k}+\Delta\mathbf{k}) = \varepsilon_n(\mathbf{k}) + \sum_i \frac{\partial \varepsilon_n(\mathbf{k})}{\partial k_i}\Delta k_i + \frac{1}{2}\sum_{ij}\frac{\partial^2 \varepsilon_n(\mathbf{k})}{\partial k_i \partial k_j}\Delta k_i \Delta k_j + \ldots, \tag{8.21}$$

we obtain from Eqs. (8.12), (8.17), (8.19), and (8.21),

$$\frac{\partial \varepsilon_n(\mathbf{k})}{\partial \mathbf{k}} = \hbar \langle u_{n\mathbf{k}} | \mathbf{v}_\mathbf{k} | u_{n\mathbf{k}} \rangle, \tag{8.22}$$

which can be rewritten in the alternate form (Problem 8.2)

$$\frac{\partial \varepsilon_n(\mathbf{k})}{\partial \mathbf{k}} = \hbar \langle \psi_{n\mathbf{k}} | \mathbf{v} | \psi_{n\mathbf{k}} \rangle = \hbar \mathbf{v}_{n\mathbf{k}}. \tag{8.23}$$

Thus, the mean velocity of a Bloch electron is given by

$$\mathbf{v}_{n\mathbf{k}} = \frac{1}{\hbar}\frac{\partial \varepsilon_n(\mathbf{k})}{\partial \mathbf{k}}. \tag{8.24}$$

It is interesting to note that one can derive a similar expression for the velocity of electrons in a periodic potential and an external field. We will show by using quasiclassical dynamics that

$$\mathbf{v}_\mathbf{k} = \frac{\hbar \mathbf{k}}{m} = \frac{1}{\hbar}\frac{\partial \varepsilon(\mathbf{k})}{\partial \mathbf{k}}. \tag{8.25}$$

It is also interesting to note that the group velocity of a wave packet moving freely in space is given by the same expression as Eq. (8.25).

8.4 QUASICLASSICAL DYNAMICS

The classical Hamilton equations are expressed as[11]

$$\dot{q}_k = \frac{\partial H}{\partial p_k}, \; \dot{p}_k = -\frac{\partial H}{\partial q_k}, \tag{8.26}$$

where H is the classical Hamiltonian function, and q_k and p_k are the generalized coordinates and momenta, respectively. In the quantum mechanical formulation, the Hamiltonian operator is obtained by replacing the classical momentum $\mathbf{p}$ by $-i\hbar\nabla$. In an external electrostatic field, the Hamiltonian operator can be written as

$$\hat{H} = \hat{H}_0 + U(\mathbf{r}) = \hat{\varepsilon}(-i\nabla) + U(\mathbf{r}), \tag{8.27}$$

where $\hat{H}_0$ is the Hamiltonian operator in the perfect lattice, $U(\mathbf{r})$ is the perturbing potential of the external field, and $\hat{\varepsilon}(-i\nabla)$ is the equivalent Hamiltonian operator. If we reverse these steps by replacing $-i\nabla$ by $\mathbf{p}/\hbar$, the classical Hamiltonian function is given by

$$H(\mathbf{r},\mathbf{p}) = \varepsilon(\mathbf{p}/\hbar) + U(\mathbf{r}). \tag{8.28}$$

From Eqs. (8.26) and (8.28), we obtain

$$\dot{\mathbf{r}} = \mathbf{v} = \frac{\partial H}{\partial \mathbf{p}} = \frac{\partial \varepsilon(\mathbf{p}/\hbar)}{\partial \mathbf{p}} = \frac{1}{\hbar}\frac{\partial \varepsilon(\mathbf{k})}{\partial \mathbf{k}}, \tag{8.29}$$

where

$$\mathbf{p} = \hbar \mathbf{k}. \tag{8.30}$$

It may be noted that $\hbar\mathbf{k}$ is the crystal momentum and not the actual momentum of an electron, as was obtained in the Sommerfeld model for free electrons. This is a consequence of the fact that in the Bloch formulation, the effect of the periodic potential has been already included.

The second Hamiltonian equation in Eq. (8.26) can be written as

$$\dot{\mathbf{p}} = \hbar\dot{\mathbf{k}} = -\frac{\partial H}{\partial \mathbf{r}} = -\nabla U(\mathbf{r}). \tag{8.31}$$

Because $U(\mathbf{r})$ is the potential energy of the electron in an external field, $\hbar\dot{\mathbf{k}}$ is essentially the force $\mathbf{F}$ acting on the Bloch electron. According to Newton's law,

$$\mathbf{F} = m\frac{d\mathbf{v}}{dt} = \hbar\dot{\mathbf{k}}, \tag{8.32}$$

where m is the mass of the electron.

8.5 EFFECTIVE MASS

It is often convenient to define an effective mass to describe the motion of Bloch electrons in an external field. Later, we will use the same concept to describe the motion of electrons and holes in a semiconductor. If we assume that the external field (electrostatic) acting on the electrons is weak and the change in $\mathbf{k}$ is slow, we can write

$$\frac{d}{dt}\mathbf{v}_{n\mathbf{k}} = \sum_{ij} \hat{\epsilon}_i \frac{\partial v^i_{n\mathbf{k}}}{\partial k_j}\frac{\partial k_j}{\partial t}, \tag{8.33}$$

where $\hat{\epsilon}_i$ is one of the three unit vectors $\hat{x}, \hat{y}$, and $\hat{z}$. We can also introduce the effective mass tensor $\mathbf{M}_n^{-1}(\mathbf{k})$ (it is actually an inverse effective mass tensor) of the Bloch electrons (it is important to note that the effect of the periodic potential of the lattice has already been included in the Bloch formulation) in analogy with Eq. (8.32) as

$$\frac{d}{dt}\mathbf{v}_{n\mathbf{k}} = \hbar\mathbf{M}_n^{-1}(\mathbf{k})\cdot\dot{\mathbf{k}}. \tag{8.34}$$

From Eqs. (8.24), (8.33), and (8.34), we obtain

$$[\mathbf{M}_n^{-1}(\mathbf{k})]_{ij} = \frac{1}{\hbar^2}\frac{\partial^2 \varepsilon_n}{\partial k_i \partial k_j}. \tag{8.35}$$

We can derive an expression for the effective mass tensor from the results of the perturbation theory. From Eqs. (8.20) and (8.21), we obtain

$$\varepsilon_n^{(2)}(\mathbf{k}) = \frac{\hbar^2}{2m}(\nabla\mathbf{k})^2 + \sum_{n'\neq n} \frac{|<u_{n\mathbf{k}}|\frac{\hbar^2}{m}\Delta\mathbf{k}\cdot(-i\nabla+\mathbf{k})|u_{n'\mathbf{k}}>|^2}{\varepsilon_n(\mathbf{k})-\varepsilon_{n'}(\mathbf{k})}$$
$$= \frac{1}{2}\sum_{ij}\frac{\partial^2\varepsilon_n(\mathbf{k})}{\partial k_i \partial k_j}\Delta k_i \Delta k_j. \tag{8.36}$$

From Eqs. (8.35) and (8.36), we obtain (Problem 8.4)

$$[\mathbf{M}_n^{-1}(\mathbf{k})]_{ij} = \frac{1}{m}\delta_{ij} + \frac{\hbar^2}{m^2}\sum_{n'\neq n}\frac{<n\mathbf{k}|-i\nabla_i|n'\mathbf{k}><n'\mathbf{k}|-i\nabla_j|n\mathbf{k}> + c.c.}{\varepsilon_n(\mathbf{k})-\varepsilon_{n'}(\mathbf{k})}. \tag{8.37}$$

Here, the summation is over all band indices n' except n. The inverse mass tensor plays an important role in the formulation of dynamics of Bloch electrons. As we will see, the inverse effective mass tensor can be either positive or negative. Because the concept of a negative effective mass associated with the negatively charged Bloch electrons is contrary to our physical understanding, we will also introduce the concept of "holes," which are positively charged particles associated with positive effective mass. The holes play a very important role in semiconductors. In addition, the mystery of a positive charge in the Hall coefficient (in the free electron model, $R_H = -nec$), in the measurement of the Hall effect in certain metals such as aluminum in high magnetic fields, is also explained by the concept of holes.

8.6 BLOCH ELECTRONS IN EXTERNAL FIELDS

In an external electric field **E**, the force acting on a free electron is given by

$$\hbar\dot{\mathbf{k}} = -e\mathbf{E}. \tag{8.38}$$

The effect of a magnetic field **B** is included by assuming that the Lorentz force equation is valid, i.e.,

$$\hbar\dot{\mathbf{k}} = -e\left(\mathbf{E} + \frac{1}{c}\mathbf{V}\times\mathbf{B}\right). \tag{8.39}$$

The proof of an equivalent result for Bloch electrons is much more difficult for a magnetic field. We will first derive an equivalent result for Bloch electrons in a static electric field and then generalize the results by using a more rigorous method. In a static electric field **E**, the general expression

$$\mathbf{E}(\mathbf{r},t) = -\frac{1}{c}\frac{\partial\mathbf{A}(\mathbf{r},t)}{\partial t} - \nabla\phi(\mathbf{r},t) \tag{8.40}$$

reduces to

$$\mathbf{E} = -\nabla\phi(\mathbf{r}). \tag{8.41}$$

Here, $\mathbf{A}(\mathbf{r}, t)$ is the vector potential, and $\phi(\mathbf{r})$ is the scalar potential. We showed in Eq. (8.24) that for Bloch electrons,

$$\dot{\mathbf{r}} = \mathbf{v}_n(\mathbf{k}) = \frac{1}{\hbar}\frac{\partial \varepsilon_n(\mathbf{k})}{\partial t}. \tag{8.42}$$

The wave packet moves in such a way that the energy

$$\varepsilon = \varepsilon_n(\mathbf{k}(t)) - e\phi(\mathbf{r}(t)) = \text{constant}. \tag{8.43}$$

Because the energy ε of the wave packet is constant, we obtain, by taking the time derivative of the energy,

$$\frac{\partial \varepsilon}{\partial t} = \frac{\partial \varepsilon_n(\mathbf{k})}{\partial t} \cdot \dot{\mathbf{k}} - e\nabla\phi \cdot \dot{\mathbf{r}} = 0. \tag{8.44}$$

From Eqs. (8.42) and (8.44), we obtain

$$\mathbf{V}_n(\mathbf{k}) \cdot [\hbar\dot{\mathbf{k}} - e\nabla\phi] = 0. \tag{8.45}$$

From Eq. (8.45), we obtain the desired result

$$\hbar\dot{\mathbf{k}} = e\nabla\phi = -e\mathbf{E}. \tag{8.46}$$

However, there are problems associated in this derivation because the periodic boundary conditions that were implicitly assumed in this derivation (by assuming that the eigenfunctions of the Hamiltonian are the Bloch functions) are no longer valid. The electrostatic potential $U(\mathbf{r})$, due to the uniform electric field $\mathbf{E}$, grows linearly in space because $U(\mathbf{r}) = -\mathbf{E}\cdot\mathbf{r}$. Therefore, when a metal is placed in an electric field, the surface charges build up and cancel the interior field. Thus, the basic assumption that the Bloch formulation is still valid is questionable, and the periodicity associated with our formulation of the problem in a Brillouin zone with a band index n and crystal momentum $\mathbf{k}$ is invalid. We will derive Eq. (8.46) by using a second method where the periodicity of the potential is not lost. However, we will first use the previous derivation to justify (in a hand-waving way) the effect of a magnetic field.

It may be noted that the results of Eq. (8.45) are valid if any term perpendicular to $\mathbf{V}_n(\mathbf{k})$ is added in the square bracket. For example, anticipating that the Lorentz equations are valid, we can rewrite Eq. (8.45) as

$$\mathbf{V}_n(\mathbf{k}) \cdot [\hbar\dot{\mathbf{k}} - e\nabla\phi + \frac{e}{c}\mathbf{V}_n(\mathbf{k}) \times \mathbf{B}] = 0. \tag{8.47}$$

Because $e\nabla\phi = -e\mathbf{E}$ (from Eq. 8.46), Eq. (8.47) leads to the desired semiclassical expression for the Bloch electron in an electric field $\mathbf{E}$ and magnetic field $\mathbf{B}$,

$$\hbar\dot{\mathbf{k}} = -e\mathbf{E} - \frac{e}{c}\mathbf{V}_n(\mathbf{k}) \times \mathbf{B}. \tag{8.48}$$

We note that Eq. (8.48) is in no way a rigorous derivation of the motion of the Bloch electrons in an electric and magnetic field.

8.6.1 Time Evolution of Bloch Electrons in an Electric Field

If we use the expression for the electric field $\mathbf{E}$ through a time-dependent vector potential $\mathbf{A}$ that increases linearly with time and set the scalar potential $\phi(\mathbf{r})$ to zero, i.e.,

$$\mathbf{A} = -c\mathbf{E}t, \ \phi(\mathbf{r}) = 0, \tag{8.49}$$

we verify that

$$\mathbf{E} = -\frac{1}{c}\frac{\partial \mathbf{A}}{\partial t} - \nabla\phi = \mathbf{E}. \tag{8.50}$$

Such a vector potential $\mathbf{A}$, which allows the electric field $\mathbf{E}$ to coexist with the periodic boundary conditions (in a ring-shaped crystal) is shown in Figure 8.1.

The use of a time-dependent vector potential instead of a scalar potential allows us to use the periodic boundary conditions. We can generalize this method to a Bloch state under the influence of a time-dependent electric field, $E(t)$, turned on at $t=0$.[6]

The time-dependent Schrodinger equation for an electron in a Bloch state under the influence of an electric field, $E(t)$, turned on at $t=0$, is given by

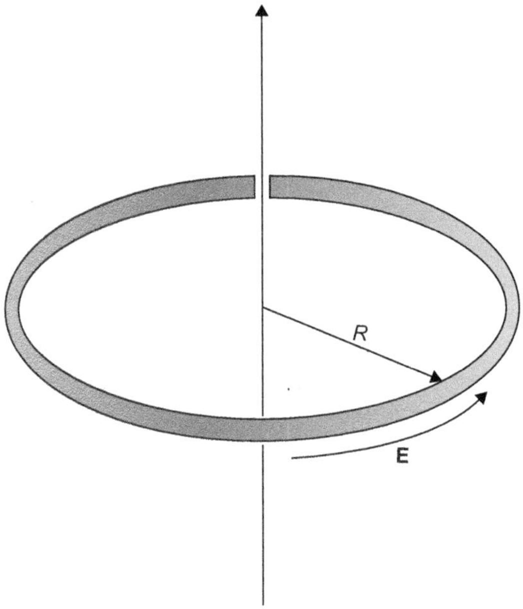

FIGURE 8.1

The vector potential $\mathbf{A} = -c\mathbf{E}t$ can be established in a ring-shaped crystal by changing the magnetic flux at a uniform rate $B_z = 2\pi RcEt\delta(\mathbf{r})$ through an infinite solenoid inside the ring.

$$H\psi(\mathbf{r},t) = \left[\frac{\left(\mathbf{p} + (e/c)\mathbf{A}\right)^2}{2m} + V(\mathbf{r})\right]\psi = i\hbar\frac{\partial\psi}{\partial t}, \tag{8.51}$$

where

$$\mathbf{A} = -c\int_0^t E(t')dt', \tag{8.52}$$

$V(\mathbf{r})$ is the lattice periodic potential, and the time-dependent electric field is included as a vector potential, as described earlier. We can obtain $\psi(\mathbf{r},t)$ from $\psi(\mathbf{r},0)$ by using eigenfunction expansion of which the elements are

$$\left[\frac{\left(\mathbf{p} + (e/c)\mathbf{A}\right)^2}{2m} + V(\mathbf{r})\right]\phi_i'(\mathbf{r},t) = \varepsilon_i(t)\phi_i'(\mathbf{r},t). \tag{8.53}$$

For each t, the Hamiltonian is invariant under a crystal lattice translation because $\mathbf{A}$ is independent of $\mathbf{r}$ for a homogeneous electric field, and $V(\mathbf{r})$ is periodic. If we substitute

$$\phi_i'(\mathbf{r},t) = e^{-ie\mathbf{A}\cdot\mathbf{r}/\hbar c}\phi_i(\mathbf{r},t), \tag{8.54}$$

in Eq. (8.53), we obtain

$$\left[\frac{p^2}{2m} + V(\mathbf{r})\right]\phi_i = \varepsilon_i\phi_i. \tag{8.55}$$

The solutions of Eq. (8.55) are the Bloch functions and energy band functions of the unperturbed crystal,

$$\phi_i = \phi_{n\mathbf{k}}(\mathbf{r}) \tag{8.56}$$

and

$$\varepsilon_i = \varepsilon_n(\mathbf{k}). \tag{8.57}$$

From Eqs. (8.54) and (8.56), we obtain

$$\phi_i'(\mathbf{r},t) = e^{-ie\mathbf{A}\cdot\mathbf{r}/\hbar c}\phi_{n\mathbf{k}}(\mathbf{r}). \tag{8.58}$$

Because H is invariant under a lattice translation, the allowed values of $\mathbf{k}$ are defined using periodic boundary conditions on the ϕ_i'. We obtain

$$\mathbf{k} - \frac{e\mathbf{A}}{\hbar c} = \sum_i \frac{n_i}{N_i}\mathbf{K}_i, \tag{8.59}$$

where $\mathbf{K}_i (i = 1, 2, 3)$ are the primitive reciprocal-lattice translation vectors, N_i are the number of cells in the i direction, and $-N_i/2 < n_i \le N_i/2$. For the periodic boundary condition to be satisfied, $\mathbf{k}$ must be a function of t and

$$\hbar\dot{\mathbf{k}} = \frac{e}{c}\dot{\mathbf{A}} = \frac{e}{c}(-c\mathbf{E}(t)) = -e\mathbf{E}(t). \tag{8.60}$$

From Eq. (8.59) and the time dependence of $\mathbf{A}(t)$, it follows that the Brillouin zone describing the allowed $\mathbf{k}(t)$ values is time dependent. The periodic boundary conditions that lead to Brillouin zone boundaries move in time. Thus, the wave vector does not undergo an Umklapp process (or U-process) back to the other side of the Brillouin zone, but continues with continuous values of $\mathbf{k}(t)$ in the time-dependent zone. Having considered the general case of a time-dependent electric field, if we restrict to a time-independent electric field,

$$\hbar\dot{\mathbf{k}} = -e\mathbf{E} \tag{8.61}$$

and

$$\mathbf{A} = -c\mathbf{E}t. \tag{8.62}$$

Thus, the **k** vectors are time dependent and should be written as $\mathbf{k}(t)$. From Eqs. (8.55) and (8.56), we also obtain the Bloch equation

$$\phi_{n\mathbf{k}(t)}(\mathbf{r}) = e^{i\mathbf{k}(t)\cdot\mathbf{r}} u_{n\mathbf{k}(t)}(\mathbf{r}). \tag{8.63}$$

The functions defined in (8.53) and (8.58) are also known as Houston functions. It may be noted that the Houston functions are not exact solutions of the Schrodinger equation because the eigenvalues are time dependent. The amplitude of Houston states with nearby **k** becomes nonzero and grows, a behavior typical of wave packets. In addition, the electron can jump from one band to another, which is known as Zener tunneling and which we will discuss later.

We note that if we use the Houston functions instead of the Bloch functions, we obtain expressions for the effective Hamiltonians by a using a similar technique followed earlier (Problem 8.5). The effective Hamiltonian for $U_{n\mathbf{k}}(\mathbf{r}, \mathbf{E}, t)$ is

$$H_{eff} = \frac{1}{2m}(-i\hbar\nabla + \hbar\mathbf{k} - e\mathbf{E}t)^2 + V(\mathbf{r}). \tag{8.64}$$

One important aspect to note is that the electric field **E** is incorporated through $\mathbf{k}(t)$ as described in Eq. (8.61).

When the electron jumps from one band to another because of a strong electric field, it is called *electric breakdown* or the *Zener breakdown.* It is easier for the electron to jump to a neighboring band in a magnetic field. This is known as *magnetic breakdown.* We will discuss these breakdowns later. We will describe the Zener breakdown after discussing an alternate (but essentially flawed) derivation of the motion of Bloch electrons in an electric and magnetic field. The controversy in this derivation is that one cannot use Bloch functions for Hamiltonians that are not periodic.

8.6.2 Alternate Derivation for Bloch Functions in an External Electric and Magnetic Field

The Hamiltonian in an electric and magnetic field can be written as

$$H = \frac{1}{2m}\left(\mathbf{p} + \frac{e\mathbf{A}}{c}\right)^2 + V(\mathbf{r}) + e\mathbf{E}\cdot\mathbf{r}. \tag{8.65}$$

In a small interval dt, the Bloch function $\psi = \psi_n(\mathbf{k}_0, \mathbf{r})$ at $t = 0$ will change by

$$\psi_n(\mathbf{k}, \mathbf{r}, dt) = e^{-i/\hbar H dt}\,\psi_n(\mathbf{k}_0, \mathbf{r}) \approx \left(1 - \frac{i}{\hbar}Hdt\right)\psi_n(\mathbf{k}_0, \mathbf{r}). \tag{8.66}$$

If we operate with the translation operator, with properties described in Eq. (4.47), we obtain

$$\hat{T}(\mathbf{R}_i)\psi = \hat{T}(\mathbf{R}_i)\left(1 - \frac{i}{\hbar}Hdt\right)\psi_n(\mathbf{k}_0, \mathbf{r}), \tag{8.67}$$

which can be written as

$$\hat{T}(\mathbf{R}_i)\psi = \left(1 - \frac{i}{\hbar}Hdt\right)\hat{T}(\mathbf{R}_i)\psi_n - \frac{i}{\hbar}dt[T(\mathbf{R}_i), H]\psi_n. \tag{8.68}$$

It can be shown that (Problem 8.6)

$$[\hat{T}(\mathbf{R}_i), H] = \left(\vec{\xi} \cdot \mathbf{R}_i + \frac{e^2 B^2}{2m} R_{ix}^2 \right) \hat{T}(\mathbf{R}_i), \tag{8.69}$$

where

$$\vec{\xi} = \frac{e}{c}\mathbf{v} \times \mathbf{B} + e\mathbf{E} + \frac{e}{c}\dot{\mathbf{A}}. \tag{8.70}$$

We neglect the last term in Eq. (8.69) in a weak magnetic field approximation and obtain

$$\hat{T}(\mathbf{R}_i)\psi = \left[1 - \frac{i}{\hbar}(H + \vec{\xi} \cdot \mathbf{R}_i)dt\right] \hat{T}(\mathbf{R}_i)\psi_n = (\ldots)e^{i\mathbf{k}_0 \cdot \mathbf{R}_i}\psi_n = e^{i\mathbf{k} \cdot \mathbf{R}_i}\psi. \tag{8.71}$$

Here,

$$\mathbf{k} = \mathbf{k}_0 - \frac{\vec{\xi}}{\hbar} dt \tag{8.72}$$

or

$$\hbar\dot{\mathbf{k}} = -\vec{\xi}. \tag{8.73}$$

If we neglect the last term in Eq. (8.70), we obtain

$$\hbar\dot{\mathbf{k}} = -e\left(\mathbf{E} + \frac{\mathbf{v} \times \mathbf{B}}{c}\right). \tag{8.74}$$

Eq. (8.74) is the standard Lorentz force equation.

8.6.3 Motion in an Applied DC Field

Eq. (8.61) can be rewritten in the alternate form

$$\mathbf{k}(t) = \mathbf{k}(0) - \frac{e\mathbf{E}t}{\hbar}, \tag{8.75}$$

from which we obtain

$$\mathbf{v}_n(\mathbf{k}(t)) = \mathbf{v}_n\left(\mathbf{k}(0) - \frac{e\mathbf{E}t}{\hbar}\right). \tag{8.76}$$

We note from Eq. (8.42) that $\mathbf{v}_n(\mathbf{k})$ is periodic in the reciprocal lattice vector $\mathbf{K}$ and when $\mathbf{E} \| \mathbf{K}$, Eq. (8.76) is oscillatory. This oscillatory behavior in one dimension is shown in Figure 8.2. The velocity, which is linear in k near the band minimum, reaches a maximum and then decreases to zero at the Brillouin zone boundary. It is surprising

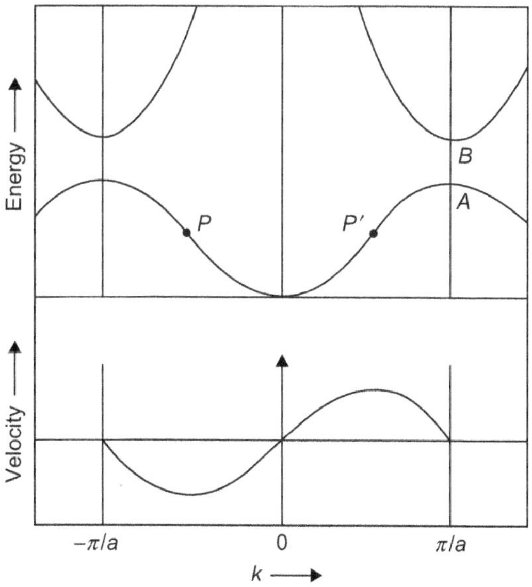

FIGURE 8.2

$v(k)$ and $\varepsilon(k)$ versus k in one dimension.

to note that in the region in which the velocity decreases with increasing k, the acceleration of the electron is opposite to the external applied field.

We first note the periodic behavior of the $\varepsilon(k)$ versus k curve in Figure 8.2. The Bloch electrons have positive effective masses at $k = 0$ and negative effective masses at $k = \pm\pi/a$. The oscillatory behavior of the velocity of the Bloch electron with k (and with time that is related to k through Eq. 8.69) is a consequence of the effect of the periodic potential that exerts an additional force. Hypothetically, the external DC electric field could induce alternating current, but collisions with phonons and impurities present in a crystal do not allow any such possibility. In fact, in the absence of damping, a perfect crystal with a periodic potential would not have any resistivity at zero temperature.

8.7 BLOCH OSCILLATIONS

The Bloch oscillations result from the fact that $\varepsilon_{n\mathbf{k}}$ is a periodic function of $\mathbf{k}$. For example, we consider the energy of a band in one dimension using the tight-binding method. From Eq. (5.13), we obtain the expression for energy of a one-dimensional crystal of lattice constant a (in the x direction) as

$$\varepsilon(k) = \varepsilon_a - \beta - 2\gamma \cos ka, \tag{8.77}$$

where the constants were defined in Eq. (5.13). When there is an external electric field E parallel to the linear chain of the lattice (in the x direction), we have from Eq. (8.61)

$$\hbar\dot{k} = -eE \tag{8.78}$$

and

$$k = -eEt/\hbar. \tag{8.79}$$

From Eqs. (8.42), (8.77), and (8.79), we have

$$\frac{dr}{dt} = \dot{r} = \frac{1}{\hbar}\frac{\partial\varepsilon}{\partial t} = -\frac{2a\gamma}{\hbar}\sin\left(\frac{aeEt}{\hbar}\right). \tag{8.80}$$

Integrating with respect to t, we obtain

$$r = \frac{2\gamma}{eE}\cos\left(\frac{aeEt}{\hbar}\right). \tag{8.81}$$

Eq. (8.81) implies that when the external electric field is very large, the Bloch electrons would oscillate around a mean position. Instead of being a good conductor, a metal would become an insulator. However, it can be easily shown that a small amount of damping can destroy this oscillation.

The existence of such Bloch oscillations was experimentally observed by Dahan et al (Ref. 3). Dahan et al. prepared ultracold cesium

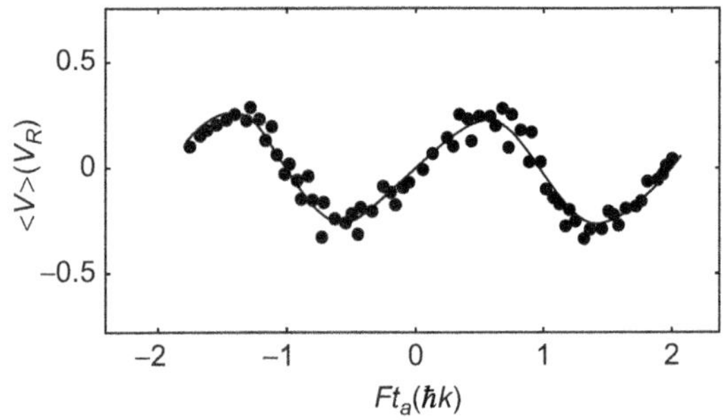

FIGURE 8.3

Bloch oscillations of cesium atoms in an optical periodic potential driven by a constant external force observed experimentally. The negative values of Ft_a were measured by changing the sign of F.

Reproduced from Dahan et al.[3] with the permission of the American Physical Society.

atoms in the ground-state energy band of the potential induced by an optical standing wave. The periodic potential results from the light shift of the ground state of cesium atoms illuminated by a laser standing wave. The constant external force was mimicked by introducing a tunable frequency difference $\delta\nu(t)$ between two counter propagating laser waves creating the optical potential. The reference frame in which the optical potential is stationary moves with velocity $\delta\nu(t)\lambda/2$. For a linear variation of time in $\delta\nu(t)$, a constant inertial force $F = -ma = -m\lambda\frac{d}{dt}\delta\nu(t)/2$ is exerted on the atoms in this frame. They observed Bloch oscillations of the atoms driven by this constant inertial force shown in Figure 8.3. The recoil velocity of the cesium atom was 0.35 cm/s, and the acceleration was $\pm 0.85\,\text{m/s}^2$. Figure 8.3 shows the results for potential depth $U_0 = 4.4\,E_R$.

8.8 HOLES

We can write the expression for the contribution of all the electrons in a given band to the current density as

$$\mathbf{j} = -e \int_{occupied} \frac{d\mathbf{k}}{4\pi^3}\mathbf{v}(\mathbf{k}). \tag{8.82}$$

Because the current in a completely filled band is zero, we can also write

$$\mathbf{j} = e \int_{unoccupied} \frac{d\mathbf{k}}{4\pi^3}\mathbf{v}(\mathbf{k}). \tag{8.83}$$

If only one electron is missing from the occupied band, the current is

$$\mathbf{j} = e\mathbf{v}(\mathbf{k}). \tag{8.84}$$

The absence of an electron in state **k** from an otherwise filled band is known as a "hole" in state **k**. There are a couple of important physical properties to note about a hole:

1. The hole is a positive charge, as evidenced by Eq. (8.84).
2. The velocity of the "wave packet" of the hole will be the same as the velocity of the electrons in either side of the hole, i.e.,

$$\mathbf{v}_{n\mathbf{k}} = \frac{1}{\hbar}\frac{\partial\varepsilon_n(\mathbf{k})}{\partial\mathbf{k}}. \tag{8.85}$$

There are several ways in which a hole can be formed in a filled band. An electron can absorb a photon and can be excited to the empty band above the filled band. Similarly, an electron can be thermally excited at room temperature to the empty band above the filled band provided the energy gap is small, as is the case for intrinsic semiconductors like Si or Ge. In fact, "holes" play a significant role in semiconductors, which is the topic of Chapter 9. The formation of a hole by the absorption of a photon by an electron in state **k** to an empty band is shown in Figure 8.4.

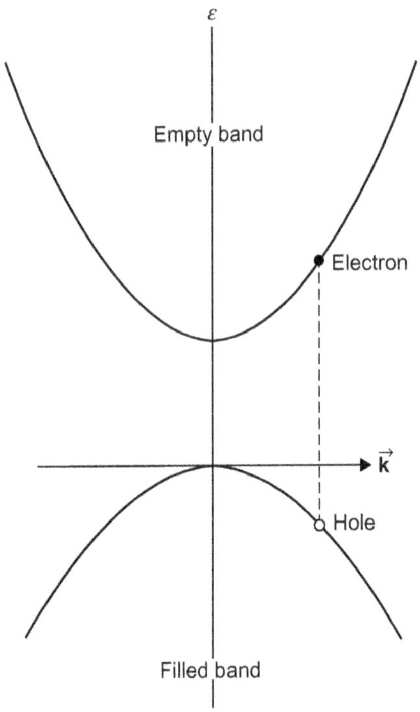

FIGURE 8.4

An electron of vector **k** absorbs a photon and is excited to the empty band, leaving behind a positively charged "hole" of the same vector **k** in the filled band.

Because the unoccupied levels normally lie near the top of the band and the band energy $\varepsilon_n(\mathbf{k})$ has its maximum value at $\mathbf{k}_m$, we can expand $\varepsilon_n(\mathbf{k})$ about $\mathbf{k}_m$ assuming that $\mathbf{k}$ is sufficiently close to $\mathbf{k}_m$. We obtain (assuming that $\mathbf{k}_m$ is a point of high symmetry)

$$\varepsilon_n(\mathbf{k}) \approx \varepsilon_n(\mathbf{k}_m) + a(\mathbf{k} - \mathbf{k}_m) - b(\mathbf{k} - \mathbf{k}_m)^2 + \ldots \tag{8.86}$$

However, the linear term on the right side will vanish because $\varepsilon_n(\mathbf{k}_m)$ is a maximum and the coefficient of b will be negative. From Eqs. (8.85) and (8.86), we obtain

$$\mathbf{v}_n(\mathbf{k}) = -\frac{2}{\hbar} b\mathbf{k} \tag{8.87}$$

and

$$\frac{d}{dt}\mathbf{v}_n(\mathbf{k}) = \mathbf{a} = -\frac{2}{\hbar} b\dot{\mathbf{k}}. \tag{8.88}$$

Comparing Eq. (8.88) with Eq. (8.34) for the effective mass tensor, we obtain

$$\hbar \mathbf{M}_n^{-1}(\mathbf{k}) \cdot \dot{\mathbf{k}} = -\frac{2}{\hbar} b\dot{\mathbf{k}} = \mathbf{a}. \tag{8.89}$$

Thus, the inverse mass tensor is negative if $\mathbf{k}$ is near a band maximum. Because a negative mass is physically unacceptable, a *hole* can be considered as a *positive charge* with a *positive effective mass.*

Thus, the general expression of the equation of motion can be written as

$$\mathbf{M}_n(\mathbf{k}) \cdot \mathbf{a} = \hbar \dot{\mathbf{k}}. \tag{8.90}$$

From Eqs. (8.74) and (8.90), we obtain

$$\mathbf{M}_n(\mathbf{k}) \cdot \mathbf{a} = \mp e(\mathbf{E} + \frac{1}{c}\mathbf{v}_n(\mathbf{k}) \times \mathbf{B}). \tag{8.91}$$

It is easier to understand the motion of the electrons and holes if we consider only the electric field $\mathbf{E}$. In that case, Eq. (8.91) can be rewritten as

$$\mathbf{M}_n(\mathbf{k}) \cdot \mathbf{a} = \mp e\mathbf{E}. \tag{8.92}$$

As shown in Figure 8.5, the symbol ∘ indicates the position of the hole in **k** space. However, the behavior of the hole, which is governed by its **k** vector, is marked ⊗. In an electric field, the electron and hole states in the same band, which are marked (•) and (∘), respectively, always move rigidly like beads on a string. However, *d***k**/*dt* for these two particles, which have opposite charges, are of opposite signs. For the holes, *d***k**/*dt* is *always positive,* and the points ⊗ move in the same direction as the applied electric field. Because the electrons have negative charge, the electron state (•) moves opposite to the direction of the electric field.

The positive and negative effective mass (in units of mass of cesium atom) as a function of the potential depth, when $k=0$ and $k=K$, respectively (in a simulated periodic potential), were demonstrated in the experiment described previously. The experimental results are shown in Figure 8.6.

It is often convenient to describe the physical behavior in terms of the hole states, especially when a Brillouin zone is nearly full. An example is shown in Figure 8.7, which shows a plane

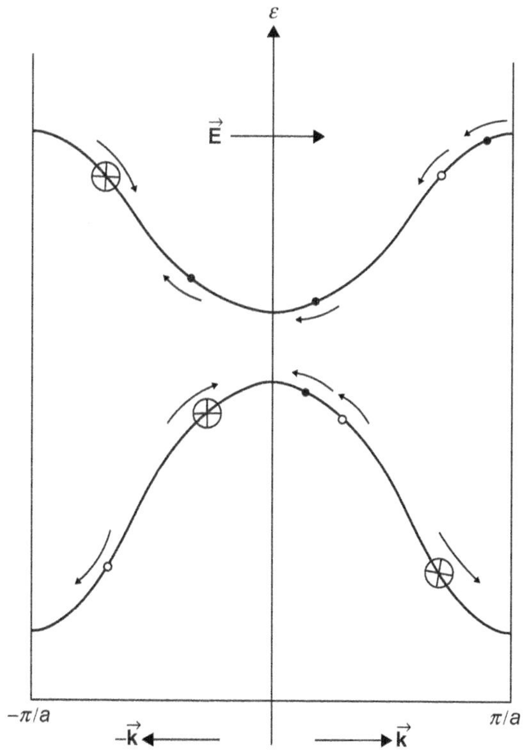

FIGURE 8.5

The electrons (•) and hole (∘) states always move together under the action of an electric field **E,** as shown by arrows. The behavior of the holes is governed by its **k** vector and is marked ⊗.

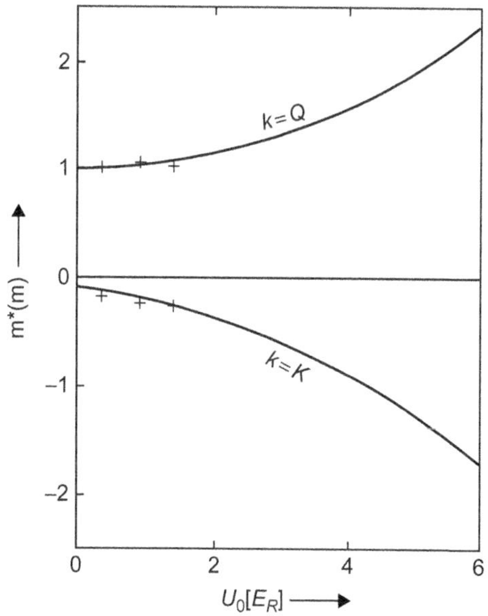

FIGURE 8.6

Effective masses m^* for $k=0$ and $k=K$ (in units of cesium atomic mass) versus potential depth U_0.

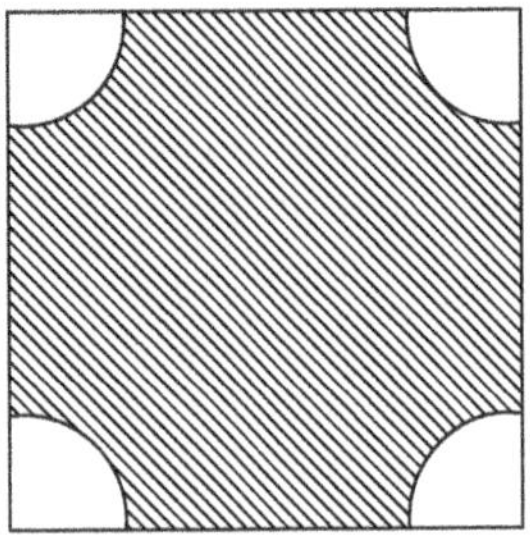

FIGURE 8.7

Occupied states in the first zone shown by a plane section perpendicular to (001).

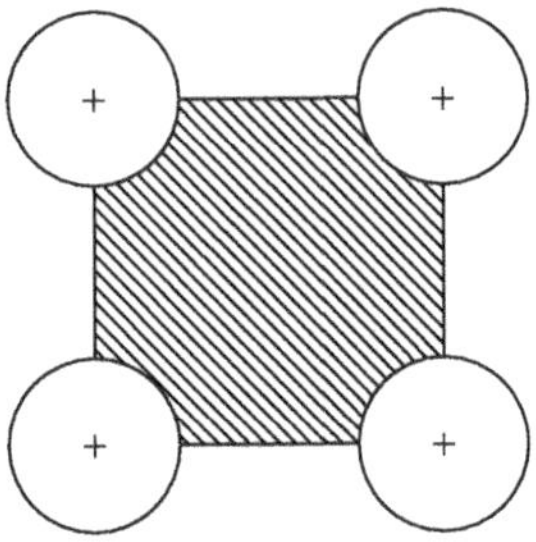

FIGURE 8.8

The circular areas of hole states at the zone corners (in two dimensions). They are spheres in three dimensions.

section perpendicular to (001) in a cubic zone structure where the occupied states are only in the first zone.

We note that the empty states of the nearly filled zone are in the corners of the zone. These empty states form spheres of hole states in the repeated zone scheme in the three-dimensional **k** space, a two-dimensional equivalent of which is shown in Figure 8.8. It is much easier to deal with these spheres of positive charges (holes) than the electrons in the Brillouin zone. The Fermi surface of divalent metals has small sections of holes in the first Brillouin zone as a consequence of electrons spilling over into the second Brillouin zone. The electrons in the second zone are ellipses of occupied states (ellipsoids in three dimensions), as shown in Figure 8.9.

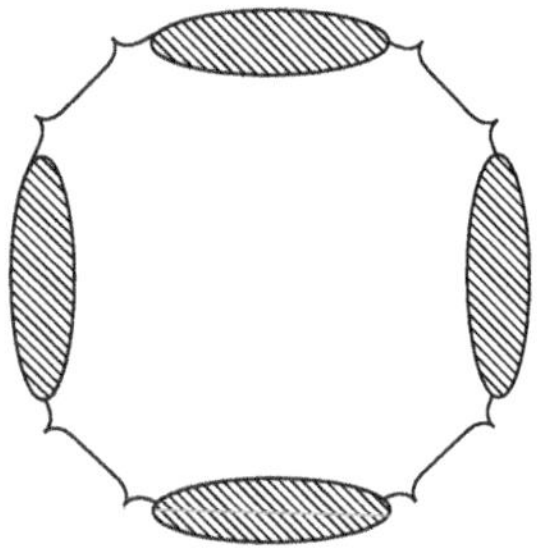

FIGURE 8.9

The electrons of the second zone in the Fermi surface of a divalent metal in the repeated zone scheme. The major portion of the first zone is filled with electrons, but these states are replaced by the hole states.

8.9 ZENER BREAKDOWN (APPROXIMATE METHOD)

We will consider the Zener breakdown (Ref. 12) in an electric field in two alternate ways. First, we will consider a simple and traditional method that is very controversial because the fundamental postulate on which the Bloch functions and the concept of the Brillouin zones have been built—i.e., the potential is periodic and the crystal has symmetry in the sense that one can consider the opposite ends to be equivalent—is lost in an external electric field. In fact, as we have seen, the wave vector **k** becomes time dependent, and the Brillouin zone boundary keeps moving with time. Nevertheless, we will first discuss the simple case neglecting these objections. Later, we will discuss the more rigorous theory of Zener breakdown (Ref. 12).

When we consider the motion of a Bloch electron (in one dimension) in an external electric field **E** along the x-axis, there are two ways in which we can represent the motion of the electron. The motion along the **k** space is shown in Figure 8.10 in the repeated zone scheme. These are the familiar Bloch oscillations in which the electron travels from 0 to A and then to B and C, and so on. The zone boundaries A and C are equivalent, and one can describe the oscillatory motion of the electron by stating that the electron has jumped from C to A.

However, the path of the electron in the external electric field in real space (in one dimension) shown in Figure 8.11 appears very different. The electron slows down as it moves from 0 to A, where it has a Bragg reflection. It cannot go forward because there is an energy gap at A, and the electron is forbidden to move in that region. The electron reverses direction until it reaches B and then accelerates again until it reaches the zone boundary at C.

In addition, in a strong electric field E, the bands are tilted as shown in Figure 8.12.

An electron moving from P to Q will be reflected back into the band, or it can move from Q to R by crossing the energy gap ε_g if the electric field is strong enough to satisfy the condition $\varepsilon_g = eEd$, where $d =$ QR. Thus, we obtain

$$d \approx \frac{\varepsilon_g}{eE}. \tag{8.93}$$

It may be noted that QR is a forbidden region, and the electron has to tunnel through this region. This tunneling problem was originally solved by Zener by using WKB approximation. However, this is a semiclassical derivation with a lot of controversies because the periodicity of the Bloch functions is lost in an electric field. We will first outline a brief derivation of tunneling

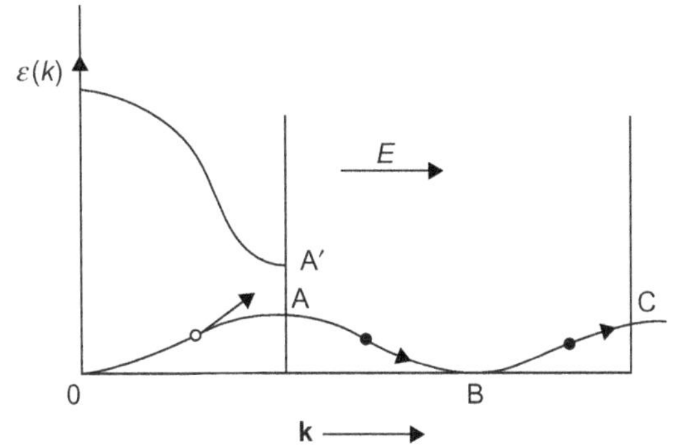

FIGURE 8.10

Electron trajectory in **k** space.

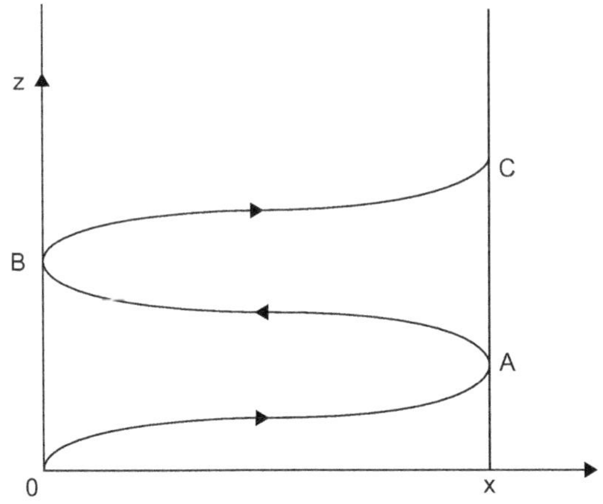

FIGURE 8.11

Path of electron in real space in an electric field E along the x axis.

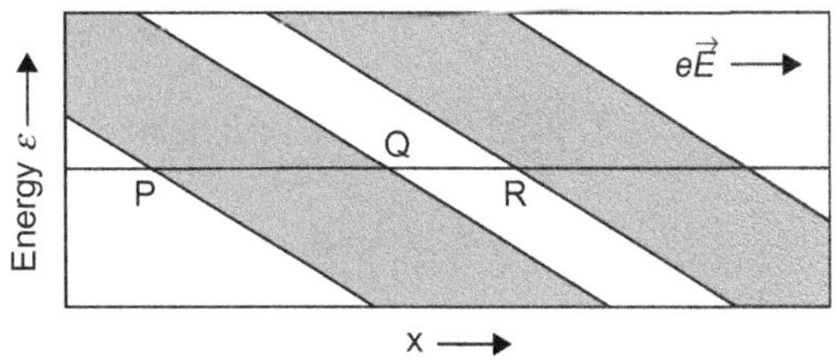

FIGURE 8.12

The energy bands in the $\varepsilon - x$ diagram are tilted in an applied electric field.

using the WKB approximation and then derive the effect of electric field on the Bloch functions for derivation of a more rigorous formula.

The tunneling of an electron in a square potential barrier is shown in Figure 8.13. This is solved by the WKB approximation.[4]

The approximate solution of the Schrodinger equation in the WKB approximation is

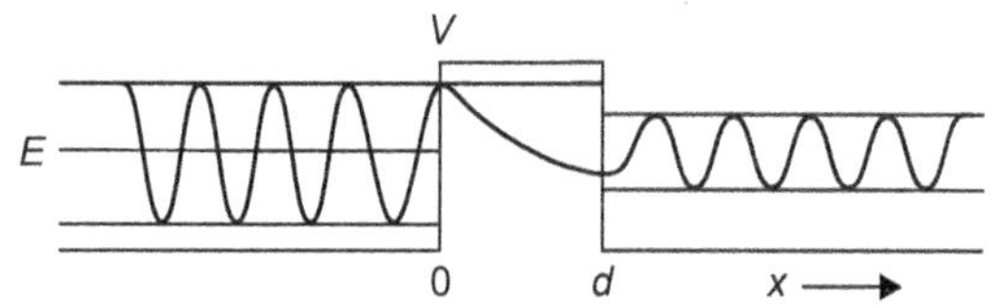

FIGURE 8.13

Square potential barrier. Penetration of a barrier by a wave function $\psi(x)$ when $V > d$.

$$\psi(x) = Ae^{i/\hbar \int^{x} dx \sqrt{2m(\varepsilon - V)}}. \tag{8.94}$$

If $V - \varepsilon = u$, where u is positive, an electron wave function that starts on the left side of Figure 8.13 and travels distance d through the gap will have a transmission coefficient[7]

$$T = e^{-2d\sqrt{2mu/\hbar^2}}. \tag{8.95}$$

However, in the case of a Bloch electron moving on an electric field, if ε_v and ε_c are the energies of the valence and conduction bands, respectively, while $\varepsilon - \varepsilon_v$ is negative in the energy gap, $\varepsilon_c - \varepsilon = 0 = \varepsilon_v - \varepsilon$ at the zone boundaries, and once the electron tunnels to the conduction band, $\varepsilon_v - \varepsilon$ is positive. If we include these specific conditions for tunneling through the bands, a rough estimate of the Zener tunneling problem can be made by rewriting the WKB approximation, and the probability for an electron to tunnel from one band to another is of the form

$$T = e^{2i/\hbar \int_0^d dx \sqrt{2m\sqrt{(\varepsilon_c - \varepsilon)(\varepsilon_v - \varepsilon)}}}. \tag{8.96}$$

Eq. (8.96) can be rewritten in the alternate form (in a very rough approximation)

$$T \approx e^{-2d\sqrt{\frac{2m\varepsilon_g}{\hbar^2}}}. \tag{8.97}$$

From Eqs. (8.87) and (8.97), we obtain the expression for tunneling probability as

$$T \approx e^{-\frac{2\varepsilon_g}{eE}\sqrt{\frac{2m\varepsilon_g}{\hbar^2}}}. \tag{8.98}$$

The probability of the electron to tunnel through the barrier obtained in Eq. (8.98) is a *very rough estimate* of the tunneling problem.

8.10 RIGOROUS CALCULATION OF ZENER TUNNELING

We expanded $\psi(\mathbf{r},t)$—in Eqs. (8.53) and (8.58)—in terms of the Houston states

$$\psi(\mathbf{r},t)=\sum_i a_i(t)\phi_i'(\mathbf{r},t), \tag{8.99}$$

which can be rewritten in the alternate form from Eq. (8.58)

$$\psi(\mathbf{r},t)=\sum_{n'\mathbf{k}'} a_{n'\mathbf{k}'}(t)e^{-ie\mathbf{A}\cdot\mathbf{r}/\hbar c}\phi_{n'\mathbf{k}'(t)}(\mathbf{r}), \tag{8.100}$$

$$H\psi(\mathbf{r},t)=\sum_{n'\mathbf{k}'} a_{n'\mathbf{k}'}(t)\varepsilon_{n'}(\mathbf{k}'(t))e^{i(\mathbf{k}'-e\mathbf{A}/\hbar c)\cdot\mathbf{r}}U_{n'\mathbf{k}'(t)}(\mathbf{r}). \tag{8.101}$$

Here, we replaced $a_i(t)$ by $a_{n'\mathbf{k}'}(t)$ and replaced the summation over i by $n'k'$ and expressed the Houston functions as

$$\phi_{n\mathbf{k}(t)}(\mathbf{r})=e^{i\mathbf{k}\cdot\mathbf{r}}U_{n\mathbf{k}(t)}(\mathbf{r}). \tag{8.102}$$

We also obtain

$$i\hbar\frac{\partial}{\partial t}\psi(\mathbf{r},t)=i\hbar\sum_{n'\mathbf{k}'}\frac{\partial a_{n'\mathbf{k}'}(t)}{\partial t}e^{i\left(\mathbf{k}'-\frac{e\mathbf{A}}{\hbar c}\right)\cdot\mathbf{r}}U_{n'\mathbf{k}'(t)}(\mathbf{r})+i\hbar\sum_{n'\mathbf{k}'}a_{n'\mathbf{k}'}(t)e^{i\left(\mathbf{k}'-\frac{e\mathbf{A}}{\hbar c}\right)\cdot\mathbf{r}}\frac{\partial}{\partial t}U_{n'\mathbf{k}'}(\mathbf{r}). \tag{8.103}$$

Here, we used the condition that $\mathbf{k}'-\frac{e\mathbf{A}}{\hbar c}$ is time independent from Eq. (8.59).

Further,

$$\frac{\partial}{\partial t}U_{n\mathbf{k}(t)}(\mathbf{r})=\frac{\partial}{\partial k_x}U_{n\mathbf{k}}\frac{dk_x}{dt}=-\frac{eE}{\hbar}\frac{\partial}{\partial k_x}U_{n\mathbf{k}}. \tag{8.104}$$

Substituting Eqs. (8.101) through (8.103) in Eq. (8.51), multiplying both sides by $e^{-i\left(\mathbf{k}-\frac{e\mathbf{A}}{\hbar c}\right)\cdot\mathbf{r}}U^*_{n\mathbf{k}}(\mathbf{r})$, using $\mathbf{E}=\hat{i}E(t)$, integrating over the volume of the crystal, and using the orthonormal conditions of the Bloch functions (Problem 8.9), we obtain

$$\varepsilon_n(\mathbf{k},t)a_{n\mathbf{k}(t)}=i\hbar\frac{\partial a_{n\mathbf{k}(t)}}{\partial t}-eE(t)\sum_{n'}A_{nn'}(\mathbf{k}(t))a_{n'\mathbf{k}(t)}, \tag{8.105}$$

where

$$A_{nn'}(\mathbf{k}(t))\equiv -i\int U^*_{n\mathbf{k}(t)}\frac{\partial}{\partial k_x}U_{n'\mathbf{k}(t)}d\tau. \tag{8.106}$$

From Eq. (8.105), it follows that the coefficients $a_{n\mathbf{k}(t)}$ are coupled only to $a_{n'\mathbf{k}(t)}$, i.e., coupled to the same $\mathbf{k}$. Therefore, for a given $\mathbf{k}$, we can write

$$a_{n\mathbf{k}(t)}=\alpha_n(t)e^{-\frac{i}{\hbar}\int_0^t \varepsilon_n(\mathbf{k}(t')dt'}. \tag{8.107}$$

Substituting Eq. (8.107) in Eq. (8.105), we obtain

$$\dot{\alpha}_n(t) = \frac{ieE(t)}{\hbar} \sum_{n'} \alpha_{n'}(t) A_{nn'}(\mathbf{k}(t)) \exp\left[\frac{i}{\hbar}\int_0^t [\varepsilon_n(\mathbf{k}(t')) - \varepsilon_{n'}(\mathbf{k}(t'))]dt'\right]. \tag{8.108}$$

This general expression can be used for derivation of Wannier–Stark ladders for time-dependent E as well as for Zener tunneling between the valence and conduction bands.

At $t = 0$, before the field E is turned on, $\mathbf{k}(t) = \mathbf{k}(t = 0)$, the electron state can be described by a Bloch wave in band n with $\mathbf{k}(t) = \mathbf{k}(t = 0)$; then at $t = 0$,

$$\alpha_{n'} = \delta_{nn'}. \tag{8.109}$$

After a time $T = -\frac{\hbar K}{eE}$, where T is the period of one Bloch oscillation, which is a sufficiently short time so that $\alpha_{n'} \ll 1$, $n' \neq n$, we can substitute Eq. (8.109) in Eq. (8.108) to obtain

$$\alpha_{n'}(t) = \int_0^t \frac{ieE(t')}{\hbar} A_{n'n}(\mathbf{k}(t')) \exp\left[\frac{i}{\hbar}\int_0^{t'} [\varepsilon_{n'}(\mathbf{k}(t'')) - \varepsilon_n(\mathbf{k}(t''))]dt''\right] dt'. \tag{8.110}$$

We consider the case in which E is constant and is in the direction of the reciprocal-lattice vector K (which is along the x axis) and use $\hbar\dot{k} = -eE$ to change the variables

$$dt = -\frac{\hbar}{eE} dk. \tag{8.111}$$

From Eqs. (8.110) and (8.111), it can be shown that the transmission probability per period T,

$$P_{nn'} = |\alpha_{n'}(T)|^2 = \left| \int_{-K/2}^{K/2} A_{n'n}(\mathbf{k}) \exp\left[-\frac{i}{eE}\int_0^{k_x} [\varepsilon_{n'}(k_x', \mathbf{k}_\perp) - \varepsilon_n(k_x', \mathbf{k}_\perp)]dk_x'\right] dk_x \right|^2. \tag{8.112}$$

Any calculation of $P_{nn'}$ requires actual band calculations for the particular metal or the semiconductor to calculate $A_{n'n}(\mathbf{k})$, and evaluation of the integrals requires the use of a computer. However, in what follows, we will make several drastic approximations to derive an expression for the Zener tunneling between a valence band and a conduction band. In the process, we will explain why the Zener tunneling between the two bands is much easier for semiconductors than for metals.

To be able to consider Zener tunneling, we consider two parabolic bands that are of the form

$$\varepsilon_{1k} = \varepsilon_v - \frac{\hbar^2 k^2}{2m_v^*} \tag{8.113}$$

and

$$\varepsilon_{2k} = \varepsilon_c + \frac{\hbar^2 k^2}{2m_c^*}. \tag{8.114}$$

Thus,

$$\varepsilon_{2k} - \varepsilon_{1k} = \varepsilon_g + \frac{\hbar^2 k^2}{2m^*}, \tag{8.115}$$

where $\varepsilon_g = \varepsilon_c - \varepsilon_v$, and m^* is the reduced mass defined by

$$\frac{1}{m^*} = \frac{1}{m_v^*} + \frac{1}{m_c^*}. \tag{8.116}$$

With these conditions, Eq. (8.105) can be rewritten as

$$\varepsilon_{1\mathbf{k}(t)} a_{1\mathbf{k}(t)} = i\hbar \frac{\partial a_{1\mathbf{k}(t)}}{\partial t}, \tag{8.117}$$

from which we obtain

$$a_{1\mathbf{k}(t)} = \exp\left[-\frac{i}{\hbar}\int_0^t dt' \varepsilon_{1\mathbf{k}}(t')\right]. \tag{8.118}$$

If the time interval is short such that $\alpha_{n'} \ll 1$, $n' \neq n$, we can substitute Eq. (8.118) in Eq. (8.108) to obtain an expression for the rate of tunneling from band 1 to band 2 after time t as

$$\alpha_2(t) = \int_0^t \frac{ieE}{\hbar} A_{21}(\mathbf{k}(t'))\exp\left[\frac{i}{\hbar}\int_0^{t'} [\varepsilon_{2\mathbf{k}(t'')} - \varepsilon_{1\mathbf{k}(t'')}]dt''\right] dt'. \tag{8.119}$$

For a linear lattice, it can be shown that apart from the oscillatory terms in k, $iA_{21} \approx a$, where a is the lattice constant. Further, if τ is the time needed for k to move by a reciprocal lattice vector $K = 2\pi/a$,

$$\alpha_2(\tau) \approx a \int_0^{2\pi/a} dk \exp\left[\frac{-i}{eE}\int_0^k dk'(\varepsilon_{2k'} - \varepsilon_{1k'})\right]. \tag{8.120}$$

Here, we used the relation $\hbar\dot{k} = -eE$ to change the variables $dt = -\frac{\hbar}{eE}dk$. From Eqs. (8.115) and (8.120), we obtain

$$\alpha_2(\tau) \approx a \int_0^{2\pi/a} dk \exp\left[\frac{-i}{eE}\int_0^k dk'\left(\varepsilon_g + \frac{\hbar^2 k'^2}{2m^*}\right)\right]. \tag{8.121}$$

The integral is impossible to perform exactly, but an approximate result achieved by using the method of steepest descent yields the expression for the rate of tunneling:

$$\alpha_2(\tau) \approx a \exp\left[-\frac{2\varepsilon_g^{3/2}}{3eE}\sqrt{\frac{2m^*}{\hbar^2}}\right]. \tag{8.122}$$

The probability of tunneling, which is given by $|\alpha_2(\tau)|^2$, is much smaller in metals than in semiconductors because both the band gaps and the effective masses are much smaller for semiconductors.

8.11 ELECTRON–PHONON INTERACTION

The electron–phonon interaction process is basically the absorption (annihilation) or emission of a phonon $(\mathbf{q}, \lambda)$ with a simultaneous change of the electron states from $|\mathbf{k}, \sigma>$ to $|\mathbf{k} \pm \mathbf{q}, \sigma>$. Here, σ is the spin index and $\lambda = x, y, z$ directions. These two processes are shown as (a) and (b) in Figure 8.14 (the spin index σ and λ are not shown).

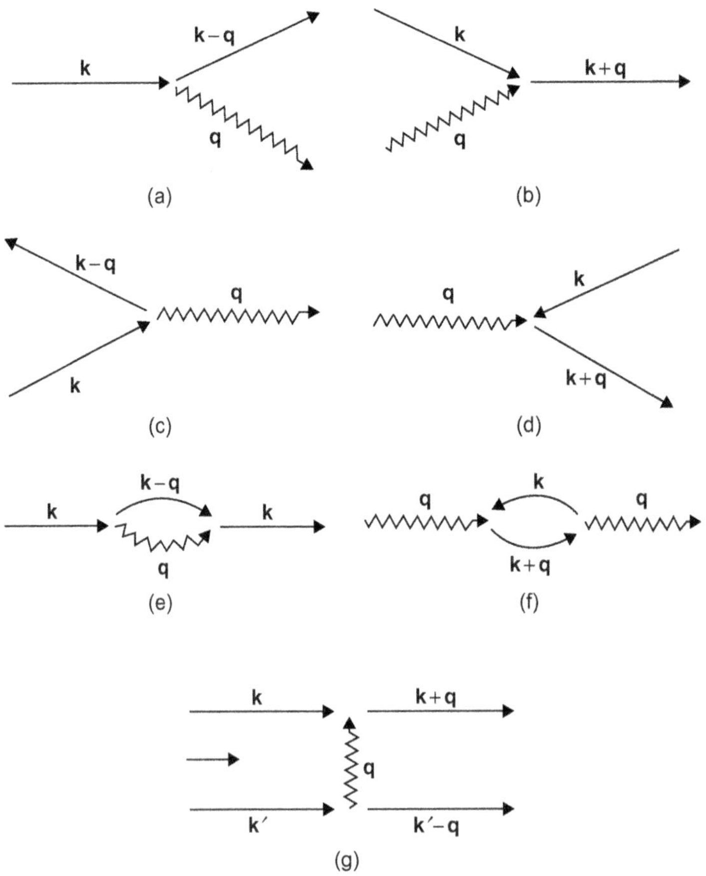

FIGURE 8.14

Graphs for various types of electron–phonon interaction. The Umklapp process is not shown here.

Process (a) is called phonon emission, and process (b) is called phonon absorption. In processes (c) and (d), the time axis runs from left to right. The electrons are assumed to run backward in time, and holes run forward in time. Process (c) describes the recombination of an electron-hole pair where a phonon is emitted. Process (d) describes the generation of an electron-hole pair by a phonon. One can describe processes (a) through (d) by using first-order perturbation theory. The conservation laws for the sum of the wave vectors (momentum) and energy are obeyed.

Processes (e) through (g) are due to contributions of perturbation calculations of higher order. Process (e) describes the emission and reabsorption of a virtual phonon. Process (f) describes the emission and reabsorption of a virtual electron-hole pair. Process (g) describes the electron–electron interaction of a virtual phonon. In these graphs, the intermediate states are not stationary states of the system. The laws of conservation of energy do not apply to these intermediate states because these processes occur in a very short time. The uncertainty relation between energy and time holds for these virtual states. The electron–electron interaction by exchange of virtual phonons will be neglected in the following discussions.

The Hamiltonian of a crystal can be written as

$$H_c = H_{el} + H_{ion} + H_{el-ion}. \tag{8.123}$$

Here, the contribution due to exchange is neglected. Using the Born–Oppenheimer (adiabatic) approximation discussed in Chapter 7, we can separate the movement of the electrons in a stationary lattice and the movement of the ions in a uniform space of electrons. The Hamiltonian H_{ion} is considered separately and is used to study lattice dynamics, which we considered in Chapter 2. We write

$$H_{el-ion} = H^0_{el-ion} + H_{el-ph}, \tag{8.124}$$

where H^0_{el-ion} describes the electron interaction with the periodic potential in the one-electron band model.

We express

$$H_e = H_{el} + H^0_{el-ion}. \tag{8.125}$$

Thus, we can write the Schrodinger equation for the electrons—from Eqs. (8.123) through (8.125)—as

$$H\psi = (H_e + H_{el-ph})\psi = E\psi. \tag{8.126}$$

We also note that we can write

$$H_{el-ion} = \sum_{j,i,\alpha} V(\mathbf{r}_j - \mathbf{R}_{i\alpha}(t)), \tag{8.127}$$

where $\mathbf{r}_j$ is the position of an electron, and $\mathbf{R}_{i\alpha}(t) = \mathbf{R}_i + \mathbf{d}_\alpha + \mathbf{u}_{\alpha i}(t)$ (see Figure 2.1 in Chapter 2). Here, the position vector of the ion is the sum of the equilibrium position of the αth ion in the ith Wigner–Seitz cell, and $\mathbf{u}_{\alpha i}(t)$ is the instantaneous deviation of the αth ion from the equilibrium position.

We assume that the interaction potential depends only on the electron–ion separation (Nordheim's rigid ion model). We can expand the potential

$$V_\alpha(\mathbf{r}_j - \mathbf{R}_{i\alpha} - \mathbf{u}_{\alpha i}) \approx V_\alpha(\mathbf{r}_j - \mathbf{R}_{i\alpha}) - \sum_{\alpha ij} \mathbf{u}_{\alpha i} \cdot \nabla V_\alpha(\mathbf{r}_j - \mathbf{R}_{i\alpha}). \tag{8.128}$$

From Eqs. (8.22), (8.127), and (8.128), we obtain

$$H_{el-ph} = -\sum_{\alpha ij} \mathbf{u}_{\alpha i} \cdot \nabla V_\alpha(\mathbf{r}_j - \mathbf{R}_{i\alpha}). \tag{8.129}$$

In Eq. (2.159), we also derived (note that we have changed the notation from **k** to **q** because we will use **k** for electrons and generalized to a lattice with a basis, and from n to α)

$$\mathbf{u}_{\alpha i} = \frac{1}{\sqrt{M_\alpha N}} \sum_{\mathbf{q}\lambda} \hat{\epsilon}_{\alpha\lambda}(\mathbf{q}) \hat{Q}_{\mathbf{q}\lambda} e^{i\mathbf{q}\cdot\mathbf{R}_i}, \tag{8.130}$$

where the expression for the normal coordinate $Q_{\mathbf{q}\lambda}$ was derived in Eq. (2.182) as

$$\hat{Q}_{\mathbf{q}\lambda} = \left(\frac{\hbar}{2\omega_{\mathbf{q}\lambda}}\right)^{\frac{1}{2}} (\hat{a}_{\mathbf{q}\lambda} + \hat{a}^\dagger_{-\mathbf{q}\lambda}). \tag{8.131}$$

From Eqs. (8.129) through (8.131), we obtain

$$H_{el-ph} = -\sum_{\alpha ij} \frac{1}{\sqrt{M_\alpha N}} \sum_{\mathbf{q}\lambda} \hat{Q}_{\mathbf{q}\lambda} e^{i\mathbf{q}\cdot\mathbf{R}_i} \hat{\epsilon}_{\alpha\lambda}(\mathbf{q}) \cdot \nabla V_\alpha(\mathbf{r}_j - \mathbf{R}_{i\alpha}). \tag{8.132}$$

We will now convert quantum mechanical equations for H_e from the **r**-representation into the occupation number representation. We assume that the Hamiltonian is a sum of one-electron operators

$$\hat{H} = \sum_i \hat{H}(\mathbf{r}_i, s_i). \tag{8.133}$$

We recall that, for fermions, we derived in Chapter 2 (Eq. 2.110) that

$$\hat{c}^\dagger_{\mathbf{k}} \hat{c}_{\mathbf{k}} = n_{\mathbf{k}} \tag{8.134}$$

(In this chapter, we use $\hat{c}^\dagger_{\mathbf{k}}$ and $\hat{c}_{\mathbf{k}}$ for creation and annihilation operators for electrons to distinguish from the phonon operators.)

Similarly, from Eqs. (2.99) through (2.101) and (2.109), we derived

$$[c_{\mathbf{k}}, c^\dagger_{\mathbf{k}}]_+ = \delta_{\mathbf{k},\mathbf{k}'}, \tag{8.135}$$

$$c^\dagger_{\mathbf{k}} |n_1, n_2, \ldots, n_{\mathbf{k}} \ldots, > = \sqrt{1 - n_{\mathbf{k}}} (-1)^m |n_1, n_2, \ldots, n_{\mathbf{k}} + 1, \ldots, > \tag{8.136}$$

and

$$c_{\mathbf{k}} |n_1, n_2, \ldots, n_{\mathbf{k}} \ldots, > = \sqrt{n_{\mathbf{k}}} (-1)^m |n_1, n_2, \ldots, n_{\mathbf{k}} - 1, \ldots, >, \tag{8.137}$$

where

$$m = \sum_{j<\mathbf{k}} n_i. \tag{8.138}$$

We also showed in Chapter 7 (Eq. 7.18) that the wave function is a Slater determinant, which can be written in the form

$$\Psi(\mathbf{r}_1 s_1, \mathbf{r}_2 s_2, \ldots, \mathbf{r}_N s_N) = \frac{1}{\sqrt{N!}} \sum_n (-1)^n \psi_{n_1}(\mathbf{r}_1 s_1)\psi_{n_2}(\mathbf{r}_2 s_2) \ldots \psi_{n_N}(\mathbf{r}_N s_N), \tag{8.139}$$

where the sum n is over all permutations n of $1 \ldots N$.

We can rewrite Eq. (8.139) in the alternate form

$$\Psi = \frac{1}{\sqrt{N!}} \sum_P P(-1)^P \psi_\alpha(\mathbf{r}_1 s_1)\psi_\beta(\mathbf{r}_2 s_2) \ldots \psi_\omega(\mathbf{r}_N s_N), \tag{8.140}$$

where the sum P is over all permutations of the indices $\alpha, \beta, \ldots,$ etc.

From Eq. (8.140), we can write

$$H\Psi(\mathbf{r}_1 s_1, \mathbf{r}_2 s_2, \ldots, \mathbf{r}_N s_N) = \frac{1}{\sqrt{N!}} H \sum_P P(-1)^P \psi_\alpha(\mathbf{r}_1 s_1)\psi_\beta(\mathbf{r}_2 s_2) \ldots \psi_\omega(\mathbf{r}_N s_N). \tag{8.141}$$

From Eqs. (8.133) and (8.141), we obtain

$$H\Psi = \frac{1}{\sqrt{N!}} \sum_i \sum_P P(-1)^P \psi_\alpha(\mathbf{r}_1 s_1)\psi_\beta(\mathbf{r}_2 s_2) \ldots H(\mathbf{r}_i s_i)\psi_\lambda(\mathbf{r}_i s_i) \ldots \psi_\omega(\mathbf{r}_N s_N). \tag{8.142}$$

If the Hamiltonian does not involve the spin explicitly (Eq. 7.20),

$$\int \psi_\lambda^*(\mathbf{r}_i s_i) H(\mathbf{r}_i) \psi_{\lambda'}(\mathbf{r}_i s_i') = \int \phi_\lambda^*(\mathbf{r}_i) H(\mathbf{r}_i) \phi_{\lambda'}(\mathbf{r}_i) \delta_{s_i, s_i'} \tag{8.143}$$

because

$$\psi_\lambda(\mathbf{r}_j s_j) = \phi_\lambda(\mathbf{r}_j)\chi_\lambda(s_j). \tag{8.144}$$

We can write the Hamiltonian as a sum of operators on single particles,

$$H = \sum_i H(\mathbf{r}_i s_i) \equiv \sum_i H_i. \tag{8.145}$$

Because ψ_λ forms a complete set, we can rewrite Eq. (8.145) in a more convenient way by writing

$$\hat{H} = \sum_{i\lambda\lambda'} |\psi_{\lambda'}(i)><\psi_{\lambda'}(i)|\hat{H}_i|\psi_\lambda(i)>|<\psi_\lambda(i)|. \tag{8.146}$$

The one-particle equivalent of Eq. (8.146) is

$$H(\mathbf{r}_i s_i)\psi_\lambda(\mathbf{r}_i s_i) = \sum_{\lambda'} \psi_{\lambda'}(\mathbf{r}_i s_i) <\lambda'|H_i|\lambda>, \tag{8.147}$$

which has i-independent matrix elements. Here, the spin index s is dropped from λ because the Hamiltonian H does not explicitly contain spin terms, whereas the summation over λ' explicitly contains spin summation.

We define the operator

$$\hat{A}_{\lambda'\lambda} = \sum_i |\psi_{\lambda'}(i)><\psi_\lambda(i)|. \tag{8.148}$$

The operator $\hat{A}_{\lambda'\lambda}$ searches for each electron (one at a time) in state ψ_λ and moves it to $\psi_{\lambda'}$. From Eqs. (8.146) and (8.148), we obtain

$$\hat{H} = \sum_{\lambda\lambda'} \hat{A}_{\lambda'\lambda} <\psi_{\lambda'}(1)|\hat{H}_1|\psi_\lambda(1)>, \tag{8.149}$$

where the label 1 is used instead of i because the matrix elements of a one-particle operator do not depend on which particle is involved.

We first consider a wave function Ψ and wish to evaluate

$$\sum_{i=1}^{N} <\Psi_a|\psi_{\lambda'}(i)><\psi_\lambda(i)|\Psi_b>. \tag{8.150}$$

We consider one term

$$<\Psi_a|\psi_{\lambda'}(1)><\psi_\lambda(1)|\Psi_b>. \tag{8.151}$$

This term is nonzero only if in $|\Psi_b>$, ψ_λ is unoccupied and $\psi_{\lambda'}$ is occupied while in $|\Psi_a>$, ψ_λ is unoccupied and $\psi_{\lambda'}$ is occupied, and otherwise Ψ_a and Ψ_b are identical. Thus, one has to permute $\psi_{\lambda'}$ past all the states in the ordering to obtain $|\psi_{\lambda'}(1)>$, and one obtains a factor of $(-1)^{m_{\lambda'}}$. Similarly, one obtains a factor $(-1)^{m_\lambda}$ to permute ψ_λ past all the states below it to obtain $|\psi_\lambda(1)>$. Here,

$$m_\lambda = \sum_{i<\lambda} n_i \text{ and } m_{\lambda'} = \sum_{i<\lambda'} n_i. \tag{8.152}$$

Thus, we obtain

$$\sum_{i=1}^{N} <\Psi_a|\psi_{\lambda'}(i)><\psi_\lambda(i)|\Psi_b> = (-1)^{m_\lambda + m_{\lambda'}}, \tag{8.153}$$

if it is not zero. Thus, in the occupation number representation,

$$\hat{A}_{\lambda'\lambda}|n_1, n_2, n_{\lambda'} \ldots n_{\lambda'} \ldots> = (-1)^{m_{\lambda'}}(-1)^{m_\lambda}\sqrt{(1-n_{\lambda'})n_\lambda}\,|n_1, n_2, \ldots, n_{\lambda'+1} \ldots n_{\lambda-1} \ldots>. \tag{8.154}$$

From Eqs. (8.149) and (8.154), we obtain (Problem 8.11)

$$\hat{H}|n_1, n_2, \ldots, n_1, n_2, \ldots, n_{\lambda'}, \ldots, n_\lambda, \ldots>$$

$$= \sum_{\lambda,\lambda'} \sqrt{n_\lambda(1-n_{\lambda'})}(-1)^{m_\lambda + m_{\lambda'}} <\lambda'|H_1|\lambda>|n_1, n_2, \ldots, n_{\lambda'+1}, \ldots, n_{\lambda-1}, \ldots> \text{ if } \lambda \neq \lambda', \tag{8.155}$$

$$= \sum_\lambda n_\lambda <\lambda|H_1|\lambda>|n_1, n_2, \ldots, n_\lambda \ldots>, \text{ if } \lambda = \lambda'. \tag{8.156}$$

Here, the terms $\sum\sqrt{n_\lambda(1-n_\lambda)}$ and n_λ occur due to the property of the fermions that n_λ can be either zero or 1.

From Eqs. (8.152) through (8.154), the operator

$$\hat{A}_{\lambda'\lambda} = \hat{c}^{\dagger}_{\lambda'}\hat{c}_{\lambda}, \tag{8.157}$$

where the creation and annihilation operators for fermions were defined in Eqs. (8.136) and (8.137). From Eqs. (8.149), (8.153), and (8.155) through (8.157), we obtain by replacing λ and λ' by $\mathbf{k}s$ and $\mathbf{k}'s$,

$$H|n_1, n_2, \ldots, n_{\mathbf{k}} \ldots> = \sum_{\mathbf{kk}'s} <\mathbf{k}'s|H_1(\mathbf{r})|\mathbf{k}s> c^{\dagger}_{\mathbf{k}'s}c_{\mathbf{k}s}|n_1, n_2, \ldots, n_{\mathbf{k}'} \ldots n_{\mathbf{k}} \ldots>. \tag{8.158}$$

The Hamiltonian H in the occupation number representation becomes

$$H = \sum_{\mathbf{k},\mathbf{k}',s} <\mathbf{k}'s|H_1(\mathbf{r})|\mathbf{k}s> c^{\dagger}_{\mathbf{k}'s}c_{\mathbf{k}s}, \tag{8.159}$$

where

$$<\mathbf{k}'s|H_1(\mathbf{r})|\mathbf{k}s> = \int \psi^*_{\mathbf{k}'s}(\mathbf{r})H_1(\mathbf{r})\psi_{\mathbf{k}s}(\mathbf{r})d\mathbf{r}. \tag{8.160}$$

Here, we dropped the index i in $\mathbf{r}_i$. We note that because spin is conserved in $\mathbf{k} \to \mathbf{k}'$ transition, we have to carry out only one spin summation. The matrix elements on the right side of Eq. (8.160) are formed from the Bloch functions.

To calculate H_{el-ph} from Eqs. (8.132) and (8.160), we expand $V_\alpha(\mathbf{r}-\mathbf{R}_{i\alpha})$ in a Fourier series

$$V_\alpha(\mathbf{r}-\mathbf{R}_i) = \sum_{\vec{K}} e^{i\vec{K}\cdot(\mathbf{r}-\mathbf{R}_i)}V_{\alpha\vec{K}}, \tag{8.161}$$

from which we obtain

$$<\mathbf{k}'s|\nabla V_\alpha|\mathbf{k}s> = \sum_{\vec{K}} e^{-i\vec{K}\cdot\vec{R}_i}V_{\alpha\vec{K}}i\vec{K} <\mathbf{k}'s|e^{i\vec{K}\cdot\vec{r}}|\mathbf{k}s>. \tag{8.162}$$

From Eqs. (8.132) and (8.162), the sum over i contains the factor

$$\sum_i e^{i(\vec{q}-\vec{K})\cdot\vec{R}_i} = N\sum_{\vec{K}_l} \delta_{\vec{K},\vec{q}+\vec{K}_l}. \tag{8.163}$$

Thus, in the sum over $\vec{K}$ in Eq. (8.162), only the term $\vec{K} = \vec{q} + \vec{K}_l$ survives. If we write the Bloch functions $|\mathbf{k}> \equiv e^{i\mathbf{k}\cdot\mathbf{r}}u_{n\mathbf{k}}(\mathbf{r})$ (note that we have also used bold letters as vectors) in Eq. (8.162), the integrand I contains the term

$$I = \int d\mathbf{r}\, e^{i(\mathbf{k}+\mathbf{q}+\mathbf{K}_l-\mathbf{k}')\cdot\mathbf{r}}u^*_{n'\mathbf{k}'}(\mathbf{r})u_{n\mathbf{k}}(\mathbf{r}). \tag{8.164}$$

Because $u^*_{n'\mathbf{k}'}(\mathbf{r})u_{n\mathbf{k}}(\mathbf{r})$ is a periodic function in $\mathbf{R}$,

$$I = \int d\mathbf{r}\, u^*_{n'\mathbf{k}'}(\mathbf{r})u_{n\mathbf{k}}(\mathbf{r})\delta_{\mathbf{k}',\mathbf{k}+\mathbf{q}+\mathbf{K}_l}. \tag{8.165}$$

From Eqs. (8.132), (8.159), and (8.162) through (8.165), and letting $n' = n$ (because the electron remains in the same band $n = n'$ when making the transition $\mathbf{k}$ to $\mathbf{k}'$ except for the U-process to be discussed later), we obtain

$$\begin{aligned} H_{el-ph} = &- \sum_{\mathbf{k}\mathbf{K}_l\mathbf{q}\alpha\lambda s} i\sqrt{\frac{N}{M_\alpha}}\, V_{\alpha,\mathbf{q}+\mathbf{K}_l}(\mathbf{q}+\mathbf{K}_l)\cdot\hat{\in}_{\alpha\lambda}\left(\frac{\hbar}{2\omega_{\mathbf{q}\lambda}}\right) \\ &\times \int u^*_{n,\mathbf{q}+\mathbf{K}_l+\mathbf{k}}(\mathbf{r})u_{n\mathbf{k}}(\mathbf{r})d\mathbf{r}\,(\hat{a}^\dagger_{-\mathbf{q}\lambda}+\hat{a}_{\mathbf{q}\lambda})\hat{c}^\dagger_{\mathbf{k}+\mathbf{q}+\mathbf{K}_l,s}c_{\mathbf{k},s}. \end{aligned} \tag{8.166}$$

We note that if $\mathbf{k}$ and $\mathbf{q}$ are added vectorially, the resultant vector can lie outside the first Brillouin zone (where $\mathbf{k}, \mathbf{k}', \mathbf{q}$, and $\mathbf{q}'$ lie in the reduced zone scheme) in the repeated zone scheme, so that $\mathbf{k}' = \mathbf{k}+\mathbf{q}+\mathbf{K}_l$. The sum over $\mathbf{K}_l$ in Eq. (8.164) is thus reduced to one term. If $\mathbf{K}_l = 0$, the transition is called a normal process (N-process) while if $\mathbf{K}_l \neq 0$, the transition is called the Umklapp process (U-process). (Umklapp is a German word for "flopping over.")

To simplify further, we assume that we restrict ourselves to Bravais lattices so that there is only one atom in the Wigner–Seitz cell. Thus, we can omit the index α and λ and count the different acoustic branches since the optical branches do not exist. We also restrict the derivation to the N-process such that $\mathbf{K}_l = 0$. We also assume that phonons are either longitudinal or transverse so that $\hat{\in}_\lambda$ is either parallel or perpendicular to $\mathbf{q}$. With these assumptions, Eq. (8.166) can be written as

$$H_{el-ph} = -\sum_{\mathbf{k}\mathbf{q}\lambda s} i\sqrt{\frac{\hbar N}{2M\omega_{\mathbf{q}\lambda}}}\, V_{\mathbf{q}}\mathbf{q}\cdot\hat{\in}_\lambda \times \int d\mathbf{r}\, u^*_{n,\mathbf{k}+\mathbf{q}}(\mathbf{r})u_{n\mathbf{k}}(\mathbf{r})\,(\hat{a}^\dagger_{-\mathbf{q}\lambda}+\hat{a}_{\mathbf{q}\lambda})\hat{c}^\dagger_{\mathbf{k}+\mathbf{q},s}c_{\mathbf{k},s}. \tag{8.167}$$

Because $\hat{\in}_\lambda \perp \mathbf{q}$ for *transverse phonons*, $\hat{\in}_\lambda \cdot \mathbf{q} = 0$. Thus, only *longitudinal acoustic phonons* are coupled to the electrons. Eq. (8.167) can be rewritten in the alternate form

$$H_{el-ph} = \sum_{\mathbf{k}\mathbf{q}s} B_{\mathbf{k}\mathbf{q}}(\hat{a}^\dagger_{-\mathbf{q}}+\hat{a}_{\mathbf{q}})\hat{c}^\dagger_{\mathbf{k}+\mathbf{q},s}\hat{c}_{\mathbf{k},s}. \tag{8.168}$$

We can calculate the probability of transition of an electron from state $|\mathbf{k}>$ into state $|\mathbf{k}+\mathbf{q}>$. Because the spin is unchanged in this transition, it can be ignored.

From Fermi's "golden rule" (originally derived by Dirac from perturbation theory), the transition probability

$$P(i \to f) = \frac{2\pi}{\hbar}|<f|H_{el-ph}|i>|^2\,\delta(\varepsilon_f-\varepsilon_i). \tag{8.169}$$

Here, the initial states are $|i>$ and the final states $|f>$ are characterized by the occupation numbers $n_{\mathbf{k}}$ and $n_{\mathbf{k}+\mathbf{q}}$ of the electron state and $n_{\mathbf{q}}$ and $n_{-\mathbf{q}}$ of the phonon states involved in the transition

$$|n_{\mathbf{k}+\mathbf{q}}, n_{\mathbf{k}};\, n_{\mathbf{q}}, n_{-\mathbf{q}}>. \tag{8.170}$$

If we consider the transition involving the absorption of a phonon, we apply the operator $\hat{c}^{\dagger}_{\mathbf{k+q}}\hat{c}_{\mathbf{k}}a_{\mathbf{q}}$ to Eq. (8.170). It can be shown that (Problem 8.12)

$$<n_{\mathbf{k+q}}+1, n_{\mathbf{k}}-1; n_{\mathbf{q}}-1|\hat{c}^{\dagger}_{\mathbf{k+q}}\hat{c}_{\mathbf{k}}\hat{a}_{\mathbf{q}}|n_{\mathbf{k+q}}, n_{\mathbf{k}}; n_{\mathbf{q}}> = \sqrt{(1-n_{\mathbf{k+q}})n_{\mathbf{k}}n_{\mathbf{q}}}. \tag{8.171}$$

The matrix elements vanish for all cases except when $n_{\mathbf{k}} = 1$; $n_{\mathbf{k+q}} = 0$. The energy relations for the absorption of a phonon are

$$\varepsilon_f - \varepsilon_i = \varepsilon(\mathbf{k}+\mathbf{q}) - \varepsilon(\mathbf{k}) - \hbar\omega_{\mathbf{q}}. \tag{8.172}$$

Similarly, the transitions involving emission of phonons can be written as (Problem 8.13)

$$<n_{\mathbf{k+q}}+1, n_{\mathbf{k}}-1; n_{-\mathbf{q}}+1|\hat{c}^{\dagger}_{\mathbf{k+q}}\hat{c}_{\mathbf{k}}\hat{a}^{\dagger}_{-\mathbf{q}}|n_{\mathbf{k+q}}, n_{\mathbf{k}}; n_{-\mathbf{q}}> = \sqrt{(1-n_{\mathbf{k+q}})n_{\mathbf{k}}(n_{-\mathbf{q}}+1)}, \tag{8.173}$$

and

$$\varepsilon_f - \varepsilon_i = \varepsilon(\mathbf{k}+\mathbf{q}) - \varepsilon(\mathbf{k}) + \hbar\omega_{\mathbf{q}}. \tag{8.174}$$

From Eqs. (8.168) through (8.174), we obtain

$$\begin{aligned} P(\mathbf{k} \rightarrow \mathbf{k}+\mathbf{q}) = \frac{2\pi}{\hbar}|B_{\mathbf{kq}}|^2 n_{\mathbf{k}}(1-n_{\mathbf{k+q}})\{n_{\mathbf{q}}\delta(\varepsilon(\mathbf{k}+\mathbf{q}) - \varepsilon(\mathbf{k}) - \hbar\omega_{\mathbf{q}}) \\ + (n_{-\mathbf{q}}+1)\delta(\varepsilon(\mathbf{k}+\mathbf{q}) - \varepsilon(\mathbf{k}) + \hbar\omega_{\mathbf{q}})\}. \end{aligned} \tag{8.175}$$

In this case, $n_{\mathbf{k}} = 1$, $n_{\mathbf{k+q}} = 0$, and the matrix element vanishes in all other cases. However, the factor $(1-n_{\mathbf{k+q}})n_{\mathbf{k}}$ is retained because if and when we consider a large number of states instead of the transition probability from *one occupied state to one empty state*, the Fermi and Bose distributions have to be used for electrons and phonons if the system is in equilibrium.

PROBLEMS

8.1. Derive Eqs. (8.19) and (8.20) from Eq. (8.16).

8.2. We can write

$$(-i\nabla + \mathbf{k})u_{n\mathbf{k}}(\mathbf{r}) = (-i\nabla + \mathbf{k})e^{-i\mathbf{k}\cdot\mathbf{r}}\psi_{n\mathbf{k}}(\mathbf{r}). \tag{1}$$

Show that

$$(-i\nabla + \mathbf{k})e^{-\mathbf{k}\cdot\mathbf{r}}\psi_{n\mathbf{k}}(\mathbf{r}) = e^{-\mathbf{k}\cdot\mathbf{r}}(-i\nabla)\psi_{n\mathbf{k}}(\mathbf{r}). \tag{2}$$

Hence, show that

$$<u_{n\mathbf{k}}|(-i\nabla + \mathbf{k})|u_{n\mathbf{k}}> = <\psi_{n\mathbf{k}}|-i\nabla|\psi_{n\mathbf{k}}>. \tag{3}$$

8.3. Show that the inverse effective mass tensor can be written as

$$[\mathbf{M}_n^{-1}(\mathbf{k})]_{ij} = \frac{1}{\hbar^2}\frac{\partial^2 \varepsilon_n}{\partial k_i \partial k_j}. \tag{1}$$

8.4. Using Eqs. (8.35) and (8.36), show that

$$[\mathbf{M}_n^{-1}(\mathbf{k})]_{ij} = \frac{1}{m}\delta_{ij} + \frac{\hbar^2}{m^2}\sum_{n'\neq n}\frac{<n\mathbf{k}|-i\nabla_i|n'\mathbf{k}><n'\mathbf{k}|-i\nabla_j|n\mathbf{k}> + c.c.}{\varepsilon_n(\mathbf{k}) - \varepsilon_{n'}(\mathbf{k})}. \tag{1}$$

8.5. Show that if in Eq. (8.53), we can write

$$\phi_i'(\mathbf{r}, t) = e^{i\mathbf{k}\cdot\mathbf{r}} u_{n\mathbf{k}}(\mathbf{r}, \mathbf{E}, t), \tag{1}$$

the effective Hamiltonian for $u_{n\mathbf{k}}(\mathbf{r}, \mathbf{E}, t)$ is

$$\frac{1}{2m}(-i\hbar\nabla + \hbar\mathbf{k} - e\mathbf{E}t)^2 + V(\mathbf{r}). \tag{2}$$

8.6. Show that

$$[\hat{T}(\mathbf{R}_i), H] = (\vec{\xi}\cdot\mathbf{R}_i + \frac{e^2B^2}{2m}R_{ix}^2)\hat{T}(\mathbf{R}_i). \tag{1}$$

8.7. The eigenfunction $\varepsilon_n(\mathbf{k})$ is periodic in **k** space and can be expanded as

$$\varepsilon_n(\mathbf{k}) = \sum_m C_{nm} e^{i\mathbf{k}\cdot\mathbf{R}_m}. \tag{1}$$

If **k** is replaced by $-i\nabla$, show that

$$\varepsilon_n(-i\nabla)\psi_n(\mathbf{k}, \mathbf{r}) = \sum_m C_{nm} e^{i\mathbf{R}_m\cdot\nabla}\psi_n(\mathbf{k}, \mathbf{r}) = \varepsilon_n(\mathbf{k})\psi_n(\mathbf{k}, \mathbf{r}). \tag{2}$$

If we write the time-dependent Schrodinger equation as

$$\left[-\frac{\hbar^2}{2m}\nabla^2 + V(\mathbf{r}) - e\phi\right]\psi = i\hbar\dot{\psi}, \tag{3}$$

and represent the electron as a wave packet constructed from all the Bloch states of all bands,

$$\psi = \sum_{n\mathbf{k}} A_n(\mathbf{k}, t)\psi_n(\mathbf{k}, \mathbf{r}), \tag{4}$$

show that

$$[E_n(-i\nabla) - e\phi]\psi = i\hbar\dot{\psi}. \tag{5}$$

8.8. If we substitute $\phi_i'(\mathbf{r},t) = e^{ie\mathbf{A}\cdot\mathbf{r}/\hbar c}\phi_i(\mathbf{r},t)$ in

$$\left[\frac{\left(\mathbf{p}+(e/c)\mathbf{A}\right)^2}{2m}+V(\mathbf{r})\right]\phi_i'(\mathbf{r},t)=\varepsilon_i(t)\phi_i'(\mathbf{r},t), \tag{1}$$

show that Eq. (1) reduces to

$$\left[\frac{p^2}{2m}+V(\mathbf{r})\right]\phi_i=\varepsilon_i\phi_i. \tag{2}$$

8.9. Derive the expression

$$\varepsilon_n(\mathbf{k},t)a_{n\mathbf{k}(t)}=i\hbar\frac{\partial a_{n\mathbf{k}(t)}}{\partial t}-eE(t)\sum_{n'}A_{nn'}(\mathbf{k}(t))a_{n'\mathbf{k}(t)}. \tag{1}$$

8.10. Use the expression

$$\dot{\alpha}_n(t)=\frac{ieE(t)}{\hbar}\sum_{n'}\alpha_{n'}(t)A_{nn'}(\mathbf{k}(t))\exp\left[\frac{i}{\hbar}\int_0^t[\varepsilon_n(\mathbf{k}(t'))-\varepsilon_{n'}(\mathbf{k}(t'))]dt'\right], \tag{1}$$

which is a general expression used for derivation of Wannier–Stark ladders for time-dependent E as well as for Zener tunneling between the valence and conduction bands.

At $t=0$, before the field E is turned on, $\mathbf{k}(t)=\mathbf{k}(t=0)$, the electron state can be described by a Bloch wave in band n with $\mathbf{k}(t)=\mathbf{k}(t=0)$, then at $t=0$,

$$\alpha_{n'}=\delta_{nn'}. \tag{2}$$

After a time $T=-\frac{\hbar K}{eE}$, where T is the period of one Bloch oscillation, which is a sufficiently short time so that $\alpha_{n'}\ll 1$, $n'\neq n$, show that we can substitute Eq. (2) in Eq. (1) to obtain

$$\alpha_{n'}(t)=\int_0^t\frac{ieE(t')}{\hbar}A_{n'n}(\mathbf{k}(t'))\exp\left[\frac{i}{\hbar}\int_0^{t'}[\varepsilon_{n'}(\mathbf{k}(t''))-\varepsilon_n(\mathbf{k}(t''))]dt''\right]dt'. \tag{3}$$

8.11. Using Eqs. (8.149) and (8.154), show that

$$\hat{H}|n_1,n_2,\ldots,n_1,n_2,\ldots,n_{\lambda'},\ldots,n_\lambda,\ldots>$$
$$=\sum_{\lambda,\lambda'}\sqrt{n_\lambda(1-n_{\lambda'})}\,(-1)^{m_\lambda+m_{\lambda'}'}<\lambda'|H_1|\lambda>|n_1,n_2,\ldots,n_{\lambda'+1},\ldots n_{\lambda-1},\ldots>\text{if } \lambda\neq\lambda', \tag{1}$$

$$=\sum_\lambda n_\lambda<\lambda|H_1|\lambda>|n_1,n_2,\ldots,n_\lambda\ldots>,\text{if } \lambda=\lambda'. \tag{2}$$

8.12. By using the expressions derived in Eqs. (2.75) and (2.76) for bosons,

$$a_{\mathbf{q}}^{\dagger}|n_1, n_2, \ldots, n_{\mathbf{q}}, \ldots > = \sqrt{n_{\mathbf{q}}+1}\,|n_1, n_2, \ldots, n_{\mathbf{q}}+1, \ldots >, \quad (1)$$

$$a_{\mathbf{q}}|n_1, n_2, \ldots, n_{\mathbf{q}}, \ldots > = \sqrt{n_{\mathbf{q}}}\,|n_1, n_2, \ldots, n_{\mathbf{q}}-1, \ldots >, \quad (2)$$

and from Eqs. (8.133) through (8.136) for fermions, show that

$$< n_{\mathbf{k+q}}+1, n_{\mathbf{k}}-1; n_{\mathbf{q}}-1 > |\hat{c}_{\mathbf{k+q}}^{\dagger}\hat{c}_{\mathbf{k}}a_{\mathbf{q}}|n_{\mathbf{k+q}}, n_{\mathbf{k}}; n_{\mathbf{q}} > = \sqrt{(1-n_{\mathbf{k+q}})n_{\mathbf{k}}n_{\mathbf{q}}}. \quad (3)$$

8.13. Using the properties of the creation and annihilation operators for bosons and fermions outlined in Problem 8.12, show that

$$< n_{\mathbf{k+q}}+1, n_{\mathbf{k}}-1; n_{-\mathbf{q}}+1|\hat{c}_{\mathbf{k+q}}^{\dagger}\hat{c}_{\mathbf{k}}\hat{a}_{-\mathbf{q}}^{\dagger}|n_{\mathbf{k+q}}, n_{\mathbf{k}}; n_{-\mathbf{q}} > = \sqrt{(1-n_{\mathbf{k+q}})n_{\mathbf{k}}(n_{-\mathbf{q}}+1)}. \quad (1)$$

References

1. Aschcroft NW, Mermin ND. *Solid state physics*. New York: Brooks/Cole; 1976.
2. Cllaway J. *Quantum theory of the solid state*. New York: Academic Press; 1976.
3. Dahan BM, Peik E, Reichel J, Castin Y, Solomon C. Bloch oscillations of aroms in an optical potential. *Phys Rev Lett* 1996;**76**:4508.
4. Goswami A. *Quantum mechanics*. Dubuque: William C. Brown; 1997.
5. Kittel C. *Quantum theory of solids*. New York: John Wiley & Sons; 1987.
6. Kreiger JB, Iafrate GJ. Time evolution of Bloch electrons in homogeneous electric field. *Phys Rev B* 1986;**33**:5494.
7. Liboff R. *Quantum mechanics*. Reading, MA: Addison-Wesley; 1980.
8. Madelung O. *Introduction to solid state theory*. New York: Springer-Verlag; 1978.
9. Marder MP. *Condensed matter physics*. New York: John Wiley & Sons; 2000.
10. Shoenberg D. *Magnetic oscillations in metals*. Cambridge: Cambridge University Press; 1984.
11. Symon KR. *Mechanics*. Reading, MA: Addison-Wesley; 1971.
12. Zener C. Non-adiabatic crossing of energy levels. *Proc R Soc Lond* 1932;**A137**:696.
13. Ziman JM. *Principles of the theory of solids*. Cambridge: Cambridge University Press; 1972.

CHAPTER

Semiconductors

9

CHAPTER OUTLINE

9.1 Introduction 275
9.2 Electrons and Holes 278
9.3 Electron and Hole Densities in Equilibrium 279
9.4 Intrinsic Semiconductors 283
9.5 Extrinsic Semiconductors 284
9.6 Doped Semiconductors 285
9.7 Statistics of Impurity Levels in Thermal Equilibrium 288
9.7.1 Donor Levels 288
9.7.2 Acceptor Levels 288
9.7.3 Doped Semiconductors 289
9.8 Diluted Magnetic Semiconductors 290
9.8.1 Introduction 290
9.8.2 Magnetization in Zero External Magnetic Field in a DMS 291
9.8.3 Electron Paramagnetic Resonance Shift 291
9.8.4 $\vec{k} \cdot \vec{\pi}$ Model 295
9.9 Zinc Oxide 296
9.10 Amorphous Semiconductors 296
9.10.1 Introduction 296
9.10.2 Linear Combination of Hybrids Model for Tetrahedral Semiconductors 297
Problems 300
References 303

9.1 INTRODUCTION

In Chapter 4 (section 4.9), we discussed that by using elementary band theory, crystalline solids can be divided into three major categories: metals, insulators, and homogeneous semiconductors. The metals are good conductors (with the exception of the divalent metals) because either the conduction band is half-filled (monovalent or trivalent metals), or there is significant overlap between the valence and conduction band (divalent metals). The crystalline solids with four valence electrons per unit cell can either be an insulator or a homogeneous semiconductor, depending on the energy gap between the valence band and the conduction band. The energy gap E_g is defined as the energy between the bottom of the lowest-filled band(s) and the top of the highest-filled bands(s). In the case of both insulators and semiconductors, the lowest unoccupied band is known as the conduction band, and the highest occupied

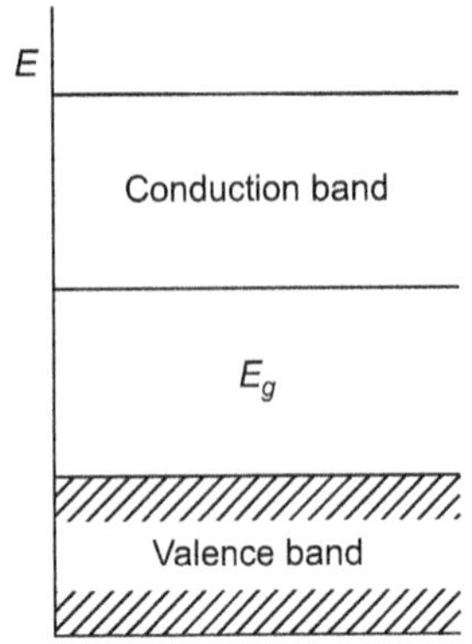

FIGURE 9.1

Schematic diagram of the valence and conduction bands.

band is known as the valence band. A schematic diagram of the valence and conduction bands, which are the energy bands in the reduced zone scheme, is shown in Figure 9.1.

At $T = 0$, the valence band is full, and the conduction band is empty for both insulators and semiconductors. Thus, the conductivity of both types of solids is zero because no carriers would be available in either of the bands to be excited by an external electric field unless the electric field (DC) is sufficiently large to cause Zener tunneling (Chapter 8) or the frequency ω of the AC electric field is such that $\hbar\omega > E_g$. At $T \neq 0$, a few electrons would be thermally excited to the conduction band, leaving behind a few positively charged holes in the valence band (as we will show, the probability of such transition is $e^{-E_g/2K_BT}$). If the energy gap is large so that very few electrons are thermally excited from the valence band to the conduction band at the room temperature, a negligible number of carriers in either of the bands would be available to conduct electricity, and essentially no current would be generated. This type of solid, which carries no current in an electric field, is known as an insulator. However, if the energy gap is small enough, a significant number of electrons are thermally excited at room temperature to the conduction band(s), leaving an equal number of positively charged "holes" at the top of the valence band(s). Thus, there are both "positively" charged carriers in the valence band and "negatively" charged carriers in the conduction band to conduct electricity and generate a perceptible current when an external field is applied. These solids are known as homogeneous semiconductors. The homogeneous semiconductors are also known as intrinsic semiconductors to distinguish them from impurity (doped) semiconductors. Thus, the distinction between the semiconductors and insulators depends essentially on the magnitude of the energy gap, and as a rule of the thumb, solids with $E_g < 2\,\text{eV}$ are semiconductors, whereas solids with $E_g > 2\,\text{eV}$ are insulators. However, most semiconductors have a much smaller energy gap.

There are two types of homogeneous semiconductors: the semiconducting elements and the semiconducting compounds. The most popular and widely used semiconducting elements are Si and Ge, both of which belong to column IV of the periodic table and crystallize in the diamond structure. The Bravais lattice of the diamond structure has a basis of two atoms, each of which has 8 *sp* electron states, but only 4 of these are occupied by electrons. The Brillouin zones have to accommodate 16 electron states, but only 8 electrons (from the two atoms) fill them. Therefore, the band structure has 8 subbands, 4 of which are completely filled and the other 4 are completely empty at 0° K. The energy gap of Si is 1.11 eV and that of Ge is 0.74 eV. These are also known as indirect semiconductors because the bottom of the conduction band does not lie directly above the top of the valence band, and thus, they have an indirect energy gap. The absorption of a photon creates an electron-hole pair, in which both the energy and momentum have to be conserved, but the momentum of a photon is negligible compared to that of the electrons. Thus, neither Si nor Ge is a good material for most optical applications because the two bands are not directly above each other. However, Si and Ge are extensively used in electronics because they can be easily doped with impurities. The other elemental semiconductors

Table 9.1 Comparison of Calculated Bandwidth with Photoemission Data for the Homopolar Materials (Energy in eV)

	Quasiparticle Theory	Expt.
Diamond	23.0	24.2 ± 1
Si	12.0	12.5 ± 0.6
Ge	12.8	12.9 ± 0.2

Reproduced from Louie[8] with the permission of Elsevier.

from column IV of the periodic table are gray tin, which has a very small energy gap (0.1 eV), red phosphorus, boron (1.5 eV), selenium, and tellurium (0.35 eV), which are solids with complex crystal structures. These semiconductors are neither used in optical applications nor in electronics. An example of the compound semiconductors of column IV of the periodic table is SiC, which has an indirect energy gap of 2.2 eV.

An example of the calculated and experimental bandwidth for the homopolar materials Si, Ge (both semiconductors), and diamond (insulator) is shown in Table 9.1.

The III–V semiconductors are crystals composed from columns III and V of the periodic table and have zincblende structure with predominantly covalent bonding. In the III–V zincblende semiconductors such as GaAs and InSb, Ga and In have three outer electrons, whereas As and Sb have five outer electrons. Ga or In occupies all the A sites in the diamond structure, whereas As or Sb occupies all the B sites (see Figure 6.4 in Chapter 6). The most popular of these is GaAs, which has a direct energy gap of 1.43 eV and thus facilitates the absorption of photons creating electron-hole pairs. GaAs is therefore widely used in optical applications. InSb has a direct energy gap of 0.18 eV. The other III–V semiconductors are GaN, GaSb, InP, and InAs, which have direct energy gaps of 3.44 eV, 0.7 eV, 1.34 eV, and 0.36 eV, respectively.

The essential features of the band structure of Ge are shown in Figure 9.2. In Ge, which

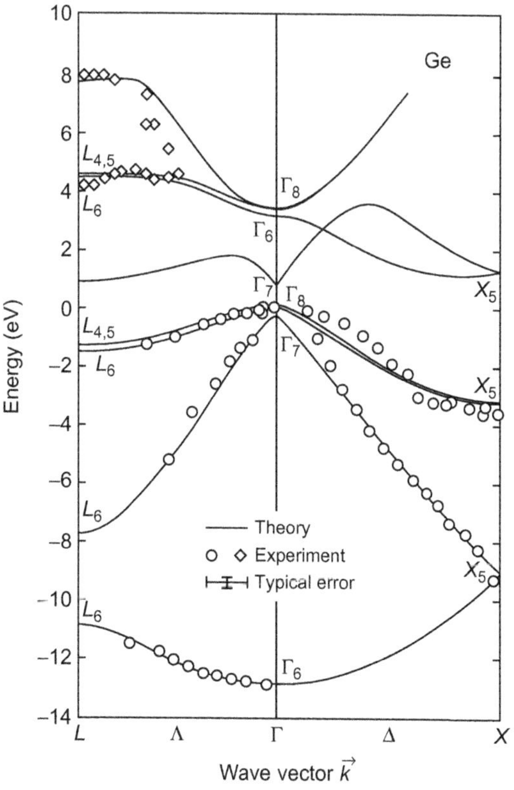

FIGURE 9.2

Calculated quasiparticle energies of Ge versus direct (○) and inverse (◇) photoemission data.

Reproduced from Louie[8] with permission of Elsevier.

has an indirect energy gap, the top of the valence band is at Γ, but the bottom of the conduction band is at L. In contrast, GaAs (not shown in the figure), which has a direct energy gap, both the top of the valence band and the bottom of the conduction band are at Γ.

9.2 ELECTRONS AND HOLES

In an intrinsic semiconductor at room temperature, the carriers are the electrons excited to the bottom of the conduction band from the top of the valence band. Because the filled valence band has no net charge (there are as many electrons as positively charged ions), the absence of an electron creates a net positive charge in the band. In addition, we consider the effective inverse mass tensor of an electron introduced in Eq. (8.35):

$$[\mathbf{M}_n^{-1}(\mathbf{k})]_{ij} = \frac{1}{\hbar^2}\frac{\partial^2 \varepsilon_n}{\partial k_i \partial k_j}. \tag{9.1}$$

We also derived an expression for the mass tensor in Eq. (8.37) by using perturbation theory,

$$[\mathbf{M}_n^{-1}(\mathbf{k})]_{ij} = \frac{1}{m}\delta_{ij} + \frac{\hbar^2}{m^2}\sum_{n'\neq n}\frac{<n\mathbf{k}|-i\nabla_i|n'\mathbf{k}><n'\mathbf{k}|-i\nabla_j|n\mathbf{k}> + c.c.}{\varepsilon_n(\mathbf{k}) - \varepsilon_{n'}(\mathbf{k})}. \tag{9.2}$$

At the top of the valence band, the inverse effective mass tensor can either be positive or negative. Because the concept of a negative mass is contrary to our physical understanding, the empty state at the top of the valence band can be considered as a positively charged "hole" with a positive effective mass. This concept of the positively charged holes is very important in formulating a theory for the semiconductors. We introduce the general definition of the effective mass tensor for both electrons and holes,

$$[\mathbf{M}_n^{-1}(\mathbf{k})]_{ij} = \pm\frac{1}{\hbar^2}\frac{\partial^2 \varepsilon_n}{\partial k_i \partial k_j}, \tag{9.3}$$

where the positive sign is for electrons and the negative sign is for holes. Because the bottom of the conduction band is at ε_c and the top of the valence band is at ε_v, we can express the energy of the electrons $(\varepsilon_e(\mathbf{k}))$ and holes $(\varepsilon_h(\mathbf{k}))$ as

$$\varepsilon_e(\mathbf{k}) = \varepsilon_c + \frac{\hbar^2}{2}\sum_{ij} k_i(\mathbf{M}_e^{-1})_{ij}k_j \tag{9.4}$$

and

$$\varepsilon_h(\mathbf{k}) = \varepsilon_v - \frac{\hbar^2}{2}\sum_{ij} k_i(\mathbf{M}_h^{-1})_{ij}k_j. \tag{9.5}$$

However, there are two types of effective hole masses (light holes and heavy holes) for Si, Ge, and GaAs. At the valence band maximum, there are two degenerate bands at Γ, while the third band (there are three degenerate bands in the absence of spin) is lowered due to spin-orbit interaction. The band with the low curvature results in holes with large effective mass (heavy holes), whereas the band with the high curvature has holes with small effective mass (light holes), which is evident from Eq. (9.3).

One can write the effective mass tensor in terms of a set of orthogonal principal axes,

$$\varepsilon_e(\mathbf{k}) = \varepsilon_c + \frac{\hbar^2}{2}\sum_i (\mathbf{M}_e^{-1})_{ii} k_i^2 \tag{9.6}$$

and

$$\varepsilon_h(\mathbf{k}) = \varepsilon_v - \frac{\hbar^2}{2}\sum_i (\mathbf{M}_h^{-1})_{ii} k_i^2. \tag{9.7}$$

We can redefine the electron and hole masses by rewriting Eqs. (9.6) and (9.7) as

$$\varepsilon_e(\mathbf{k}) = \varepsilon_c + \frac{\hbar^2}{2}\sum_{i=1}^{3} \frac{1}{m_i^e} k_i^2 \tag{9.8}$$

and

$$\varepsilon_h(\mathbf{k}) = \varepsilon_v - \frac{\hbar^2}{2}\sum_{i=1}^{3} \frac{1}{m_i^h} k_i^2. \tag{9.9}$$

9.3 ELECTRON AND HOLE DENSITIES IN EQUILIBRIUM

To determine the number of carriers in each band in a semiconductor, we will modify the expression between the electron density and the density of states derived in Eq. (3.59), which was obtained for free electrons,

$$n = \int_{-\infty}^{\infty} g(\varepsilon) f(\varepsilon) d\varepsilon. \tag{9.10}$$

Here, $f(\varepsilon)$ is the Fermi distribution function, which for electrons is

$$f_e(\varepsilon) = \frac{1}{e^{(\varepsilon-\mu)/k_B T} + 1}. \tag{9.11}$$

In the case of free electrons, the chemical potential μ at $T=0$ is equal to the Fermi energy ε_F, which is defined as the energy at the boundary between the filled and the empty states. In the case of semiconductors, the filled and empty states are separated by an energy gap E_g. Thus, we can argue that the chemical potential μ, known as the Fermi level in the case of semiconductors, lies somewhere between the energy gap. Later, we will show that for an intrinsic semiconductor, the Fermi level lies exactly at the middle of the gap at $T=0$. In addition, we are considering the density of electrons in the conduction band for which Eq. (9.10) is modified as

$$n_c(T) = \int_{\varepsilon_c}^{\infty} g_c(\varepsilon) f_e(\varepsilon) d\varepsilon, \tag{9.12}$$

where $g_c(\varepsilon)$ is the density of states of the electrons in the conduction band. We can write a similar expression for the holes in the valence band except that the distribution function for a hole can be written as

$$f_h(\varepsilon) = 1 - \frac{1}{e^{(\varepsilon-\mu)/k_BT}+1} = \frac{1}{e^{(\mu-\varepsilon)/k_BT}+1}. \tag{9.13}$$

In addition, the density of states for holes lies below the valence band edge. Thus, the density of holes in the valence band can be written as

$$p_v(T) = \int_{-\infty}^{\varepsilon_v} g_v(\varepsilon) f_h(\varepsilon) d\varepsilon, \tag{9.14}$$

where $g_v(\varepsilon)$ is the density of states of the holes in the valence band. For semiconductors, the conduction bands are nearly empty, and the valence bands are nearly full. We assume that the band shapes are nearly parabolic as in the case of free electrons. Therefore, we use the expression for the density of states for a free electron gas derived in Eq. (3.56) and suitably modify it for semiconductors by substituting effective masses m_n^* and m_p^* for the free electron mass m, and the fact that the energy of the electrons is $\varepsilon \geq \varepsilon_c$ and the energy of the holes is $|\varepsilon| \leq \varepsilon_v$ (we note that the energy of the holes is negative),

$$\begin{aligned} g_c(\varepsilon) &= \frac{\sqrt{2m_n^{*3}\varepsilon'}}{\pi^2\hbar^3}\eta_c, \quad \varepsilon' > \varepsilon_c \\ &= 0, \qquad\qquad\qquad \varepsilon' \leq \varepsilon_c, \end{aligned} \tag{9.15}$$

where

$$\varepsilon' \equiv \varepsilon - \varepsilon_c, \tag{9.16}$$

and

$$\begin{aligned} g_v(\varepsilon) &= \frac{\sqrt{2m_p^{*3}|\varepsilon''|}}{\pi^2\hbar^3}, \quad |\varepsilon''| > \varepsilon_v \\ &= 0, \qquad\qquad\qquad |\varepsilon''| \leq \varepsilon_v, \end{aligned} \tag{9.17}$$

where

$$|\varepsilon''| \equiv |\varepsilon - \varepsilon_v|. \tag{9.18}$$

Here, η_c is the number of symmetrically equivalent minima in the conduction band (six for Si and eight for Ge). m_n^* and m_p^* are the effective mass of electrons and holes that are obtained from the relation

$$m_n^* = (m_1^e m_2^e m_3^e)^{1/2} \tag{9.19}$$

and

$$m_p^{*3/2} = (m_{pl}^*)^{3/2} + (m_{ph}^*)^{3/2}, \tag{9.20}$$

where

$$m_{pl}^* = (m_1^{lh} m_2^{lh} m_3^{lh})^{1/2} \tag{9.21}$$

and

$$m_{ph}^* = (m_1^{hh} m_2^{hh} m_3^{hh})^{1/2}. \tag{9.22}$$

Here, m_{pl}^* and m_{ph}^* are the effective masses of the light and heavy holes defined in Eqs. (9.15) and (9.17), respectively. We note that in Eq. (9.17), the energy of the holes is zero at the valence band and negative downwards.

For most semiconductors, the following approximations can be easily made:

$$\varepsilon_c - \mu \gg k_B T \tag{9.23}$$

and

$$\mu - \varepsilon_v \gg k_B T. \tag{9.24}$$

The semiconductors for which the approximations (9.23) and (9.24) are valid are known as nondegenerate semiconductors, whereas those for which these approximations are not valid are known as degenerate semiconductors and one has to use Eqs. (9.11) and (9.13) for $f_e(\varepsilon)$ and $f_h(\varepsilon)$, respectively. For nondegenerate semiconductors, we can rewrite Eqs. (9.11) and (9.13) as

$$f_e(\varepsilon) \approx e^{-(\varepsilon-\mu)/k_B T} \approx e^{[-\varepsilon_c-(\varepsilon'-\mu)]/k_B T} \approx e^{-\varepsilon_c/k_B T} f_e(\varepsilon') \tag{9.25}$$

and

$$f_h(\varepsilon) \approx e^{(\varepsilon-\mu)/k_B T} \approx e^{[\varepsilon_v+(|\varepsilon-\varepsilon_v|-\mu)]/k_B T} \approx e^{\varepsilon_v/k_B T} f_h(|\varepsilon''|). \tag{9.26}$$

The density of carriers (in the conduction and valence bands) is obtained from the relations,

$$n_c(T) = \int_0^\infty g_c(\varepsilon') f_e(\varepsilon') d\varepsilon' \tag{9.27}$$

and

$$p_v(T) = \int_0^\infty g_v(|\varepsilon''|) f_h(|\varepsilon''|) d|\varepsilon''|. \tag{9.28}$$

From Eqs. (9.15), (9.25), and (9.27), the expression for $n_c(T)$ is obtained as

$$n_c(T) = \eta_c \frac{\sqrt{2m_n^{*3}}}{\pi^2 \hbar^3} e^{(\mu-\varepsilon_c)/k_B T} \int_0^\infty \varepsilon'^{1/2} e^{-\varepsilon'/k_B T} d\varepsilon'. \tag{9.29}$$

From Eqs. (9.17), (9.26), and (9.28), the expression for $p_v(T)$ is obtained as

$$p_v(T) = \frac{\sqrt{2m_p^{*3}}}{\pi^2 \hbar^3} e^{(\varepsilon_v-\mu)/k_B T} \int_0^\infty |\varepsilon''|^{1/2} e^{-|\varepsilon''|/k_B T} d|\varepsilon''|. \tag{9.30}$$

It can be easily shown that

$$\int_0^\infty \varepsilon^{1/2} e^{-\varepsilon/k_B T} d\varepsilon = \frac{1}{2}(k_B T)^{3/2} \pi^{1/2}. \tag{9.31}$$

From Eqs. (9.29) and (9.31), we obtain

$$n_c(T) = \aleph_c(T) e^{(\mu - \varepsilon_c)/k_B T} \tag{9.32}$$

where

$$\aleph_c(T) = 2\eta_c \left(\frac{m_n^* k_B T}{2\pi\hbar^2} \right)^{3/2}. \tag{9.33}$$

$\aleph_c(T)$ can be expressed numerically as

$$\aleph_c(T) = 2.51 \eta_c \left(\frac{m_n^*}{m} \right)^{3/2} \left(\frac{T}{300\ \text{K}} \right)^{3/2} \times 10^{19} \text{cm}^{-3}. \tag{9.34}$$

Similarly, from Eqs. (9.30) and (9.31), we obtain

$$p_v(T) = \wp_v(T)\, e^{(\varepsilon_v - \mu)/k_B T}, \tag{9.35}$$

where

$$\wp_v(T) = 2 \left(\frac{m_p^* k_B T}{2\pi\hbar^2} \right)^{3/2}. \tag{9.36}$$

$\wp_v(T)$ can be expressed numerically as

$$\wp_v(T) = 2.51 \left(\frac{m_p^*}{m} \right)^{3/2} \left(\frac{T}{300\ \text{K}} \right)^{3/2} 10^{19} \text{cm}^{-3}. \tag{9.37}$$

From Eqs. (9.32) and (9.35), we obtain

$$n_c(T) p_v(T) = \aleph_c(T) \wp_v(T) e^{-E_g/k_B T}, \tag{9.38}$$

where E_g is the energy gap between the conduction and valence band,

$$E_g = \varepsilon_c - \varepsilon_v. \tag{9.39}$$

Eq. (9.38) is known as the *law of mass action*. It states that at a given temperature, one can obtain the density of one type of carrier if one knows the density of the other type of carrier, the effective masses of both carriers (including that of light and heavy holes), the number of symmetrically equivalent minima in the conduction band, and the energy gap of the semiconductor.

We note that Eqs. (9.32) and (9.35) are valid for both intrinsic (pure) and extrinsic (semiconductors with natural or doped impurities) semiconductors. We will first consider the case of intrinsic semiconductors.

9.4 INTRINSIC SEMICONDUCTORS

In an intrinsic semiconductor, because the number of electrons $n_c(T)$ in the conduction band is equal to the number of holes $p_v(T)$ in the valence band, we can express n_i, the number of carriers in each band from Eq. (9.38), as

$$n_i(T) = \sqrt{n_c(T)p_v(T)} = \sqrt{\aleph_c(T)\wp_v(T)}\, e^{-E_g/2k_BT}. \tag{9.40}$$

From Eqs. (9.34), (9.37), and (9.40), we obtain

$$n_i(T) = 2.51\eta_c^{1/2}\left(\frac{m_n^* m_p^*}{m^2}\right)^{3/4}\left(\frac{T}{300\text{ K}}\right)^{3/2} e^{-E_g/2k_BT}\,10^{19}\text{ cm}^{-1}. \tag{9.41}$$

The chemical potential of an intrinsic semiconductor is obtained from the equality relation

$$n_i(T) = n_c(T) = p_v(T), \tag{9.42}$$

and from Eqs. (9.32) and (9.35),

$$\aleph_c(T)e^{(\mu_i-\varepsilon_c)/k_BT} = \wp_v(T)e^{(\varepsilon_v-\mu_i)/k_BT} \tag{9.43}$$

or

$$\mu = \mu_i = \varepsilon_v + \frac{1}{2}E_g + \frac{1}{2}k_BT\ln\frac{\wp_v(T)}{\aleph_c(T)}. \tag{9.44}$$

From Eqs. (9.33), (9.36), and (9.44), we obtain

$$\mu_i = \varepsilon_v + \frac{1}{2}E_g + \frac{3}{4}k_BT\ln\left(\frac{m_p^*}{m_n^*}\right) - \frac{1}{2}k_BT\ln\eta_c. \tag{9.45}$$

From Eq. (9.45), at $T=0$, μ_i has a simple form,

$$\mu_i = \varepsilon_v + \frac{1}{2}E_g, \tag{9.46}$$

which is precisely at the center of the energy gap. Even at reasonable temperatures, because $m_p^* \approx m_n^*$ for most semiconductors, μ_i is close to the center of the energy gap. This is shown in Figure 9.3.

From Eq. (9.32) and in analogy with the free electron model, we can write the conductivity σ_i of an intrinsic semiconductor as

$$\sigma_i = \frac{n_i e^2\tau_e}{m_n^*} + \frac{p_i e^2\tau_h}{m_p^*}, \tag{9.47}$$

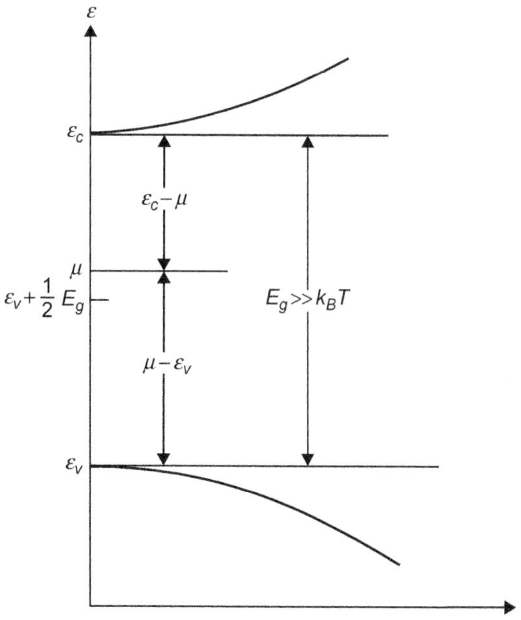

FIGURE 9.3

The chemical potential for an intrinsic semiconductor lies within the energy gap at $T \neq 0$.

where τ_e and τ_h are the relaxation times for electrons and holes. Eq. (9.47) can be rewritten in the alternate and more familiar form

$$\sigma_i = n_i e\mu_e + p_i e\mu_h, \tag{9.48}$$

where μ_e and μ_h are the mobility of the electrons and holes, which is the velocity (always defined as positive) of the carriers in unit electric field.

9.5 EXTRINSIC SEMICONDUCTORS

If there are impurities present in a semiconductor, which contribute a significant number of electrons to the conduction band or holes to the valence band, the semiconductor is known as an extrinsic semiconductor. This can be obtained either by doping (which is the base of modern electronics) or by contamination. In either case,

$$\Delta n(T) = n_c(T) - p_v(T) \neq 0. \tag{9.49}$$

From Eqs. (9.38) and (9.40), we have

$$n_c(T)p_v(T) = n_i(T)^2, \tag{9.50}$$

which can be rewritten in the alternate form (dropping the T in the bracket),

$$(p_v + \Delta n)p_v = n_i^2, \tag{9.51}$$

or

$$p_v^2 + \Delta n p_v - n_i^2 = 0. \tag{9.52}$$

Solving the quadratic equation, we have

$$p_v = -\frac{1}{2}\Delta n + \frac{1}{2}\left[(\Delta n)^2 + 4n_i^2\right]^{\frac{1}{2}}. \tag{9.53}$$

Following a similar procedure, we obtain

$$n_c = \frac{1}{2}\Delta n + \frac{1}{2}\left[(\Delta n)^2 + 4n_i^2\right]^{\frac{1}{2}}. \tag{9.54}$$

We dropped the negative sign before the square root in Eqs. (9.53) and (9.54) because both p_v and n_c are positive. We also obtain from Eqs. (9.32), (9.35), and (9.42),

$$n_c = n_i e^{(\mu-\mu_i)/k_B T} \tag{9.55}$$

and

$$p_v = p_i e^{-(\mu-\mu_i)/k_B T} = n_i e^{-(\mu-\mu_i)/k_B T}. \tag{9.56}$$

From Eqs. (9.49), (9.55), and (9.56), we have

$$n_c - p_v = n_i\left[e^{(\mu-\mu_i)/k_BT} - e^{-(\mu-\mu_i)/k_BT}\right] = 2n_i \sinh\left[(\mu-\mu_i)/k_BT\right]. \tag{9.57}$$

From Eqs. (9.49) and (9.57), we obtain

$$\frac{\Delta n}{n_i} = 2\sinh\left[(\mu-\mu_i)/k_BT\right], \tag{9.58}$$

which can be expressed in the alternate form

$$\mu = \mu_i + k_BT\sinh^{-1}\left[\frac{\Delta n}{2n_i}\right]. \tag{9.59}$$

Because μ_i is at the center of the energy gap, Δn must exceed n_i by many orders of magnitude before the chemical potential μ violates the condition of "nondegeneracy" stated earlier (Eqs. 9.23 and 9.24). The exception is in a region of "extreme extrinsic behavior."

We also note from Eqs. (9.53) and (9.54), if $|\Delta n| \gg n_i$, and if Δn is positive,

$$n_c \approx \Delta n \text{ and } p_v \approx n_c\left(\frac{n_i}{\Delta n}\right)^2, \tag{9.60}$$

in which case, $n_c \gg p_v$ and the semiconductor is called n-type. If Δn is negative,

$$p_v \approx |\Delta n| \text{ and } n_c \approx p_v\left(\frac{n_i}{|\Delta n|}\right)^2, \tag{9.61}$$

in which case, $p_v \gg n_c$ and the semiconductor is called p-type. These types of semiconductors are known as doped semiconductors and are the base on which modern electronics is built.

9.6 DOPED SEMICONDUCTORS

Doped impurities, which contribute additional electrons to the conduction band such that the semiconductor becomes n-type, are known as donors, whereas those that contribute holes to the valence band to make the semiconductor p-type are known as acceptors. The simplest example is that when in a group IV semiconductor such as a crystal of pure Si, an Si atom is replaced with a P atom, and there is an extra electron (because phosphorous has five valence electrons) that does not participate in the covalent bonds of Si. A deliberate substitution of quite a large number of P in Si is called doping, and because the excess electrons that are originally bound to their parent phosphorus atoms eventually end up in the conduction band at room temperature (because, as we will show, this binding energy is small), the semiconductor is called the n-type. A schematic diagram of this type of doping is shown in Figure 9.4.

The problem of the substitutional impurity can be simplified by ignoring the structural difference between silicon and phosphorus ion cores. In addition, as shown in Figure 9.5, the extra electron, which is bound to the parent phosphorus atom, can essentially be considered as a particle of charge $-e$ and mass m^* moving in the presence of an attractive center of charge e in a medium of dielectric constant $\in$. This problem is equivalent to that of the ground state of a hydrogen atom with two modifications: the dielectric constant $\in_0$ of a vacuum (in a hydrogen atom) is replaced by the dielectric constant $\in$ of the semiconductor, and the free electron mass (m_e) is replaced by the effective mass m_e^*.

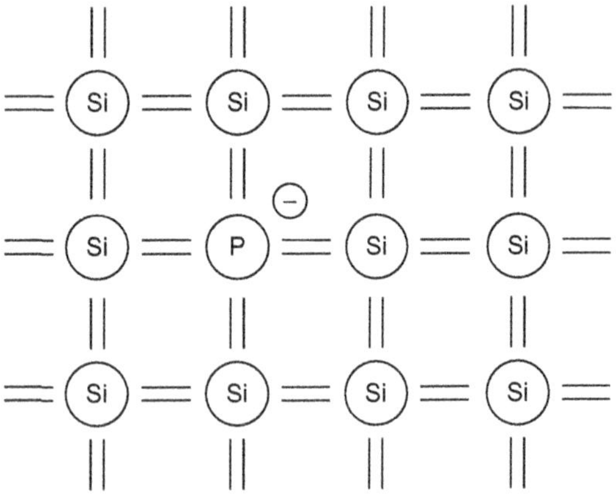

FIGURE 9.4

Phosphorus atom in a pure silicon crystal (donor impurity).

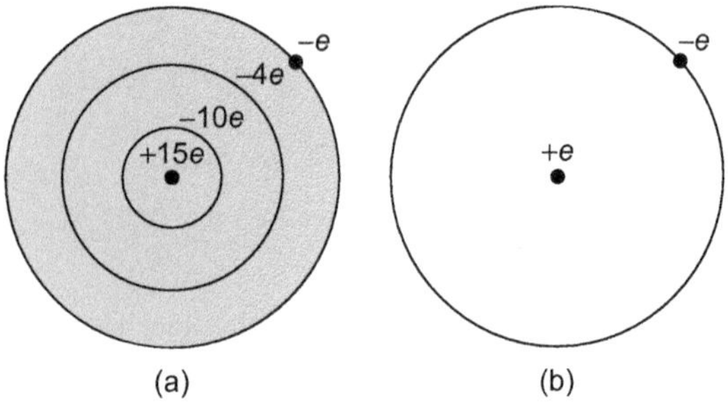

FIGURE 9.5

(a) The electrons in a phosphorus impurity. (b) The 10 inner electrons and the 4 valence electrons, which participate in the covalent bonding, screen the nucleus, which has an effective attractive charge *e*.

The binding energy of the donor electron is given by

$$\varepsilon = -\frac{m_e^* e^4}{2(4\pi \in \hbar)^2}, \tag{9.62}$$

and the radius of the first Bohr orbit is given by

$$r = \frac{m}{m_e^*} \in a_0, \tag{9.63}$$

where a_0 is the first Bohr radius and m is the free electron mass. Because the dielectric constant $\in$ is large ($\sim$20) and the effective mass m_e^* is small ($\sim$0.1 m), $r \sim 100$ A°. This justifies the use of a semiclassical model, and the binding energy is obtained from the expression

$$\varepsilon \approx \frac{m_e^*}{m} \frac{1}{\in^2} \times 13.6 \text{ eV}. \tag{9.64}$$

Using the same arguments, one can easily show that the binding energy of the donor electron is $\sim -0.01\,5$ eV. The bound impurity level is formed relative to the energy of the conduction band, and hence, the binding energy, which is much smaller than the energy gap (($E_g \approx 1.14$ eV for Si), is measured relative to these levels. At room temperature, the donor impurity is ionized, and the electron jumps to the conduction band. When a large number of such extra electrons are introduced in the conduction band by doping Si with P (or other elements from group V), the crystal is known as an *n*-type semiconductor.

It is easy to make the same argument for acceptor impurities by substituting an Si atom with an element from group III in a silicon crystal. An example of substituting an Si atom with an Al atom in a silicon crystal is shown in Figure 9.6.

As shown in Figure 9.6, there is a deficiency of one electron in the formation of a covalent bond that requires four electrons while aluminum has only three valence electrons. Thus, a hole is created that is initially bound to the parent aluminum atom. This bound state is known as an acceptor state with a role reversal of a hydrogen-like atom in the sense that there is a net charge of '$-e$' and mass m at the center and hole with charge 'e' and effective mass m_h^* orbiting around it. One can easily show that the radius of the first Bohr orbit is given by

$$r_h = \frac{m}{m_h^*} \in a_0. \tag{9.65}$$

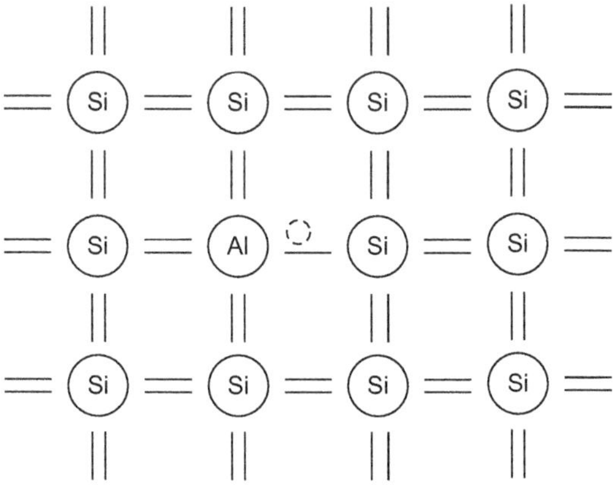

FIGURE 9.6

An example of acceptor impurity: aluminum atom in a silicon crystal.

The binding energy of the ground state is nearly the same for both electrons and holes, i.e., on the order of -0.015 eV. However, the binding energy of the donor levels is measured relative to the conduction band, and the binding energy of the acceptor levels is measured relative to the valence band (the zero of the valence band is at the top of the band, and the energy of the holes is measured positive downwards). Thus, both the donor and the acceptor levels are formed in the energy gap, as shown in Figure 9.7.

However, at room temperature, the thermal energy is sufficient for both the electrons and holes to fall into the conduction and valence bands, respectively.

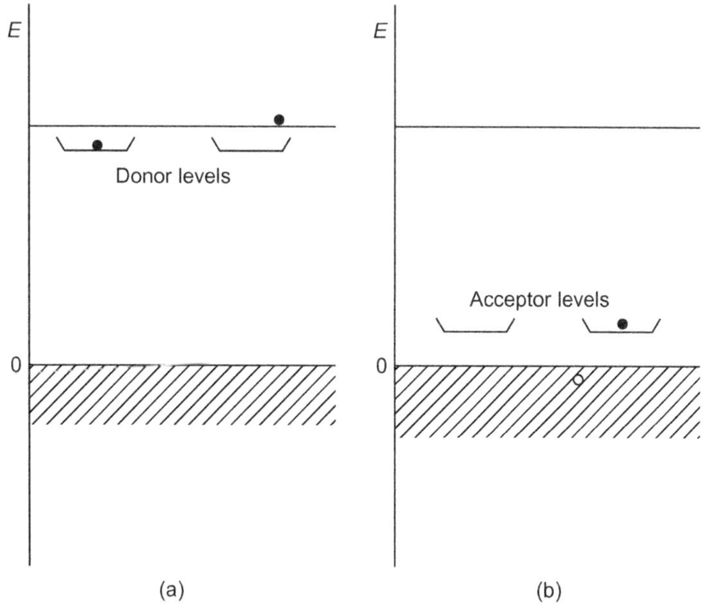

FIGURE 9.7

Donor and acceptor levels for (a) n-type and (b) p-type semiconductors.

9.7 STATISTICS OF IMPURITY LEVELS IN THERMAL EQUILIBRIUM

9.7.1 Donor Levels

For simplicity, we consider a semiconductor with donor states that have a binding energy ε_d located just below the bottom of the conduction band. Ignoring electron–electron interaction, the three possibilities for the donor level are either the level could be empty, or the donor could trap one electron of either spin (up or down). The donor level cannot bind two electrons of opposite spin at the same time. Using the Fermi statistics (in a grand canonical ensemble), we obtain f_d, the mean number of electrons in the donor level,

$$f_d = \frac{\sum_i n_i e^{-\beta(\varepsilon_i - \mu N_i)}}{\sum_i e^{-\beta(\varepsilon_i - \mu N_i)}}, \tag{9.66}$$

which can be written in the form

$$f_d = \frac{0 + 2e^{-\beta(\varepsilon_d - \mu)}}{1 + 2e^{-\beta(\varepsilon_d - \mu)}}. \tag{9.67}$$

A more appropriate way to write Eq. (9.67) is

$$f_d = \frac{1}{1 + \frac{1}{2} e^{\beta(\varepsilon_d - \mu)}}. \tag{9.68}$$

If the density of donors per unit volume is N_d, we can express n_d, the number density of electrons bound to the donor sites $(n_d = N_d f_d)$, as

$$n_d = \frac{N_d}{1 + \frac{1}{2} e^{\beta(\varepsilon_d - \mu)}}. \tag{9.69}$$

9.7.2 Acceptor Levels

The statistics of the holes can be simplified by considering the holes as the "absence" of electrons. An acceptor level, which is placed at an energy ε_a above the valence band, can be either singly or doubly occupied but cannot be empty. An acceptor impurity is essentially a fixed, negatively charged $(-e)$ attractive center superimposed on an unaltered host atom, which can weakly bind one hole. The binding energy of the hole is $\varepsilon_a - \varepsilon_v$, and a second electron moves into the acceptor level when the hole is "ionized." There is little probability that two holes would be localized in the presence of the acceptor impurity because the Coulomb repulsion between two holes would be very large. This scenario corresponds to no electrons in the acceptor level. In addition, we must account for the fact that the valence maximum is four-fold degenerate (including spin degeneracy). Thus, we obtain an expression for f_d', the occupation probability of electrons in the acceptor level,

$$f_d' = \frac{4e^{\beta\mu} + 2e^{-\beta(\varepsilon_a - 2\mu)}}{4e^{\beta\mu} + e^{-\beta(\varepsilon_a - 2\mu)}} = \frac{1 + \frac{1}{2} e^{\beta(\mu - \varepsilon_a)}}{1 + \frac{1}{4} e^{\beta(\mu - \varepsilon_a)}}. \tag{9.70}$$

The average number of holes in the acceptor level is the difference between the maximum number of electrons the level can hold (two) and the mean number of electrons in the level, which can be rewritten in the alternate form (for the mean number of holes)

$$f_a = 2 - f_a' = 2 - \frac{1 + \frac{1}{2}e^{\beta(\mu-\varepsilon_a)}}{1 + \frac{1}{4}e^{\beta(\mu-\varepsilon_a)}} = \frac{1}{1 + \frac{1}{4}e^{\beta(\mu-\varepsilon_a)}}. \tag{9.71}$$

If the semiconductor is doped with N_a acceptor impurities per unit volume, $p_a(p_a = N_a f_a)$, the number of holes in the acceptor levels is

$$p_a = \frac{N_a}{1 + \frac{1}{4}e^{\beta(\mu-\varepsilon_a)}}. \tag{9.72}$$

9.7.3 Doped Semiconductors

We assume that a semiconductor is doped with N_d donor impurities and N_a acceptor impurities (per unit volume) and assume that $N_d \geq N_a$. Thus, N_a of the N_d electrons, which are supplied by the donor atoms, will drop from the donor levels into the acceptor levels. The eventual scenario is that in the ground state ($T=0$), both the valence band and the acceptor levels are filled. In addition, $N_d - N_a$ of the donor levels are filled, but the conduction bands are empty. However, at finite temperature, the number of empty (electron) levels is $p_a + p_v$ in the valence band and the acceptor levels. So, we can write

$$N_d - N_a = n_c + n_d - p_v - p_a, \tag{9.73}$$

where n_c and n_d are the number of electrons in the conduction band and donor levels. To simplify the formulation, we assume (known as the conditions of nondegeneracy) that

$$\varepsilon_d - \mu \gg k_B T \tag{9.74}$$

and

$$\mu - \varepsilon_a \gg k_B T. \tag{9.75}$$

From Eqs. (9.69), (9.70), (9.74), and (9.75), we conclude that the impurities are essentially ionized and hence $n_d \ll N_d$ and $p_a \ll N_a$. Therefore, Eq. (9.74) can be approximated as

$$N_d - N_a \approx n_c - p_v \approx \Delta n. \tag{9.76}$$

From Eqs. (9.53), (9.54), and (9.76), we obtain

$$n_c = \frac{1}{2}[(N_d - N_a)] + \frac{1}{2}[(N_d - N_a)^2 + 4n_i^2]^{\frac{1}{2}} \tag{9.77}$$

and

$$p_v = \frac{1}{2}[(N_a - N_d)] + \frac{1}{2}[(N_d - N_a)^2 + 4n_i^2]^{\frac{1}{2}}. \tag{9.78}$$

In addition, Eq. (9.59) can be rewritten in the alternate form

$$\mu = \mu_i + k_B T \sinh^{-1}(|N_d - N_a|/2n_i). \tag{9.79}$$

Thus, $|N_d - N_a|$ must exceed n_i by many orders of magnitude before the conditions of nondegeneracy outlined in Eqs. (9.74) and (9.75) are violated. From Eqs. (9.69) and (9.72), we obtain $n_d \approx N_d$ and $p_a \approx N_a$, thereby ensuring that most of the impurities are fully ionized. From Eq. (9.77), we obtain in the extrinsic regime ($n_i \ll |N_d - N_a|$), when $N_d \gg N_a$,

$$n_c \approx N_d - N_a \tag{9.80}$$

and

$$p_v \approx \frac{n_i^2}{N_d - N_a}. \tag{9.81}$$

Eqs. (9.80) and (9.81) state that the number of donors is essentially equal to the number of mobile electrons while the number of holes in the valence band is very small. In the other extrinsic regime, when $N_a \gg N_d$, we obtain from Eq. (9.78),

$$p_v \approx N_a - N_d \tag{9.82}$$

and

$$n_c \approx \frac{n_i^2}{N_a - N_d}. \tag{9.83}$$

Eqs. (9.82) and (9.83) state that the number of acceptors is essentially equal to the number of mobile holes in the valence band while the number of electrons in the conduction band is very small.

9.8 DILUTED MAGNETIC SEMICONDUCTORS

9.8.1 Introduction

The diluted magnetic semiconductors (DMSs), which are ternary or quaternary alloys, in which a part of the nonmagnetic cations of the host material has been substituted by magnetic ions, have attracted considerable attention in recent years. In theory, any semiconductor with a fraction of its constituent ions replaced by ions bearing a net magnetic moment is considered as a DMS. However, in practice, the majority of DMS involve Mn^{2+} ions embedded in various $A^{II}B^{VI}$ or $A^{IV}B^{VI}$ hosts because Mn^{2+} can be incorporated in the host without affecting the crystallographic quality of the resulting material. In addition, Mn^{2+} possesses a relatively large magnetic moment characteristic of a half-filled d-shell ($S = 5/2$). Further, because Mn^{2+} is electrically neutral in $A^{II}B^{VI}$ or $A^{IV}B^{VI}$ hosts, it does not have accepting or donating centers. The Mn containing IV–VI DMSs can be divided into two groups, according to their carrier concentrations. The magnetic behavior of the first group (charge–carrier concentration, or p-type, range 10^{17}–10^{19}cm^{-3}) can be ascribed to antiferromagnetic interactions of the superexchange type between the Mn ions. This group has a spin-glass phase at low temperatures. The magnetic behavior of the second group (charge-carrier concentration, or p, of 10^{21}cm^{-3}) has a ferromagnetic ordering, induced by RKKY interactions (an indirect exchange mechanism in which the interaction between the magnetic ions is mediated by the itinerant charge carriers, or p) at low temperatures.

The magnetic properties of $Sn_{1-x}Mn_xTe$ and $Pb_{0.28-x}Sn_{0.72}Mn_xTe$ (which can be considered as partly like semimetals and partly like semiconductors) are very diverse. The ferromagnetic, spin-glass (SG) and reentrant-spin-glass phases have been observed, but the transition to a paramagnetic phase has not been observed at $T > 1.5$ K. The occurrence of these magnetic phases depends not only on the Mn concentration but also on the concentration of free carriers. The significance of carrier-mediated magnetism in a DMS is mainly due to the search for materials suitable for information processing and storage (in the light of developments in the rapidly emerging area of spintronics, which will be discussed in Chapter 11).

Some experimental results of the three-dimensional (T, x, p) magnetic phase diagram for $Sn_{1-x}Mn_xTe$ are shown in Figure 9.8.

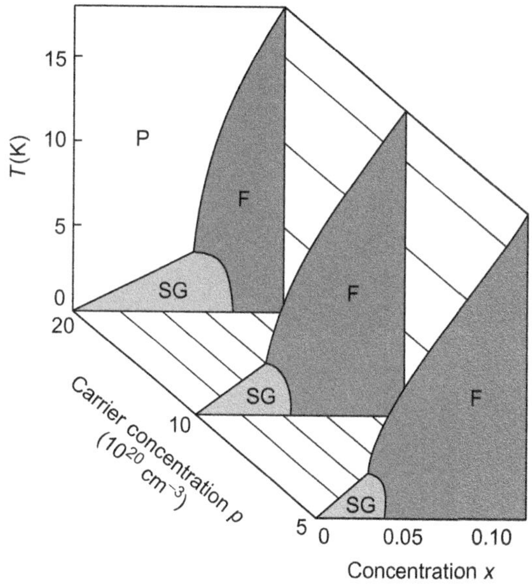

FIGURE 9.8

(T, x, p) phase diagram for $Sn_{1-x}Mn_xTe$. F = ferromagnetic phase; P = paramagnetic phase; SG = spin-glass phase.

Reproduced from Vennix et al.,[18] with the permission of the American Physical Society.

In subsequent experiments, the magnetic phase of the samples included magnetization, AC susceptibility, specific heat, and neutron diffraction. In Figure 9.9, the magnetic phase is displayed as a function of both the Mn concentration and the carrier concentration. The carrier concentration (the theoretical calculation of which is described later) has a significant effect on the magnetic phase of the material. At a carrier concentration of $p_c = 3 \times 10^{20}\,\text{cm}^{-3}$ $(T > 1.5\text{ K})$, there is an abrupt change from the paramagnetic to ferromagnetic phase. At high carrier concentrations $(p \gg p_c)$, there is a gradual transition to an intermediate reentrant-spin-glass phase and eventually to the spin-glass state. The location of these transitions is shifted to higher carrier concentrations if the Mn concentration is increased.

9.8.2 Magnetization in Zero External Magnetic Field in a DMS

Another interesting property of some DMSs is that magnetization in a zero external magnetic field can be induced by optically excited carriers in $Hg_{1-x}Mn_xTe$ and $Cd_{1-x}Mn_xTe$. The observed magnetization is due to the orientation of the Mn ions caused by the spin dynamics of the magnetic response on a picosecond time scale and is the result of spin-polarized electrons. This can be calculated by the spin-EPR shift, of which the theory is briefly described.

9.8.3 Electron Paramagnetic Resonance Shift

The electron paramagnetic resonance (EPR) shift is a measure of the internal field created at the magnetic ion site in magnetic materials or materials with magnetic impurities by the partial

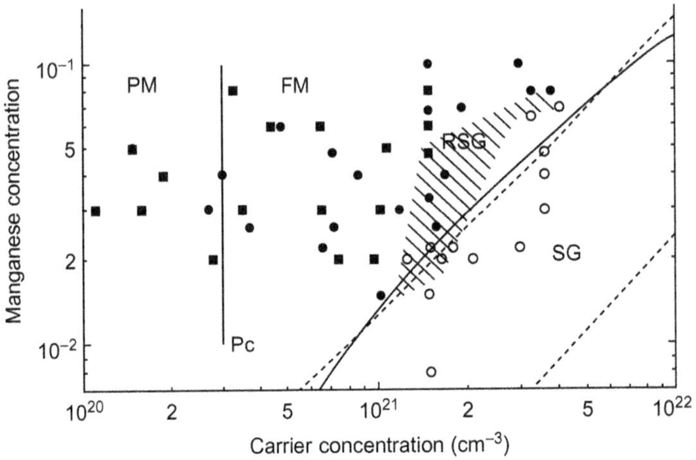

FIGURE 9.9

Magnetic-phase diagram of $Pb_{0.28-x}Sn_{0.72}Mn_xTe$ (squares) and $Sn_{1-x}Mn_xTe$ (circles). PM = paramagnetic, FM = ferromagnetic, RSG = reentrant spin glass, SG = spin glass.

Reproduced from Eggenkamp et al.,[5] with the permission of the American Physical Society.

polarization of electrons and/or carriers in an applied magnetic field. It can be shown that the contribution to the spin-EPR shift at the Mn^{2+} ion on the *j*th site of a DMS ($P_j^{\mu\nu}$) is given by the expression

$$P_j^{\mu\nu} = -\frac{\partial^2 \Omega}{\partial B^\mu \partial M_j^\nu}\bigg|_{\vec{B}\to 0, \vec{M}_j \to 0}, \tag{9.84}$$

where Ω is the thermodynamic potential and is given by

$$\Omega = \frac{1}{\beta} Tr \ln(-\tilde{G}_{\xi_l}). \tag{9.85}$$

Here, $\vec{M}_j$ is the local moment at the *j*th site, and $\tilde{G}$ is related to the one-particle Green's function (Eq. 7.219) in the presence of a periodic potential $V(\vec{r})$, spin-orbit interaction, applied magnetic field $\vec{B}$, and local magnetic moment. *Tr* involves summation of over both imaginary frequencies and one-particle states, forming a complete orthonormal set. The one-particle Green's function $G(\vec{r}, \vec{r}', \vec{B}, \vec{M}_j, \xi_l)$ satisfies the equation

$$(\xi_l - H)G(\vec{r}, \vec{r}', \vec{B}, \vec{M}_j, \xi_l) = \delta(\vec{r} - \vec{r}'), \tag{9.86}$$

where the complex energy

$$\xi_l = \frac{(2l+1)i\pi}{\beta} + \mu, \quad l = 0, \pm 1, \pm 2, \ldots, \tag{9.87}$$

and in a symmetric gauge,

$$G(\vec{r}, \vec{r}', \vec{B}, \vec{M}_j, \xi_l) = e^{i\vec{h}\cdot\vec{r}\times\vec{r}'}\tilde{G}(\vec{r}, \vec{r}', \vec{B}, \vec{M}_j, \xi_l), \tag{9.88}$$

where

$$\vec{h} = \frac{e\vec{B}}{2\hbar c}. \tag{9.89}$$

Here, $\widetilde{G}$ is the Green's function that satisfies the lattice translational symmetry, and $e^{i\vec{h}\cdot\vec{r}\times\vec{r}'}$ is the Peierls phase factor. The phase factor has the effect of translating the origin of the vector potential.

Recently, Trellakis[16] used a singular gauge transformation based on a lattice of magnetic flux lines, an equivalent quantum system with a periodic vector potential. However, his theory is beyond the scope of this book. The one-particle Hamiltonian is

$$\hat{H} = \frac{1}{2m}(\vec{p} + \frac{e}{c}\vec{A})^2 + V(\vec{r}) + \frac{\hbar}{4m^2c^2}\vec{\sigma}\cdot\vec{\nabla}V\times(\vec{p} + \frac{e}{c}\vec{A}) + \frac{1}{2}g_0\mu_0\vec{B}\cdot\vec{\sigma} + \hat{H}_I, \tag{9.90}$$

where

$$\hat{H}_I = \frac{1}{2g_J\mu_0}\sum_{\vec{j}}\vec{M}_j\cdot\vec{\sigma}\Im(\vec{r} - \vec{R}_j). \tag{9.91}$$

Here, $\vec{A}$ is the magnetic vector potential, $\Im(\vec{r} - \vec{R}_j)$ is the strength of the exchange interaction between the conduction electrons and/or the carriers and the local moment at the jth site, and the other symbols have their usual meaning. One can write the equation of motion in a representation defined by the periodic part $u_{\vec{k}\rho}(\vec{r})$ of the Bloch function $\psi_{\vec{k}\rho}(\vec{r})$, where $\vec{k}$ is the reduced wave vector and ρ is the spin index.

It has been shown (Misra et al.,[11] p. 1903) that in this representation, Eqs. (9.86), (9.88), (9.90), and (9.91) can be rewritten as

$$[\xi_l - H(\vec{\kappa})]\widetilde{G}(\vec{k}, \xi_l) = I, \tag{9.92}$$

where

$$H(\vec{\kappa}) = H_0(\vec{k}) + H'(\vec{\kappa}), \tag{9.93}$$

$$H_0(\vec{k}) = \frac{1}{2m}(\vec{p} + \hbar\vec{k})^2 + V + \frac{\hbar^2}{4m^2c^2}\vec{\sigma}.\vec{\nabla}V\times(\vec{p} + \hbar\vec{k}), \tag{9.94}$$

$$H'(\vec{\kappa}) = -i\frac{\hbar}{m}h_{\alpha\beta}\pi^{\alpha}\nabla_k^{\beta} + \frac{1}{2}g_0\mu_0\sigma^{\mu}H^{\mu} + \frac{1}{2\mu_0}\sum_j\frac{1}{g_j}M_j^{\nu}\sigma^{\nu}\Im, \tag{9.95}$$

$$\vec{\kappa} = \vec{k} + i\vec{h}\times\vec{\nabla}_k, \tag{9.96}$$

$$\vec{\pi} = \vec{p} + \frac{\hbar}{4mc^2}\vec{\sigma}\times\vec{\nabla}V, \tag{9.97}$$

$h_{\alpha\beta} = \in_{\alpha\beta\mu}h^{\mu}$, where $\in_{\alpha\beta\mu}$ is the antisymmetric tensor of the third rank, and we follow Einstein summation convention. It may be noted that $\vec{\kappa}$ (Eq. 9.96) is the Misra–Roth operator. Eq. (9.85) can be further simplified by writing the frequency summation as

$$\Omega = -\frac{1}{2\pi i}tr\oint_c \phi(\xi)\widetilde{G}(\xi)d\xi, \tag{9.98}$$

where

$$\phi(\xi) = -\frac{1}{\beta}\ln[1+e^{-\beta(\mu-\xi)}] \tag{9.99}$$

and the contour c encircles the imaginary axis in a counterclockwise direction. Eq. (9.92) can be solved by using a perturbation expansion of $\widetilde{G}(\vec{k},\xi)$,

$$\widetilde{G}(\vec{k},\xi) = \widetilde{G}_0(\vec{k},\xi) + \widetilde{G}_0(\vec{k},\xi)H'\widetilde{G}_0(\vec{k},\xi) + \widetilde{G}_0(\vec{k},\xi)H'\widetilde{G}_0(\vec{k},\xi)H'\widetilde{G}_0(\vec{k},\xi). \tag{9.100}$$

Here, the terms only up to second order are retained because the EPR shift independent of the applied field and the local moment is calculated. In Eq. (9.100), $\widetilde{G}_0(\vec{k},\xi)$ satisfies the equation

$$[\xi - H_0(\vec{k})]\widetilde{G}_0(\vec{k},\xi) = I \tag{9.101}$$

and is diagonal in the basis $u_{\vec{k}\rho}(\vec{r})$. Using the identity in Eq. (9.100),

$$\nabla_k^\alpha \widetilde{G}_0(\vec{k},\xi) = \frac{\hbar}{m}\widetilde{G}_0(\vec{k},\xi)\pi^\alpha \widetilde{G}_0(\vec{k},\xi), \tag{9.102}$$

we obtain

$$\widetilde{G}(\vec{k},\xi) = \sum_j M_j^\nu B^\mu \frac{1}{2g_j}\begin{bmatrix} (\widetilde{G}_0\sigma^\mu\widetilde{G}_0\sigma^\nu\Im\widetilde{G}_0 + \widetilde{G}_0\sigma^\nu\Im\widetilde{G}_0\sigma^\mu\widetilde{G}_0) \\ -\frac{i}{m}\varepsilon_{\alpha\beta\mu}(\widetilde{G}_0\pi^\alpha\widetilde{G}_0\pi^\beta\widetilde{G}_0\sigma^\nu\Im\widetilde{G}_0 + \widetilde{G}_0\sigma^\nu\Im\widetilde{G}_0\pi^\alpha\widetilde{G}_0\pi^\beta\widetilde{G}_0) \end{bmatrix}. \tag{9.103}$$

Here, $\widetilde{G}_0$ is the compact form of $\widetilde{G}_0(\vec{k},\xi)$. Eq. (9.98) is evaluated by using Eqs. (9.99) and (9.103). Expressing the contributions of the two terms in Eq. (9.103) to Ω as

$$\Omega = \Omega_1 + \Omega_2, \tag{9.104}$$

we obtain

$$\Omega_1 = \sum_j H^\mu M_j^\nu \sum_{n,\vec{k},\rho,\rho'} \left[\frac{1}{2g_j}\sigma^\nu\Im\right]_{n\rho,n\rho'} \sigma^\mu_{n\rho',n\rho} f'\left(E_{n\vec{k}\rho}\right) \tag{9.105}$$

and

$$\Omega_2 = \sum_j H^\mu M_j^\nu \sum_{n,m,\vec{k},\rho,\rho',\rho'' m\neq n} \frac{i}{m}\varepsilon_{\alpha\beta\mu}\frac{\left[\frac{1}{2g_j}\sigma^\nu\Im\right]_{n\rho,n\rho'} \pi^\alpha_{n\rho',m\rho''}\pi^\beta_{m\rho'',n\rho}}{E_{mn}} f'\left(E_{n\vec{k}\rho}\right), \tag{9.106}$$

where

$$E_{mn} \equiv E_{m\vec{k}} - E_{n\vec{k}} \tag{9.107}$$

and

$$H_0 u_{n\vec{k}\rho}(\vec{r}) = E_{n\vec{k}} u_{n\vec{k}\rho}(\vec{r}). \tag{9.108}$$

The matrix elements are of the type

$$A_{n\rho,n\rho'} \equiv \int u^*_{n\vec{k}\rho}(\vec{r})Au_{n\vec{k}\rho}(\vec{r})d\vec{r}, \tag{9.109}$$

and $f'(E_{n\vec{k}})$ is the first derivative of the Fermi function. The expressions for Ω_1 and Ω_2 derived in Eqs. (9.105) and (9.106) are substituted in Eq. (9.84), and the complex spin-orbit terms are neglected. After considerable algebra, we obtain the expression for the spin contribution to the EPR shift at the jth site,

$$P^{\nu\mu}_{js} = -\frac{1}{2}\sum_{n\vec{k}\rho\rho'}\left(\frac{1}{2g_j}\sigma^{\nu}\Im\right)_{n\rho,n\rho'} g^{\mu}_{nn}(\vec{k})\sigma^{\mu}_{n\rho',n\rho}f'\left(E_{n\vec{k}}\right), \tag{9.110}$$

where the effective g factor $g^{\mu}_{nn}(\vec{k})$ is defined as

$$g^{\mu}_{nn}(\vec{k})\sigma^{\mu}_{n\rho',n\rho} = g_0\sigma^{\mu}_{n\rho',n\rho} + \frac{2i}{m}\epsilon_{\alpha\beta\mu}\sum_{m\neq n,\rho''}\frac{\pi^{\alpha}_{n\rho',m\rho''}\pi^{\beta}_{m\rho'',n\rho}}{E_{mn}}. \tag{9.111}$$

Eq. (9.110) can be used to calculate the EPR shift of any diluted magnetic semiconductor of which the band structure is known.

A brief discussion of the calculation of the EPR shift from Eq. (9.110) by using the $\vec{k}\cdot\vec{\pi}$ model is explained in the next section.

9.8.4 $\vec{k}\cdot\vec{\pi}$ Model

The $\vec{k}\cdot\vec{p}$ model of Luttinger and Kohn[9] (p. 869) was modified by Tripathi et al.[17] (p. 3091) to include the effect of magnetic fields. Their method of calculation is known as the $\vec{k}\cdot\vec{\pi}$ model.

In the Luttinger–Kohn $(\vec{k}\cdot\vec{p})$ model, a complete orthonormal set of functions,

$$\chi_j(\vec{k},\vec{r}) = e^{i(\vec{k}-\vec{k}_0)\cdot\vec{r}}\psi_j(\vec{k}_0,\vec{r}), \tag{9.112}$$

is defined, where $\vec{k}_0$ is a reference point in the Brillouin zone at which the energy bands and wave functions $\psi_j(\vec{k}_0,\vec{r})$ have been determined. The unknown wave function is expanded as

$$\psi_n(\vec{k},\vec{r}) = \sum_j A_{nj}(\vec{k})\chi_j(\vec{k},\vec{r}). \tag{9.113}$$

In the $\vec{k}\cdot\vec{\pi}$ model, the effective equation of motion is

$$\sum_j\left[\left(E_j(\vec{k}_0) - E_n(\vec{k}) + \frac{\hbar^2(k^2-k_0^2)}{2m}\right)\delta_{jl} + \frac{\hbar}{m}(\vec{k}-\vec{k}_0)\cdot\vec{\pi}_{lj}\right]A_{nj}(\vec{k}) = 0, \tag{9.114}$$

where

$$\vec{\pi}_{lj} = \frac{(2\pi)^3}{\Omega_c}\int_{cell} d^3r\, u^*_l(\vec{k}_0,\vec{r})\vec{\pi}\, u_j(\vec{k}_0,\vec{r}). \tag{9.115}$$

There is one equation for each value of the band index l, and the condition for this infinite set of simultaneous, linear, and homogeneous equations to have a nontrivial solution is that the determinant of the coefficients should vanish. A general element of the determinant has the form $H_{jl} - E(\vec{k})\delta_{jl}$, with

$$H_{jl} = \left(E_j(\vec{k}_0) + \frac{\hbar^2}{2m}(k^2 - k_0^2) \right)\delta_{jl} + \frac{\hbar}{m}(\vec{k} - \vec{k}_0) \cdot \vec{\pi}_{lj}. \tag{9.116}$$

One can use Eq. (9.116) to determine the spin contribution to the EPR shift (P_s) of any diluted magnetic semiconductor. An example of using this method to determine the EPR shift of $Pb_{1-x}Mn_xTe$ as a function of carrier concentration can be found in Das et al.[4]

9.9 ZINC OXIDE

The semiconductor ZnO has attracted widespread attention in recent years for its optoelectronic properties and, more recently, for the possibility of finding room-temperature ferromagnetism when doped with magnetic and nonmagnetic impurities. The electronic structure of ZnO has been studied for the past 50 years. However, in spite of extensive energy band calculations, there still exists a controversy with regard to the valence band ordering in ZnO. The wurtzite ZnO conduction band is mainly constructed from the *s*-like state having Γ_7 symmetry, whereas the valence band is *p*-like, which is split into three bands due to the crystal field and spin-orbit interactions. A schematic picture of the band diagram is given in Figure 9.10. By treating the wurtzite energy levels as a perturbation over those of the zincblende, a formula has been derived for the valence band mixing, the extent of which is controlled by the relative magnitudes of the spin-orbit and crystal field splittings.

No satisfactory theory has yet been proposed for the possibility of finding room-temperature ferromagnetism in ZnO, when doped with magnetic and nonmagnetic impurities. Some of the fascinating theoretical models include the $\mathbf{k} \cdot \mathbf{p} + \mathrm{U}$ formalism, where U is the many-body Hubbard term.

9.10 AMORPHOUS SEMICONDUCTORS

9.10.1 Introduction

In recent years, amorphous semiconductors have attracted considerable attention. The amorphous structures can be obtained by rapid cooling from the melt or by evaporation onto a cooled substrate. They can also be obtained by sputtering the components on to a cooled substrate. The evaporation or sputtering of Si or Ge onto substrates held below 300°C produces amorphous films, which are stable up to 425°C. In the amorphous (a) state, the sp^3 covalent bonds are strong in Si and Ge. However, although these bonds exist between nearest neighbors, the

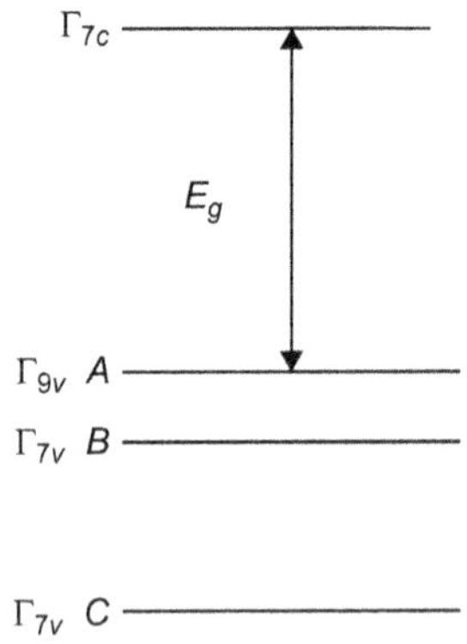

FIGURE 9.10

Schematic picture of valence band ordering for the wurtzite ZnO.

bonding is not perfect. There is four-fold coordination over small regions of the solid, but the tetrahedral symmetry is lost over extended regions. The bond angles are distorted from the ideal value, and many atoms can have only three neighbors. One can consider the system as a random network of imperfectly bonded atoms in which even small voids can exist.

Recent experiments of amorphous semiconductors exhibit many interesting properties. Tetrahedral and quasitetrahedral materials exhibit an unusual enhancement of diamagnetic susceptibility, χ (450% for Si, 270% for Ge, and 150% for $CdGeAs_2$) in the amorphous (a) phase relative to the crystalline (c) phase but no change in dielectric susceptibility (χ_e). In contrast, in chalcogenides, there is very little change in χ in either phase but appreciable reduction of χ_e in the (a) phase. These unusual properties are related to the nature of the chemical bonding and the presence or absence of long-range order. The chemical bond approach to the study of electronic properties of solids is much simpler than the band theory; emphasizes the bond aspect of the crystal structure; and is valuable in studying chemical trends such as covalency, polarity, and metallicity. This approach has a certain degree of flexibility and is well suited to the study of amorphous as well as periodic ones. In the chemical bond approach, each atom in the amorphous system has four nearest neighbors, but the bonds are distorted in bond lengths and directions. Thus, one starts with a crystalline phase and considers the covalent bonds that give rise to bonding and antibonding orbitals, leading to valence and conduction bands. Then distortion in both bond lengths and angles is introduced, and the random network model of Polk[13] (p. 365) is used to derive appropriate expressions for physical properties of amorphous structures.

The electron energy levels of amorphous systems can be derived from a modified bond orbital (LCH) theory typical of the crystalline phase. Even if they are randomly arranged, the covalent bonds give rise to some system of bonding and antibonding levels, which leads to valence and conduction bands. However, due to distortion of the bonds, the bands become broader. We first consider a linear combination of hybrids (LCH) model for tetrahedral semiconductors.

9.10.2 Linear Combination of Hybrids Model for Tetrahedral Semiconductors

We describe a linear combination of hybrids (LCH) model for a chemical bond approach to tetrahedral semiconductors (Sahu and Misra,[14] p. 6795). In this model, the basis set for the valence bands is a linear combination of sp^3 hybrids forming a bond, in which their relative phase factors, which have been neglected in the usual chemical bond models, adapted from the bond orbital models for molecules (Sukhatme and Wolff,[15] p. 1369; Chadi et al.,[2] p. 1372), have been included properly. A basis set for the conduction bands, which are orthogonal to the valence band functions, has also been constructed. It can be shown that the basic assumption of the earlier chemical bond models—i.e., that the localized functions have the character of chemical bonds—is equivalent to ignoring the relative Bloch phase factor $e^{i\mathbf{k}\cdot\mathbf{d}_j}$ (where $\mathbf{d}_j$ is a bond length) between the hybrids forming a bond. However, because $\mathbf{d}_j - \mathbf{d}_{j'} (j \neq j')$ is a lattice vector, these relative phase factors play an important role in solids, unlike the case of molecules where it can be neglected.

We will now discuss the LCH model in detail. In the zincblende structure, each atom is surrounded tetrahedrally by four identical atoms, which may be of the second type. The primitive cell contains two basic atoms at site i, with four sp^3 hybrids $h_j^1(\mathbf{r} - \mathbf{R}_i)$ pointing from atom I to the nearest neighbors (atom II) along the directions $j (j = 1, \ldots, 4)$ and four other sp^3 hybrids $h_j^2(\mathbf{r} - \mathbf{R}_i - \mathbf{d}_j)$ pointing from these nearest neighbors to atom I. Here, $\mathbf{R}_i$ is a lattice vector for site i and locates atoms of type I (one of the atomic sites I is chosen as the origin) and $\mathbf{d}_j$ is a nearest-neighbor vector joining atom I with atom II. The hybrids can be expressed as

$$h_j^1(\mathbf{r}-\mathbf{R}_i) = \frac{1}{2}[s_1 + \sqrt{3}(\xi_j^x p_{x1} + \xi_j^y p_{y1} + \xi_j^z p_{z1})] \tag{9.117}$$

and

$$h_j^2(\mathbf{r}-\mathbf{R}_i-\mathbf{d}_j) = \frac{1}{2}[s_2 - \sqrt{3}(\xi_j^x p_{x2} + \xi_j^y p_{y2} + \xi_j^z p_{z2})], \tag{9.118}$$

where $\vec{\xi}_1 = \frac{1}{\sqrt{3}}(1,1,1)$; $\vec{\xi}_2 = \frac{1}{\sqrt{3}}(1,\bar{1},\bar{1})$; $\vec{\xi}_3 = \frac{1}{\sqrt{3}}(\bar{1},1,\bar{1})$; and $\vec{\xi}_4 = \frac{1}{\sqrt{3}}(\bar{1},\bar{1},1)$ are the atomic orbitals at sites I and II, respectively. The Bloch-type tight-binding sums for valence-band basis functions are constructed by taking a linear combination of the hybrids forming a bond

$$\chi_j^v(\mathbf{r},\mathbf{k}) = \sum_i f_j^v(\mathbf{k}) e^{i\mathbf{k}\cdot\mathbf{R}_i}[h_j^1(\mathbf{r}-\mathbf{R}_i) + \lambda h_j^2(\mathbf{r}-\mathbf{R}_i-\mathbf{d}_j)e^{i\mathbf{k}\cdot\mathbf{d}_j}], \tag{9.119}$$

where

$$f_j^v(\mathbf{k}) = [N(1+\lambda^2+2\lambda S\cos\mathbf{k}\cdot\mathbf{d}_j)]^{-1/2}. \tag{9.120}$$

Here, S is the overlap integral, and $\lambda^2/(1+\lambda^2)$ is the probability of the electron being around atom II for III–V semiconductors and is related to Coulson's ionicity (Coulson et al.,[3] p. 357; Nucho et al.,[12] p. 1843) by the expression

$$f_c = \frac{(1-S)^{1/2}(1-\lambda^2)}{(1+\lambda^2+2\lambda S)}. \tag{9.121}$$

We note that the LCH model discussed here is different from the usual bond-orbital model[12] in the sense that a relative phase factor $e^{i\mathbf{k}\cdot\mathbf{d}_j}$ between the two hybrids forming a bond has been included to properly account for the origin.

The basis functions for the conduction band $\chi_j^c(\mathbf{r},\mathbf{k})$ are obtained by constructing functions orthogonal to $\chi_j^v(\mathbf{r},\mathbf{k})$:

$$\chi_j^c(\mathbf{r},\mathbf{k}) = \sum_i f_j^c(\mathbf{k}) e^{i\mathbf{k}\cdot\mathbf{R}_i}[(\lambda + Se^{i\mathbf{k}\cdot\mathbf{d}_j})h_j^1(\mathbf{r}-\mathbf{R}_i) - (\lambda S + e^{i\mathbf{k}\cdot\mathbf{d}_j})h_j^2(\mathbf{r}-\mathbf{R}_i-\mathbf{d}_j)], \tag{9.122}$$

where

$$f_j^c(\mathbf{k}) = \left[\frac{\lambda + Se^{-i\mathbf{k}\cdot\mathbf{d}_j}}{N(1-S^2)(1+\lambda^2+2\lambda S\cos\,\mathbf{k}\cdot\mathbf{d}_j)(\lambda+Se^{i\mathbf{k}\cdot\mathbf{d}_j})}\right]^{1/2}. \tag{9.123}$$

The Bloch eigenfunctions for the valence and conduction bands are

$$\psi_n(\mathbf{r},\mathbf{k}) = \sum_j \alpha_{jn}^v(\mathbf{k})\chi_j^v(\mathbf{r},\mathbf{k}) \tag{9.124}$$

and

$$\psi_m(\mathbf{r},\mathbf{k}) = \sum_j \alpha_{jm}^c(\mathbf{k})\chi_j^c(\mathbf{r},\mathbf{k}), \tag{9.125}$$

where

$$\sum_n \alpha^v_{jn}(\mathbf{k})\alpha^{v\dagger}_{nj'}(\mathbf{k}) = \delta_{jj'} \tag{9.126}$$

and

$$\sum_m \alpha^c_{jm}(\mathbf{k})\alpha^{c\dagger}_{mj'}(\mathbf{k}) = \delta_{jj'}. \tag{9.127}$$

The conduction and valence bands for the tetrahedral semiconductors as well as the expression for the energy gap can be obtained by using these Bloch functions in the same way they were derived earlier in Chapter 4.

The physical properties of model amorphous semiconductors (a-phase), such as magnetic susceptibility and dielectric constant, are obtained by introducing disorder in the bond angles and the bond lengths. The nearest-neighbor coordinations of the a-phase are not supposed to change appreciably from the c-phase. However, the bands become broader due to distortion of the bonds. In addition, the disorder causes a spread of the energy levels into the energy gap region. These are known as tail states, which arise due to the distorted bonds. These tail states are nonconducting and localized. A schematic picture of these tail states is shown in Figure 9.11.

In addition, an amorphous tetrahedral semiconductor has imperfectly coordinated atoms, and thereby, the bonds are uncompensated. These are known as "dangling" bonds, each of which has limited degeneracy. A dangling bond has an empty state and an electron associated with it. Because these bonds are uncompensated, both the electron and the empty state are localized at 0° K. Thus, the immobile states are separate but overlapping, and the distributions are obtained within the energy gap. In fact, the concentration of the dangling bonds is on the order of 10^{25} m^{-3}. Due to this unusually high concentration of the dangling bonds, each of which produces a localized electron and a localized empty state, the Fermi level (μ) is at the center of the energy gap. Thus, the amorphous semiconductors essentially become insensitive to doping. The "band" picture of an amorphous tetrahedral semiconductor is shown in Figure 9.12.

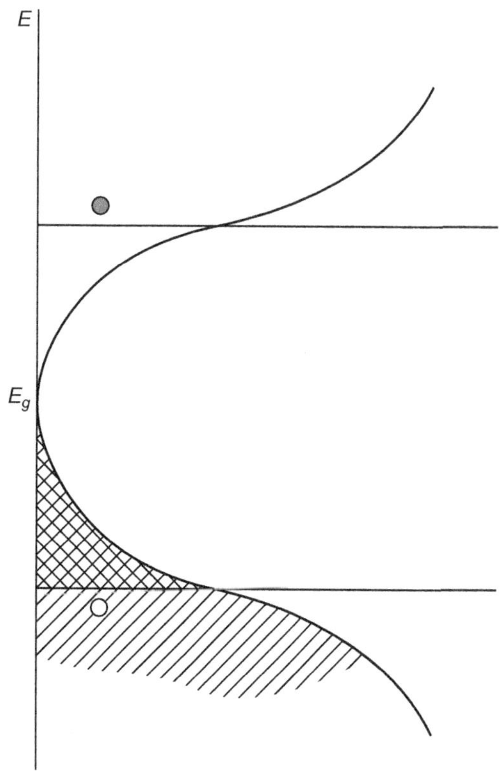

FIGURE 9.11

The nonconducting tail states in a distorted tetrahedral semiconductor arise from the distorted bonds.

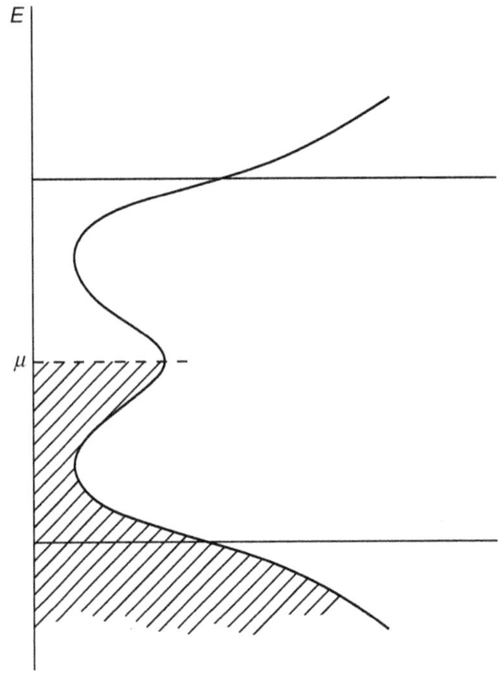

FIGURE 9.12

The band picture of amorphous tetrahedral semiconductors including tail states.

PROBLEMS

9.1. The effective mass of electrons and holes in a semiconductor can be expressed as

$$[\mathbf{M}_n^{-1}(\mathbf{k})]_{ij} = \pm\frac{1}{\hbar^2}\frac{\partial^2 \varepsilon_n}{\partial k_i \partial k_j}, \tag{1}$$

where the positive sign is for electrons and the negative sign is for holes. Because the bottom of the conduction band is at ε_c and the top of the valence band is at ε_v, show that the energy of the electrons ($\varepsilon_e(\mathbf{k})$) and holes ($\varepsilon_h(\mathbf{k})$) can be written as

$$\varepsilon_e(\mathbf{k}) = \varepsilon_c + \frac{\hbar^2}{2}\sum_{ij} k_i(\mathbf{M}_e^{-1})_{ij}k_j \tag{2}$$

and

$$\varepsilon_h(\mathbf{k}) = \varepsilon_v - \frac{\hbar^2}{2}\sum_{ij} k_i(\mathbf{M}_h^{-1})_{ij}k_j. \tag{3}$$

9.2. Show that the density of states in the conduction and valence bands can be expressed as

$$\begin{aligned} g_c(\varepsilon) &= \frac{\sqrt{2m_n^{*3}\varepsilon'}}{\pi^2\hbar^3}\eta_c, \quad \varepsilon > \varepsilon_c \\ &= 0, \qquad\qquad \varepsilon \leq \varepsilon_c, \end{aligned} \tag{1}$$

where

$$\varepsilon' \equiv \varepsilon - \varepsilon_c \tag{2}$$

and

$$g_v(\varepsilon) = \frac{\sqrt{2m_p^{*3/2}|\varepsilon''|}}{\pi^2\hbar^3}, \quad |\varepsilon''| > \varepsilon_v$$
$$= 0, \quad |\varepsilon''| \leq \varepsilon_v, \tag{3}$$

where

$$|\varepsilon''| \equiv |\varepsilon - \varepsilon_v|. \tag{4}$$

Here, η_c is the number of symmetrically equivalent minima in the conduction band (six for Si and eight for Ge). m_n^* and m_p^* are the density of states' effective mass of electrons and holes that are obtained from the relation

$$m_n^* = \left(m_1^e m_2^e m_3^e\right)^{1/2} \tag{5}$$

and

$$m_p^{*3/2} = \left(m_{pl}^*\right)^{3/2} + \left(m_{ph}^*\right)^{3/2}, \tag{6}$$

where

$$m_{pl}^* = \left(m_1^{lh} m_2^{lh} m_3^{lh}\right)^{1/2} \tag{7}$$

and

$$m_{ph}^* = \left(m_1^{hh} m_2^{hh} m_3^{hh}\right)^{1/2}. \tag{8}$$

Here, m_{pl}^* and m_{ph}^* are the effective masses of the light and heavy holes.

9.3. From Eq. (9.32) and in analogy with the free electron model, one can write the conductivity σ_i of an intrinsic semiconductor as

$$\sigma_i = \frac{n_i e^2 \tau_e}{m_n^*} + \frac{p_i e^2 \tau_h}{m_p^*}, \tag{1}$$

where τ_e and τ_h are the relaxation times for electrons and holes. Show that Eq. (9.48) can be written in the alternate and more familiar form

$$\sigma_i = n_i e \mu_e + p_i e \mu_h, \tag{2}$$

where μ_e and μ_h are the mobility of the electrons and holes, which is the velocity (always defined as positive) of the carriers in a unit electric field.

9.4. We have derived an expression for the valence band

$$p_v(T) = \frac{\sqrt{2m_p^{*3}}}{\pi^2\hbar^3} e^{(\varepsilon_v - \mu)/k_B T} \int_0^\infty |\varepsilon''|^{1/2} e^{-|\varepsilon''|/k_B T} d|\varepsilon''|. \tag{1}$$

It can be easily shown that

$$\int_0^\infty \varepsilon^{1/2} e^{-\varepsilon/k_B T} d\varepsilon = \frac{1}{2}(k_B T)^{3/2}\pi^{1/2}. \tag{2}$$

From Eqs. (1) and (2), show that

$$p_v(T) = \wp_v(T) e^{(\varepsilon_v - \mu)/k_B T}, \tag{3}$$

where

$$\wp_v(T) = 2\left(\frac{m_p^* k_B T}{2\pi\hbar^2}\right)^{3/2}. \tag{4}$$

$\wp_v(T)$ can be expressed numerically as

$$\wp_v(T) = 2.51\left(\frac{m_p^*}{m}\right)^{3/2}\left(\frac{T}{300\text{ K}}\right)^{3/2} 10^{19}\text{ cm}^{-3}. \tag{5}$$

9.5. Substitute the expressions for Ω_1 and Ω_2 derived in Eqs. (9.105) and (9.106) in Eq. (9.84) and neglect the complex spin-orbit terms. Show that the expression for the spin contribution to the EPR shift at the jth site is

$$P_{js}^{\nu\mu} = -\frac{1}{2}\sum_{n\vec{k}\rho\rho'}\left(\frac{1}{2g_j}\sigma^\nu \Im\right)_{n\rho,n\rho'} g_{nn}^\mu(\vec{k})\sigma_{n\rho',n\rho}^\mu f'(E_{n\vec{k}}), \tag{1}$$

where the effective g factor $g_{nn}^\mu(\vec{k})$ is defined as

$$g_{nn}^\mu(\vec{k})\sigma_{n\rho',n\rho}^\mu = g_0\sigma_{n\rho',n\rho}^\mu + \frac{2i}{m}\epsilon_{\alpha\beta\mu}\sum_{m\neq n,\rho''}\frac{\pi_{n\rho',m\rho''}^\alpha \pi_{m\rho'',n\rho}^\beta}{E_{mn}}. \tag{2}$$

9.6. If the Bloch-type tight-binding sums for valence-band basis functions are constructed by taking a linear combination of the hybrids forming a bond

$$\chi_j^v(\mathbf{r},\mathbf{k}) = \sum_i f_j^v(\mathbf{k}) e^{i\mathbf{k}\cdot\mathbf{R}_i}[h_j^1(\mathbf{r}-\mathbf{R}_i) + \lambda h_j^2(\mathbf{r}-\mathbf{R}_i-\mathbf{d}_j)e^{i\mathbf{k}\cdot\mathbf{d}_j}], \tag{1}$$

where

$$f_j^v(\mathbf{k}) = [N(1+\lambda^2+2\lambda S\cos\,\mathbf{k}\cdot\mathbf{d}_j)]^{-1/2}, \tag{2}$$

show that the basis functions for the conduction band $\chi_j^c(\mathbf{r},\mathbf{k})$ are obtained by constructing functions orthogonal to $\chi_j^v(\mathbf{r},\mathbf{k})$:

$$\chi_j^c(\mathbf{r},\mathbf{k}) = \sum_i f_j^c(\mathbf{k}) e^{i\mathbf{k}\cdot\mathbf{R}_i}[(\lambda + S e^{i\mathbf{k}\cdot\mathbf{d}_j})h_j^1(\mathbf{r}-\mathbf{R}_i) - (\lambda S + e^{i\mathbf{k}\cdot\mathbf{d}_j})h_j^2(\mathbf{r}-\mathbf{R}_i-\mathbf{d}_j)], \tag{3}$$

where

$$f_j^c(\mathbf{k}) = \left[\frac{\lambda + S e^{-i\mathbf{k}\cdot\mathbf{d}_j}}{N(1-S^2)(1+\lambda^2+2\lambda S\cos\,\mathbf{k}\cdot\mathbf{d}_j)(\lambda + S e^{i\mathbf{k}\cdot\mathbf{d}_j})}\right]^{1/2}. \tag{4}$$

9.7. Show that the Bloch eigenfunctions for the valence and conduction bands

$$\psi_n(\mathbf{r},\mathbf{k}) = \sum_j \alpha^v_{jn}(\mathbf{k})\chi^v_j(\mathbf{r},\mathbf{k}) \tag{1}$$

and

$$\psi_m(\mathbf{r},\mathbf{k}) = \sum_j \alpha^c_{jm}(\mathbf{k})\chi^c_j(\mathbf{r},\mathbf{k}) \tag{2}$$

are orthonormal. Here, α's are elements of (4×4) unitary matrices

$$\sum_n \alpha^v_{jn}(\mathbf{k})\alpha^{v\dagger}_{nj'}(\mathbf{k}) = \delta_{jj'} \tag{3}$$

and

$$\sum_m \alpha^c_{jm}(\mathbf{k})\alpha^{c\dagger}_{mj'}(\mathbf{k}) = \delta_{jj'}. \tag{4}$$

References

1. Ashcroft NW, Mermin ND. *Solid state physics.* New York: Brooks/Cole; 1976.
2. Chadi DJ, White RM, Harrison WA. Theory of the magnetic susceptibility of tetrahedral semiconductors. *Phys Rev Lett* 1975;**35**:1372.
3. Coulson CA, Redei LB, Stocker D. The electronic properties of tetrahedral intermetallic compounds I. charge distribution. *Proc R Soc* 1962;**270**:357.
4. Das RK, Tripathi GS, Misra PK. Theory of the spin EPR shift: application to Pb1-xMnxTe. *Phys Rev B* 2005;**72**:035216.
5. Eggenkamp PJT, Swegten HJM, Story T, Litvinov VI, Swuste CHW, de Jonge WJM. Calculations of the ferromagnet-to-spin-glass transition in diluted magnetic systems with RKKY interaction. *Phys Rev* 1995; **51**:15250.
6. Furdyna JK, Kossut J, eds. *Diluted magnetic semiconductors.* Boston: Academic Press; 1988.
7. Kittel C. *Introduction to solid state physics.* New York: John Wiley & Sons; 1976.
8. Louie SG. Quasiparticle excitations and photoemission. *Stud Surf Sci Catal* 1992;**74**:32.
9. Luttinger JM, Kohn W. Motion of electrons and holes in perturbed periodic fields. *Phys Rev* 1969;**97**:869.
10. Marder MP. *Condensed matter physics.* New York: John Wiley & Sons; 2000.
11. Misra SK, Misra PK, Mahanti SD. Many-body theory of magnetic susceptibility of electrons in solids. *Phys Rev B* 1982;**26**:1903.
12. Nucho RN, Ramos JG, Wolff PA. Chemical-bond approach to the magnetic susceptibility of semiconductors. *Phys Rev B* 1978;**17**:1843.
13. Polk DE. Structural model for amorphous silicon and germanium. *J Non-Cryst Solids* 1971;**5**:365.
14. Sahu T, Misra PK. Magnetic susceptibility of tetrahedrally coordinated solids. *Phys Rev B* 1982;**26**:6795.
15. Sukhatme VP, Wolff PA. Chemical-bond approach to the magnetic susceptibility of tetrahedral semiconductors. *Phys Rev Lett* 1975;**35**:1369.
16. Trellakis A. Nonperturbative solution for Bloch electrons in constant magnetic fields. *Phys Rev Lett* 2003;**91**:056405.
17. Tripathi GS, Das LK, Misra PK, Mahanti SD. Theory of spin-orbit and many-body effects of the Knight shift. *Phys Rev B* 1982;**25**:3091.
18. Vennix CWHM, Frikkee E, Eggenkamp PJT, Swagten HJM, Kopinga K, de Jonge WJM. Neutron-diffraction study of the carrier-concentration-induced ferromagnet-to-spin-glass transition in the diluted magnetic semiconductor Sn1-xMnxTe. *Phys Rev B* 1993;**48**:3770.

CHAPTER

Electronics

10

CHAPTER OUTLINE

10.1 Introduction 305
10.2 p-n Junction 306
10.2.1 Introduction 306
10.2.2 p-n Junction in Equilibrium 307
10.3 Rectification by a p-n Junction 311
10.3.1 Equilibrium Case 311
10.3.2 Nonequilibrium Case ($V \neq 0$) 313
10.4 Transistors 318
10.4.1 Bipolar Transistors 318
10.4.2 Field-Effect Transistor 319
10.4.3 Single-Electron Transistor 321
10.5 Integrated Circuits 325
10.6 Optoelectronic Devices 325
10.7 Graphene 329
10.8 Graphene-Based Electronics 332
Problems 333
References 336

10.1 INTRODUCTION

The importance of semiconductors is due to their use in modern electronics. The original semiconductor device was the Schottky diode, constructed from a metal-semiconductor interface, which was used instead of a thermoionic diode for rectification. The Schottky diode was constructed by placing a metal whisker against an n-doped semiconductor crystal (the p-doped semiconductor-metal contact does not have any rectifying properties). The work function of the semiconductor is less than that of the metal. Because of the higher chemical potential, the electrons rush from the semiconductor to the metal, and the voltage of the metal is lowered until further motion of charges is prevented by electrostatic forces. The increase in voltage $(-eV(x))$ of the semiconductor compensates for the difference in the chemical potential μ, which is equal to the Fermi energy ε_F of the metal. This also adds an electrostatic potential $-eV(x)$ to the conduction and valence band levels, which are bent by the potentials formed across the junction. This scenario is shown in Figure 10.1.

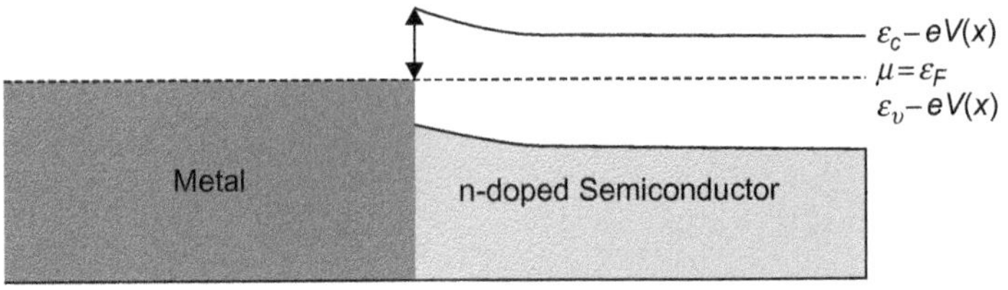

FIGURE 10.1

The charges move from the n-doped semiconductor to the metal, and both the valence and conduction bands are increased by $-eV(x)$. The electrons flow to the metal when the voltage of the metal is raised relative to the semiconductor by eV_A.

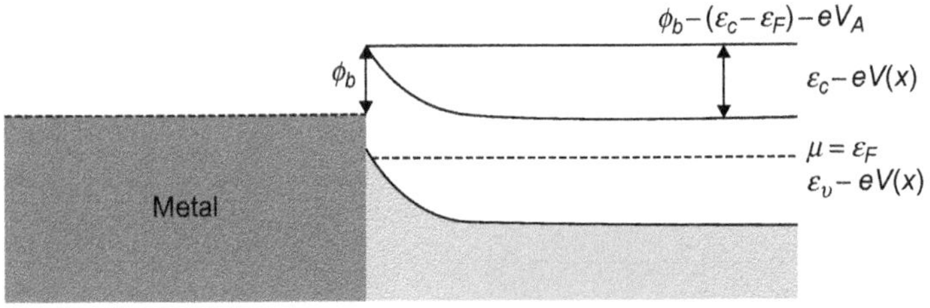

FIGURE 10.2

There is no flow of current when V_A is lowered. The height of the barrier is ϕ_b.

The effect of an external voltage V_A (which can be positive or negative) can be understood in a simple manner. When V_A is positive, the electrons flow from the semiconductor into the metal because the barrier is lowered. However, when V_A is negative, there is no change in the barrier, and hence, the current does not flow. This rectifying effect of the Schottky diode is shown in Figure 10.2.

In practice, it is difficult to construct an ideal Schottky diode because semiconductor surfaces acquire oxide layers after cleaving, and it is virtually impossible to produce atomic flat surfaces. The age of modern electronics was started with the creation of the p-n junction in a semiconductor.

10.2 p-n JUNCTION

10.2.1 Introduction

A simple p-n junction is fabricated by taking a single intrinsic semiconductor such as Si in which donor impurities are introduced in one region and acceptor impurities are introduced into another. It is easier to visualize the p-n junction as equivalent to two Si crystals doped with donor (n-type) and acceptor (p-type) impurities joined together with polished surfaces. In practice, this is not a good idea because the surface effects would interfere with the properties of the semiconductor. These semiconductors have different chemical potentials, $\mu(n)$ and $\mu(p)$, respectively. Initially, most of the conduction electrons are at the n-type, and the holes are at the p-type semiconductor. This scenario is shown in Figure 10.3a. However, when the two materials are joined together, some of the conduction electrons will diffuse to the p-type material while some of the holes will diffuse to the n-type material. When an electron

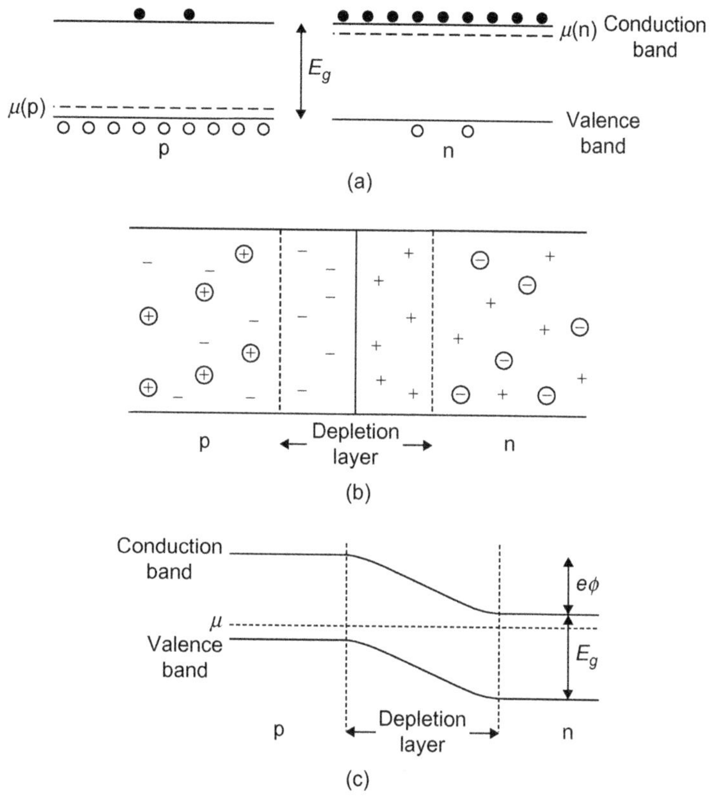

FIGURE 10.3

(a) The electrons (solids) and holes (open circles) in p- and n-type semiconductors. (b) When the p- and n-type regions are in contact, there are practically no carriers in the depletion region. The acceptor and donor ions are represented by – and + and electrons and holes are represented by circles with ⊖ and ⊕ signs. (c) The distribution of charge at a p-n junction produces a contact potential $\phi(x)$ across the junction.

diffuses to the p-type layer, it will recombine with a hole due to the large concentration of holes in that layer. A similar situation will arise when a hole diffuses to the n-type layer. As more and more electrons and holes diffuse to the p- and n-type layers, the number of carriers in the junction will be greatly reduced. Thus, the layer on either side of the junction has much lower concentration of carriers than the rest of the semiconductor. This is known as the depletion layer, which is shown in Figure 10.3b and usually has an approximate length of 10^2 Å. We note that the donor and acceptor ions are much too heavy to diffuse like the electrons and holes. Thus, the p-n junction is formed in the semiconductor with a common chemical potential μ. This is shown in Figure 10.3c.

10.2.2 p-n Junction in Equilibrium

We assume that the impurity concentration varies along the x-axis only in a small region around $x = 0$. If the "abrupt junction," which is defined as the region about $x = 0$ where the impurity concentrations change, is narrow compared with the "depletion layer" in which the carrier densities are not uniform, the donor impurities dominate at positive x, and the acceptor impurities dominate at negative x (see Figure 10.4).

The depletion layer, which extends from x_p to x_n, is shown in Figure 10.5. The abrupt junction is at $x = 0$. The region to the left of x_p is the p-doped mobile holes, and the region to right of x_n is the n-doped mobile electrons (note that x_p is negative). In the beginning, when the p-n junction is formed, the carrier concentration would be such that there would be charge neutrality everywhere in the crystal. However, the concentration of the electrons in the n-side would be very high and that of the holes in the p-side would be very low. Thus, the electrons would diffuse to the p-side, and the holes would diffuse to the n-side. The charge transfer will eventually build up an electric field that will prevent further diffusion. In fact, the effect of the field cancels the effect of the diffusion.

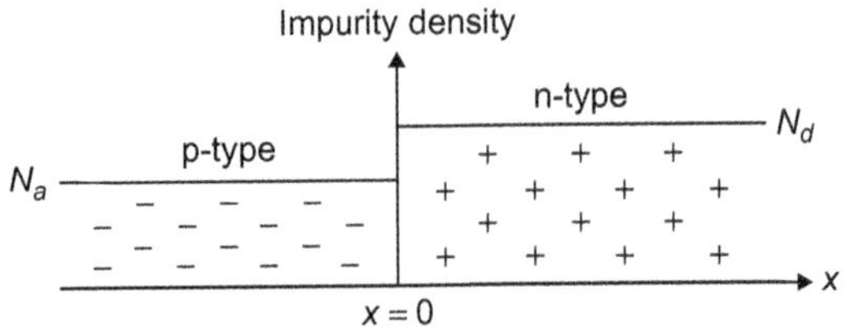

FIGURE 10.4

"Abrupt" p-n junction in which the impurity-concentration change is narrow compared with the depletion layer shown in Figure 10.5.

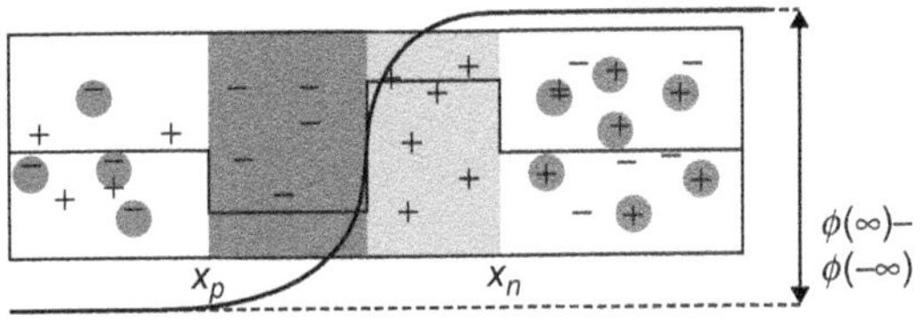

FIGURE 10.5

The depletion layer has length between 10^2 Å and 10^4 Å extending from x_p to x_n.

To derive expressions for the carrier densities at a position x at temperature T in the presence of a potential $\phi(x)$, one can use Eqs. (9.32) and (9.35) subject to the semiclassical condition that each one-electron energy level is shifted by $-e\phi(x)$. Thus, one obtains

$$\varepsilon_c \to \varepsilon_c - e\phi(x) \tag{10.1}$$

and

$$\varepsilon_v \to \varepsilon_v - e\phi(x). \tag{10.2}$$

From Eqs. (9.32) and (10.1), we obtain

$$n_c(x) = \aleph_c(T)e^{-\beta(\varepsilon_c - e\phi(x) - \mu)}, \tag{10.3}$$

and from Eqs. (9.35) and (10.2), we obtain

$$p_v(x) = \wp_v(T)e^{\beta(\varepsilon_v - e\phi(x) - \mu)}, \tag{10.4}$$

where $\beta \equiv 1/k_BT$. If we define the electrochemical potential $\mu_e(x)$ as

$$\mu_e(x) = \mu + e\phi(x), \tag{10.5}$$

Eqs. (10.3) and (10.4) can be rewritten in the alternate form,

$$n_c(x) = \aleph_c(T)e^{-\beta(\varepsilon_c-\mu_e(x))} \tag{10.6}$$

and

$$p_v(x) = \wp_v(T)e^{\beta(\varepsilon_v-\mu_e(x))}. \tag{10.7}$$

The total charge density must vanish in both the limits $x\to-\infty$ and $x\to\infty$, which requires that $N_d = n_c(\infty)$ and $N_a = p_v(-\infty)$. From Eqs. (10.3) through (10.5), one can also derive (Problem 10.1)

$$\mu_e(\infty)-\mu_e(-\infty) = e\Delta\phi = E_g + \frac{1}{\beta}\ln\left[\frac{N_d N_a}{\aleph_c(T)\wp_v(T)}\right], \tag{10.8}$$

where

$$\Delta\phi \equiv \phi(\infty)-\phi(-\infty). \tag{10.9}$$

The effect of the internal potential $\phi(x)$ on the electron and hole densities of a p-n junction is shown in Figure 10.6, in which the electrochemical potential $\mu_e(x)$ is plotted along the p-n junction. The carrier densities at a point x are obtained by using $\mu_e(x)$ as the equivalent chemical potential.

The alternate method of representing the carrier densities at any point x is shown in Figure 10.7. Here, $\varepsilon_d(x) = \varepsilon_d - e\phi(x)$ $\varepsilon_a(x) = \varepsilon_a - e\phi(x)$ and μ is the constant chemical potential.

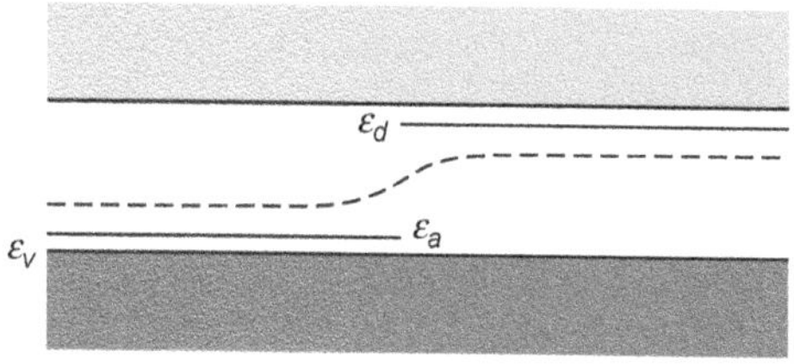

FIGURE 10.6

ε_c and ε_v are conduction and valence bands; ε_d and ε_a are donor and acceptor levels.

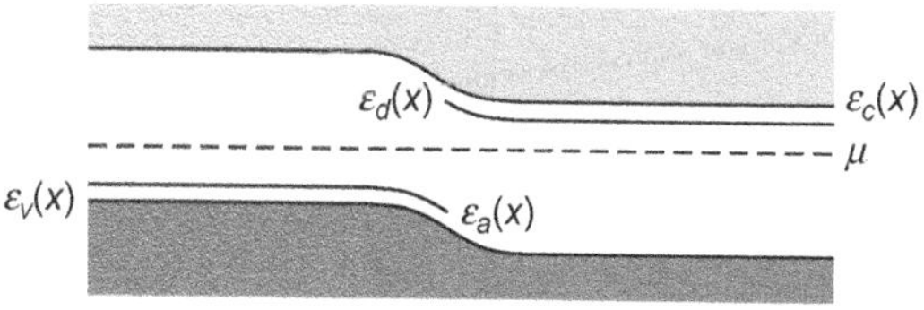

FIGURE 10.7

Carrier densities at a point x obtained by using a fixed chemical potential μ.

To calculate $\phi(x)$, one uses the Poisson's equation in one dimension,

$$\frac{\partial^2\phi}{\partial x^2} = -\frac{4\pi\rho(x)}{\epsilon}, \tag{10.10}$$

where the charge density $\rho(x)$ is obtained from

$$\rho(x) = N_d(x) - n_c(x) - N_a(x) + p_v(x), \tag{10.11}$$

and ϵ is the dielectric constant. It can be easily shown that (Problem 10.2)

$$n_c(x) = N_d e^{-e\beta[\phi(\infty)-\phi(x)]} \tag{10.12}$$

and

$$p_v(x) = N_a e^{e\beta[\phi(-\infty)-\phi(x)]}. \tag{10.13}$$

It can also be shown that (Problem 10.3; except at the boundaries of the depletion layer), outside the depletion layer,

$$\rho(x) = 0, \tag{10.14}$$

and inside the depletion layer,

$$\rho(x) = e[N_d(x) - N_a(x)]. \tag{10.15}$$

The solution for $\phi(x)$ is obtained in Problem 10.4,

$$\phi(x) = \begin{cases} \phi(-\infty), & x < x_p \\ \phi(-\infty) + \left(\frac{2\pi e N_a}{\epsilon}\right)(x - x_p)^2, & 0 > x > x_p \\ \phi(\infty) - \left(\frac{2\pi e N_d}{\epsilon}\right)(x - x_n)^2, & 0 < x < x_n \\ \phi(\infty), & x > x_n \end{cases}. \tag{10.16}$$

In addition, both $\phi(x)$ and $\phi'(x)$ must be continuous at $x = 0$. From the condition $\phi(x)|_{x=-\epsilon} = \phi(x)|_{x=\epsilon}$, where $\epsilon \to 0$, we obtain

$$\phi(0) = \phi(\infty) - \frac{2\pi e N_d}{\epsilon} x_n^2 = \phi(-\infty) + \frac{2\pi e N_a}{\epsilon} x_p^2, \tag{10.17}$$

which can be rewritten in the alternate form

$$\Delta\phi = \frac{2\pi e(N_a x_p^2 + N_d x_n^2)}{\epsilon}. \tag{10.18}$$

From the condition $\phi'(x)|_{x=-\epsilon} = \phi'(x)|_{x=\epsilon}$, we obtain

$$N_d x_n = -N_a x_p. \tag{10.19}$$

From Eqs. (10.18) and (10.19), we obtain expressions for the lengths x_n and x_p,

$$x_{n,p}(0) = \pm\left[\frac{\epsilon (N_a/N_d)^{\pm 1} \Delta\phi}{2\pi e(N_a + N_d)}\right]^{1/2}, \tag{10.20}$$

where $x_{n,p}(0)$ denotes the values of x_n and x_p when the external potential $V = 0$. Because N_a and N_d are of the order of $10^{18}\ \text{cm}^{-3}$ and $e\Delta\phi$ is of the order of 0.1 eV, the depletion layer is of the order of 10^2 to 10^4 Å. It is important to note that because the depletion layer has no mobile charges, the resistance of this region is considerably higher than that of the doped regions. It is equivalent to a series circuit in which a high resistance is sandwiched between two low resistances. When there is no external potential ($V = 0$), the p-n junction is known as the unbiased junction. The potential $\phi(x)$ for the unbiased junction is plotted versus the position x in Figure 10.8.

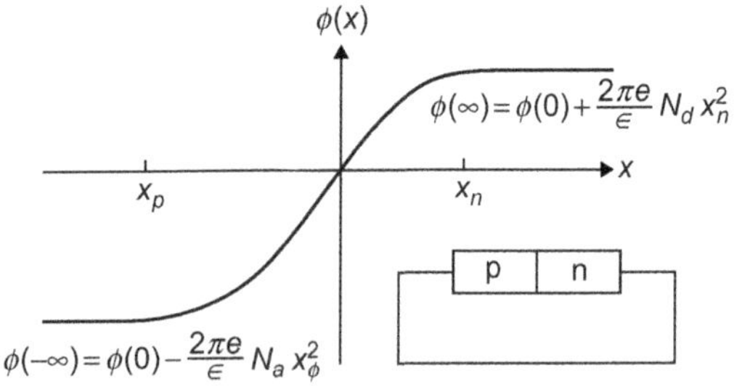

FIGURE 10.8

The potential for the unbiased junction $\phi(x)$ plotted against position x. The depletion layer is from $x = x_p$ to $x = x_n$ (note that x_p is negative).

10.3 RECTIFICATION BY A p-n Junction

10.3.1 Equilibrium Case

When the external potential $V=0$, the change in potential from the p-side to the n-side of the depletion layer is given by $\Delta\phi$, the expression for which is obtained in Eq. (10.8). When the external potential $V \neq 0$, the potential drop (or gain) $\Delta\phi'$ will occur mostly across the high-resistance region, which is the depletion layer. Thus, $\Delta\phi'$ is given by

$$\Delta\phi' = \Delta\phi - V. \tag{10.21}$$

The size of the layer, which extends from x_p to x_n, would accordingly change because in Eq. (10.20), $\Delta\phi$ would be replaced by $\Delta\phi'$ in the expressions for x_n and x_p. From Eqs. (10.20) and (10.21), we obtain

$$x_{n,p}(V) = \pm\left[\frac{\in(N_a/N_d)^{\pm 1}\Delta\phi'}{2\pi e(N_a+N_d)}\right]^{1/2}. \tag{10.22}$$

Eq. (10.22) can be rewritten in the alternate form, by using the expressions for $x_{n,p}(0)$ and $\Delta\phi'$ from Eqs. (10.20) and (10.21),

$$x_{n,p}(V) = x_{n,p}(0)\left[1-\frac{V}{\Delta\phi}\right]^{1/2}. \tag{10.23}$$

This scenario is shown in Figures 10.9 and 10.10.

We will first discuss the scenario when there is no external potential ($V=0$) in the p-n junction. Due to the thermal excitation of electrons in the valence band on the n-side of the depletion layer, holes that move at random are generated. These holes are the "minority carriers" in the n-side compared to the electrons that are the "majority carriers," but they play an important role. The holes are swept away from the n-side to the p-side as soon as they wander into the depletion layer because there is a strong electric field $\Delta\phi$ across the layer. The current due to the motion of the holes from the n-side to the p-side is known as the hole generation current, $j_{hg}(0)$. In addition, the holes

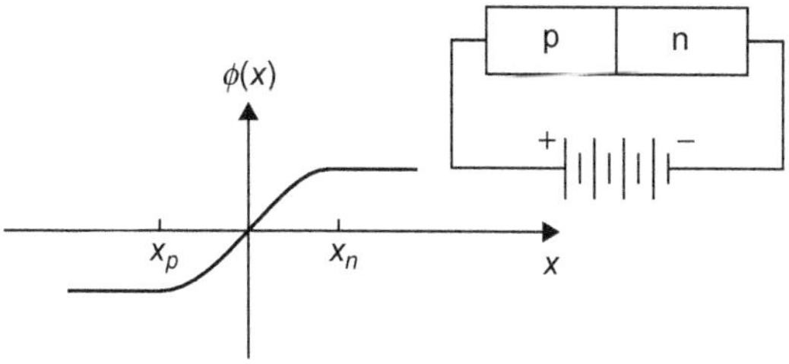

FIGURE 10.9

The potential $\phi(x)$ versus x when $V>0$ (forward bias). $\Delta\phi' = \phi(\infty)-\phi(-\infty)-V$.

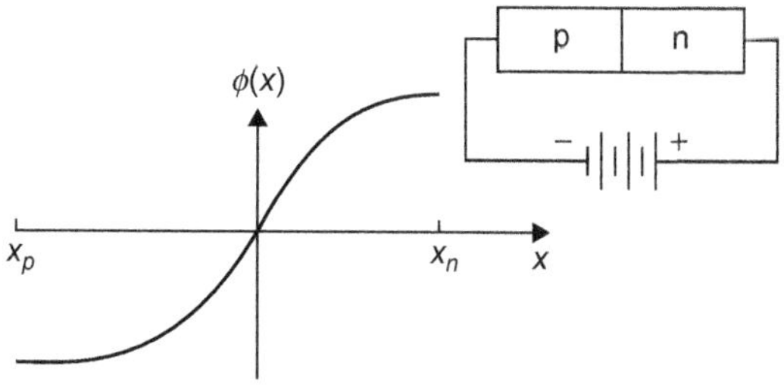

FIGURE 10.10

The potential $\phi(x)$ versus x when $V<0$ (reverse bias). $\Delta\phi' = \phi(\infty)-\phi(-\infty)+V$.

that are thermally generated and are the majority carriers in the p-side flow to the n-side of the junction. The motion of these holes, which are positively charged, is opposed by the electric field of the depletion layer, but those holes of which the thermal energy is greater than the potential barrier would reach the n-side and would recombine with the electrons. The current due to the motion of the holes from the p-side to the n-side of the depletion layer is known as the hole recombination current, $j_{hr}(0)$. Because there is no external potential, the sum of the two currents is zero. Thus, we obtain

$$j_{hg}(0) + j_{hr}(0) = 0. \tag{10.24}$$

If we write $j_{hg} = e\aleph_{hg}$ and $j_{hr} = e\aleph_{hr}$, where $\aleph_{hg}$ and $\aleph_{hr}$ are the number density of holes generated in the n-side and the p-side of the depletion layer, respectively, it follows that at $V = 0, \aleph_{hg}(0) = \aleph_{hr}(0)$ and Eq. (10.24) is satisfied because the positively charged holes flow in opposite directions. We also note that $\aleph_{hr}$ is approximately the same as N_a. Therefore, when an external potential V is applied across the p-n junction,

$$\aleph_{hr} \propto e^{-e\beta\phi'}. \tag{10.25}$$

From Eqs. (10.21) and (10.25), we obtain

$$\aleph_{hr}(V) = \aleph_{hr}(0)\, e^{\beta eV}. \tag{10.26}$$

In contrast, the external potential V would have no effect on the number of holes crossing the depletion layer. Thus, we obtain

$$\aleph_{hg}(V) = \aleph_{hg}(0) = \aleph_{hg}. \tag{10.27}$$

The net number of holes moving across the depletion layer is given by

$$\aleph_h(V) = \aleph_{hr}(V) + \aleph_{hg}(V). \tag{10.28}$$

The expression for the hole current density, $j_h = e\aleph_h(V)$, is obtained from Eqs. (10.26) through (10.28),

$$j_h = j_{hr} - j_{hg} = e\aleph_{hg}(e^{\beta eV} - 1). \tag{10.29a}$$

We can make a similar argument for the motion of the electrons that are the majority carriers in the n-side and the minority carriers in the p-side. However, the electrons are negatively charged particles, and therefore, only those electrons that have enough thermal energy to cross the potential barrier can move from the n-side to the p-side. The current produced by the motion of the negatively charged electrons is known as the electron generation current $j_{eg}(0)$. In contrast, the electrons that reach the edge of the p-side of the p-n junction will be swept across the depletion layer and will eventually recombine with the holes in the n-side. This is known as the electron recombination current, $j_{er}(0)$. We can write

$$j_{eg}(0) + j_{er}(0) = 0. \tag{10.29b}$$

If we write $j_{eg} = -e\aleph_{eg}$ and $j_{er} = -e\aleph_{er}$, where $\aleph_{eg}$ and $\aleph_{er}$ are the number current density of electrons generated in the p-side and the n-side, respectively, it follows that at $V = 0$, $\aleph_{eg}(0) = \aleph_{er}(0)$ and

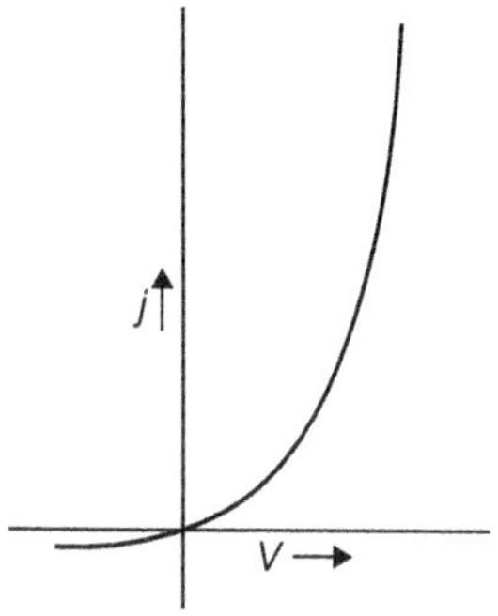

FIGURE 10.11

The $j(V)$ characteristic of the p-n junction.

Eq. (10.25) is satisfied because the electrons flow in opposite directions. We also note that $\aleph_{er}$ is approximately the same as N_d. Therefore, we can express

$$\aleph_{er}(V) = \aleph_{er}(0)e^{\beta eV}. \tag{10.30}$$

However, the external potential does not make any difference in the motion of the electrons generated in the p-side and moving across the depletion layer to the n-side. Thus, we obtain

$$\aleph_{eg}(V) = \aleph_{eg}(0) = \aleph_{eg}. \tag{10.31}$$

The net number of electrons moving across the depletion layer is given by

$$\aleph_e(V) = \aleph_{er}(V) + \aleph_{eg}(V). \tag{10.32}$$

From Eqs. (10.30) through (10.32), we obtain

$$j_e(V) = j_{er}(V) - j_{eg}(V) = e\aleph_{eg}(e^{\beta eV} - 1). \tag{10.33}$$

The total electrical current density is obtained from Eqs. (10.29) and (10.33),

$$j(V) = j_h(V) + j_e(V) = e(\aleph_{hg} + \aleph_{eg})(e^{\beta eV} - 1). \tag{10.34}$$

Eq. (10.34) is the characteristic property of rectifiers and is at the heart of modern transistors. A schematic diagram of the $j(V)$ characteristic for the p-n junction is shown in Figure 10.11.

10.3.2 Nonequilibrium Case ($V \neq 0$)

In the preceding discussion, we considered the equilibrium case ($V = 0$) and then introduced the external potential V in an ad hoc manner to explain the elementary theory of rectification by a p-n junction. We will now consider the nonequilibrium problem ($V \neq 0$). The nonequilibrium problem is the direct generalization of the equilibrium problem ($V = 0$). When $V \neq 0$, one has to solve the Boltzmann equation for semiconductors. In the relaxation-time approximation, the Boltzmann equation for semiconductors is different from that of metals because of the presence of electrons and holes. In the relaxation-time approximation, the Boltzmann equation for electrons can be written as (in analogy with Eqs. 6.73 and 6.74)

$$\frac{\partial g}{\partial t} = -\dot{\mathbf{r}} \cdot \frac{\partial g}{\partial \mathbf{r}} - \dot{\mathbf{k}} \cdot \frac{\partial g}{\partial \mathbf{k}} - \frac{g - f}{\tau_n}, \tag{10.35}$$

where g is the density of particles at position $\mathbf{r}$. In the presence of weak applied fields, g is replaced by $g_{\mathbf{rk}}(t)$, which is the occupation number of electrons at position $\mathbf{r}$, which have a wave vector $\mathbf{k}$ at time t. It can be easily shown from Eq. (6.83) that at constant temperature

$$g_{\mathbf{rk}} \approx f_{\mathbf{rk}} - e\tau_n \frac{\partial f}{\partial \mu} \mathbf{v_k} \cdot \mathbf{E}. \tag{10.36}$$

The electron density n_c at position $\mathbf{r}$ in the conduction band can be written as

$$n_c = \frac{1}{4\pi^3}\int g_{r\mathbf{k}} d\mathbf{k}, \tag{10.37}$$

while the equilibrium density of electrons in the conduction band can be written as

$$n_c^0 = \frac{1}{4\pi^3}\int f\, d\mathbf{k}. \tag{10.38}$$

Integrating Eq. (10.35) over $d\mathbf{k}$, we obtain from Eqs. (10.35) through (10.38) (see Problem 10.6)

$$\frac{\partial n_c}{\partial t} = -\frac{\partial}{\partial \mathbf{r}} \cdot <\dot{\mathbf{r}}> + \frac{n_c^0 - n_c}{\tau_n}. \tag{10.39}$$

Here, n_c^0 is the equilibrium density of the electrons in the conduction band, and $<\dot{\mathbf{r}}>$ is $\mathbf{v_k}$ averaged over the Brillouin zone,

$$<\dot{\mathbf{r}}> = \frac{1}{4\pi^3 n_c}\int d\mathbf{k}\; g_{r\mathbf{k}}\mathbf{v_k} = \frac{1}{4\pi^3 n_c}\int d\mathbf{k}\left[f - \tau_n \mathbf{v_k}\cdot\left\{e\mathbf{E}\frac{\partial f}{\partial \mu} + \frac{\partial f}{\partial \mathbf{r}}\right\}\right]\mathbf{v_k}. \tag{10.40}$$

Because $\int f d\mathbf{k}$ vanishes by symmetry, $\partial f/\partial\mu = \beta f$ and $f \approx g$, Eq. (10.40) can be rewritten as

$$<\dot{\mathbf{r}}> \approx \frac{1}{4\pi^3 n_c}\int d\mathbf{k}\left[-\tau_n \mathbf{v_k}\cdot\left\{e\mathbf{E}\beta g + \frac{\partial g}{\partial \mathbf{r}}\right\}\right]\mathbf{v_k}. \tag{10.41}$$

We define the mobility μ_n and the diffusion constant D_n as

$$\mu_n = \frac{e\beta}{3}<\tau_n v_k^2> \tag{10.42}$$

and

$$D_n = \frac{1}{3}<\tau_n v_k^2> = \frac{\mu_n}{e\beta}. \tag{10.43}$$

From Eqs. (10.41) through (10.43), we obtain

$$<\dot{\mathbf{r}}> = -\mu_n \mathbf{E} - \frac{D_n}{n_c}\frac{\partial n_c}{\partial \mathbf{r}}. \tag{10.44}$$

The electron current is given by

$$\mathbf{j}_e = -en_c<\dot{\mathbf{r}}> = en_c\mu_n\mathbf{E} + eD_n\vec{\nabla}n_c. \tag{10.45}$$

Similarly, the hole current can be obtained as

$$\mathbf{j}_h = ep_v\mu_p\mathbf{E} - eD_p\vec{\nabla}p_v. \tag{10.46}$$

From Eq. (10.39) and its analogous equations for holes, Eqs. (10.45) and (10.46), we obtain

$$\frac{\partial n_c}{\partial t} = \frac{1}{e}\vec{\nabla}\cdot\mathbf{j}_e + \frac{n_c^0 - n_c}{\tau_n} \tag{10.47}$$

and

$$\frac{\partial p_v}{\partial t} = -\frac{1}{e}\vec{\nabla} \cdot \mathbf{j}_h + \frac{p_v^0 - p_v}{\tau_p}. \tag{10.48}$$

Because we have assumed that the carriers in the doped semiconductor primarily move in the x direction, Eqs. (10.45) and (10.46) can be rewritten as

$$j_e = en_c\mu_n E + eD_n\frac{dn_c}{dx} \tag{10.49}$$

and

$$j_h = ep_v\mu_p E - eD_p\frac{dp_v}{dx}. \tag{10.50}$$

Similarly, Eqs. (10.47) and (10.48) can be rewritten as

$$\frac{\partial n_c}{\partial t} = \frac{1}{e}\frac{\partial j_e}{\partial x} + \frac{n_c^0 - n_c}{\tau_n} \tag{10.51}$$

and

$$\frac{\partial p_v}{\partial t} = -\frac{1}{e}\frac{\partial j_h}{\partial x} + \frac{p_v^0 - p_v}{\tau_p}. \tag{10.52}$$

We note that

$$n_i^2 = n_c^0 p_v = p_v^0 n_c. \tag{10.53}$$

In the steady state, $\frac{\partial n_c}{\partial t} = \frac{\partial p_v}{\partial t} = 0$. Thus, in the steady state, we can rewrite Eqs. (10.51) and (10.52) as

$$\frac{1}{e}\frac{\partial j_e}{\partial x} + \frac{n_c^0 - n_c}{\tau_n} = 0 \tag{10.54}$$

and

$$-\frac{1}{e}\frac{\partial j_h}{\partial x} + \frac{p_v^0 - p_v}{\tau_p} = 0. \tag{10.55}$$

In the regions, where $E \to 0$, the majority carrier density is constant. In these regions, from Eqs. (10.49) and (10.50), we obtain

$$j_e(x_p) \approx eD_n\frac{dn_c}{dx}\Big|_{x=x_p} \tag{10.56}$$

and

$$j_h(x_n) \approx -eD_p\frac{dp_v}{dx}\Big|_{x=x_n}. \tag{10.57}$$

Eqs. (10.56) and (10.57) essentially state that in the steady state, the minority carrier drift current can be ignored compared to the minority carrier diffusion current. From Eqs. (10.54) through (10.57), we obtain

$$D_n \frac{d^2 n_c}{dx^2} + \frac{n_c^0 - n_c}{\tau_n} = 0 \tag{10.58}$$

and

$$D_p \frac{d^2 p_v}{dx^2} + \frac{p_v^0 - p_v}{\tau_p} = 0. \tag{10.59}$$

It can be easily shown (Problem 10.7) that in the n-side of the depletion layer, for $x \geq x_n$,

$$p_v(x) - p_v(\infty) = [p_v(x_n) - p_v(\infty)]e^{-(x-x_n)/d_p}, \tag{10.60}$$

and in the p-side of the depletion layer, for $x \leq x_p$,

$$n_c(x) - n_c(-\infty) = [n_c(x_p) - n_c(-\infty)]e^{(x-x_p)/d_n}, \tag{10.61}$$

where the diffusion lengths for holes in the n-doped region and electrons in the p-doped region are defined as

$$d_p = (D_p \tau_p)^{1/2} \text{ and } d_n = (D_n \tau_n)^{1/2}. \tag{10.62}$$

In fact, we can define two diffusion regions, which extend over a distance of the order of the diffusion length in either side of the depletion layer, sandwiched between the depletion layer and the homogeneous regions. The diffusion regions do not exist when $V = 0$.

Thus, we have three regions when $V \neq 0$. The difference between the three regions is that in the depletion layer, the electric field, the space charge, and the carrier density gradients are large (Eq. 10.23), but there is no current in the equilibrium case ($V = 0$) because the drift and the diffusion currents are equal and opposite for both electrons and holes. When $V \neq 0$, there is a net current in the depletion layer due to the difference between the drift and diffusion currents of each carrier type, which are large. In the diffusion regions, the drift current of the majority carrier density is quite large, whereas that of the minority carrier is small. However, the diffusion current due to both the carriers is appreciable. In contrast, the only current in the homogeneous regions is the majority carrier drift current. These regions are shown in Figure 10.12.

We will now calculate the total current j flowing in the p-n junction for a given value of V. We make the assumption that the passage of carriers across the depletion layer is so swift that the generation and recombination of electrons and holes within the layer are negligible. In the steady state, the total current $j = j_e + j_h$ of electrons and holes will be constant. Thus, j_e and j_h can be evaluated separately at any arbitrary point in the depletion layer. The electron current j_e can easily be calculated at the boundary between the depletion layer and the diffusion region on the p-side (x_p), and the hole current j_h can easily be calculated at the boundary between the two regions on the n-side

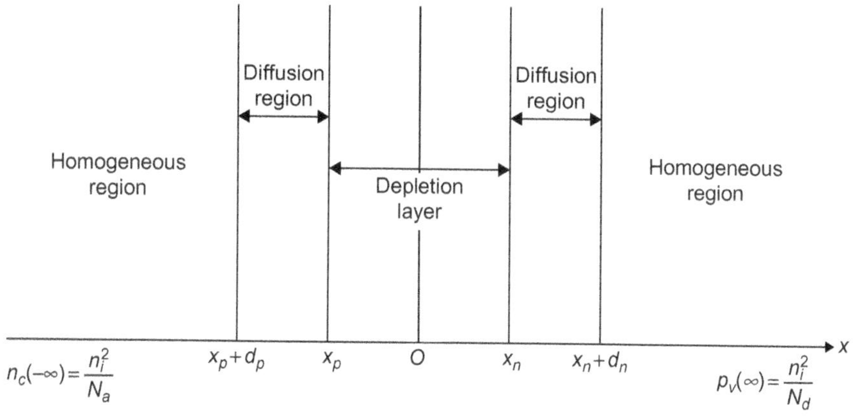

FIGURE 10.12

The depletion layer, the two diffusion and homogeneous regions when $V \neq 0$.

(x_n) in Figure 10.12. From Eq. (10.60) and Problem 10.4, we obtain the solution of the diffusion equations

$$p_v(x) = \frac{n_i^2}{N_d} + [p_v(x_n) - \frac{n_i^2}{N_d}]e^{-(x-x_n)/d_p}, \quad x \geq x_n \tag{10.63}$$

and

$$n_c(x) = \frac{n_i^2}{N_a} + \left[n_c(x_p) - \frac{n_i^2}{N_d}\right]e^{(x-x_p)/d_n}, \quad x \leq x_p. \tag{10.64}$$

Note that both x and x_p are negative in Eq. (10.64). From Eqs. (10.56), (10.57), (10.63), and (10.64), we obtain expressions for the minority carrier currents at the edges of the depletion layer ($x = x_n$ and $x = x_p$),

$$j_e(x_p) = \frac{eD_n}{d_n}\left[n_c(x_p) - \frac{n_i^2}{N_a}\right] \tag{10.65}$$

and

$$j_h(x_n) = \frac{eD_p}{d_p}\left[p_v(x_n) - \frac{n_i^2}{N_d}\right]. \tag{10.66}$$

From Problem 10.8,

$$n_c(x) \approx N_d e^{e\beta[V(x)-V(x_n)]} \tag{10.67}$$

and

$$p_v(x) \approx N_a e^{-e\beta[V(x)-V(x_p)]}. \tag{10.68}$$

From Eqs. (10.67), (10.68), and Problem 10.9,

$$n_c(x_p) = N_d e^{e\beta(V-\Delta\phi)} = \frac{n_i^2}{N_a} e^{e\beta V} \tag{10.69}$$

and

$$p_v(x_n) = N_a e^{e\beta(V-\Delta\phi)} = \frac{n_i^2}{N_d} e^{e\beta V}. \tag{10.70}$$

From Eqs. (10.65) and (10.69),

$$j_e(x_p) = en_i^2 \frac{D_n}{d_n N_a}(e^{e\beta V} - 1), \tag{10.71}$$

and from Eqs. (10.66) and (10.70),

$$j_h(x_n) = en_i^2 \frac{D_p}{d_p N_d}(e^{e\beta V} - 1). \tag{10.72}$$

The total current, $j = j_e(x_p) + j_h(x_n)$, is the sum of Eqs. (10.71) and (10.72),

$$j = en_i^2\left(\frac{D_n}{d_n N_a} + \frac{D_p}{d_p N_d}\right)(e^{e\beta V} - 1). \tag{10.73}$$

Eq. (10.73) is the ideal diode, or Shockley, equation. We note that because N_a and N_d appear in the denominator, the heavily doped side acts as a short circuit, whereas the current flow is in the lightly doped side.

10.4 TRANSISTORS

10.4.1 Bipolar Transistors

In the bipolar transistor, three alternately doped layers of semiconductor (n-p-n or p-n-p) are sandwiched. We will discuss the n-p-n bipolar transistor in the following section. The middle layer (p) is the base that is very narrow, and the other two layers (n) are known as the emitter and the collector. Energy band diagrams of an n-p-n bipolar transistor with no applied voltage and with an applied voltage, which makes the emitter negative with respect to the collector, are shown in Figure 10.13.

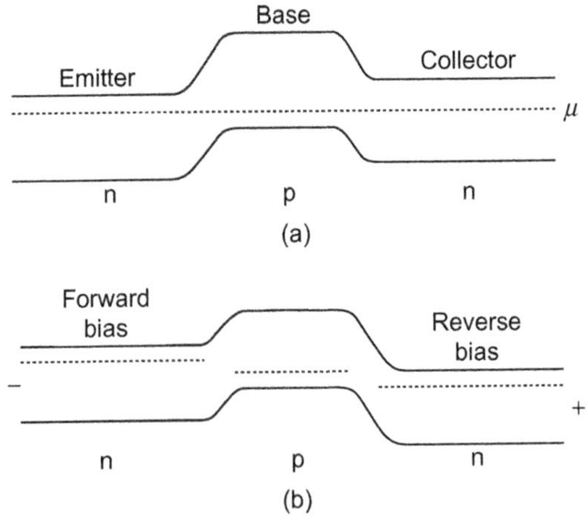

FIGURE 10.13

(a) An n-p-n bipolar transistor with no applied voltage. (b) The same transistor with the emitter negative with respect to the collector.

In Figure 10.13b, a voltage is applied such that the emitter is negative with respect to the collector. In such a scenario, the emitter-base junction is forward biased, whereas the base-collector junction is reverse biased. In a p-n junction, each time an electron recombines with a hole in the base region, an acceptor ion is left behind that accumulates as more electrons flow through the junction and ultimately inhibits the flow of electrons. However, in a transistor, due to the contact with the base region, a current of holes flows into the base. The base remains electrically neutral because the number of holes drawn into the base compensates those that are lost due to recombination. The way in which a bipolar transistor acts as an amplifier is shown in Figure 10.14.

When a small input of hole current i_B is supplied to the base of the transistor, the base is positively charged, due to which the potential barrier between the base and the emitter is lowered. This results in an increase in the flow of electrons from the emitter to the base. Even if only a small percentage of the conduction electrons (from the emitter) recombine with the holes in the base while most of them flow into the collector, $i_E \gg i_B$, so that base remains neutral. Thus, the flow of current due to the electrons in the collector, i_C, is nearly equal to i_E, and the current gain in the amplifier, α, which is defined as the ratio of change in the output current Δi_C to the change in the input current Δi_B, is given by

$$\alpha = \frac{\Delta i_C}{\Delta i_B}. \tag{10.74}$$

Thus, a signal can be easily amplified by the bipolar transistor because a small change in the base current produces a very large change in the collector current.

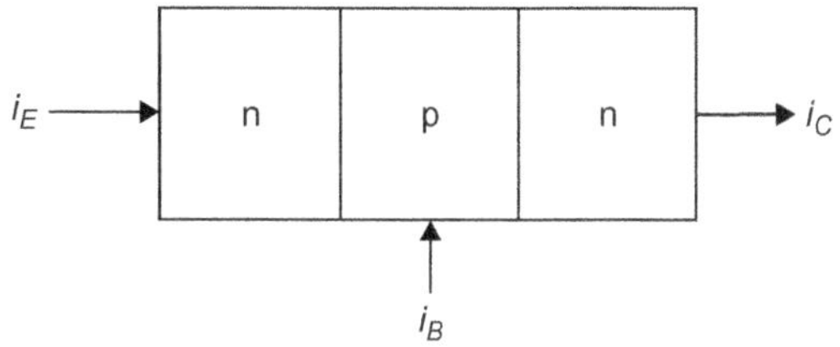

FIGURE 10.14

Amplification of current in a bipolar transistor.

10.4.2 Field-Effect Transistor

The most common field-effect transistor (FET) is the metal-oxide-semiconductor FET, or the MOSFET. The structure of an n-type MOSFET is shown in Figure 10.15. In a MOSFET, a metal contact above the gate region is separated from the p-type semiconductor (Si) substrate by a thin oxide layer (usually SiO_2), which is an insulator.

When a positive voltage is applied to the metal contact above the gate region, the holes in the p-type semiconductor are repelled from the surface, whereas the electrons, which are the minority carriers, are attracted to the surface. This scenario is shown in Figure 10.16.

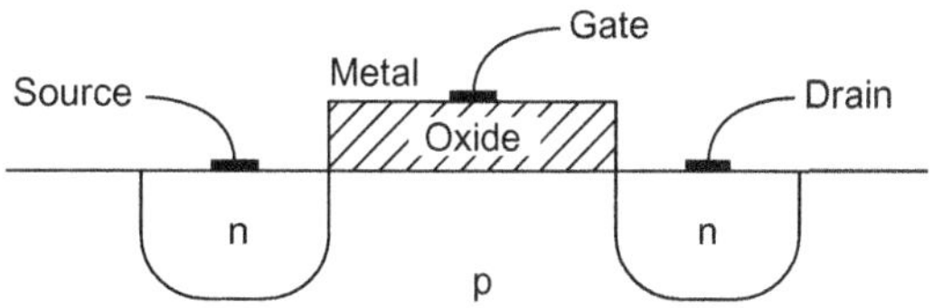

FIGURE 10.15

The structure of an n-type MOSFET.

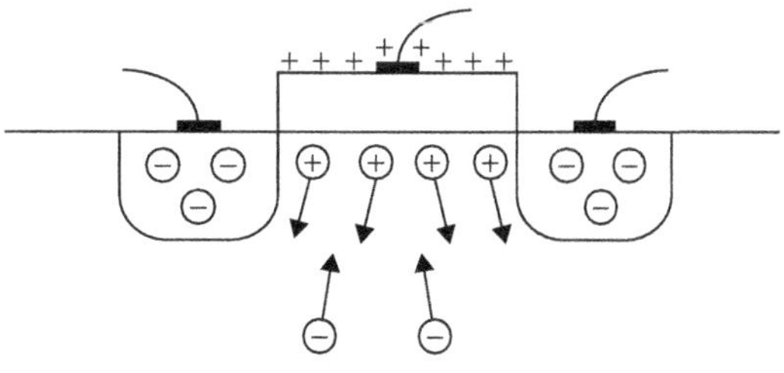

FIGURE 10.16

Electrons are attracted to the surface when a positive voltage is applied to the gate.

As the voltage is increased, more and more electrons, which are the minority carriers in the p-type semiconductor, are attracted to the surface. When the voltage increases beyond a threshold value, the number of electrons becomes greater than the number of holes near the surface even though it is a p-type semiconductor. This is known as the inversion layer, and the semiconductor behaves as if it is an n-type. This scenario is shown in Figure 10.17.

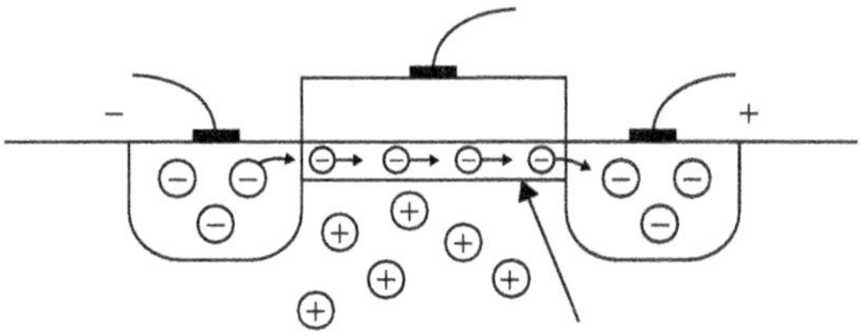

FIGURE 10.17

Inversion layer near the surface, when the gate voltage is greater than the threshold voltage.

Figure 10.17 explains how the MOSFET acts as a switch. The inversion layer allows the conduction electrons to flow from the source to the drain. The device is on when the current flows and off when the flow of current is stopped by decreasing the gate voltage below the threshold voltage. In such a situation, there is no inversion layer. An explanation of how the inversion layer is obtained is shown in Figure 10.18.

The Fermi energy is close to the valence band edge in the p-type semiconductor when it is well below the interface with the oxide layer. However, the electron energies are lowered near the surface of the semiconductor because there is a positive charge on the other side of the oxide layer, and both the conduction and valence bands are bent. If the Fermi energy is closer to the conduction band than to the valence band in this region, as shown in Figure 10.18, the n-type behavior is restricted to this region. Because the semiconductor displays n-type behavior, the probability of inversion depends on the degree of band bending.

The switching time of a transistor is defined as the minimum time period over which a transistor can be switched from off to on and again to the off position. The most important factor in designing integrated circuits is that a transistor must be in the on state until an electron (or hole) moves from the source to the drain. If it does not reach the drain before the transistor reaches the off position, no current will reach the drain, and the electronic device would not be switched on for any length of time. The MOSFET is eminently suitable to reduce the switching time in a

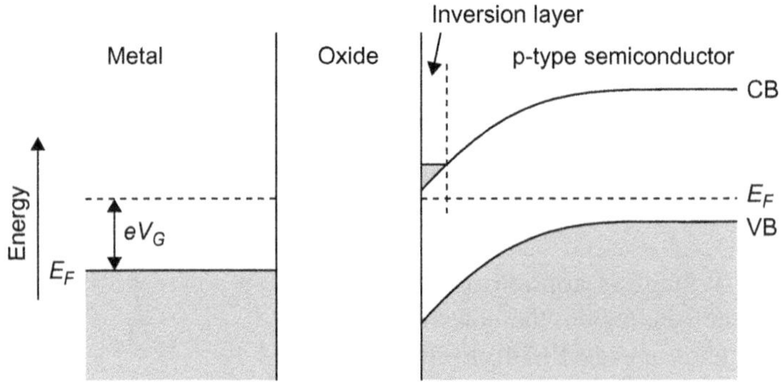

FIGURE 10.18

The band picture of the inversion layer in the MOSFET.

transistor. It may be noted that there are p-type MOSFETs, but because the mobility of the holes is much less than the mobility of the conduction electrons, the n-type MOSFETs can switch much faster than the p-type.

10.4.3 Single-Electron Transistor

As we have noted, a MOSFET is essentially a parallel-plate capacitor, one plate of which is a metal and the other of which is silicon, a semiconductor. SiO_2 is the insulator of the capacitor. Kastner et al.[7] designed a single-electron transistor (SET) that had two metal gates. A schematic diagram of their device is shown in Figure 10.19.

The electron gas, which is formed in the p-type Si by the positively biased upper gate, which is continuous as in the conventional MOSFET, is confined by the lower gate, which has a narrow gap of about 70 nm width. The narrow channel is approximately 20 nm wide and 1–10 μm long. The surface electrons are isolated from the bulk by the p-n junctions. The electron-rich region is called an inversion layer. When the voltage in the upper gate is positive with respect to the semiconductor while the bottom gate is neutral or negative, electrons are added to the semiconductor only under the gap in the bottom gate. Figure 10.19b shows how the electron gas is confined to move in one direction.

If V_g is the gate voltage, it can be easily shown that

$$e\Delta(N/L) = (C/L)\Delta V_g, \qquad (10.75)$$

where C is the capacitance, L is the length, e is the electronic charge, and N is the number of electrons added. It was observed that when conductance g was plotted against the electron density (N/L), oscillations of the conductance were periodic. This is shown in Figure 10.20.

From Eq. (10.75) and Figure 10.20, it was observed that the transistor's conductance oscillated as a function of the number of electrons per unit length. In fact, there is a special length L_0, such

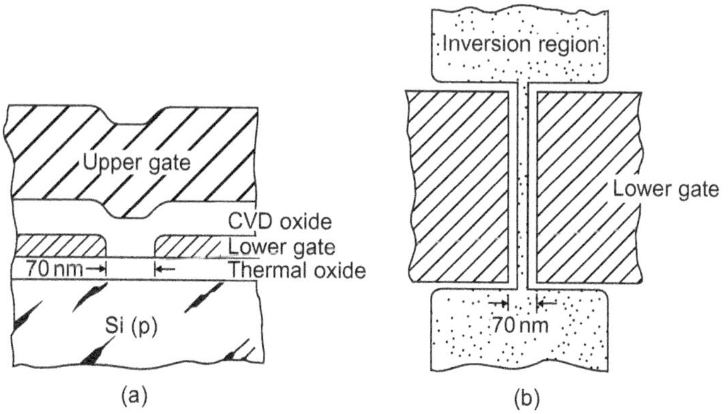

FIGURE 10.19

Schematic diagram of the (a) cross-section and (b) top view of the silicon transistor with a continuous upper gate and a gap in the lower gate.

Reproduced from Kastner et al.[7] with the permission of the American Physical Society.

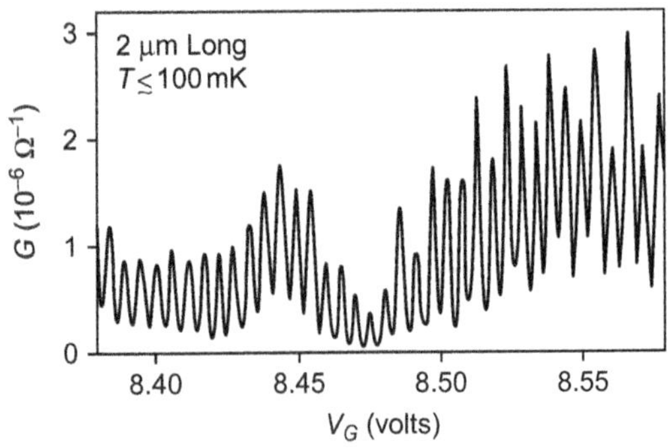

FIGURE 10.20

Conductance C versus gate voltage V_g in single-transistor Si devices.

Reproduced from Kastner et al.[7] with the permission of the American Physical Society.

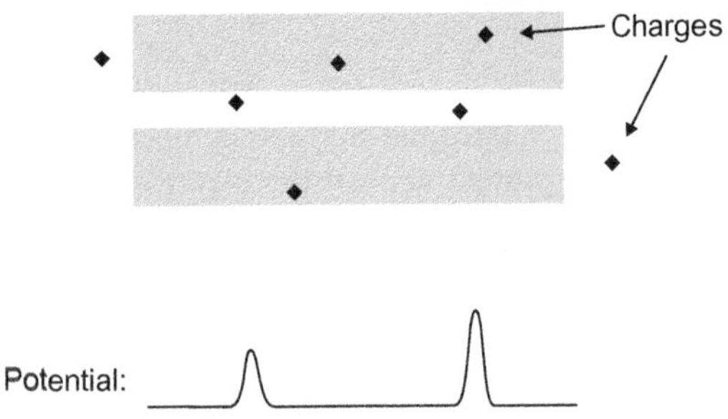

FIGURE 10.21

Narrow channel through which the electrons move. The metal gate with the gap is shown by the shaded area. The random distribution of charges near the interface is shown by black diamonds. The bottom sketch shows the electrostatic potential resulting from such impurities versus position along the channel.

Reproduced from Kastner et al.[7] with the permission of the American Physical Society.

that the voltage difference for one period is that necessary to add some fixed number of electrons to some fixed length,

$$e\Delta N = L_0(C/L)\Delta V_g. \tag{10.76}$$

In a separate experiment, it was also observed that the period was not correlated to the channel length. The sensitivity of the conductance fluctuations to thermal cycling lead to the conclusion that the period was determined by the distribution of charges at the Si-SiO_2 interfaces. These charges are always present even at densities of approximately 10^{10} per cm^2 for the best interfaces. It was calculated that for the transistor that had $2\,\mu m \times 1\,\mu m$ in size, there were approximately two charges adjacent to it. In fact, each sample had random distribution of charges that changed each time the sample was warmed to the room temperature. Kastner et al.[7] guessed that the charges create potential barriers along the length of the transistor, as shown schematically in Figure 10.21. They postulated that each period of conductance oscillations corresponded to the addition of one electron to the distance between the charges.

The proof that the period of oscillations did not correspond to the addition of two electrons due to spin degeneracy or four electrons due to the energy-band structure of Si depended on the measurement of the length L_0 defined in Eq. (10.76). The measurement of L_0 was facilitated by the discovery of small transistors in GaAs by Meirav et al.[13] Their device is shown in Figure 10.22.

Using molecular-beam epitaxy (MBE), a layer of AlGaAs, which has larger band gap than GaAs, is grown on a heavily doped GaAs crystal (n^+). Then a layer of pure GaAs is grown where the electrons accumulate. The density of the electrons is controlled by the positive voltage applied to the n^+ substrate. A metal gate, which is negatively biased so that the electrons are repelled from it, is deposited on the top by electron-beam lithography. The negative bias on the top gates creates a potential barrier for the electrons moving down the narrow channel.

The advantage of the GaAs-AlGaAs interface is that the density of charges near the semiconductor and insulator is smaller than the Si-SiO_2 interface. Figure 10.23 shows the conductance as a function of gate voltage V_g for two devices with different length.

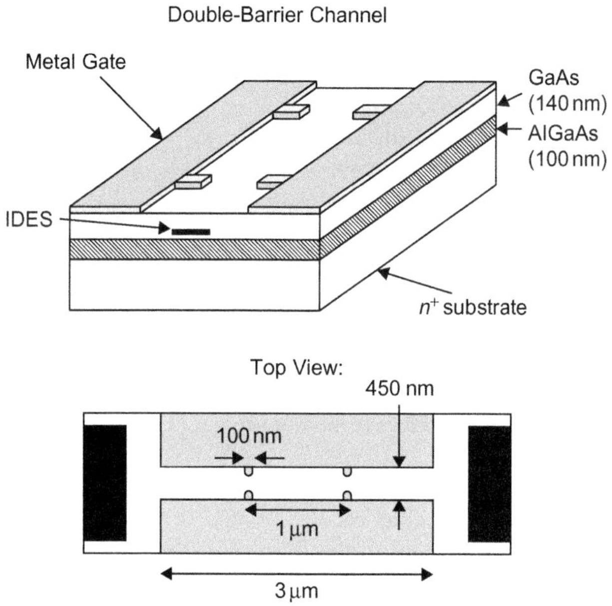

FIGURE 10.22

The double-barrier channel shows a one-dimensional electron gas (1DES) or narrow two-dimensional gas forms at the top of the GaAs-AlGaAs interface. The density is controlled by the substrate voltage V_g. The top view shows the top metal gate structure, which has a narrow channel with two constrictions.

Reproduced from Kastner et al.[7] with the permission of the American Physical Society.

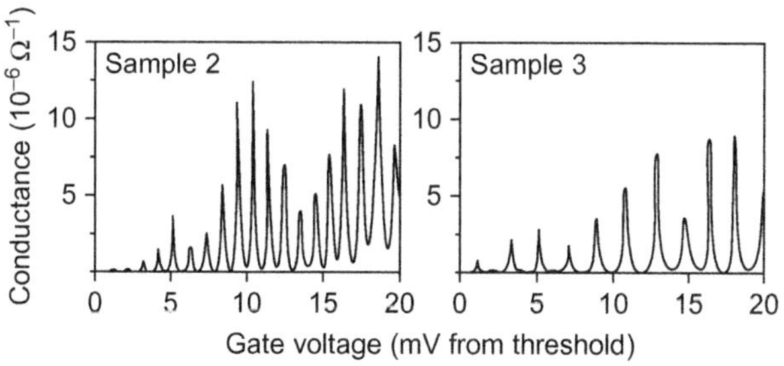

FIGURE 10.23

Conductance as a function of V_g for $L_0 = 0.8$ μm (sample 2) and $L_0 = 0.6$ μm (sample 3).

Reproduced from Kastner et al.[7] with the permission of the American Physical Society.

Each period corresponds to the addition of the same number of electrons to the regions between constrictions. The voltage to add one electron is larger for a shorter segment because the capacitance is smaller. However, it is necessary to know the absolute capacitance of a segment to be able to estimate whether the number of electrons added per period is one or two. The charge density was obtained by solving the Poisson equation and estimated by a method similar to the Thomas–Fermi approximation. The capacitance was calculated at each gate voltage by integrating the charge contained in the region between the two constrictions. The capacitance was compared with the measured period. It was conclusively proved that the period of oscillations corresponds to the addition of one electron per oscillation. Thus, the one-electron transistor turns on and off again each time an electron is added.

We note from Figure 10.23 that the conductance of a small region of electron gas, separated by tunnel junctions from its leads, oscillates with density, which is explained by the simple Coulomb blockade model. A schematic picture of the tunnel barrier with a metal particle is shown in Figure 10.24.

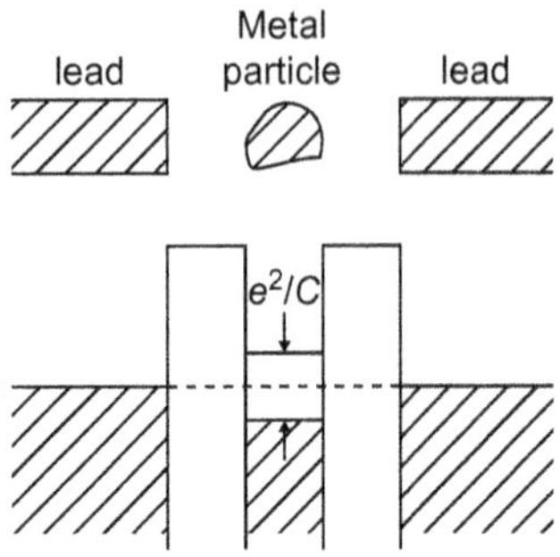

FIGURE 10.24

A Coulomb blockade system. Electrons tunnel from one lead onto a small metal particle and then to the other lead.

Reproduced from Kastner et al.[7] with the permission of the American Physical Society.

Due to the Coulomb interaction between the electrons in the metal particle, the electrons cannot tunnel from one plate of the capacitor to the other plate through the metal particle. For current to flow, an electron or hole of charge $q(\pm e)$ has to be added to the particle. This costs an energy $q^2/2C$, where C is the capacitance between the particle and the rest of the system. Thus, there is an energy gap in the single-particle density of states. If we consider the tunneling of an electron or a hole, the gap width is e^2/C. This is shown in Figure 10.24.

Because the potential difference between the gate and the electron gas is V_g, the isolated region of the transistor has an electrostatic energy

$$\varepsilon = \frac{q^2}{2C} - qV_g, \tag{10.77}$$

where $-qV_g$ is the attractive interaction between the positively charged gate and the charge in the isolated region, and $q^2/2C$ is the repulsive term between two charges in the isolated region. We note that Eq. (10.77) can be written in the alternate form

$$\varepsilon = (q - q_0)^2/2C, \tag{10.78}$$

where $q_0 = CV_g$ and a constant $q_0^2/2C$ has been added to ε. Any value of q_0 can be added to minimize the energy, but because the charge is quantized, only discrete values of energy are possible for a given q_o. Thus, either $q_0 = ne$ or $q_0 = (n + 1/2)e$, where n is an integer. These two cases are illustrated in Figure 10.25.

We note that when $q_0 = (n + 1/2)e$, the states where $q = ne$ and $q = (n + 1)e$ are degenerate. The energy gap in the tunneling density of states disappears because the charge fluctuates between the two values even at zero temperature. Because the conductance is thermally activated at all values of the gate voltage except those for $q_0 = (n + 1/2)e$, it has sharp peaks at low temperatures. The change in voltage to alter q_0 from $(n + 1/2)e$ to $(n + 3/2)e$ is $\Delta V_g = e/C$, which is the period in V_g. The activation energy at the minimum is $e^2/2C$.

To summarize, the number of electrons on the isolated segment is quantized because the time for tunneling onto and off the segment is long. The Coulomb interactions and the quantization of

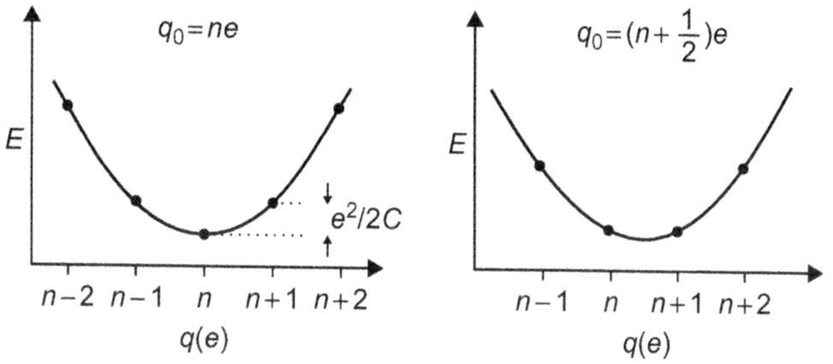

FIGURE 10.25

Energy versus charge for $q_0 = ne$ and $q_0 = (n+1/2)e$. Because the charge is quantized, the allowed values of energy are shown by solid circles.

Reproduced from Kastner et al.[7] with the permission of the American Physical Society.

charge suppress all charge fluctuations at zero temperature for all values of q_0 not equal to $(n+1/2)e$ and at that value the charge fluctuates only by one electron. This device is called a Coulomb island because the Coulomb interaction suppresses the charge fluctuations.

10.5 INTEGRATED CIRCUITS

An integrated circuit consists of capacitors, resistors, transistors, and other metallic connections required for a complete electrical circuit. The most popular electrical circuits are the MOSFET circuits because the switching time can be easily reduced. In addition, the transistors switch faster if the devices are made smaller. However, it is important to keep in check the power dissipated by a transistor so that the amount of heat produced in an integrated circuit can be kept under control.

Due to the improvement of the technology in building integrated circuits, primarily due to the decrease in the individual devices as well as in the increase in the area of the circuit, there has been a rapid growth in the number of transistors on an integrated circuit since the first such circuit was fabricated in 1961 with only four transistors. At present, a typical integrated circuit has about 80 million transistors. The single most important criterion is to keep in check the enormous heat produced by such circuits.

10.6 OPTOELECTRONIC DEVICES

Semiconductors can be utilized for a variety of optical properties. The simplest way to consider the role of a photon is that if it has energy $\in = h\nu$ greater than the band gap, it can excite an electron from the valence band to the conduction band, leaving behind a hole in the valence band. It may be noted that both the laws of conservation of momentum and energy have to be obeyed by the photons and electrons involved in the collision process where the energy is absorbed from the photon by the electron. This is usually possible only in direct band-gap semiconductors such as GaAs where the minimum of the conduction band is directly above the maximum of the valence band. This process, shown in Figure 10.26a, is known as photoconductivity because a beam of light with appropriate frequency can produce a large number of electrons and holes, thereby significantly

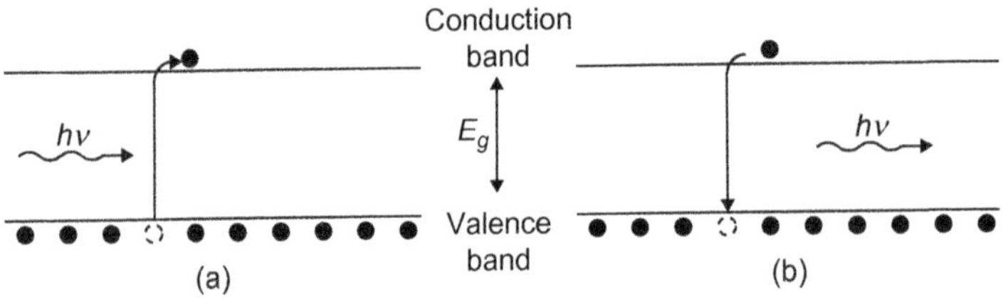

FIGURE 10.26

(a) A photon excites an electron from the valence band to the conduction band. (b) An electron in the conduction band recombines with a hole and produces a photon.

increasing the conductivity. In the reverse process, shown in Figure 10.26b, an electron in the conduction band recombines with a hole in the valence band and produces a photon of energy $\in = h\nu$. This is the basis of both the light-emitting diode (LED) and the semiconductor laser. As emphasized earlier, only semiconductors of which the minimum of the conduction band lies directly above the conduction band can be used for such a process.

A light-emitting diode constitutes a forward-biased p-n junction made from a direct band-gap material. The band gap of the semiconductor determines the wavelength of the emitted light. One can obtain light across most of the visible spectrum by using GaAsP if one varies the phosphorus content of the alloy semiconductor.

A photodiode, shown in Figure 10.27, is used to detect light. One uses a reverse-biased p-n junction. In this junction, when a valence electron in the depletion region on the p-type region absorbs a photon of the appropriate frequency, it is excited to the conduction band, where it is the minority carrier. The strong electric field sweeps it across the depletion region to the n-type region, and hence, the electron contributes to the drift current. The magnitude of this current is proportional to the intensity of light. This process can be used to construct a solar cell.

A laser (light amplification by stimulation of radiation) that produces coherent light is different from an LED in the sense that although the production of photons in an LED by the recombination of electrons and holes is a spontaneous process, in a laser the recombination of one-electron-hole pair triggers similar events in other electrons and holes, simultaneously producing a large number of photons of the same frequency. This is known as stimulated emission. The original diode laser, shown in Figure 10.28, had more electrons at the bottom of the conduction band than at the top of the valence band, which is obtained in a heavily doped p-n junction.

In Figure 10.28, the population inversion exists only near the middle of the depletion region even though the recombination events take place in a wide region. Consequently, most of these events have spontaneous emission rather than stimulated emission as required in lasers. The efficiency of the lasing process is low. Because a major proportion of the electrons and holes are lost due to spontaneous emission, to replenish their loss, a large current must flow through the junction in a continuous basis. This large current produces so

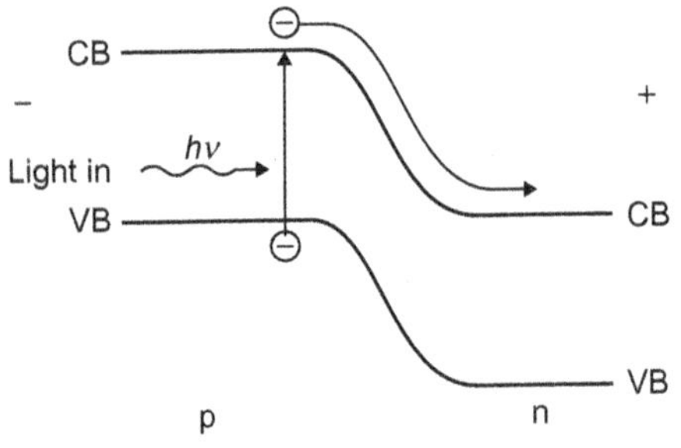

FIGURE 10.27

A reverse-biased p-n junction used as a photodetector.

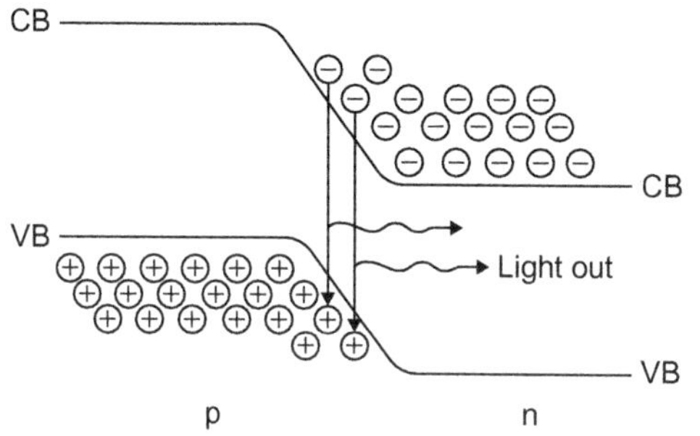

FIGURE 10.28

Schematic diagram of a diode laser.

much heat that the diode laser can be used for only short bursts or pulses.

The efficiency of the diode laser is very low. However, the efficiency of the laser is considerably improved by using two heterostructures instead of a p-n junction. A schematic diagram of the double-heterostructure laser is shown in Figure 10.29.

In a double-heterostructure laser, the conduction electrons and holes are confined in the GaAs layer. Due to the fact that both the conduction electrons and holes collect in the same region, the proportion of stimulated emission is much larger than a diode laser. In view of the above, this type of laser does not get heated very quickly, and therefore, it can be used for a longer period.

Semiconducting lasers are very widely used in communication systems using optical fibers. For best results, the wavelength is matched to the performance of the fiber. An InGaAsP laser, which is widely used in optical communication systems, is shown in Figure 10.30.

Recent studies on semiconductor lasers has shown that a semiconductor laser generates a number-phased squeezed state rather than a squeezed state, primarily due to the fact that a semiconductor laser is pumped by a shot-noise-free electric current. When the squeezed state becomes phase coherent with an independent local laser oscillator, the squeezed light can be detected by optical homodyne detectors and used for various interferometric measurements. Thus, a phase-coherent squeezed state can be generated by injection locking the squeezed slave laser with an external master laser.

A setup for generating a squeezed vacuum state by destructively interfering with an amplitude-squeezed state from an injection-locked slave laser with strong coherent light from a master laser is shown in Figure 10.31. A constant-current-driven semiconductor laser, which is denoted as a *slave laser,* is injection locked by an external *master laser*. Thus, the two signals, which are phase coherent, are combined at a high transmission mirror where the coherent excitation of the squeezed

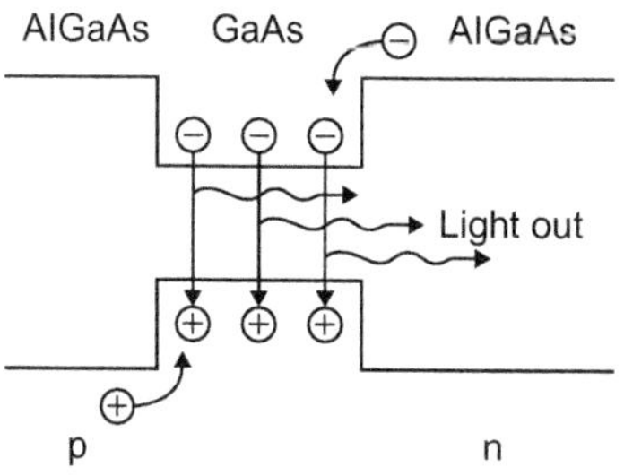

FIGURE 10.29

Schematic diagram of a double-heterostructure laser.

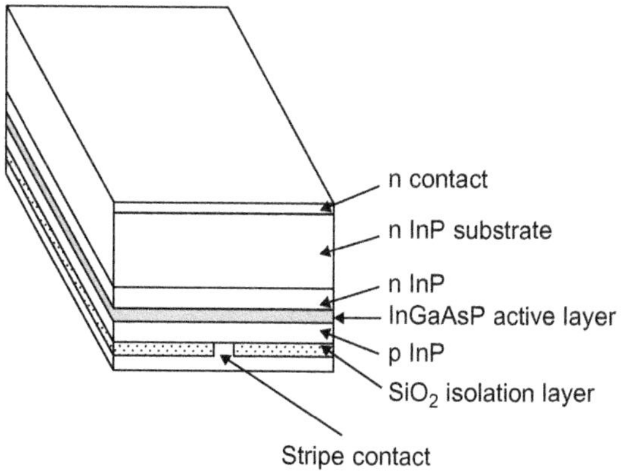

FIGURE 10.30

Schematic diagram of an InGaAsP laser.

output signal from the injection-locked slave laser is canceled by the destructive interference with the master laser signal. The squeezed output signal of the slave laser is not degraded because the noise of the master laser signal is attenuated by the mirror.

The actual experimental setup is shown in Figure 10.32. The master laser is an AlGaAs single-mode high-power semiconductor laser, and the slave laser is a single-mode low-power GaAs

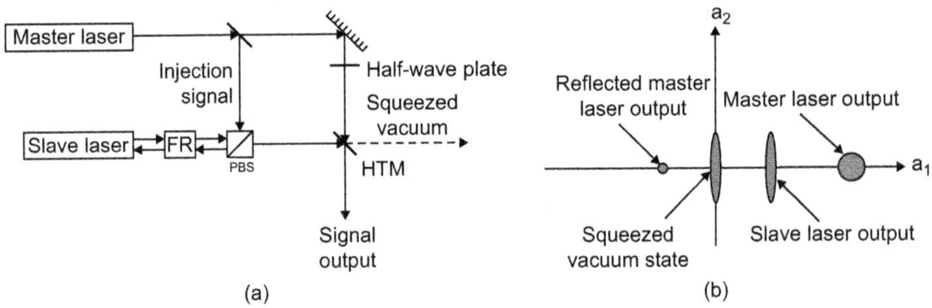

FIGURE 10.31

Squeezed-vacuum-state generation by mixing an amplitude-squeezed state with a coherent state. FR, Faraday rotator; PBS, polarization beam splitter, HTM, high-transmission mirror.

Reproduced from Yamamoto et al.[6] with the permission of Elsevier.

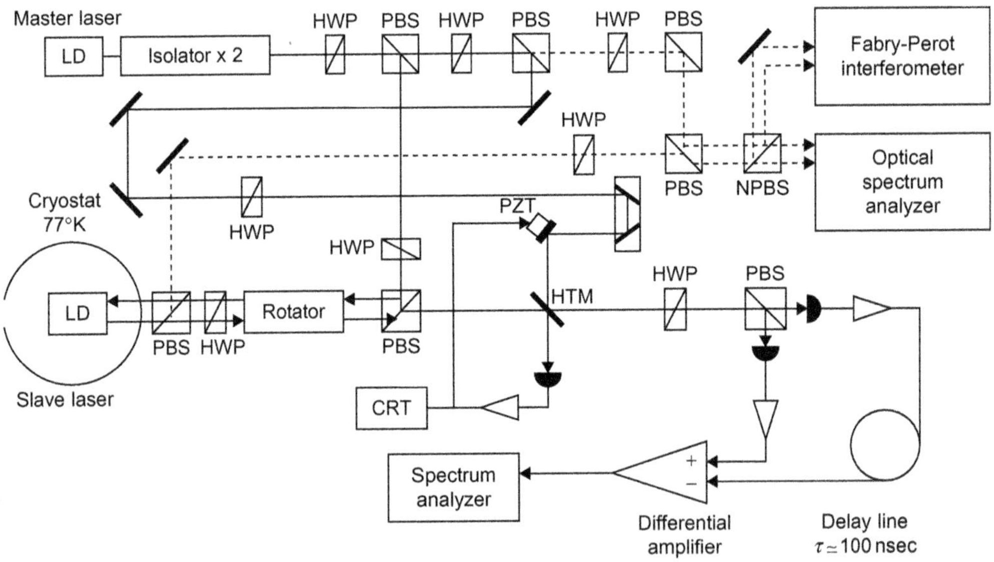

FIGURE 10.32

The experimental setup for the squeezed-vacuum-state generation by a semiconductor laser system; HWP, half-wave plate; PBS, polarization beam splitter; NPBS, nonpolarization bean splitter; HTM, high-transmission mirror; PZT, pizeo translator.

Reproduced from Yamamoto et al.[6] with the permission of Elsevier.

transverse-junction strip semiconductor laser with antireflection coating (~10%) on the front facet and high-reflection coating (~90%) on the rear facet.

10.7 GRAPHENE

Graphene is a flat monolayer of carbon atoms that is tightly packed into a two-dimensional honeycomb lattice. Graphene can be wrapped up into zero-dimensional fullerenes or rolled into one-dimensional nanotubes. It can also be stacked into three-dimensional graphite. Figure 10.33 shows the different ways in which graphene can be wrapped or stacked.

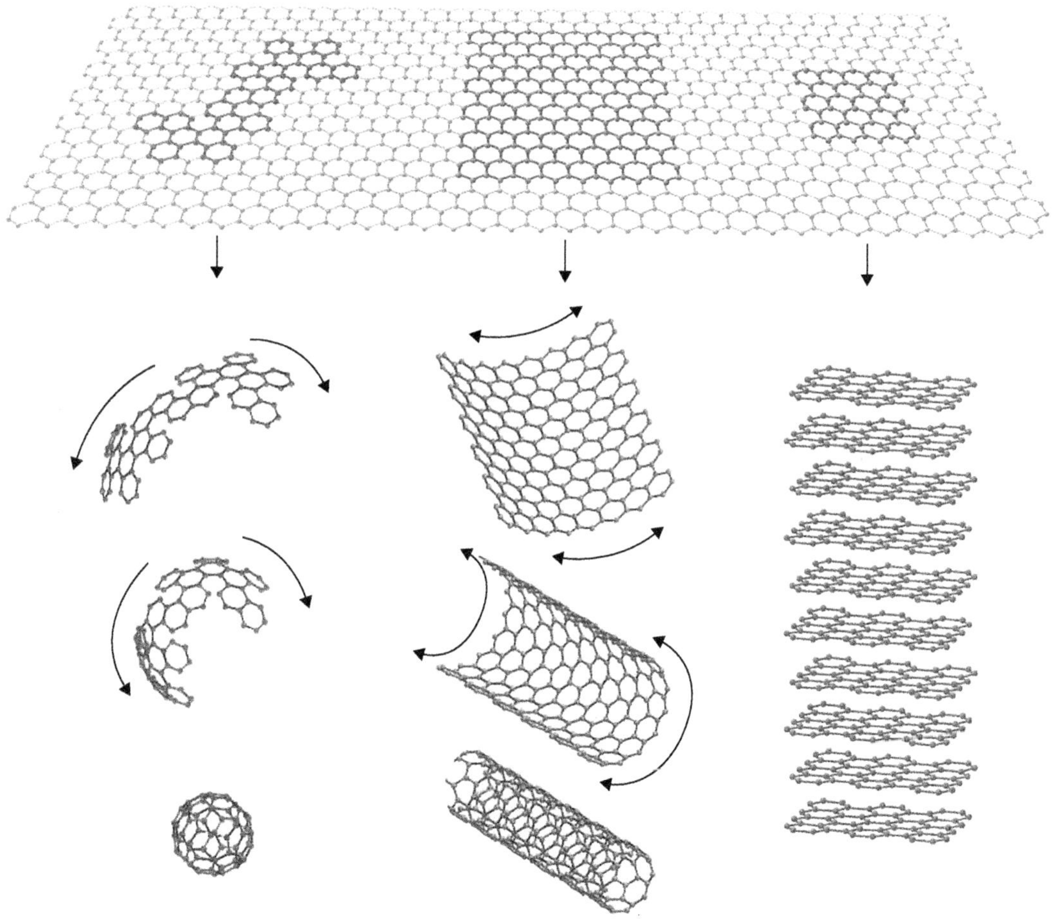

FIGURE 10.33

Graphene is a 2D carbon material that can be wrapped into 0D buckyballs, rolled into 1D nanotubes, or stacked into 3D graphites.

Reproduced from Geim[5] with the permission of MacMillan Publishers Ltd. Copyright 2011.

Graphene and its bilayer are both zero-gap semiconductors (which can also be considered as zero-overlap semimetals) with one type of electron and one type of hole. The conduction and valence bands start overlapping for three or more layers. Thus, single-, double-, and few- (3- to 9-) layer graphene can be considered as three types of 2D crystals. In these crystals, the charge carriers can travel thousands of interatomic distances.

Single- and few-layer graphenes have been grown epitaxially by chemical vapor deposition of hydrocarbons on metal substrates and thermal decomposition of SiC. Few-layer graphene obtained on SiC shows high-mobility charge carriers. Although epitaxial growth of graphene is the viable route for experimental applications, current experiments mostly use samples obtained by micromechanical cleavage of bulk graphite, which provides high-quality graphene crystallites up to 100 μm in size. Graphene becomes visible in an optical microscope if placed on an Si wafer with an appropriate thickness of SiO_2. This is due to feeble interference-like contrast with an empty wafer. These high-quality graphene crystallites are shown in Figure 10.34.

There is a pronounced ambipolar electric field effect in graphene that is shown in Figure 10.35.

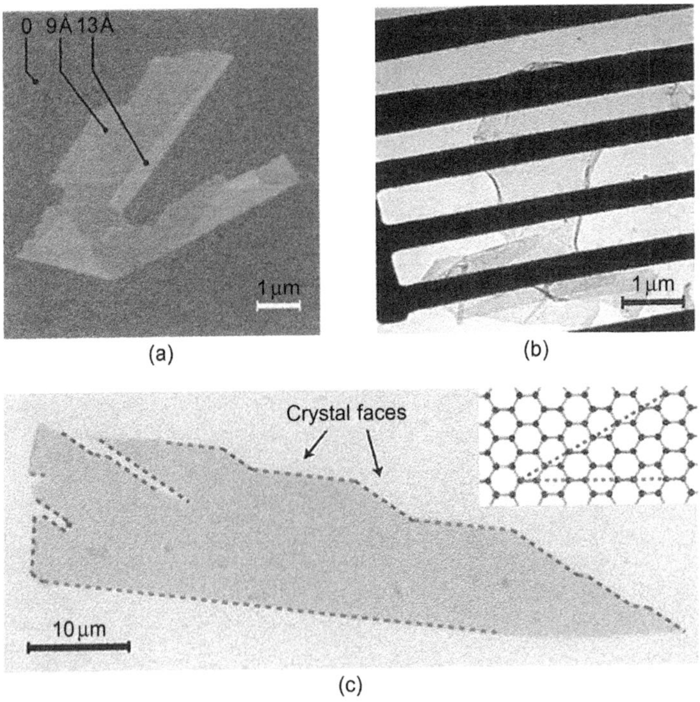

FIGURE 10.34

(a) Graphene visualized by atomic force microscopy. The folded region exhibiting a relative height of ≈ 4 Å indicates that it is a single crystal. (b) Transmission electron microscopy image of a graphene sheet freely suspended on a micrometer-sized metallic scaffold. (c) Scanning electron microscopy of relatively large graphene crystal, which shows that most of the crystal's faces are zigzag and armchair edges.

Reproduced from Geim[5] with the permission of MacMillan Publishers Ltd. Copyright 2011.

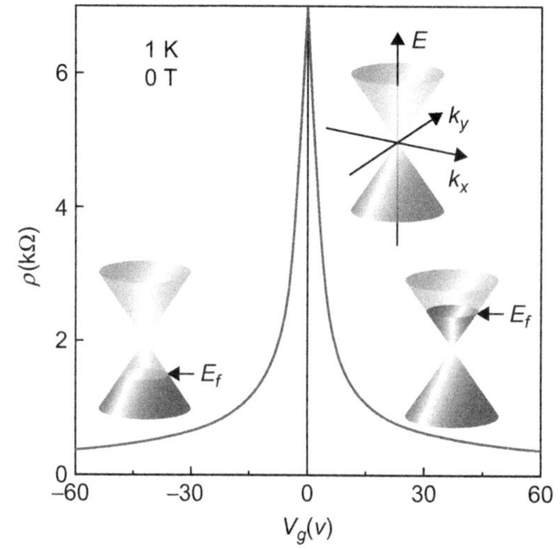

FIGURE 10.35

Ambipolar electric field effect in single-layer graphene.

Reproduced from Geim[5] with the permission of MacMillan Publishers. Copyright 2011.

The insets in Figure 10.35 show the conical low-energy spectrum $E(\mathbf{k})$, which shows the change in the Fermi energy E_F with gate voltage V_g. Positive V_g induces electrons and negative V_g induces holes. The concentration $n = \alpha V_g$, where $\alpha = 7.2 \times 10^{10}\ \text{cm}^{-2}\text{V}^{-1}$ for field-effect devices with a 300 nm SiO_2 layer used as a dielectric. The rapid decrease in resistivity ρ on adding charge carriers indicates their high mobility and does not change significantly when the temperature is increased to 300° K. The charge carriers in graphene can be tuned continuously between electrons and holes in concentrations $n \sim 10^{13}\ \text{cm}^{-2}$ and mobilities $\mu \geq 15{,}000\ \text{cm}^2\ \text{V}^{-1}\ \text{s}^{-1}$.

A unique property of graphene is that the interaction of the electrons with the periodic potential of the honeycomb lattice gives rise to quasiparticles that are described at low energies E by a (2 + 1)–dimensional Dirac equation with an effective speed of light $v_F = 10^6\ \text{m}^{-1}\text{s}^{-1}$. These quasiparticles are essentially electrons that have lost their rest mass m_e and are called massless Dirac fermions.

The conical sections of the energy spectrum for $|E| < 1$ eV are a consequence of the fact that graphene is a zero-gap semiconductor. The low-E quasiparticles within each valley can be described by the Dirac-like Hamiltonian, which is a direct consequence of graphene's crystal symmetry. A detailed analysis of the physical properties of graphene, including the origin of Dirac fermions, is given in Chapter 18 (on novel materials). The Hamiltonian H can be written as (for details, see Chapter 18):

$$H = \hbar v_F \begin{pmatrix} 0 & k_x - ik_y \\ k_x + ik_y & 0 \end{pmatrix} = \hbar v_F\ \vec{\sigma} \cdot \mathbf{k}. \tag{10.79}$$

Here, $\mathbf{k}$ is the quasiparticle momentum, $\vec{\sigma}$ is the 2D Pauli matrix (but pseudospin rather than real spin), and the Fermi velocity v_F, which is independent of k, plays the role of the speed of light. The honeycomb lattice of graphene is made up of two equivalent carbon sublattices, A and B. The cosine-like energy bands associated with the sublattices give rise to the conical sections of the energy spectrum for $|E| < 1$ eV (Figure 10.33) because they intersect at zero E near the edges of the Brillouin zone.

In addition to the unique feature of the band structure $E = \hbar v_F k$, the electronic states at zero E, where bands intersect, are a mixture of states of different sublattices. One has to use two-component wave functions (spinors) by requiring an index to identify sublattices A and B, to account for their relative contributions in the makeup of the quasiparticles. Thus, in Eq. (10.79), $\vec{\sigma}$ is referred as the pseudospin. The real spin of the electrons is described by additional terms in the Hamiltonian. However, the pseudospin effects, which are inversely proportional to the speed of light c (note that v_F plays the role of c) in quantum electrodynamics (QED), dominate the effects due to real spin because $c/v_F \approx 300$.

In QED, chirality (which is positive or negative for electrons or holes, respectively) is defined as the projection of $\vec{\sigma}$ on the direction of motion **k**. In graphene, the intricate connection of chirality is a consequence of the fact that k electrons and $-k$ hole states originate from the same carbon sublattices. Both chirality and pseudospin are conserved in graphene.

10.8 GRAPHENE-BASED ELECTRONICS

The high mobility (μ) of the charge carriers of graphene does not decrease even in the highest-field-induced concentrations and remains unchanged by chemical doping. Thus, there can be ballistic transport on a submicrometer scale at room temperature. The switching time is reduced due to the large value of v_F and low-resistance contacts without a Schottky barrier. The on–off ratios, which are comparatively low for graphene because of poor conductivity ($\approx$100), do not create any problem for high-frequency applications. In fact, it has been shown that transistors can be operational at THz frequencies.

The fact that graphene remains metallic even at the neutrality point creates a problem for mainstream logic applications. However, it has been shown that significant semiconductor gaps, $\Delta E(\approx 0.3\ eV)$, can be induced in bilayer graphene, which can be used in tunable infrared lasers and detectors. ΔE can also be induced in single-layer graphene by spatial confinement or lateral-superlattice potential. If graphene is epitaxially grown on top of crystals with matching lattices such as BN or SiC, such superlattice effects are likely to occur.

It can be shown that the confinement gap for graphene is

$$\Delta E(eV) \approx \alpha \hbar v_F / d \approx 1/d\ (\text{nm}), \tag{10.80}$$

where the coefficient $\alpha \approx 0.5$ for Dirac fermions. For room-temperature operations, $d \approx 10$ nm, which is achievable with the rapidly advancing Si-based technology. However, hitherto, no technique has been found to make anisotropic etching of graphene to make devices with crystallographic-defined faces in order to avoid irregular edges. Electronic states associated with short irregular edges in short channels usually induce a significant sample-dependent conductance, whereas those associated in long channels lead to additional scattering.

Graphene can also be used as a conductive sheet where the different nanometer-size structures can be carved to make a single-electron transistor circuit. Graphene nanostructures are stable down to nanometer sizes that allow the exploration of a region between SET and molecular electronics. Figure 10.36 shows an SET made from graphene by using electron-beam lithography and dry etching.

Figure 10.36b shows that for a minimum feature size of $\approx$10 nm, the combined Coulomb and confinement gap reaches >3 k_BT. This would allow a SET-like circuitry operational at room temperature, and resistive barriers can be used to induce a Coulomb blockade.

There are two impediments for growth of graphene electronics. High-quality wafers suitable for industrial applications are yet to be developed despite the recent progress in epitaxial growth of graphene. In addition, individual features in graphene devices have to be controlled to provide reasonably accurate reproducibility in their properties.

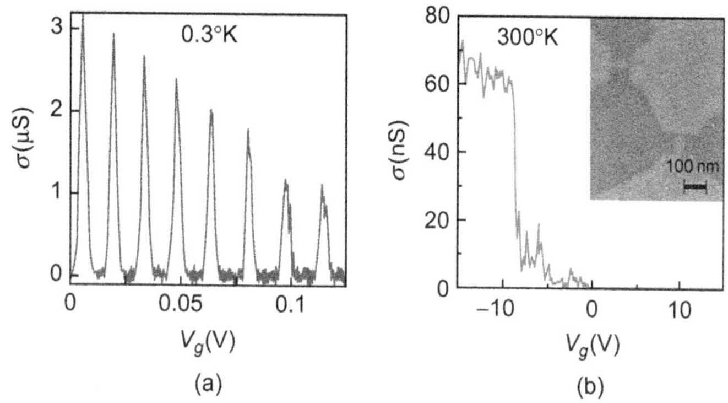

FIGURE 10.36

(a) Coulomb blockade in large quantum dots (diameter ≈0.25 μm) at low temperatures. Narrow constrictions in graphene serve as quantum barriers. (b) Here, 10 nm–scale graphene structures are stable under ambient conditions and survive thermal cycling to liquid-helium temperature. The inset shows a scanning electron micrograph of two graphene dots of ≈40 nm in diameter with narrower (<10 nm) constrictions.

Reproduced from Geim[5] with the permission of MacMillan Publishers Ltd. Copyright 2011.

PROBLEMS

10.1. In Eqs. (10.3) and (10.4), we derived

$$n_c(x) = \aleph_c(T)e^{-\beta(\varepsilon_c - e\phi(x) - \mu)} \tag{1}$$

and

$$p_v(x) = \wp_v(T)e^{\beta(\varepsilon_v - e\phi(x) - \mu)}. \tag{2}$$

Because $N_d = n_c(\infty)$ and $N_a = p_v(-\infty)$, show that

$$e\phi(\infty) - e\phi(-\infty) = E_g + \frac{1}{\beta}\ln\left[\frac{N_d N_a}{\aleph_c(T)\wp_v(T)}\right]. \tag{3}$$

10.2. From the preceding expressions, show that

$$n_c(x) = N_d e^{-e\beta[\phi(\infty) - \phi(x)]} \tag{1}$$

and

$$p_v(x) = N_a e^{e\beta[\phi(-\infty) - \phi(x)]}. \tag{2}$$

10.3. Within the boundaries of the depletion layer ($x = x_p$ and $x = x_n$), $n_c \ll N_d$ and $p_v \ll N_a$. Outside the depletion layer, $n_c = N_d$ *on the n-side* and $p_v = N_a$ *on the p-side*. Neglecting the boundary effects, show that, outside the depletion layer,

$$\rho(x) = 0, \tag{1}$$

and, inside the depletion layer,

$$\rho(x) = e[N_d(x) - N_a(x)]. \tag{2}$$

10.4. In the "abrupt junctions," one can approximate

$$N_d(x) = \begin{Bmatrix} N_d, & x>0 \\ 0, & x<0 \end{Bmatrix} \tag{1}$$

and

$$N_a(x) = \begin{Bmatrix} N_a, & x<0 \\ 0, & x>0 \end{Bmatrix}. \tag{2}$$

Show that the one-dimensional Poisson's equation (Eq. 10.10) can be written as

$$\frac{\partial^2\phi}{\partial x^2} = \begin{Bmatrix} 0, & x<x_p \\ -\frac{4\pi e N_a}{\epsilon}, & 0>x>x_p \\ \frac{4\pi e N_d}{\epsilon}, & 0<x<x_n \\ 0, & x>x_n \end{Bmatrix}. \tag{3}$$

By integrating Eq. (3), show that

$$\phi(x) = \begin{Bmatrix} \phi(-\infty), & x<x_p \\ \phi(-\infty) + \left(\frac{2\pi e N_a}{\epsilon}\right)(x-x_p)^2, & 0>x>x_p \\ \phi(\infty) - \left(\frac{2\pi e N_d}{\epsilon}\right)(x-x_n)^2, & 0<x<x_n \\ \phi(\infty), & x>x_n \end{Bmatrix}. \tag{4}$$

10.5. From Eq. (10.6) and (10.7), prove the law of mass action,

$$n_c(x,T)p_v(x,T) = \aleph_c(T)\wp_v(T)e^{-\beta E_g}. \tag{1}$$

Since, for an intrinsic semiconductor, $n_c(x,T) = p_v(x,T) = n_i(T)$, and from Eq. (10.12), $n_c(\infty) = N_d$ (the n-side of the depletion layer), using Eq. (1), show that

$$p_v(\infty) \approx \frac{n_i^2(T)}{N_d}, \tag{2}$$

where

$$n_i(T) = n_i(0)e^{-\beta E_g/2}. \tag{3}$$

Similarly, because $p_v(-\infty) = N_a$ in the p-side of the depletion layer, show that

$$n_c(-\infty) \approx \frac{n_i^2(T)}{N_a}. \tag{4}$$

10.6. It has been shown that

$$\frac{\partial g}{\partial t} = -\dot{\mathbf{r}} \cdot \frac{\partial g}{\partial \mathbf{r}} - \dot{\mathbf{k}} \cdot \frac{\partial g}{\partial \mathbf{k}} - \frac{g - f}{\tau_n}, \tag{1}$$

where g is the density of particles at position $\mathbf{r}$. In the presence of weak-applied fields, g is replaced by $g_{\mathbf{rk}}(t)$, which is the occupation number of electrons at position $\mathbf{r}$, which have a wave vector $\mathbf{k}$ at time t. It can be easily shown from Eq. (6.83) that at constant temperature

$$g_{\mathbf{rk}} \approx f_{\mathbf{rk}} - e\tau_n \frac{\partial f}{\partial \mu} \mathbf{v_k} \cdot \mathbf{E}. \tag{2}$$

The electron density n_c at position $\mathbf{r}$ in the conduction band can be written as

$$n_c = \frac{1}{4\pi^3} \int g_{r\mathbf{k}} d\mathbf{k}, \tag{3}$$

while the equilibrium density of electrons in the conduction band can be written as

$$n_c^0 = \frac{1}{4\pi^3} \int f d\mathbf{k}. \tag{4}$$

Integrating Eq. (1) over $d\mathbf{k}$, show from Eqs. (1) through (4),

$$\frac{\partial n_c}{\partial t} = -\frac{\partial}{\partial \mathbf{r}} \cdot <\dot{\mathbf{r}}> + \frac{n_c^0 - n_c}{\tau_n}. \tag{5}$$

Here, n_c^0 is the equilibrium density of the electrons in the conduction band and $<\dot{\mathbf{r}}>$ is $\mathbf{v_k}$ averaged over the Brillouin zone,

$$<\dot{\mathbf{r}}> = \frac{1}{4\pi^3 n_c} \int d\mathbf{k}\, g_{\mathbf{rk}} \mathbf{v_k} = \frac{1}{4\pi^3 n_c} \int d\mathbf{k} \left[f - \tau_n \mathbf{v_k} \cdot \left\{ e\mathbf{E} \frac{\partial f}{\partial \mu} + \frac{\partial f}{\partial \mathbf{r}} \right\} \right] \mathbf{v_k}. \tag{6}$$

10.7. It has been shown that in the steady state, for the minority carriers,

$$D_n \frac{d^2 n_c}{dx^2} + \frac{n_c^0 - n_c}{\tau_n} = 0 \tag{1}$$

and

$$D_p \frac{d^2 p_v}{dx^2} + \frac{p_v^0 - p_v}{\tau_p} = 0. \tag{2}$$

Show that in the n-side of the depletion layer, for $x \geq x_n$,

$$p_v(x) - p_v(\infty) = [p_v(x_n) - p_v(\infty)] e^{-(x - x_n)/d_p}, \tag{3}$$

and in the p-side of the depletion layer, for $x \leq x_p$,

$$n_c(x) - n_c(-\infty) = [n_c(x_p) - n_c(-\infty)] e^{(x - x_p)/d_n}, \tag{4}$$

where the diffusion lengths for holes and electrons are defined as

$$d_p = (D_p\tau_p)^{1/2} \text{ and } d_n = (D_n\tau_n)^{1/2}. \tag{5}$$

10.8. From Eqs. (10.49) and (10.50), we have

$$j_e = en_c\mu_n E + eD_n \frac{dn_c}{dx} \tag{1}$$

and

$$j_h = ep_v\mu_p E - eD_p \frac{dp_v}{dx}. \tag{2}$$

Show that

$$n_c(x) = N_d e^{\beta[V(x)-V(x_n)]}\left[1 + \frac{j_e}{eN_d D_n}\int_{x_n}^{x} dx' e^{-\beta[V(x')-V(x_n)]}\right] \tag{3}$$

and

$$p_v(x) = N_a e^{-\beta[V(x)-V(x_p)]}\left[1 - \frac{j_h}{eN_a D_p}\int_{x_p}^{x} dx' e^{\beta[V(x')-V(x_p)]}\right]. \tag{4}$$

Hence, show that the second term in the square bracket can be neglected compared to the first term in both Eqs. (3) and (4).

10.9. We derived in Chapter 9 that

$$n_i^2 = \aleph_c(T)\wp_v(T)e^{-\beta E_g}. \tag{1}$$

We also derived in Eq. (10.8),

$$e\Delta\phi = E_g + \frac{1}{\beta}\ln\left[\frac{N_d N_a}{\aleph_c(T)\wp_v(T)}\right]. \tag{2}$$

Show from Eqs. (1) and (2) that

$$n_i^2 = N_d N_a e^{-e\beta\Delta\phi}. \tag{3}$$

References

1. Ashcroft NW, Mermin ND. *Solid state physics*. New York: Brooks/Cole; 1976.
2. Bardeen J. Surface states and rectification at a metal semi-conductor contact. *Phys Rev* 1947;**71**:717.
3. Bardeen J, BrattinWH. The transistor, a semi-conductor triode. *Phys Rev* 1948;**74**:230.

4. Cohen ML, Chelikosky JR. *Electronic structure and optical properties of semiconductors.* Berlin: Springer-Verlag; 1989.
5. Geim AK, Novoselov KS. The rise of graphene. *Nat Mater* 2007;**6**:183.
6. Yamamoto Y, Inoue S, Bjork G, Heitmann H, Matinaga F. In: Kapon E, editor. *Semiconductor lasers,* vol. 1. Amsterdam: Elsevier; 1999:361.
7. Kastner MA. The single-electron transistor. *Rev Mod Phys* 1992;**64**:849.
8. Kimmerling LC, Kolenbrander KD, Michel J, Palm J. Light emission from silicon. *Solid State Phys Adv Res Appl* 1996;**50**:333.
9. Kittel C. *Introduction to solid state physics.* New York: John Wiley & Sons; 1976.
10. Klingston CF. *Semiconductor optics.* Berlin: Springer-Verlag; 1995.
11. Maiman TH. Stimulated optical radiation in ruby. *Nature* 1960;**147**:493.
12. Marder MP. *Condensed matter physics.* New York: John Wiley & Sons; 2000.
13. Meirav U, Kastner MA, Heiblum M, Wind SJ. *Phys Rev B* 1989;**40**:5871.

CHAPTER

Spintronics

11

CHAPTER OUTLINE

11.1 Introduction ... 339
11.2 Magnetoresistance ... 340
11.3 Giant Magnetoresistance ... 340
11.3.1 Metallic Multilayers ... 340
11.4 Mott's Theory of Spin-Dependent Scattering of Electrons ... 342
11.5 Camley–Barnas Model ... 345
11.6 CPP-GMR ... 348
11.6.1 Introduction ... 348
11.6.2 Theory of CPP-GMR of Multilayered Nanowires ... 350
11.7 MTJ, TMR, and MRAM ... 352
11.8 Spin Transfer Torques and Magnetic Switching ... 356
11.9 Spintronics with Semiconductors ... 357
11.9.1 Introduction ... 357
11.9.2 Theory of an FM-T-N Junction ... 358
11.9.3 Injection Coefficient ... 361
Problems ... 364
References ... 367

11.1 INTRODUCTION

As we noted in Chapter 10 on electronics, charges are manipulated by electric fields, but spins are generally ignored. Magnetic recording and other techniques use the spin only through the magnetization of a ferromagnet. When giant magnetoresistance (GMR) of the magnetic multilayers was discovered in 1988, an efficient control of the electrons was achieved through the orientation of their magnetization by acting on their spin. The application of the GMR to the read heads of the hard discs contributed significantly to the quick rise in the density of stored information. This led to the extension of hard disc technology and consumer electronics.

A large number of phenomena related to the control and manipulation of spin currents were developed through a new area of physics/materials science called spintronics. The rapidly growing area of research in spintronics includes such phenomena as spin transfer, molecular spintronics, spintronics with semiconductors, and single-electron spintronics.

11.2 MAGNETORESISTANCE

In Chapter 3, we discussed the Hall effect by considering a conductor in the shape of a rod that has a rectangular cross-section, which is placed under a magnetic field **B** in the z direction. There is a longitudinal electric field E_x. The electric and magnetic fields are so adjusted that the current cannot flow out of the rod in the y direction ($j_y = 0$). The charges pile up on the surface of the sample, thereby setting up an electric field E_y. This field, which nullifies the Lorentz force (Eq. 3.90), is known as the Hall field. The Hall coefficient R_H, which defines the size of the carrier, is defined as $R_H = E_y/Bj_x$. The component of the resistivity tensor ρ, ρ_{yx}, is the Hall resistance defined as $\rho_{yx} = E_y/j_x$, whereas the diagonal component ρ_{xx} is the magnetoresistance, defined as

$$\rho_{xx} = \frac{E_x}{j_x}, \tag{11.1}$$

and the magnetoresistance ratio is defined as

$$\frac{\Delta\rho}{\rho_0} = \frac{\rho - \rho_0}{\rho_0}, \tag{11.2}$$

where ρ and ρ_0 are the resistivities (along a given direction) in the presence and absence of a magnetic field, respectively. The dependence of the electrical resistance of the material on an applied magnetic field, usually perpendicular to the direction of the current, represents this effect.

11.3 GIANT MAGNETORESISTANCE

11.3.1 Metallic Multilayers

A typical multilayer unit is shown in Figure 11.1. In a magnetic multilayer system, one of the metals (A) is magnetic, whereas the other (B) is nonmagnetic, which is referred to as the spacer layer. When the thickness of the spacer layer is varied, there are oscillations in the magnetic coupling between the magnetic layers. The thickness of these thin films can vary from a few tenths of a nanometer to tens of nanometers. The magnetization directions of the ferromagnetic layers are coupled to each other through an exchange interaction. The sign of this coupling oscillates as the thickness of the spacer layer is varied. The best multilayer samples have around 30 periods of oscillations.

Some metallic multilayers exhibit drastic changes in magnetoresistance and, hence, the name giant magnetoresistance (GMR). The drastic change of magnetoresistance in metallic multilayers has resulted in a new definition of the magnetoresistance ratio, MR, as

$$MR = \frac{R_{AP} - R_P}{R_P} \times 100, \tag{11.3}$$

where R_P and R_{Ap} are the resistances of the parallel and antiparallel magnetic configurations, respectively.

In the mid-1980s, it became possible to develop molecular-beam epitaxy (MBE) and other techniques to fabricate multilayers composed of very thin individual layers. The explanation of the

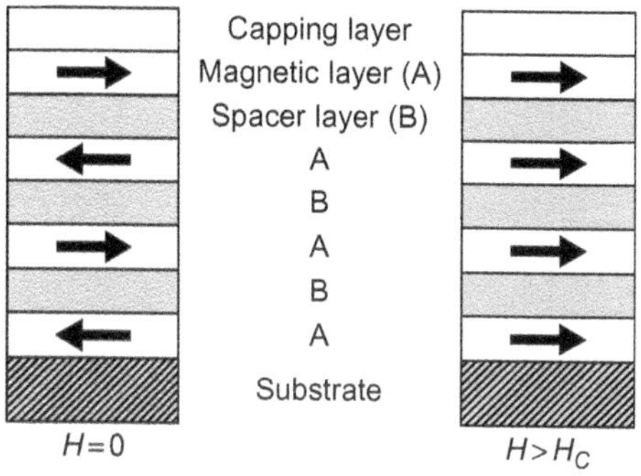

FIGURE 11.1

A multilayer structure and the changes in magnetization directions with an applied magnetic field **H**. The parallel magnetizations lead to a large magnetoresistance compared to that of an antiparallel structure.

Reproduced from Fernando[6] with the permission of Elsevier.

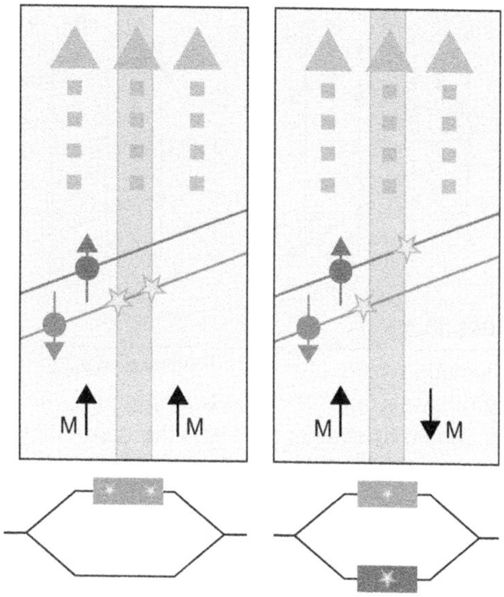

FIGURE 11.2

Spin-dependent electron scattering and redistribution of scattering events upon anti-alignment of magnetization.

Reproduced from Grunberg[11] with the permission of the American Physical Society.

GMR effect includes spin-dependent electron scattering and redistribution of scattering events due to anti-alignment of magnetizations. This scenario is shown in Figure 11.2.

The microscopic explanation for the GMR effect is that the scattering rates of the electrons depends on the orientation (parallel or antiparallel) of the electron spins with respect to the local magnetizations. Figure 11.2 shows two structures, one with parallel and the other with antiparallel alignment with the magnetization. The electrons with spin parallel to the local magnetization in their random walk are not scattered in the ideal situation. In Figure 11.2, only one passage from the left to the right is shown as a sample of the electron motion. A short circuit is caused by the electrons that are not scattered. However, when the electron with spin-up enters the layer where the magnetization has been turned around, its spin is opposite to the local magnetization. The bottom diagram in Figure 11.2 shows the increase in resistivity due to the removal of the short circuit. In practice, although both types of electrons are scattered, the resistivity due to the antiparallel magnetization is more than that due to parallel magnetization.

In Figure 11.3, the arrows represent the majority spin direction in the magnetic layers. In the ferromagnetic (F) configuration, the spin+ ($s_z = 1/2$) electrons are weakly scattered everywhere, which gives a short circuit effect and hence a small resistivity. In the antiferromagnetic (AF) configuration, each spin direction is scattered at every second magnetic layer, and the resistivity is higher because there is no short circuit effect.

The short circuit effect for the ferromagnetic (F) configuration is visually explained in Figure 11.4. In the parallel (P) configuration, majority electrons in the (+) channel ($s_z = 1/2$) experience little or

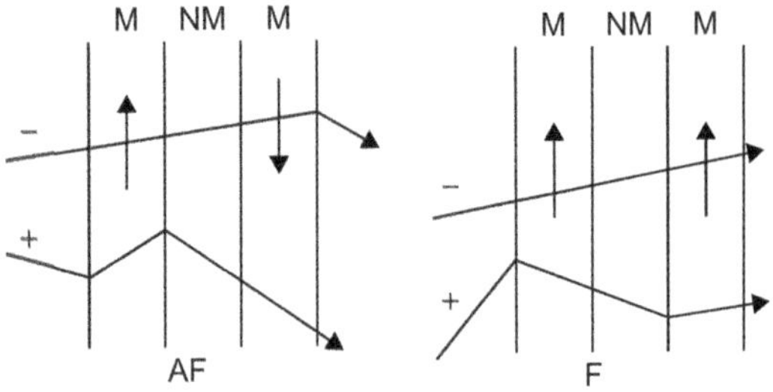

FIGURE 11.3

Schematic picture of the electron trajectories. The electron trajectories between two scatterings are represented by straight lines and the scatterings by abrupt change in direction. The + and – are for electron spins $s_z = 1/2$ and $s_z = -1/2$, respectively.

Reproduced from Fert et al.[10] with the permission of Elsevier.

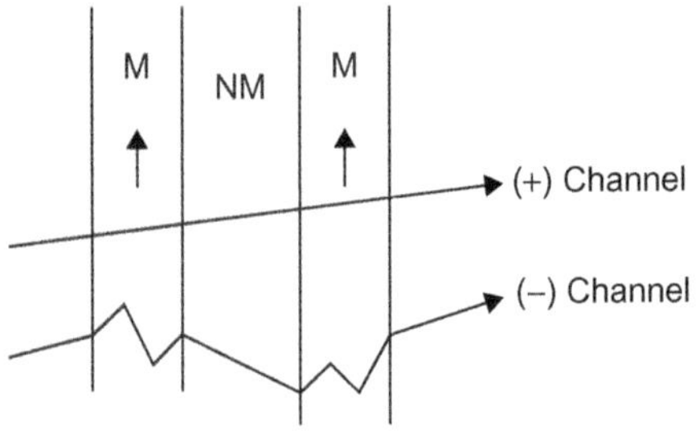

FIGURE 11.4

Schematic diagram of the short circuit effect.

Reproduced from Fernando[6] with the permission of Elsevier.

no resistance, and hence, a short circuit effect occurs. In the antiparallel (AP) configuration (not shown), electrons in the (+) and (–) ($s_z = -1/2$) channels will experience a significant resistance when going through the slab with opposite magnetization with no short circuit effects.

11.4 MOTT'S THEORY OF SPIN-DEPENDENT SCATTERING OF ELECTRONS

A simple model of a scattering of electrons, which was given by Mott[15] will be presented here. It was modified later by more complex models. We consider a ferromagnetic *d*-band metal with a magnetization $M(T)$ at temperature T. The schematic density of states representing *s*-, *p*-, and *d*-bands of a transition metal is shown in Figure 11.5.

If M_0 is the saturation magnetization at $T=0$, and $\beta = M(T)/M_0$, a fraction $(1-\beta)/2$ of the unoccupied *d* states will have their spins parallel to M, and a fraction $(1+\beta)/2$ will have their spins antiparallel. If we define $\rho_1(E)$ and $\rho_2(E)$ as the respective density of states, then they will have a parabolic form,

$$\rho_1(E) = a\sqrt{(E_1 - E)} \tag{11.4}$$

and

$$\rho_2(E) = a\sqrt{(E_2 - E)}, \tag{11.5}$$

where E_1 and E_2 are the highest energies the two spins can have. However, above the Curie temperature, where the system becomes paramagnetic

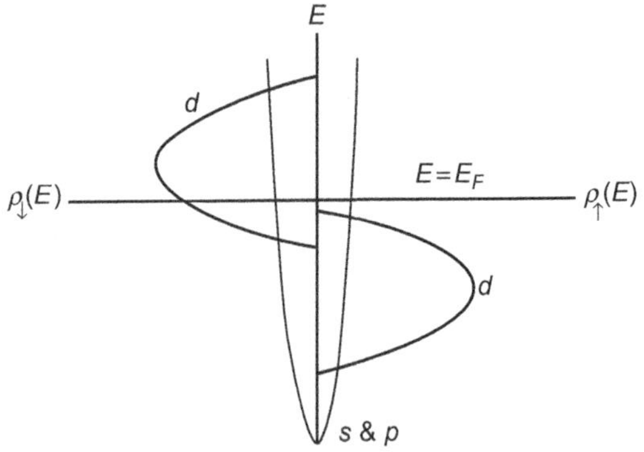

FIGURE 11.5

Density of states plotted against *E* for *s*, *p*, and *d* states of a transition metal.

and the number of parallel spins equals the number of antiparallel spins, $E_1 = E_2 = E_0$. It can be shown that the corresponding relaxation times are related to the density of states as

$$\frac{1}{\tau_1} \propto \rho_1(E) + \delta \tag{11.6}$$

and

$$\frac{1}{\tau_2} \propto \rho_2(E) + \delta, \tag{11.7}$$

where δ is a contribution from spin-independent scattering. From Eqs. (11.4) through (11.7), it can be shown that (Problem 11.1)

$$\frac{1}{\tau_1} = A(T/M_i\theta^2)\left[\frac{\sqrt{(E_1 - E)}}{\sqrt{(E_0 - \zeta_0)}} + \alpha\right] \tag{11.8}$$

and

$$\frac{1}{\tau_2} = A(T/M_i\theta^2)\left[\frac{\sqrt{(E_2 - E)}}{\sqrt{(E_0 - \zeta_0)}} + \alpha\right], \tag{11.9}$$

where $\alpha \approx 1/4$, θ is the Debye temperature, and A is a constant. If ζ_0 represents the energy at $T=0$ and $\beta = 0$, and ζ_0' represents the energy at the highest occupied state (Fermi energy) at $T=0$ when the states are split, because the density of states is proportional to $n_0^{1/3}$, we can write (Problem 11.2)

$$\sqrt{\frac{E_1 - \zeta_0'}{E_0 - \zeta_0}} = (1-\beta)^{1/3} \tag{11.10}$$

and

$$\sqrt{\frac{E_2 - \zeta_0'}{E_0 - \zeta_0}} = (1+\beta)^{1/3}. \tag{11.11}$$

One can write the expression for conductivity by using a Drude-type formula, but including the energy dependence of the relaxation time,

$$\sigma = -Ne^2/m \int \frac{(\tau_1 + \tau_2)}{2} \frac{\partial f}{\partial E} dE. \tag{11.12}$$

The resistivity $\rho(\beta, T) = 1/\sigma$ is obtained by approximating the partial derivative of the integrand in Eq. (11.12) to be nonzero only when $E = \zeta_0'$. We obtain, after some algebra (Problem 11.3),

$$\rho(\beta, T) = A(T/m\theta^2)\left[\frac{1}{(1-\beta)^{1/3} + \alpha} + \frac{1}{(1+\beta)^{1/3} + \alpha}\right]^{-1}. \tag{11.13}$$

Because $\alpha \simeq 0.25$, the ratio of the resistances between the saturated magnetization case ($\beta = 1$) and the paramagnetic case ($\beta = 0$) is given by

$$\frac{\rho(\beta = 1, T)}{\rho(\beta = 0, T)} \simeq 0.34. \tag{11.14}$$

If we define the resistances for up and down channels as $R^{\uparrow}$ and $R^{\downarrow}$, then the resistances R_P and R_{AP} for the parallel and antiparallel alignments of spins relative to a given magnetic layer are given by

$$R_P = (1/R^{\uparrow} + 1/R^{\downarrow})^{-1} \tag{11.15}$$

and

$$R_{AP} = (R^{\uparrow} + R^{\downarrow})/4, \tag{11.16}$$

when the mean path of the electrons is much higher than the repeat length of the multilayer. From Eqs. (11.3), (11.15), and (11.16), we obtain

$$\mathrm{MR} = \frac{(1-\eta)^2}{4\eta} \times 100\%, \tag{11.17}$$

where

$$\eta = \frac{R_{\downarrow}}{R_{\uparrow}}. \tag{11.18}$$

Because η is a positive parameter, it follows from Eq. (11.17) that $R_{AP} > R_P$. This is a very simplified explanation of the higher resistance encountered in the antiparallel case.

When an applied field changes an alignment from antiferromagnetic (AF) to ferromagnetic (F) alignment, the difference in resistivity is the largest. The AF alignment is usually provided by interlayer exchange or by coercivities of successive magnetic layers, by pinning the magnetization using an antiferromagnetic material in direct contact, known as exchange biasing. If GMR is obtained by exchange biasing, it is called a spin-valve system.

The first discovery of GMR was done by Baibich et al.[1] on Fe/Cr magnetic multilayers, in which it is possible to switch the relative orientation in adjacent magnetic layers from antiparallel to parallel by applying a magnetic field. The resistivity is strongly enhanced in the antiparallel magnetic configuration of two adjacent layers A and B because the electrons in each channel are slowed down at every second magnetic layer. There is no such enhancement in layers A and B in the parallel magnetic configuration because the electrons can go easily through all the magnetic layers, and the short circuit through this channel leads to a small resistance. This opens up the possibility of switching between high and low resistivity states by changing the relative orientation of the magnetizations of A and B layers from parallel to antiparallel. Similar GMR effects were discovered by Binash et al.[2] in Fe/Cr/Fe trilayers. These effects are shown later in Figure 11.9. Camley and Barnas[3] presented a theoretical description of the GMR effects in Fe/Cr/Fe trilayers by calculating the resistivity using Boltzmann transport equations, with spin-dependent scattering at the interface.

11.5 CAMLEY–BARNAS MODEL

The Camley–Barnas model[3] includes the idea that there is an antiferromagnetic coupling between Fe films due to the intervening Cr films. At the zero field, the resulting magnetic moments of neighboring Fe films are antiparallel to each other. However, in a strong external magnetic field, all the magnetic moments of the Fe films can be forced to lie in the same direction. Earlier experimental results suggested that the resistance is the largest when the magnetic moments in neighboring Fe films are antiparallel and smallest when they are parallel. Further, multilayer structures with thin Fe films have a much larger magnetoresistance than a single sandwich structure of Fe/Cr/Fe. In addition, the magnetoresistance is increased by a factor of 2 or 3 when the temperature is changed from room temperature to that of liquid He. Camley and Barnas interpreted these results to conclude that a spin-dependent scattering is responsible for the observed effects and the relative lengths of the mean-free path are more important than the thickness of the various films.

To develop a theory for the multilayer, Camley and Barnas first considered a single sandwich structure of Fe/Cr/Fe, as shown in Figure 11.6, and computed the conductivity by using the Boltzmann equation.

In Figure 11.6, the dashed line in the center of the Cr films is the position at which the change in axis of quantization for the electron spin is calculated. In each region, the Boltzmann equation reduces to a differential equation that depends on the coordinate z only,

$$\frac{\partial g}{\partial z} + \frac{g}{\tau v_z} = \frac{eE}{mv_z}\frac{\partial f_0}{\partial v_x}, \tag{11.19}$$

where f_0 is the equilibrium distribution function, τ is the relaxation time, E is the external field in the x direction, and g is the correction to the distribution function due to scattering.

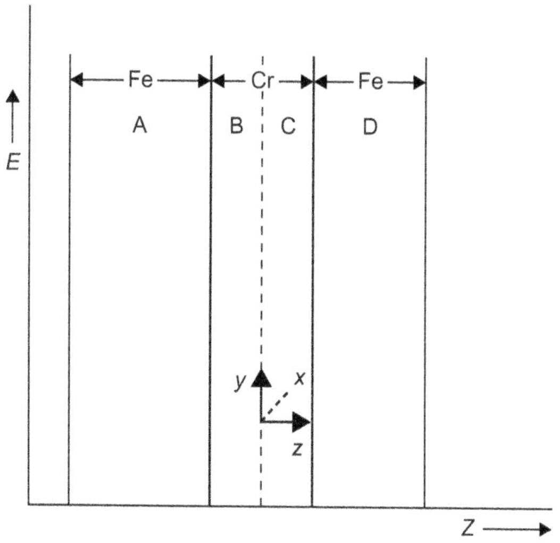

FIGURE 11.6

The geometry of the Fe/Cr/Fe sandwich structure.

In the region A, the contributions to g for regions for spin-up or spin-down electrons moving to the right (positive v_z) or left (negative v_z) are

$$g = g_{A+\uparrow}(v_z, z) + g_{A+\downarrow}(v_z, z) \tag{11.20}$$

and

$$g = g_{A-\uparrow}(v_z, z) + g_{A-\downarrow}(v_z, z). \tag{11.21}$$

From Eqs. (11.19) through (11.21), it can be easily shown that (Problem 11.4)

$$g_{A\pm\uparrow}(v_z, z) = \frac{eE\tau}{m}\frac{\partial f_0}{\partial v_x}\left[1 + A_{\pm\uparrow}\exp\left(\frac{\mp z}{\tau|v_z|}\right)\right] \tag{11.22}$$

and

$$g_{A\pm\downarrow}(v_z, z) = \frac{eE\tau}{m}\frac{\partial f_0}{\partial v_x}\left[1 + A_{\pm\downarrow}\exp\left(\frac{\mp z}{\tau|v_z|}\right)\right]. \tag{11.23}$$

The coefficients $A_{\pm\uparrow(\downarrow)}$ and similar coefficients for the regions B, C, and D are determined through the boundary conditions. The distribution function g for an electron leaving the surface (at the outer surfaces of the sandwich) is equal to the distribution function g for an electron of the same spin striking the surface multiplied by a specular scattering event R_0. Thus, we obtain

$$g_{A+\uparrow} = R_0 g_{A-1} \text{ at } z = -b \tag{11.24}$$

and

$$g_{D-\uparrow} = R_0 g_{D+\uparrow} \text{ at } z = +b. \tag{11.25}$$

Similar equations are obtained for down-spins.

Camley and Barnas[3] assumed a model system of two equivalent simple metals that have the same Fermi energies, mean-free path values, and so on. They neglected the angular dependence of scattering and assumed that there is only transmission or diffusive scattering at the Fe/Cr interfaces. They further assumed that the scattering at the outer boundaries is purely diffusive (the reflection coefficients are zero). If the transmission coefficients are $T_\uparrow$ for up-spins and $T_\downarrow$ for down-spins, for up-spins at $z = -a$,

$$g_{A-\uparrow(\downarrow)} = T_{\uparrow(\downarrow)} g_{B-\uparrow(\downarrow)} \tag{11.26}$$

and

$$g_{B+\uparrow(\downarrow)} = T_{\uparrow(\downarrow)} g_{A+\uparrow(\downarrow)}. \tag{11.27}$$

A similar set of equations holds for the $z = +a$ interface. Similar boundary conditions are obtained for $g_{B-\uparrow(\downarrow)}$ and $g_{C-\uparrow(\downarrow)}$, except that one has to account for the fact that the magnetic moments in the two films are in different directions by an angle θ, which is the angle between the magnetization vectors in the two Fe films. We define the transmission coefficients,

$$T_{\uparrow\uparrow} = T_{\downarrow\downarrow} = \cos^2(\theta/2) \tag{11.28}$$

and

$$T_{\uparrow\downarrow} = T_{\downarrow\uparrow} = \sin^2(\theta/2), \tag{11.29}$$

where $T_{\uparrow\uparrow}$ is the probability of an electron of spin-up (in layer A, with respect to the magnetization) at $z = -0$ to continue as a spin-up (in layer D, with respect to magnetization) at $z = 0$, and the other symbols are defined similarly. We obtain

$$g_{B-\uparrow(\downarrow)} = g_{C-\uparrow(\downarrow)} \cos^2(\theta/2) + g_{C-\downarrow(\uparrow)} \sin^2(\theta/2) \tag{11.30}$$

and

$$g_{C+\uparrow(\downarrow)} = g_{B+\uparrow(\downarrow)} \cos^2(\theta/2) + g_{B+\downarrow(\uparrow)} \sin^2(\theta/2). \tag{11.31}$$

The current density at different fields are obtained by using the expression

$$J(z) = \int v_x g(v_x, z)\, d^3v. \tag{11.32}$$

The current of the whole structure is obtained by integrating $J(z)$ over the coordinate z. Camley and Barnas defined the diffusive scattering parameter,

$$D_{\uparrow} = 1 - T_{\uparrow}, \tag{11.33}$$

and the asymmetry in up-spin and down-spin scattering,

$$N = D_{\uparrow}/D_{\downarrow}. \tag{11.34}$$

Their results are plotted in Figure 11.7.

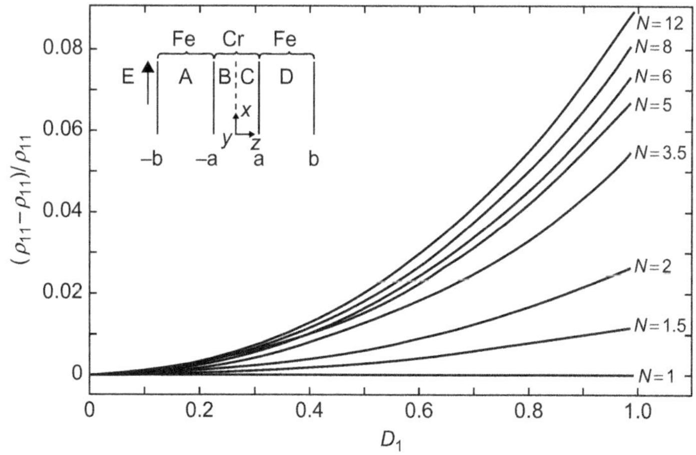

FIGURE 11.7

The maximum normalized change in resistance as a function of D_1 and N. The inset shows the geometry shown in Figure 11.6.

Reproduced from Camley and Barnas[3] with the permission of the American Physical Society.

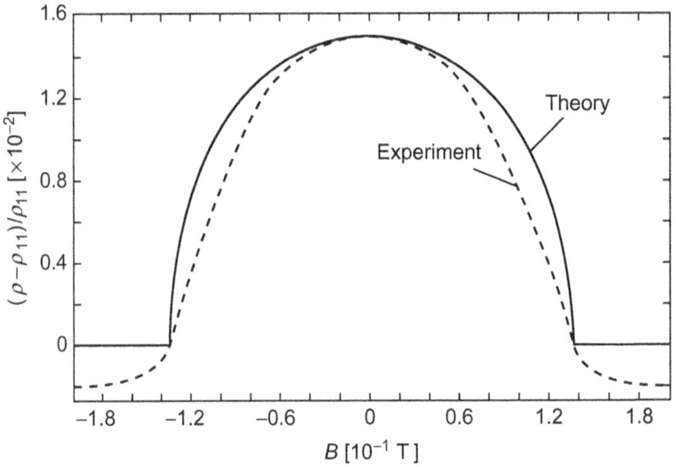

FIGURE 11.8

The percentage change of resistance as a function of an applied field ($N = 6$, $D_\uparrow = 0.48$). The experimental results are from Binash et al.[2] from Figure 11.9b.

Reproduced from Camley and Barnas[3] with the permission of the American Physical Society.

The calculations shown in Figure 11.7 were made to show how the magnetoresistance effect depends on $D_\uparrow$ and N for a (120 Å Fe)/(10 Å Cr)/(120 Å Fe) sandwich structure with a mean-free path of $\lambda = 180$ Å. The theoretical value of the percentage change in resistance as a function of an applied field for a (120 Å Fe)/(10 Å Cr)/(120 Å Fe) sandwich structure is shown in Figure 11.8.

In the preceding calculations, the mean-free path is $\lambda = 180$ Å, and the angle θ between the magnetizations in the two Fe films was calculated by minimizing the sum of the exchange, anisotropy, and Zeeman energies for the Fe/Cr/Fe sandwich.

The first observation of giant magnetoresistance was discovered by Baibich et al.[1] and is shown in Figure 11.9a. It may be noted that if we define $\mathrm{MR} = (R_{AP} - R_P)/R_P \times 100$, MR = 85% for the (Fe 3 nm/Cr 0.9 nm) multilayer. The experimental results of Binash et al.[2] for Fe/Cr/Fe trilayers is shown in Figure 11.9b. A schematic diagram of the electrons in parallel (low resistance) and antiparallel (high resistance) spin configurations is shown in Figure 11.9c.

As we noted earlier, GMR can be obtained in the current in plane (CIP) and current perpendicular to plane (CPP) geometry. At present, the CIP configuration is used for most sensor applications, and in most experiments the current flows in the CIP geometry. However, Pratt et al.[17] showed that for the Ag/Co multilayers, CPP-MR is several times larger than the CIP-GMR. There has been considerable interest in CPP-MR, which is more likely to be used in future sensor applications.

11.6 CPP-GMR

11.6.1 Introduction

The first experiment of CPP-GMR was done by sandwiching a magnetic multilayer between superconducting electrodes. This restricted the use of such multilayers only at very low temperatures. However, Fert and Piraux[9] as well as other groups have fabricated magnetic nanowires by electrodepositing into

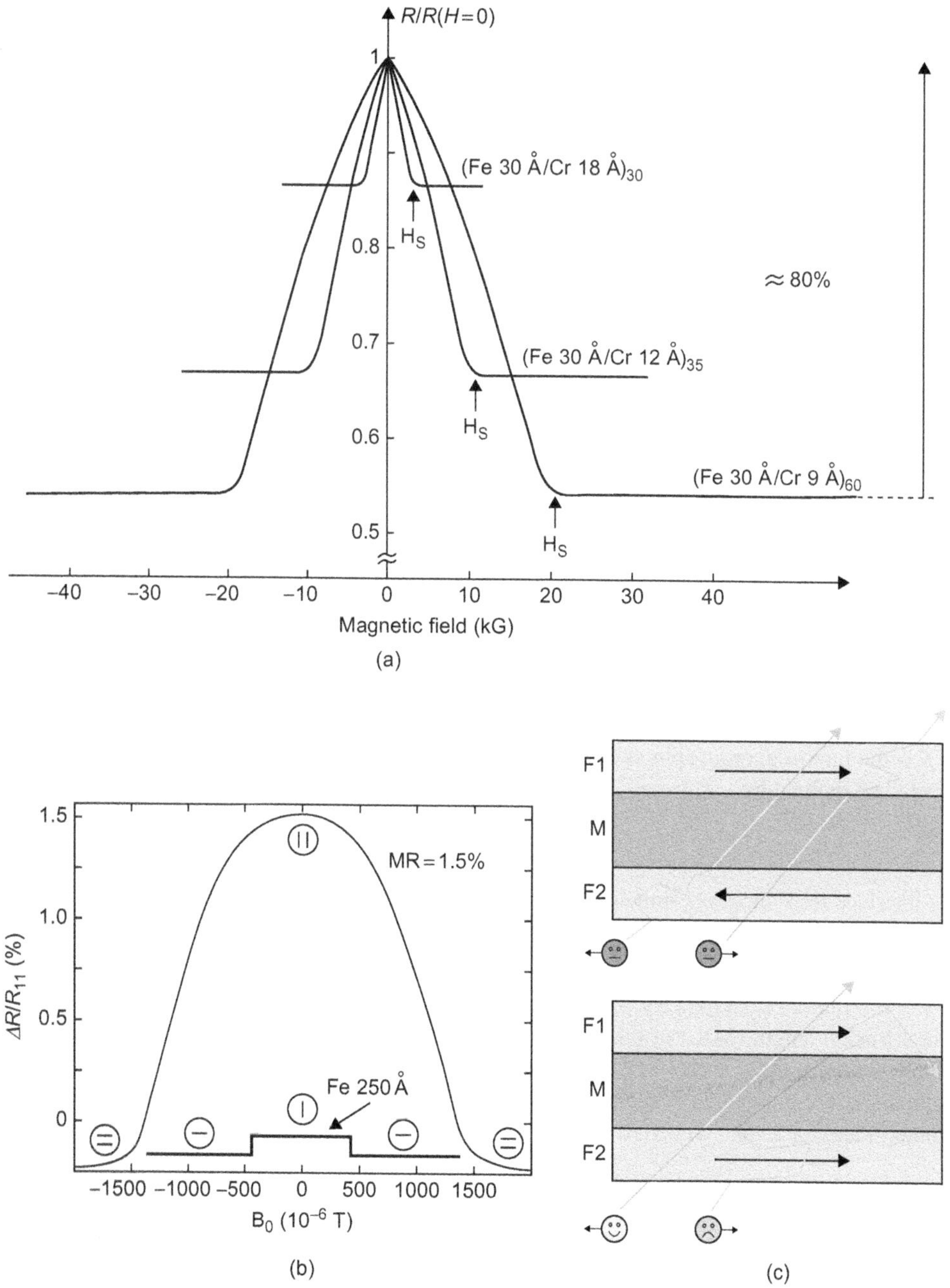

FIGURE 11.9

(a) GMR in Fe/Cr(001) multilayers. MR = 85% for the (Fe 3 nm/Cr 0.9 nm) multilayer. (b) GMR in Fe/Cr/Fe trilayers. (c) Schematic of the mechanism of GMR.

Reproduced from Fert[7] with the permission of American Physical Society.

pores of membranes. These nanowires usually have a diameter in the range 30–500 nm for a length of 10 μm. The multilayered nanowires can be composed of a stack of layers of different metals with thicknesses in the nanometer range.

The general technique for fabrication of magnetic multilayered nanowires in nanoporous polymer membranes (Figure 11.10b) consists of a pulse-plating method in which two metals are deposited from a single solution by switching between the deposition potential of two constituents. These types of multilayers include Co/Cu, NiFe/Cu, Ni/Cu, and Fe/Cu. Other types of multilayered nanowires such as Ni/NiO/Co heterostructures and Co/Pb multilayers have been grown by electrodeposition. We will now summarize the theory of the perpendicular magnetoresistance in magnetic multilayers (CPP-MR) derived by Valet and Fert[22] (the V–F model) and in magnetic nanowires by Fert and Piraux.[9]

11.6.2 Theory of CPP-GMR of Multilayered Nanowires

The V–F theory uses the Boltzmann equation to calculate the transport properties of magnetic multilayers for currents perpendicular to the layer. Their model takes into account both volume and interface scattering and includes a spin-lattice relaxation term τ_{sf}, which describes the relaxation of spin accumulation by spin-flip scattering. The data on the perpendicular magnetoresistance can be used to separate the volume and interface scattering. The notations generally used in the calculation of the magnetoresistance of a multilayer are as follows. The spin ↑ (majority) and spin ↓ (minority) resistivities of the ferromagnetic metal are defined by

$$\rho^F_{\uparrow(\downarrow)} = 2\rho^*_F[1-(+\beta)], \tag{11.35}$$

where β is the bulk-scattering spin asymmetry coefficient. The (equal) spin resistivities of the nonmagnetic metal are defined by

$$\rho^N_{\uparrow(\downarrow)} = 2\rho^*_N, \tag{11.36}$$

and the interface resistances per unit area, $r_\uparrow$ and $r_\downarrow$, are defined by

$$r_{\uparrow(\downarrow)} = 2r^*_b[1-(+)\gamma], \tag{11.37}$$

where γ is the interface asymmetry coefficient, t_N and t_f are defined as the thickness, and l^N_{sf} and l^F_{sf} as the spin diffusion length (SDL) of the nonmagnetic and ferromagnetic layers, respectively. In the long SDL limit, where $t_N \ll l^N_{sf}$ and $t_F \ll l^F_{sf}$, Valet and Fert[22] showed that the resistances R^P (parallel) and R^{AP} (antiparallel) of a unit area of the multilayer (which is also valid for nanowires for a true antiparallel configuration and for a state with zero net magnetization in a volume of the cube of the SDL) are

$$R^{AP} = N(\rho^*_F t_F + \rho^*_N t_N + 2r^*_b) \tag{11.38}$$

and

$$R^P = R^{AP} - \frac{\{\beta\rho^*_F t_F + 2\gamma r^*_b\}^2 N2}{R^{AP}}, \tag{11.39}$$

where N is the number of periods.

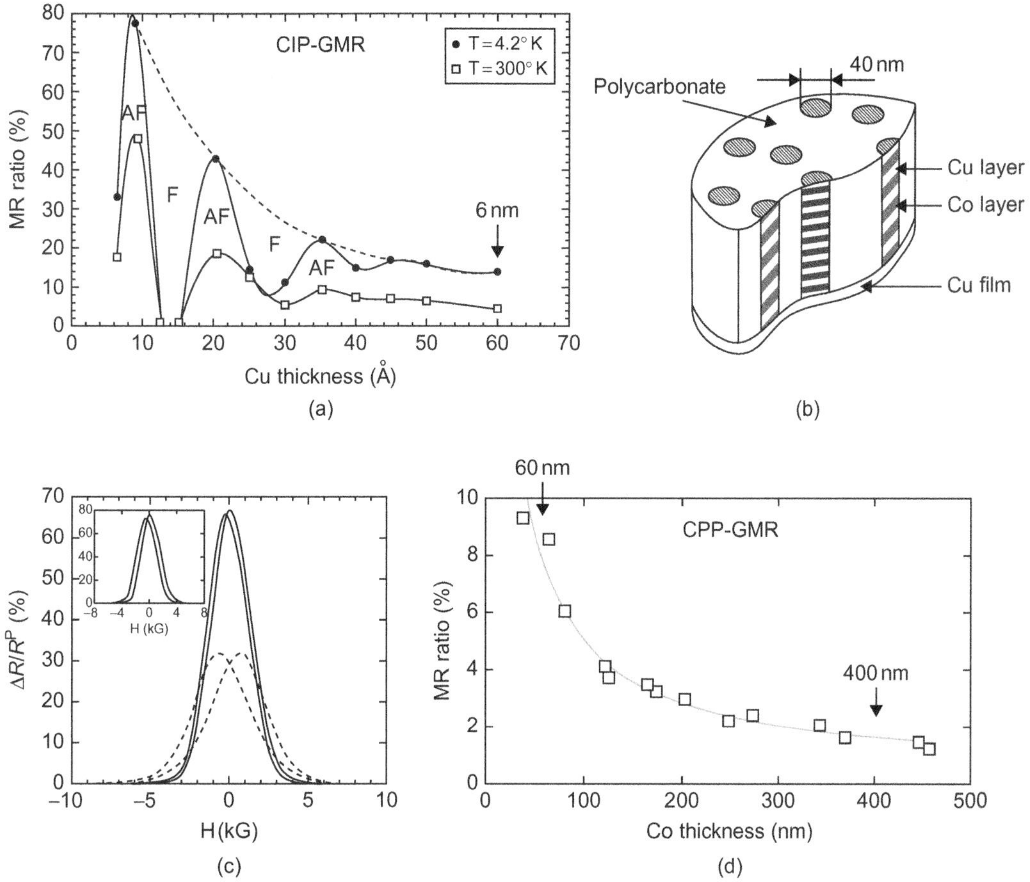

FIGURE 11.10

(a) Variation of the GMR ratio of Co/Cu multilayers as a thickness of Cu in the CIP geometry. (b) Schematic of an array of multilayered nanowires in nanoporous track-etched polymer membranes. (c) The variation of the CPP-GMR ($\Delta R/R^P$) at 77° K for $Ni_{80}Fe_{20}$(12 nm)/Cu(4 nm) (solid lines) and Co(10 nm)/Cu(5 nm) (dotted lines) multilayered nanowires as a function of H. The inset shows the same plot at 4.2° K for the $Ni_{80}Fe_{20}$/Cu sample. (d) Variation of the CPP-GMR ratio of Co/Cu multilayered nanowires as a thickness of the Co layers.

Reproduced from Fert[7] with the permission of the American Physical Society.

From Eqs. (11.38) and (11.39), it can be shown that (Problem 11.5)

$$\left(\frac{R^{AP}-R^{P}}{R^{AP}}\right)^{-1/2} = \frac{\rho_F^* t_F + 2r_b^*}{\beta\rho_F^* t_F + 2\gamma r_b^*} + \frac{\rho_N^* t_N}{\beta\rho_F^* t_F + 2\gamma r_b^*}. \tag{11.40}$$

The magnetoresistance properties of multilayered nanowires helps considerably in the understanding of the CPP-GMR and the determination of the spin diffusion length. In addition, in current-induced switching (or reversal) of magnetization, nanowires can be used. They are ideal for high-injection density currents in order to probe the change in spin configuration of multilayers.

In Figure 11.10a, the large and oscillatory GMR effects in Co/Cu, which were discovered simultaneously by Mosca et al.[13] and Parkin et al.[16], are shown. In fact, these effects became the archetypical GMR system. In 1991, Dieny et al.[5] observed the GMR in spin valves, i.e., trilayered structures in which the magnetization of one magnetic layer is pinned by coupling with the antiferromagnetic layer while the magnetization by the second layer is free. A small magnetic field can reverse the magnetization of the free layer; this concept is now used in most applications in spin valves.

Figure 11.10b shows a schematic picture of an array of multilayered nanowires in nanoporous track-etched polymer membranes and the technique of fabrication, the theory of which was described earlier. Due to the spin accumulation effects that occur in the CPP geometry, the length scale of the spin transport becomes the long spin diffusion length. In fact, the CPP-GMR has demonstrated the spin accumulation effects that determine the propagation of a spin-polarized current through magnetic and nonmagnetic materials. The CPP-GMR plays an important role in all recent developments of spintronics.

As one can see in Figures 11.10c and d, in the CPP, the GMR is not only higher than in CIP, but also subsists in multilayers with relatively thick layers, up to the micron range. The CPP-GMR has demonstrated the spin accumulation effects that govern the propagation of a spin-polarized current through a series of magnetic and nonmagnetic materials. Thus, it plays an important role in the development and future use of spintronics. The key mechanism driving a spin-polarized current at a large distance from the interface is the diffusion current induced by the accumulation of spins at the magnetic–nonmagnetic interface. For example, spin-polarized currents can be transported in long carbon nanotubes because the SDL is quite large beyond the micron range.

When an electron flux crosses the interface between a ferromagnetic and a nonmagnetic material, far from the magnetic side, the current is large in one of the spin channels. However, when the flux is far from the interface on the other side, the current is equally distributed in both channels. This scenario is shown in Figure 11.11.

Figure 11.11a shows the spin-up and spin-down currents far from an interface between ferromagnetic and nonmagnetic conductors outside the spin accumulation zone. Figure 11.11b shows the splitting of the chemical potentials $E_{F\uparrow}$ and $E_{F\downarrow}$ at the interface. When the current travels through the spin-accumulation zone, it is polarized due to the inversion of the spin accumulation and opposite spin flips. The spin flips control the gradual depolarization of the current due to the left and the right. Figure 11.11c shows the variation of the current spin polarization on both sides of metal/metal, where there is a balance between the spin flips on both sides, and the metal/nonmagnetic semiconductor (without spin-dependent interface resistance), where the spin flips dominate in the left side. Thus, the current is nearly depolarized when it enters the semiconductor. One can introduce a tunnel junction, which results in a discontinuity of the spin accumulation at the interface, thereby increasing the depolarization from the metallic to the semiconductor side. However, due to the large tunnel resistances, it is difficult to efficiently transform the spin information into an electrical signal.

11.7 MTJ, TMR, AND MRAM

There have been significant advances in the research on the tunneling magnetoresistance (TMR) of magnetic tunnel junctions (MTJ). The MTJs are tunnel junctions with ferromagnetic electrodes of which the resistances are different for parallel and antiparallel configurations. In addition, the two

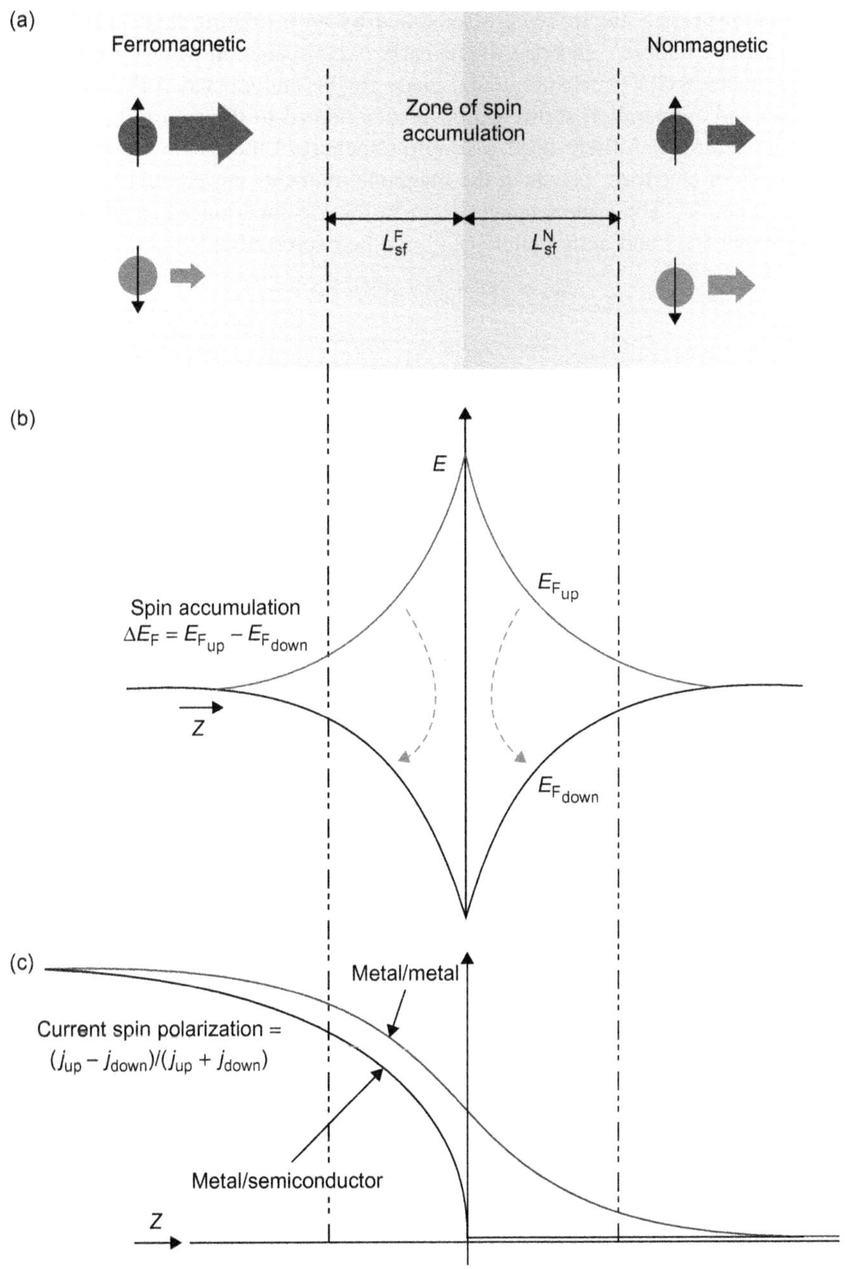

FIGURE 11.11

Schematic representation of the spin accumulation at an interface between a ferromagnetic and a nonmagnetic layer.

Reproduced from Fert[7] with the permission of the American Physical Society.

metallic layers that are called electrodes are separated by an insulating layer thin enough to allow some tunneling current. Julliere[13] did one of the early experiments on MTJ by using the Fe-Ge-Co system. The conductance G(V) measurements, made for Fe and Co when the average magnetizations were parallel and antiparallel, showed a difference related to the spin polarizations of the tunneling conduction electrons. Julliere used a simple (stochastic) model for tunneling electrons and denoted the fractions of electrons, of which the magnetic moments are parallel to the magnetizations in Fe and Co, as a and a'. The conductance of the Fe-Ge-Co junction when the magnetizations in Fe and Co are parallel (G_p) and antiparallel (G_{AP}) can be expressed as[21]

$$G_p \propto [aa' + (1-a)(1-a')] \tag{11.41}$$

and

$$G_{ap} \propto [a(1-a') + a'(1-a)]. \tag{11.42}$$

Assuming that the spin is conserved,

$$\text{TMR} = \frac{G_P - G_{ap}}{G_{ap}} = \frac{2PP'}{(1-PP')}, \tag{11.43}$$

where the conduction electron spin polarization of the two ferromagnetic metals are[13]

$$P = 2a - 1 \quad \text{and} \quad P' = 2a' - 1. \tag{11.44}$$

The original value measured by Julliere was 14%, mainly due to interface roughness. Recent experiments indicate that in MTJs consisting of ferromagnetic amorphous CoFeB and MgO, the TMR can be as large as 500%. In fact, it appears that MgO is crucial for achieving large TMR values.

The polarization of tunneling electrons depends on the barrier height of the insulator in addition to the polarization of the ferromagnets. Slonczewski[19] made the simple assumption that the ferromagnets had parabolic but spin-split bands that are separated by an insulating barrier. The schematic picture is shown in Figure 11.12. A schematic picture of the one- and two-band parabolic models is shown in Figure 11.13.

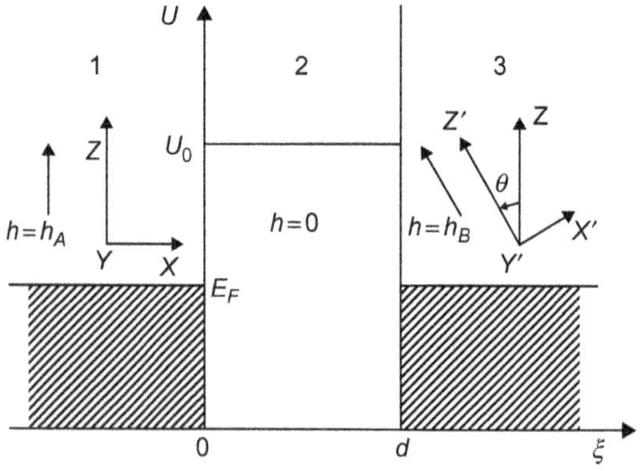

FIGURE 11.12

Schematic picture of an insulator between two different types of ferromagnets.

Reproduced from Slonczewski[19] with the permission of the American Physical Society.

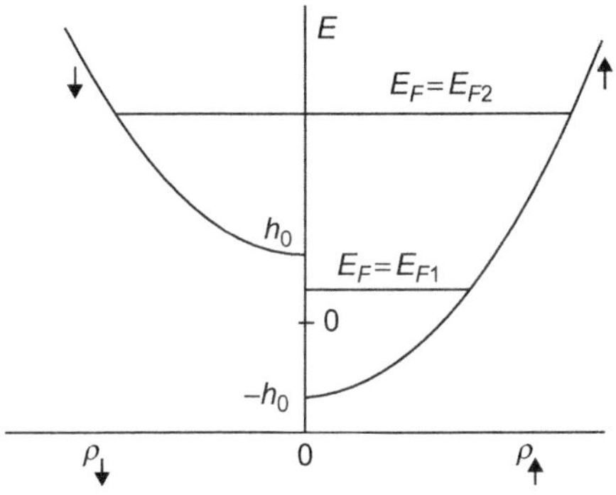

FIGURE 11.13

Density of spin-up ($\rho_\uparrow$) and spin-down ($\rho_\downarrow$) electrons. The positions of the Fermi energy for one-band (E_{F1}) and two-band (E_{F2}) models of a ferromagnet are shown schematically.

Reproduced from Slonczewski[19] with the permission of the American Physical Society.

When the barrier has low energy and thin size, the wave function of the two ferromagnets has to be matched at the interfaces. Slonczewski[19] showed that P', the polarization of the tunneling electrons, depends on the polarization of the ferromagnets as well as on the barrier height,

$$P' = \left[\frac{k^\uparrow - k^\downarrow}{k^\uparrow + k^\downarrow}\right]\left[\frac{\kappa^2 - k^\uparrow k^\downarrow}{\kappa^2 + k^\uparrow k^\downarrow}\right], \tag{11.45}$$

where $k^\uparrow$ and $k^\downarrow$ are the Fermi wave vectors for the up- and down-spin bands, and

$$\hbar\kappa = \sqrt{[2m(V_b - E_F]}, \tag{11.46}$$

where V_b is the barrier height, and E_F is the Fermi level of the ferromagnet. This barrier-dependent factor can vary from -1 to 1. It may be noted that for parabolic, free electron bands, the polarization associated with the ferromagnets is

$$P = \left[\frac{k^\uparrow - k^\downarrow}{k^\uparrow + k^\downarrow}\right]\left[\frac{\rho^\uparrow - \rho^\downarrow}{\rho^\uparrow + \rho^\downarrow}\right]. \tag{11.47}$$

Thus, Eqs. (11.46) and (11.47) change Julliere's result obtained in Eq. (11.44), especially when the tunnel barrier is high. The signs of the polarization can even change in certain cases.

The magnetic random access memory (MRAM) is built from the concept of MTJs, as shown in Figure 11.14.

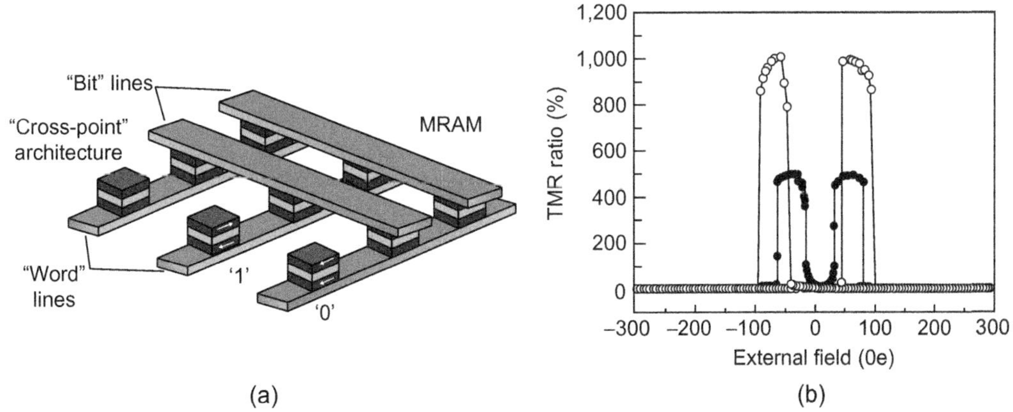

FIGURE 11.14

Principles of the magnetic random access memory (MRAM) in the basic "cross-point" structure.

Reproduced from Fert[7] with the permission of the American Physical Society.

In Figure 11.14a, the binary information "0" and "1" is recorded on the two opposite orientations of the magnetization of the MTJs, which are connected to two crossing points of two perpendicular arrays of parallel conducting lines. The current pulses are sent through one line of each array for "writing," and only at the crossing point, the resulting magnetic field is high enough to orient the magnetization of the free layer. The resistance between the two lines between the addressed cell is measured for "reading." Figure 11.14b shows the TMR = $(R_{max} - R_{min})/R_{min}$ for the stack $(Co_{25}Fe_{75})_{80}B_{20}$(4 nm)/MgO(2.1 nm)/$(Co_{25}Fe_{75})_{80}B_{20}$(4.3 nm) annealed at 475°C after growth. The measurements were done at room temperature (closed circles) and low temperatures (open circles). In the first MRAMs, the memory cells are MTJs with an alumina barrier. The "word" and "bit" lines generate magnetic fields that switch the magnetic configuration. The future of the technology of computers is based on the ST-RAM, which is based on MgO tunnel junctions and switching by spin transfer.

11.8 SPIN TRANSFER TORQUES AND MAGNETIC SWITCHING

A spin current, $\overleftrightarrow{\mathbf{Q}}$, which consists of moving spins, can be written as

$$\overleftrightarrow{\mathbf{Q}} = (\hbar/2)P\hat{\mathbf{s}} \otimes \mathbf{j}, \tag{11.48}$$

where P is the polarization (scalar). The spin and current densities can be written as

$$\vec{\mathbf{s}}(\mathbf{r}) = \sum_{i\sigma\sigma'} \psi^*_{i\sigma}(\mathbf{r})\hat{\mathbf{S}}_{\sigma,\sigma'}\psi^*_{i\sigma'}(\mathbf{r}) \tag{11.49}$$

and

$$\overleftrightarrow{\mathbf{Q}}(\mathbf{r}) = \sum_{i\sigma\sigma'} \mathrm{Re}(\psi^*_{i\sigma}(\mathbf{r})\hat{\mathbf{S}}_{\sigma,\sigma'} \otimes \hat{\mathbf{v}}\psi^*_{i\sigma}(\mathbf{r})), \tag{11.50}$$

where $\hat{\mathbf{S}}_{\sigma,\sigma'}$ and $\hat{\mathbf{v}}$ are spin and velocity operators. The continuity equation, which expresses the conservation of number of electrons, is given by

$$\frac{\partial n}{\partial t} = -\nabla \cdot \mathbf{j}, \tag{11.51}$$

where n is the number density, and $\mathbf{j}$ is the current density. It can be shown that there is a similar equation for the spin density, $\mathbf{s}$, which consists of an extra term arising due to the noncommutivity of spin density with magnetocrystalline anisotropy,

$$\frac{\partial \mathbf{s}}{\partial t} = -\nabla \cdot \overleftrightarrow{\mathbf{Q}} + \mathbf{n}_{ext}, \tag{11.52}$$

where $\nabla \cdot \overleftrightarrow{\mathbf{Q}} = \partial_k Q_{ik}$. Here, $\mathbf{n}_{ext}$ is the external torque density that rotates the spins. Eq. (11.52) is equivalent to the continuity equation stated in Eq. (11.51) with additional terms $\mathbf{n}_{ext}$. In general, there are two contributions to the spin current. If the direction of magnetization is nonuniform in a ferromagnet, the left and right spin currents do not cancel, and the gradient of this current gives rise to a spin torque $\mathbf{n}_{ex} = -\nabla \cdot \overleftrightarrow{\mathbf{Q}}_{ex}$. There is also a second contribution to the spin current at the interfaces due to the imbalance in the population of the spin states at or near the Fermi energy. It can be

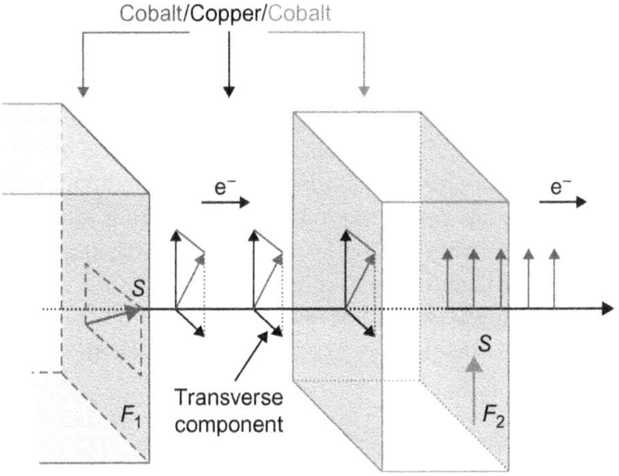

FIGURE 11.15

Illustration of the spin-transfer concept. The spin-transfer torque described previously acts on F_2.

Reproduced from Fert[7] with the permission of the American Physical Society.

shown that the spin current along the direction of the magnetization is conserved. Further, the reflected and transmitted components do not (generally) have any transverse components. Therefore, the main mechanism for transfer of angular momentum is the exchange interaction experienced by the electrons in the ferromagnet, which exerts a torque on the spin current. The scattering of spin at a ferromagnetic interface is a growing area of research, and we have considered only an elementary discussion of this topic.

To summarize, in spin-transfer phenomena, the magnetic moments of a ferromagnet are manipulated by the transfer of spin-angular momentum from a spin-polarized current without the application of a magnetic field. The transfer of a transverse spin current to a "free" magnetic layer F_2 is due to the torque acting on the magnetic moment, as briefly described previously. The spin torque can induce irreversible switching of the magnetic moment or can generate precession of the moment in the microwave frequency range in the presence of a magnetic field in a second regime.

This scenario is schematically shown in Figure 11.15. The first magnetic layer, F, prepares a spin-polarized current with an obliquely oriented spin polarization with respect to the magnetization axis of a second layer, F_2. When this current goes through F_2, the exchange interaction aligns its spin polarization along the magnetization axis. Because the exchange interaction is spin conserving, the transverse spin polarization lost by the current is transferred to the total spin of F_2. This can lead to a magnetic switching of the F_2 layer or to magnetic oscillations in the microwave frequency range.

11.9 SPINTRONICS WITH SEMICONDUCTORS

11.9.1 Introduction

Spintronics with semiconductors has the tremendous advantage of combining the potential of the magnetic materials such as the control of current by spin manipulation, nonvolatility, etc., with the potential of the semiconductors (control of current by gate, coupling with optics, etc.). Datta and Das[4] proposed the concept of a spin-effect transistor (spin FETs) based on spin transport in semiconductor lateral channels between spin-polarized sources and drains with control of the spin transmission by a field-effective gate. The spin precession can be controlled by spin-orbit coupling. The three ingredients for a spin transistor are (1) long relaxation time of a semiconductor, (2) gate voltage control of the spin-orbit coupling, and (3) high spin injection coefficients. Optical experiments have established that electron spins of semiconductors have long relaxation time. Modulation of the spin-orbit splitting at the Fermi level by gate voltage has been reported for both electrons and holes for a variety of semiconductors. However, the spin injection from a ferromagnetic (FM) source into a semiconductor is a very difficult task and yields a maximum of 1% spin polarization.

The conductivity mismatch between an FM metal emitter and a semiconductor is the primary reason for the difficulty in spin injection. It has been shown that if we define the spin injection coefficient in a diffusive regime, γ, as

$$\gamma \propto \sigma_N / \sigma_F, \tag{11.53}$$

where σ_N and σ_F are the conductivities of a normal (N) and FM contacts, then

$$\sigma_N / \sigma_F \gg 1 \tag{11.54}$$

if N is a paramagnetic metal, and

$$\sigma_N / \sigma_F \ll 1 \tag{11.55}$$

if N is a semiconductor. This explains why the spin injection from an FM source into a paramagnetic metal is very efficient, whereas the same from an FM source to a semiconductor is practically impossible. Rashba[18] showed that tunnel contacts (T) can significantly increase spin injection and solve the problem of the mismatch between the conductivities of a ferromagnetic metal (FM) and a semiconductor (N) microstructure. The tunnel resistance, r_c, should be larger than the competing "effective resistances" making the total contact resistance,

$$r_c \geq L_F / \sigma_F, \quad \min\{L_N, w\} / \sigma_N, \tag{11.56}$$

where L_F and L_N are spin diffusion lengths in the FM and N conductors and w is the width of the N conductor.

11.9.2 Theory of an FM-T-N Junction

Rashba considered a semi-infinite FM ($x < 0$) and N($x > 0$) and assumed that the T contact, at $x = 0$, has different spin conductivities, $\Sigma_\uparrow$ and $\Sigma_\downarrow$, for up- and down-spins, and there is no spin relaxation in it. The currents $j_{\uparrow,\downarrow}(x)$ carried by up- and down-spins can be written as

$$j_{\uparrow,\downarrow}(x) = \sigma_{\uparrow,\downarrow}\, \xi'_{\uparrow,\downarrow}(x), \tag{11.57}$$

where $\xi'_{\uparrow,\downarrow}(x)$ is the space derivative of the electrochemical potentials $\xi_{\uparrow,\downarrow}(x)$, which are related to the nonequilibrium parts $n_{\uparrow,\downarrow}(x)$ of the electron concentrations and $\varphi_F(x)$ in the FM region,

$$\xi_{\uparrow,\downarrow}(x) = (eD_{\uparrow,\downarrow} / \sigma_{\uparrow,\downarrow}) n_{\uparrow,\downarrow}(x) - \varphi_F(x). \tag{11.58}$$

Here, $D_{\uparrow,\downarrow}$ are diffusion coefficients, and $\sigma_{\uparrow,\downarrow}$ are conductivities of up- and down-spin electrons. To maintain the charge neutrality under the spin injection conditions,

$$n_{\uparrow}(x) + n_{\downarrow}(x) = 0. \tag{11.59}$$

The continuity equation is

$$j'_{\uparrow}(x) = en_{\uparrow}(x)/\tau_s^F, \tag{11.60}$$

where τ_s^F is the spin-relaxation time. Because the charge is conserved,

$$J = j_{\uparrow}(x) + j_{\downarrow}(x) = \text{constant}. \tag{11.61}$$

Introducing the notations

$$\xi_F(x) = \xi_{\uparrow}(x) - \xi_{\downarrow}(x) \tag{11.62}$$

and

$$j_F(x) = j_{\uparrow}(x) - j_{\downarrow}(x), \tag{11.63}$$

one can show (Problem 11.7) that the diffusion equation can be written as

$$D_F \xi''_F(x) = \xi_F(x)/\tau_s^F, \tag{11.64}$$

where

$$D_F = (\sigma_{\downarrow} D_{\uparrow} + \sigma_{\uparrow} D_{\downarrow})/\sigma_F \tag{11.65}$$

and

$$\sigma_F = \sigma_{\uparrow} + \sigma_{\downarrow}. \tag{11.66}$$

From Eqs. (11.57) through (11.66), it can be shown that (Problem 11.8)

$$\varphi'_F(x) = [(D_{\uparrow} - D_{\downarrow})/D_F](\sigma_{\uparrow}\sigma_{\downarrow}/\sigma_F^2)\xi'_F(x) - J/\sigma_F. \tag{11.67}$$

At $T = 0$, the Einstein relations are

$$e^2 D_{\uparrow,\downarrow} = \sigma_{\uparrow,\downarrow}/\rho_{\uparrow,\downarrow}, \tag{11.68}$$

where $\rho_{\uparrow,\downarrow}$ are the densities of states at the Fermi level. One can show (Problem 11.9) from Eqs. (11.65) through (11.68) that

$$e^2 D_F = (\sigma_{\uparrow}\sigma_{\downarrow}/\sigma_F)(\rho_F/\rho_{\uparrow}\rho_{\downarrow}) \tag{11.69}$$

and

$$(\rho_{\downarrow}\sigma_{\uparrow} - \rho_{\uparrow}\sigma_{\downarrow})/\rho_F\sigma_F = [(\Delta\sigma/\sigma_F) - (\Delta\rho/\rho_F)]/2. \tag{11.70}$$

It can be easily shown (Problem 11.10) from Eqs. (11.67), (11.69), and (11.70) that

$$\varphi'_F(x) = [(\Delta\sigma/\sigma_F) - (\Delta\rho/\rho_F)]\xi'_F(x)/2 - J/\sigma_F. \tag{11.71}$$

From Eqs. (11.57) and (11.71), it is easy to show that

$$j_F(x) = 2(\sigma_\uparrow\sigma_\downarrow/\sigma_F)\xi'_F(x) + (\Delta\sigma/\sigma_F)J. \tag{11.72}$$

Eqs. (11.64), (11.65), (11.71), and (11.72) are a complete system of bulk equations for the F region. One can also show (Problem 11.11) that

$$\xi_\uparrow(x) + \xi_\downarrow(x) = -[2\varphi_F(x) + (\Delta\rho/\rho_F)\xi_F(x)] \tag{11.73}$$

and

$$n_\uparrow(x) = (\rho_\uparrow\rho_\downarrow/\rho_F)\xi_F(x). \tag{11.74}$$

The equations for the N region can be obtained from the equations for the F region by substituting the following in Eqs. (11.64), (11.67), and (11.72):

$$\sigma_\uparrow = \sigma_\downarrow = \sigma_N/2, \tag{11.75}$$

$$\Delta\rho = \Delta\sigma = 0, \tag{11.76}$$

and

$$D_N = D_\uparrow = D_\downarrow. \tag{11.77}$$

We obtain

$$\begin{aligned} D_N\xi_N(x) &= \xi_N(x)/\tau_s^N, \\ \varphi'_N(x) &= -J/\sigma_N, \end{aligned} \tag{11.78}$$

and

$$j_N(x) = \sigma_N\xi'_N(x)/2. \tag{11.79}$$

In Eqs. (11.77) through (11.79), the symbol N in the prefix has been used for the N region, instead of the prefix F, which was used in the F region in the previous equations. Because there is no spin relaxation at the interface $x=0$, the boundary conditions are $j_\uparrow(x)$ is continuous at $x=0$ and hence $j_F(0) = j_N(0)$. Substituting these in Eqs. (11.72) and (11.79), we obtain

$$\sigma_N\xi'_N(0) - 4(\sigma_\uparrow\sigma_\downarrow/\sigma_F)\xi'_F(0) = 2(\Delta\sigma/\sigma_F)J. \tag{11.80}$$

The currents $j_{\uparrow,\downarrow}(0)$ are related to the conductivities of the T contact,

$$j_{\uparrow,\downarrow}(0) = \Sigma_{\uparrow,\downarrow}(\xi^N_{\uparrow,\downarrow} - \xi_{\uparrow,\downarrow}). \tag{11.81}$$

Using Eq. (11.62) and its equivalent for ξ_N, we can rewrite Eq. (11.81) as

$$\xi_N(0) - \xi_F(0) = -2(\Delta\Sigma/\Sigma)r_cJ + 2r_cj(0), \tag{11.82}$$

where

$$\Delta\Sigma = \Sigma_\uparrow - \Sigma_\downarrow,\ \Sigma = \Sigma_\uparrow + \Sigma_\downarrow \text{ and } r_c = \Sigma/4\Sigma_\uparrow\Sigma_\downarrow. \tag{11.83}$$

From Eq. (11.73), its equivalent for φ_N and Eq. (11.83), we obtain

$$(\varphi_F(0) - \varphi_N(0)) + \frac{\Delta\rho}{2\rho_F}\xi_F(0) = r_c J - \frac{\Delta\Sigma}{\Sigma} r_c j(0). \tag{11.84}$$

Eq. (11.84) implies that even when $r_c = 0$, because $\Delta\rho \neq 0$, there is a finite potential drop at the interface, $[\varphi_F(0) - \varphi_N(0)] \propto J$.

11.9.3 Injection Coefficient

The solutions of Eqs. (11.64) and (11.78) are

$$\xi_F(x) = Ae^{-L_F x} \tag{11.85}$$

and

$$\xi_N(x) = Be^{-L_N x}, \tag{11.86}$$

where

$$L_F = (D_F \tau_s^F)^{1/2} \tag{11.87}$$

and

$$L_N = (D_F \tau_s^N)^{1/2}. \tag{11.88}$$

Here, L_F and L_N are known as the diffusion lengths. Thus, we obtain

$$\xi_N'(0) = -\xi_N(0)/L_N = 2\gamma J/\sigma_N \tag{11.89}$$

and

$$\xi_F'(0) = \xi_F(0)/L_F. \tag{11.90}$$

The injection coefficient is defined as

$$\gamma = j(0)/J. \tag{11.91}$$

If we eliminate $\xi_F(0)$ from Eqs. (11.80) and (11.82), we obtain (Problem 11.14)

$$\gamma = [r_F(\Delta\sigma/\sigma_F) + r_c(\Delta\Sigma/\Sigma)]/r_{FN}. \tag{11.92}$$

Here,

$$r_{FN} = r_F + r_N + r_c, \tag{11.93}$$

$$r_F = L_F \sigma_F / 4\sigma_\uparrow \sigma_\downarrow, \tag{11.94}$$

and

$$r_N = L_N/\sigma_N. \tag{11.95}$$

Eq. (11.93) shows that r_c, r_F, and r_N are connected in series. We note that if $r_F = r_N$, $\gamma \sim 1$ if and only if $r_c \geq r_N$, a criterion that can be satisfied for narrow tunnel junctions of the atomic scale. If $r_c \gg r_N$, r_F, $r_{FN} \sim r_c$ from Eq. (11.93). Using this approximation, we obtain (from Eq. 11.92) the injection coefficient $\gamma \approx \Delta\Sigma/\Sigma$. The contact completely determines γ in this regime. Thus, the spin injection coefficient is controlled by the element of an FM-T-N junction having the largest effective resistance.

There are alternate possibilities of spintronics with semiconductors based on the use of ferromagnetic semiconductors such as $Ga_{1-x}Mn_xAs$, where $x \ll 1$. There is a good possibility of controlling the ferromagnetic properties with a gate voltage as well as having large TMR effects.

An alternate approach is the spin accumulation effect due to spin-orbit coupling or anomalous scattering mechanisms. When a spin-unpolarized current flows in a metal, the spin-orbit interaction produces asymmetric scattering of the conduction electrons so that spin-up electrons have a larger probability to be scattered to the right compared to spin-down electrons, and spin-down electrons would tend to scatter to the left more than spin-up electrons. This results in a spin current that is generated in a direction transverse to the direction of the flow of current. If a spin-polarized current is present in a semiconductor, a Hall-like effect can be induced by the spin-orbit coupling without an external field. Zhang[23] derived an expression for the spin Hall effect (SHE) using a semiclassical Boltzmann equation and extending it to the case where the spin diffusion effect is finite. He showed that when the

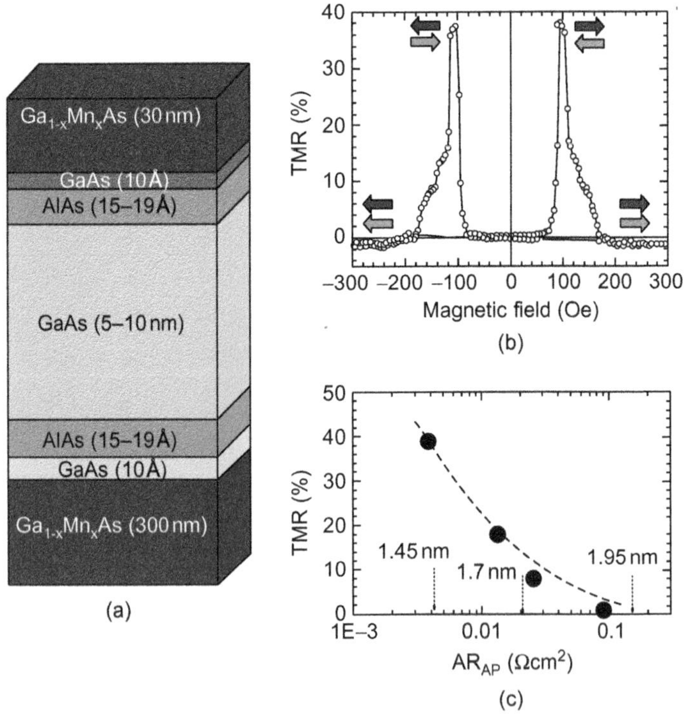

FIGURE 11.16

Spintronics of semiconductors illustrated by experimental results.

Reproduced from Fert[7] with the permission of the American Physical Society.

formulation is applied to certain metals and semiconductors, the magnitude of the spin Hall voltage is much larger than that of magnetic multilayers. Because SHE is also found in nonmagnetic metals, further research in this area is very active.

Spintronics with semiconductors is shown in Figure 11.16. The structure, shown in Figure 11.16a, is composed of a GaAs layer that is separated from the GaMnAs source and drain by tunnel barriers of AlAs. Figure 11.16b shows the MR curve at 4.2° K, which shows a difference in resistance of 40% between the parallel and antiparallel magnetic configuration between the source and the drain. Figure 11.16c shows the MR ratio as a function of the resistance of the tunnel barriers.

It may be noted that spintronics is a very rapidly growing area of research. Recently, large GMR- and TMR-like effects were predicted in carbon-based molecules. In fact, due to small spin-orbit coupling, carbon molecules have a long spin lifetime. Recent experiments on carbon nanotubes between a ferromagnetic source and drain made of the metallic manganite $La_{2/3}Sr_{1/3}MnO_3$ (LSMO) have been encouraging (see Figure 11.17).

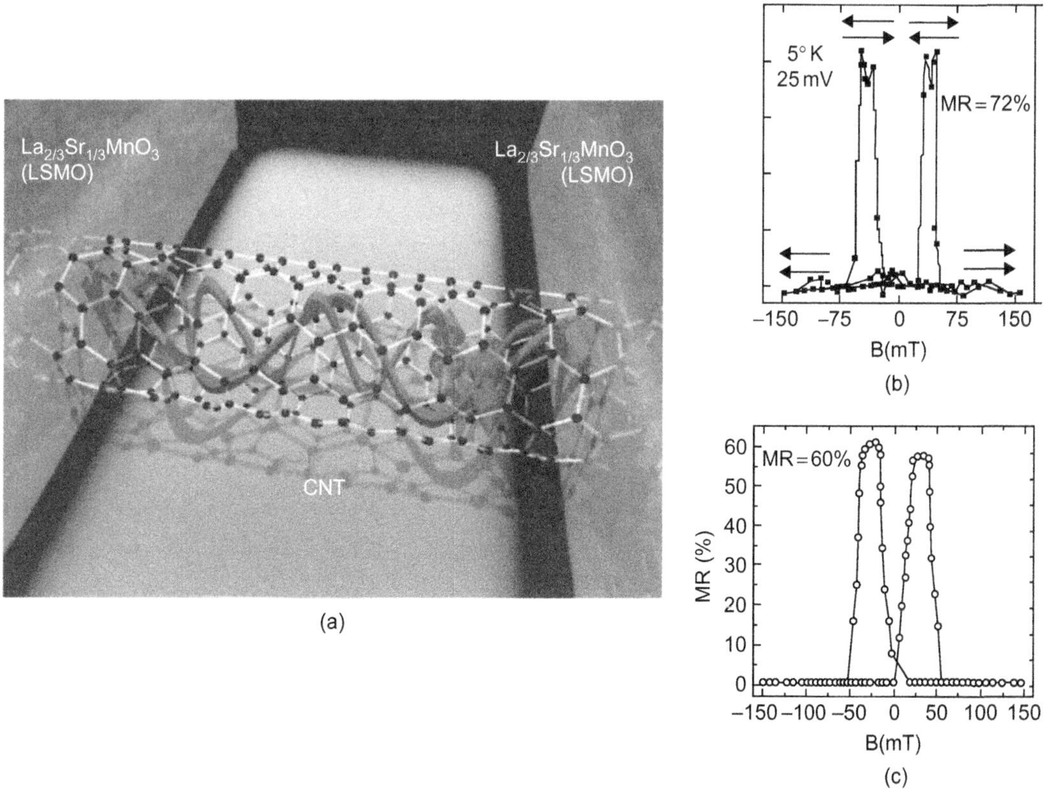

FIGURE 11.17

(a) Sketch of spintronics with LSMO molecules. Magnetoresistance experimental results of Heuso et al.[12] at 4.2°K on carbon nanotubes between electrodes made of LSMO are shown in (b) and (c). A contrast of 72% and 60% is obtained for the parallel (high field) and antiparallel (peaks) magnetic configurations of the source and drain.

Reproduced from Fert[7] with the permission of the American Physical Society.

The results indicate that the resistances of the parallel and antiparallel configurations are between 60% and 70%, which is larger than that obtained for semiconductor channels. In addition, their high Fermi velocity, which is responsible for short dwell time, is an advantage over semiconductors. The research is very active in the general area of spintronics and, in particular, on graphene-based devices.

In view of the above, any review of the general area of spintronics would very quickly become outdated. Therefore, this chapter may be considered as an introduction to the topic.

PROBLEMS

11.1. By using Mott's two-current model, in which the spin transport in a metal is perceived as being due to two independent spin channels, and by using Eqs. (11.4) through (11.7), show that the energy-dependent relaxation times can be expressed as

$$\frac{1}{\tau_1} = A(T/M_i\theta^2)\left[\frac{\sqrt{(E_1 - E)}}{\sqrt{(E_0 - \varsigma_0)}} + \alpha\right] \qquad (1)$$

and

$$\frac{1}{\tau_2} = A(T/M_i\theta^2)\left[\frac{\sqrt{(E_2 - E)}}{\sqrt{(E_0 - \varsigma_0)}} + \alpha\right], \qquad (2)$$

where $\alpha \approx 1/4$ and θ is the Debye temperature.

11.2. Show that

$$\sqrt{\frac{E_1 - \zeta_0'}{E_0 - \zeta_0}} = (1 - \beta)^{1/3} \qquad (1)$$

and

$$\sqrt{\frac{E_2 - \zeta_0'}{E_0 - \zeta_0}} = (1 + \beta)^{1/3}, \qquad (2)$$

where ζ_0' represents the energy of the highest occupied state (Fermi energy) at $T = 0$ when the states are split and ζ_0 when $\beta = 0$.

11.3. By using Eqs. (11.8) through (11.12), derive the equation for resistivity (Eq. (11.13),

$$\rho(\beta, T) = A(T/m\theta^2)\left[\frac{1}{(1-\beta)^{1/3} + \alpha} + \frac{1}{(1+\beta)^{1/3} + \alpha}\right]^{-1}. \qquad (1)$$

11.4. From Eqs. (11.19) through (11.21), show that

$$g_{A\pm\uparrow}(v_z, z) = \frac{eE\tau}{m}\frac{\partial f_0}{\partial v_x}\left[1 + A_{\pm\uparrow}\exp\left(\frac{\mp z}{\tau|v_z|}\right)\right] \tag{1}$$

and

$$g_{A\pm\downarrow}(v_z, z) = \frac{eE\tau}{m}\frac{\partial f_0}{\partial v_x}\left[1 + A_{\pm\downarrow}\exp\left(\frac{\mp z}{\tau|v_z|}\right)\right]. \tag{2}$$

11.5. We obtained in Eqs. (11.38) and (11.39)

$$R^{AP} = N(\rho_F^* t_F + \rho_N^* t_N + 2r_b^*) \tag{1}$$

and

$$R^P = R^{AP} - \frac{\{\beta\rho_F^* t_F + 2\gamma r_b^*\}^2 N^2}{R^{AP}}. \tag{2}$$

Show that

$$\left(\frac{R^{AP} - R^P}{R^{AP}}\right)^{-1/2} = \frac{\rho_F^* t_F + 2r_b^*}{\beta\rho_F^* t_F + 2\gamma r_b^*} + \frac{\rho_N^* t_N}{\beta\rho_F^* t_F + 2\gamma r_b^*}. \tag{3}$$

11.6. Using a simple model, we have shown in Eqs. (11.41) and (11.42) that

$$G_p \propto [aa' + (1-a)(1-a')] \tag{1}$$

and

$$G_{ap} \propto [a(1-a') + a'(1-a)]. \tag{2}$$

Assuming that the spin is conserved, show that

$$\text{TMR} = \frac{G_P - G_{ap}}{G_{ap}} = \frac{2PP'}{(1-PP')}, \tag{3}$$

where the conduction electron spin polarization of the two ferromagnetic metals is

$$P = 2a - 1 \quad \text{and} \quad P' = 2a' - 1. \tag{4}$$

11.7. Introducing the notations

$$\xi_F(x) = \xi_\uparrow(x) - \xi_\downarrow(x) \tag{1}$$

and

$$j_F(x) = j_\uparrow(x) - j_\downarrow(x), \tag{2}$$

show that the diffusion equation can be rewritten as

$$D_F \xi''_F(x) = \xi_F(x)/\tau_s^F, \tag{3}$$

where

$$D_F = (\sigma_\downarrow D_\uparrow + \sigma_\uparrow D_\downarrow)/\sigma_F \tag{4}$$

and

$$\sigma_F = \sigma_\uparrow + \sigma_\downarrow. \tag{5}$$

11.8. From Eqs. (11.57) through (11.66), show that

$$\varphi'_F(x) = [(D_\uparrow - D_\downarrow)/D_F](\sigma_\uparrow \sigma_\downarrow/\sigma_F^2)\xi'_F(x) - J/\sigma_F. \tag{1}$$

11.9. By using Eqs. (11.65) through (11.68), show that

$$e^2 D_F = (\sigma_\uparrow \sigma_\downarrow/\sigma_F)(\rho_F/\rho_\uparrow \rho_\downarrow) \tag{1}$$

and

$$(\rho_\downarrow \sigma_\uparrow - \rho_\uparrow \sigma_\downarrow)/\rho_F \sigma_F = [(\Delta\sigma/\sigma_F) - (\Delta\rho/\rho_F)]/2, \tag{2}$$

where

$$\Delta\sigma = \sigma_\uparrow - \sigma_\downarrow, \tag{3}$$

$$\Delta\rho = \rho_\uparrow - \rho_\downarrow, \tag{4}$$

and

$$\rho_F = \rho_\uparrow + \rho_\downarrow. \tag{5}$$

11.10. Show from Eqs. (11.67), (11.69), and (11.70) that

$$\varphi'_F(x) = [(\Delta\sigma/\sigma_F) - (\Delta\rho/\rho_F)]\xi'_F(x)/2 - J/\sigma_F. \tag{1}$$

11.11. From Eqs. (11.64), (11.65), (11.71), and (11.72), show that

$$\xi_\uparrow(x) + \xi_\downarrow(x) = -[2\phi_F(x) + (\Delta\rho/\rho_F)\xi_F(x)] \tag{1}$$

and

$$n_\uparrow(x) = (\rho_\uparrow \rho_\downarrow/\rho_F)\xi_F(x). \tag{2}$$

11.12. Using Eq. (11.62) and its equivalent for ξ_N, show that Eq. (11.81) can be rewritten as

$$\xi_N(0) - \xi_F(0) = -2(\Delta\Sigma/\Sigma) r_c J + 2 r_c j(0), \tag{1}$$

where

$$\Delta\Sigma = \Sigma_{\uparrow} - \Sigma_{\downarrow},\ \Sigma = \Sigma_{\uparrow} + \Sigma_{\downarrow} \text{ and } r_c = \Sigma/4\Sigma_{\uparrow}\Sigma_{\downarrow}. \tag{2}$$

Here, r_c is the effective contact resistance.

11.13. From Eq. (11.73), its equivalent for φ_N, and Eq. (11.83), show that

$$(\varphi_F(0) - \varphi_N(0)) + \frac{\Delta\rho}{2\rho_F}\xi_F(0) = r_c J - \frac{\Delta\Sigma}{\Sigma} r_c j(0). \tag{1}$$

11.14. The injection coefficient is defined as

$$\gamma = j(0)/J. \tag{1}$$

If one eliminates $\xi_F(0)$ from Eqs. (11.80) and (11.82), show that

$$\gamma = \left[r_F(\Delta\sigma/\sigma_F) + r_c(\Delta\Sigma/\Sigma)\right]/r_{FN}. \tag{2}$$

Here,

$$r_{FN} = r_F + r_N + r_c, \tag{3}$$

$$r_F = L_F\sigma_F/4\sigma_{\uparrow}\sigma_{\downarrow}, \tag{4}$$

and

$$r_N = L_N/\sigma_N. \tag{5}$$

References

1. Baibich MN, Roole JM, Fert A, Van Dare FN, Petroff F, Etienne P, et al. Giant magnetoresistance of (001)Fe/(001) Cr magnetic superlattices. *Phys Rev Lett* 1988;**61**:2472.
2. Binash G, Grunberg P, Sauerbech F, Zonn W. Enhanced magnetoresistance in layered magnetic structures with antiferromagnetic interlayer exchange. *Phys Rev B* 1989;**39**:4828.
3. Camley RE, Barnas J. Theory of giant magnetoresistance effects in magnetic-layered structures with antiferromagnetic coupling. *Phys Rev Lett* 1989;**63**:662.
4. Datta S, Das B. Electronic analog of the electro-optic modulator. *Appl Phys Lett* 1990;**56**:665.
5. Dieny B, Sperosu VS, Parkin SSP, Gurney BA, Wilhoit DR, Maurri D. Giant magnetoresistance in soft ferromagnetic multilayers. *Phys Rev B* 1991;**43**:1297.
6. Fernando GW. Metallic multilayers and their applications. In: Misra PK, editor. *Handbook of metal physics (series)*. Amsterdam: Elsevier; 2008.
7. Fert A. Nobel lecture: origin, development, and future of spintronics. *Rev Mod Phys* 2008;**80**:1517.
8. Fert A, Valet T, Barnas J. Perpendicular magnetoresistance in magnetic multilayers. Theoretical model and discussions (invited). *J Appl Phys* 1994;**75**:6693.
9. Fert A, Piraux L. Magnetic nanowires. *J Magn Mag Mater* 1999;**200**:338.

10. Fert A, Gruenberg P, Barthelemey A, Petroff F, Zinn W. Layered magnetic structures: interlayer exchange coupling and giant magnetoresistance. *J Magn Mag Mater* 1995;**140–144**:1.
11. Grunberg PA. Nobel lecture: from spin waves to magnetoresistance and beyond. *Rev Mod Phys* 2008;**80**:1531.
12. Heuso LE, Pruneda JM, Ferrari V, Burnell G, Valdes-Herrere JP, Simmons BD, et al. Transformation of spin information into large electrical signals using carbon nanotubes. *Nature (London)* 2007;**445**:410.
13. Julliere M. Tunneling between ferromagnetic films. *Phys Lett A* 1975;**54**:225.
14. Mosca DH, Petroff F, Fert A, Schroeder PA, Pratt Jr. Jr WP, Laloee T. Oscillatory interlayer coupling and giant magnetoresistance in Co/Cu multilayers. *J Mag Mater* 1991;**94**:L1.
15. Mott N. The resistance and thermoelectric properties of the transition metals. *Proc Roy Soc* 1936;**156**:368.
16. Parkin SSP, Bhadra R, Roche KP. Oscillatory magnetic exchange coupling through thin copper layers. *Phys Rev Lett* 1991;**66**:2152.
17. Pratt Jr. Jr WP, Lee SF, Slaughter JM, Laloee R, Schroeder PA, Bass J. Perpendicular giant magnetoresistance of Ag/Co multilayers. *Phys Rev Lett* 1992;**66**:3060.
18. Rashba EI. Theory of electrical spin injection: tunnel contacts as a solution of the conducting mismatch process. *Phys Rev* 2000;**62**:R 16267.
19. Slonczewski JC. Conductance and exchange coupling of two ferromagnets separated by a tunneling barrier. *Phys Rev B* 1989;**39**:6995.
20. Stiles MD. (a) Exchange coupling in magnetic heterostructures. *Phys Rev B* 1993;**48**:7238; (b) Oscillatory exchange coupling in Fe/Cr multilayers. *Phys Rev B* 1996;**54**:14679.
21. Tedrow PM, Meservy R. Spin-dependent tunneling into ferromagnetic Nickel. *Phys Rev Lett* 1971;**26**:192.
22. Valet T, Fert A. Theory of the perpendicular magnetoresistance in magnetic multilayers. *Phys Rev B* 1993;**48**:7099.
23. Zhang S. Spin hall effect on the presence of spin diffusion. *Phys Rev Lett* 2000;**85**:393.

CHAPTER

Diamagnetism and Paramagnetism 12

CHAPTER OUTLINE

12.1 Introduction 370
12.2 Atomic (or Ionic) Magnetic Susceptibilities 371
12.2.1 General Formulation 371
12.2.2 Larmor Diamagnetism 372
12.2.3 Hund's Rules 373
12.2.4 Van Vleck Paramagnetism 374
12.2.5 Landé g Factor 375
12.2.6 Curie's Law 377
12.3 Magnetic Susceptibility of Free Electrons in Metals 378
12.3.1 General Formulation 378
12.3.2 Landau Diamagnetism and Pauli Paramagnetism 380
12.3.3 De Haas–van Alphen Effect 383
12.4 Many-Body Theory of Magnetic Susceptibility of Bloch Electrons in Solids 388
12.4.1 Introduction 388
12.4.2 Equation of Motion in the Bloch Representation 388
12.4.3 Thermodynamic Potential 390
12.4.4 General Formula for χ 390
12.4.5 Exchange Self-Energy in the Band Model 393
12.4.6 Exchange Enhancement of χ_s 394
12.4.7 Exchange and Correlation Effects on χ_o 395
12.4.8 Exchange and Correlation Effects on χ_{so} 396
12.5 Quantum Hall Effect 396
12.5.1 Introduction 396
12.5.2 Two-Dimensional Electron Gas 396
12.5.3 Quantum Transport of a Two-Dimensional Electron Gas in a Strong Magnetic Field 397
12.5.4 Quantum Hall Effect from Gauge Invariance 400
12.6 Fractional Quantum Hall Effect 400
Problems 401
References 407

12.1 INTRODUCTION

The magnetization density of a quantum-mechanical system is defined as

$$M^{\nu} = -\frac{1}{V}\frac{\partial \Omega}{\partial B^{\nu}}, \tag{12.1}$$

where Ω is the thermodynamic potential, V is the volume of the solid, and B^{ν} is the νth component of the magnetic induction **B**, the microscopic field perceived by the nuclei or the electrons due to an external magnetic field **H**. The magnetic susceptibility is defined as

$$\chi^{\mu\nu} = \lim_{B \to 0} \frac{\partial M^{\nu}}{\partial B^{\mu}} = -\frac{1}{V} \lim_{B \to 0} \frac{\partial^2 \Omega}{\partial B^{\mu} \partial B^{\nu}}. \tag{12.2}$$

It may be noted that the magnetic susceptibility is a tensor, whereas the magnetization is a vector. If a system has positive magnetic susceptibility, it is known as paramagnetic, and if it has negative magnetic susceptibility, it is known as diamagnetic. In a linear medium, $\mathbf{B} = \mu \mathbf{H}$, where μ is known as the magnetic permeability. Because an alternate definition of susceptibility is $\chi^{\mu\nu} = \lim_{H \to 0} \frac{\partial M^{\nu}}{\partial H^{\mu}}$, a linear system is paramagnetic if $\mu > 1$ and diamagnetic if $\mu < 1$.

There exists a lot of confusion in the literature as to when to use **H** and when to use **B**. **H** is the external applied magnetic field usually produced by external currents. Therefore, experiments that control external currents control **H** more directly than **B**, and hence, it is appropriate to consider **H** as the experimentally applied field. However, when this external magnetic field is applied to the system through external currents, the microscopic field perceived by the nuclei or the electrons is the magnetic induction **B**. Thus, the microscopic Hamiltonian should be written in terms of **B**, whereas the experimental results should be expressed in terms of **H**.

In this chapter, we will first calculate by using approximate methods, the atomic susceptibility, the susceptibility of insulators with filled shells that leads to Larmor diamagnetism, and the susceptibility of a collection of magnetic ions with partially filled shells that leads to paramagnetism. Then we will calculate the magnetic susceptibility of free electrons in metals that leads to Pauli paramagnetism and Landau diamagnetism, as well as the de Haas–van Alphen effect. Finally, we will outline the many-body theory of magnetic susceptibility of Bloch electrons in a magnetic field. We will express the total magnetic susceptibility as a sum of the contributions of the orbital (χ_o), spin (χ_s), and spin-orbit interactions (χ_{so}), and discuss the effects of exchange and correlation on each of these terms. Here, the spin susceptibility χ_s includes the effect of spin-orbit interaction on the spin, whereas χ_{so} is the contribution to magnetic susceptibility from the effect of spin-orbit coupling on the orbital motion of Bloch electrons. In the usual derivations of many-body theories of magnetic susceptibility, attention is focused either on the orbital part or on the spin part of the Hamiltonian, and the effects of spin-orbit coupling are accounted for in χ_o through the modifications of the Bloch functions and in χ_s by replacing the free electron g factor by the effective g factor. In this process, χ_{so}, which is of the same order as χ_s for solids with large effective g factors, has been neglected.

12.2 ATOMIC (OR IONIC) MAGNETIC SUSCEPTIBILITIES

12.2.1 General Formulation

It can be easily shown that the classical Hamiltonian for an electron in a magnetic field $\mathbf{B}$ is given by

$$H = \frac{1}{2m}\left(\mathbf{p} + \frac{e\mathbf{A}(\mathbf{r})}{c}\right)^2, \tag{12.3}$$

where $\mathbf{p}$ is the classical momentum, and $\mathbf{A}(\mathbf{r})$ is the vector potential obtained from the relation $\mathbf{B} = \nabla \times \mathbf{A}$ and $\nabla \cdot \mathbf{A} = 0$. In quantum mechanics, the vector $\mathbf{p}$ is replaced by the operator $-i\hbar\nabla$. Thus, the Hamiltonian for a free electron with spin operator $\hat{\mathbf{s}}$ in a magnetic field $\mathbf{B}$ is given by

$$\hat{H} = \frac{1}{2m}\left(-i\hbar\nabla + \frac{e\mathbf{A}(\mathbf{r})}{c}\right)^2 + g_0\mu_B \mathbf{B}\cdot\hat{\mathbf{s}} = \frac{1}{2m}\left(\hat{\mathbf{p}} + \frac{e\mathbf{A}(\mathbf{r})}{c}\right)^2 + g_0\mu_B \mathbf{B}\cdot\hat{\mathbf{s}}, \tag{12.4}$$

where $\hat{\mathbf{p}}$ and $\hat{\mathbf{s}}$ are the momentum and spin operators. The Hamiltonian of an atom (or ion) in a uniform magnetic field $\mathbf{B}$ (in the z direction) can be written as

$$\hat{H} = \frac{1}{2m}\sum_i \left(\hat{\mathbf{p}}_i + \frac{e\mathbf{A}(\mathbf{r}_i)}{c}\right)^2 + g_0\mu_B B \sum_i \hat{s}_z^i, \tag{12.5}$$

where $\mathbf{r}_i, \mathbf{p}_i$, and $\hat{\mathbf{s}}_i$ are the position, momentum, and spin operator of each electron in the atom (or ion), and $\mathbf{A}(\mathbf{r}_i)$ is the vector potential such that

$$\mathbf{B} = \nabla \times \mathbf{A} \text{ and } \nabla \cdot \mathbf{A} = 0. \tag{12.6}$$

Here, $g_0 \approx 2.0023$ and $\mu_B = e\hbar/2mc$. The first term in the Hamiltonian in Eq. (12.5), in which the sum is taken over all the electrons in the atom, is due to $\hat{T}$, the kinetic energy operator of the electrons, which includes the interaction of the orbital magnetic momentum with the magnetic induction. This interaction is accounted for by including the vector potential $\mathbf{A}(\mathbf{r}_i)$ in the momentum operator $\mathbf{p}_i$ of the electron at $\mathbf{r}_i$. The second term in Eq. (12.5) is due to the interaction of $\mathbf{B}$ with the electron spin and is a consequence of Dirac's relativistic theory. We will assume that

$$\mathbf{A} = -\frac{1}{2}\mathbf{r} \times \mathbf{B}, \tag{12.7}$$

which satisfies both conditions in Eq. (12.6). Thus, the kinetic energy operator, $\hat{T}$, in Eq. (12.5) can be written as

$$\hat{T} = \frac{1}{2m}\sum_i \hat{\mathbf{p}}_i^2 + \mu_B \hat{\mathbf{L}}\cdot\mathbf{B} + \frac{e^2}{8mc^2}B^2\sum_i (x_i^2 + y_i^2), \tag{12.8}$$

where $\hat{\mathbf{L}}$ is the total electronic orbital angular momentum operator,

$$\hat{\mathbf{L}} = \frac{1}{\hbar}\sum_i \mathbf{r}_i \times \hat{\mathbf{p}}_i. \tag{12.9}$$

From Eqs. (12.5), (12.8), and (12.9), we obtain (Problem 12.1)

$$\hat{H} = \frac{1}{2m}\sum_i \hat{\mathbf{p}}_i^2 + \mu_B(\hat{\mathbf{L}} + 2\hat{\mathbf{S}})\cdot\mathbf{B} + \frac{e^2}{8mc^2}B^2\sum_i(x_i^2 + y_i^2), \tag{12.10}$$

where we have approximated $g_0 \approx 2$, and

$$\hat{\mathbf{S}} = \sum_i \hat{\mathbf{s}}_i. \tag{12.11}$$

We can rewrite Eq. (12.10) in the alternate form

$$\hat{H} = \hat{H}_0 + \Delta\hat{H}, \tag{12.12}$$

where $\hat{H}_0$ is the Hamiltonian in the absence of the magnetic field, and $\Delta\hat{H}$ is the perturbation due to the magnetic field,

$$\Delta\hat{H} = \mu_B(\hat{\mathbf{L}} + 2\hat{\mathbf{S}})\cdot\mathbf{B} + \frac{e^2B^2}{8mc^2}\sum_i(x_i^2 + y_i^2). \tag{12.13}$$

If the states $|n\rangle$ are the orbital states of the electrons in the absence of the magnetic field, we obtain, by using second-order perturbation theory,

$$E_n = E_n^0 + \Delta E_n, \tag{12.14}$$

where

$$E_n^0 = \langle n|H_0|n\rangle \tag{12.15}$$

and

$$\Delta E_n = \langle n|\Delta\hat{H}|n\rangle + \sum_{n'\neq n}\frac{|\langle n|\Delta\hat{H}|n'|^2\rangle}{E_n - E_{n'}}. \tag{12.16}$$

From Eqs. (12.13) and (12.16), retaining terms through those quadratic in $\mathbf{B}$, we obtain

$$\Delta E_n = \mu_B\mathbf{B}\cdot\langle n|\hat{\mathbf{L}} + 2\hat{\mathbf{S}}|n\rangle + \sum_{n'\neq n}\frac{|\langle n|\mu_B\mathbf{B}\cdot(\hat{\mathbf{L}} + 2\hat{\mathbf{S}})|n'\rangle|^2}{E_n - E_{n'}} + \frac{e^2B^2}{8mc^2}\langle n|\sum_i(x_i^2 + y_i^2)|n\rangle. \tag{12.17}$$

The magnetic susceptibility of atoms, ions, and molecules is obtained from Eq. (12.17). Usually, the first term on the right side of the equation is the dominant term unless it vanishes. As we will see, the first term will be zero for either closed shells or shells with one electron short of being half filled ($J = 0$). However, for magnetic ions, the first term is much larger than the second and third terms, which can be neglected. The collection of ions becomes paramagnetic in such cases.

12.2.2 Larmor Diamagnetism

If a solid has atoms or ions where all the electronic shells are filled, each atom has zero orbital and spin angular momentum in its ground state. Therefore,

$$\hat{\mathbf{J}}|0\rangle = \hat{\mathbf{L}}|0\rangle = \hat{\mathbf{S}}|0\rangle = 0. \tag{12.18}$$

In the ground state, the third term in Eq. (12.17) is the only term that contributes to the magnetic susceptibility,

$$\Delta E_0 = \frac{e^2 B^2}{8mc^2} <0|\sum_i (x_i^2 + y_i^2)|0> = \frac{e^2 B^2}{12mc^2} <0|\sum_i r_i^2|0>, \tag{12.19}$$

which follows from the spherical symmetry of the atom or ion. Assuming that the free energy is equal to the ground-state energy (which is true if and only if $J = 0$), the magnetic susceptibility of the solid of N such ions in a volume V (in a semiclassical approximation) is given by

$$\chi \approx -\frac{\partial^2 \Delta E_0}{\partial B^2} = -\frac{N}{V} <0|\sum_i r_i^2|0>. \tag{12.20}$$

Eq. (12.20) is known as Larmor diamagnetic susceptibility or sometimes as Langevin susceptibility. It is valid only for solids composed of atoms or ions of filled shells.

We will now discuss Hund's rules, which are needed to discuss atoms or ions with partially filled shells.

12.2.3 Hund's Rules

Hund's rules, which are valid for atoms or ions with partially filled shells, were obtained from the analysis of atomic spectra as well as by rigorous theoretical calculations. They are valid for incomplete shells of which the one-electron levels are characterized by orbital angular momentum l. The shell would have $2(2l+1)$ one-electron levels (including spin). If n is the number of electrons in the shell,

$$0 < n < 2(2l+1). \tag{12.21}$$

The degeneracy of these levels is lifted by electron–electron interaction as well as by the spin-orbit interaction.

We will first discuss Russsell–Saunders coupling, which states that the Hamiltonian of the atom or ion (with partially filled shells) commutes with the total angular momentum $\hat{\mathbf{J}} = \hat{\mathbf{L}} + \hat{\mathbf{S}}$ and with the total electronic orbital and spin angular momentum $\hat{\mathbf{L}}$ and $\hat{\mathbf{S}}$ (provided the spin-orbit coupling is not too large such that it can be considered as a small perturbation). Thus, the partially filled shells can be indexed by the quantum numbers J, J_z, L, L_z, S, and S_z. This indexing is based on the fact that the eigenvalues of the operators $\hat{\mathbf{J}}^2$, $\hat{\mathbf{J}}_z$, $\hat{\mathbf{L}}^2$, $\hat{\mathbf{L}}_z$, $\hat{\mathbf{S}}^2$, and $\hat{\mathbf{S}}_z$ are $J(J+1)$, J_z, $L(L+1)$, L_z, $S(S+1)$, and S_z.

Hund's rules are as follows:

1. In an incomplete shell, the electrons that lie lowest in energy have the largest total spin S, which is consistent with the exclusion principle. The largest value S can have is equal to the largest magnitude that S_z can have. When $n \le 2l+1$, each electron can have parallel spin without multiple occupation of any one-electron level in the shell, provided each electron has a different value of l_z. Thus, when $n \le 2l+1$, $S = \frac{1}{2}n$. The exclusion principle requires that when $n > 2l+1$, the spin of each additional electron has spin opposite to the first $2l+1$ electrons. Therefore, S is reduced by half a unit from its maximum vale of $l + \frac{1}{2}$ for each electron after the first $2l+1$ electrons.

2. Once S has been determined, as per Hund's first rule, the total angular momentum L of the electrons of the lowest-lying states has the largest value. For example, the first electron in the shell is in the level $|l_z| = l$, which is its maximum value. The second electron is in the level $|l_z| = l-1$. When the electrons are half filled, $L = l + (l-1) + \ldots [L-(n-1)] = 0$. The spin of the electrons that fill the second half of the shell are opposite to those in the first half, and using the same arguments, $L = 0$.
3. The values of L and S or the states of lowest energy are obtained from the first two rules. However, there are $(2L+1)(2S+1)$ possible states that are degenerate. The degeneracy of these states is lifted by spin-orbit coupling, which is of the form $\lambda(\hat{\mathbf{L}} \cdot \hat{\mathbf{S}})$. The total angular momentum J can take on all integral values between $|L-S|$ and $L+S$. Spin-orbit coupling favors minimum J if λ is positive (for shells that are less than half filled) and maximum J if λ is negative (for shells that are more than half filled). Thus, in the ground state,

$$\begin{aligned} J &= |L-S|, \quad n \leq (2l+1), \\ J &= L+S, \quad n \geq (2l+1) \cdot \end{aligned} \tag{12.22}$$

We note that when the shell is half full, $L = 0$ and there is no jump in J because $J = S$.

We further note that Hund's rules apply to partially filled d and f shells but not to partially filled p shells, which contain valence electrons and broaden into bands in the solid.

12.2.4 Van Vleck Paramagnetism

We consider the susceptibility of insulators containing ions with a partially filled shell. First, we consider the case where $J = 0$ (shells that are one electron short of being half filled). The first term in Eq. (12.17) still vanishes, as in the case of a filled shell. However, the second and third terms contribute to the shift in the ground-state energy, and we obtain

$$\Delta E_0 = -\sum_n \frac{|<0|\mu_B \mathbf{B} \cdot (\hat{\mathbf{L}} + 2\hat{\mathbf{S}})|n>|^2}{E_n - E_0} + \frac{e^2 B^2}{8mc^2} \sum_i <0|(x_i^2 + y_i^2)|0>. \tag{12.23}$$

Because $J = 0$, assuming as before that the free energy is equal to the ground-state energy, the magnetic susceptibility is given by

$$\begin{aligned} \chi &\approx -\frac{N}{V}\frac{\partial^2 E_0}{\partial B^2} \\ &\approx -\frac{N}{V}\left[-2\mu_B^2 \sum_n \frac{|<0|\hat{\mathbf{L}}_z + 2\hat{\mathbf{S}}_z|n>|^2}{E_n - E_0} + \frac{e^2}{4mc^2} <0|\sum_i (x_i^2 + y_i^2)|0> \right]. \end{aligned} \tag{12.24}$$

The first term in Eq. (12.24), which is positive, is known as Van Vleck paramagnetism, and the second term is the Larmor diamagnetism derived for ions of filled shells. Thus, the magnetic susceptibility of ions with a shell one electron short of being half filled is obtained by the sum of the two terms. The Van Vleck paramagnetism also exists in molecules that have a more complex structure than single atoms or ions.

It may be noted that the basic assumption made in these derivations is that the ground state is occupied with appreciable probability in thermal equilibrium, and hence, the free energy is equal to the ground-state energy. However, if the next state is close to the $J = 0$ ground state, the free energy is not just the ground-state energy, and the derivation of the formula for magnetic susceptibility is much more complicated.

12.2.5 Landé g Factor

If $J \neq 0$ for the shell, the first term for the shift in energy in Eq. (12.17) becomes the dominant term compared to the other two terms, which yield the Larmor diamagnetism and Van Vleck paramagnetism. If we ignore the last two terms in Eq. (12.17), we have to consider the matrix elements' dominant term

$$\Delta E_n \approx \mu_B B <n|\hat{L}_z + 2\hat{S}_z|n>. \tag{12.25}$$

Here, we have assumed that $\mathbf{B} = B\hat{\mathbf{z}}$. The ground state $|n>$ is $(2J+1)$-fold degenerate in the absence of the magnetic field. The matrix elements can be evaluated by diagonalizing and evaluating the matrix elements of the $(2J+1)$-dimensional square matrix

$$<JLSJ_z|\hat{L}_z + 2\hat{S}_z|JLSJ_z'>, \tag{12.26}$$

where

$$J_z, J_z' = -J, \ldots, J. \tag{12.27}$$

To evaluate these matrix elements, we use the Wigner–Eckart theorem according to which the matrix elements of any vector operator $\hat{\mathbf{A}}$ in the $(2J+1)$-dimensional space of eigenstates of $\mathbf{J}^2$ and $\mathbf{J}_z$ with a given value of J are proportional to the matrix elements of $\hat{\mathbf{J}}$ itself:

$$<JLSJ_z|\hat{\mathbf{A}}|JLSJ_z'> = g(JLS) <JLSJ_z|\hat{\mathbf{J}}|JLSJ_z'>. \tag{12.28}$$

The proportionality constant $g(JLS)$ depends on $\hat{\mathbf{A}}$ but does not depend on the values of J_z and J_z'. Applying the Wigner–Eckart theorem to magnetism,

$$<JLSJ_z|\hat{L}_z + 2\hat{S}_z|JLSJ_z'> = g(JLS) <JLSJ_z|\hat{J}_z|JLSJ_z'> = g(JLS) J_z \delta_{J_z J_z'}. \tag{12.29}$$

Here, $g(JLS)$ is known as the Landé g factor. The matrix is diagonal in the states of definite J_z, and the ground state $|n>$, which is $(2J+1)$-fold degenerate, is split into states with definite values of J_z. From Eqs. (12.25) and (12.29), we obtain

$$\Delta E_{J_z} \approx g(JLS)\, \mu_B B J_z\, \delta_{J_z J_z'}. \tag{12.30}$$

The energies are uniformly split by $g(JLS)\mu_B B$. The Landé g factor can be evaluated by writing, in analogy with Eq. (12.28),

$$<JLSJ_z|(\hat{\mathbf{L}} + 2\hat{\mathbf{S}})|JLSJ_z'> = g(JLS) <JLSJ_z|\hat{\mathbf{J}}|JLSJ_z'>. \tag{12.31}$$

We can also rewrite Eq. (12.31) in the alternate form (Problem 12.2)

$$<JLSJ_z|\hat{\mathbf{L}} + 2\hat{\mathbf{S}}|J'L'S'J_z'> = g(JLS) <JLSJ_z|\hat{\mathbf{J}}|J'L'S'J_z'>, \tag{12.32}$$

because both matrix elements vanish unless $J = J'$, $L = L'$, and $S = S'$. By using the completeness relation,

$$\sum_{J''L''S''J''} |J''L''S''J_z''> <J''L''S''J_z''| = 1, \tag{12.33}$$

we can rewrite Eq. (12.32) in the alternate form

$$\begin{aligned} &\sum_{J''L''S''J_z''} <JLSJ_z|(\hat{\mathbf{L}}+2\hat{\mathbf{S}})|J''L''S''J_z''>\cdot<J''L''S''J_z''|\hat{\mathbf{J}}|J'L'S'J_z'> \\ &= g(JLS) \sum_{J''L''S''J_z''} <JLSJ_z|\hat{\mathbf{J}}|J''L''S''J_z''>\cdot<J''L''S''J_z''|\hat{\mathbf{J}}|J'L'S'J_z'> \cdot \end{aligned} \tag{12.34}$$

Because the sum over Eq. (12.34) is taken over a complete set, from Eqs. (12.31), (12.32), and (12.34), we obtain (Problem 12.3)

$$<JLSJ_z|(\hat{\mathbf{L}}+2\hat{\mathbf{S}})\cdot\hat{\mathbf{J}}|JLSJ_z'> = g(JLS) <JLSJ_z|\hat{\mathbf{J}}^2|JLSJ_z'>. \tag{12.35}$$

We have, from the relation (Problem 12.4),

$$\hat{\mathbf{J}} = \hat{\mathbf{L}} + \hat{\mathbf{S}}, \tag{12.36}$$

$$(\hat{\mathbf{L}}+2\hat{\mathbf{S}})\cdot(\hat{\mathbf{L}}+\hat{\mathbf{S}}) = \hat{\mathbf{L}}^2 + 2\hat{\mathbf{S}}^2 + \hat{\mathbf{L}}\cdot\hat{\mathbf{J}} + 2\hat{\mathbf{S}}\cdot\hat{\mathbf{J}} = \frac{1}{2}[3\hat{\mathbf{J}}^2 - \hat{\mathbf{L}}^2 + \hat{\mathbf{S}}^2]. \tag{12.37}$$

From Eqs. (12.35) through (12.37), we obtain

$$g(JLS) = \frac{[3J(J+1) - L(L+1) + S(S+1)]}{2J(J+1)}. \tag{12.38}$$

We note that for the $(2J+1)$-dimensional set of states that make up the degenerate atomic ground state in the zero field, Eq. (12.32) can often be rewritten without the state vectors as

$$\hat{\mathbf{L}} + 2\hat{\mathbf{S}} = g(JLS)\,\mathbf{J}, \tag{12.39}$$

as long as the matrix elements are diagonal in $J, L,$ and S. If the splitting between the zero-field atomic ground-state multiplet and the first excited multiplet is large compared with k_BT, then the $(2J+1)$ states in the ground-state multiplet will contribute significantly to the free energy. In this case, we can rewrite the first term of Eq. (12.32) by using Eq. (12.39), as the interaction $(-\vec{\mu}\cdot\mathbf{B})$ of the field $\mathbf{B}$ with a magnetic moment,

$$\vec{\mu} = -g(JLS)\mu_B\mathbf{J}. \tag{12.40}$$

The magnetic susceptibility has to be obtained from the free energy because the free energy cannot be equated with the ground-state energy in this case. When $\mathbf{B} \to 0$, the splitting of the $(2J+1)$ lowest-lying states would be small compared with k_BT. As we will see, the magnetic susceptibility of a collection of magnetic ions is paramagnetic and leads to Curie's law.

12.2.6 Curie's Law

To calculate the Helmholtz free energy F, we assume that only the lowest-lying spin multiplet contributes to the statistical mechanical sums. Thus, only the lowest $2J+1$ states are thermally excited with appreciable probability. The Helmholtz free energy F, for a single ion in magnetic field $\mathbf{B}$, is given by

$$e^{-\beta F} = \sum_{J_z=-J}^{J} e^{-\beta g(JLS)\mu_B B J_z} = \sum_{J_z=-J}^{J} e^{-\beta \gamma B J_z}, \tag{12.41}$$

where

$$\gamma = g(JLS)\mu_B. \tag{12.42}$$

By summing over the geometric series, we obtain

$$e^{-\beta F} = \frac{e^{\beta\gamma B(J+\frac{1}{2})} - e^{-\beta\gamma B(J+\frac{1}{2})}}{e^{\beta\gamma B/2} - e^{-\beta\gamma B/2}}. \tag{12.43}$$

The magnetization of N ions in a volume V is defined as

$$M = -\frac{N}{V}\frac{\partial F}{\partial B}. \tag{12.44}$$

From Eqs. (12.43) and (12.44), we obtain (Problem 12.5)

$$M = \frac{N}{V}\gamma J B_J(\beta\gamma J B), \tag{12.45}$$

where the Brillouin function $B_J(x)$ is defined as

$$B_J(x) = \frac{(2J+1)}{2J}\coth\left(\frac{2J+1}{2J}x\right) - \frac{1}{2J}\coth\left(\frac{1}{2J}x\right). \tag{12.46}$$

If $\beta\gamma B \ll 1$, $x \ll 1$, and

$$\coth x \approx \frac{1}{x} + \frac{1}{3}x + O(x^3), \tag{12.47}$$

from Eqs. (12.46) and (12.47), we obtain

$$B_J(x) \approx \frac{J+1}{3J}x \approx \frac{J+1}{3}\beta\gamma B \approx \frac{J+1}{3}g(JLS)\beta\mu_B B. \tag{12.48}$$

From Eqs. (12.45) and (12.48), we obtain

$$M = \frac{N}{V}\frac{\beta g^2\mu_B^2}{3}BJ(J+1). \tag{12.49}$$

The magnetic susceptibility is

$$\chi = \frac{\partial M}{\partial B} = \frac{N}{V}\frac{(g\mu_B)^2}{3}\frac{J(J+1)}{k_B T}. \tag{12.50}$$

This is known as Curie's law and characterizes paramagnetic systems with "permanent moments" of which the alignment is opposed by thermal disorder but favored by the magnetic field. We note that Curie's law, as derived here, is valid as long as the system of magnetic ions is considered as noninteracting.

We can rewrite Curie's law in the alternate form

$$\chi = \frac{1}{3}\frac{N}{V}\frac{\mu_B^2 p^2}{k_B T}, \tag{12.51}$$

where the "effective Bohr magneton number" p is given by

$$p = g(JLS)\sqrt{J(J+1)}. \tag{12.52}$$

12.3 MAGNETIC SUSCEPTIBILITY OF FREE ELECTRONS IN METALS

12.3.1 General Formulation

The diamagnetic and paramagnetic susceptibilities of free electrons in a metal were calculated by Landau[4] and Pauli, respectively. The diamagnetic susceptibility arises out of the orbital motion of the electrons, and the paramagnetic susceptibility arises out of the realignment of spins in a magnetic field. It can be shown that Pauli paramagnetism is three times the value of Landau diamagnetism with the signs reversed. In addition, there is an oscillatory contribution to the magnetic susceptibility, which is known as the de Haas–van Alphen effect[5]. The de Haas–van Alphen effect is an effective tool to measure the contours of the Fermi surface of a metal.

The spin-orbit interaction is not included in the following derivation for magnetic susceptibility of free electrons. We will first calculate the magnetic susceptibility of a free electron gas confined in a box.

For a free electron gas, one can write the magnetization density M as

$$M = -\frac{1}{V}\frac{\partial F}{\partial B}, \tag{12.53}$$

where F is the free energy

$$F = N\mu - \frac{1}{\beta}\sum_{\mathbf{k}} \ln Z_{\mathbf{k}}, \tag{12.54}$$

$\beta = 1/k_B T$, N is the total number of electrons, μ is the chemical potential, and the partition function is

$$Z_{\mathbf{k}} = 1 + e^{-\beta(E(\mathbf{k})-\mu)}. \tag{12.55}$$

The Schrodinger equation for free electrons in a magnetic field $\mathbf{B}$ (neglecting spin) is given by

$$H\psi(\mathbf{r}) = \frac{1}{2m}\left(-i\hbar\nabla + \frac{e\mathbf{A}}{c}\right)^2\psi(\mathbf{r}) = E\psi(\mathbf{r}). \tag{12.56}$$

We assume that the magnetic field is along the z direction, $\mathbf{B} = B\hat{z}$, in which case $\mathbf{A} = Bx\hat{y}$ (from $\mathbf{B} = \nabla \times \mathbf{A}$). This is also known as the Landau gauge. In the Landau gauge, we can write

$$H = \frac{1}{2m}[p_x^2 + p_z^2 + (p_y + m\omega_c x)^2], \tag{12.57}$$

where the cyclotron frequency

$$\omega_c = eB/2mc. \tag{12.58}$$

If we make the substitution

$$p_y = \hbar k_y,\ p_z = \hbar k_z, \text{ and } x = -\frac{\hbar}{m\omega_c}k_y + q = x_0 + q, \tag{12.59}$$

Eq. (12.57) can be rewritten in the alternate form

$$H = \frac{1}{2m}(p_x^2 + m^2\omega_c^2 q^2) + \frac{\hbar^2}{2m}k_z^2. \tag{12.60}$$

We consider the electrons to be confined in a rectangular parallelepiped of sides L_x, L_y, and L_z with the Born–von Karman periodic boundary conditions. The solution of Eq. (12.56), which is a plane wave $e^{i\mathbf{k}\cdot\mathbf{r}}$ when $\mathbf{A} = 0$, is modified (by using Eq. 12.60) as

$$\psi(\mathbf{r}) = e^{i(k_y y + k_z z)}\phi(x). \tag{12.61}$$

Substituting Eqs. (12.60) and (12.61) in (12.56), we obtain

$$-\frac{\hbar^2}{2m}\phi''(x) + \frac{m\omega_c^2}{2}(x - x_0)^2\phi(x) = \left(E - \frac{\hbar^2 k_z^2}{2m}\right)\phi(x), \tag{12.62}$$

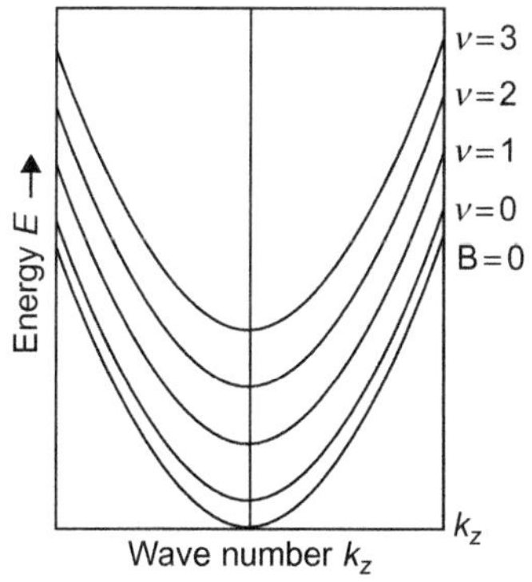

FIGURE 12.1

Sketch of the one-dimensional energy sub-bands in the free electron model.

where Eq. (12.62) is the Schrodinger equation of the one-dimensional oscillator centered at x_0. The eigenvalues are

$$E_v = \frac{\hbar^2 k_z^2}{2m} + \left(v + \frac{1}{2}\right)\hbar\omega_c, \quad v = 0,1,2,\ldots. \tag{12.63}$$

Thus, the energy of the electron is the sum of the kinetic energy of its undisturbed motion along the z direction, and the quantized energy of the oscillatory motion in the plane orthogonal to the field direction. This part of the energy is the contribution due to the orbital motion. The energy of the one-dimensional sub-bands is plotted as a function of the wave number k_z in Figure 12.1.The wave function in Eq. (12.61) depends on k_y and k_z directly and on k_z and v indirectly through $\phi(x)$.

Because one can choose any k_y for a particular value of k_z and v, the state k_y is degenerate. Further, x_0 lies in the range

$$-L_x/2 < x_0 < L_x/2. \tag{12.64}$$

If we neglect spin, the states occur on the k_y axis in the intervals of $2\pi/L_y$. One can write

$$-\frac{1}{2\hbar} m\omega_c L_y < k_y < \frac{1}{2\hbar} m\omega_c L_x. \tag{12.65}$$

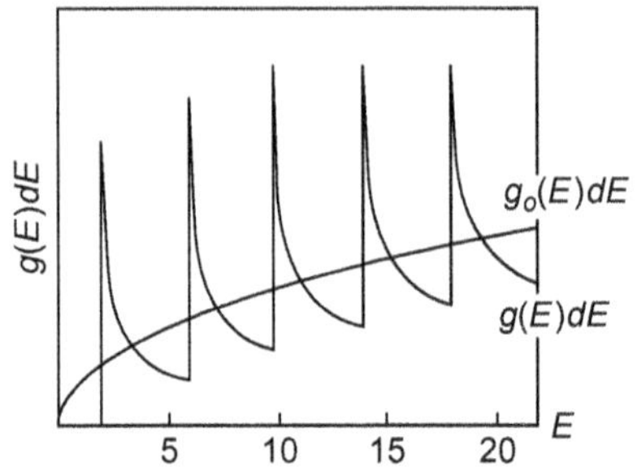

FIGURE 12.2

Sketch of the density of states for $\mathbf{B} = 0$ and $\mathbf{B} \neq 0$.

Thus, k_y can have $(L_y/2\pi)(m\omega_c L_x/\hbar)$ different values. Furthermore, in the range dk_z, the z-component of k will have $(L_z/2\pi)dk_z$ different values. The density of states is obtained by multiplying by a factor of 2 for spin and dividing by the volume of the cube $(L_x L_y L_z)$,

$$g(v, k_z)dk_z = \frac{2}{(2\pi)^2} \frac{m\omega_c}{\hbar} dk_z. \tag{12.66}$$

From Eqs. (12.63) and (12.66), we obtain (Problem 12.6)

$$g(E, v)\, dE = \frac{2}{(2\pi)^2} \frac{h\omega_c}{2} \left(\frac{2m}{\hbar^2}\right)^{3/2} \left(E - \left(v + \frac{1}{2}\right)\hbar\omega_c\right)^{-1/2} dE. \tag{12.67}$$

The total density of states is obtained by summation over all bands that lie "below" the energy E,

$$g(E)\, dE = \sum_{v=0}^{v'} g(E, v)\, dE. \tag{12.68}$$

It can be easily shown (Problem 12.7) that Eq. (12.68) has the same density of states as the zero-field case. The average density of states is unaffected by a magnetic field. The states are redistributed due to the magnetic field **B**, which pulls together a large number of states into a single level (see Figure 12.2).

The contribution due to spin is $\pm(g/2)\mu_B B$, depending on its direction, where the Bohr magneton $\mu_B = e\hbar/2mc$ and the g factor for free electrons is $g = 2$. If we include the contribution of spin, the energy E is modified as

$$E_\pm = \frac{\hbar^2 k_z^2}{2m} + (2v+1)\mu_B B \pm \frac{g}{2}\mu_B B = E \pm \mu_B B. \tag{12.69}$$

12.3.2 Landau Diamagnetism and Pauli Paramagnetism

From Eqs. (12.53) through (12.55) and (12.69), we obtain the expression for magnetization per unit volume (Problem 12.8),

$$M = -\frac{d}{dB}\left\{ n\mu - \frac{1}{\beta}\int_0^\infty g(E_+)\ \ln\ [e^{\beta(\mu-E)} + 1]\, dE_+ - \frac{1}{\beta}\int_0^\infty g(E_-)\ \ln\ [e^{\beta(\mu-E)} + 1]\, dE_- \right\}. \tag{12.70}$$

In Eq. (12.70), the summation over the energy states is replaced by an integration because the energy levels are very close together. Here, n is the electron concentration, and μ is the chemical potential. We have

$$n = \int_0^\infty n(E)dE = \int_0^\infty f(E)g(E)dE, \tag{12.71}$$

where $f(E)$ is the Fermi function,

$$f(E) = [1 + e^{\beta(E(\mathbf{k})-\mu)}]^{-1}. \tag{12.72}$$

The density of states in **k** space is given by

$$g(E)dE = \frac{1}{2\pi^2}\left(\frac{2m}{\hbar^2}\right)^{3/2} E^{1/2}\, dE. \tag{12.73}$$

From Eqs. (12.71) through (12.73), we obtain

$$n = 2\left(\frac{mk_BT}{2\pi\hbar^2}\right)^{3/2} \frac{2}{\sqrt{\pi}} F(\beta\mu), \tag{12.74}$$

where the Fermi integral $F(x)$ is given by

$$F(x) = \int_0^\infty \frac{y^{1/2}}{1+e^{(y-x)}}\, dy \cdot \tag{12.75}$$

It can be shown (Problem 12.9) that

$$F(x) \approx \frac{\sqrt{\pi}}{2} e^x, \quad \text{for} \quad x<0$$

and

$$F(x) \approx \frac{2}{3} x^{3/2}, \quad \text{for} \quad x>0. \tag{12.76}$$

Using a similar expansion for Eq. (12.70) and for low temperatures, in which the first term is retained, one can show that (Problem 12.10)

$$M = \frac{3n\mu_B^2 B}{2E_F}\left[1 - \frac{1}{3} + \frac{\pi k_B T}{\mu_B B}\left(\frac{E_F}{\mu_B B}\right)^{1/2} \sum_{v=1}^{\infty} \frac{(-1)^v}{\sqrt{v}} \cos(\pi v) \frac{\sin\left(\frac{\pi}{4} - \frac{\pi v E_F}{\mu_B B}\right)}{\sinh \frac{\pi^2 v k_B T}{\mu_B B}}\right]. \tag{12.77}$$

Here, E_F is the Fermi energy (value of μ at $T=0$). The magnetic susceptibility is obtained from the expression

$$\chi = \frac{\partial M}{\partial B} \cdot \tag{12.78}$$

For most cases, M is linear in B for attainable field strengths, and the definition reduces to

$$\chi \approx \frac{M}{B}. \tag{12.79}$$

From Eqs. (12.77) and (12.79), we obtain (retaining only the first term in the series expansion),

$$\chi = \chi_P + \chi_L + \chi_{dH-vA}, \tag{12.80}$$

where χ_P is the Pauli spin paramagnetism,

$$\chi_P = \frac{3n\mu_B^2}{2E_F}, \tag{12.81}$$

χ_L is the Landau diamagnetism,

$$\chi_L = -\frac{n\mu_B^2}{2E_F}, \tag{12.82}$$

and χ_{dH-vA} is the de Haas–van Alphen effect, which is oscillatory with a period $\frac{2\mu_B}{E_F}$. This is an additional diamagnetic term given by

$$\chi_{dH-vA} \approx \frac{3n\pi}{2\beta}\left(\frac{\mu_B}{E_F B}\right)^{1/2}\left(\frac{\sin\left(\frac{\pi}{4}-\frac{\pi E_F}{\mu_B B}\right)}{\sinh\left(\frac{\pi^2}{\mu_B \beta B}\right)}\right). \tag{12.83}$$

We note from Eq. (12.83) that because the sin term is oscillatory,

$$\sin\left(\frac{\pi}{4}-\frac{\pi E_F}{\mu_B B}\right) = \sin\left(\frac{\pi}{4}-\frac{\pi E_F}{\mu_B B}+2n\pi\right), \tag{12.84}$$

χ_{dH-vA} is oscillatory whenever

$$\frac{\pi E_F}{\mu_B B} = \frac{\pi E_F}{\mu_B B}+2n\pi \tag{12.85}$$

or

$$\frac{1}{B} = \frac{1}{B}+\frac{2n\mu_B}{E_F}. \tag{12.86}$$

χ_{dH-vA} is periodic in $1/B$ with a temperature-independent period $2\mu_B/E_F$. However, these oscillations can be observed only at low temperatures and high magnetic fields. It will be shown that the condition for such oscillations is

$$\hbar\omega_c = 2\mu_B B \gg \frac{1}{\beta}. \tag{12.87}$$

Otherwise, the distribution of electrons in the region of E_F is widely spread, and the oscillations are spread out.

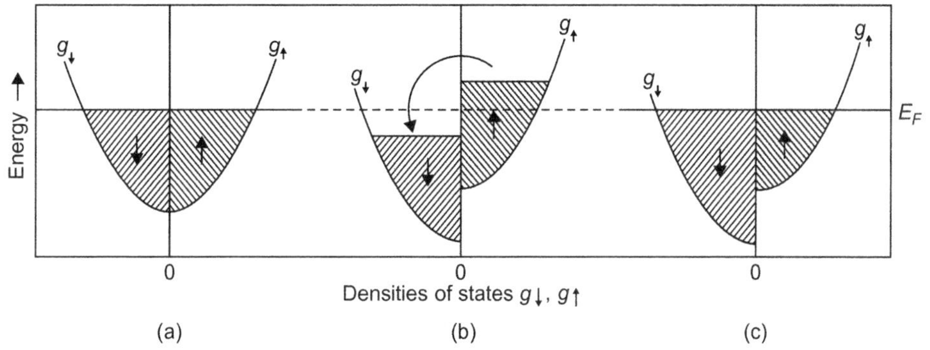

FIGURE 12.3

Sketch of densities of states $g_\downarrow$ *and* $g_\uparrow$ as a function of energy.

The origin of each of the previous three contributions to the magnetic susceptibility of a free electron gas can be explained in simple manner. The origin of the Pauli spin paramagnetism is the additional contribution $\pm\mu_B B$ to the kinetic energy of the electrons due to the magnetic field, depending on their spin directions. This is shown in Figure 12.3.

In Figure 12.3a, the occupied states for both spins are the same in the absence of a magnetic field. The highest occupied states for each spin have the energy equal to the Fermi energy E_F. In Figure 12.3b, the states of opposite spins are shifted in a magnetic field B. However, the "occupied" states above E_F with spin "up" flow to the "unoccupied" states with spin "down" until the states are filled. The imbalance of the density of states in a magnetic field between the two spin states contributes to the positive Pauli paramagnetism.

The Landau diamagnetic term represents the orbital quantization of the electrons in a magnetic field. It is one-third the contribution of the Pauli paramagnetism and is of the opposite (negative) sign. The theory of diamagnetic susceptibility of metals was derived by Misra and Roth (Ref. 12). The theory of magnetic susceptibility of Bloch electrons was derived by Misra and Kleinman (Ref. 13).

12.3.3 De Haas–van Alphen Effect

The de Haas–van Alphen effect[5], which has been used extensively to investigate the Fermi surfaces of metals, can be explained by first assuming that the Fermi level is approximately constant as B is varied. The validity of this assumption can be seen in Figure 12.4. We note $G(E) \sim G_0(E)$ as the energy increases. Here, $G(E)$ and $G_0(E)$ are the number of states below the energy E in the presence and absence of the magnetic field. Because the number of states below the Fermi level E_F is half of the number of electrons at $T=0$ (each state can have two electrons), E_F is approximately constant as B is varied.

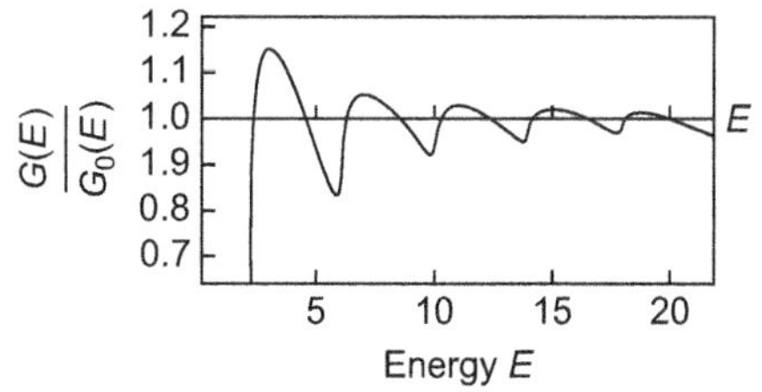

FIGURE 12.4

The ratio of the number of states $G(E)$ below E in a magnetic field to the number $G_0(E)$ in the absence of a magnetic field.

If we consider the case at $T=0$, all states are filled up to the Fermi level E_F, and all states above it are empty. If we consider a plane slab in **k** space of thickness δk_z at k_z, the number of allowed values of k_z in the range δk_z is $(L_z/2\pi)\,\delta k_z$. The total degeneracy of the state ν (neglecting spin) per unit volume in the slice δk_z is (from Eq. 12.66)

$$\frac{m\omega_c}{4\pi^2\hbar}\delta k_z = \xi B. \tag{12.88}$$

Thus, the degeneracy parameter (apart from spin), which is defined as the degeneracy per unit magnetic field per unit volume, is

$$\xi = \frac{e\delta k_z}{4\pi^2\hbar c}. \tag{12.89}$$

We consider the de Haas–van Alphen effect for a free electron gas at absolute zero. Figure 12.5 shows the spectrum of the Landau levels when the energy E is plotted versus B.

At $T=0$, all levels in the slice δk_z will be filled (Figure 12.5) for which

$$(\nu+\frac{1}{2})\hbar\omega_c + \frac{\hbar^2 k_z^2}{2m} \le E_F, \tag{12.90}$$

which can be rewritten in the alternate form,

$$(\nu+\frac{1}{2})\hbar\omega_c \le \varepsilon_F', \tag{12.91}$$

where

$$\varepsilon_F' = E_F - \frac{\hbar^2 k_z^2}{2m}. \tag{12.92}$$

At $T=0$, if ν' is the highest filled level, n, the number of electron states in the slice of thickness δk_z is (from Eq. 12.89 and the fact that $\nu = 0$ is a filled level)

$$n = (\nu'+1)\xi B. \tag{12.93}$$

When B is increased, n increases linearly with B until ν' coincides with ε_F'. As B is further increased, all the electrons in ν' will have energy greater than ε_F', and hence, they will empty out into the orbits in other slices with different values of k_z and ε_F'. This oscillatory evacuation occurs when

$$(\nu'+\frac{1}{2}) = \frac{\varepsilon_F'}{\hbar\omega_c} = \frac{mc\varepsilon_F'}{e\hbar}\frac{1}{B}. \tag{12.94}$$

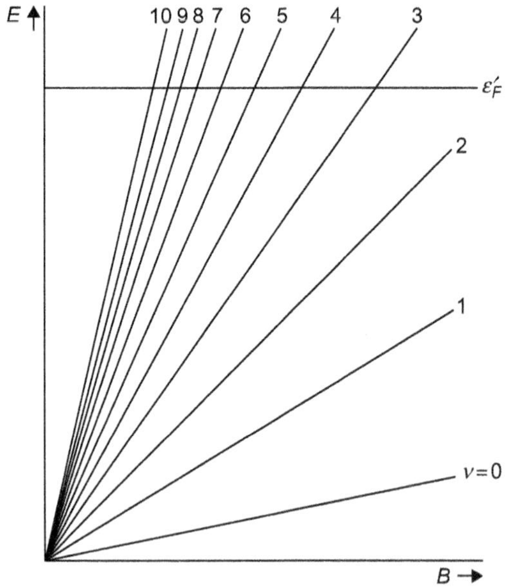

FIGURE 12.5

Sketch of the spectrum of Landau levels versus B.

Hence, the population δn is approximately a periodic function of $1/B$, with period $\frac{e\hbar}{mc\varepsilon'_F}$. The energy of the electrons in the slice δk_z in a magnetic field B such that the population n_0 is

$$E_0 = \hbar\xi B\omega_c \sum^{v'} \left(v + \frac{1}{2}\right) + n_0 \frac{\hbar^2 k_z^2}{2m} = \frac{1}{2}\hbar\xi B\omega_c (v' + 1)^2 + n_0 \frac{\hbar^2 k_z^2}{2m}. \tag{12.95}$$

From Eqs. (12.89) and (12.95), we obtain

$$E_0 = \frac{1}{2}\frac{\hbar\omega_c}{\xi B} n_0^2 + n_0 \frac{\hbar^2 k_z^2}{2m}, \tag{12.96}$$

where $n_0 = (v' + 1)\xi B$ from Eq. (12.89) for this value of B. For a nearby B that has the same value of v' for the highest filled state,

$$E = \frac{1}{2}\frac{\hbar\omega_c}{\xi B} n^2 + n \frac{\hbar^2 k_z^2}{2m} + (n_0 - n)\varepsilon_F, \tag{12.97}$$

where $(n_0 - n)\varepsilon_F$ is the change in energy arising out of the transfer of $n - n_0$ electrons at the Fermi level while the first two terms are the energy of the electrons in the slice δk_z. From Eqs. (12.92), (12.96), and (12.97), we obtain

$$\delta E = E - E_0 = \frac{\mu}{\xi}(n^2 - n_0^2) + (n_0 - n)\varepsilon'_F, \tag{12.98}$$

where

$$\mu = \frac{e\hbar}{2mc}. \tag{12.99}$$

Further, from Eq. (12.94),

$$\varepsilon'_F \approx (n_0/\xi B)\hbar\omega_c = 2\mu n_0/\xi. \tag{12.100}$$

From Eqs. (12.98) and (12.100), we obtain

$$\delta E = \left(\frac{\mu}{\xi}\right)(n - n_0)^2. \tag{12.101}$$

From Eq. (12.101), δM, the magnetization of the slice at $T = 0$,

$$\delta M = -\frac{\partial E}{\partial B} = -\frac{2\mu}{\xi}(n - n_0)\frac{dn}{dB}. \tag{12.102}$$

From Eqs. (12.93), (12.94), and (12.100), we have

$$\frac{dn}{dB} \approx \varepsilon'_F\left(\frac{\xi}{2\mu B}\right). \tag{12.103}$$

From Eqs. (12.102) and (12.103), we obtain

$$\delta M \approx -\frac{\varepsilon'_F}{B}(n - n_0). \tag{12.104}$$

As B is increased, $n-n_0$ oscillates with extrema $\pm\frac{1}{2}\xi B$ because the population of the level v' varies between ξB and 0. Thus, the magnetization varies between $\mp\frac{1}{2}\xi\varepsilon_F'$. This oscillation of the magnetization as a periodic function of B is known as the de Haas–van Alphen effect, a more rigorous analysis of which was presented earlier in this section. The de Haas–van Alphen oscillations are sensitive probes of the geometrical property of the Fermi surface. The areas of the extremals of the electron orbits can be determined from the period of oscillation.

Because the de Haas–van Alphen effect is an important tool and is widely used to experimentally measure the contours of the Fermi surface of crystalline solids, we will present the sequence of events that occur when a magnetic field is applied to a free electron gas in a rectangular parallelepiped, the Fermi surface of which is a sphere prior to the application of a magnetic field.

Figure 12.6 shows a plane section in the **k** space. The electron states are uniformly distributed in the (k_x, k_y) plane. The magnetic field **B** is directed normally into the plane section. The **k** states are subjected to a Lorenz force, and all the **k** states rotate with the cyclotron frequency about an axis through the origin and parallel to the field direction.

The magnetic field causes a redistribution of the **k** states that lie on rings that correspond to the energies

$$E_v = \left(v+\tfrac{1}{2}\right)\hbar\omega_c = \frac{\hbar^2 k_v^2}{2m}. \tag{12.105}$$

These rings are shown in Figure 12.7. Because the magnetic field **B** is in the z direction, k_z is not affected by it. Thus, in the **k** space, the representative points lie on cylinders (Landau cylinders) of which the cross-sections are the Landau rings.

At $T=0$, the **k** states are within the Fermi sphere of radius k_F. Thus, the Landau cylinders, shown in Figure 12.8, are either partially occupied or fully empty.

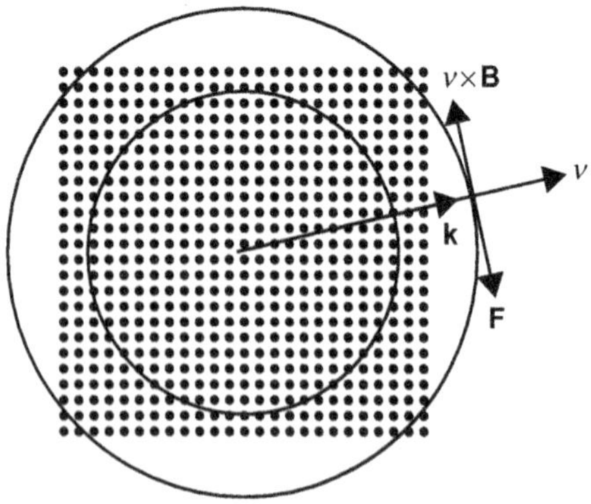

FIGURE 12.6

A magnetic field **B** is directed normally into the plane section of **k** space.

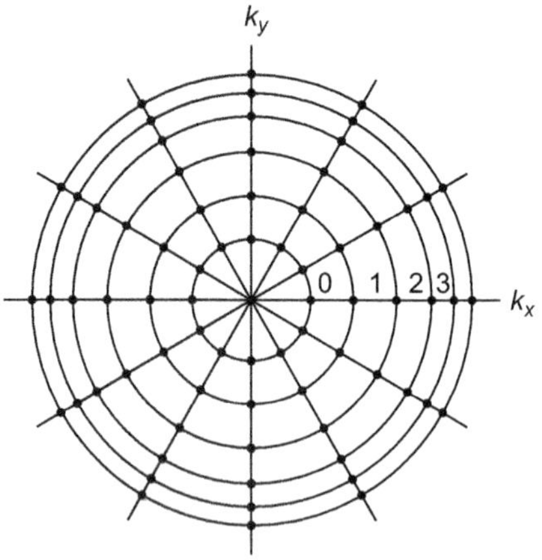

FIGURE 12.7

The Landau levels have a concentric circular form in the (k_x, k_y) plane but have cylindrical surfaces in the **k** space.

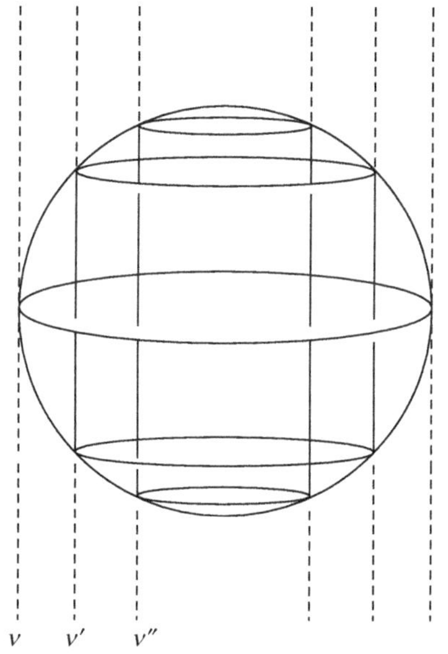

FIGURE 12.8

Landau cylinders. As the magnetic field **B** is increased, the cylinders expand until they become empty when they cross the Fermi sphere.

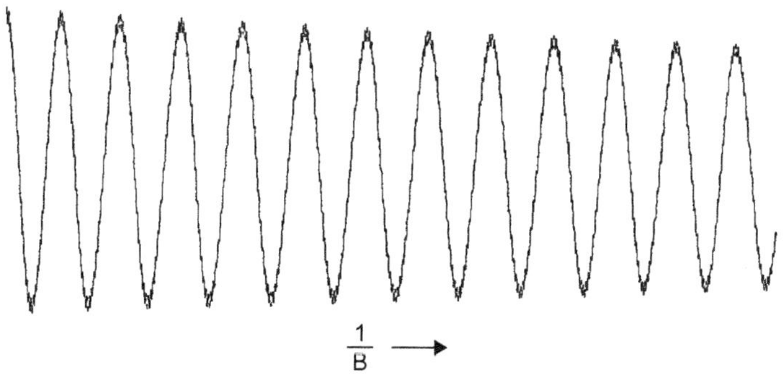

FIGURE 12.9

The de Haas–van Alphen oscillations for a metal with complex structure with decreasing magnetic field.

Figure 12.9 shows the de Haas–van Alphen oscillations for metals with complicated crystal structures as **B** is decreased. Usually, the Fermi surface of such metals has a thin neck and a thick belly. The large-scale oscillations, of which the amplitude decreases with decreasing **B**, are due to the extremal orbits around the thin neck, whereas the barely resolved small-scale oscillations are due to the extremal orbits of the thick belly.

12.4 MANY-BODY THEORY OF MAGNETIC SUSCEPTIBILITY OF BLOCH ELECTRONS IN SOLIDS

12.4.1 Introduction

Misra et al. (Ref. 14) derived an expression for the total magnetic susceptibility (χ) of Bloch electrons in solids (including spin-orbit interaction) that includes both many-body and interband effects.[14] They used a finite-temperature Green's function formalism to express the thermodynamic potential $\Omega(T, V, \mu, \mathbf{B})$ for an interacting electron system in the presence of a periodic potential $V(\mathbf{r})$, spin-orbit interaction, and external magnetic field $\mathbf{B}$ in terms of the exact one-particle propagator G. They showed that the total magnetic susceptibility of a nonferromagnetic solid can be expressed as the sum of three terms,

$$\chi = \chi_o + \chi_s + \chi_{so}, \tag{12.106}$$

where χ_o is the orbital contribution, χ_s is the spin contribution, and χ_{so} is the contribution of the spin-orbit coupling on the orbital motion of the Bloch electrons. An important aspect of their derivation is the analysis of exchange and correlation effects on each of these terms that have been explicitly calculated.

Misra et al. (Ref. 14) have shown that if they make a simple approximation for the self-energy, their expression for χ_o is reduced to the earlier results. If they make drastic assumptions while solving the matrix integral equations for the field-dependent part of the self-energy, their expression for χ_s is equivalent to the earlier results for the exchange-enhanced χ_s but with the g factor replaced by the effective g factor, a result that has been intuitively used but not yet rigorously derived. An important aspect of their derivation is the analysis of exchange and correlation effects on χ_{so} that are more subtle and cannot be included in an intuitive way.

12.4.2 Equation of Motion in the Bloch Representation

The exact one-particle propagator G satisfies the equation

$$(\xi_l - \hat{H})G(\mathbf{r}, \mathbf{r}', \xi_l) + \int d\mathbf{r}''\hat{\Sigma}(\mathbf{r}, \mathbf{r}'', \xi_l)G(\mathbf{r}'', \mathbf{r}', \xi_l) = \delta(\mathbf{r} - \mathbf{r}'), \tag{12.107}$$

where $\hat{\Sigma}$ is the exact proper self-energy operator, ξ_l is the complex energy,

$$\xi_l = \frac{(2l+1)\pi i}{\beta} + \mu, \quad l = 0, \pm 1, \pm 2, \ldots, \tag{12.108}$$

and $\hat{H}$ is the Hamiltonian of the Bloch electron in a magnetic field,

$$\hat{H} = \frac{1}{2m}\left(\mathbf{p} + \frac{e\mathbf{A}}{c}\right)^2 + \frac{\hbar^2}{4m^2c^2}\vec{\sigma} \cdot \vec{\nabla} V \times \left(\mathbf{p} + \frac{e\mathbf{A}}{c}\right) + V(\mathbf{r}) + \frac{\hbar^2}{8m^2c^2}\nabla^2 V + \frac{1}{2} g\mu_B \mathbf{B} \cdot \vec{\sigma}. \tag{12.109}$$

In the absence of the magnetic field, both G and $\hat{\Sigma}$ have the symmetry

$$G(\mathbf{r} + \mathbf{R}, \mathbf{r}' + \mathbf{R}, \xi_l) = G(\mathbf{r}, \mathbf{r}', \xi_l) \tag{12.110}$$

and

$$\hat{\Sigma}(\mathbf{r}+\mathbf{R},\, \mathbf{r}'+\mathbf{R},\, \xi_l) = \hat{\Sigma}(\mathbf{r},\, \mathbf{r}',\, \xi_l). \tag{12.111}$$

The vector potential in the Hamiltonian destroys this symmetry. It can be shown that in a symmetric gauge ($\mathbf{A} = \frac{1}{2}\, \mathbf{B}\times\mathbf{r}$), both $\hat{G}$ and $\hat{\Sigma}$ can be written as the product of a Peierls phase factor and a part that has the preceding symmetry,

$$G(\mathbf{r},\, \mathbf{r}',\, \xi_l,\, \mathbf{h}) = e^{i\mathbf{h}\cdot\mathbf{r}\times\mathbf{r}'}\, \widetilde{G}(\mathbf{r},\, \mathbf{r}',\, \xi_l,\, \mathbf{h}) \tag{12.112}$$

and

$$\hat{\Sigma}(\mathbf{r},\, \mathbf{r}',\, \xi_l,\, \mathbf{h}) = e^{i\mathbf{h}\cdot\mathbf{r}\times\mathbf{r}'}\, \widetilde{\Sigma}(\mathbf{r},\, \mathbf{r}',\, \xi_l,\, \mathbf{h}), \tag{12.113}$$

where

$$\mathbf{h} = \frac{e\mathbf{B}}{2\hbar c}. \tag{12.114}$$

Substituting Eqs. (12.112) and (12.113) in Eq. (12.108), commuting the differential operator through the Peierls phase factor, and then multiplying the left side by $e^{-i\mathbf{h}\cdot\mathbf{r}\times\mathbf{r}'}$, we obtain (Problem 12.11)

$$\begin{aligned}&\Big(\xi_l-\frac{1}{2m}[\mathbf{p}+\hbar\mathbf{h}\times(\mathbf{r}-\mathbf{r}')]^2-\frac{\hbar}{4m^2c^2}\vec{\sigma}\cdot\vec{\nabla}V\times[\mathbf{p}+\hbar\mathbf{h}\times(\mathbf{r}-\mathbf{r}')]-V(\mathbf{r})-\frac{\hbar^2}{8m^2c^2}\nabla^2V\\&-\frac{1}{2}\,g\mu_B\mathbf{B}\cdot\vec{\sigma}\Big)\widetilde{G}(\mathbf{r},\mathbf{r}',\xi_l,\mathbf{h})-\int d\mathbf{r}''e^{i\mathbf{h}\cdot(\mathbf{r}'\times\mathbf{r}+\mathbf{r}\times\mathbf{r}''+\mathbf{r}''\times\mathbf{r}')}\widetilde{\Sigma}(\mathbf{r},\mathbf{r}'',\xi_l,\mathbf{h})\widetilde{G}(\mathbf{r}'',\mathbf{r}',\xi_l,\mathbf{h})=\delta(\mathbf{r}-\mathbf{r}')\cdot\end{aligned} \tag{12.115}$$

One can write the equation of motion in the Bloch representation, i.e., in terms of the basis functions,

$$\psi_{n\mathbf{k}\rho}(\mathbf{r}) = e^{i\mathbf{k}\cdot\mathbf{r}}U_{n\mathbf{k}\rho}(\mathbf{r}), \tag{12.116}$$

where $U_{n\mathbf{k}\rho}(\mathbf{r})$ is a periodic two-component function, n is the band index, $\mathbf{k}$ is the reduced wave vector, and ρ is the spin index. Using the Bloch representation, one can show that Eq. (12.115) can be rewritten as (Problem 12.12)

$$[\xi_l-\hat{H}(\vec{\kappa},\xi_l)]\widetilde{G}(\mathbf{k},\xi_l)=I, \tag{12.117}$$

where

$$\hat{H}(\vec{\kappa},\xi_l)=\frac{1}{2m}(\mathbf{p}+\hbar\vec{\kappa})^2+V(\mathbf{r})+\frac{\hbar}{4m^2c^2}\vec{\sigma}\cdot\vec{\nabla}V\times(\mathbf{p}+\hbar\vec{\kappa})+\frac{\hbar^2}{8m^2c^2}\nabla^2V+\frac{1}{2}\,g\mu_B\mathbf{B}\cdot\vec{\sigma}+\Sigma(\vec{\kappa},\xi_l) \tag{12.118}$$

and

$$\vec{\kappa}=\mathbf{k}+i\mathbf{h}\times\nabla_{\mathbf{k}}. \tag{12.119}$$

12.4.3 Thermodynamic Potential

The grand partition function of a system is defined as

$$Z_G = Tr\{e^{-\beta(\hat{H}-\mu\hat{N})}\}, \tag{12.120}$$

where μ is the chemical potential, and $\hat{N}$ is the operator giving the number of particles. If we write Z_G in the form

$$Z_G = e^{-\beta\Omega(T,V,\mu,\mathbf{B})}, \tag{12.121}$$

all the thermodynamic properties may be derived from $\Omega(T, V, \mu, \mathbf{B})$, which is called the thermodynamic potential. It can be easily shown that the mean energy E is given by

$$\Omega = E - \mu\overline{N} - TS, \tag{12.122}$$

where $\overline{N}$ is the mean number of particles, and S is the entropy. Because $S \rightarrow 0$ when $T \rightarrow 0$, the last term in Eq. (12.122) is neglected.

It can be shown that the thermodynamic potential for an interacting system is given by

$$\Omega = \frac{1}{\beta} Tr\ \ln(-\widetilde{G}_{\xi_l}) + \frac{1}{\beta}[-Tr\widetilde{\Sigma}_{\xi_l}\widetilde{G}_{\xi_l} + \phi(\widetilde{G}_{\xi_l})] \equiv \Omega_{qp} + \Omega_{corr}. \tag{12.123}$$

Here, $\widetilde{G}_{\xi_l} \equiv \widetilde{G}(\xi_l)$ and $\widetilde{\Sigma}_{\xi_l} \equiv \widetilde{\Sigma}(G_{\xi_l})$ are the one-particle Green's function and the proper self-energy, respectively. ξ_l stands for the imaginary frequencies, Ω_{qp} is the contribution from quasiparticles, and Ω_{corr} describes the corrections from electron correlations. The functional

$$\phi(\widetilde{G}_{\xi_l}) = \lim_{\lambda\rightarrow 1} Tr\sum_n \frac{\lambda^n}{2n} \widetilde{\Sigma}^{(n)}(\widetilde{G}_{\xi_l})\widetilde{G}_{\xi_l}, \tag{12.124}$$

where Tr involves summation over both the imaginary frequencies and one-particle states, and $\Sigma^{(n)}(\widetilde{G}_{\xi_l})$ is the nth-order self-energy part, where only the interaction parameter λ occurring explicitly in Eq. (12.123) is used to determine the order. In fact, $\phi(\widetilde{G}_{\xi_l})$ is defined through the decomposition of $\Sigma^{(n)}(\widetilde{G}_{\xi_l})$ into skeleton diagrams. There are $2n$ $\widetilde{G}_{\xi_l}$ lines for the nth-order diagrams in $\phi(\widetilde{G}_{\xi_l})$. Differentiating $\phi(\widetilde{G}_{\xi_l})$ with respect to $\widetilde{G}_{\xi_l}$ has the effect of "opening" any of the $2n$ lines of the nth-order diagram, and each will give the same contribution when Tr is taken.

12.4.4 General Formula for χ

The magnetic susceptibility is calculated from the expression

$$\chi^{\mu\nu} = -\frac{1}{V}\lim_{\mathbf{B}\rightarrow 0} \frac{\partial^2\Omega}{\partial B^\mu \partial B^\nu}. \tag{12.125}$$

From Eqs. (12.121) through (12.123), it can be shown that[14]

$$\chi^{\mu\nu} = \frac{1}{V\beta}\left[-\frac{\partial^2}{\partial B^\mu \partial B^\nu} Tr\ \ln(-\widetilde{G}_{\xi_l}) + Tr\frac{\partial^2\widetilde{\Sigma}_{\xi_l}}{\partial B^\mu \partial B^\nu}\widetilde{G}_{\xi_l} + Tr\frac{\partial\widetilde{\Sigma}_{\xi_l}}{\partial B^\mu}\frac{\partial\widetilde{G}_{\xi_l}}{\partial B^\nu}\right]_{B\rightarrow 0}. \tag{12.126}$$

We can expand

$$\widetilde{\Sigma}(\vec{\kappa}, \mathbf{B}, \xi_l) = \widetilde{\Sigma}(\mathbf{k}, \mathbf{B}, \xi_l) - ih_{\alpha\beta}\frac{\partial\widetilde{\Sigma}(\mathbf{k}, \mathbf{B}, \xi_l)}{\partial k^{\alpha}}\nabla^{\beta}_{\mathbf{k}} - \frac{1}{2}h_{\alpha\beta}h_{\gamma\delta}\frac{\partial^2\widetilde{\Sigma}}{\partial k^{\alpha}\partial k^{\gamma}}\nabla^{\beta}_{\mathbf{k}}\nabla^{\delta}_{\mathbf{k}} + \ldots \tag{12.127}$$

and

$$\widetilde{\Sigma}(\mathbf{k}, \mathbf{B}, \xi_l) = \Sigma^0(\mathbf{k}, \xi_l) + B^{\mu}\Sigma^{1,\mu}(\mathbf{k}, \xi_l) + B^{\mu}B^{\nu}\Sigma^{2,\mu\nu}(\mathbf{k}, \xi_l) + \ldots, \tag{12.128}$$

where

$$h_{\alpha\beta} = \epsilon_{\alpha\beta\gamma}\, h^{\gamma}, \tag{12.129}$$

$\epsilon_{\alpha\beta\gamma}$ is the antisymmetric tensor of the third rank, and we follow the Einstein summation convention. From Eqs. (12.118), (12.119), (12.127), and (12.128) (Problem 12.13), we obtain

$$\hat{H}(\vec{\kappa}, \xi_l) = \hat{H}_0(\mathbf{k}, \xi_l) + \hat{H}'(\mathbf{k}, \xi_l), \tag{12.130}$$

where

$$\hat{H}_0(\mathbf{k}, \xi_l) = \frac{1}{2m}(\mathbf{p} + \hbar\mathbf{k})^2 + V(\mathbf{r}) + \Sigma^0(\mathbf{k}, \xi_l) + \frac{\hbar^2}{8mc^2}\nabla^2 V + \frac{\hbar}{4m^2c^2}\vec{\sigma}\cdot\vec{\nabla}V \times (\mathbf{p} + \hbar\mathbf{k}) \tag{12.131}$$

and

$$\begin{aligned}\hat{H}'(\mathbf{k}, \xi_l) = &-ih_{\alpha\beta}\prod{}^{\alpha}\nabla^{\beta}_{\mathbf{k}} + \frac{1}{2}\, g\mu_B B^{\mu}\sigma^{\mu} + B^{\mu}\Sigma^{1,\mu}(\mathbf{k}, \xi_l) - ih_{\alpha\beta}B^{\mu}\frac{\partial\Sigma^{1,\mu}}{\partial k^{\alpha}}\nabla^{\beta}_{\mathbf{k}}\\ &-\frac{1}{2}\, h_{\alpha\beta}h_{\gamma\delta}\left[\frac{\hbar^2}{m}\delta_{\alpha\gamma} + \frac{\partial^2\Sigma^0}{\partial k^{\alpha}\partial k^{\gamma}}\right]\nabla^{\beta}_{\mathbf{k}}\nabla^{\delta}_{\mathbf{k}} + B^{\mu}B^{\nu}\Sigma^{2,\mu\nu}(\mathbf{k}, \xi_l),\end{aligned} \tag{12.132}$$

where the terms up to the second order in the magnetic field are retained, and $\overrightarrow{\prod}/\hbar$ is the velocity operator,

$$\overrightarrow{\prod} = \frac{\hbar}{m}(\vec{p} + \hbar\mathbf{k}) + \frac{\hbar^2}{4m^2c^2}\vec{\sigma}\times\vec{\nabla}V + \nabla_{\mathbf{k}}\Sigma^0(\mathbf{k}, \xi_l). \tag{12.133}$$

We can make a perturbation expansion

$$\widetilde{G}(\mathbf{k}, \xi_l) = G_0(\mathbf{k}, \xi_l) + G_0(\mathbf{k}, \xi_l)H'G_0(\mathbf{k}, \xi_l) + G_0(\mathbf{k}, \xi_l)H'G_0(\mathbf{k}, \xi_l)H'G_0(\mathbf{k}, \xi_l) + \ldots, \tag{12.134}$$

where

$$G_0(\mathbf{k}, \xi_l) = \frac{1}{\xi_l - H_0(\mathbf{k}, \xi_l)}, \tag{12.135}$$

and only terms up to the second order in the magnetic field are retained. It can be shown that (Problem 12.14)

$$\nabla^{\alpha}_{\mathbf{k}}G_0(\mathbf{k}, \xi_l) = G_0(\mathbf{k}, \xi_l)\prod{}^{\alpha}G_0(\mathbf{k}, \xi_l) \tag{12.136}$$

and

$$\nabla_{\mathbf{k}}^{\alpha}\nabla_{\mathbf{k}}^{\gamma}G_0(\mathbf{k},\xi_l) = G_0\left[\frac{\hbar^2}{m}\delta_{\alpha\gamma} + X^{\alpha\gamma}\right]G_0 + G_0\Pi^{\alpha}G_0\Pi^{\gamma}G_0 + G_0\Pi^{\gamma}G_0\Pi^{\alpha}G_0, \tag{12.137}$$

where

$$X^{\alpha\gamma} = \nabla_{\mathbf{k}}^{\alpha}\left[\frac{\hbar^2}{4m^2c^2}(\vec{\sigma}\times\vec{\nabla}V)^{\gamma} + \nabla_{\mathbf{k}}^{\gamma}\Sigma^0(\mathbf{k},\xi_l)\right]. \tag{12.138}$$

After considerable algebra (for details, see Misra et al.[14]), the general expression for the total magnetic susceptibility of nonferromagnetic solids (including exchange and correlation effects) is obtained as

$$\chi^{\mu\nu} = \chi_0^{\mu\nu} + \chi_s^{\mu\nu} + \chi_{so}^{\mu\nu}, \tag{12.139}$$

where

$$\begin{aligned}
\chi_0^{\mu\nu} = \sum_{\mathbf{k}}(1+\delta_{\mu\nu})\Bigg\{&\frac{e^2\epsilon_{\alpha\beta\mu}\epsilon_{\gamma\delta\nu}}{48\hbar^2c^2}\nabla_{\mathbf{k}}^{\alpha}\nabla_{\mathbf{k}}^{\gamma}E_n\,\nabla_{\mathbf{k}}^{\beta}\nabla_{\mathbf{k}}^{\delta}E_n f'(E_n)\\
&+\Bigg[\frac{e^2\epsilon_{\alpha\beta\mu}\epsilon_{\gamma\delta\nu}}{4\hbar^2c^2}\Big(-\frac{2\hbar^2}{m}\frac{\Pi^{\alpha}_{np,mp'}\Pi^{\gamma}_{mp',np}}{E^2_{mn}}\delta_{\beta\delta} + 2\frac{\Pi^{\alpha}_{np,mp'}\Pi^{\gamma}_{mp',np''}\Pi^{\beta}_{np'',qp'''}\Pi^{\delta}_{qp''',np}}{E^2_{mn}E_{qn}}\\
&-2\frac{\Pi^{\alpha}_{np,mp'}\Pi^{\gamma}_{mp',qp''}\Pi^{\beta}_{qp'',lp'''}\Pi^{\delta}_{lp''',np}}{E_{ln}E_{qn}E_{mn}} - \frac{\Pi^{\alpha}_{np,np}\Pi^{\gamma}_{np,mp'}\Pi^{\delta}_{mp',qp''}\Pi^{\beta}_{qp'',np}}{E_{mn}E^2_{qn}}\\
&+\frac{\Pi^{\alpha}_{np,np}\Pi^{\beta}_{np,mp'}\Pi^{\gamma}_{mp',qp''}\Pi^{\delta}_{qp'',np}}{E^2_{mn}E_{qn}} - \frac{\Pi^{\beta}_{np,mp'}X^{\alpha\gamma}_{mp',qp''}\Pi^{\delta}_{qp'',np}}{E_{mn}E_{qn}}\\
&+\frac{\Pi^{\beta}_{np,mp'}\Pi^{\delta}_{mp',qp''}X^{\alpha\gamma}_{qp'',np}}{E_{mn}E_{qn}} + \frac{X^{\alpha\gamma}_{np,mp'}\Pi^{\beta}_{mp',qp''}\Pi^{\delta}_{qp'',np}}{E_{mn}E_{qn}} - \frac{X^{\alpha\gamma}_{np,np}\Pi^{\beta}_{np,mp'}\Pi^{\delta}_{mp',np}}{E^2_{mn}}\Bigg]\\
&+\frac{ie}{4\hbar c}\epsilon_{\alpha\beta\nu}\left[\frac{\Pi^{\beta}_{np,mp'}Y^{\alpha\mu}_{mp',np}}{E_{mn}} - \frac{Y^{\alpha\mu}_{np,mp'}\Pi^{\beta}_{mp',np}}{E_{mn}}\right]f(E_n)\Bigg\},
\end{aligned} \tag{12.140}$$

where

$$E_{mn} \equiv E_m - E_n, \tag{12.141}$$

and repeated indices means summation over band and spin. Similarly, one can show[14] that the effective Pauli spin susceptibility, including the exchange and correlation effects is,

$$\chi_s^{\mu\nu} = -\frac{1}{8}(1+\delta_{\mu\nu})\mu_B^2\sum_{n,\mathbf{k},p,p'} g^{\nu}_{nn}(\mathbf{k})\sigma^{\nu}_{np,np'}\left(g^{\mu}_{nn}(\mathbf{k})\sigma^{\mu}_{np',np} + \frac{2}{\mu_B}\Sigma^{1,\mu}_{np',np}\right)f'(E_n), \tag{12.142}$$

where the effective g matrix is defined as

$$g^{\nu}_{nn}(\mathbf{k})\sigma^{\nu}_{np,np'} = \frac{ie}{\mu_B\hbar c}\epsilon_{\alpha\beta\nu}\sum_{m,p''}\frac{\Pi^{\alpha}_{np,mp''}\Pi^{\beta}_{mp'',np'}}{E_{mn}} + g\sigma^{\nu}_{np,np'}. \tag{12.143}$$

The additional spin-orbit contribution to the magnetic susceptibility is[14]

$$
\chi_{so}^{\mu\nu} = \sum_{\mathbf{k}} (1+\delta_{\mu\nu}) \Bigg[\frac{e^2 \epsilon_{\alpha\beta\mu} \epsilon_{\gamma\delta\nu}}{2\hbar^2 c^2} \frac{\Pi^{\alpha}_{n\rho,m\rho'} \Pi^{\beta}_{m\rho',n\rho''} \Pi^{\gamma}_{n\rho'',q\rho'''} \Pi^{\delta}_{q\rho''',n\rho}}{E_{mn}^2 E_{qn}}
$$

$$
+ \frac{ieg\mu_B}{4\hbar c} \epsilon_{\alpha\beta\nu} \Bigg\{ -3 \frac{J^{\mu}_{n\rho,n\rho'} \Pi^{\alpha}_{n\rho',m\rho''} \Pi^{\beta}_{m\rho'',n\rho}}{E_{mn}^2} + \frac{\Pi^{\alpha}_{n\rho,m\rho'} \Pi^{\beta}_{m\rho',q\rho''} J^{\mu}_{q\rho'',n\rho}}{E_{qn} E_{mn}}
$$

$$
+ \frac{\Pi^{\alpha}_{n\rho,m\rho'} J^{\mu}_{m\rho',q\rho''} \Pi^{\beta}_{q\rho'',n\rho}}{E_{qn} E_{mn}} + \frac{J^{\mu}_{n\rho,m\rho'} \Pi^{\alpha}_{m\rho',q\rho''} \Pi^{\beta}_{q\rho'',n\rho}}{E_{qn} E_{mn}} + \frac{\Pi^{\alpha}_{n\rho,n\rho} J^{\mu}_{n\rho,m\rho''} \Pi^{\beta}_{m\rho'',n\rho}}{E_{mn}^2}
$$

$$
- \frac{\Pi^{\alpha}_{n\rho,n\rho} \Pi^{\beta}_{n\rho,m\rho'} J^{\mu}_{m\rho',n\rho}}{E_{mn}^2} \Bigg\} + \frac{1}{8} g^2 \mu_B^2 \Bigg\{ \frac{\sigma^{\mu}_{n\rho,m\rho'} F^{\nu}_{m\rho',n\rho}}{E_{mn}} + \frac{F^{\mu}_{n\rho,m\rho'} \sigma^{\nu}_{m\rho',n\rho}}{E_{mn}} \Bigg\} \Bigg] f(E_n),
$$

(12.144)

where

$$
\vec{J} = \vec{\sigma} + \frac{1}{g\mu_B} \Sigma^1,
$$

(12.145)

$$
Y^{\mu\nu} = \frac{\partial \Sigma^{1,\nu}}{\partial k^{\mu}},
$$

(12.146)

and

$$
F^{\nu} = \sigma^{\nu} + \frac{2}{g\mu_B} \Sigma^{1,\nu}.
$$

(12.147)

12.4.5 Exchange Self-Energy in the Band Model

The exchange contribution to the self-energy is local in $\mathbf{r}$ space. In the simple static screening approximation, the self-energy is independent of ξ_l. Neglecting the field dependence of screening as well as that of $v_{eff}(\mathbf{r}, \mathbf{r}')$,

$$
\widetilde{\Sigma}(\mathbf{r}, \mathbf{r}') = -\frac{1}{\beta} \sum_{\xi_l} v_{eff}(\mathbf{r}, \mathbf{r}') \widetilde{G}(\mathbf{r}, \mathbf{r}', \xi_l),
$$

(12.148)

$\widetilde{\Sigma}$ and $\widetilde{G}$ can be expanded in terms of Bloch states as follows:

$$
\widetilde{\Sigma}(\mathbf{r}, \mathbf{r}') = \sum_{n,m,\mathbf{k},\rho,\rho'} \widetilde{\Sigma}_{n\rho,m\rho'}(\mathbf{k}) \psi_{n\mathbf{k}\rho}(\mathbf{r}) \psi^*_{m\mathbf{k}\rho'}(\mathbf{r}')
$$

(12.149)

and

$$
\widetilde{G}(\mathbf{r}, \mathbf{r}') = \sum_{n,m,\mathbf{k},\rho,\rho'} \widetilde{G}_{n\rho,m\rho'}(\mathbf{k}) \psi_{n\mathbf{k}\rho}(\mathbf{r}) \psi^*_{m\mathbf{k}\rho'}(\mathbf{r}') \cdot
$$

Substituting Eqs. (12.149) and (12.150) in (12.148), we obtain

$$\begin{aligned}\sum_{n,m,\rho,\rho'} \widetilde{\Sigma}_{n\rho,m\rho'}(\mathbf{k})\psi_{n\mathbf{k}\rho}(\mathbf{r})\psi^*_{m\mathbf{k}\rho'}(\mathbf{r}') \\ = -\frac{1}{\beta}\sum_{\xi_l,p,q,\mathbf{k}',\overline{\rho},\overline{\rho}'} v_{eff}(\mathbf{r},\mathbf{r}')\widetilde{G}_{p\overline{\rho},q\overline{\rho}'}(\mathbf{k}',\xi_l)\psi_{p\mathbf{k}'\overline{\rho}}(\mathbf{r})\psi^*_{q\mathbf{k}'\overline{\rho}'}(\mathbf{r}')\cdot\end{aligned} \tag{12.151}$$

If the effective electron–electron interaction is spin independent, then $\rho=\overline{\rho}$, $\rho'=\overline{\rho}'$, and we obtain

$$\overline{\Sigma}_{n\rho,m\rho'}(\mathbf{k}) = -\frac{1}{\beta}\sum_{\mathbf{k}',\xi_l,p,q} <nm|v_{eff}(\mathbf{k},\mathbf{k}')|pq>_{\rho\rho'}\widetilde{G}_{p\rho,q\rho'}(\mathbf{k}',\xi_l), \tag{12.152}$$

where (Problem 12.15)

$$<nm|v_{eff}(\mathbf{k},\mathbf{k}')|pq>_{\rho\rho'} = \int \psi^*_{n\mathbf{k}\rho}(\mathbf{r})\psi_{m\mathbf{k}\rho'}(\mathbf{r}')v_{eff}(\mathbf{r},\mathbf{r}')\psi_{p\mathbf{k}'\rho}(\mathbf{r})\psi^*_{q\mathbf{k}'\rho'}(\mathbf{r}')d\mathbf{r}d\mathbf{r}'. \tag{12.153}$$

Eq. (12.153) is the exchange self-energy in the band model. One can obtain Σ^0, Σ^1, Σ^2, and so on (defined in Eq. 12.128), by expanding $\widetilde{G}$. We make the further approximation

$$<nn|v_{eff}(\mathbf{k},\mathbf{k}')|pq>_{\rho\rho'} \approx <nn|v_{eff}(\mathbf{k},\mathbf{k}'|pp>\delta_{pq} = v_{np}(\mathbf{k},\mathbf{k}')\delta_{pq}. \tag{12.154}$$

From Eqs. (12.152) and (12.154),

$$\widetilde{\Sigma}_{n\rho,n\rho'}(\mathbf{k}) = -\frac{1}{\beta}\sum_{\mathbf{k}',\xi_l,p} v_{np}(\mathbf{k},\mathbf{k}')\widetilde{G}_{p\rho,p\rho'}(\mathbf{k}',\xi_l)\cdot \tag{12.155}$$

Substituting only the value of the first-order terms of **B** occurring in $\widetilde{G}$ from Eq. (12.134) on the right side of Eq. (12.155), and neglecting the terms proportional to f, we obtain (Problem 12.16)

$$\Sigma^{1,\mu}_{n\rho,n\rho'}(\mathbf{k}) \simeq -\sum_{m\mathbf{k}'} v_{nm}(\mathbf{k},\mathbf{k}')\Sigma^{1,\mu}_{m\rho,m\rho'}(\mathbf{k}')f'_m(\mathbf{k}') - \frac{1}{2}\mu_B\sum_{m\mathbf{k}'} v_{nm}(\mathbf{k},\mathbf{k}')g^{\mu}_{mm}(\mathbf{k}')\sigma^{\mu}_{m\rho,m\rho'}f'_m(\mathbf{k}'). \tag{12.156}$$

Similarly, to calculate $\Sigma^1_{n\rho,m\rho'}(\mathbf{k})$, we assume

$$<nm|v_{eff}(\mathbf{k},\mathbf{k}')|pq>_{\rho\rho'} = \overline{v}_{nm}(\mathbf{k},\mathbf{k}')\delta_{np}\delta_{mq}. \tag{12.157}$$

From Eqs. (12.152) and (12.157), we obtain

$$\widetilde{\Sigma}_{n\rho,m\rho'}(\mathbf{k}) = -\frac{1}{\beta}\sum_{\mathbf{k}',\xi_l} \overline{v}_{nm}(\mathbf{k},\mathbf{k}')\widetilde{G}_{n\rho,m\rho'}(\mathbf{k}',\xi_l)\cdot \tag{12.158}$$

12.4.6 Exchange Enhancement of χ_s

We will first discuss how $\chi^{\mu\mu}_s$ gets exchange enhanced. One can rewrite Eq. (12.142) in the alternate form

$$\chi^{\mu\mu}_s = \chi^{\mu\mu}_{0,s} + \chi^{\mu\mu}_{1\cdot s}, \tag{12.159}$$

where

$$\chi_{0,s}^{\mu\mu} = -\frac{1}{4}\mu_B^2 \sum_{n,\mathbf{k},\rho,\rho'} g_{nn}^{\mu}\sigma_{n\rho,n\rho'}^{\mu} g_{nn}^{\mu}\sigma_{n\rho',n\rho}^{\mu} f'(E_n) \tag{12.160}$$

is the effective Pauli spin susceptibility for noninteracting Bloch electrons, and

$$\chi_{1,s}^{\mu\mu} = -\frac{1}{2}\mu_B \sum_{n,\mathbf{k},\rho,\rho'} g_{nn}^{\mu}\Sigma_{n\rho,n\rho'}^{1,\mu}\sigma_{n\rho',n\rho}^{\mu} f'(E_n) \tag{12.161}$$

is the contribution due to exchange and correlation. If we consider the individual band enhancement and neglect interband interactions in the expression for $\Sigma_{n\rho,n\rho'}^{1,\mu}$ in Eq. (12.156), make an average exchange enhancement ansatz, and assume $v_{nm} \simeq v_{nn}\delta_{nm}$, which is equivalent to the assumption that $\Sigma^{1,\mu}$ is independent of **k**, we obtain

$$\Sigma_{n\rho,n\rho'}^{1,\mu} = \frac{1}{2}\frac{\alpha_n}{1-\alpha_n}\mu_B g_{nn}^{\mu}\sigma_{n\rho,n\rho'}^{\mu}, \tag{12.162}$$

where

$$\alpha_n = -\sum_{\mathbf{k}',m} v_{nm}(\mathbf{k},\mathbf{k}')f'(E_m(\mathbf{k}')). \tag{12.163}$$

From Eqs. (12.159) through (12.162), we obtain

$$\chi_s^{\mu\mu} = \sum_n \frac{\chi_{0s,n}^{\mu}}{(1-\alpha_n)}, \tag{12.164}$$

where $\chi_{0s\text{-}n}^{\mu\mu}$ is the contribution to effective Pauli susceptibility for each band. Eq. (12.164) is known as the Stoner enhancement, which was obtained by making drastic assumptions while solving the matrix integral equations for $\Sigma_{n\rho,n\rho'}^{1,\mu}$. However, the neglect of coupling of interband terms, i.e., coupling between $\Sigma_{n\rho,n\rho'}^{1,\mu}$, might be too drastic for systems such as Be, Cd, and so on. It can be easily shown[14] that even in a simple two-band model, the exchange enhancement of χ_s is quite different from the simple form obtained from Eq. (12.164).

12.4.7 Exchange and Correlation Effects on χ_o

The exchange and correlation effects on χ_o arc very complicated. We consider only the first term of Eq. (12.140),

$$\chi_0^{\mu\nu} \approx \chi_{LP}^{qp} = \sum_{\mathbf{k}}(1+\delta_{\mu\nu})\frac{e^2\epsilon_{\alpha\beta\mu}\epsilon_{\gamma\delta\nu}}{48\hbar^2c^2}\nabla_{\mathbf{k}}^{\alpha}\nabla_{\mathbf{k}}^{\gamma}E_n\,\nabla_{\mathbf{k}}^{\beta}\nabla_{\mathbf{k}}^{\delta}E_n f'(E_n), \tag{12.165}$$

which is the familiar Landau–Peierls susceptibility for quasiparticles (χ_{LP}^{qp}), because the energy in the Landau–Peierls term is the quasiparticle energy. This is the well-known Sampson–Seitz prescription that had been stated without proof. If we include the effects of electron–electron interaction through an effective mass and ignore the band effects, we obtain the well-known results for χ_o,[21]

$$\chi_{LP}^{qp} = \frac{\chi_{LP}}{1+A_1/3}, \tag{12.166}$$

where A_1 is the Fermi-liquid parameter. The second through fifth terms in Eq. (12.140) are corrections to the Landau–Peierls term, which are zero for free electrons, but are of the same order as χ_{LP} for band electrons even in the absence of electron–electron interactions.[12] Therefore, while one considers many-body effects on χ_o, it is wrong to consider only χ_{LP}^{qp} as was done in earlier calculations.

12.4.8 Exchange and Correlation Effects on χ_{so}

The effect of electron–electron interactions is different on the various terms in χ_{so} in Eq. (12.144). A discussion of these effects is beyond the scope of this book except to note that even in the absence of exchange and correlation effects, the contributions of χ_{so} are of the same order as χ_s for some metals and semiconductors.

12.5 QUANTUM HALL EFFECT

12.5.1 Introduction

von Klitzing et al.[20] first observed the integer quantum Hall effect (QHE) in a two-dimensional electron gas formed by an inversion layer at an Si/SiO_2 interface (discussed in Chapter 11). We note that a two-dimensional electron gas can be formed at the semiconductor surface if the electrons are fixed close to the surface by an external electric field in either Silicon MOSFETS or GaAs-$Al_xGa_{1-x}As$ heterostructures (Chapter 11). In fact, a two-dimensional electron gas is essential for the observation of the quantum Hall effect. In addition to the quantum phenomena connected with the confinement of electrons within a two-dimensional layer, the Landau quantization of the electron motion in a strong magnetic field is necessary for the interpretation of the quantum Hall effect. We will discuss the integer quantum Hall effect in detail and briefly mention the fractional quantum Hall effect, discovered soon after, because a detailed discussion involving many-body theory is beyond the scope of this book.

12.5.2 Two-Dimensional Electron Gas

The energy of mobile electrons in semiconductors can be written as

$$E = \frac{\hbar^2}{2m^*}(k_x^2 + k_y^2 + k_z^2). \tag{12.167}$$

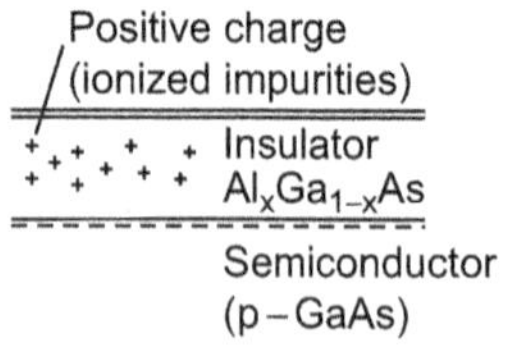

FIGURE 12.10

Two-dimensional electron gas formed close to the semiconductor surface of GaAs-$Al_xGa_{1-x}As$ heterostructures by an external electric field along the *z* direction.

When the energy for the motion in the z direction is fixed by using a triangular potential with an infinite barrier at the surface ($z = 0$) and with a constant electric field F_s for $z \geq 0$ (z is positive downward), one obtains a quasi-two-dimensional electron gas (as shown in Figure 12.10).

The electrons are confined close to the surface due the electrostatic field F_s normal to the interface originating from the positive charges, which causes a drop in the electron potential toward the surface.

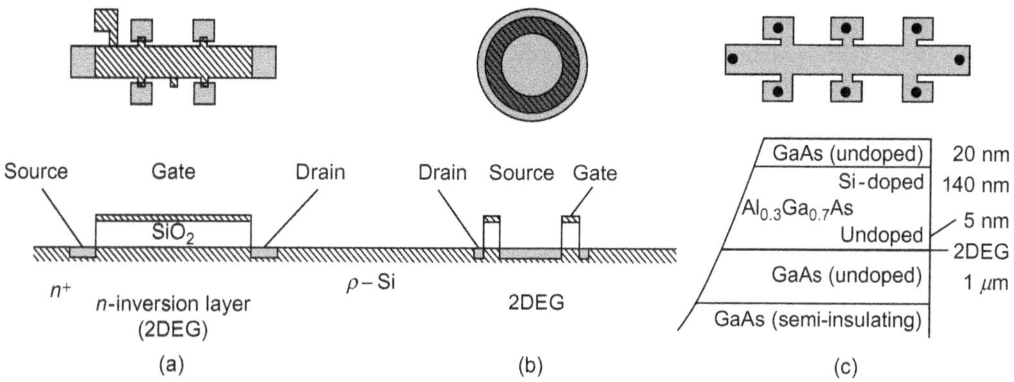

FIGURE 12.11

Typical geometries and cross-sections of devices used in the quantum Hall experiments. From left to right: (a) Long silicon device (Hall geometry) with potential probes (for R_x measurements) and Hall probes (for R_H measurements): typical length: 0.5 mm. (b) Circular MOS device for σ_{xy} measurements. (c) Cross-section and top view of a GaAs–$Al_{0.3}Ga_{0.7}As$ heterostructure with Hall geometry.

Reproduced from von Klitzing[19] with the permission of Elsevier.

The energy of the carriers is grouped into electric sub-bands corresponding to the quantized levels for the motion in the z direction provided the potential well is small compared to the de Broglie wavelength of the electrons. At very low temperatures (T < 4° K), if the carrier densities of the two-dimensional electron gas are small such that only the lowest electric sub-band E_0 is occupied with electrons (the electric quantum limit that occurs when the Fermi energy E_F is small compared with the sub-band separation $E_1 - E_0$), the energy spectrum can be written as

$$E = E_0 + \frac{\hbar^2 k_{||}^2}{2m^*}, \tag{12.168}$$

where $k_{||}$ is a wave vector within the plane of the electron gas.

The experimental arrangement for QHE measurements is shown in Figure 12.11. For measurement of current, heavily doped n^+ contacts are used as current contacts and potential probes at the semiconductor surface.

The resistivity component ρ_{xx} is directly proportional to $\sigma_{xx} = \sigma_{yy}$, $\sigma_{xy} = -\sigma_{yx}$. Hence, $\rho_{xx} = \sigma_{xx}/(\sigma_{xx}^2 + \sigma_{yy}^2)$. This means that the condition $\sigma_{xx} = 0$ (fully occupied Landau levels) leads to $\rho_{xx} = 0$. Thus, a correct value for the quantized Hall resistance is $R_H = h/e^2 n$, which is expected only under the condition $\rho_{xx} = 0$.

12.5.3 Quantum Transport of a Two-Dimensional Electron Gas in a Strong Magnetic Field

When a strong magnetic field **B** is applied such that B_z is normal to the interface, the two-dimensional electrons move in cyclotron orbits parallel to the surface. The energy levels can be expressed as

$$E_n = E_0 + \left(n + \frac{1}{2}\right)\hbar\omega_c + gs\mu_B B, \tag{12.169}$$

where ω_c is the cyclotron frequency ($\omega_c = -eB/m^*$), and s is the spin quantum number, $s = \pm\frac{1}{2}$. Laughlin (Ref. 8) derived an expression of a 2DEG in a strong magnetic field by considering the isotropic effective-mass Hamiltonian

$$\hat{H} = \frac{1}{2m^*}\left[\mathbf{p} + \frac{e}{c}\mathbf{A}\right]^2 - eE_0 y, \tag{12.170}$$

where the y coordinate is related to the vector potential (in the Landau gauge),

$$\mathbf{A} = By\hat{x}. \tag{12.171}$$

From Eqs. (12.170) and (12.171), the wave functions are given by

$$\psi_{k,n} = e^{ikx}\phi_n(y - y_0), \tag{12.172}$$

where ϕ_n is the solution of the harmonic-oscillator equations

$$\frac{1}{2m^*}\left[p_y^2 + (\frac{e}{c}B)^2 y^2\right]\phi_n = (n + \frac{1}{2})\hbar\omega_c\phi_n \tag{12.173}$$

and

$$y_0 = \frac{1}{\omega_c}\left[\frac{\hbar k}{m^*} - \frac{cE_0}{B}\right]. \tag{12.174}$$

The energy of the state is

$$E_{n,k} = (n + \frac{1}{2})\hbar\omega_c - eE_0 y_0 + \frac{1}{2}m^*(cE_0 B)^2. \tag{12.175}$$

The y_0 are changed by a vector potential increment $\Delta A\hat{x}$ only through the location of their centers,

$$y_0 \to y_0 - \Delta A/B. \tag{12.176}$$

It is obvious from Eqs. (12.175) and (12.176) that the energy changes linearly with ΔA.

From Eq. (12.174), one can obtain the degeneracy factor for each Landau level that is given by the number of center coordinates y_0 (note that $y - y_0$ is a good quantum number) in the sample. For a two-dimensional electron gas confined in a device of dimensions L_x, L_y,

$$\Delta y_0 = \frac{1}{\omega_c}\frac{\hbar\Delta k}{m^* c} = \frac{-\hbar}{eB}\Delta k = \frac{-\hbar}{eB}\frac{2\pi}{L_x} = \frac{-h}{eBL_x}. \tag{12.177}$$

The degeneracy factor is given by

$$N_0 = \frac{L_y}{\Delta y_0} = \frac{-L_x L_y eB}{h}, \tag{12.178}$$

(note that e is negative) which is the same as the number of the flux quanta in the sample. The degeneracy factor per unit area (from Eq. 12.178) is

$$N = \frac{N_0}{L_x L_y} = \frac{-eB}{h}. \tag{12.179}$$

Thus, the degeneracy factor for each Landau level is independent of the effective mass and other semiconductor parameters.

It can be shown that the Hall voltage U_H of a two 2DEG with a surface carrier density n_s ($n_s = nN$, when n energy levels are fully occupied) is

$$U_H = -\frac{B}{n_s e} I, \tag{12.180}$$

where I is the current in the sample. From Eqs. (12.179) and (12.180), the Hall resistance

$$R_H = \frac{U_H}{I} = -\frac{B}{n_s e} = -\frac{B}{nNe} = \frac{h}{ne^2}, \tag{12.181}$$

where $n = 1, 2, 3, \ldots$. Thus, whenever

$$n = -\frac{n_s h}{eB} \tag{12.182}$$

is an integer (by adjusting the magnetic field B and the density of states n_s), the Hall resistance is quantized. When the condition outlined in Eq. (12.182) is satisfied, there is no current flow in the direction of the electric field, and hence, the conductivity $\sigma_{xx} = 0$. The electrons move like free particles perpendicular to the electric field. The quantized plateaus in the Hall resistance are shown in Figure 12.12.

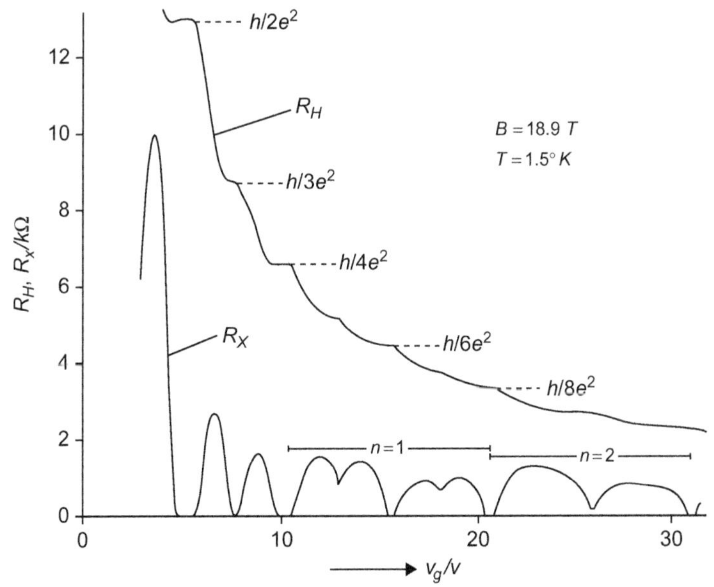

FIGURE 12.12

Gate voltage dependence of the Hall resistance R_H and resistivity R_x at $B = 18.9$ T for a long silicon MOS device at $B = 18.9$ T and $T = 1.5$°K.

Reproduced from von Klitzing[19] with the permission of Elsevier.

12.5.4 Quantum Hall Effect from Gauge Invariance

Laughlin (Refs. 7-9) considered a two-dimensional metallic loop (see Figure 12.13) pierced by a magnetic field B normal to its surface. A voltage U_H is applied between the two edges of the ring. When $\sigma_{xx} = 0$, the energy is conserved and Faraday's law of induction can be written as

$$I = c\frac{\partial E}{\partial \phi}, \qquad (12.183)$$

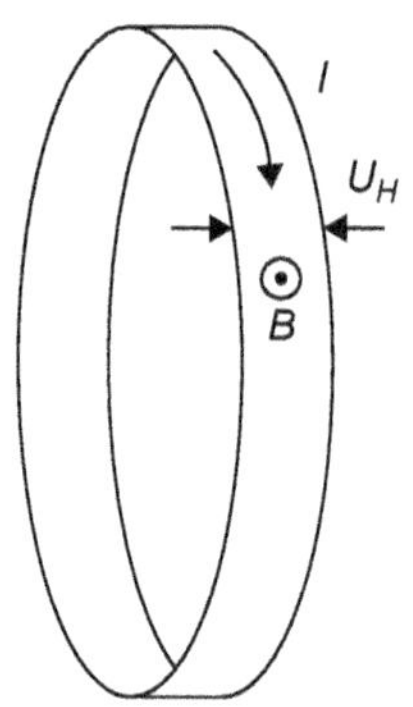

FIGURE 12.13

Quantized Hall resistance for a model two-dimensional metallic loop.

where E is the total energy of the system, and ϕ is the magnetic flux threading the loop. If $\phi \to \phi + \Delta\phi$, where $\Delta\phi \equiv \phi_0 = -hc/e$ is a flux quantum, the wave function enclosing the flux changes by a factor 2π, which implies $k \to k + (2\pi)/L$, where L is the circumference of the ring.

The change in energy ΔE when the states are transported from one edge to the other is

$$\Delta E = -neU_H, \qquad (12.184)$$

where n corresponds to the filled Landau levels. From Eqs. (12.183) and (12.184), we obtain an expression for the dissipationless Hall current and the Hall voltage,

$$I = c\frac{\partial E}{\partial \phi} = c\frac{\Delta E}{\Delta \phi} = -\frac{cneU_H}{\phi_0} = \frac{ne^2 U_H}{h}. \qquad (12.185)$$

The quantized Hall resistance is obtained from the expression

$$R_H = \frac{U_H}{I} = \frac{h}{ne^2}. \qquad (12.186)$$

12.6 FRACTIONAL QUANTUM HALL EFFECT

There have been numerous papers on the fractional quantum Hall effect. It is generally interpreted on the elementary excitations of quasiparticles with a charge $e/3$, $e/5$, $e/7$, and so on. A typical experimental result is shown in Figure 12.14.

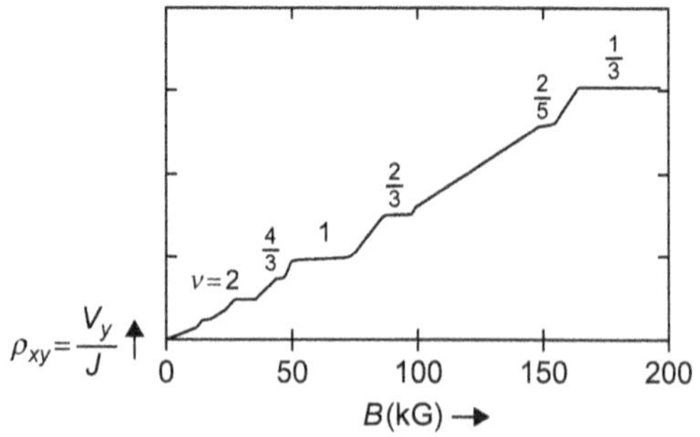

FIGURE 12.14

Schematic diagram of fractional quantum Hall effect in GaAs/$Ga_{1-x}Al_xAs$ heterostructure.

PROBLEMS

12.1. The Hamiltonian of an atom (or ion) in a uniform magnetic field **B** (in the z direction) can be written as

$$H = \frac{1}{2m}\sum_i \left(\mathbf{p}_i + \frac{e\mathbf{A}(\mathbf{r}_i)}{c}\right)^2 + g_0\mu_B B\sum_i \hat{\mathbf{s}}_z^i. \tag{1}$$

If

$$\mathbf{A}(\mathbf{r}_i) = -\frac{1}{2}\,\mathbf{r}\times\mathbf{B}, \tag{2}$$

show that Eq. (1) can be rewritten in the alternate form

$$\hat{H} = \frac{1}{2m}\sum_i \hat{\mathbf{p}}_i^2 + \mu_B(\hat{\mathbf{L}} + g_0\hat{\mathbf{S}})\cdot\mathbf{B} + \frac{e^2}{8mc^2}B^2\sum_i (x_i^2 + y_i^2). \tag{3}$$

12.2. We have derived

$$<JLSJ_z|(\hat{\mathbf{L}} + 2\hat{\mathbf{S}})|JLSJ_z'> = g(JLS)<JLSJ_z|\,\hat{\mathbf{J}}\,|JLSJ_z'>. \tag{1}$$

Show that Eq. (1) can be rewritten in the alternate form

$$<JLSJ_z|\hat{\mathbf{L}} + 2\hat{\mathbf{S}}|J'L'S'J_z'> = g(JLS)<JLSJ_z|\,\hat{\mathbf{J}}\,|J'L'S'J_z'>. \tag{2}$$

12.3. We have proved that

$$\begin{aligned}&\sum_{J''L''S''J_z''} <JLSJ_z|(\hat{\mathbf{L}} + 2\hat{\mathbf{S}})|J''L''S''J_z''>\cdot<J''L''S''J_z''|\,\hat{\mathbf{J}}\,|J'L'S'J_z'>\\ &= g(JLS)\sum_{J''L''S''J_z''} <JLSJ_z|\,\hat{\mathbf{J}}\,|J''L''S''J_z''>\cdot<J''L''S''J_z''|\,\hat{\mathbf{J}}\,|J'L'S'J_z'>.\end{aligned} \tag{1}$$

Because the sum over Eq. (1) is taken over a complete set, using Eqs. (12.31) and (12.32), show that Eq. (1) can be rewritten in the alternate form

$$<JLSJ_z|(\hat{\mathbf{L}} + 2\hat{\mathbf{S}})\cdot\hat{\mathbf{J}}|JLSJ_z'> = g(JLS)<JLSJ_z|\,\hat{\mathbf{J}}^2|JLSJ_z'>. \tag{2}$$

12.4. Using the standard relation for the total angular momentum of an electron in an atom,

$$\hat{\mathbf{J}} = \hat{\mathbf{L}} + \hat{\mathbf{S}}, \tag{1}$$

show that

$$(\hat{\mathbf{L}} + 2\hat{\mathbf{S}})\cdot(\hat{\mathbf{L}} + \hat{\mathbf{S}}) = \hat{\mathbf{L}}^2 + 2\hat{\mathbf{S}}^2 + \hat{\mathbf{L}}\cdot\hat{\mathbf{J}} + 2\hat{\mathbf{S}}\cdot\hat{\mathbf{J}} = \frac{1}{2}[3\hat{\mathbf{J}}^2 - \hat{\mathbf{L}}^2 + \hat{\mathbf{S}}^2]. \tag{2}$$

12.5. The Helmholtz free energy is obtained from

$$e^{-\beta F} = \frac{e^{\beta\gamma B\left(J+\frac{1}{2}\right)} - e^{-\beta\gamma B\left(J+\frac{1}{2}\right)}}{e^{\beta\gamma B/2} - e^{-\beta\gamma B/2}}. \tag{1}$$

The magnetization of N ions in a volume V is defined as

$$M = -\frac{N}{V}\frac{\partial F}{\partial B}. \tag{2}$$

From Eqs. (1) and (2), show that

$$M = \frac{N}{V}\gamma J B_J(\beta\gamma J B), \tag{3}$$

where the Brillouin function $B_J(x)$ is defined by

$$B_J(x) = \frac{(2J+1)}{2J}\coth\left(\frac{2J+1}{2J}x\right) - \frac{1}{2J}\coth\left(\frac{1}{2J}x\right). \tag{4}$$

12.6. From the equations

$$E_\nu = \frac{\hbar^2 k_z^2}{2m} + \left(\nu + \frac{1}{2}\right)\hbar\omega_c, \quad \nu = 0, 1, 2, \ldots \tag{1}$$

and

$$g(\nu, k_z)\,dk_z = \frac{2}{(2\pi)^2}\frac{m\omega_c}{\hbar}\,dk_z, \tag{2}$$

show that

$$g(E, \nu)\,dE = \frac{2}{(2\pi)^2}\frac{\hbar\omega_c}{2}\left(\frac{2m}{\hbar^2}\right)^{3/2}\left(E - \left(\nu + \frac{1}{2}\right)\hbar\omega_c\right)^{-1/2} dE. \tag{3}$$

12.7. In the limiting case, B → 0, the summation

$$g(E)dE = \sum_0^{\nu'} g(E, \nu)\,dE \tag{1}$$

can be replaced by an integration because the sub-bands of different quantum numbers move very close to each other. By carrying out this integration (using Eq. 3 from Problem 12.6), and substituting $B = 0$, show that

$$g(E)dE = \frac{1}{2\pi^2}\left(\frac{2m}{\hbar^2}\right)^{3/2} E^{1/2} dE, \tag{2}$$

which is the density of states of free electrons.

12.8. Show from Eqs. (12.53) through (12.55) and (12.69) that the expression for magnetization per unit volume can be written as

$$M = -\frac{d}{dB}\left\{ n\mu - \frac{1}{\beta}\int_0^\infty g(E_+)\ \ln[e^{\beta(\mu-E)}+1]\,dE_+ - \frac{1}{\beta}\int_0^\infty g(E_-)\ \ln[e^{\beta(\mu-E)}+1]\,dE_- \right\}. \quad (1)$$

12.9. The Fermi integral $F(x)$ is given by

$$F(x) = \int_0^\infty \frac{y^{1/2}}{1+e^{(y-x)}}\,dy\cdot \quad (1)$$

Show that

$$F(x) \approx \frac{\sqrt{\pi}}{2}e^x, \quad \text{for } x<0$$

and

$$F(x) \approx \frac{2}{3}x^{3/2}, \quad \text{for } x>0. \quad (2)$$

12.10. The magnetization per unit volume is given by (Problem 12.8)

$$M = -\frac{d}{dB}\left\{ n\mu - \frac{1}{\beta}\int_0^\infty g(E_+)\ \ln[e^{\beta(\mu-E)}+1]\,dE_+ - \frac{1}{\beta}\int_0^\infty g(E_-)\ \ln[e^{\beta(\mu-E)}+1]\,dE_- \right\}. \quad (1)$$

By retaining the first term in the expansion similar to the one outlined in the text, show that at very low temperatures

$$M = \frac{3n\mu_B^2 B}{2E_F}\left[1-\frac{1}{3}+\frac{\pi k_B T}{\mu_B B}\left(\frac{E_F}{\mu_B B}\right)^{1/2}\sum_{\nu=1}^{\infty}\frac{(-1)^\nu}{\sqrt{\nu}}\cos(\pi\nu)\frac{\sin\left(\frac{\pi}{4}-\frac{\pi\nu E_F}{\mu_B B}\right)}{\sinh\frac{\pi^2\nu k_B T}{\mu_B B}}\right], \quad (2)$$

where E_F is the Fermi energy (the value of μ at $T = 0$).

12.11. Substituting Eqs. (12.112) and (12.113) in Eq. (12.108), commuting the differential operator through the Peierls phase factor, and then multiplying the left side of the equation by $e^{-i\mathbf{h}\cdot\mathbf{r}\times\mathbf{r}'}$, show that

$$\begin{aligned}&(\xi_l - \frac{1}{2m}[\mathbf{p}+\hbar\mathbf{h}\times(\mathbf{r}-\mathbf{r}')]^2 - \frac{\hbar}{4m^2c^2}\vec{\sigma}\cdot\vec{\nabla}V\times[\mathbf{p}+\hbar\mathbf{h}\times(\mathbf{r}-\mathbf{r}')] - V(\mathbf{r}) - \frac{\hbar^2}{8m^2c^2}\nabla^2 V \\ &- \frac{1}{2}g\mu_B\mathbf{B}\cdot\vec{\sigma})\widetilde{G}(\mathbf{r},\mathbf{r}',\xi_l,\mathbf{h}) - \int d\mathbf{r}''e^{i\mathbf{h}\cdot(\mathbf{r}'\times\mathbf{r}+\mathbf{r}\times\mathbf{r}''+\mathbf{r}''\times\mathbf{r}')}\widetilde{\Sigma}(\mathbf{r},\mathbf{r}'',\xi_l,\mathbf{h})\widetilde{G}(\mathbf{r}'',\mathbf{r}',\xi_l,\mathbf{h}) \\ &= \delta(\mathbf{r}-\mathbf{r}')\cdot\end{aligned} \quad (1)$$

12.12. Using the Bloch representation, $\psi_{n\mathbf{k}\rho}(\mathbf{r}) = e^{i\mathbf{k}\cdot\mathbf{r}}U_{n\mathbf{k}\rho}(\mathbf{r})$, show that Eq. (1) in Problem 12.11 can be rewritten as

$$\sum_{n'',\rho'',\mathbf{k}',\mathbf{k}''}\int d\mathbf{r}d\mathbf{r}'d\mathbf{r}''e^{-i\mathbf{k}\cdot\mathbf{r}}U^*_{n\mathbf{k}\rho}(\mathbf{r})\begin{pmatrix}\xi_l-\frac{1}{2m}[\mathbf{p}+\hbar\mathbf{h}\times(\mathbf{r}-\mathbf{r}')]^2-\frac{\hbar}{4m^2c^2}\vec{\sigma}\cdot\vec{\nabla}V\times[\mathbf{p}+\hbar\mathbf{h}\times(\mathbf{r}-\mathbf{r}')]\\ -V(\mathbf{r})-\frac{\hbar^2}{8m^2c^2}\nabla^2V-\frac{1}{2}\,g\mu_B\mathbf{B}\cdot\vec{\sigma}\end{pmatrix}$$

$$\times e^{-\mathbf{k}''\cdot(\mathbf{r}-\mathbf{r}'')}U_{n''\mathbf{k}''\rho''}(\mathbf{r})U^*_{n''\mathbf{k}''\rho''}(\mathbf{r}'')\widetilde{G}(\mathbf{r}'',\,\mathbf{r}',\,\xi_l,\,\mathbf{h})U_{n'\mathbf{k}'\rho'}(\mathbf{r}')e^{i\mathbf{k}'\cdot\mathbf{r}'}$$

$$+\sum_{n'',\rho'',\mathbf{k}',\mathbf{k}''}\int d\mathbf{r}d\mathbf{r}'d\mathbf{r}''d\mathbf{r}'''e^{-i\mathbf{k}\cdot\mathbf{r}}U^*_{n\mathbf{k}\rho}(\mathbf{r})e^{i\mathbf{h}\cdot(\mathbf{r}'\times\mathbf{r}+\mathbf{r}\times\mathbf{r}''+\mathbf{r}''\times\mathbf{r}')}\widetilde{\Sigma}(\mathbf{r},\,\mathbf{r}'')e^{i\mathbf{k}''\cdot(\mathbf{r}''-\mathbf{r}''')}$$

$$U_{n''\mathbf{k}''\rho''}(\mathbf{r}'')U^*_{n''\mathbf{k}''\rho''}(\mathbf{r}''')\times\widetilde{G}(\mathbf{r}''',\,\mathbf{r}',\,\xi_l,\,\mathbf{h})U_{n'\mathbf{k}'\rho'}(\mathbf{r}')e^{i\mathbf{k}'\cdot\mathbf{r}'}=\delta_{nn'}\delta_{\rho\rho'}\,. \tag{1}$$

By introducing a change of variables $\mathbf{R}_1=\mathbf{r}''-\mathbf{r}'$, and $\mathbf{R}_2=\frac{1}{2}\,(\mathbf{r}'+\mathbf{r}'')$ in the first term, $\mathbf{R}_1=\mathbf{r}-\mathbf{r}''$, $\mathbf{R}_2=\frac{1}{2}\,(\mathbf{r}+\mathbf{r}'')$, $\mathbf{R}_3=\mathbf{r}'''-\mathbf{r}'$, and $\mathbf{R}_4=\frac{1}{2}\,(\mathbf{r}'+\mathbf{r}''')$ in the second term, and by using partial integration of the type

$$\begin{aligned}&\sum_{\mathbf{k}''}(\mathbf{r}-\mathbf{r}')e^{i\mathbf{k}''\cdot(\mathbf{r}-\mathbf{r}')}e^{i\mathbf{k}''\cdot(\mathbf{r}'-\mathbf{r}'')}U_{n''\mathbf{k}''\rho''}(\mathbf{r})U^*_{n''\mathbf{k}''\rho''}(\mathbf{r}'')\\ &\qquad=\sum_{\mathbf{k}''}e^{i\mathbf{k}''\cdot(\mathbf{r}-\mathbf{r}')}i\nabla_{\mathbf{k}''}e^{i\mathbf{k}''\cdot(\mathbf{r}'-\mathbf{r}'')}U_{n''\mathbf{k}''\rho''}(\mathbf{r})U^*_{n''\mathbf{k}''\rho''}(\mathbf{r}''),\end{aligned} \tag{2}$$

show that Eq. (1) can be rewritten in the form

$$\sum_{n'',\rho''}[\xi_l-\hat{H}(\vec{\kappa}',\,\xi_l)]_{n\mathbf{k}\rho,n''\mathbf{k}\rho''}\widetilde{G}_{n''\mathbf{k}\rho'',n'\mathbf{k}'\rho'}(\mathbf{k}',\,\xi_l)|_{\mathbf{k}'=\mathbf{k}}=\delta_{nn'}\delta_{\rho\rho'}, \tag{3}$$

where

$$\vec{\kappa}=\mathbf{k}+i\mathbf{h}\times\nabla_{\mathbf{k}}. \tag{4}$$

$$\hat{H}(\vec{\kappa},\xi_l)=\frac{1}{2m}(\mathbf{p}+\hbar\vec{\kappa})^2+V(\mathbf{r})+\frac{\hbar}{4m^2c^2}\vec{\sigma}\cdot\vec{\nabla}V\times(\mathbf{p}+\hbar\vec{\kappa})+\frac{\hbar^2}{8m^2c^2}\nabla^2V+\frac{1}{2}\,g\mu_B\mathbf{B}\cdot\vec{\sigma}+\widetilde{\Sigma}(\vec{\kappa},\,\xi_l), \tag{5}$$

$$\widetilde{\Sigma}_{n\mathbf{k}\rho,n''\mathbf{k}\rho''}(\vec{\kappa}',\xi_l)=\int d\mathbf{r}d\mathbf{r}'U^*_{n''\mathbf{k}\rho''}(\mathbf{r})e^{-i\vec{\kappa}'\cdot(\mathbf{r}-\mathbf{r}')}\widetilde{\Sigma}(\mathbf{r},\,\mathbf{r}',\,\xi_l)U_{n''\mathbf{k}\rho''}(\mathbf{r}'), \tag{6}$$

and

$$\widetilde{G}_{n''\mathbf{k}\rho'',n'\mathbf{k}\rho'}(\mathbf{k}',\xi_l)=\int d\mathbf{r}d\mathbf{r}'U^*_{n''\mathbf{k}\rho''}(\mathbf{r})\widetilde{G}(\mathbf{r},\,\mathbf{r}',\,\xi_l)e^{-i\mathbf{k}'\cdot(\mathbf{r}-\mathbf{r}')}U_{n'\mathbf{k}\rho'}(\mathbf{r}'). \tag{7}$$

Hence, show that because the $U_{n\mathbf{k}\rho}$'s form a complete set of functions, Eq. (3) can be rewritten in the alternate form

$$[\xi_l - \hat{H}(\vec{\kappa}, \xi_l)]\widetilde{G}(\mathbf{k}, \xi_l) = I. \quad (8)$$

12.13. From Eqs. (12.118), (12.119), (12.127), and (12.128), show that

$$\hat{H}(\vec{\kappa}, \xi_l) = \hat{H}_0(\mathbf{k}, \xi_l) + \hat{H}'(\mathbf{k}, \xi_l), \quad (1)$$

where

$$\hat{H}_0(\mathbf{k}, \xi_l) = \frac{1}{2m}(\mathbf{p} + \hbar\mathbf{k})^2 + V(\mathbf{r}) + \Sigma^0(\mathbf{k}, \xi_l) + \frac{\hbar^2}{8mc^2}\nabla^2 V + \frac{\hbar}{4m^2c^2}\vec{\sigma}\cdot\vec{\nabla}V \times (\mathbf{p} + \hbar\mathbf{k}), \quad (2)$$

and

$$\begin{aligned}\hat{H}'(\mathbf{k}, \xi_l) = &-ih_{\alpha\beta}\prod{}^{\alpha}\nabla^{\beta}_{\mathbf{k}} + \frac{1}{2}g\mu_B B^{\mu}\sigma^{\mu} + B^{\mu}\Sigma^{1,\mu}(\mathbf{k}, \xi_l) - ih_{\alpha\beta}B^{\mu}\frac{\partial\Sigma^{1,\mu}}{\partial k^{\alpha}}\nabla^{\beta}_{\mathbf{k}} \\ &-\frac{1}{2}h_{\alpha\beta}h_{\gamma\delta}\left[\frac{\hbar^2}{m}\delta_{\alpha\gamma} + \frac{\partial^2\Sigma^0}{\partial k^{\alpha}\partial k^{\gamma}}\right]\nabla^{\beta}_{\mathbf{k}}\nabla^{\delta}_{\mathbf{k}} + B^{\mu}B^{\nu}\Sigma^{2,\mu\nu}(\mathbf{k}, \xi_l),\end{aligned} \quad (3)$$

where the terms up to the second order in the magnetic field are retained and $\vec{\prod}/\hbar$ is the velocity operator,

$$\vec{\prod} = \frac{\hbar}{m}(\vec{p} + \hbar\mathbf{k}) + \frac{\hbar^2}{4m^2c^2}\vec{\sigma}\times\vec{\nabla}V + \nabla_{\mathbf{k}}\Sigma^0(\mathbf{k}, \xi_l). \quad (4)$$

12.14. The temperature Green's function operator is defined as

$$G_0(\mathbf{k}, \xi_l) = \frac{1}{\xi_l - H_0(\mathbf{k}, \xi_l)}, \quad (1)$$

where $H_0(\mathbf{k}, \xi_l)$ is defined in Eq. (12.131). Show that

$$\nabla^{\alpha}_{\mathbf{k}}G_0(\mathbf{k}, \xi_l) = G_0(\mathbf{k}, \xi_l)\prod{}^{\alpha}G_0(\mathbf{k}, \xi_l) \quad (2)$$

and

$$\nabla^{\alpha}_{\mathbf{k}}\nabla^{\gamma}_{\mathbf{k}}G_0(\mathbf{k}, \xi_l) = G_0\left[\frac{\hbar^2}{m}\delta_{\alpha\gamma} + X^{\alpha\gamma}\right]G_0 + G_0\prod{}^{\alpha}G_0\prod{}^{\gamma}G_0 + G_0\prod{}^{\gamma}G_0\prod{}^{\alpha}G_0, \quad (3)$$

where

$$X^{\alpha\gamma} = \nabla^{\alpha}_{\mathbf{k}}\left[\frac{\hbar^2}{4m^2c^2}(\vec{\sigma}\times\vec{\nabla}V)^{\gamma} + \nabla^{\gamma}_{\mathbf{k}}\Sigma^0(\mathbf{k}, \xi_l)\right]. \quad (4)$$

12.15. We have derived

$$\sum_{n,m,\rho,\rho'} \widetilde{\Sigma}_{n\rho,m\rho'}(\mathbf{k})\psi_{n\mathbf{k}\rho}(\mathbf{r})\psi^*_{m\mathbf{k}\rho'}(\mathbf{r}') = -\frac{1}{\beta}\sum_{\xi_l,p,q,\mathbf{k}',\overline{\rho},\overline{\rho}'} v_{eff}(\mathbf{r},\mathbf{r}')\widetilde{G}_{p\overline{\rho},q\overline{\rho}'}(\mathbf{k}',\xi_l)\psi_{p\mathbf{k}'\overline{\rho}}(\mathbf{r})\psi^*_{q\mathbf{k}'\overline{\rho}'}(\mathbf{r}')\cdot \quad (1)$$

If the effective electron–electron interaction is spin independent, then $\rho = \overline{\rho}$, $\rho' = \overline{\rho}'$. Show that

$$\widetilde{\Sigma}_{n\rho,m\rho'}(\mathbf{k}) = -\frac{1}{\beta}\sum_{\mathbf{k}',\xi_l,p,q} <nm|v_{eff}(\mathbf{k},\mathbf{k}')|pq>\rho\rho'\,\widetilde{G}_{p\rho,q\rho'}(\mathbf{k}',\xi_l), \quad (2)$$

where

$$<nm|v_{eff}(\mathbf{k},\mathbf{k}')|pq>\rho\rho' = \int \psi^*_{n\mathbf{k}\rho}(\mathbf{r})\psi_{m\mathbf{k}\rho'}(\mathbf{r}')v_{eff}(\mathbf{r},\mathbf{r}')\psi_{p\mathbf{k}'\rho}(\mathbf{r})\psi^*_{q\mathbf{k}'\rho'}(\mathbf{r}')d\mathbf{r}d\mathbf{r}'. \quad (3)$$

12.16. From Eqs. (12.152) and (12.154), we have obtained

$$\widetilde{\Sigma}_{np,np'}(\mathbf{k}) = -\frac{1}{\beta}\sum_{\mathbf{k}',\xi_l,p} v_{np}(\mathbf{k},\mathbf{k}')\widetilde{G}_{pp,pp'}(\mathbf{k}',\xi_l)\cdot \quad (1)$$

Substituting only the value of the first-order terms of **B** occurring in $\widetilde{G}$ from Eq. (12.134) on the right side of Eq. (1), and neglecting the terms proportional to f, show that

$$\Sigma^{1,\mu}_{np,np'}(\mathbf{k}) \simeq -\sum_{m\mathbf{k}'} v_{nm}(\mathbf{k},\mathbf{k}')\Sigma^{1,\mu}_{mp,mp'}(\mathbf{k}')f'_m(\mathbf{k}') - \frac{1}{2}\mu_0\sum_{m\mathbf{k}'} v_{nm}(\mathbf{k}')\sigma^{\mu}_{mp,mp'}f'_m(\mathbf{k}). \quad (2)$$

12.17. The isotropic effective-mass Hamiltonian (Eq. 12.170) is given by

$$\hat{H} = \frac{1}{2m^*}\left[\mathbf{p} + \frac{e}{c}\mathbf{A}\right]^2 - eE_0 y, \quad (1)$$

where the y coordinate is related to the vector potential (in the Landau gauge),

$$\mathbf{A} = By\hat{x}. \quad (2)$$

From Eqs. (1) and (2), show that the wave functions are given by

$$\psi_{k,n} = e^{ikx}\phi_n(y - y_0), \quad (3)$$

where ϕ_n is the solution of the harmonic-oscillator equations

$$\frac{1}{2m^*}\left[p_y^2 + \left(\frac{e}{c}B\right)^2 y^2\right]\phi_n = \left(n + \frac{1}{2}\right)\hbar\omega_c\phi_n \quad (4)$$

and

$$y_0 = \frac{1}{\omega_c}\left[\frac{\hbar k}{m^*} - \frac{cE_0}{B}\right]. \quad (5)$$

References

1. Aschroft NW, Mermin ND. *Solid state physics*. New York: Brooks/Cole; 1976.
2. Harrison WA. *Solid state theory*. New York: McGraw-Hill; 1969.
3. Kittel C. *Introduction to solid state physics*. New York: John Wiley & Sons; 1976.
4. Landau LD. Diamagnetism of metals. *Z Physik* 1930;**64**:629.
5. Landau LD. On the de Haas-van Alphen effect. *Proc Roy Soc (London)* 1939;**A170**:363.
6. Landau LD, Lifshitz LM. *Statistical physics part 1*. Oxford: Pergamon Press; 1980.
7. Laughlin RB. Quantized Hall conductivity in two dimensions. *Phys Rev B* 1981;**23**:5632.
8. Laughlin RB. Impurities and edges in the quantum Hall effect *Surf. Sci* 1982;**22**:113.
9. Laughlin RB. Anomalous quantum Hall effect: An incompressible quantum fluid with fractionally charged excitations. *Phys Rev Lett* 1983;**50**:1395.
10. Madelung O. *Introduction to solid state theory*. New York: Springer-Verlag; 1978.
11. Marder MP. *Condensed matter physics*. New York: John Wiley & Sons; 2000.
12. Misra PK, Roth LM. Theory of diamagnetic susceptibility of metals. *Phys Rev* 1969;**177**:1089.
13. Misra PK, Kleinman L. Theory of magnetic susceptibility of Bloch electrons. *Phys Rev B* 1972;**5**:4581.
14. Misra SK, Misra PK, Mahanti SD. Many-body theory of magnetic susceptibility of electrons in solids. *Phys Rev B* 1982;**26**:1903.
15. Myers HP. *Introduction to solid state physics*. London: Taylor & Francis; 1990.
16. Prang RE, Girvin SM. *The quantum hall effect*. New York: Springer-Verlag; 1987.
17. ter Haar D, editor. Collected papers of *L.D. Landau*. New York: Gordon and Breach; 1965.
18. van Vleck JH. *Theory of electric and magnetic susceptibilities*. Oxford: Oxford University Press; 1932.
19. von Klitzing K. The quantized Hall effect. *Rev Mod Phys* 1986;**58**:519.
20. von Klitzing K, Dorda G, Pepper M. New method for high-accuracy determination of the fine structure constant based on quantized Hall resistance. *Phys Rev Lett* 1980;**45**:494.
21. White RM. *Quantum theory of magnetism*. New York: McGraw-Hill; 1970.
22. Ziger HJ, Pratt GW. *Magnetic interactions in solids*. Oxford: Oxford University Press; 1970.
23. Ziman JM. *Principles of the theory of solids*. Cambridge: Cambridge University Press; 1972.

CHAPTER

Magnetic Ordering

13

CHAPTER OUTLINE

13.1 Introduction 410
13.2 Magnetic Dipole Moments 411
13.3 Models for Ferromagnetism and Antiferromagnetism 412
13.3.1 Introduction 412
13.3.2 Heitler–London Approximation 412
13.3.3 Spin Hamiltonian 414
13.3.4 Heisenberg Model 416
13.3.5 Direct, Indirect, and Superexchange 416
13.3.6 Spin Waves in Ferromagnets: Magnons 417
13.3.7 Schwinger Representation 417
13.3.8 Application to the Heisenberg Hamiltonian 418
13.3.9 Spin Waves in Antiferromagnets 421
13.4 Ferromagnetism in Solids 422
13.4.1 Ferromagnetism Near the Curie Temperature 422
13.4.2 Comparison of Spin-Wave Theory with the Weiss Field Model 424
13.4.3 Ferromagnetic Domains 425
13.4.4 Hysteresis 426
13.4.5 Ising Model 427
13.5 Ferromagnetism in Transition Metals 427
13.5.1 Introduction 427
13.5.2 Stoner Model 428
13.5.3 Ferromagnetism in Fe, Co, and Ni from Stoner's Model and Kohn–Sham Equations 430
13.5.4 Free Electron Gas Model 431
13.5.5 Hubbard Model 433
13.6 Magnetization of Interacting Bloch Electrons 434
13.6.1 Introduction 434
13.6.2 Theory of Magnetization 434
13.6.3 The Quasiparticle Contribution to Magnetization 435
13.6.4 Contribution of Correlations to Magnetization 436
13.6.5 Single-Particle Spectrum and the Criteria for Ferromagnetic Ground State 437
13.7 The Kondo Effect 439
13.8 Anderson Model 439

13.9 The Magnetic Phase Transition 440
13.9.1 Introduction 440
13.9.2 The Order Parameter 441
13.9.3 Landau Theory of Second-Order Phase Transitions 441
Problems 443
References 448

13.1 INTRODUCTION

The phenomena of magnetic ordering such as ferromagnetism, antiferromagnetism, and ferrimagnetism are very complex. Ferromagnets have been known to exist for thousands of years in the shape of loadstones. However, ferromagnetism in transition metals, which is one of the most important as well as complex phenomena in physics, remains one of the major unsolved problems in solid state physics and is not well understood compared to most other physical properties. We will discuss some of the important models used in the theory of magnetic ordering used in solids before discussing some specific cases, including that of ferromagnetism in transition metals.

There are three different types of "magnetic ordering," known as ferromagnetism, antiferromagnetism, and ferrimagnetism. The ferromagnetic solid has a nonvanishing magnetic moment even in the absence of an external magnetic field. This is known as "spontaneous magnetization," which is a result of parallel orientation of the individual magnetic moments that must be due to interactions between these moments. In a ferromagnet, because all the individual moments are aligned in the same direction, there is a net total moment even in the absence of a magnetic field. As the temperature is increased, these orientations become gradually disordered, and at a critical temperature known as the Curie temperature, the spontaneous magnetization vanishes. This situation is far more complicated than the simple model used at $T=0$. In the preceding discussion, we tacitly assumed that the electrons of an ion in a lattice are tightly bound so that each ion has a net magnetic moment.

In some other types of solids, the ions of the nearest neighbors have antiparallel spin. The ground state consists of two sublattices of identical ions having opposite spin directions. This is known as an antiferromagnet, in which the two sublattices have mutually compensating magnetic moments. Thus, the net magnetic moment of an antiferromagnet is zero.

In ferrimagnets, there are usually two types of basis atoms and therefore two sublattices because the ions in the individual sublattices are different. The individual ions in each sublattice will possess a magnetic moment, and each sublattice will have a net magnetic moment in the ground state. The total moment in the ground state will be the vector sum of the moments of the sublattices. For sublattices with opposite magnetic moments, the net magnetic moment will be the difference between the two moments. Such types of solids are known as ferrimagnets. These differences are illustrated in Figure 13.1.

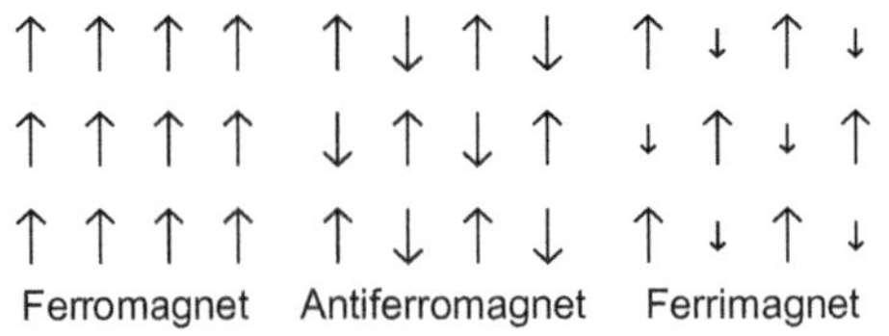

FIGURE 13.1

Ferromagnetic, antiferromagnetic, and ferrimagnetic states.

13.2 MAGNETIC DIPOLE MOMENTS

The simplest problem involves magnetic dipole moments and how nearby magnetic dipoles interact with each other even though such interaction does not lead to ferromagnetism.

The magnetic dipole moment **m** of a current distribution **j** can be defined as

$$\mathbf{m} = \int d\mathbf{r} \, \frac{1}{2c} \mathbf{r} \times \mathbf{j}(\mathbf{r}). \tag{13.1}$$

We note that **m** will be independent of the origin provided $\int d\mathbf{r} \, \mathbf{j}(\mathbf{r}) = 0$, which implies that the current distribution is over a closed loop and vanishes except at the origin. The Lorentz force on a current distribution is

$$\mathbf{F} = \frac{1}{c} \int d\mathbf{r} \, \mathbf{j}(\mathbf{r}) \times \mathbf{B}(\mathbf{r}). \tag{13.2}$$

Because $\mathbf{j}(\mathbf{r})$ vanishes except at the origin, we can expand $\mathbf{B}(\mathbf{r})$ in a Taylor series and write

$$\mathbf{F} = \frac{1}{c} \int d\mathbf{r} \, \mathbf{j}(\mathbf{r}) \times [\mathbf{B}(0) + (\mathbf{r} \cdot \vec{\nabla})\mathbf{B}(0) + \cdots]. \tag{13.3}$$

Using vector identities, we can easily show from Eqs. (13.1) through (13.3) that (Problem 13.1)

$$\mathbf{F} = (\mathbf{m} \times \vec{\nabla}) \times \mathbf{B} = \vec{\nabla}(\mathbf{m} \cdot \mathbf{B}), \tag{13.4}$$

where $\mathbf{B} \equiv \mathbf{B}(0)$. Thus, the potential energy U of a dipole in an external magnetic field is

$$U = -\mathbf{m} \cdot \mathbf{B}. \tag{13.5}$$

When two dipoles are close to each other, they interact with each other's magnetic fields.

The induction **B** produced by a magnetic dipole of moment $\mathbf{m}_1$ at a distance **r**, where a second dipole of moment $\mathbf{m}_2$ is located, is given by

$$\mathbf{B} = \vec{\nabla}\left[\mathbf{m}_1 \cdot \vec{\nabla} \frac{1}{r}\right] = \frac{3\hat{\mathbf{r}}(\mathbf{m}_1 \cdot \hat{\mathbf{r}}) - \mathbf{m}_1}{r^3}. \tag{13.6}$$

From Eqs. (13.5) and (13.6), we obtain the expression for the direct dipolar interaction energy of two magnetic dipoles separated by **r**,

$$U = \frac{1}{r^3}[\mathbf{m}_1 \cdot \mathbf{m}_2 - 3(\mathbf{m}_1 \cdot \hat{\mathbf{r}})(\mathbf{m}_2 \cdot \hat{\mathbf{r}})]. \tag{13.7}$$

One can calculate the energy scale for the dipole interaction from Eq. (13.7), which is of the order of 10^{-4} eV (equivalent to $1°$ K) for solids in which the distance between magnetic moments is of the order of 2 Å, while the electrostatic energy difference between two atomic states is on the order of 0.1–1.0 eV. Thus, dipolar interactions, which are important in explaining the phenomenon of ferromagnetic domains, are not the source of magnetic interaction responsible for ferromagnetism.

13.3 MODELS FOR FERROMAGNETISM AND ANTIFERROMAGNETISM

13.3.1 Introduction

Various models describe ferromagnetism and antiferromagnetism. The direct exchange model assumes that the electrons of a lattice ion, which contribute to the magnetic moment, are tightly bound so that the ions can be assumed to be isolated. However, the nearest neighbors are sufficiently close enough for a significant exchange interaction between them. The spins of nearest neighbors in the Bravais lattice of a ferromagnet (where electrons are tightly bound in the lattice ion) are aligned parallel in the ground state by the exchange interaction. In the absence of such interactions, the moments would be thermally disordered due to random orientations, and there would be no magnetic moment.

In the superexchange model, an exchange between magnetic ions occurs over large distances in an insulator, in which a paramagnetic ion between them facilitates the interaction. An example is MnO, in which two metallic ions (Mn) with unfilled *d*-shells are linked by an oxygen atom that has two *p*-electrons with spins in opposite directions. Each *d*-electron will interact with one of the two *p*-electrons, and because the two *p*-electrons are linked by the Pauli principle, there is an effective interaction between the two *d*-electrons that is known as the superexchange.

In the indirect exchange (RKKY) model (Refs. 10, 23, 31), the localized spins of a lattice ion interact with the conduction electrons of a metal. Essentially, the electrons mediate between the interaction of the lattice ions. This ion–ion interaction via conduction electrons plays a major role in rare-earth metals, which have a variety of ordered magnetism.

In the transition metals, which form the most significant group of ferromagnetic metals, the spins of itinerant electrons give rise to ferromagnetism. The *d*- and *s*-bands are only partly filled, and hence, there is a superposition of 3*d* and 4*s* bands. We will discuss thc occurrence of ferromagnetism in transition metals as a special category.

13.3.2 Heitler–London Approximation

The Heitler–London (Ref. 6) approximation is designed to describe the interaction between two spins arising from Coulomb forces between the two electrons of two adjacent atoms, as shown in Figure 13.2. It was originally designed to explain the bonding of a hydrogen molecule and is also an example of the valence bond method. We further note that the Heitler–London theory does not yield good results for the binding energy of the hydrogen molecule. The reason is that when two protons are close, the system looks like the helium atom with $Z=2$, whereas in the Heitler–London theory, the electrons are in *s* states for $Z=1$. However, it features the essential ingredients of the exchange

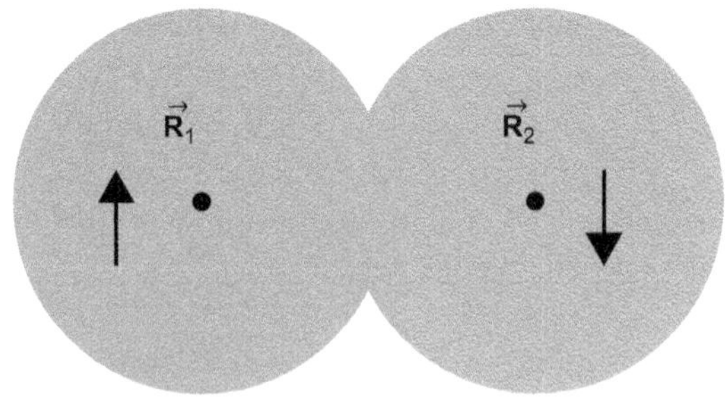

FIGURE 13.2

The overlapping wave functions of electrons (of opposite spin) of two adjacent hydrogen atoms. The protons are located at $\vec{\mathbf{R}}_1$ and $\vec{\mathbf{R}}_2$. The arrows are the directions of the spins of the electrons.

interaction used in the theory of magnetism in insulators, and hence, we will discuss the theory in some detail.

When the two hydrogen atoms ($i = 1$ and $i = 2$) are infinitely apart, the Schrodinger equation for each atom can be written as

$$\left[-\frac{\hbar^2\nabla_i^2}{2m} - \frac{e^2}{|\mathbf{r}_i - \mathbf{R}_i|}\right]\phi_i(\mathbf{r}_i) = \varepsilon_0\phi_i(\mathbf{r}_i), \tag{13.8}$$

which is the spatial part of the wave function. The wave function of the two atoms is just the product of the individual atomic wave functions. When the two atoms come much closer, there will be considerable overlap, and a molecule is formed. The individual atomic wave functions will not be orthogonal. In fact,

$$\int \phi_1^*(\mathbf{r})\phi_2(\mathbf{r})d\mathbf{r} = I, \tag{13.9}$$

where I is the overlap integral, which can be positive or negative. The Hamiltonian can be written as

$$\hat{H} = \frac{-\hbar^2\nabla_1^2}{2m} - \frac{\hbar^2\nabla_2^2}{2m} - \frac{e^2}{|\mathbf{r}_1 - \mathbf{R}_1|} - \frac{e^2}{|\mathbf{r}_1 - \mathbf{R}_2|} - \frac{e^2}{|\mathbf{r}_2 - \mathbf{R}_1|} - \frac{e^2}{|\mathbf{r}_2 - \mathbf{R}_2|} + \frac{e^2}{|\mathbf{R}_1 - \mathbf{R}_2|} + \frac{e^2}{|\mathbf{r}_1 - \mathbf{r}_2|}. \tag{13.10}$$

Here, the electrostatic interactions between the two electrons and the two protons, with all permutations, have been included in the Hamiltonian of the molecule. It may be noted that the wave function of the molecule is no longer the product of the individual atomic functions. The spin component of the wave function has to be included such that the total wave function will be antisymmetric when the particle numbers are exchanged. The spin operators commute with the spatial part of the Hamiltonian. The spin eigenfunctions are the eigenfunctions of the commuting operators $\hat{S}^2$ and $\hat{S}_z$ and form either one singlet or three triplet states. The spin singlet state of the molecule is given by

$$\chi_s = \frac{1}{\sqrt{2}}[\chi_\uparrow(1)\chi_\downarrow(2) - \chi_\downarrow(1)\chi_\uparrow(2)], \tag{13.11}$$

which is antisymmetric, and the three-spin triplet states are

$$\begin{aligned}\chi_t &= \chi_\uparrow(1)\chi_\uparrow(2),\\ &= \frac{1}{\sqrt{2}}[\chi_\uparrow(1)\chi_\downarrow(2) + \chi_\downarrow(2)\chi_\uparrow(1)],\\ &= \chi_\downarrow(1)\chi_\downarrow(2),\end{aligned} \tag{13.12}$$

which are all symmetric. Here, the symbols in the parentheses denote the spinors for the electrons at $\mathbf{r}_1$ and $\mathbf{r}_2$. Thus, the normalized spatial part of the wave functions of the singlet and triplet spin states are

$$\psi_{s,t}(\mathbf{r}_1, \mathbf{r}_2) = [2(1 \pm I^2)]^{-\frac{1}{2}}[\phi_1(\mathbf{r}_1)\phi_2(\mathbf{r}_2) \pm \phi_2(\mathbf{r}_1)\phi_1(\mathbf{r}_2)], \tag{13.13}$$

where the s, t symbols as well as the $+(-)$ signs are for the singlet (triplet) spin states.

From Eqs. (13.8) through (13.10) and Eq. (13.13), we obtain (Problem 13.3)

$$\varepsilon_{s,t} = <\hat{H}>_{s,t} = 2\varepsilon_0 + \frac{e^2}{R_{12}} + e^2\int d\mathbf{r}_1 d\mathbf{r}_2\, \psi_{s,t}^*(\mathbf{r}_1, \mathbf{r}_2)\left[\frac{1}{|\mathbf{r}_1 - \mathbf{r}_2|} - \frac{1}{|\mathbf{r}_1 - \mathbf{R}_2} - \frac{1}{|\mathbf{r}_2 - \mathbf{R}_1|}\right]\psi_{s,t}(\mathbf{r}_1, \mathbf{r}_2), \tag{13.14}$$

where

$$\frac{1}{R_{12}} \equiv \frac{1}{|\mathbf{R}_1 - \mathbf{R}_2|}. \tag{13.15}$$

We define the Coulomb and exchange integrals as

$$V_c(R_{12}) = e^2 \int d\mathbf{r}_1 d\mathbf{r}_2 |\phi_1(\mathbf{r}_1)|^2 |\phi_2(\mathbf{r}_2)|^2 \left[\frac{1}{|\mathbf{r}_1 - \mathbf{r}_2|} - \frac{1}{|\mathbf{r}_1 - \mathbf{R}_1|} - \frac{1}{|\mathbf{r}_2 - \mathbf{R}_2|}\right] \tag{13.16}$$

and

$$V_{ex}(R_{12}) = e^2 \int d\mathbf{r}_1 d\mathbf{r}_2 \phi_1^*(\mathbf{r}_1)\phi_2^*(\mathbf{r}_2) \left[\frac{1}{|\mathbf{r}_1 - \mathbf{r}_2|} - \frac{1}{|\mathbf{r}_1 - \mathbf{R}_2|} - \frac{1}{|\mathbf{r}_2 - \mathbf{R}_1|}\right] \phi_2(\mathbf{r}_1)\phi_1(\mathbf{r}_2). \tag{13.17}$$

From Eqs. (13.14) through (13.17), it can be shown that (Problem 13.4)

$$\varepsilon_s = 2\varepsilon_0 + \frac{e^2}{R} + \frac{V_c(R_{12}) + V_{ex}(R_{12})}{(1 + I^2)} \tag{13.18}$$

and

$$\varepsilon_t = 2\varepsilon_0 + \frac{e^2}{R} + \frac{V_c(R_{12}) - V_{ex}(R_{12})}{(1 - I^2)}. \tag{13.19}$$

From Eqs. (13.18) and (13.19), we obtain

$$\varepsilon_t - \varepsilon_s = \frac{2(I^2 V_e - V_s)}{1 - I^4} = -J. \tag{13.20}$$

Heitler and London (Ref. 6) found that J is negative, which implies that $\varepsilon_t > \varepsilon_s$. Because the singlet state is of lower energy, the spins of the two atoms are in opposite directions, which is an example of two-atom antiferromagnetism.

13.3.3 Spin Hamiltonian

Dirac and Heisenberg argued that the original Hamiltonian in Eq. (13.10) acts only on the spatial degrees of freedom and yields two eigenvalues, ε_s and ε_t (one singlet and three degenerate triplet states, depending on whether the spins of the two electrons are parallel or antiparallel). They showed that the same results can be obtained by considering a Hamiltonian that involves only the spin degrees of freedom and that is more useful in the study of magnetism.

The actual wave function of the Heitler–London problem (Ref. 6) is the product of the spin (Eqs. 13.11 or 13.12) and spatial part of the wave function (Eq. 13.13), i.e.,

$$\Psi_s = \psi_s \chi_s \tag{13.21}$$

and

$$\Psi_t = \psi_t \chi_t. \tag{13.22}$$

It is much more convenient to represent the Hamiltonian in a spin representation as long as the energy in the singlet (ε_s) and triplet (ε_t) states derived in Eqs. (13.18) and (13.19) by using the spatial Hamiltonian of the two-electron system are the same in the new representation. This is possible because the coupling in the spin Hamiltonian depends only on the relative orientation of the two spins but not on their directions on $\mathbf{R}_1 - \mathbf{R}_2$. If there are dipolar interactions or spin-orbit coupling, which breaks the rotational symmetry of the Hamiltonian in the Heitler–London model, one must add additional terms to the spin Hamiltonian. Thus, the spin Hamiltonian can be written as

$$\hat{H}^{spin} = \lambda_1 + \lambda_2 \hat{\mathbf{S}}_1 \cdot \hat{\mathbf{S}}_2, \tag{13.23}$$

where λ_1 and λ_2 are such that

$$\hat{H}^{spin} \Psi_s = \varepsilon_s \Psi_s \tag{13.24}$$

and

$$\hat{H}^{spin} \Psi_t = \varepsilon_t \Psi_t. \tag{13.25}$$

We also note that for a two-electron system,

$$\hat{\mathbf{S}}_1 \cdot \hat{\mathbf{S}}_2 = \frac{1}{2}[\hat{S}_1^+ \hat{S}_2^- + \hat{S}_2^+ \hat{S}_1^-] + \hat{S}_1^z \hat{S}_2^z, \tag{13.26}$$

where

$$\hat{S}^{\pm} = \hat{S}_x \pm i\hat{S}_y. \tag{13.27}$$

It can be easily shown from Eqs. (13.11), (13.12), and (13.24) through (13.26) that (Problem 13.5)

$$\hat{\mathbf{S}}_1 \cdot \hat{\mathbf{S}}_2 \chi_s = -\frac{3}{4}\chi_s \tag{13.28}$$

and

$$\hat{\mathbf{S}}_1 \cdot \hat{\mathbf{S}}_2 \chi_t = \frac{1}{4}\chi_t. \tag{13.29}$$

From Eqs. (13.24) through (13.29), we obtain

$$\hat{H}^{spin} = \frac{1}{4}(\varepsilon_s + 3\varepsilon_t) - (\varepsilon_s - \varepsilon_t)\hat{\mathbf{S}}_1 \cdot \hat{\mathbf{S}}_2, \tag{13.30}$$

such that

$$\hat{H}^{spin} \Psi_s = \varepsilon_s \Psi_s \tag{13.31}$$

and

$$\hat{H}^{spin} \Psi_t = \varepsilon_t \Psi_t \tag{13.32}$$

for each of the three triplet states. Comparing Eqs. (13.23) and (13.30), we obtain

$$\lambda_1 = \frac{1}{4}(\varepsilon_s + 3\varepsilon_t) \tag{13.33}$$

and

$$\lambda_2 = (\varepsilon_s - \varepsilon_t). \tag{13.34}$$

From Eqs. (13.18) through (13.20), (13.23), (13.33), and (13.34), we obtain

$$\hat{H}^{spin} = 2\varepsilon_0 + \frac{e^2}{R} + \frac{1}{4}\left[\frac{V_c(R_{12}) + V_{ex}(R_{12})}{1+l^2} + 3\frac{V_c(R_{12} - V_{ex}(R_{12})}{1-l^2}\right] - J\hat{\mathbf{S}}_1 \cdot \hat{\mathbf{S}}_2. \quad (13.35)$$

By shifting the zero of energy, we can rewrite Eq. (13.35) in the convenient form

$$\hat{H}^{spin} = -J\hat{\mathbf{S}}_1 \cdot \hat{\mathbf{S}}_2. \quad (13.36)$$

Because $J = \varepsilon_s - \varepsilon_t$ (from Eq. 13.20), a positive value of J means that $\varepsilon_t < \varepsilon_s$. This implies that the two spins are aligned in the same direction (ferromagnetism). Similarly, a negative value of J implies that the spins are aligned in the opposite direction. This alternate change in the direction of the spins is known as antiferromagnetism.

13.3.4 Heisenberg Model

Heisenberg (Ref. 5) extended Eq. (13.36) for the general case of a large collection of magnetic ions placed in a lattice. He postulated that the Hamiltonian can be written as

$$\hat{H}^{spin} = -\sum_{ij}{}' J_{ij}\hat{\mathbf{S}}_i \cdot \hat{\mathbf{S}}_j, \quad (13.37)$$

where the exchange integral J_{ij} is a function of the positions of the lattice sites $\mathbf{R}_i$ and $\mathbf{R}_j$ (note that $\mathbf{R}_i \neq \mathbf{R}_j$). We further note that the exchange integral J_{ij}, which is a function of $\mathbf{R}_i - \mathbf{R}_j$, is large only for one- or two-lattice spacing. It is not possible to "derive" the Heisenberg Hamiltonian, but there have been many attempts to "derive" it or at least make reasonable attempts to justify it. Nevertheless, it remains the starting point for the theory of ferromagnetism and antiferromagnetism.

13.3.5 Direct, Indirect, and Superexchange

The Heisenberg model is the result of the direct overlap of wave functions of two magnetic ions. This is due to the direct exchange interaction between localized spins of nearest neighbors, and the basic assumption is that the electrons (of a lattice ion) are tightly bound such that the ion can be considered as isolated, but the nearest neighbors are close enough for exchange interaction to occur.

However, there are other mechanisms for exchange. In indirect exchange, the localized spins of the lattice ions interact with the conduction electrons of a metal. Thus, the information on the spin over a given ion is passed on by an electron to another ion. Hence, the interaction between the two ions is mediated by conduction electrons. This indirect ion–ion interaction is known as the RKKY interaction (Refs. 10, 20, 23) and plays a major role in the rare-earth metals (Tm and Gd).

In the superexchange mechanism in an insulator, the exchange between magnetic ions often occurs over large distances. A paramagnetic ion (an ion with closed electronic shells) between the two magnetic ions facilitates the interaction. For example, in MnO, two metallic ions with unfilled d-shells are linked by an oxygen atom. Each d-electron interacts with one of the two p-electrons of the spin-saturated outermost electron pair of the oxygen atom. There is an effective interaction between the two d-electrons, known as superexchange, because the two spins of the two p-electrons are linked by the Pauli principle. These three types of interactions are shown schematically in Figure 13.3.

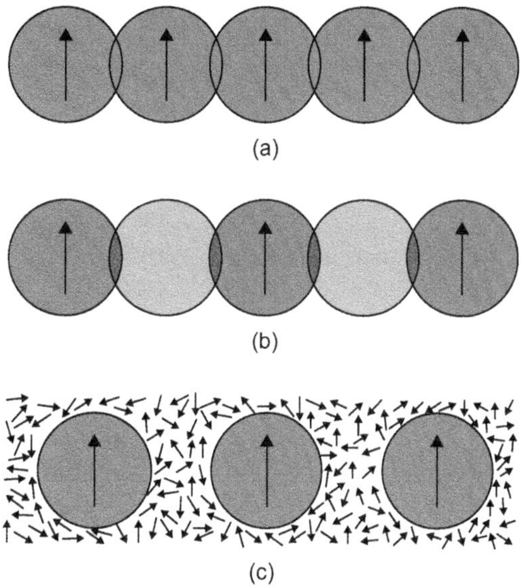

FIGURE 13.3

(a) Direct exchange, (b) superexchange, and (c) indirect exchange.

13.3.6 Spin Waves in Ferromagnets: Magnons

For ferromagnets, the ground state of the Heisenberg model is such that all the J_{ij} are positive and all the spins point in the same direction. It can be easily shown from Eq. (13.35) that if all the spins point in the same direction (for instance, z), the ground-state energy is

$$<\uparrow\uparrow\uparrow\uparrow \cdots\cdot |\hat{H}|\uparrow\uparrow\uparrow\uparrow \cdots\cdot ...> = -\sum_{<ii'>}{}' \frac{J_{ii'}}{4}. \tag{13.38}$$

The ferromagnetic state of the ground state is degenerate. Even though the spins have to point in the same direction, there is no preference for a particular direction like z direction. The low-energy excitons can be constructed by slowly twisting the local spin orientation while propagating through the crystal. These propagations are known as spin waves.

13.3.7 Schwinger Representation

We note that in an excited state, the spin of one or more ions at a lattice site is reversed from ↑ to ↓. Each time such a spin reversal occurs, the net spin change is 1. Thus, the appropriate operators to represent such changes are Bose operators. Schwinger (Ref. 24) proposed a formal theory of spin waves by using a representation in which the spins are represented by Bose operators even though the Heisenberg Hamiltonian is a product of Fermion operators. In this representation, the Bose operators are defined as

$$\hat{S}^{\alpha} = \frac{1}{2}\sum_{ij} \hat{a}_i^{\dagger}\sigma_{ij}^{\alpha}\hat{a}_j, \tag{13.39}$$

where σ^{α} are the Pauli spin matrices ($\alpha = x, y, z$). It is easy to show from Eq. (13.39) that (Problem 13.6)

$$\begin{aligned} \hat{S}^x &= \frac{1}{2}(\hat{a}_1^{\dagger}\hat{a}_2 + \hat{a}_2^{\dagger}\hat{a}_1), \\ \hat{S}^y &= \frac{i}{2}(\hat{a}_2^{\dagger}\hat{a}_1 - \hat{a}_1^{\dagger}\hat{a}_2), \text{ and} \\ \hat{S}^z &= \frac{1}{2}(\hat{a}_1^{\dagger}\hat{a}_1 - \hat{a}_2^{\dagger}\hat{a}_2). \end{aligned} \tag{13.40}$$

It can be easily shown from Eq. (13.40) that

$$[\hat{S}^x, \hat{S}^y] = i\hat{S}^z. \tag{13.41}$$

If one introduces the operators

$$\hat{S}^{\pm} = \hat{S}^x \pm i\hat{S}^y, \tag{13.42}$$

it can be shown from Eqs. (13.40) and (13.42) that

$$\hat{S}^{+} = \hat{a}_1^{\dagger}\hat{a}_2 \tag{13.43}$$

and

$$\hat{S}^{-} = \hat{a}_2^{\dagger}\hat{a}_1. \tag{13.44}$$

Because the total spin is $2S$, one can write

$$\hat{a}_1^{\dagger}\hat{a}_1 + \hat{a}_2^{\dagger}\hat{a}_2 = 2S. \tag{13.45}$$

Through use of the Holstein–Primakoff transformation (Ref. 7) (which is an unusual transformation to be used for operators, but seems to be valid), Eq. (13.45) becomes

$$\hat{a}_2 = \sqrt{2S - \hat{a}_1^{\dagger}\hat{a}_1}. \tag{13.46}$$

From Eqs. (13.40) and (13.46), we obtain

$$\begin{aligned} \hat{S}^{+} &= \hat{a}_1^{\dagger}\sqrt{2S - \hat{a}_1^{\dagger}\hat{a}_1}, \\ \hat{S}^{-} &= \sqrt{2S - \hat{a}_1^{\dagger}\hat{a}_1\hat{a}_1}, \text{ and} \\ \hat{S}^{z} &= (\hat{a}_1^{\dagger}\hat{a}_1 - S). \end{aligned} \tag{13.47}$$

One can easily show that (Problem 13.7)

$$[\hat{S}^{+}, \hat{S}^{-}] = 2\hat{S}^z. \tag{13.48}$$

13.3.8 Application to the Heisenberg Hamiltonian

If we assume that the direct exchange interaction between the nearest neighbors is dominant, for simple lattices, Eq. (13.37) can be rewritten as

$$\hat{H} = -J\sum_{i,\delta}{}' \hat{\mathbf{S}}_i \cdot \hat{\mathbf{S}}_{i+\delta}, \tag{13.49}$$

where $\mathbf{R}_j = \mathbf{R}_i + \mathbf{R}_\delta$, $\mathbf{R}_\delta\,(\delta = 1, 2, ..., z)$ is a vector to the nearest neighbor of the ith ion and $J_{i,i+\delta} = J$ for all δ. If J is positive in Eq. (13.49), it leads to ferromagnetism. The ion spins are assumed to be so aligned that in the ground state their z-components have the maximum values S. If $|S>_n$ represents the spin of the nth ion in state s, the ground state can be written as

$$\Phi_0 = \Pi_n |S>_n. \tag{13.50}$$

Eq. (13.49) can be rewritten in the alternate form

$$\hat{H} = -J\sum_{i,j}{}' \left[\hat{S}_{iz}\hat{S}_{jz} + \frac{1}{2}(\hat{S}_i^+\hat{S}_j^- + \hat{S}_i^-\hat{S}_j^+)\right], \tag{13.51}$$

where $j = i + \delta$. Thus, we obtain

$$\hat{H}\Phi_0 = \varepsilon_0\Phi_0 = (-S^2J\sum_{i,i+\delta} 1)\Phi_0 = -JS^2zN\Phi_0, \tag{13.52}$$

where N is the total number of sites, z is the coordination number of each site, and the application of S^+ to a function with maximum spin leads to zero. Here, N is the total number of ions. Thus, we obtain the expression for the energy of the ground state for a ferromagnet,

$$\varepsilon_0 = -JNzS^2. \tag{13.53}$$

However, if we consider the state Φ_m in which the mth spin is reduced by 1, the new state can be written as

$$\Phi_m = \hat{S}_m^-\Pi_n|S>_n. \tag{13.54}$$

From Eqs. (13.51) and (13.54), we can write

$$\hat{H}\Phi_m = -J\sum_{ij}{}' \left[\hat{S}_i^z\hat{S}_j^z\hat{S}_m^- + \frac{1}{2}(\hat{S}_i^+\hat{S}_j^-\hat{S}_m^- + \hat{S}_i^-\hat{S}_j^+\hat{S}_m^-)\right]\Phi_0. \tag{13.55}$$

It can be easily shown by using the commutation relation of the spin operators (Problem 13.8) that

$$\hat{H}\Phi_m = E_0\Phi_m + 2JS\sum_{\mathbf{R}_\delta}(\Phi_m - \Phi_{m+\delta}). \tag{13.56}$$

Thus, Φ_m is not an eigenstate of the spin Hamiltonian. In fact, the deviation of spin at one site spreads into the neighboring sites, creating spin waves. To solve this, we take S to be large and expand in powers of $1/S$. For example, we obtain from Eqs. (13.47) and (13.51)

$$\begin{aligned}\hat{H} = -J\sum_{i,j}\Big[&(S - \hat{a}_i^\dagger\hat{a}_i)(S - \hat{a}_j^\dagger\hat{a}_j) + \frac{1}{2}\hat{a}_i^\dagger\sqrt{2S - \hat{a}_i^\dagger\hat{a}_i}\sqrt{2S - \hat{a}_j^\dagger\hat{a}_j}\hat{a}_j \\ &+ \frac{1}{2}\hat{a}_j^\dagger\sqrt{2S - \hat{a}_j^\dagger\hat{a}_j}\sqrt{2S - \hat{a}_i^\dagger\hat{a}_i\hat{a}_i}\Big],\end{aligned} \tag{13.57}$$

where $j = i + \delta$, δ being the nearest neighbors. For large values of S, wc can write

$$\hat{a}_i = \sqrt{S}b_i + (\hat{a}_i - \sqrt{S}b_i), \tag{13.58}$$

where $(\hat{a}_i - \sqrt{S}b_i)$ is very small, and b_i is determined by minimizing the Hamiltonian. The constants b_i are determined by minimizing the Hamiltonian while the series in $1/S$ is obtained by expanding the remainder. Thus, Eq. (13.57) can be rewritten as

$$\hat{H} = -J\sum_{ij} S^2\left[\frac{1}{2}(b_i^*b_j + b_jb_i^*)\sqrt{2 - |b_i|^2}\sqrt{2 - |b_j|^2} + (1 - |b_i|)^2(1 - |b_j|)^2\right]. \tag{13.59}$$

Here, it is tacitly assumed that i and j are nearest pairs. To obtain the ground-state energy, we minimize the Hamiltonian with respect to all values of b_i and b_j. The only possible way in which the ground-state energy ε_0 obtained from Eq. (13.59) will be the same as obtained in Eq. (13.53) is to assume $b_i = b_j = b$ (Problem 13.9). In fact, if the spins rotate in any direction as long as they all point together, the ground-state energy is independent of b, which can therefore be chosen as $b = 0$. The operator $\hat{a}$ is treated as small (Eq. 13.58) if we continue with the expansion of the Hamiltonian.

From Eqs. (13.53) and (13.59), we can write $\hat{H}$ (up to the first order in S), treating $\hat{a}$ as small, $b = 0$, and multiplying the second term by 2 because <ij> is a sum over nearest-neighbor pairs and each pair appears twice (Problem 13.10),

$$\hat{H} = -JNzS^2 - 2JS \sum_{<ij>} (\hat{a}_i^\dagger \hat{a}_j + \hat{a}_j^\dagger \hat{a}_i - \hat{a}_i^\dagger \hat{a}_i - \hat{a}_j^\dagger \hat{a}_j). \tag{13.60}$$

We make a Fourier transformation

$$\hat{a}_i = \frac{1}{\sqrt{N}} \sum_{\mathbf{k}} \hat{b}_{\mathbf{k}} e^{-i\mathbf{k}\cdot\mathbf{R}_i}, \tag{13.61}$$

where **k** (for cyclic boundary conditions) is limited to the N values inside a Brillouin zone in the **k** space. From Eqs. (13.60) and (13.61), we obtain

$$\hat{H} = -JNzS^2 + \sum_{\mathbf{k}} \hbar\omega_{\mathbf{k}} \hat{n}_{\mathbf{k}}, \tag{13.62}$$

where

$$\hat{n}_{\mathbf{k}} = \hat{b}_{\mathbf{k}}^\dagger \hat{b}_{\mathbf{k}} \tag{13.63}$$

and

$$\hbar\omega_{\mathbf{k}} = 2JS \sum_l (1 - \cos \mathbf{k} \cdot \mathbf{R}_l). \tag{13.64}$$

Eq. (13.64) gives the spin-wave dispersion relation. Here, $\mathbf{R}_l$ are the nearest neighbors. The first term in Eq. (13.64) is the energy of the ground state, the second term is the energy contained in the magnons, and $n_{\mathbf{k}} = \hat{b}_{\mathbf{k}}^\dagger \hat{b}_{\mathbf{k}}$ is the magnon particle number. The magnon energy can be expressed as

$$E = \sum_{\mathbf{k}} \frac{\hbar\omega_{\mathbf{k}}}{e^{\beta\hbar\omega_{\mathbf{k}}} - 1}, \tag{13.65}$$

where $\beta = 1/k_B T$. In the isotropic case, $\hbar\omega_{\mathbf{k}} \propto k^2 = \gamma k^2$ (from Eq. 13.65), and we obtain

$$E = \frac{V}{4\pi^3} \int_0^{k_{max}} \frac{4\pi\gamma k^4 dk}{e^{\gamma k^2/k_B T} - 1}. \tag{13.66}$$

At very low temperatures, when only a few magnons are excited, the upper limit of the integration can be considered as infinity. Thus, Eq. (13.66) can be rewritten as

$$E = \frac{\gamma V}{\pi^2} \left(\frac{k_B T}{\gamma}\right)^{5/2} \int_0^\infty \frac{x^4 dx}{e^{x^2} - 1} = \alpha T^{5/2}, \tag{13.67}$$

where α is a constant. Thus, the specific heat is proportional to $T^{3/2}$, which agrees with the experimental results.

13.3.9 Spin Waves in Antiferromagnets

When J is negative in Eq. (13.49), the neighboring spins are aligned in opposite directions. If all the ions are of the same type, this leads to antiferromagnetism. In antiferromagnets, the neighboring spins are antiparallel at zero temperature; this is known as a Néel state (Ref. 21). The ground state of an antiferromagnet is schematically illustrated in Figure 13.4.

↑	↓	↑	↓	↑	↓
A	B	A	B	A	B
↓	↑	↓	↑	↓	↑
B	A	B	A	B	A
↑	↓	↑	↓	↑	↓
A	B	A	B	A	B

FIGURE 13.4

The spins are divided into two interpenetrating sublattices A and B in the Néel state.

The spin states can be described as two interpenetrating sublattices, A and B. Each spin state in A has spins in the B sublattice as the nearest neighbors. One way of solving the complicated problem is as follows. The spin Hamiltonian in the antiferromagnetic state can be obtained from that of the ferromagnetic state by rotating all the spin operators on the B sublattice by 180° about the x-axis, while keeping the spin operators intact in the A sublattice. Thus, $x \to x$, $y \to -y$, and $z \to -z$. If all the $\mathbf{R}_i$ vectors are in the A sublattice and the $\mathbf{R}_j$ vectors are in the B sublattice, the spin operators defined in Eqs. (13.40) and (13.42) for ferromagnets can be rewritten for antiferromagnets as

$$\hat{S}_j^z \to -\hat{S}_j^z \tag{13.68}$$

and

$$\hat{S}_j^{\pm} \to \hat{S}_j^{\mp}. \tag{13.69}$$

The Hamiltonian in Eq. (13.51) for ferromagnets can be rewritten for antiferromagnets as (Problem 13.11)

$$\hat{H} = 2|J| \sum_{<ij>} \left[-\hat{S}_{iz}\hat{S}_{jz} + \frac{1}{2}(\hat{S}_i^+\hat{S}_j^+ + \hat{S}_i^-\hat{S}_j^-)\right], \tag{13.70}$$

where the factor of 2 is multiplied because each neighboring pair <ij> appears twice in the summation over i and j. Using a procedure adapted to obtain Eq. (13.57) with the modifications for antiferromagnets (Eqs. 13.68 through 13.70), we obtain (Problem 13.12)

$$\begin{aligned}\hat{H} = 2|J| \sum_{<ij>} \Bigg[&-(S-\hat{a}_i^\dagger\hat{a}_i)(S-\hat{a}_j^\dagger\hat{a}_j) + \frac{1}{2}\hat{a}_i^\dagger\sqrt{2S-\hat{a}_i^\dagger\hat{a}_i}\,\hat{a}_j^\dagger\sqrt{2S-\hat{a}_j^\dagger\hat{a}_j}\\ &+\frac{1}{2}\sqrt{2S-\hat{a}_i^\dagger\hat{a}_i}\,\hat{a}_i\sqrt{2S-\hat{a}_j^\dagger\hat{a}_j}\,\hat{a}_j\Bigg].\end{aligned} \tag{13.71}$$

We follow a $1/S$ expansion method similar to the procedure outlined in Eq. (13.57) and from Eqs. (13.70) and (13.71), we obtain (Problem 13.13)

$$\hat{H} = -|J|NzS^2 + 2|J|S \sum_{<i,j>} [\hat{a}_i^\dagger\hat{a}_i + \hat{a}_j^\dagger\hat{a}_j + \hat{a}_i^\dagger\hat{a}_j^\dagger + \hat{a}_i\hat{a}_j], \tag{13.72}$$

where N is the number of lattice sites, and z is the number of nearest neighbors of each site. We define the operators

$$\hat{a}_i = \frac{1}{\sqrt{N}}\sum_{\mathbf{k}} e^{i\mathbf{k}\cdot\mathbf{R}_i}\hat{b}_{\mathbf{k}} \tag{13.73}$$

and

$$\hat{a}_i^\dagger = \frac{1}{\sqrt{N}} \sum_{\mathbf{k}} e^{-i\mathbf{k}\cdot\mathbf{R}_i} \hat{b}_{\mathbf{k}}^\dagger. \tag{13.74}$$

Substituting Eqs. (13.73) and (13.74) in Eq. (13.72), we obtain (Problem 13.14)

$$\hat{H} = -|J|NzS^2 + |J|S| \sum_{\mathbf{k},l} [2\hat{b}_{\mathbf{k}}^\dagger \hat{b}_{\mathbf{k}} + (\hat{b}_{\mathbf{k}}^\dagger \hat{b}_{-\mathbf{k}}^\dagger + \hat{b}_{\mathbf{k}} \hat{b}_{-\mathbf{k}}) \cos(\mathbf{k}\cdot\mathbf{R}_l)], \tag{13.75}$$

where $\mathbf{R}_l$ are nearest-neighbor vectors. To diagonalize the Hamiltonian, we introduce two new operators through the transformation,

$$\hat{b}_{\mathbf{k}} = (\sinh \beta_{\mathbf{k}})\hat{c}_{-\mathbf{k}}^\dagger + (\cosh \beta_{\mathbf{k}})\hat{c}_{\mathbf{k}}, \tag{13.76}$$

where β is real. Substituting Eq. (13.76), we can show that the Hamiltonian in Eq. (13.75) is diagonalized provided (Problem 13.14)

$$\tanh 2\beta_{\mathbf{k}} = -\frac{1}{z} \sum_l \cos(\mathbf{k}\cdot\mathbf{R}_l). \tag{13.77}$$

Substituting Eqs. (13.76) and (13.77) in Eq. (13.75), we obtain

$$\hat{H} = -N|J|zS(S+1) + 2|J|zS \sum_{\mathbf{k}} \left(\hat{b}_{\mathbf{k}}^\dagger \hat{b}_{\mathbf{k}} + \frac{1}{2}\right) \sqrt{1 - \tanh^2 2\beta_{\mathbf{k}}}. \tag{13.78}$$

Thus, the ground-state energy is given by

$$\varepsilon_0 = -N|J|zS(S+1) + |J|zS \sum_{\mathbf{k}} \sqrt{1 - \frac{1}{z^2}\left(\sum_l \cos \mathbf{k}\cdot\mathbf{R}_l\right)^2}, \tag{13.79}$$

and the energy of a magnon of wave number $\mathbf{k}$ is given by

$$\varepsilon_{\mathbf{k}} = 2|J|zS \sqrt{1 - \frac{1}{z^2}\left(\sum_l \cos \mathbf{k}\cdot\mathbf{R}_l\right)^2}. \tag{13.80}$$

13.4 FERROMAGNETISM IN SOLIDS

13.4.1 Ferromagnetism Near the Curie Temperature

The spontaneous magnetization of a ferromagnet vanishes above the Curie temperature. This phenomenon can be explained by the exchange interaction by using the molecular field approximation. The Hamiltonian for the exchange interaction in the presence of an external field is given by

$$\hat{H} = -\sum_{ij}{}' J_{ij} \hat{\mathbf{S}}_i \cdot \hat{\mathbf{S}}_j - g\mu_B \mathbf{B} \cdot \sum_{i=1}^{N} \hat{\mathbf{S}}_i. \tag{13.81}$$

In the mean-field approximation, of which the validity is in general questionable, one of the operators appearing in the exchange integral is replaced by the mean value, so we can rewrite Eq. (13.81) as

$$\hat{H} = -\sum_{i=1}^{N}(g\mu_B \mathbf{B} + \sum_{j=1, j\neq i}^{N} J_{ij} <\hat{S}_j>) \cdot \hat{S}_i, \tag{13.82}$$

which we can rewrite in the alternate form

$$\hat{H} = -\sum_{i=1}^{N} g\mu_B(\mathbf{B} + \mathbf{B}_{\mathrm{M}}). \tag{13.83}$$

Here, $\mathbf{B}_{\mathrm{M}}$ is equivalent to the internal field originally introduced by Weiss to explain ferromagnetism. Assuming nearest-neighbor interaction between the spins, we obtain from Eqs. (13.82) and (13.83)

$$\mathbf{B}_{\mathrm{M}} = \frac{zJ}{g\mu_B} <\hat{S}_j>. \tag{13.84}$$

If we further assume that $<\hat{S}_j>$ is in the same direction as the magnetization $\mathbf{M}$,

$$\mathbf{M} = Ng\mu_B <\hat{S}_j> \tag{13.85}$$

and

$$\mathbf{B}_{\mathrm{M}} = \lambda \mathbf{M}, \tag{13.86}$$

where λ is the constant originally introduced by Weiss, known as the Weiss constant. From Eqs. (13.84) through (13.86), the relation between the Weiss constant λ and the exchange integral J is given by

$$\lambda = \frac{zJ}{Ng^2\mu_B^2}. \tag{13.87}$$

We obtain an expression for the spontaneous magnetic moment $\mathbf{M}$ for ferromagnetic solids ($\mathbf{B} = 0$) by following a procedure similar to that used earlier for the derivation of M for paramagnetic solids in Eq. (12.45),

$$M = NgS\mu_B B_s\left(\frac{g\mu_B S\lambda M}{k_B T}\right), \tag{13.88}$$

where the Brillouin function $B_s(x)$ was defined in Eq. (12.46),

$$B_S(x) = \frac{(2S+1)}{2S} \coth\left(\frac{2S+1}{2S}x\right) - \frac{1}{2S} \coth\frac{x}{2S}. \tag{13.89}$$

We note that for ferromagnetic solids, J is replaced with S in Eqs. (12.45) and (12.46). In addition, the external field $\mathbf{B} = 0$. We can easily check the accuracy of Eq. (13.89) by noting that when $T = 0$, coth $x = 1$ and $B_S = 1$. From Eqs. (13.88) and (13.89), we obtain the expression for saturation magnetization,

$$M = Ng\mu_B S. \tag{13.90}$$

When the temperature increases, the spins become randomly oriented, and the spontaneous magnetization gradually decreases until it disappears. The critical temperature T_C at which the spontaneous magnetization disappears is obtained from the following approximation for the Brillouin function $B_S(x)$. When $x \to 0$, which occurs when $M \to 0$,

$$B_S(x) \approx \frac{S+1}{S}\frac{x}{3} - \frac{(2S+1)^4 - 1}{(2S)^4}\frac{x^3}{45}. \tag{13.91}$$

Substituting Eq. (13.91) in (13.88), we obtain $(T \to T_C)$,

$$M \approx Ng^2\mu_B^2 S(S+1)\lambda M/3k_B T_C, \tag{13.92}$$

from which we obtain the expression for the critical temperature, known as the Curie temperature,

$$T_C = \frac{Ng^2\mu_B^2 S(S+1)\lambda}{3k_B}. \tag{13.93}$$

In the paramagnetic phase, the temperature dependence of magnetization can be described by the Curie law, provided the internal field is included along with the external field,

$$\mathbf{M} = \frac{C}{T}(\mathbf{B} + \lambda\mathbf{M}) \tag{13.94}$$

or

$$\mathbf{M} = \frac{C}{T - \lambda C}\mathbf{B}. \tag{13.95}$$

Further, we can write

$$\mathbf{M} = \chi\mathbf{B} = \frac{C}{T - T_C}\mathbf{B}, \tag{13.96}$$

from which we obtain the expression for the paramagnetic susceptibility χ,

$$\chi = \frac{C}{T - T_C}, \tag{13.97}$$

where T_C is the Curie temperature. From Eqs. (13.95) and (13.96), the Curie constant to be inserted in Eq. (13.97) is given by

$$C = \frac{T_C}{\lambda}, \tag{13.98}$$

where an expression for T_C was obtained in Eq. (13.93).

13.4.2 Comparison of Spin-Wave Theory with the Weiss Field Model

Both the spin-wave theory and the Weiss field model use the concept of direct exchange interaction between localized spins of nearest neighbors. The basic assumption in both models is that the electrons of a lattice ion contributing to the magnetic moment are tightly bound for the ions to be isolated, but the nearest neighbors are sufficiently close for a significant interaction to arise. However, the results obtained by using the two models differ significantly at low and high temperatures. For example, from Eqs. (13.88) and (13.93), it can be easily shown that (Problem 13.15)

$$\frac{M(T)}{M(0)} = 1 - \frac{1}{S} e^{-3T_C/(S+1)T}, \qquad (13.99)$$

while the experimental results agree with the $T^{3/2}$ law obtained by using spin-wave theory.

This result suggests that for temperatures near the ground state ($T = 0$), the spin-wave theory can be derived by using the method of elementary excitations rather than using other approximations. At higher temperatures, it is more appropriate to use semiclassical methods such as the Weiss field model, which can be justified by the general exchange-interaction concept. However, many aspects of the behavior of a ferromagnet near the Curie temperature can be understood by using the concept of magnons. Even at temperatures above the Curie temperature, paramagnons can be used to understand the properties in the paramagnetic region. The temperature dependence of the spontaneous magnetization, obtained from Eq. (13.88), is shown in Figure 13.5.

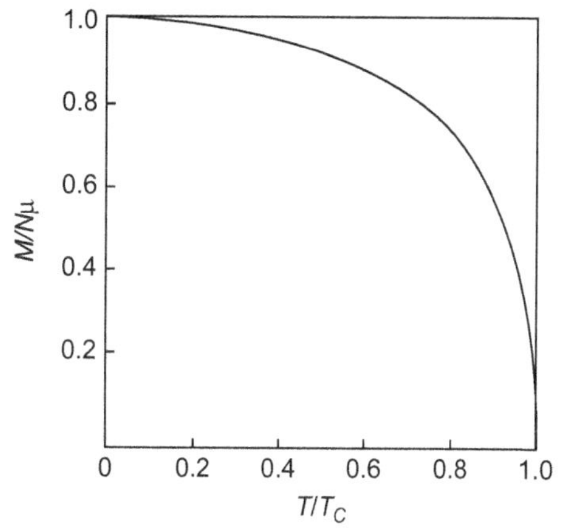

FIGURE 13.5

Saturation magnetization as a function of temperature.

13.4.3 Ferromagnetic Domains

It is a common experience to note that a ferromagnetic material is not necessarily magnetized when the temperature is lower than the Curie temperature. However, it is strongly attracted by magnetic fields and can be "magnetized" by stroking it with a "permanent magnet." The basic question is how the atomic magnetic dipoles are aligned below the Curie temperature and yet produce zero magnetization.

The key to explain this phenomenon is the fact that we have only considered the Heisenberg Hamiltonian while deriving an expression for the magnetization and neglected the magnetic dipolar interaction between the spins introduced in Eq. (13.7). The main reason for omitting the latter is that the dipolar coupling between nearest neighbors is much smaller than the exchange coupling. However, the exchange interaction is very short-ranged and decreases exponentially with spin separation in a ferromagnetic insulator. In contrast, the dipolar interaction falls off as the inverse cube of the separation. The magnetic configuration of a macroscopic sample depends on both interactions, especially when a large number of spins is involved where the dipolar energies that have been hitherto neglected become quite significant.

It can be easily shown that the dipolar energy can be substantially reduced by dividing the ferromagnetic solid into uniformly magnetized domains of much smaller size and of which the magnetization vectors point in different directions. The concept of ferromagnetic domains was introduced by Weiss. Figure 13.6 shows a schematic diagram of the domains in a ferromagnet (Ref. 7). The atomic dipoles are aligned in the same direction within a domain but have no common link with the neighboring domains.

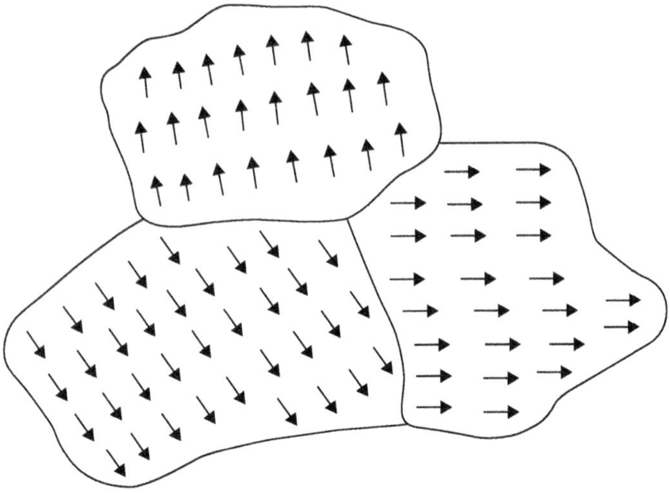

FIGURE 13.6

Ferromagnetic domains. The dimensions are 10 μm to 100 μm.

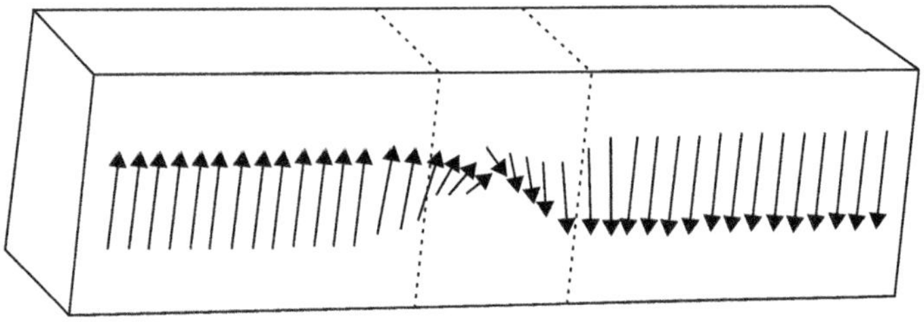

FIGURE 13.7

Schematic diagram of the change in orientation of magnetic dipoles in neighboring domains.

Figure 13.7 shows schematically a change in orientation of magnetic dipoles, in which each dipole is slightly misaligned with the neighboring dipole. The change in the orientation of the dipoles from the direction of one domain to that of the neighboring domain takes place over a distance of a few hundred atomic spaces. This narrow region between the adjacent domains is called a Bloch wall.

13.4.4 Hysteresis

Figure 13.8 shows the magnetization curve that describes the process by which a ferromagnetic material can be converted from a nonmagnetic state to a ferromagnetic state with the application of a reasonably small magnetic field. This curve is known as the hysteresis curve (Ref. 27). The magnetization curve is plotted as B versus H, where H is the external magnetic field and the magnetic induction, $B = H + 4\pi M$. The external field H is applied to an initially unmagnetized sample until the magnetization reaches the saturation value.

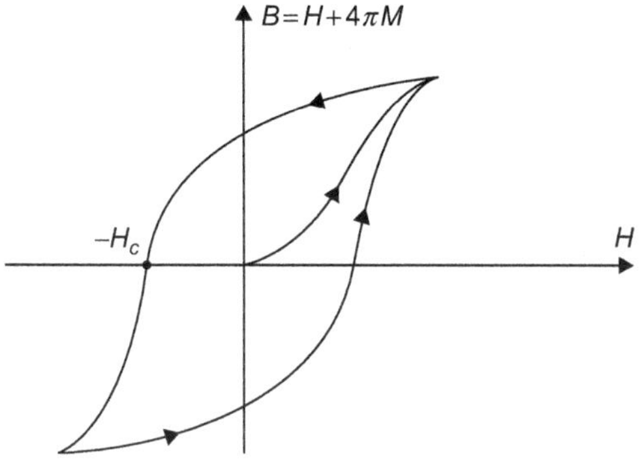

FIGURE 13.8

The hysteresis curve.

After the sample has become fully magnetized, the field is subsequently reduced, and the magnetization decreases to a constant value when the external field is zero. To return the sample to its original state (i.e., to demagnetize it), a magnetic field has to be applied in the opposite direction until $B = 0$ when $H = -H_c$. Usually, this is considered as the definition of the coercive force.

13.4.5 Ising Model

The Ising model is a very simplified version of the Heisenberg model. In this model, the Hamiltonian is written as

$$\hat{H} = -J\sum_{i,j}{}'\hat{S}_{iz}\hat{S}_{jz} - g\mu_B B\sum_{i=1}^{N}\hat{S}_{iz}. \tag{13.100}$$

Thus, the terms S^+ and S^- are essentially dropped from the Heisenberg model, and the magnetic field is taken in the z direction. Because all $\hat{S}_{iz}$ commute, the Hamiltonian $\hat{H}$ is diagonal in the representation in which each $\hat{S}_{iz}$ is diagonal. Hence, all the eigenfunctions and the eigenvalues of the Hamiltonian are known. Thus, the Ising model is very convenient in describing the statistical theory of phase transitions—for example, in describing the system when the ferromagnetic state is changed to a paramagnetic state at the Curie temperature. However, in spite of these simplifications, only by using a two-dimensional Ising model for simple lattices (i.e., square, triangular, honeycomb) can one calculate the exact free energy in zero magnetic field and the spontaneous magnetization. We will not discuss this model in more detail.

13.5 FERROMAGNETISM IN TRANSITION METALS

13.5.1 Introduction

The problem of the origin of ferromagnetism in some transition metals, of which the common feature is that they have narrow unfilled d-bands ($3d$) as well as filled s-bands (s^2), remains one of the major unsolved problems in solid state physics. There have been many theoretical explanations, but no theory can satisfactorily explain why Fe, Co, and Ni are ferromagnetic metals while a large number of transition elements with narrow unfilled d-bands are not ferromagnetic; i.e., they do not have spontaneous magnetization. For example, the configurations of these elements are

$$\text{Fe}([\text{Ar}]3d^6 4s^2), \quad \text{Co}([\text{Ar}]3d^7 4s^2), \quad \text{Ni}([\text{Ar}]3d^8 4s^2).$$

These three elements (Fe, Co, and Ni) are ferromagnetic metals, whereas Mn and Pd become ferromagnetic under certain conditions.

There have been many attempts to solve this problem of band ferromagnetism (Ref. 4). It is indeed easy to understand the origin of an atomic magnetic moment arising out of an intra-atomic exchange but difficult to explain the cooperative interaction that couples the moments on different atoms. Such cooperative phenomena require a band model of ferromagnetism. However, one of the essential features of the band model is to delocalize the moment from the atom. Thus, the Heisenberg model is inappropriate for use in ferromagnetism in transition metals. We will discuss the existing theories, with the cautionary note that not one of them is adequate in explaining band ferromagnetism.

The magnetic moments of transition metals can indeed be calculated with reliable accuracy by using a combination of band calculations and density functional theory. However, this approach does not provide a specific model for band ferromagnetism.

13.5.2 Stoner Model

Stoner (Ref. 26) considered a collective electron model in which there is an interaction term between the pairs of electrons of opposite spin. Thus,

$$H_i = \frac{U}{N} \sum_{\mathbf{k},\mathbf{k}'} n_{\mathbf{k}\uparrow} n_{\mathbf{k}'\downarrow}, \tag{13.101}$$

where $n_{\mathbf{k}\uparrow(\downarrow)}$ is the occupation number of the states $|\mathbf{k}\uparrow(\downarrow)>\cdot$ Each pair of electrons of opposite spin contributes a positive "exchange" energy U/N. It should be made clear at this point that the nature of this positive "exchange" energy between opposite spins in the same d-shell was neither explained by Stoner nor has it been explained by any other group since then. In a later section, we will present an alternate interpretation for U. The contributions from electrons of the same spin in the d-shell are included in the definition of zero of energy and therefore not counted in Eq. (13.101). If we define the total number of electrons n_σ per atom of spin σ, the energy of an electron of $\uparrow(\downarrow)$ spin is

$$\varepsilon_{\mathbf{k}\uparrow(\downarrow)} = \varepsilon(\mathbf{k}) \mp \mu_B B + U n_{\downarrow(\uparrow)}. \tag{13.102}$$

The number of electrons in each state is given by two Fermi–Dirac distributions with the same chemical potential μ. Thus, we have

$$n_{\uparrow(\downarrow)} = \int_0^\infty D(\varepsilon) f^0(\varepsilon_{\mathbf{k}\uparrow(\downarrow)}) d\varepsilon, \tag{13.103}$$

where $f^0(\varepsilon_{\mathbf{k}\uparrow(\downarrow)})$ is the Fermi–Dirac distribution function, and $D(\varepsilon)$ is the density of states per spin. The chemical potential is the same for the two electron distributions. To satisfy this, the following conditions have to be met:

$$n<\mu> = \mu_B \int_0^\infty \{f^0(\varepsilon_{\mathbf{k}\uparrow}) - f^0(\varepsilon_{\mathbf{k}\downarrow})\} D(\varepsilon) d\varepsilon \tag{13.104}$$

and

$$n = \int_0^\infty \{f^0(\varepsilon_{\mathbf{k}\uparrow}) + f^0(\varepsilon_{\mathbf{k}\downarrow})\} D(\varepsilon) d\varepsilon. \tag{13.105}$$

The magnetic moment is given by

$$M = \mu_B (n_\uparrow - n_\downarrow). \tag{13.106}$$

From Eqs. (13.102), (13.103), and (13.106), we obtain

$$M = \mu_B \int_0^\infty \{ f^0(\varepsilon - \mu_B B + U n_\downarrow) - f^0(\varepsilon + \mu_B B + U n_\uparrow) \} d\varepsilon, \tag{13.107}$$

which can be written in the alternate form in the limit of $B \to 0, T \to 0$,

$$M \approx [MU + 2\mu_B^2 B] \int_0^\infty \left(-\frac{\partial f^0}{\partial \varepsilon} D(\varepsilon) d\varepsilon \right). \tag{13.108}$$

At $T \to 0$, we have, for any function $F(\varepsilon)$,

$$-\int_0^\infty \frac{\partial f^0}{\partial \varepsilon} F(\varepsilon) d\varepsilon = F(\varepsilon_F). \tag{13.109}$$

From Eqs. (13.108) and (13.109), we obtain

$$M \approx [(MU + 2\mu_B^2 B) D(\varepsilon_F)]. \tag{13.110}$$

Eq. (13.110) can be rewritten in the alternate form

$$M \approx \frac{2\mu_B^2 B D(\varepsilon_F)}{1 - U D(\varepsilon_F)}. \tag{13.111}$$

The magnetic susceptibility is easily obtained from Eq. (13.111),

$$\chi = \frac{M}{B} = \frac{2\mu_B^2 D(\varepsilon_F)}{1 - U D(\varepsilon_F)}. \tag{13.112}$$

When the "exchange field" is sufficiently large so that

$$U D(\varepsilon_F) > 1, \tag{13.113}$$

Eq. (13.112) is unstable. In the Stoner model, this leads to a transition to ferromagnetism. In addition, from Eq. (13.106), to have a permanent magnetic moment,

$$n_\uparrow \gg n_\downarrow. \tag{13.114}$$

However, Stoner's model does not explain this large difference between $n_\uparrow$ and $n_\downarrow$. In addition, Stoner's model neither explains the origin of the positive "exchange interaction" U nor does it explain why only Fe, Co, and Ni are ferromagnetic metals among such a large number of elements in the transition group with narrow d-bands.

The typical density of states of iron and the other transition metals in the same group with $3d$ narrow bands and $4s$ wide bands is shown in Figure 13.9. Because the Fermi energy lies inside the $3d$ band, the density of states at the Fermi energy is very high.

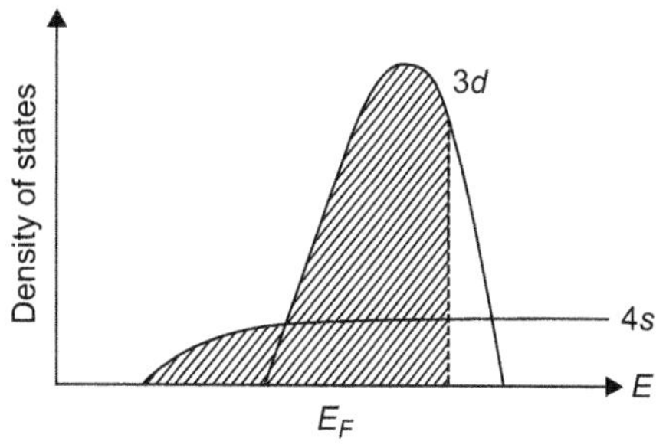

FIGURE 13.9

Schematic diagram of density of states of $3d$ and $4s$ bands in iron and other transition metals.

13.5.3 Ferromagnetism in Fe, Co, and Ni from Stoner's Model and Kohn–Sham Equations

In the $3d$ electrons in bulk transition metals, there is a significant overlap between neighboring atoms. The $4s$ electrons form broad free-electron-like bands of width 20–30 eV, whereas the $3d$ electron bandwidths are typically 5–10 eV. Due to hybridization and crystal environment, one electron is transferred from s- to d-bands. Thus, there are 7, 8, and 9 electrons per atom, respectively, in the d-bands in Fe, Co, and Ni. The predicted spin moments from Hund's rule are $\mu_{spin}(\text{Fe}) = 3\,\mu_B$, $\mu_{spin}(\text{Co}) = 2\,\mu_B$, and $\mu_{spin}(\text{Ni}) = \mu_B$. However, the actual values are noninteger and smaller because of partial delocalization; i.e., $\mu_{spin}(\text{Fe}) = 2.12\ \mu_B$, $\mu_{spin}(\text{Co}) = 1.58\,\mu_B$, and $\mu_{spin}(\text{Ni}) = 0.56\,\mu_B$.

Due to the large density of states, there are many unoccupied states just above the Fermi energy, which allows the promotion of electrons from minority spin to majority spin states at a modest energy cost. We have obtained the Stoner criterion for ferromagnetic susceptibility,

$$UD(\varepsilon_F) > 1. \tag{13.115}$$

The parameter U can be evaluated by perturbation theory based on non-spin-polarized solutions of the Kohn–Sham equations of density functional theory described in detail in Section 7.8.

Janak[9] showed that for infinitesimal Stoner splitting, one can write

$$U = \int d\mathbf{r}\, \gamma^2(\mathbf{r})|K(\mathbf{r}), \tag{13.116}$$

where $\gamma(\mathbf{r})$ is essentially a normalized local density of states at the Fermi energy,

$$\gamma(\mathbf{r}) = \sum_i \frac{\delta(\varepsilon_F - \varepsilon_i)|\psi_i(\mathbf{r})|^2}{D(\varepsilon_F)}, \tag{13.117}$$

and ε_i and $\psi_i(\mathbf{r})$ are the self-consistent energies and wave functions of the Kohn–Sham equations. $K(\mathbf{r})$ is a kernel giving the exchange-correlation enhancement of the field due to magnetization defined by

$$\left\{\frac{\delta^2 E_{xc}[nm]}{\delta m(\mathbf{r})\delta m(\mathbf{r}')}\right\}_{m=0} = 2K(\mathbf{r})\delta(\mathbf{r}-\mathbf{r}'), \tag{13.118}$$

where $E_{xc}[n;m]$ is the exchange-correlation functional, which is defined in Eq. (7.165) in the local density approximation of Kohn and Sham. Here, n and m are defined in the usual manner,

$$n_p(\mathbf{r}) = \sum_i \theta(\varepsilon_F - \varepsilon_{ip})|\psi_{ip}|^2, \tag{13.119}$$

$$n(\mathbf{r}) = n_\uparrow(\mathbf{r}) + n_\downarrow(\mathbf{r}), \tag{13.120}$$

and

$$m(\mathbf{r}) = n_\uparrow(\mathbf{r}) - n_\downarrow(\mathbf{r}). \tag{13.121}$$

Janak (Ref. 9) calculated the values of both U and $D(\varepsilon_F)$ as functions of atomic number Z. The exchange-correlation-enhanced spin susceptibilities of 32 elements from Li through In were calculated using the spin-polarized exchange-correlation functional of von Barth and Hedin. The methods of Janak's calculation are summarized in his paper, and the results are tabulated in Table 1 of the same paper [note that his results for $D(\varepsilon_f)$ are for both spins]. He has also plotted U and $D(\varepsilon_F)$ as a function of atomic number Z in Figure 13.1 of his paper. In Figure 13.10, we have plotted $1-UD(\varepsilon_F)$ versus Z, where $D(\varepsilon_F)$ is the density of states per spin. We note that according to the data, both Fe (Z = 26) and Ni (Z = 28) are ferromagnetic according to the Stoner criterion while $UD(\varepsilon_F) = 0.97$ for Co (Z = 27). In contrast, calculations by Gunnarson (Ref. 4) indicate that $UD(\varepsilon_F)$ is in the range 1.6–1.8 for Co.

To summarize, the ferromagnetism of Fe, Co, and Ni can be explained by using density-functional theory through a combination of band theory and Stoner criterion, but we still lack a fundamental theory of ferromagnetism in transition metals.

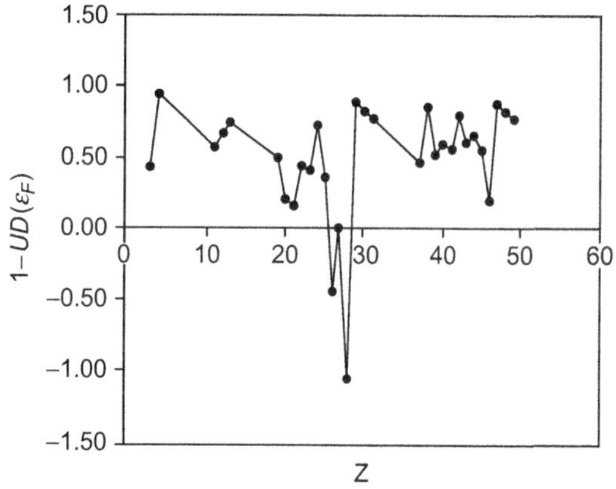

FIGURE 13.10

Using data from Figure 1 of Janak (Ref. 9), $1-UD(\varepsilon_F)$ is plotted against Z (by Binns C and Blackman JA, p.231 of Ref. 19.).

Reproduced from Misra[19] with permission of Elsevier.

13.5.4 Free Electron Gas Model

It is well known that the free electron gas model does not explain the ferromagnetism in metals. However, one can derive conditions for the onset of ferromagnetism using the free electron gas model without using spin-dependent interactions. The free electron gas model provides a simple theory that is grossly inadequate but nevertheless provides a glimpse into the much harder and yet unsolved problem. It deals with the itinerant aspect of exchange through the use of the Hartree–Fock approximation of the uniform electron gas.

We derived in Eq. (7.44) an expression for the ground-state energy of the free electrons,

$$E = N\frac{e^2}{2a_0}\left[\frac{3}{5}(k_F a_0)^2 - \frac{3}{2\pi}(k_F a_0)\right], \tag{13.122}$$

where a_0 is the Bohr radius, the first term is the total kinetic energy, and the second term is the exchange energy that is between the electrons of the same spin.

Eq. (13.122) was derived with the assumption that each occupied one-electron state was occupied with two electrons of opposite spin. However, if it so happens that there is a spin imbalance, then each one-electron state has k less than some $k_\uparrow$ with spin-up electrons and similarly $k < k_\downarrow$ for some spin-down electrons. Because the exchange interaction is between electrons of the same spin, we obtain from Eq. (13.122) an equation for each spin population,

$$E_{(\uparrow,\downarrow)} = N_{(\uparrow,\downarrow)}\frac{e^2}{2a_0}\left[\frac{3}{5}(k_{(\uparrow,\downarrow)}a_0)^2 - \frac{3}{2\pi}(k_{(\uparrow,\downarrow)}a_0)\right]. \tag{13.123}$$

The total energy

$$E = E_\uparrow + E_\downarrow, \tag{13.124}$$

and the total number of electrons

$$N = N_\uparrow + N_\downarrow = V\left(\frac{k_\uparrow^3 + k_\downarrow^3}{6\pi^2}\right) = V\frac{k_F^3}{3\pi^2}. \tag{13.125}$$

If $N_\uparrow > N_\downarrow$, the ground state will have a nonvanishing magnetic density

$$M = -g\mu_B \frac{N_\uparrow - N_\downarrow}{V}, \tag{13.126}$$

which leads to a ferromagnetic electron gas. In the other extreme, if

$$\begin{aligned} N_\downarrow &= N, \\ N_\uparrow &= 0, \\ E &= E_\downarrow, \\ k_\downarrow &= 2^{1/3} k_F \text{ (from Eq. 13.125)}. \end{aligned} \tag{13.127}$$

From Eqs. (13.124), (13.125), and (13.127), we obtain

$$E_\downarrow = N\left[\frac{3}{5}2^{2/3}(k_F a_0)^2 - \frac{3}{2\pi}2^{1/3}(k_F a_0)\right]. \tag{13.128}$$

Comparing Eqs. (13.122) with (13.128), we note that the energy of the fully magnetized state is lower than the energy of the unmagnetized state when the exchange energy dominates the kinetic energy. It can be shown from Eq. (13.128) that when $E = E_\downarrow$, transition to a fully magnetized state occurs, in which case

$$k_F a_0 = \frac{5}{2\pi}\frac{1}{2^{1/3}+1}. \tag{13.129}$$

We earlier defined the electron density r_s,

$$r_s = \left(\frac{3V}{4\pi N}\right)^{1/3} = \frac{1.92}{k_F}. \tag{13.130}$$

From Eqs. (13.129) and (13.130), the condition for transition from a nonmagnetic to a ferromagnetic state is

$$\frac{r_s}{a_0} = \frac{2\pi}{5}(2^{1/3}+1)\left(\frac{9\pi}{4}\right)^{1/3} = 5.45. \tag{13.131}$$

The only metal with such a low conduction electron density is cesium. There are some metallic compounds that also satisfy the criteria that $r_s/a_0 > 5.45$.

However, none of these are ferromagnetic metals, even though their band structures are reasonably described by the free electron model. Thus, one has to know the specific features of the band structure and the itinerant exchange interactions to be able to account for magnetic ordering. The starting point is the fact that the d-electrons form a narrow band. Next, we will briefly discuss a few models for the theory of ferromagnetic metals.

13.5.5 Hubbard Model

Hubbard (Ref. 8) proposed a simple model that yields both bandlike and localized behavior in suitable limits. He proposed an extension of the tight-binding model by adding a term that includes an energy penalty U for any atomic site occupied by more than one electron. Hubbard's main argument was that because the off-diagonal elements of the Coulomb interaction were much smaller than the diagonal elements, the Hamiltonian can be approximated as

$$\hat{H} = \sum_{<ij>,\rho} -t[\hat{a}_{i\rho}^{\dagger}\hat{a}_{j\rho} + \hat{a}_{j\rho}^{\dagger}\hat{a}_{i\rho}] + U\sum_{i} \hat{a}_{i\uparrow}^{\dagger}\hat{a}_{i\uparrow}\hat{a}_{i\downarrow}^{\dagger}\hat{a}_{i\downarrow}, \tag{13.132}$$

where the sum $<ij>$ is taken over nearest-neighbor pairs. Here, the off-diagonal states have nonvanishing matrix elements t, known as the hopping parameter, between those pairs of states that differ only by a single electron that has moved from an ion to one of its neighbors without change of spin. This set of terms leads to a tight-binding model of one-electron Bloch levels. Even this oversimplified Hamiltonian can be solved only by using mean-field theory.

Eq. (13.132) can be rewritten as

$$\hat{H} = \sum_{<ij>,\rho} -t[\hat{a}_{i\rho}^{\dagger}\hat{a}_{j\rho} + \hat{a}_{j\rho}^{\dagger}\hat{a}_{i\rho}] + U\sum_{i} \hat{n}_{i\uparrow}\hat{n}_{i\downarrow}. \tag{13.133}$$

If we write

$$\hat{n}_{i\rho} = n_{\rho} + (\hat{n}_{i\rho} - n_{\rho}), \tag{13.134}$$

and consider the second term in Eq. (13.134) as small, Eq. (13.133) can be rewritten as

$$\hat{H} = \sum_{<ij>,\rho} -t[\hat{a}_{i\rho}^{\dagger}\hat{a}_{j\rho} + \hat{a}_{j\rho}^{\dagger}\hat{a}_{i\rho}] + U\sum_{i} (\hat{n}_{i\uparrow}n_{\downarrow} + n_{\uparrow}\hat{n}_{i\downarrow} - n_{\uparrow}n_{\downarrow}). \tag{13.135}$$

Making a Fourier transformation of the type

$$\hat{a}_{i\rho} = \frac{1}{\sqrt{N}}\sum_{\mathbf{k}} e^{i\mathbf{k}\cdot\mathbf{R}_i}\hat{b}_{\mathbf{k}\rho}, \tag{13.136}$$

from Eqs. (13.135) and (13.136), we obtain (Problem 13.17)

$$\hat{H} = \sum_{\mathbf{k},\mathbf{R}_l,\rho} -t\hat{b}_{\mathbf{k}\rho}^{\dagger}\hat{b}_{\mathbf{k}\rho}\cos \mathbf{R}_l\cdot\mathbf{k} + U\sum_{\mathbf{k}} (\hat{n}_{\mathbf{k}\uparrow}n_{\downarrow} + n_{\uparrow}\hat{n}_{\mathbf{k}\downarrow} - n_{\uparrow}n_{\downarrow}), \tag{13.137}$$

where $\mathbf{R}_l$ are nearest-neighbor vectors. The Hamiltonian in Eq. (13.137) becomes diagonal in $\mathbf{k}$ space. Specific calculations for a ferromagnetic solid (Ref. 30) can be made depending on the lattice structure and the dimension. In addition to the difficulties in making specific calculations in the Hubbard model, its major drawback is that it was designed for solutions that lead to ferromagnetism in metals.

We will present an alternate model for ferromagnetism in the transition metals starting with a theory of magnetization of interacting Bloch electrons.

13.6 MAGNETIZATION OF INTERACTING BLOCH ELECTRONS

13.6.1 Introduction

Recently, Tripathi and Misra[29] derived a theory of magnetization of interacting electrons in the presence of a periodic potential, spin-orbit interaction, and an applied magnetic field in the paramagnetic limits. Starting from a thermodynamic potential, which includes both the quasiparticle and correlation contributions, they showed that the modifications brought about by the electron–electron interactions for the magnetization in the quasiparticle approximation are precisely canceled by the contributions due to electron correlations. The magnetization is expressed as a product of the spin density and effective g factor, due mainly to the spin-orbit interaction. Tripathi and Misra (Ref. 29) also showed the importance of the self-energy corrections in the single-particle energy spectrum. By considering a variant of the Hubbard Hamiltonian in the momentum space, their theory can predict whether or not the ground state of the interacting electron system is magnetic. Tripathi and Misra's model for ferromagnetism appears as a variant of the Stoner model but from a very different perspective.

13.6.2 Theory of Magnetization

The magnetization of an interacting electron system is given by

$$M^{\nu} = -\frac{\partial \Omega}{\partial B^{\nu}}, \tag{13.138}$$

where Ω is the thermodynamic potential, and in Eq. (12.123), Misra et al. (Ref. 18) showed

$$\Omega = \Omega_{qp} + \Omega_{corr}, \tag{13.139}$$

where

$$\Omega_{qp} = \frac{1}{\beta} Tr \ln(-\widetilde{G}_{\xi_l}) = -\frac{1}{2i\pi} Tr \oint_C F(\xi)\widetilde{G}(\xi)d\xi, \tag{13.140}$$

$$F(\xi) = -\frac{1}{\beta} \ln\left[1 + e^{-\beta(\xi-\mu)}\right], \tag{13.141}$$

and the contour encircles the imaginary axis in a counterclockwise direction and Tr involves summation over one-particle states.[18] Misra et al.[18] derived (Eq. 12.117)

$$[\xi_l - \hat{H}(\vec{\kappa}, \xi_l)]\hat{G}(\mathbf{k}, \xi_l) = I, \tag{13.142a}$$

where

$$\vec{\kappa} = \mathbf{k} + i\mathbf{h} \times \nabla_{\mathbf{k}}, \tag{13.142b}$$

(from Eq. 12.119) and $\mathbf{h} = \frac{e\mathbf{B}}{2\hbar c}$ (from Eq. 12.114).

Misra et al. (Ref. 18) also derived (Eqs. 12.130 through 12.132)

$$\hat{H}(\vec{\kappa}, \xi_l) = \hat{H}_0(\mathbf{k}, \xi_l) + \hat{H}'(\mathbf{k}, \xi_l), \tag{13.143}$$

where

$$\hat{H}_0(\mathbf{k},\xi_l) = \frac{1}{2m}(\mathbf{p}+\hbar\mathbf{k})^2 + V(\mathbf{r}) + \Sigma^0(\mathbf{k},\xi_l) + \frac{\hbar^2}{8mc^2}\nabla^2 V + \frac{\hbar}{4m^2c^2}\vec{\sigma}\cdot\vec{\nabla}V\times(\mathbf{p}+\hbar\mathbf{k}), \quad (13.144)$$

and retaining terms up to the first order in the magnetic field in Eq. (12.132),

$$H'(\mathbf{k},\xi_l) = -ih_{\alpha\beta}\Pi^\alpha\nabla^\beta_{\mathbf{k}} + \tfrac{1}{2}g\mu_B B^\mu\sigma^\mu + B^\mu\Sigma^{1,\mu}(\mathbf{k},\xi_l). \quad (13.145)$$

In Eq. (13.145), $h_{\alpha\beta} = \epsilon_{\alpha\beta\mu}h^\mu$, where $\epsilon_{\alpha\beta\mu}h^\mu$, where $\epsilon_{\alpha\beta\mu}$ is the antisymmetric tensor of the third rank and we follow Einstein summation convention.

Here, $\vec{\Pi}$ was defined in Eq. (12.133),

$$\vec{\Pi} = \frac{\hbar}{m}(\mathbf{p}+\hbar\mathbf{k}) + \frac{\hbar^2}{4m^2c^2}\vec{\sigma}\times\vec{\nabla}V + \nabla_{\mathbf{k}}\Sigma^0(\mathbf{k},\xi_l). \quad (13.146)$$

In addition, from Eqs. (12.134) through (12.136), Misra et al. (Ref. 18) derived

$$\widetilde{G}(\mathbf{k},\xi_l) = G_0(\mathbf{k},\xi_l) + G_0(\mathbf{k},\xi_l)\hat{H}'G_0(\mathbf{k},\xi_l) + \cdots \quad (13.147)$$

and

$$\nabla^\alpha_{\mathbf{k}}G_0(\mathbf{k},\xi_l) = G_0(\mathbf{k},\xi_l)\Pi^\alpha G_0(\mathbf{k},\xi_l), \quad (13.148a)$$

where (from Eq. 12.135)

$$G_0(\mathbf{k},\xi_l) = \frac{1}{\xi_l - \hat{H}_0(\mathbf{k},\xi_l)}. \quad (13.148b)$$

From Eqs. (13.145) through (13.148), it can be shown that (Problem 13.18)

$$\widetilde{G}(\mathbf{k},\xi_l) = G_0 - G_0\left[ih_{\alpha\beta}\Pi^\alpha\nabla^\beta_{\mathbf{k}} - \frac{1}{2}g\mu_B B^\nu F^\nu\right]G_0, \quad (13.149)$$

where

$$F^\nu = \sigma^\nu + \frac{2}{g\mu_B}\widetilde{\Sigma}^{1,\nu} \quad (13.150)$$

was defined in Eq. (12.147) and is the renormalized spin vertex in the presence of the electron–electron interaction.

13.6.3 The Quasiparticle Contribution to Magnetization

From Eqs. (12.123) and (13.149), one can obtain the quasiparticle contribution to the thermodynamic potential, after evaluating the trace and contour integration (Problem 13.19) as

$$\Omega_{qp} = \frac{1}{2}\mu_B B^\nu \sum_{n\mathbf{k}\rho m\rho',m\neq n}\left[gF^\nu_{n\rho,n\rho'} + \frac{ie}{\hbar c\mu_B}\epsilon_{\alpha\beta\nu}\frac{\Pi^\alpha_{n\rho,m\rho'}\Pi^\beta_{m\rho',n\rho}}{E_{mn}}\right]f(E_{n\mathbf{k}\rho}), \quad (13.151)$$

where $f(E_{n\mathbf{k}\rho})$ is the Fermi distribution function for an electron in band n and spin ρ,

$$\Pi^\alpha_{n\rho,m\rho'} = \int_{cell} d^3r\, u^*_{n\mathbf{k}\rho}\Pi^\alpha u_{m\mathbf{k}\rho'} \quad (13.152)$$

and

$$E_{mn} = E_m(\mathbf{k}) - E_n(\mathbf{k}). \tag{13.153}$$

Using

$$M^{\nu}_{qp} = -\frac{\partial \Omega_{qp}}{\partial B^{\nu}}, \tag{13.154}$$

and Eqs. (13.150) and (13.151), Tripathi and Misra (Ref. 29) obtain

$$M^{\nu}_{qp} = -\frac{1}{2}\mu_B \sum_{n\mathbf{k}\rho\rho'} \left[g^{\nu}_{nn}(\mathbf{k})\sigma^{\nu}_{n\rho,n\rho'} + \frac{2}{\mu_B}\widetilde{\Sigma}^{1,\nu}_{n\rho,n\rho'} \right] f(E_{n\mathbf{k}\rho}), \tag{13.155}$$

where

$$g^{\nu}_{nn}(\mathbf{k})\sigma^{\nu}_{n\rho,n\rho'} = g\sigma^{\nu}_{n\rho,n\rho'} + \frac{ie}{\hbar c \mu_B} \in_{\alpha\beta\nu} \sum_{m\neq n,\rho'} \frac{\Pi^{\alpha}_{n\rho,m\rho'}\Pi^{\beta}_{m\nu',n\rho}}{E_{mn}}, \tag{13.156}$$

and is the effective g factor in the presence of spin-orbit interaction. In the absence of spin-orbit interactions, the second term in Eq. (13.156) vanishes, and the effective g factor reduces to the free electron g factor. In the absence of many-body interactions, the second term in the square brackets of Eq. (13.155) vanishes. However, Tripathi and Misra (Ref. 29) showed that the inclusion of M_{corr} would precisely cancel this term, and the final expression for magnetization would be free from any explicit many-body corrections—a surprising result that we will discuss in detail at the end of this section.

13.6.4 Contribution of Correlations to Magnetization

In Eq. (12.123), Misra et al. (Ref. 18) derived

$$\Omega_{corr} = \frac{1}{\beta}\left[-Tr\, \widetilde{\Sigma}_{\xi_l}\widetilde{G}_{\xi_l} + \phi(\widetilde{G}_{\xi_l}) \right]. \tag{13.157}$$

The contribution of correlation to magnetization is (Ref. 29)

$$M^{\nu}_{corr} = -\frac{\partial \Omega_{corr}}{\partial B^{\nu}} = \frac{1}{\beta} Tr \frac{\partial \widetilde{\Sigma}_{\xi_l}}{\partial B^{\nu}} \widetilde{G}_{\xi_l}. \tag{13.158}$$

Simplifying Eq. (13.158) with the use of Eqs. (12.127), (12.128), (13.155), and the Luttinger–Ward (Ref. 15) prescription for frequency summation, we obtain

$$\frac{1}{\beta}\sum_{\xi_l} \frac{1}{(\xi_l - \hat{H}_0)^m} = \frac{1}{2i\pi} Tr \int_{\Gamma_0} \frac{1}{(\xi - \hat{H}_0)^m} f(\xi)\, d\xi, \tag{13.159}$$

where Γ_0 encircles the real axis in a clockwise direction, and $f(\xi)$ is the Fermi distribution function obtained (Problem 13.20) after evaluating the *tr* over a complete set of single particle states,

$$M^{\nu}_{corr} = \sum_{n\mathbf{k}\rho} \widetilde{\Sigma}^{1,\nu}_{n\rho,n\rho'} f(E_{n\mathbf{k}\rho}) \cdot \tag{13.160}$$

From Eqs. (13.155) and (13.160),

$$M^{v} = -\frac{1}{2}\mu_B \sum_{n\mathbf{k}\rho\rho'} g^{v}_{nn}(\mathbf{k})\sigma^{v}_{n\rho,n\rho'} f(E_{n\mathbf{k}\rho}). \tag{13.161}$$

Thus, the renormalization of the spin vertex by the electron–electron interactions in the quasiparticle contribution is precisely canceled by M^{v}_{corr}. It may appear surprising that although both the spin contribution to the susceptibility (Ref. 18) (see Chapter 12) and spin Knight shift (Ref. 28) are exchange enhanced by electron–electron interactions, the self-energy corrections do not explicitly appear in the expression for magnetization. The reason is that the magnetic susceptibility and Knight shift arise from second-order effects in the magnetic field, where both the spin vertices are renormalized, and the renormalization of only one of the vertices is canceled due to the electron correlations. The renormalization of the other spin vertex, in a Hartree–Fock analysis, contributes to the exchange enhancement of the spin contributions to the susceptibility (Chapter 12) and Knight shift. In contrast, the single-particle energies appearing in Eqs. (13.144) and (13.145) depend on electron–electron interaction and the magnetic field through the self-energy. In the next section, we will show the consequences of these self-energy corrections on the single-particle spectrum in predicting whether or not the ground state of the interacting system is magnetic.

Assuming the effective magnetic field to be in the z direction and considering a single band, Tripathi and Misra (Ref. 29) obtained a tractable expression by averaging over the g matrix and denoting it as the effective g factor g_{eff}. They showed that Eq. (13.161) can be expressed as

$$M = -\frac{1}{2}\mu_B g_{eff} \sum_{\mathbf{k}\rho\rho'} \sigma_{\rho\rho'} f(E_{\mathbf{k}\rho}). \tag{13.162}$$

Here, ρ and ρ' take both $\uparrow$ and $\downarrow$ states. Eq. (13.162), therefore, can be rewritten as

$$M = -\frac{1}{2}\mu_B g_{eff}(n_\uparrow - n_\downarrow), \tag{13.163}$$

where

$$n_{\uparrow(\downarrow)} = \sum_{\mathbf{k}} f(E_{\mathbf{k}\uparrow(\downarrow)}). \tag{13.164}$$

13.6.5 Single-Particle Spectrum and the Criteria for Ferromagnetic Ground State

The single-particle energies $E_{\mathbf{k}\rho}$ appearing in Eq. (13.154) are the eigenvalues of the field-independent Hamiltonian described in Eq. (13.144). If the spin-orbit interaction is neglected, ρ becomes a pure spin state, and $E_{\mathbf{k}\rho}$ can be obtained by the following procedure. The many-body Hamiltonian can be written as

$$\hat{H}_0 = \sum_{\mathbf{k}\rho} \varepsilon_{\mathbf{k}} c^{\dagger}_{\mathbf{k}\rho} c_{\mathbf{k}\rho} + \frac{1}{2}\sum_{\mathbf{q}\mathbf{k}\mathbf{k}'\rho\rho'} V(\mathbf{q}) c^{\dagger}_{\mathbf{k}+\mathbf{q},\rho} c^{\dagger}_{\mathbf{k}'-\mathbf{q},\rho'} c_{\mathbf{k}',\rho'} c_{\mathbf{k},\rho}, \tag{13.165}$$

where $c^{\dagger}_{\mathbf{k}\rho}$ and $c_{\mathbf{k}\rho}$ are the creation and annihilation operators for an electron with wave vector $\mathbf{k}$ and spin ρ, and $V(\mathbf{q})$ is the Fourier transformation of the Coulomb interaction.

$E_{\mathbf{k}\rho}$ can be obtained by using the mean-field approximation,

$$E_{\mathbf{k}\rho} = \frac{1}{N}\left\langle \left\{ \left[c_{\mathbf{k}\rho}, \hat{H}_0\right], c_{\mathbf{k}\rho}^{\dagger} \right\} \right\rangle. \tag{13.166}$$

From Eqs. (13.165) and (13.166), we obtain

$$E_{\mathbf{k}\rho} = \varepsilon_{\mathbf{k}} - \frac{1}{N}\sum_{\mathbf{q}} V(\mathbf{q}) f(\varepsilon_{\mathbf{k}-\mathbf{q},\rho}) + \frac{1}{N} V(0) \sum_{\mathbf{k}} [f(\varepsilon_{\mathbf{k}\rho}) + f(\varepsilon_{\mathbf{k},-\rho})]. \tag{13.167}$$

Because V is assumed to be a constant U in the approximation of the averaged Coulomb interaction ansatz, we can rewrite Eq. (13.167) as

$$E_{\mathbf{k}\rho} = \varepsilon_{\mathbf{k}} + \frac{1}{N} U \sum_{\mathbf{k}} f(\varepsilon_{\mathbf{k},-\rho}) \equiv \varepsilon_{\mathbf{k}} + U n_{-\rho}, \tag{13.168}$$

where

$$n_{-\rho} = \frac{1}{N}\sum_{\mathbf{k}} f(\varepsilon_{\mathbf{k},-\rho}). \tag{13.169}$$

Eq. (13.168) is the Hartree–Fock representation of the single-particle spectrum. If we write $n = n_\uparrow + n_\downarrow$ and $m = n_\uparrow - n_\downarrow$ where n is the number of electrons/atoms and m is the average magnetization per atom in a Bohr magneton, the total energy per atom is

$$E = \frac{1}{N}\sum_{\mathbf{k}\rho} \varepsilon_{\mathbf{k}} f(E_{\mathbf{k}\rho}) + U n_\uparrow n_\downarrow. \tag{13.170}$$

The nonmagnetic solution in the ground state is obtained from the criteria

$$\begin{aligned} f(E_{\mathbf{k}\rho}) &= 1 \text{ if } E_{\mathbf{k}\rho} < E_F \\ &= 0 \text{ if } E_{\mathbf{k}\rho} > E_F. \end{aligned} \tag{13.171}$$

The Fermi energy E_F is determined from the condition that one has the right number (n) of electrons per atom. To study the stability of this state, let us transfer some electrons in an energy range δE below E_F from the down-spin states to the up-spin states. The change in the kinetic energy is

$$\Delta T = D(E_F)(\delta E^2), \tag{13.172}$$

where $D(E_F)$ is the density of states per atom per spin at the Fermi energy. The change in the interaction energy is

$$\Delta E_{int} = U\left(\frac{n}{2} + D\delta E\right)\left(\frac{n}{2} - D\delta E\right) - \frac{1}{4} U n^2 \equiv -U D^2(E_F)(\delta E^2). \tag{13.173}$$

From Eqs. (13.172) and (13.173), the change in the total energy is

$$\Delta E = D(E_F)(\delta E)^2 [1 - U D(E_F)]. \tag{13.174}$$

The nonmagnetic state would be stable if $UD(\varepsilon_F) < 1$ and the condition for the ferromagnetic stability is $UD(\varepsilon_F) > 1$. This appears similar to the Stoner condition for ferromagnetic stability except that here U is well defined (Eq. 13.168) instead of an arbitrary positive attractive interaction between electrons of opposite spin in an atom. It can be shown that the preceding conditions could also be obtained if one considers the second-order terms in the field in the thermodynamic potential and calculates the magnetic susceptibility.

13.7 THE KONDO EFFECT

In 1936, de Haas and van den Berg observed in experiments with gold (probably containing a small amount of iron impurities) that electrical resistivity dropped as the normal temperature decreased until about 8° K, and below that it increased. This observation was in contrast to the theory of electrical resistivity of normal metals (nonsuperconductors) where the resistance is supposed to fall with temperature as the atomic vibrations decrease and, below about 10° K, remain constant. This low-temperature residual resistivity is due to scattering by defects in the material.

Kondo (Ref. 14) used second-order perturbation theory to calculate the scattering of the conduction electrons of a noble metal by magnetic defects, using a Heisenberg model to describe the interaction between conduction electron spins and the spin moments and the defects. Kondo observed that spin-flip processes can occur in the second order, which results in a resistance that increases logarithmically when the temperature is lowered. These results are valid above a temperature T_K (known as the Kondo temperature), but the perturbation theory breaks down for $T < T_K$. In 1963, Nagaoka[20] published a self-consistent treatment that yielded solutions both above and below the Kondo temperature and reproduced the resistance minimum.

13.8 ANDERSON MODEL

In 1961, Anderson (Ref. 1) proposed a model for the Hamiltonian of a single magnetic impurity to a Fermi sea of electrons,

$$H = \sum_{\mathbf{k}\sigma} \varepsilon_{\mathbf{k}} c_{\mathbf{k}\sigma}^{\dagger} c_{\mathbf{k}\sigma} + \varepsilon_d \sum_{\sigma} d_{\sigma}^{\dagger} d_{\sigma} + \sum_{\mathbf{k}\sigma} (V_{d\mathbf{k}} d_{\sigma}^{\dagger} c_{\mathbf{k}\sigma} + V_{d\mathbf{k}}^{*} c_{\mathbf{k}\sigma}^{\dagger} d_{\sigma}) + U d_{\uparrow}^{\dagger} d_{\uparrow} d_{\downarrow}^{\dagger} d_{\downarrow}. \tag{13.175}$$

The first term is the kinetic energy of the electrons, where $c_{\mathbf{k}\sigma}^{\dagger}$ and $c_{\mathbf{k}\sigma}$ are the creation and annihilation operators for electrons with spin σ in state $\mathbf{k}$. The magnetic state is represented by a single state with energy ε_d and electron operators $d_{\sigma}^{\dagger}$ and d_{σ}. The third term represents the $s-d$ hybridization of the d-level with the conduction band. The fourth term represents the cost, in Coulomb energy, of double occupancy of the d-level. U is assumed to be large and $\varepsilon < \varepsilon_F$, where ε_F is the Fermi energy. In the absence of the hybridization term, we have a single occupied (magnetic) d state. The Anderson model does not attempt to describe the details of the atomic $3d$ orbitals, but it captures the essential physics of the Kondo problem.

The Anderson Hamiltonian leads to the spin exchange process. The electron can tunnel from the localized impurity state to an unoccupied state just above the Fermi energy, and another electron from the Fermi energy sea replaces it. If these two electrons have opposite spin, it is known as a spin-flip process. An alternate spin-flip process involves double occupancy of the impurity state, with a cost of Coulomb energy. When many such processes are taken together, a new many-body process known as the Kondo resonance is generated. The Kondo resonance, which has a spin-singlet state, has a width of $k_B T_K$ and is pinned at the Fermi level, as schematically shown in Figure 13.11.

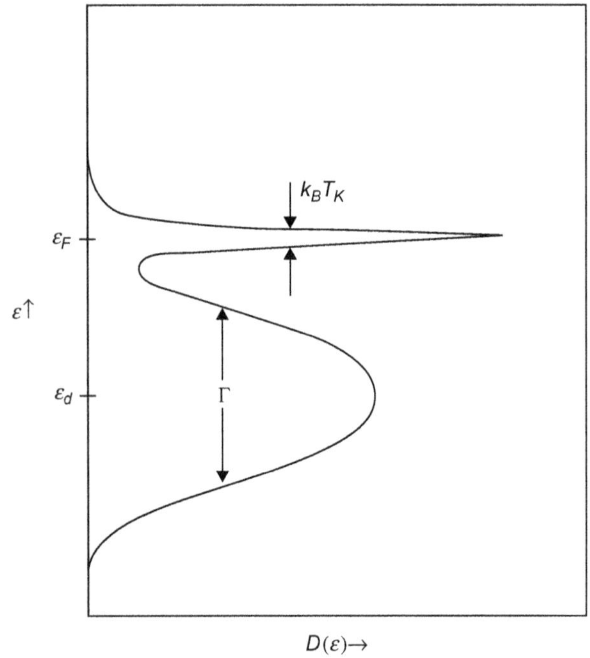

FIGURE 13.11

Schematic diagram of Kondo resonance of width $k_B T_K$. The impurity state is broadened to width Γ by hybridization with the conduction electrons.

The impurity state at ε_d is shifted and broadened through hybridization with electron states of the Fermi sea. This is known as a *d*-resonance. The net effect is an alignment of the spins of the conduction electrons near the impurity atom so that the local moment is screened. It has been shown by scaling laws and parameters of the Anderson model that

$$k_B T_K = \frac{1}{2}(\Gamma U)^{\frac{1}{2}} \exp\left[\pi\varepsilon_d(\varepsilon_d + U)/\Gamma U\right]. \tag{13.176}$$

13.9 THE MAGNETIC PHASE TRANSITION

13.9.1 Introduction

There are several types of phase transitions, but the most thoroughly investigated type is the paramagnetic to ferromagnetic phase transition. When the temperature is lowered through a critical temperature T_c of the system, a long-range order appears that was absent above T_c. The order, which usually has an infinite derivative at T_c, rapidly increases below T_c. In the absence of external fields, the ordering is not completely determined, and the transition is characterized by divergence of certain quantities like the magnetic susceptibility and the specific heat.

13.9.2 The Order Parameter

In the case of magnetization, the order parameter $\sigma(\mathbf{r})$, which describes the sudden appearance of order in the phase transition, is defined as

$$\sigma(\mathbf{r}) = M_z/M_0, \tag{13.177}$$

where M_z is the magnetization that defines the z-axis, and M_0 is the maximum possible value of magnetization,

$$M_0 = Ng\beta S. \tag{13.178}$$

The order parameter, which is zero above but not below T_c, can continuously approach zero as $T \to T_c$ from below if the transition is not of the first order. However, it is not completely determined below T_c in the absence of external fields.

13.9.3 Landau Theory of Second-Order Phase Transitions

According to Landau theory (Ref.15), the free energy is defined as a function of the order parameter

$$G = G(T, \sigma) = \int g(\mathbf{r})d\mathbf{r}, \tag{13.179}$$

where $g(\mathbf{r})$ is the free energy density. The entropy S is given by

$$S = -\left(\frac{\partial G}{\partial T}\right)_\sigma. \tag{13.180}$$

Landau introduced a quantity $h(\mathbf{r})$, which is related to the external field $H_0(\mathbf{r})$ by

$$h(\mathbf{r}) = M_0 H(\mathbf{r}). \tag{13.181}$$

Landau assumed that $g(\mathbf{r})$ can be expanded as a power series in σ,

$$g(\mathbf{r}) = g_0(T) - h(\mathbf{r})\sigma(\mathbf{r}) + \alpha(T)[\sigma(\mathbf{r})]^2 + \beta(T)[\sigma(\mathbf{r})]^4 + \gamma(T)|\nabla\sigma(\mathbf{r})|^2 + \cdots. \tag{13.182}$$

Here, g is minimized (which determines the most probable value of σ) by requiring that

$$\delta \int g(\mathbf{r})d\mathbf{r} = 0, \tag{13.183}$$

from which we obtain

$$\int d\sigma[-h(\mathbf{r}) + 2a\sigma + 4b\sigma^3 - 2c\,\nabla^2\sigma(\mathbf{r})]d\mathbf{r} = 0. \tag{13.184}$$

Eq. (13.184) yields

$$-2c\nabla^2\sigma(\mathbf{r}) + [2a + 4b\sigma^2(\mathbf{r})]\sigma(\mathbf{r}) = h(\mathbf{r}). \tag{13.185}$$

If σ and h are assumed to be independent of $\mathbf{r}$, Eq. (13.185) yields

$$[2a + 4b\sigma^2]\sigma = h. \tag{13.186}$$

If $h = 0$, the two possible solutions are

$$\sigma = 0 \tag{13.187}$$

and

$$\sigma = \pm(-a/2b)^{1/2}. \tag{13.188}$$

From Eqs. (13.182) and (13.183), if $a > 0$, Eq. (13.187) minimizes the free energy, whereas if $a < 0$, Eq. (13.188) minimizes the free energy. The free energy should describe a system with non-zero magnetization in the absence of external fields for $T < T_c$. This suggests the assumption

$$a(T) = \alpha(T - T_c), \tag{13.189}$$

where α is a constant. From Eqs. (13.188) and (13.189), we obtain

$$\sigma(T) \approx (T_c - T)^{1/2}, \quad (T < T_c). \tag{13.190}$$

Eq. (13.190) is consistent with the molecular field theory. To obtain the reduced susceptibility, we consider $h \neq 0$. The reduced susceptibility is defined as

$$\chi_r = (\partial\sigma/\partial h)_T. \tag{13.191}$$

From Eq. (13.186) and (13.191), we obtain

$$2a\chi_r + 12\sigma^2\chi_r = 1. \tag{13.192}$$

Because σ^2 is small and, hence, can be neglected for $T > T_c$, Eq. (13.192) can be rewritten in the alternate form with the help of Eq. (13.189),

$$\chi_r = 1/2\alpha(T - T_c), \quad (T > T_c). \tag{13.193}$$

Eq. (13.193) is the Curie–Weiss law. Below T_c, from Eqs. (13.190) and (13.192), we obtain

$$\chi_r = -1/4a = -1/4\alpha(T_c - T), \quad (T < T_c). \tag{13.194}$$

At a constant external field, the specific heat is obtained from the relation

$$C_M = -T\left(\frac{\partial^2 G}{\partial T^2}\right)_h. \tag{13.195}$$

First, we consider the case of zero external field ($h = 0$). Because $\sigma = 0$ for $T > T_c$,

$$C = C_0 = -Td^2/dT^2 \int g_0(\mathbf{r})d\mathbf{r}, \quad (T > T_c). \tag{13.196}$$

For a uniform system below T_c, it can be shown that (Problem 13.22)

$$G = \int d\mathbf{r}\,[g_0 - (a^2/4b)]. \tag{13.197}$$

Thus, from Eqs. (13.195) and (13.197), the specific heat is

$$C = C_0 + T\int[(\alpha^2/2b) - [\alpha^2(T - T_c)^2/4b^2]db/dT]d\mathbf{r}, \quad (T < T_c). \tag{13.198}$$

Comparing Eqs. (13.196) and (13.198), we note that there is a finite contribution from magnetization to the specific heat just below T_c that results in a finite discontinuity to C at T_c.

PROBLEMS

13.1. Using vector identities, show from Eqs. (13.1) through (13.3) that

$$\mathbf{F} = (\mathbf{m} \times \vec{\nabla}) \times \mathbf{B} = \vec{\nabla}(\mathbf{m} \cdot \mathbf{B}),$$

where $\mathbf{B} \equiv \mathbf{B}(0)$.

13.2. Show that the induction $\mathbf{B}$ produced by a magnetic dipole of moment $\mathbf{m}_1$ at a distance $\mathbf{r}$, where a second dipole of moment $\mathbf{m}_2$ is located, is given by

$$\mathbf{B} = \vec{\nabla}\left[\mathbf{m}_1 \cdot \vec{\nabla}\frac{1}{r}\right] = \frac{3\hat{\mathbf{r}}(\mathbf{m}_1 \cdot \hat{\mathbf{r}}) - \mathbf{m}_1}{r^3}.$$

13.3. From Eqs. (13.8) through (13.10) and Eq. (13.13), show that

$$\varepsilon_{s,t} = <\hat{H}>_{s,t} = 2\varepsilon_0 + \frac{e^2}{R_{12}} + e^2 \int d\mathbf{r}_1 d\mathbf{r}_2\, \psi^*_{s,t}(\mathbf{r}_1, \mathbf{r}_2)\left[\frac{1}{|\mathbf{r}_1 - \mathbf{r}_2|} - \frac{1}{|\mathbf{r}_1 - \mathbf{R}_2|} - \frac{1}{|\mathbf{r}_2 - \mathbf{R}_1|}\right]\psi_{s,t}(\mathbf{r}_1, \mathbf{r}_2), \quad (1)$$

where

$$\frac{1}{R_{12}} \equiv \frac{1}{|\mathbf{R}_1 - \mathbf{R}_2|}. \quad (2)$$

13.4. From Eqs. (13.14) through (13.17), show that

$$\varepsilon_s = 2\varepsilon_0 + \frac{e^2}{R} + \frac{V_c(R_{12}) + V_{ex}(R_{12})}{(1 + I^2)} \quad (1)$$

and

$$\varepsilon_t = 2\varepsilon_0 + \frac{e^2}{R} + \frac{V_c(R_{12}) - V_{ex}(R_{12})}{(1 - I^2)}. \quad (2)$$

13.5. For a two-electron system, we can express

$$\hat{S}_1 \cdot \hat{S}_2 = \frac{1}{2}[\hat{S}_1^+ \hat{S}_2^- + \hat{S}_2^+ \hat{S}_1^-] + \hat{S}_1^z \hat{S}_2^z, \quad (1)$$

where

$$\hat{S}^{\pm} = \hat{S}_x \pm i\hat{S}_y. \quad (2)$$

Show that

$$\hat{S}_1 \cdot \hat{S}_2 \chi_s = -\frac{3}{4}\chi_s \quad (3)$$

and

$$\hat{S}_1 \cdot \hat{S}_2 \chi_t = \frac{1}{4}\chi_t, \quad (4)$$

where χ_s and χ_t are the spin singlet and triplet states.

13.6. In the Schwinger representation, the Bose operators are defined as

$$\hat{S}^{\alpha} = \frac{1}{2}\sum_{ii'} \hat{a}_i^{\dagger}\sigma_{ii'}^{\alpha}\hat{a}_{i'}, \tag{1}$$

where σ^{α} are the Pauli spin matrices ($\alpha = x, y, z$). Show from Eq. (1) that

$$\begin{aligned} \hat{S}^x &= \frac{1}{2}(\hat{a}_1^{\dagger}\hat{a}_2 + \hat{a}_2^{\dagger}\hat{a}_1) \\ \hat{S}^y &= \frac{1}{2}(\hat{a}_2^{\dagger}\hat{a}_1 - \hat{a}_1^{\dagger}\hat{a}_2) \\ \hat{S}^z &= \frac{1}{2}(\hat{a}_1^{\dagger}\hat{a}_1 - \hat{a}_2^{\dagger}\hat{a}_2). \end{aligned} \tag{2}$$

13.7. By using the Holstein–Primakoff transformation, we have obtained

$$\begin{aligned} \hat{S}^{+} &= \hat{a}_1^{\dagger}\sqrt{2S - \hat{a}_1\hat{a}_1} \\ \hat{S}^{-} &= \sqrt{2S - \hat{a}_1\hat{a}_1}\,\hat{a}_1 \\ \hat{S}^{z} &= (\hat{a}_1^{\dagger}\hat{a}_1 - S). \end{aligned} \tag{1}$$

Show that

$$[\hat{S}^{+}, \hat{S}^{-}] = 2\hat{S}^{z}. \tag{2}$$

13.8. In Eq. (13.54), we defined

$$\Phi_m = \hat{S}_m^{-}\Pi_n|S>_n. \tag{1}$$

From Eqs. (13.51) and (13.54), we can write

$$\hat{H}\Phi_m = -J\sum_{ij}{}' \left[\hat{S}_i^z\hat{S}_j^z\hat{S}_m^{-} + \frac{1}{2}(\hat{S}_i^{+}\hat{S}_j^{-}\hat{S}_m^{-} + \hat{S}_i^{-}\hat{S}_j^{+}\hat{S}_m^{-})\right]\Phi_0. \tag{2}$$

Show by using the commutation relation of the spin operators that

$$\hat{H}\Phi_m = E_0\Phi_m + 2JS\sum_{\mathbf{R}_l}(\Phi_m - \Phi_{m+l}). \tag{3}$$

13.9. Show that the only possible way in which the ground-state energy ε_0 obtained from Eq. (13.59) will be the same as obtained in Eq. (13.53) is to assume $b_i = b_j = b$.

13.10. From Eq. (13.53) and (13.59), show that we can write $\hat{H}$ (up to the first order in S), treating $\hat{a}$ as small, $b = 0$, and multiplying the second term by 2 because $<ij>$ is a sum over nearest-neighbor pairs (each pair appears twice),

$$\hat{H} = -JNzS^2 - 2JS\sum_{<ij>}(\hat{a}_i^{\dagger}\hat{a}_j + \hat{a}_j^{\dagger}\hat{a}_i - \hat{a}_i^{\dagger}\hat{a}_i - \hat{a}_j^{\dagger}a_j). \tag{1}$$

13.11. If all the $\mathbf{R}_i$ vectors are in the A sublattice and the $\mathbf{R}_j$ vectors are in the B sublattice, the spin operators defined in Eqs. (13.40) and (13.42) for ferromagnets can be rewritten for antiferromagnets as

$$\hat{S}_j^z \to -\hat{S}_j^z \tag{1}$$

and

$$\hat{S}_j^{\pm} \to \hat{S}_j^{\mp}. \tag{2}$$

Show that the Hamiltonian in Eq. (13.51) for ferromagnets can be rewritten for antiferromagnets as

$$\hat{H} = 2|J| \sum_{<ij>} \left[-\hat{S}_{iz}\hat{S}_{jz} + \frac{1}{2}(\hat{S}_i^+ \hat{S}_j^+ + \hat{S}_i^- \hat{S}_j^-) \right]. \tag{3}$$

13.12. Using a procedure adapted to obtain Eq. (13.57) with the modifications for antiferromagnets (Eqs. 13.68 through 13.70), show that

$$\hat{H} = 2|J| \sum_{<ij>} \left[-(S - \hat{a}_i^\dagger \hat{a}_i)(S - \hat{a}_j^\dagger \hat{a}_j) + \frac{1}{2}\hat{a}_i^\dagger \sqrt{2S - \hat{a}_i^\dagger \hat{a}_i}\, \hat{a}_j^\dagger \sqrt{2S - \hat{a}_j^\dagger \hat{a}_j} \right.$$
$$\left. + \frac{1}{2}\sqrt{2S - \hat{a}_i^\dagger \hat{a}_i}\hat{a}_i \sqrt{2S - \hat{a}_j^\dagger \hat{a}_j}\hat{a}_j \right]. \tag{1}$$

13.13. By following a $1/S$ expansion method similar to the procedure outlined in Eqs. (13.59) and Eq. (13.70), derive from Eq. (1) of Problem 13.12

$$\hat{H} = -|J|NzS^2 + 2|J|S \sum_{<i,j>} [\hat{a}_i^\dagger \hat{a}_i + \hat{a}_j^\dagger \hat{a}_j + \hat{a}_i^\dagger \hat{a}_j^\dagger + \hat{a}_i \hat{a}_j]. \tag{1}$$

13.14. We derived

$$\hat{H} = -|J|NzS^2 + |J|S| \sum_{\mathbf{k},l} [2\hat{b}_\mathbf{k}^\dagger \hat{b}_\mathbf{k} + (\hat{b}_\mathbf{k}^\dagger \hat{b}_{-\mathbf{k}}^\dagger + \hat{b}_\mathbf{k} \hat{b}_{-\mathbf{k}}) \cos(\mathbf{k} \cdot \mathbf{R}_l)], \tag{1}$$

where $\mathbf{R}_l$ are nearest-neighbor vectors. To diagonalize the Hamiltonian, we introduced two new operators through the transformation,

$$\hat{b}_\mathbf{k} = (\sinh \beta_\mathbf{k})\, \hat{c}_{-\mathbf{k}}^\dagger + (\cosh \beta_\mathbf{k})\, \hat{c}_\mathbf{k}, \tag{2}$$

where β is real. Show that by substituting Eq. (2), the Hamiltonian in Eq. (1) is diagonalized provided

$$\tanh 2\beta_\mathbf{k} = -\frac{1}{z} \sum_l \cos(\mathbf{k} \cdot \mathbf{R}_l). \tag{3}$$

13.15. Derive from Eqs. (13.88) and (13.91) that the ratio of saturation magnetization $M(T)$ at temperature T and $M(0)$ at zero temperature is

$$\frac{M(T)}{M(0)} = 1 - \frac{1}{S} e^{-3T_C/(S+1)T}. \tag{1}$$

13.16. The chemical potential is the same for the two electron distributions with up- and down-spins. Show that, to satisfy this, the following conditions have to be met:

$$n\langle\mu\rangle = \mu_B \int_0^\infty \frac{1}{2}\{f^0(\varepsilon_{\mathbf{k}\uparrow}) - f^0(\varepsilon_{\mathbf{k}\downarrow})\}D(\varepsilon)d\varepsilon \tag{1}$$

and

$$n = \int_0^\infty \frac{1}{2}\{f^0(\varepsilon_{\mathbf{k}\uparrow}) + f^0(\varepsilon_{\mathbf{k}\downarrow})\}D(\varepsilon)d\varepsilon. \tag{2}$$

13.17. We derived in Eq. (13.135)

$$\hat{H} = \sum_{\langle ij\rangle,\rho} -t[\hat{a}^\dagger_{i\rho}\hat{a}_{j\rho} + \hat{a}^\dagger_{j\rho}\hat{a}_{i\rho}] + U\sum_i(\hat{n}_{i\uparrow}n_\downarrow + n_\uparrow\hat{n}_{i\downarrow} - n_\uparrow n_\downarrow). \tag{1}$$

Making a Fourier transformation of the type

$$\hat{a}_{i\rho} = \frac{1}{\sqrt{N}}\sum_{\mathbf{k}} e^{i\mathbf{k}\cdot\mathbf{R}_i}\hat{b}_{\mathbf{k}\rho}, \tag{2}$$

show that

$$\hat{H} = \sum_{\mathbf{k},\mathbf{R}_l,\rho} -t\hat{b}^\dagger_{\mathbf{k}\rho}\hat{b}_{\mathbf{k}\rho}\cos\mathbf{k}\cdot\mathbf{R}_l + U\sum_{\mathbf{k}}(\hat{n}_{\mathbf{k}\uparrow}n_\downarrow + n_\uparrow\hat{n}_{\mathbf{k}\downarrow} - n_\uparrow n_\downarrow). \tag{3}$$

13.18. We derived in Eq. (13.145)

$$H'(\mathbf{k},\xi_l) = -ih_{\alpha\beta}\Pi^\alpha\nabla^\beta_{\mathbf{k}} + \frac{1}{2}g\mu_B B^\mu\sigma^\mu + B^\mu\Sigma^{1,\mu}(\mathbf{k},\xi_l). \tag{1}$$

Here, $\vec{\Pi}$ was defined in Eq. (12.133),

$$\vec{\Pi} = \frac{\hbar}{m}(\mathbf{p} + \hbar\mathbf{k}) + \frac{\hbar^2}{4m^2c^2}\vec{\sigma}\times\vec{\nabla}V + \nabla_{\mathbf{k}}\Sigma^0(\mathbf{k},\xi_l). \tag{2}$$

In addition, from Eqs. (13.147) and (13.148), we have

$$\widetilde{G}(\mathbf{k},\xi_l) = G_0(\mathbf{k},\xi_l) + G_0(\mathbf{k},\xi_l)\hat{H}'G_0(\mathbf{k},\xi_l) + \cdots \tag{3}$$

and

$$\nabla^\alpha_{\mathbf{k}}G_0(\mathbf{k},\xi_l) = G_0(\mathbf{k},\xi_l)\,\Pi^\alpha G_0(\mathbf{k},\xi_l). \tag{4}$$

From Eqs. (1) through (4), show that

$$\widetilde{G}(\mathbf{k},\xi_l) = G_0 - G_0\left[ih_{\alpha\beta}\Pi^\alpha\nabla^\beta_{\mathbf{k}} - \frac{1}{2}g_0\mu_B B^\nu F^\nu\right]G_0, \tag{5}$$

where

$$F^\nu = \sigma^\nu + \frac{2}{g\mu_B}\widetilde{\Sigma}^{1,\nu} \tag{6}$$

was defined in Eq. (12.147) and is the renormalized spin vertex of the electron–electron interaction.

13.19. From Eqs. (12.123) and (13.149), the quasiparticle contribution to the thermodynamic potential is obtained. After evaluating the trace and contour integration, show that

$$\Omega_{qp} = \frac{1}{2}\mu_B B^\nu \sum_{n\mathbf{k}\rho m\rho', m\neq n} \left[gF^\nu_{n\rho,n\rho'} + \frac{ie}{\hbar c\mu_B}\epsilon_{\alpha\beta\nu}\frac{\Pi^\alpha_{n\rho,m\rho'}\Pi^\beta_{m\rho',n\rho}}{E_{mn}} \right] f(E_{n\mathbf{k}\rho}), \tag{1}$$

where $f(E_{n\mathbf{k}\rho})$ is the Fermi distribution function for an electron in band n and spin ρ,

$$\Pi^\alpha_{n\rho,m\rho'} = \int_{cell} d^3r u^*_{n\mathbf{k}\rho}\Pi^\alpha u_{m\mathbf{k}\rho'} \tag{2}$$

and

$$E_{mn} = E_m(\mathbf{k}) - E_n(\mathbf{k}). \tag{3}$$

13.20. The contribution of correlation to magnetization is

$$M^\nu_{corr} = -\frac{\partial\Omega_{corr}}{\partial B^\nu} = \frac{1}{\beta}Tr\frac{\partial\widetilde{\Sigma}_{\xi_l}}{\partial B^\nu}\widetilde{G}_{\xi_l}. \tag{1}$$

Simplifying Eq. (1) with the use of Eqs. (12.127), (12.128), (13.145), (13.147), and the Luttinger–Ward (Ref. 16) prescription for frequency summation,

$$\frac{1}{\beta}\sum_{\xi_l}\frac{1}{(\xi_l - \hat{H}_0)^m} = \frac{1}{2i\pi}Tr\int_{\Gamma_0}\frac{1}{(\xi - \hat{H}_0)^m}f(\xi)\, d\xi, \tag{2}$$

where Γ_0 encircles the real axis in a clockwise direction, and $f(\xi)$ is the Fermi distribution function, show that

$$M^\nu_{corr} = \sum_{n\mathbf{k}\rho}\widetilde{\Sigma}^{1,\nu}_{n\rho,n\rho'} f(E_{n\mathbf{k}\rho}). \tag{3}$$

13.21. The many-body Hamiltonian can be written as

$$\hat{H}_0 = \sum_{\mathbf{k}\rho}\varepsilon_{\mathbf{k}}c^\dagger_{\mathbf{k}\rho}c_{\mathbf{k}\rho} + \frac{1}{2}\sum_{\mathbf{q}\mathbf{k}\mathbf{k}'\rho\rho'} V(\mathbf{q})c^\dagger_{\mathbf{k}+\mathbf{q},\rho}c^\dagger_{\mathbf{k}'-\mathbf{q},\rho'}c_{\mathbf{k}',\rho'}c_{\mathbf{k},\rho}, \tag{1}$$

where $c^\dagger_{\mathbf{k}\rho}$ and $c_{\mathbf{k}\rho}$ are the creation and annihilation operators for an electron with wave vector $\mathbf{k}$ and spin ρ, and $V(\mathbf{q})$ is the Fourier transformation of the Coulomb interaction. $E_{\mathbf{k}\rho}$ can be obtained by using the mean-field approximation,

$$E_{\mathbf{k}\rho} = \frac{1}{N}\left\langle\left\{\left[c_{\mathbf{k}\rho}, \hat{H}_0\right], c^\dagger_{\mathbf{k}\rho}\right\}\right\rangle. \tag{2}$$

From Eqs. (1) and (2), show that

$$E_{\mathbf{k}\rho} = \varepsilon_{\mathbf{k}} - \frac{1}{N}\sum_{\mathbf{q}} V(\mathbf{q}) f(\varepsilon_{\mathbf{k}-\mathbf{q},\rho}) + \frac{1}{N} V(0) \sum_{\mathbf{k}} [f(\varepsilon_{\mathbf{k}\rho}) - f(\varepsilon_{\mathbf{k},-\rho})]. \quad (3)$$

13.22. Show that for a uniform system below T_c,

$$G = \int d\mathbf{r}[g_0 - (a^2/4b)]. \quad (1)$$

Hence, show by using Eq. (13.195) that the specific heat is

$$C = C_0 + T\int [(\alpha^2/2b) - [\alpha^2 (T - T_c)^2/4b^2] db/dT] d\mathbf{r}. \quad (T < T_c) \quad (2)$$

References

1. Anderson PW. Localized magnetic states in metals. *Phys Rev* 1961;**124**:41.
2. Ashcroft NW, Mermin ND. *Solid state physics.* New York: Brooks/Cole; 1976.
3. Callaway J. *Quantum theory of the solid state.* New York: Academic Press; 1976.
4. Gunnarson O. Band model for magneticism of transition metals in the spin-density-funcional formalism. *J Phys F, Met Phys* 1976;**6**:587.
5. Heisenberg W. On the theory of ferromagnetism. *Z Physik* 1926;**49**:619.
6. Heitler W, London F. Interaction of neutral atoms and homopolar binding in quantum mechanics. *Z Physik* 1927;**44**:455.
7. Holstein T, Primakoff H. Field dependence of the intrinsic domain magnetization of a ferromagnet. *Phys Rev* 1940;**58**:1098.
8. Hubbard J. Electron correlations in narrow energy bands. *Proc R Soc Lond* 1963;**A277**:237.
9. Janak JF. Uniform susceptibilities of metallic elements. *Phys Rev B* 1976;**16**:255.
10. Kasuya T. A theory of metallic ferro- and antiferromagnetism on Zener's model. *Prog Theor Phys (Kyoto)* 1956;**16**:45.
11. Kittel C. *Introduction to solid state physics.* New York: John Wiley & Sons; 1976.
12. Kittel C. *Quantum theory of solids.* New York: John Wiley & Sons; 1987.
13. Kittel C. *Indirect Exchange Interaction in Metals. Solid State Phys* 1968;**22**:1.
14. Kondo J. Resistance minimum in dilute magnetic alloys. *Prog Theor Phys* 1964;**32**:37.
15. Landau LD. On the theory of phase transition. *Phhysikalliische Zeitschraft der Sowjetunion* 1937;**11**:26 and 545.
16. Luttinger JM, Ward JC. Ground-state energy of a many fermion system II. *Phys. Rev.* 1960;**118**:1417.
17. Marder MP. *Condensed matter physics.* New York: John Wiley & Sons; 2000.
18. Misra SK, Misra PK, Mahanti SD. Many-body theory of magnetic susceptibility of electrons in solids. *Phys Rev B* 1982;**26**:1903.
19. Misra PK, series editor, Blackman J, editor. *Metallic nanoparticles.* Amsterdam: Elsevier; 2008.
20. Nagaoka Y. Self-consistent treatment of Kondo effect in dilute alloys. *Phys Rev* 1965;**138**:A1112.
21. Neel L. Magnetic properties of ferrites: Ferromagnetism and sntiferromagnetism. *Annales de Physique* 1948;**3**:137.
22. Rado GT, Suhl H, editors. *Magnetism.* New York: Academic Press; 1963.
23. Ruderman MA, Kittel C. Indirect exchange coupling of nuclear magnetic moments by conduction electrons. *Phys Rev* 1954;**96**:99.
24. Schwinger J. *Quantum theory of angular momentum.* New York: Academic Press; 1965.

25. Stoner EC. *Magnetism in matter*. London: Methuen; 1934.
26. Stoner EC. Collective electron ferromagnetism II. Energy and specific heat. *Proc R Soc Lond A* 1939;**169**:339.
27. Stoner EC, Wohlfarth EP. A Mechanism of Magnetic Hysteresis in Heterogeneous Alloys. *Phil Trans Lond A* 1948;**240**:599.
28. Tripathi GS, Das LK, Misra PK, Mahanti SD. Theory of spin-orbit and many-body effects on the Knight shift. *Phys. Rev. B* 1982;**25**:3091.
29. Tripathi GS, Misra PK. Many-body theory of magnetization of Bloch electrons in te presence pf spin-orbit interaction. *J Magn Mag Mater* 2010;**322**:88.
30. Vosko SH, Pedrew V. Theory of the spin susceptibility of an inhomogeneous electron gas via the density functional formalism. *Can J Phys* 1975;**53**:1385.
31. Yosida K. Many-body theory of magnetization of Bloch electrons in the presence of spin-orbit interaction. *Phys Rev* 1957;**106**:893.
32. Ziman JM. *Principles of the theory of solids*. Cambridge: Cambridge University Press; 1972.

CHAPTER

Superconductivity

14

CHAPTER OUTLINE

14.1 Properties of Superconductors 452
14.1.1 Introduction 452
14.1.2 Type I and Type II Superconductors 453
14.1.3 Second-Order Phase Transition 454
14.1.4 Isotope Effect 454
14.1.5 Phase Diagram 454
14.2 Meissner–Ochsenfeld Effect 455
14.3 The London Equation 455
14.4 Ginzburg–Landau Theory 456
14.4.1 Order Parameter 456
14.4.2 Boundary Conditions 457
14.4.3 Coherence Length 457
14.4.4 London Penetration Depth 458
14.5 Flux Quantization 459
14.6 Josephson Effect 460
14.6.1 Two Superconductors Separated by an Oxide Layer 460
14.6.2 AC and DC Josephson Effects 462
14.7 Microscopic Theory of Superconductivity 462
14.7.1 Introduction 462
14.7.2 Quasi-Electrons 463
14.7.3 Cooper Pairs 464
14.7.4 BCS Theory 466
14.7.5 Ground State of the Superconducting Electron Gas 466
14.7.6 Excited States at $T = 0$ 469
14.7.7 Excited States at $T \neq 0$ 470
14.8 Strong-Coupling Theory 472
14.8.1 Introduction 472
14.8.2 Upper Limit of the Critical Temperature, T_c 472
14.9 High-Temperature Superconductors 473
14.9.1 Introduction 473
14.9.2 Properties of Novel Superconductors (Cuprates) 474

14.9.3 Brief Review of s-, p-, and d-wave Pairing 474
14.9.4 Experimental Confirmation of d-wave Pairing 476
14.9.5 Search for a Theoretical Mechanism of High T_c Superconductors 481
Problems 481
References 485

14.1 PROPERTIES OF SUPERCONDUCTORS

14.1.1 Introduction

The main source of electrical resistance in a metal is the electron–phonon interaction caused by the scattering of the electrons due to vibrations of the ions. This is the lattice excitations in bulk metals, which corresponds to small ionic vibrations ($d/L \ll 1$), where d is the amplitude of the vibrations and L is the lattice period. This is described as acoustic quanta (phonons) with energies,

$$\varepsilon_{ph}^{i} = \hbar\Omega_i(\mathbf{q}), \tag{14.1}$$

where $\mathbf{q} = \hbar\mathbf{k}$ is the momentum of the phonon, and $|\mathbf{k}| = 2\pi/\lambda$, where λ is the phonon's wavelength. Here, i corresponds to the various phonon branches (longitudinal, transverse, and optical). The electron–lattice interaction, i.e., the energy exchange between the electrons and lattice, is due to the radiation and adsorption of phonons and is known as the electron–phonon interaction. As the temperature is lowered, the amplitude of the ions becomes smaller, and the electrical resistance is reduced.

But the resistivity of a normal metal does not become zero at zero temperature because there are other sources of electrical resistance due to the presence of impurities and imperfections in the crystal structure. As the temperature is reduced, there is a residual resistivity ρ_0 at absolute zero that is approximately 1% of the resistivity of a pure sample. The temperature around which the resistivity is constant is approximately 1° K. The resistivity of a metal can be expressed as

$$\rho(T) = \rho_0 + AT^5, \tag{14.2}$$

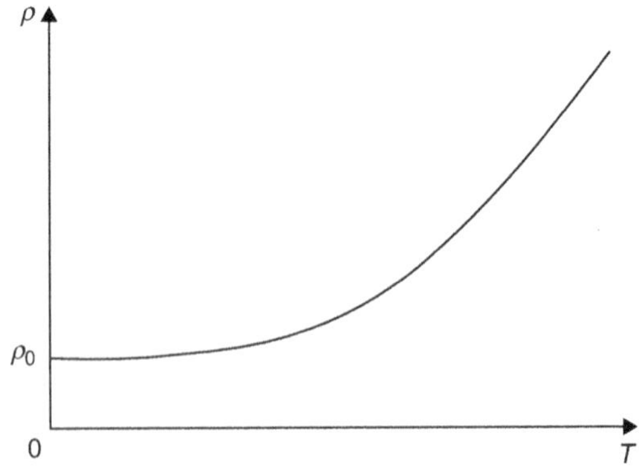

FIGURE 14.1

Variation of resistivity of a normal metal.

where ρ_0 is the residual resistivity, and AT^5 is the term arising from electron–phonon scattering. The variation of resistivity with temperature for a non-superconducting metal is shown in Figure 14.1.

The phenomenon of superconductivity was discovered by Kammerlingh Onnes in 1911.[15] He measured the resistivity of platinum and found that the resistivity followed a curve similar to the curve shown in Figure 14.1. However, when he performed a similar experiment with a sample of mercury, he found that the resistance of the sample dropped sharply to a value of zero at 4.2° K. The metals of which the resistivity becomes zero at a particular temperature are known as superconductors. The temperature at or below which a metal becomes a superconductor is called the critical temperature, T_c. Thus, there are two types of metals: some that become superconductors at or below T_c and others that continue to have resistivity even at 0° K. The variation of resistivity with temperature for a superconducting metal is shown in Figure 14.2.

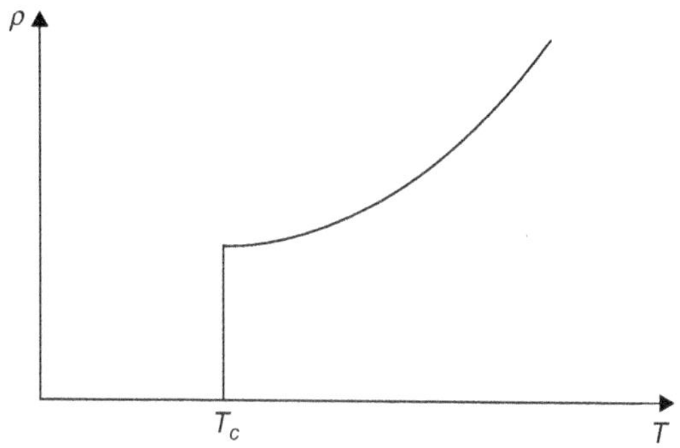

FIGURE 14.2

Variation of resistivity with temperature of a superconducting metal.

14.1.2 Type I and Type II Superconductors

The experimental results displayed in the previous section were for metals when there was no external magnetic field. However, when an external magnetic field H is applied, the superconducting state is destroyed above a critical external magnetic field H_c. This type of a superconductor is called a type I superconductor. From experiment results, one can obtain a relationship between the critical magnetic field H_T, which destroys superconductivity at a given temperature T,

$$H_T = H_c\left[1 - \left(\frac{T}{T_c}\right)^2\right]. \tag{14.3}$$

This experimental result is plotted in Figure 14.3.

One can distinguish type I and type II superconductors in the following manner. In type I superconductors, the diamagnetism grows linearly with the magnetic field until the critical value is reached. This is shown in Figure 14.4.

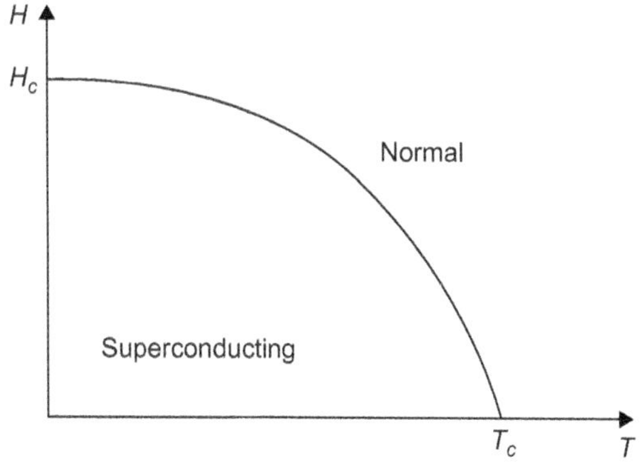

FIGURE 14.3

Variation of external magnetic field H against temperature T in a type I superconductor.

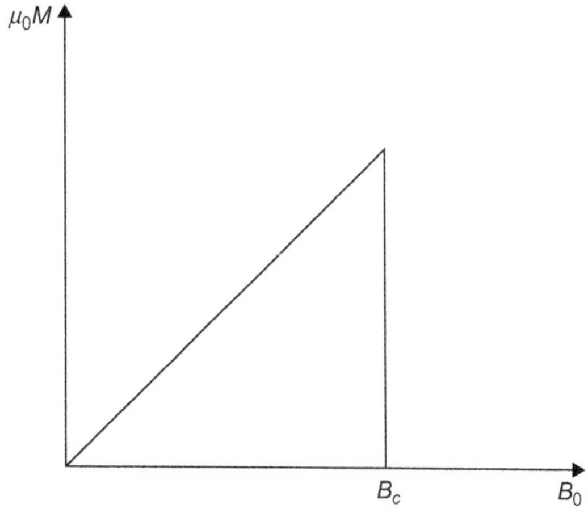

FIGURE 14.4

Type I superconductor.

In type II superconductors, the diamagnetization is linear with the external field up to a value B_{c1}, but the magnetization decreases when the flux begins to penetrate the metal. This is shown in Figure 14.5.

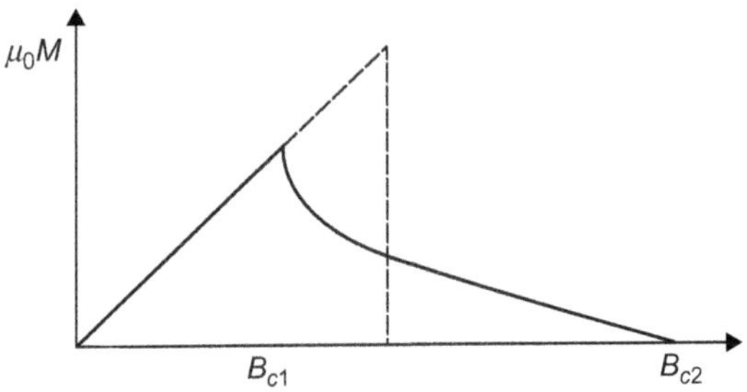

FIGURE 14.5

Type II superconductor.

14.1.3 Second-Order Phase Transition

The transition from the normal (N) state to the superconducting (S) state is known as a second-order phase transition. In the first-order phase transitions, the heat capacity is continuous, whereas in second-order phase transitions, the heat capacity is discontinuous. Figure 14.6 schematically shows the variation of heat capacity from a normal state to a superconducting state.

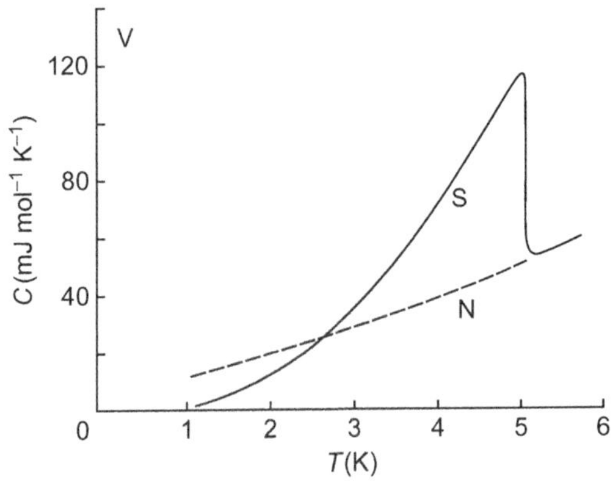

FIGURE 14.6

Second-order phase transition.

14.1.4 Isotope Effect

It was discovered in 1950 that the different isotopes of the same element possess different transition temperatures to enter the superconducting state,

$$T_c M_i^{\alpha} = \text{constant}, \tag{14.4}$$

where M_i is the isotopic mass of the same element. Usually, $\alpha \approx 1/2$. This was a strong indication that the interaction of the electrons with ions in the lattice was important for superconductivity.

14.1.5 Phase Diagram

Superconductivity is destroyed by application of a large magnetic field. In addition, superconductivity is destroyed if the current exceeds a "critical current." This is known as the Silsbee effect. To summarize, superconductivity depends on three factors: (a) the temperature T, (b) the external magnetic field H, and (c) the current density j. This type of dependence on three different experimental parameters is shown as a three-dimensional curve in Figure 14.7.

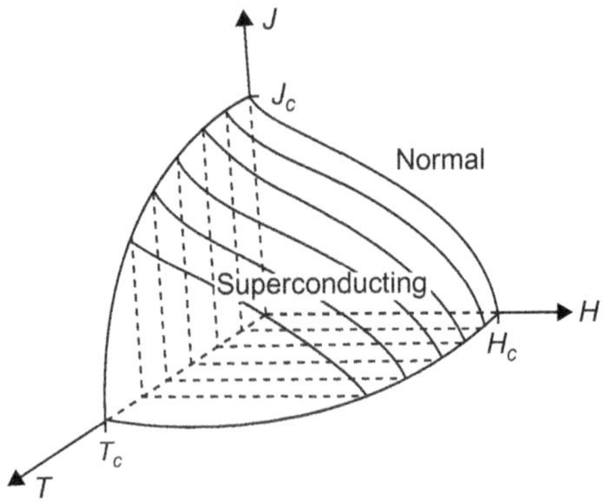

FIGURE 14.7

Critical surface separating the superconducting and normal states.

14.2 MEISSNER–OCHSENFELD EFFECT

Meissner and Ochsenfeld observed that an external magnetic field (as long as it is not too large) cannot penetrate inside a superconductor.[22] Thus, a superconductor behaves as a perfect diamagnet. If a normal metal is cooled below the critical temperature in a magnetic field, the magnetic flux is expelled abruptly. The transition to the superconducting state in a magnetic field is accompanied by the surface currents necessary to cancel the magnetic field inside the specimen. This scenario is shown in Figure 14.8.

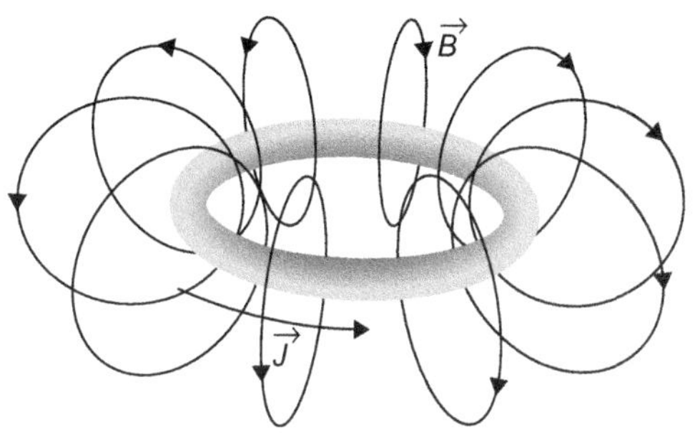

FIGURE 14.8

Flux lines cannot pass through the current loop in a superconductor.

14.3 THE LONDON EQUATION

London and London explained the Meissner–Ochsenfeld effect by adopting the two-fluid model of Gorter and Casimir.[18] The basic assumption of their model is that there are two types of electrons, $n_s(T)$ (density of superconducting electrons) and n (density of total number of conducting electrons), in a metal in the superconducting state. At temperatures $T < T_c$, only a fraction $n_s(T)/n$ of the total number of conduction electrons can carry a supercurrent. $n_s(T) \to 0$ as $T \to T_c \cdot n - n_s$ cannot carry an electric current without dissipation. In a small electric field, the normal electrons that flow parallel to the superconducting electrons are inert and therefore ignored while discussing the motion of electrons. In the presence of an electric field **E**,

$$m\dot{\mathbf{v}}_s = -e\mathbf{E}, \tag{14.5}$$

where $\mathbf{v}_s$ is the mean velocity of superconducting electrons. Because the current density $\mathbf{j} = -e\mathbf{v}_s n_s$, Eq. (14.5) can be rewritten in the alternate form

$$\frac{d\mathbf{j}}{dt} = \frac{n_s e^2}{m}\mathbf{E}. \tag{14.6}$$

From Eq. (14.6) and Faraday's law of induction,

$$\vec{\nabla} \times \mathrm{E} = -\frac{1}{c}\frac{\partial \mathbf{B}}{\partial t}, \tag{14.7}$$

we obtain

$$\frac{\partial}{\partial t}\left(\vec{\nabla} \times \mathrm{j} + \frac{n_s e^2}{mc}\mathbf{B}\right) = 0. \tag{14.8}$$

It is easy to show that Maxwell's equation (neglecting the displacement current **D** as well as replacing **H** with **B** because **j** is the mean microscopic current) can be written as (Problem 14.1)

$$\vec{\nabla} \times \mathrm{B} = \frac{4\pi}{c}\mathbf{j}. \tag{14.9}$$

Because Eqs. (14.8) and (14.9) are consistent with any static magnetic field and the Meissner–Ochsenfeld (Ref. 22) effect requires that the magnetic field be expelled in the superconducting state, London and London (Ref. 18) postulated that

$$\left(\vec{\nabla}\times \mathrm{j}+\frac{n_s e^2}{mc}\mathbf{B}\right)=0. \tag{14.10}$$

Using the vector identity,

$$\vec{\nabla}\times(\vec{\nabla}\times \mathrm{V})=\vec{\nabla}(\vec{\nabla}\cdot\mathbf{V})-\nabla^2\mathbf{V}, \tag{14.11}$$

we obtain from Eqs. (14.9) through (14.11),

$$\nabla^2\mathbf{B}=\frac{4\pi n_s e^2}{mc^2}\mathbf{B} \tag{14.12}$$

and

$$\nabla^2\mathbf{j}=\frac{4\pi n_s e^2}{mc^2}\mathbf{j}. \tag{14.13}$$

Eqs. (14.12) and (14.13) predict that the magnetic fields and currents in a superconductor can exist within a layer of thickness λ_L, known as the London penetration depth,

$$\lambda_L=\left(\frac{mc^2}{4\pi n_s e^2}\right)^{\frac{1}{2}}. \tag{14.14}$$

One can rewrite λ_L in the alternate form

$$\lambda_L=41.9\left(\frac{r_s}{a_0}\right)^{3/2}\left(\frac{n}{n_s}\right)^{1/2}\text{Å}. \tag{14.15}$$

Thus, when $T << T_c$, $(n\approx n_s)$ the surface currents that screen out the applied magnetic field occur within a layer of 10^2-10^3 Å. The magnetic field drops continuously to zero within this layer. It has indeed been observed experimentally that the field penetration is incomplete in superconducting films that are thinner than λ_L.

14.4 GINZBURG–LANDAU THEORY

14.4.1 Order Parameter

Landau and Ginzburg[17] described the superconductivity state through a position-dependent order parameter $\Psi(\mathbf{r})=|\Psi(\mathbf{r})|e^{i\phi(\mathbf{r})}$, which describes the macroscopic properties of a superfluid condensate.[17] The superfluid density $n_s(\mathbf{r})=|\Psi(\mathbf{r})|^2$, suggesting that $\Psi(\mathbf{r})$ is a wave function that vanishes at T_c. Landau and Ginzburg (Ref. 17) guessed that to be able to study the magnetic and thermodynamic properties of superconductors, the total free energy F_s with respect to its value in the normal state F_n should be

$$F_s=F_n+\int d\mathbf{r}\left[\alpha|\Psi|^2+\frac{\beta}{2}|\Psi|^4+\frac{1}{8\pi}B^2+\frac{1}{2m^*}\left|\left[\frac{\hbar}{i}\vec{\nabla}+\frac{e^*}{c}\mathbf{A}(\mathbf{r})\right]\Psi(\mathbf{r})\right|^2\right]. \tag{14.16}$$

Here, the first two terms are from Landau's general theory of second-order phase transition borrowed from their theory for liquid ^{4}He. According to their phenomenological model, one assumes that the complete ground-state wave function $\psi_N(\mathbf{r}_1......\mathbf{r}_N)$ for N electrons is known. If an extra electron is added, one knows $\psi_{N+1}(\mathbf{r}_1......\mathbf{r}_{N+1})$. We define

$$\Psi(\mathbf{r}) = \int d^N \mathbf{r} \psi_N^*(\mathbf{r}_1......\mathbf{r}_N)\psi_{N+1}(\mathbf{r}_1......\mathbf{r}_N, \mathbf{r}). \tag{14.17}$$

Far away from $\mathbf{r}$, Ψ_N and Ψ_{N+1} coincide, but in its neighborhood Ψ_{N+1} accommodates one extra particle. The last two terms are due to the fact that the superconducting electron wave function might interact with the vector potential like a single macroscopic particle of effective charge e^* (it was later found that $e^* = 2e$, *consistent with Cooper pairs* (Ref. 7)). Minimizing Eq. (14.16) with respect to $\mathbf{A}$, we obtain

$$\vec{\nabla} \times \mathbf{B} = \frac{4\pi}{c}\mathbf{j}, \tag{14.18}$$

where

$$\mathbf{j}(\mathbf{r}) = \frac{-e^*\hbar}{2im^*}\left[\Psi^* \vec{\nabla}\Psi - \Psi\vec{\nabla}\Psi^*\right] - \frac{e^{*2}}{m^*c}\mathbf{A}\Psi^*\Psi. \tag{14.19}$$

Minimizing Eq. (14.16) with respect to Ψ^*, by first integrating such that all spatial derivatives act on Ψ and then taking functional derivatives with respect to Ψ^*, we obtain (Problem 14.3)

$$\left[\alpha + \beta|\Psi|^2 + \frac{1}{2m^*}\left(\frac{\hbar}{i}\vec{\nabla} + \frac{e^*}{c}\mathbf{A}\right)^2\right]\Psi = 0. \tag{14.20}$$

Eqs. (14.19) and (14.20) are known as the Landau–Ginzburg (Ref. 17) equations.

14.4.2 Boundary Conditions

No current can flow out of the boundary when a superconductor is in contact with a vacuum. Thus, one can write

$$\hat{n} \cdot \left(\frac{\hbar}{i}\vec{\nabla} + \frac{e^*}{c}\mathbf{A}\right) = 0. \tag{14.21}$$

The additional boundary condition imposed by Landau and Ginzburg[17] is that $\Psi = 0$ was not considered as an acceptable condition, because in that case, it would be impossible to obtain solutions for thin superconducting films.

14.4.3 Coherence Length

The coherence length ξ specifies the scale for variation of the order parameter Ψ. The coherence length is defined by

$$\xi = \left[\frac{\hbar^2}{2m^*|\alpha|}\right]^{\frac{1}{2}}, \tag{14.22}$$

where α is defined in Eq. (14.20). If the external magnetic field vanishes, Eq. (14.20) can be rewritten as

$$|\Psi|^2 = \Psi_0^2 = -\frac{\alpha}{\beta} \text{ or } |\Psi|^2 = 0. \tag{14.23}$$

Because β is positive (otherwise, F is minimized when $\Psi \to \infty$), $\alpha < 0$, which corresponds to a uniform superconducting state. The free energy per unit volume is given by Eq. (14.16),

$$F_s - F_N = -\frac{\alpha^2}{2\beta}. \tag{14.24}$$

It can be easily shown that

$$\Delta F = F_N - F_s = \frac{H_c^2}{8\pi}. \tag{14.25}$$

From Eqs. (14.24) and (14.25), we obtain

$$H_c^2 = \frac{4\pi\alpha^2}{\beta}. \tag{14.26}$$

We can scale the order parameter for superconductors,

$$\psi = \frac{\Psi}{\Psi_0}. \tag{14.27}$$

From Eqs. (14.20), (14.22), and (14.27), we obtain for $\mathbf{A} = 0$,

$$-\xi^2\nabla^2\psi - \psi + \psi|\psi|^2 = 0. \tag{14.28}$$

Thus, the coherence length ξ is the characteristic scale on which ψ varies (in the absence of a magnetic field).

14.4.4 London Penetration Depth

In Eq. (14.19), in the presence of a weak magnetic field, $\Psi \approx \Psi_0$ for $x < 0$. Thus, we obtain from Eqs. (14.18) and (14.19),

$$\mathbf{j} = \frac{c}{4\pi}\vec{\nabla} \times \mathbf{B} = -\frac{e^{*2}}{m^* c}\Psi_0^2 \mathbf{A} \tag{14.29}$$

and

$$\nabla \times \nabla \times \mathbf{B} = -\frac{4\pi}{c}\frac{e^{*2}}{m^* c}\Psi_0^2 \mathbf{B} = -\frac{\mathbf{B}}{\lambda_L^2}. \tag{14.30}$$

Because $m^* = 2m$, $e^* = 2e$, and $\Psi_0^2 = -\frac{\alpha}{\beta}$, λ_L is the London penetration depth defined in Eq. (14.14). The parameter

$$\kappa = \lambda_L/\xi = \frac{m^* c}{e\hbar}\sqrt{\frac{\beta}{2\pi}} \tag{14.31}$$

is the only parameter in Landau–Ginzburg theory.

It can be easily shown that the surface energy is positive if $\kappa < 1/\sqrt{2}$ and negative if $\kappa > 1/\sqrt{2}$. The magnetic flux can easily penetrate a superconductor if the surface energy is negative and can form type II superconductors, which are interlocking regions of normal and superconducting metal. Type I superconductors are those for which $\kappa < 1/\sqrt{2}$ and the flux cannot penetrate.

Consider a magnetic field $H > H_c$ so that $\Psi = 0$ because superconductivity is destroyed. When the field is gradually lowered, considering only the first-order terms in Ψ (because $\mathbf{j} = 0$),

$$\left[\alpha + \frac{1}{2m^*}\left(\frac{\hbar}{i}\vec{\nabla} + \frac{e^*}{c}\mathbf{A}\right)^2\right]\Psi = 0. \tag{14.32}$$

Eq. (14.32) is an eigenvalue equation of a charge ($e^* = 2e$) in a constant magnetic field, the lowest energy is $\hbar\omega_c/2$, and because H_{c2} is the largest magnetic field that permits a solution,

$$\omega_c = \frac{e^* H_{c2}}{m^* c}. \tag{14.33}$$

From Eqs. (14.32) and (14.33),

$$-\alpha\Psi = \frac{\hbar e^* H_{c2}}{2m^* c}\Psi \tag{14.34}$$

or

$$|\alpha| = \frac{\hbar e H_{c2}}{m^* c}. \tag{14.35}$$

From Eqs. (14.26), (14.31), and (14.35), we obtain

$$\frac{H_{c2}}{H_c} = \sqrt{2}\kappa. \tag{14.36}$$

14.5 FLUX QUANTIZATION

In Eq. (14.18), if we substitute

$$\Psi(\mathbf{r}) = \Psi_0 e^{i\phi(\mathbf{r})}, \tag{14.37}$$

where $\phi(\mathbf{r})$ is real, we obtain from Eqs. (14.18) and (14.37)

$$\mathbf{j} = -\frac{\Psi_0^2}{m^*}\left(\frac{e^{*2}}{c}\mathbf{A} + e^*\hbar\vec{\nabla}\phi\right), \tag{14.38}$$

which can be rewritten in the alternate form

$$\vec{\nabla}\phi = -\frac{1}{\hbar}\left(\frac{m^*}{e^*\Psi_0^2}\mathbf{j} + \frac{e^*}{c}\mathbf{A}\right). \tag{14.39}$$

If we consider a superconductor in the shape of a ring (Figure 14.9), which shows a path encircling the aperture but lying well within the interior,

$$\oint \mathbf{j}\cdot d\mathbf{l} = 0. \tag{14.40}$$

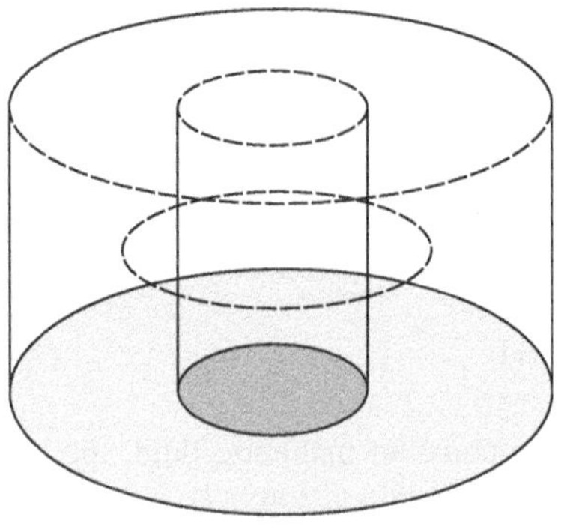

FIGURE 14.9

A path encircling the aperture ring of superconducting material.

From Eqs. (14.23) and (14.24), we obtain

$$\oint \vec{\nabla}\phi \cdot d\mathbf{l} = -\oint \frac{e^*\mathbf{A}}{\hbar c} \cdot d\mathbf{l}. \quad (14.41)$$

From Stoke's theorem,

$$\oint \mathbf{A} \cdot d\mathbf{l} = \oint \vec{\nabla} \times \mathbf{A} \cdot d\mathbf{S} = \oint \mathbf{B} \cdot d\mathbf{S} = \Phi, \quad (14.42)$$

where Φ is the flux enclosed by the ring. We note that because the magnetic field cannot penetrate a superconducting material, the enclosed flux is independent of the choice of the path. In addition, the order parameter Ψ is single-valued, and hence, its phase changes by $2\pi n$ (n is an integer) when the ring is encircled. Thus, we have

$$\oint \vec{\nabla}\phi \cdot d\mathbf{l} = 2\pi n. \quad (14.43)$$

From Eqs. (14.41) through (14.43), we obtain

$$\frac{-e^*}{\hbar c}\Phi = 2\pi n. \quad (14.44)$$

Defining a fluxoid or a flux quantum,

$$\Phi_0 \equiv \frac{hc}{2e} = 2.0679 \text{ G cm}^2, \quad (14.45)$$

we obtain from Eqs. (14.43) and (14.44)

$$|\Phi| = \frac{2ne}{e^*}\Phi_0. \quad (14.46)$$

The integrated magnetic flux that penetrates a hole through a superconductor is quantized in units of $(2e/e^*)\Phi_0$. Thus, $e^* = 2e$, confirming the fact that the microscopic theory of superconductivity involves pairing of electrons. Flux quantization was experimentally observed independently by Deaver and Fairbank[8] and Doll and Nabauer.[12]

14.6 JOSEPHSON EFFECT

14.6.1 Two Superconductors Separated by an Oxide Layer

In 1962, Josephson (Refs. 13, 14) predicted that if two superconductors were separated by a small strip of nonsuperconducting material (shown in Figure 14.10), the wave function would oscillate because it would interfere with itself. The two superconductors would interact through their small residues while decaying through the barrier. Due to tunneling, the change in energy $\Delta\varepsilon$ is given by

$$\Delta\varepsilon = \int d\mathbf{r}\, f(\mathbf{r})\left(\Psi_1^*(\mathbf{r})\Psi_2(\mathbf{r}) + \Psi_1(\mathbf{r})\Psi_2^*(\mathbf{r})\right). \quad (14.47)$$

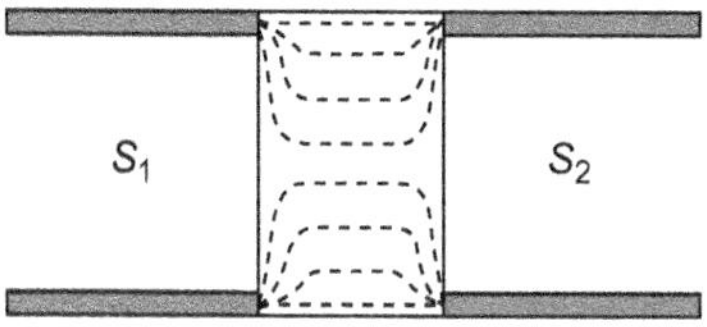

FIGURE 14.10

The oxide layer between two superconductors S_1 and S_2.

At some reference point in the bulk, if the macroscopic wave functions are Ψ_1 and Ψ_2,

$$\Delta\varepsilon = \in(\Psi_1^*\Psi_2 + \Psi_1\Psi_2^*), \tag{14.48}$$

where $\in$ is a very small quantity. The Schrodinger equations for Ψ_1 and Ψ_2 can be written as

$$i\hbar\frac{\partial\Psi_1}{\partial t} = \varepsilon_1\Psi_1 + \in\Psi_2 \tag{14.49}$$

and

$$i\hbar\frac{\partial\Psi_2}{\partial t} = \in\Psi_1 + \varepsilon_2\Psi_2. \tag{14.50}$$

The wave functions are taken of the form

$$\Psi_j = \sqrt{n_j}e^{i\phi_j}, \tag{14.51}$$

where n_j ($j = 1$ or 2) are the superconducting electron densities $n = \sqrt{n_1 n_2}$ and $n_1 \approx n_2$. Substituting Eq. (14.51) in Eqs. (14.49) and (14.50), we obtain

$$i\hbar\left(\frac{\dot{n}_1}{\sqrt{n_1}} + i\sqrt{n_1}\dot{\phi}_1\right)e^{i\phi_1} = (\varepsilon_1\sqrt{n_1}e^{i\phi_1} + \in\sqrt{n_2}e^{i\phi_2}) \tag{14.52}$$

and

$$i\hbar\left(\frac{\dot{n}_2}{\sqrt{n_2}} + i\sqrt{n_2}\dot{\phi}_2\right)e^{i\phi_2} = (\varepsilon_2\sqrt{n_2}e^{i\phi_2} + \in\sqrt{n_1}e^{i\phi_1}). \tag{14.53}$$

It can be easily shown from Eqs. (14.52) and (14.53) that

$$\dot{n}_1 = 2\frac{\in n}{\hbar}\sin(\phi_2 - \phi_1) = -\dot{n}_2 = \frac{j}{e^*} \tag{14.54}$$

and

$$\dot{\phi}_2 - \dot{\phi}_1 = \frac{1}{\hbar}(\varepsilon_1 - \varepsilon_2) = \frac{e^*}{\hbar}(V_2 - V_1). \tag{14.55}$$

We note that the phase of a wave function changes when there is a magnetic field. To make a wave function phase gauge-invariant, we have to add the line integral $\frac{e^*}{\hbar c}\int_0 d\mathbf{l}\cdot\mathbf{A}$, taken from some arbitrary reference point, to the phase. Incorporating this and taking into account the fact that $e^* = 2e$, we obtain

$$\mathbf{j} = \mathbf{j}_0 \sin\left(\phi_2 - \phi_1 + \frac{2e}{\hbar c}\int_1^2 d\mathbf{l}\cdot\mathbf{A}\right) \tag{14.56}$$

and

$$-\frac{1}{\hbar}(\varepsilon_2-\varepsilon_1)=\frac{2eV}{\hbar}=\frac{\partial}{\partial t}\left(\phi_2-\phi_1+\frac{2e}{\hbar c}\int_1^2 d\mathbf{l}\cdot\mathbf{A}\right). \tag{14.57}$$

Here, the energy difference $\varepsilon_2-\varepsilon_1$ is the difference in energies of electron pairs, and V is the difference of the voltage between the two superconductors.

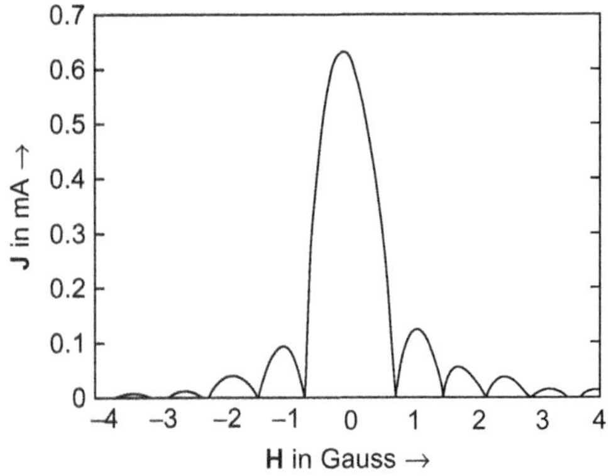

FIGURE 14.11

Maximum zero-voltage current in an Sn-SnO-Sn junction at $T = 1.9°$ K (sketch of results of R.C. Jaklevic (Ref. 23)).

14.6.2 AC and DC Josephson Effects

If we place two superconductors at different voltages in contact, then $\varepsilon_1-\varepsilon_2\neq 0$. Because $\phi_2-\phi_1$ drifts in time, n_1 and n_2 oscillate about their mean values at 484 MHz/μV. This is known as the AC Josephson effect.

However, if $\varepsilon_1-\varepsilon_2=0$, $\phi_1-\phi_2$ is independent of time. Because $\phi_1\neq\phi_2$, there is a steady current flow according to Eq. (14.56). This phenomenon is known as the DC Josephson effect. If the electrons are injected into a Josephson junction, $\phi_2-\phi_1$ adjusts to a nonzero value. The maximum value of $\phi_2-\phi_1$ is $\pi/2$. Thus, there is a current flow in the absence of any voltage difference (see Figure 14.11). This is a unique property of the superconductors and is a significant consequence of the Josephson effect.

14.7 MICROSCOPIC THEORY OF SUPERCONDUCTIVITY

14.7.1 Introduction

In Figure 8.14g, the graph shows an electron emitting a virtual phonon that is absorbed by a second electron. The lattice is deformed (polarized) in the vicinity of the first electron. A second electron near this polarized cloud experiences a force of repulsion or attraction that is independent of the Coulomb interaction between the two electrons. This is termed an effective electron–electron interaction via virtual phonons and is responsible for superconductivity under certain conditions. In the graph of Figure 8.14g, there are two possible intermediate states: electron **k** emits a phonon −**q**, which is absorbed by electron **k**′; or electron **k**′ emits a phonon **q**, which is absorbed by electron **k**.

In Eq. (8.165), we derived an expression for electron–phonon coupling for longitudinal acoustic (LA) phonons,

$$\hat{H}_{el-ph}=\sum_{\mathbf{kq}s}M_{\mathbf{q}}(\hat{a}^{\dagger}_{-\mathbf{q}}+\hat{a}_{\mathbf{q}})\hat{c}^{\dagger}_{\mathbf{k+q},s}\hat{c}_{\mathbf{k},s}. \tag{14.58}$$

We note that the direction of the spin is not changed by this interaction.

14.7.2 Quasi-Electrons

We now consider a system of electrons and phonons and write a Hamiltonian of the form

$$\hat{H} = \sum_{\mathbf{k},s} E(\mathbf{k})\hat{c}^\dagger_{\mathbf{k}s}\hat{c}_{\mathbf{k}s} + \sum_{\mathbf{q}} \hbar\omega_{\mathbf{q}}\hat{a}^\dagger_{\mathbf{q}}\hat{a}_{\mathbf{q}} + \sum_{\mathbf{kq}} M_{\mathbf{q}}(\hat{a}^\dagger_{-\mathbf{q}} + \hat{a}_{\mathbf{q}})\hat{c}^\dagger_{\mathbf{k+q},s}\hat{c}_{\mathbf{k},s} = \hat{H}_0 + \hat{H}_1, \tag{14.59}$$

where $\hat{H}_1$ is the electron–phonon interaction term. We use a canonical transformation,

$$\hat{H}_S = e^{-\hat{S}}\hat{H}e^{\hat{S}} = \hat{H}_0 + (\hat{H}_1 + [\hat{H}_0, \hat{S}]) + \frac{1}{2}[(\hat{H}_1 + [\hat{H}_0, \hat{S}], \hat{S}] + \frac{1}{2}[\hat{H}_1, \hat{S}] + \cdots. \tag{14.60}$$

If we choose $\hat{S}$ such that

$$\hat{H}_1 + [\hat{H}_0, \hat{S}] = 0, \tag{14.61}$$

the electron–phonon interaction $\hat{H}_1$ is eliminated apart from a higher-order term. We choose $\hat{S}$ of the form

$$\hat{S} = \sum_{\mathbf{kq}s'} M_{\mathbf{q}}(\alpha\hat{a}^\dagger_{-\mathbf{q}} + \beta\hat{a}_{\mathbf{q}})\hat{c}^\dagger_{\mathbf{k+q},s'}\hat{c}_{\mathbf{k},s'}. \tag{14.62}$$

From Eqs. (14.61) and (14.62), it can be easily shown that (Problem 14.5)

$$\alpha^{-1} = E(\mathbf{k}) - E(\mathbf{k+q}) - \hbar\omega_{\mathbf{q}} \tag{14.63}$$

and

$$\beta^{-1} = E(\mathbf{k}) - E(\mathbf{k+q}) + \hbar\omega_{\mathbf{q}}. \tag{14.64}$$

From Eqs. (14.60) and (14.61), the next-order interaction term is $\frac{1}{2}[\hat{H}_1, \hat{S}]$, which is a sum of terms that contain operator products of the form

$$\hat{a}^\dagger_{\pm\mathbf{q}}\hat{a}_{\pm\mathbf{q}'}\hat{c}^\dagger_{\mathbf{k'+q'},s'}\hat{c}_{\mathbf{k'},s'}\hat{c}^\dagger_{\mathbf{k+q},s}\hat{c}_{\mathbf{k},s}. \tag{14.65}$$

One can show (Problem 14.6) that because $\mathbf{q}' = -\mathbf{q}$ (from momentum conservation), only one out of all the possible combinations does not contain any phonon operators,

$$\hat{c}^\dagger_{\mathbf{k+q},s}\hat{c}^\dagger_{\mathbf{k-q},s'}\hat{c}_{\mathbf{k'},s'}\hat{c}_{\mathbf{k},s}, \tag{14.66}$$

when

$$\mathbf{k}' \neq \mathbf{k},\ \mathbf{k+q}. \tag{14.67}$$

From. Eqs. (14.60), (14.62), and (14.66), the explicit form for this interaction is (Problem 14.7)

$$\hat{H}_{eff} = \frac{1}{2}\sum_{\mathbf{k,k',q},s,s'} |M_{\mathbf{q}}|^2(\alpha - \beta)\hat{c}^\dagger_{\mathbf{k+q},s}\hat{c}^\dagger_{\mathbf{k'-q},s'}\hat{c}_{\mathbf{k'},s'}\hat{c}_{\mathbf{k},s}. \tag{14.68}$$

Substituting the values of α and β from Eqs. (14.63) and (14.64) in Eq. (14.68), we obtain (Problem 14.8)

$$<f|\hat{H}_{eff}|i> = \frac{1}{2}\sum_{\mathbf{k,k',q},s,s'} <f|V_{\mathbf{kq}}\hat{c}^\dagger_{\mathbf{k+q},s}\hat{c}^\dagger_{\mathbf{k-q},s'}\hat{c}_{\mathbf{k'}s'}\hat{c}_{\mathbf{k},s}|i>, \tag{14.69}$$

where

$$V_{\mathbf{kq}} = \frac{2|M_{\mathbf{q}}|^2 \hbar\omega_{\mathbf{q}}}{[E(\mathbf{k}+\mathbf{q}) - E(\mathbf{k})]^2 - (\hbar\omega_{\mathbf{q}})^2}, \tag{14.70}$$

and $|i>$ and $|f>$ are the initial and final states, respectively. $V_{\mathbf{kq}}$ is essentially the Fourier coefficient of the effective interaction. When $|E(\mathbf{k}+\mathbf{q}) - E(\mathbf{k})| < \hbar\omega_{\mathbf{q}}$, $V_{\mathbf{kq}}$ is negative and the interaction is attractive, whereas the interaction is repulsive when $V_{\mathbf{kq}}$ is positive.

Omitting all the terms in the transformed Hamiltonian that contains phonon creation or annihilation operators, we obtain from Eqs. (14.59) and (14.70),

$$\hat{H}_S = \sum_{\mathbf{k}s} E(\mathbf{k}) \hat{c}^{\dagger}_{\mathbf{k},s} \hat{c}_{\mathbf{k},s} + \sum_{\mathbf{k},\mathbf{k}',\mathbf{q},s,s'} |M_{\mathbf{q}}|^2 \frac{\hbar\omega_{\mathbf{q}}}{[E(\mathbf{k}+\mathbf{q}) - E(\mathbf{k})]^2 - (\hbar\omega_{\mathbf{q}})^2} \hat{c}^{\dagger}_{\mathbf{k}+\mathbf{q},s} \hat{c}^{\dagger}_{\mathbf{k}'-\mathbf{q},s'} \hat{c}_{\mathbf{k}',s'} \hat{c}_{\mathbf{k},s}. \tag{14.71}$$

14.7.3 Cooper Pairs

We consider a state containing N noninteracting electron gas, which fills the Fermi sphere in $\mathbf{k}-$ space. The ground state $|G>$ is the filled Fermi sphere. We introduce two electrons $\mathbf{k}_1$ and $\mathbf{k}_2$ into this system and take as the interaction between these electrons the positive part of $V_{\mathbf{kq}}$ (Eq. 14.70), which is repulsive. Thus, the interaction involving phonon exchange will take place only when $|E(\mathbf{k}+\mathbf{q}) - E(\mathbf{k})| \le \hbar\omega_{\mathbf{q}}$. The wave function of the electron pair can be written as

$$\psi_{12} = \sum_{\mathbf{k}_1 \mathbf{k}_2 s_1 s_2} A_{s_1 s_2}(\mathbf{k}_1, \mathbf{k}_2) \hat{c}^{\dagger}_{\mathbf{k}_1 s_1} \hat{c}^{\dagger}_{\mathbf{k}_2 s_2} |G>. \tag{14.72}$$

The summation over $\mathbf{k}_1$ and $\mathbf{k}_2$ is carried out subject to the condition that $\mathbf{K} = \mathbf{k}_1 + \mathbf{k}_2 =$ constant in order to form a state with definite momentum. This scenario is shown in Figure 14.12.

If we consider two electrons just above the Fermi sphere, an interaction will occur only when $E(\mathbf{k}_1) \le E_F + \hbar\omega_{\mathbf{q}}$. The regions in $\mathbf{k}$ space that are summed in Eq. (14.72) are shown in the shaded areas because $\mathbf{K} = \mathbf{k}_1 + \mathbf{k}_2$. These regions are at a maximum when $\mathbf{K} = 0$.

Assuming that $\mathbf{K} = 0$, and the electron spins are antiparallel, we can rewrite Eq. (14.72) as

$$\psi_{12} = \sum_{\mathbf{k}} b(\mathbf{k})\, \hat{c}^{\dagger}_{\mathbf{k},s} c_{-\mathbf{k},-s} |G>. \tag{14.73}$$

We will explain later why only antiparallel spins are included in Eq. (14.73). We make a further approximation in Eq. (14.70) by considering $V_{\mathbf{kq}}$ to be a constant in the range of attractive interaction ($V_{\mathbf{kq}} = -U$), and $V_{\mathbf{kq}} = 0$ otherwise. In fact, from Eq. (14.71), we obtain the condition

$$U \ne 0, \text{ when } |E(\mathbf{k}+\mathbf{q}) - E(\mathbf{k})| \le \hbar\omega_{\mathbf{q}}. \tag{14.74}$$

The Debye frequency ω_D is the maximum value of $\omega_{\mathbf{q}}$ in the Debye approximation.

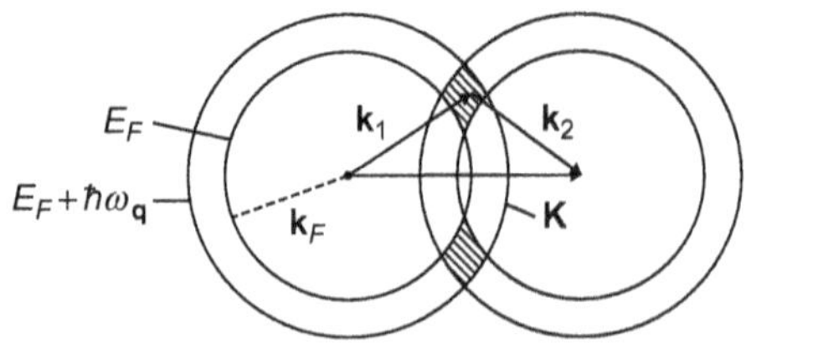

FIGURE 14.12

Here, $\mathbf{k}$ vectors of two interacting electrons ($\mathbf{K} = \mathbf{k}_1 + \mathbf{k}_2$) that lie within a shell of thickness $\hbar\omega_{\mathbf{q}}$ above E_F.

Substituting Eq. (14.74) in Eq. (14.71), we obtain

$$\hat{H} = \sum_{\mathbf{k},s} E(\mathbf{k})\hat{c}^{\dagger}_{\mathbf{k},s}\hat{c}_{\mathbf{k},s} - \frac{U}{2}\sum_{\mathbf{k},\mathbf{q},s} \hat{c}^{\dagger}_{\mathbf{k}+\mathbf{q},s}\hat{c}^{\dagger}_{-\mathbf{k}-\mathbf{q},-s}\hat{c}_{-\mathbf{k},-s}\hat{c}_{\mathbf{k},s}. \tag{14.75}$$

Here, we note that the factor ½ in the second term in Eq. (14.75) is due to the summation over only one spin index. From Eqs. (14.73) and (14.75), we obtain an expression for the energy E,

$$E = <\psi|\hat{H}|\psi> = 2\sum_{\mathbf{k}} E(\mathbf{k})|b(\mathbf{k})|^2 - U\sum_{\mathbf{k},\mathbf{q}} b^*(\mathbf{k}+\mathbf{q})b(\mathbf{k}), \tag{14.76}$$

where $b(\mathbf{k})$ is obtained by varying E subject to the condition,

$$\sum_{\mathbf{k}} |b(\mathbf{k})|^2 = 1. \tag{14.77}$$

By varying E in Eq. (14.76), we obtain

$$\frac{\partial}{\partial b^*_{\mathbf{k}'}}\left[E - \beta\sum_{\mathbf{k}'}|b(\mathbf{k}'')|^2\right] = 2E(\mathbf{k}')b(\mathbf{k}') - U\sum_{\mathbf{q}} b(\mathbf{k}'-\mathbf{q}) - \beta b(\mathbf{k}') = 0, \tag{14.78}$$

from which we have

$$[2E(\mathbf{k}) - \beta]b(\mathbf{k}) = U\sum_{\mathbf{k}'} b(\mathbf{k}'). \tag{14.79}$$

Here, β is Lagrange's parameter and $U \neq 0$ only for the energies E_F to $E_F + \hbar\omega_{\mathbf{q}}$. One can show (Problem 14.9) that if we take the complex conjugate of Eq. (14.79), multiply by $b(\mathbf{k})$, and sum over $\mathbf{k}$, we obtain an equation that agrees with Eq. (14.76) provided $\beta = E$.

Similarly, the sum over $b(\mathbf{k}')$ is over a small number of $\mathbf{k}'$, such that we can write

$$\sum_{\mathbf{k}'} b(\mathbf{k}') = B. \tag{14.80}$$

From Eqs. (14.79) and (14.80), we obtain

$$\sum_{\mathbf{k}} b(\mathbf{k}) = B = \sum_{E(\mathbf{k})} \frac{UB}{2E(\mathbf{k}) - \beta} \tag{14.81}$$

or

$$\sum_{E(\mathbf{k})} \frac{U}{2E(\mathbf{k}) - \beta} = 1. \tag{14.82}$$

We note that if we had considered parallel spins in Eq. (14.73), the constant B would have been zero because of the antisymmetry of the spatial part of the wave function. If we take the complex conjugate of Eq. (14.79), multiply by $b(\mathbf{k})$, and sum over $\mathbf{k}$, we obtain an equation that agrees with Eq. (14.76), provided $\beta = E$ (Problem 14.9). Because $U \neq 0$ only in the range E_F and $E_F + \hbar\omega_{\mathbf{q}}$, we can rewrite Eq. (14.82) in the alternate form

$$U\int_{E_F}^{E_F+\hbar\omega_{\mathbf{q}}} \frac{D(x)dx}{2x - E} = 1, \tag{14.83}$$

where $D(x)$ is the density of states. We approximate $D(x)$ by $D(E_F)$ in the narrow range of integration and obtain (Problem 14.10)

$$E = 2E_F - \frac{2\hbar\omega_{\mathbf{q}} \exp[-2/D(E_F)U]}{1 - \exp[-2/D(E_F)U]} \approx 2E_F - 2\hbar\omega_{\mathbf{q}} \exp[-2/D(E_F)U]. \tag{14.84}$$

The expression on the extreme right in Eq. (14.84) is valid for weak interactions (small U). We note that we have made the approximation $V_{\mathbf{kq}} = -U$ in the range of attractive interaction and zero elsewhere. Thus, the energy of the electron pair is less than $2E_F$ in the absence of the interaction. It can be shown that all other solutions lead to energies greater than $2E_F$. Hence, the lowest energy state of the electron pair is a bound state known as a Cooper pair (Ref. 7) that is formed from a pair of electrons with opposite spin and opposite wave vector.

The formation of a bound state by the additional electron pair means that when two electrons from states directly below the Fermi surface are excited into states above it, the energy is lowered. Thus, the filled Fermi sphere is unstable, and one can gain energy by combining electrons into Cooper pairs.

14.7.4 BCS Theory

The theory of superconductivity was proposed by Bardeen, Cooper, and Schrieffer on the basic assumption that electron–phonon interaction provides a means for creating Cooper pairs.[2] A ground state of a superconductor consists of a condensate, which is a single quantum state available only to interacting Cooper pairs. The condensate is the product of the pair wave functions, and its center of mass is stationary in the absence of electric and magnetic fields. If one wants to remove a pair of electrons from the condensate, an energy 2Δ (the binding energy of the Cooper pair) is required to "break" a pair.

The BCS model Hamiltonian (Ref. 2) can be written as

$$\hat{H}_{BCS} = \sum_{\mathbf{k},s} E_{\mathbf{k}} \hat{c}^{\dagger}_{\mathbf{k}s} \hat{c}_{\mathbf{k}s} + \sum_{\mathbf{kk}'} U_{\mathbf{kk}'} \hat{c}^{\dagger}_{\mathbf{k}\uparrow} \hat{c}^{\dagger}_{-\mathbf{k}\downarrow} \hat{c}_{-\mathbf{k}'\downarrow} \hat{c}_{\mathbf{k}'\uparrow}. \tag{14.85}$$

Bardeen et al.[2] used the grand canonical ensemble and showed that the wave function for coherent states can be written as

$$|\Phi\rangle = \prod_{\mathbf{k}} [1 + g_{\mathbf{k}} \hat{c}^{\dagger}_{\mathbf{k}\uparrow} \hat{c}^{\dagger}_{-\mathbf{k}\downarrow}] |\Theta\rangle = \hat{\Phi} |\Theta\rangle, \tag{14.86}$$

which creates all possible pairs of $2N$ particles with various weights through $g_{\mathbf{k}}$. They used a variational procedure by using the constraint that the average particle number is maintained at N. However, we will use the Bogoliubov–Valatin transformation and obtain the same result as the BCS theory.[5]

14.7.5 Ground State of the Superconducting Electron Gas

We can write the Hamiltonian in Eq. (14.75) in a simpler form,

$$H = \sum_{\mathbf{k}} E(\mathbf{k}) (\hat{c}^{\dagger}_{\mathbf{k}} \hat{c}_{\mathbf{k}} + \hat{c}^{\dagger}_{-\mathbf{k}} \hat{c}_{-\mathbf{k}}) - U \sum_{\mathbf{kk}'} \hat{c}^{\dagger}_{\mathbf{k}'} \hat{c}^{\dagger}_{-\mathbf{k}'} \hat{c}_{-\mathbf{k}} \hat{c}_{\mathbf{k}}, \tag{14.87}$$

where a positive spin is associated with the index $\mathbf{k}$, and a negative spin is associated with the index $-\mathbf{k}$. Eq. (14.87) is rearranged by introducing the following creation and annihilation operators:

$$\begin{aligned}\hat{\alpha}_{\mathbf{k}} &= u_{\mathbf{k}}\hat{c}_{\mathbf{k}} - v_{\mathbf{k}}\hat{c}^{\dagger}_{-\mathbf{k}}\\ \hat{\alpha}_{-\mathbf{k}} &= u_{\mathbf{k}}\hat{c}_{-\mathbf{k}} + v_{\mathbf{k}}\hat{c}^{\dagger}_{\mathbf{k}}\\ \hat{\alpha}^{\dagger}_{\mathbf{k}} &= u_{\mathbf{k}}\hat{c}^{\dagger}_{\mathbf{k}} - v_{\mathbf{k}}\hat{c}_{-\mathbf{k}}\\ \alpha^{\dagger}_{-\mathbf{k}} &= u_{\mathbf{k}}c^{\dagger}_{-\mathbf{k}} + v_{\mathbf{k}}\hat{c}_{\mathbf{k}}.\end{aligned} \tag{14.88}$$

In the preceding transformations, we use the conditions $u_{\mathbf{k}}^2 + v_{\mathbf{k}}^2 = 1$, $u_{\mathbf{k}} = u_{-\mathbf{k}}$, $v_{\mathbf{k}} = -v_{-\mathbf{k}}$. In addition, $u_{\mathbf{k}}$ and $v_{\mathbf{k}}$ are nonzero outside and inside the Fermi sphere, respectively. These conditions guarantee that the commutation relations for the $\hat{c}-$ operators also apply to the $\hat{\alpha}-$ operators (Problem 14.11). We avoid the confusion created by the situation that a hole in the state $\mathbf{k}$ has a momentum $-\hbar\mathbf{k}$ while an electron has a momentum $\hbar\mathbf{k}$ by defining quasi-particles which have momentum $\hbar\mathbf{k}$ both outside and inside the Fermi sphere. Since a quasi-particle of momentum $\hbar\mathbf{k}$ is created by the operator $c^{\dagger}_{\mathbf{k}}$ when it is outside the Fermi sphere and by $c_{-\mathbf{k}}$ when it is inside the Fermi sphere, we have in Eq. (14.88),

$$\begin{aligned} u_{\mathbf{k}} = 1, v_{\mathbf{k}} = 0 \quad &\text{for} \quad k > k_F\\ u_{\mathbf{k}} = 0, v_{\mathbf{k}} = 1 \quad &\text{for} \quad k < k_F.\end{aligned} \tag{14.88a}$$

We also introduce the operator $\hat{\overline{H}} = \hat{H} - E_F\hat{N}$ and the energy $\varepsilon(\mathbf{k}) = E(\mathbf{k}) - E_F$. The transition from the $\hat{c}-$ to the $\hat{\alpha}-$ operators (Bogoliubov–Valatin transformation) leads to (Problem 14.12)

$$\begin{aligned}\hat{\overline{H}} = &\sum_{\mathbf{k}} \varepsilon(\mathbf{k})[2v_{\mathbf{k}}^2 + (u_{\mathbf{k}}^2 - v_{\mathbf{k}}^2)(\hat{\alpha}^{\dagger}_{\mathbf{k}}\hat{\alpha}_{\mathbf{k}} + \hat{\alpha}^{\dagger}_{-\mathbf{k}}\hat{\alpha}_{-\mathbf{k}}) + 2u_{\mathbf{k}}v_{\mathbf{k}}(\alpha^{\dagger}_{\mathbf{k}}\hat{\alpha}^{\dagger}_{-\mathbf{k}} + \hat{\alpha}_{-\mathbf{k}}\hat{\alpha}_{\mathbf{k}})]\\ &- U\sum_{\mathbf{k}\mathbf{k}'}\{[u_{\mathbf{k}}v_{\mathbf{k}}u_{\mathbf{k}'}v_{\mathbf{k}'}(1 - \hat{\alpha}^{\dagger}_{-\mathbf{k}'}\hat{\alpha}_{-\mathbf{k}'} - \hat{\alpha}^{\dagger}_{\mathbf{k}'}\hat{\alpha}_{\mathbf{k}'})(1 - \hat{\alpha}^{\dagger}_{-\mathbf{k}}\hat{\alpha}_{-\mathbf{k}} - \hat{\alpha}^{\dagger}_{\mathbf{k}}\hat{\alpha}_{\mathbf{k}})]\\ &+ (u_{\mathbf{k}}^2 - v_{\mathbf{k}}^2)u_{\mathbf{k}'}v_{\mathbf{k}'}(1 - \hat{\alpha}^{\dagger}_{-\mathbf{k}'}\hat{\alpha}_{-\mathbf{k}'} - \hat{\alpha}^{\dagger}_{\mathbf{k}'}\hat{\alpha}_{\mathbf{k}'})(\hat{\alpha}_{-\mathbf{k}}\hat{\alpha}_{\mathbf{k}} + \hat{\alpha}^{\dagger}_{\mathbf{k}}\hat{\alpha}^{\dagger}_{-\mathbf{k}})\\ &+ (u_{\mathbf{k}}^2\hat{\alpha}_{-\mathbf{k}}\hat{\alpha}_{\mathbf{k}} - v_{\mathbf{k}}^2\hat{\alpha}^{\dagger}_{\mathbf{k}}\alpha^{\dagger}_{-\mathbf{k}})(u_{\mathbf{k}'}^2\hat{\alpha}^{\dagger}_{\mathbf{k}'}\alpha^{\dagger}_{-\mathbf{k}'} - v_{\mathbf{k}'}^2\hat{\alpha}_{-\mathbf{k}'}\hat{\alpha}_{\mathbf{k}'})\}.\end{aligned} \tag{14.89}$$

The last term of the first line in Eq. (14.89) vanishes from the condition that $u_{\mathbf{k}}$ and $v_{\mathbf{k}}$ are zero inside and outside the Fermi sphere (Eq. 14.88 (a)). Normally, the first term (energy of the filled Fermi sphere) and the second term (energy of the quasiparticles defined by $\alpha^{\dagger}_{\mathbf{k}}\alpha_{\mathbf{k}}$) remain, but in the ground state, the second term vanishes. As we will show, the terms containing the products of $\alpha^{\dagger}_{\mathbf{k}}\alpha^{\dagger}_{-\mathbf{k}}$ and $\alpha_{-\mathbf{k}}\alpha_{\mathbf{k}}$ can be eliminated by choosing different values of $u_{\mathbf{k}}$ and $v_{\mathbf{k}}$. The last term in Eq. (14.89) contributes very little to the final result and hence can be neglected.

Thus, in the ground state, Eq. (14.89) can be rewritten as

$$\overline{H}_G = 2\sum_{\mathbf{k}} \varepsilon(\mathbf{k})v_{\mathbf{k}}^2 - U\sum_{\mathbf{k},\mathbf{k}'} u_{\mathbf{k}}v_{\mathbf{k}}u_{\mathbf{k}'}v_{\mathbf{k}'} + \sum_{\mathbf{k}}\left[2u_{\mathbf{k}}v_{\mathbf{k}}\varepsilon(\mathbf{k}) - (u_{\mathbf{k}}^2 - v_{\mathbf{k}}^2)U\sum_{\mathbf{k}'} u_{\mathbf{k}'}v_{\mathbf{k}'}\right](\hat{\alpha}^{\dagger}_{\mathbf{k}}\hat{\alpha}^{\dagger}_{-\mathbf{k}} + \hat{\alpha}_{-\mathbf{k}}\hat{\alpha}_{\mathbf{k}}). \tag{14.90}$$

For the square bracket in Eq. (14.90) to vanish, we first introduce a constant,

$$\Delta = U\sum_{\mathbf{k}} u_{\mathbf{k}}v_{\mathbf{k}}, \tag{14.91}$$

and then set the condition,

$$2u_{\mathbf{k}}v_{\mathbf{k}}\varepsilon(\mathbf{k}) = \Delta(u_{\mathbf{k}}^2 - v_{\mathbf{k}}^2). \tag{14.92}$$

In addition,

$$u_{\mathbf{k}}^2 + v_{\mathbf{k}}^2 = 1. \tag{14.93}$$

From Eqs. (14.92) and (14.93), it is easy to show that

$$\xi_{\mathbf{k}} = \frac{\varepsilon(\mathbf{k})}{\sqrt{\varepsilon^2(\mathbf{k}) + \Delta^2}}, \tag{14.94}$$

$$\begin{aligned} u_{\mathbf{k}}^2 &= \frac{1}{2}(1 + \xi_{\mathbf{k}}) \\ v_{\mathbf{k}}^2 &= \frac{1}{2}(1 - \xi_{\mathbf{k}}). \end{aligned} \tag{14.95}$$

From Eqs. (14.91), (14.94), and (14.95), one can show (Problem 14.12)

$$\Delta = U\sum_{\mathbf{k}} u_{\mathbf{k}}v_{\mathbf{k}} = \frac{U}{2}\sum_{\mathbf{k}} \frac{\Delta}{\sqrt{\varepsilon^2(\mathbf{k}) + \Delta^2}}. \tag{14.96}$$

Because $\Delta \neq 0$ (no vanishing interaction), $U \neq 0$ only in the range $|\varepsilon(\mathbf{k})| \le \hbar\omega_{\mathbf{q}}$. Since the sum is over one spin direction, the density of states is $D(\varepsilon)/2$. Thus, while converting the summation over $\mathbf{k}$ to an integration, we can write Eq. (14.96) in the alternate form

$$\frac{U}{4}\int_{-\hbar\omega_{\mathbf{q}}}^{\hbar\omega_{\mathbf{q}}} \frac{D(\varepsilon)d\varepsilon}{\sqrt{\varepsilon^2 + \Delta^2}} \approx \frac{UD(E_F)}{4}\int_{-\hbar\omega_{\mathbf{q}}}^{\hbar\omega_{\mathbf{q}}} \frac{d\varepsilon}{\sqrt{\varepsilon^2 + \Delta^2}} = 1. \tag{14.97}$$

We obtain from Eq. (14.97)

$$\Delta = 2\hbar\omega_{\mathbf{q}} \exp[-2/D(E_F)U]. \tag{14.98}$$

This is precisely the binding energy of the Cooper pair. Because the energy of the filled Fermi sphere is

$$\overline{H}_0 = 2\sum_{\mathbf{k}<\mathbf{k}_F} \varepsilon(\mathbf{k}), \tag{14.99}$$

from Eqs. (14.90) and (14.99), we obtain the difference between the Hamiltonians with and without interactions (note that the coefficient of the square bracket in Eq. 14.90 is equal to zero), which, in the absence of any operators, is the energy difference between the ground state of the interacting and noninteracting electron gas:

$$E = \overline{H}_G - \overline{H}_0 = 2\sum_{\mathbf{k}} \varepsilon(\mathbf{k})v_{\mathbf{k}}^2 - 2\sum_{\mathbf{k}<\mathbf{k}_F} \varepsilon(\mathbf{k}) - U\sum_{\mathbf{k}\mathbf{k}'} u_{\mathbf{k}}v_{\mathbf{k}}u_{\mathbf{k}'}v_{\mathbf{k}'}. \tag{14.100}$$

Substituting the values of $u_{\mathbf{k}}, v_{\mathbf{k}}$, and $\xi_{\mathbf{k}}$ from Eqs. (14.94) and (14.95) in Eq. (14.100), we obtain

$$E = \sum_{\mathbf{k}<\mathbf{k}_F} |\varepsilon| \left(1 - \frac{|\varepsilon|}{\sqrt{\varepsilon^2 + \Delta^2}}\right) + \sum_{\mathbf{k}>\mathbf{k}_F} \varepsilon \left(1 - \frac{\varepsilon}{\sqrt{\varepsilon^2 + \Delta^2}}\right) - \sum_{\mathbf{k}} \frac{\Delta^2}{2\sqrt{\varepsilon^2 + \Delta^2}}. \tag{14.101}$$

When we convert the sum over $\mathbf{k}$ to an integration (with only one spin direction),

$$E = D(E_F) \int_0^{\hbar\omega_{\mathbf{q}}} \left(\varepsilon - \frac{1}{2} \frac{2\varepsilon^2 + \Delta^2}{\sqrt{\varepsilon^2 + \Delta^2}}\right) d\varepsilon. \tag{14.102}$$

After integrating Eq. (14.102) (see Problem 14.13), we obtain for weak interactions, $(\Delta \ll \hbar\omega_{\mathbf{q}})$,

$$E = \frac{D(E_F)}{2} (\hbar\omega_{\mathbf{q}})^2 \left[1 - \sqrt{1 + \left(\frac{\Delta}{\hbar\omega_{\mathbf{q}}}\right)^2}\right] \approx -\frac{D(E_F)\Delta^2}{4}. \tag{14.103}$$

E is known as the condensate energy of the ground state.

14.7.6 Excited States at $T=0$

From Eqs. (14.89) and (14.90), we can write the Hamiltonian of the excited states, $\overline{H}_e$, as

$$\overline{H}_e = \overline{H}_G + \sum_{\mathbf{k}} \left[\varepsilon(\mathbf{k})(u_{\mathbf{k}}^2 - v_{\mathbf{k}}^2) + U \sum_{\mathbf{k}'} 2u_{\mathbf{k}} v_{\mathbf{k}} u_{\mathbf{k}'} v_{\mathbf{k}'}\right] \left(\hat{\alpha}_{\mathbf{k}}^\dagger \hat{\alpha}_{\mathbf{k}} + \hat{\alpha}_{-\mathbf{k}}^\dagger \hat{\alpha}_{-\mathbf{k}}\right) + \cdots. \tag{14.104}$$

Eq. (14.104) can be expressed by using Eq. (14.91),

$$\overline{H}_e = \overline{H}_G + \sum_{\mathbf{k}} \left[\varepsilon(\mathbf{k})(u_{\mathbf{k}}^2 - v_{\mathbf{k}}^2) + 2\Delta u_{\mathbf{k}} v_{\mathbf{k}}\right] \left(\hat{\alpha}_{\mathbf{k}}^\dagger \hat{\alpha}_{\mathbf{k}} + \hat{\alpha}_{-\mathbf{k}}^\dagger \hat{\alpha}_{-\mathbf{k}}\right) + \cdots. \tag{14.105}$$

Eq. (14.105) can be rewritten with the help of Eq. (14.92),

$$\overline{H}_e = \overline{H}_G + \sum_{\mathbf{k}} \frac{2u_{\mathbf{k}} v_{\mathbf{k}}}{\Delta} \left(\varepsilon^2(\mathbf{k}) + \Delta^2\right) (n_{\mathbf{k}\uparrow} + n_{-\mathbf{k}\downarrow}). \tag{14.106}$$

From Eqs. (14.96) and (14.106), we obtain

$$E - E_0 = \sum_{\mathbf{k}} \left[\varepsilon^2(\mathbf{k}) + \Delta^2\right]^{1/2} n_{\mathbf{k}}. \tag{14.107}$$

The energy of a quasiparticle is given by

$$\overline{\varepsilon}(\mathbf{k}) = \sqrt{\varepsilon^2(\mathbf{k}) + \Delta^2}. \tag{14.108}$$

Note that here the energy $\varepsilon(\mathbf{k})$ is measured from the Fermi surface:

$$\varepsilon(\mathbf{k}) = E(\mathbf{k}) - E_F. \tag{14.109}$$

Thus, the ground and the first excited states are separated by an energy gap Δ. This is shown in Figure 14.13.

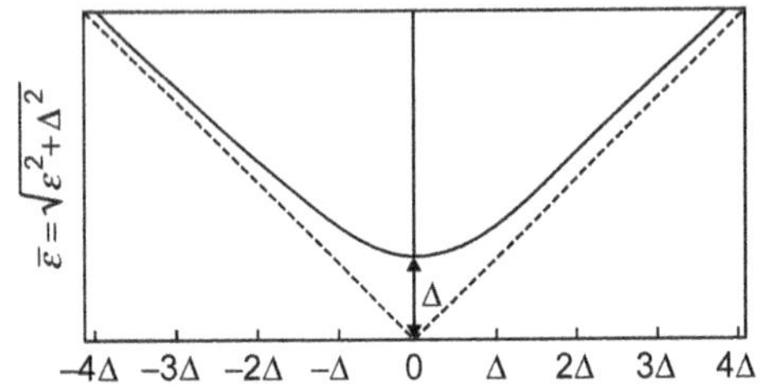

FIGURE 14.13

Energy $\bar{\varepsilon}(\mathbf{k})$ of the quasiparticles plotted as a function of $\varepsilon(\mathbf{k}) = E(\mathbf{k}) - E_F$.

Because in a scattering process, a quasiparticle is created in pairs, corresponding to the electron-hole pair of the noninteracting electron gas, the threshold energy of an excitation is 2Δ. This leads to the conclusion that the Cooper pairs must be broken up during the scattering process. Thus, both Cooper pairs and single quasiparticles are present in the excited state, and while individual particles are scattered, Cooper pairs lead to a current that flows without resistance. We now are back to a two-fluid model.

14.7.7 Excited States at $T \neq 0$

The states $\mathbf{k}\uparrow$ and $\mathbf{k}\downarrow$ are fermions, and at temperatures $T \neq 0$, replacing the particle numbers $n_{\mathbf{k}}$ by their statistical average, we obtain by taking the Fermi distribution as the occupational probability,

$$<n_{\mathbf{k}}> \equiv f_{\mathbf{k}} = \frac{1}{e^{\beta\bar{\varepsilon}(\mathbf{k})}+1}, \tag{14.110}$$

where the energy of the quasiparticles $\bar{\varepsilon}(\mathbf{k})$ replaces the usual energy difference $E-\mu$. We further note that $\bar{\varepsilon}(\mathbf{k})$ is always positive (from Eq. 14.108), except that $\varepsilon(\mathbf{k})$ is defined as $\varepsilon(\mathbf{k}) = E(\mathbf{k}) - \mu$ at $T \neq 0$. Because $n_{\mathbf{k}} \neq 0$ for $T \neq 0$, the energy gap Δ is now temperature dependent, $\Delta(T)$. Thus, the bracket in Eq. (14.90) has to be replaced by the condition

$$2u_{\mathbf{k}}v_{\mathbf{k}}\varepsilon(\mathbf{k}) - (u_{\mathbf{k}}^2 - v_{\mathbf{k}}^2)U\sum_{\mathbf{k}'} u_{\mathbf{k}'}v_{\mathbf{k}'}(1-2f_{\mathbf{k}'}) = 0. \tag{14.111}$$

Similarly, $\Delta(T)$ is defined as

$$\Delta(T) = U\sum_{\mathbf{k}'} u_{\mathbf{k}'}v_{\mathbf{k}'}(1-2f_{\mathbf{k}'}) = \frac{U}{2}\sum_{\mathbf{k}'}\frac{\Delta(T)}{\sqrt{\varepsilon^2(\mathbf{k}')+\Delta^2(T)}}(1-2f_{\mathbf{k}'}). \tag{14.112}$$

Because the critical temperature T_c is the highest one for which Eq. (14.112) has a solution for which $\Delta(T) \neq 0$, and the sum over $\mathbf{k}'$ is over one spin state, we can eliminate $\Delta(T)$ from both sides, and by using a procedure similar to Eq. (14.97), we obtain

$$\frac{UD(E_F)}{4}\int_{-\hbar\omega_{\mathbf{q}}}^{\hbar\omega_{\mathbf{q}}}\frac{d\varepsilon}{\sqrt{\varepsilon^2+\Delta^2(T)}}\left\{1-2f\left(\frac{\sqrt{\varepsilon^2+\Delta^2(T)}}{k_BT}\right)\right\} = 1. \tag{14.113}$$

Eq. (14.113) can be rewritten in the alternate form

$$\int_{-\hbar\omega_{\mathbf{q}}}^{\hbar\omega_{\mathbf{q}}}\frac{d\varepsilon}{\sqrt{\varepsilon^2+\Delta^2(T)}} - \int_{-\hbar\omega_{\mathbf{q}}}^{\hbar\omega_{\mathbf{q}}}\frac{d\varepsilon}{\sqrt{\varepsilon^2+\Delta^2(T)}}2f\left(\frac{\sqrt{\varepsilon^2+\Delta^2(T)}}{k_BT}\right) = \frac{4}{UD(E_F)}. \tag{14.114}$$

Because $\Delta(0) \ll \hbar\omega_{\mathbf{q}}$, it can be shown from Eqs. (14.98) and (14.114) that (Problem 14.14)

$$\ln \frac{\Delta(T)}{\Delta(0)} = -2\int_0^{\infty} \frac{dx}{\sqrt{x^2+1}} f\left(\sqrt{x^2+1}\frac{\Delta(T)}{\Delta(0)}\left[\frac{k_BT}{\Delta(0)}\right]^{-1}\right) = F\left(\frac{\Delta(T)}{\Delta(0)}, \frac{k_BT}{\Delta(0)}\right). \tag{14.115}$$

Because Eq. (14.115) is a function of two parameters $\Delta(T)/\Delta(0)$ and $k_BT/\Delta(0)$, T_c is calculated from the condition that $\Delta(T_c) = 0$ is linearly dependent on $\Delta(0)$. It can be shown by numerical integration that

$$k_BT_c \approx 0.57\Delta(0). \tag{14.116}$$

Eq. (14.116) can be rewritten in the alternate form

$$\frac{2\Delta(0)}{k_BT_c} \approx 3.53. \tag{14.117}$$

Approximating $\omega_{\mathbf{q}}$ by ω_D (the Debye frequency), replacing the Debye temperature $\theta_D = \hbar\omega_D/k_B$, from Eqs. (14.98) and (14.117), we obtain

$$T_c \approx 1.13\, \theta_D \exp\left[-\frac{2}{D(E_F)U}\right]. \tag{14.118}$$

[Be aware that in the notations of BCS theory, $N(E_F) = D(E_F)/2$ is the density of states per spin and $U = V$.]

The BCS theory of superconductivity (Ref. 2) can be briefly summarized as follows. When a uniform electric field is applied to a superconductor, a current is generated because all the pairs in the condensate experience the same force and move in the same direction. When a voltage difference V exists across the two ends of the superconductor, an energy $2eV$ would be gained by a Cooper pair and different parts of the condensate would have different energy, and eventually, the condensate would be lost. Thus, the condensate must move and carry current without any potential difference; i.e., it must exist as a phase-locked entity. The condensate cannot receive any energy less than $2\Delta(T)$, the energy required to break a single pair of the condensate. However, when the temperature is above zero, some pairs are broken due to thermal excitation. Because the density of Cooper pairs is large at absolute zero, the initial reduction of numbers has little effect on 2Δ, but as T approaches a critical temperature $T = T_c$, $2\Delta(T_c) = 0$ and the condensate ceases to exist. The Cooper pairs are broken into individual electrons, and superconductivity is destroyed. A diagram that displays the reduced energy gap $\Delta(T)/\Delta(0)$ as a function of the reduced temperature T/T_c is shown in Figure 14.14.

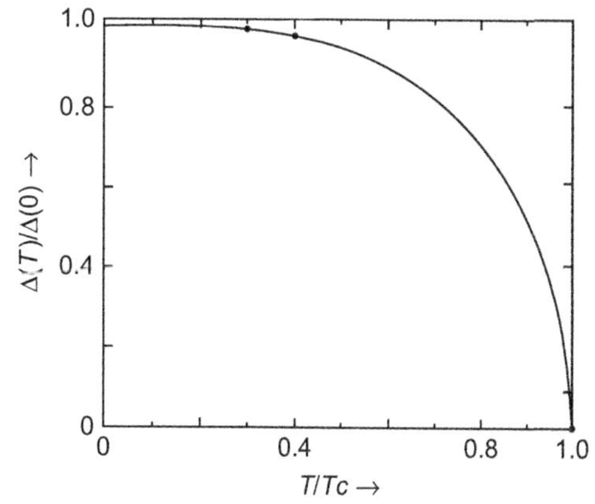

FIGURE 14.14

Reduced energy gap $\Delta(T)/\Delta(0)$ as a function of reduced temperature T/T_c.

It can also be shown that the condensation energy per unit volume is given by

$$W = -1/4\, D(E_F)\Delta^2. \tag{14.119}$$

If we distinguish the density of states in a normal conductor as $D_n(E)dE$ and in its superconducting state as $D_s(\bar{\varepsilon}(\mathbf{k}))d\bar{\varepsilon}(\mathbf{k})$, where $\bar{\varepsilon}(\mathbf{k}) = E - E_F$ (Eq. 14.108), because no states are lost,

$$D_n(E)dE = D_s(\bar{\varepsilon}(\mathbf{k}))d\bar{\varepsilon}(\mathbf{k}), \tag{14.120}$$

from which we obtain

$$D_s(\bar{\varepsilon}) = D_n(E)\frac{dE}{d\bar{\varepsilon}}. \tag{14.121}$$

Eq. (14.121) can be rewritten in the alternate form by using Eqs. (14.108) and (14.109),

$$\begin{aligned} D_s(\bar{\varepsilon}) &= D_n(E)\frac{\bar{\varepsilon}}{(\bar{\varepsilon}^2 - \Delta^2)^{1/2}}, \quad |\bar{\varepsilon}| > \Delta, \\ &= 0, \quad |\bar{\varepsilon}| < 0. \end{aligned} \tag{14.122}$$

14.8 STRONG-COUPLING THEORY

14.8.1 Introduction

The BCS theory is based on a weak coupling approximation, i.e., the electron–phonon coupling constant $\lambda \ll 1$ at $T = 0$. Here, λ reflects the strength of the electron–lattice interaction. However, for many superconductors, $\lambda \geq 1$. For example, $\lambda = 1.4$ for lead, $\lambda = 1.6$ for mercury, and $\lambda = 2.1$ for $Pb_{0.65}Bi_{0.35}$. Thus, the more universal approach of the strong-coupling theory was developed after the formulation of the BCS theory. The strong-coupling theory is based on the Green's function method of the many-body theory, and we will introduce only the main results of this theory as derived by McMillan[21] and modified by Dynes (Ref. 11).

14.8.2 Upper Limit of the Critical Temperature, T_c

The electron–phonon coupling constant can be written as

$$\lambda = 2\int \alpha^2(\Omega)F(\Omega)\Omega^{-1}d\Omega, \tag{14.123}$$

where Ω is the phonon frequency (we note that $\hbar\Omega_D = k_B\theta_D$), $F(\Omega)$ is the phonon density of states, and $\alpha^2(\Omega)$ is a measure of the phonon-frequency-dependent electron–phonon interaction. The characteristic phonon frequency $\tilde{\Omega}$ is defined as

$$\tilde{\Omega} = <\Omega^2>^{1/2}, \tag{14.124}$$

where the average is determined by

$$<f(\Omega)> = \frac{2}{\lambda}\int f(\Omega)\alpha^2(\Omega)F(\Omega)d\Omega. \tag{14.125}$$

McMillan (Ref. 21) introduced a convenient expression for the coupling constant, λ,

$$\lambda = \nu<I^2>/M\overline{\Omega}^2, \tag{14.126}$$

where ν is the bulk density of states,

$$\nu = m^* p_F/2\pi^2, \tag{14.127}$$

and $<I^2>$ contains the average value of the electron–phonon matrix element I. The expression in the strong-coupling limit, for the critical temperature T_c, obtained by McMillan[21] (later modified by Dynes (Ref. 11)) is given by

$$T_c = \frac{\theta_D}{1.2}\exp\left[-\frac{1.04(1+\lambda)}{\lambda - \mu^*(1+0.62\lambda)}\right], \tag{14.128}$$

and μ^* can be expressed as the Coulombic pseudopotential ($\varepsilon_0 \approx E_F$),

$$\mu^* = V_c[1 + V_c \ln(\varepsilon_0/\widetilde{\Omega})]^{-1}. \tag{14.129}$$

When we use a rough estimate of upper limit, each of these values is $\theta_D \approx 400°$ K, $\lambda \approx 0.8$, and $\mu^* \approx 0.1$. In Eq. (14.128), we obtain the upper limit for $T_c \approx 30°$ K in the strong coupling limit. In fact, until 1986, Nb_3Ge was the material that had the highest critical temperature, $T_c = 23°$ K.

14.9 HIGH-TEMPERATURE SUPERCONDUCTORS

14.9.1 Introduction

A new class of superconducting materials, the high T_c copper oxides (often called cuprates), was discovered by Bednroz and Muller in 1986.[3] They found that $La_{1.85}Ba_{0.15}CuO_4$ became superconducting at a critical temperature of $T_c \approx 30°$ K. The atomic structure of this remarkable superconductor is shown in Figure 14.15.

$La_{1.85}Ba_{0.5}CuO_4$ is very different from traditional superconductors in the sense that it is not a conventional metal but brittle ceramic, which is an antiferromagnetic insulator, carefully doped so as to produce metallic and superconducting phases. The discovery of $La_{1.85}Ba_{0.15}CuO_4$ set out a frenzy to discover more high T_c superconductors. The synthesis of the similar compound $La_{1.82}Sr_{0.18}CuO_4$ moved the transition temperature close to $T_c \approx 40°$ K. The first high-temperature superconductor of which the critical temperature exceeded the liquid nitrogen temperature is

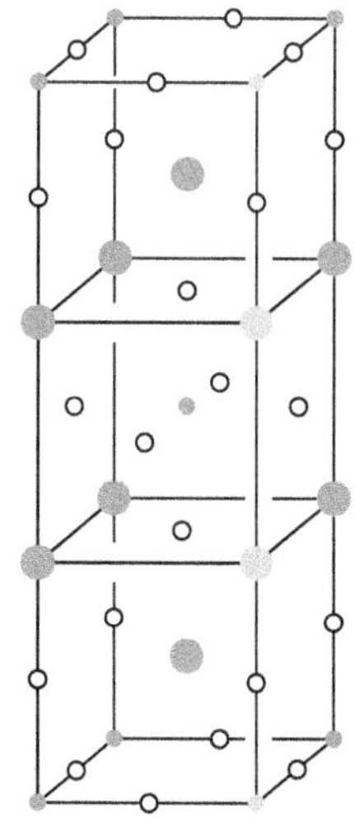

FIGURE 14.15

The atomic structure of $La_{1.85}Ba_{0.15}CuO_4$.

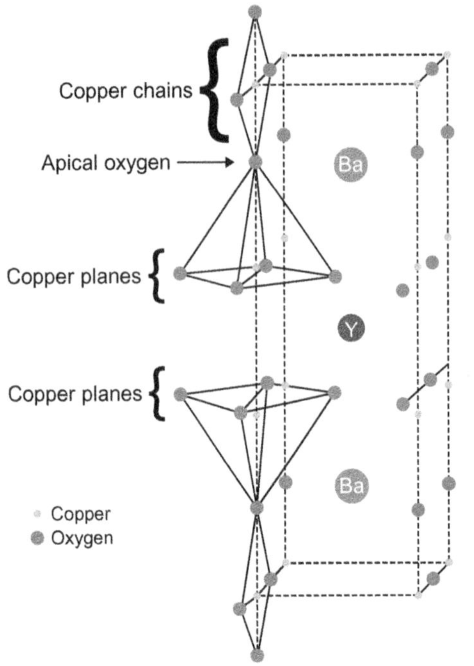

FIGURE 14.16

Structure of the YBCO compound.

Reproduced from Kresin and Wolf[16] with the permission of the American Physical Society.

$YBa_2Cu_3O_{7-x}$ ($x = 0.1$) with $T_c \approx 93°$ K, which was discovered by Wu et al.[29] in February 1987. In fact, it was soon found that $YBa_2Cu_3O_{7-x}$ is superconducting in the orthogonal structure ($0 \le x \le 0.6$) although it is not superconducting in the tetragonal structure ($x > 0.6$). This is the most studied high T_c compound. The orthogonal YBCO structure is shown in Figure 14.16. The highest observed value of T_c is 150° K for the $HgBa_2Ca_2Cu_3O_{8+x}$ compound under pressure.

14.9.2 Properties of Novel Superconductors (Cuprates)

All cuprates have a layered structure. The main structural unit typical for the whole family is the Cu-O plane (see Figure 14.16), where the pairing originates and the charge reservoir is located. The YBCO compound contains the Cu-O chains, and the change in the oxygen content in the chain layers leads to charge transfer between these two subsystems. The charge transfer occurs through the apical oxygen ion located between the chains and the planes, as shown in Figure 14.16.

Whereas the undoped parent compounds of the cuprates are insulators, the novel superconductors are doped materials. Doping leads to conductivity and, for larger concentration, to superconductivity. The doping is provided either by changing the oxygen content or by chemical substitution (La $\rightarrow$ Sr substitution in $La_{2-x}Sr_xCuO_4$). The dopants create electrons, whereas the holes are produced by doping, which removes electrons. For example, some cuprates like YBCO contain carriers that are holes, whereas other cuprates like Nd-Ce-Cu-O contain carriers that are electrons. The maximum value of $T_c \equiv T_c^{max}$ is obtained for a characteristic value n_m of the carrier concentration. $T_c < T_c^{max}$ for both $n < n_m$ (the underdoped region) and $n > n_m$ (the overdoped region).

14.9.3 Brief Review of *s*-, *p*-, and *d*-wave Pairing

Next, we will review the three possible choices for pairing—the *s*-, *p*-, and *d*-wave pairings—and discuss the differences between these types of pairings. The Landau–Ginzburg (Ref. 17) order parameter $\Psi(\mathbf{r})$ (defined in Eq. 14.17) becomes nonzero only in the presence of superconductivity. Under a gauge transformation,

$$\mathbf{A} \rightarrow \mathbf{A} + \vec{\nabla}\phi, \tag{14.130}$$

$$\Psi \rightarrow \Psi e^{-2ie\phi/\hbar c}. \tag{14.131}$$

In fact, Gorkov showed that the pair potential $\Delta_{\mathbf{r}}$ (used in Bogoliubov theory of superconductivity) is proportional to the order parameter $\Psi(\mathbf{r})$ and also transforms as

$$\Delta_{\mathbf{r}} \rightarrow \Delta_{\mathbf{r}} e^{-2ie\phi/\hbar c}. \tag{14.132}$$

The implicit assumption in the BCS theory of superconductivity was that the effective potential U was isotropic and $\Psi(\mathbf{r})$ depended only on the center-of-mass coordinate $\mathbf{r} = (\mathbf{r}+\mathbf{r}')/2$. This is known as s-wave superconductivity. However, in general, the order parameter is a function of $\Psi(\mathbf{r}, \mathbf{r}')$, and one can define $\mathbf{R} = \mathbf{r} - \mathbf{r}'$. When Ψ is independent of the direction of $\mathbf{R}$, the superconductor is called s-wave. When Ψ decays rapidly as a function of $\mathbf{R}$, Ψ has the symmetry of $\vec{x} \cdot \mathbf{R}$ (proportional to $\cos\theta$, as in the case of superfluid ^{3}He); this is called p-wave. When Ψ depends on the direction of $\mathbf{R}$, the symmetry is $(\vec{x} \cdot \mathbf{R})^2 - (\vec{y} \cdot \mathbf{R})^2$, which is proportional to $\cos 2\theta$ and is known as d-wave.

One can also gain insight into the nature of the pair-condensate state based on symmetry considerations. For example, the parity of a superconductor with inversion symmetry can be specified using the Pauli exclusion principle. Because the crystal structures of bulk superconductors are all characterized by a center of inversion, they can be classified by the parity of the pair state. The spin-triplet state (total spin $S = 1$) has a superconducting order parameter (gap function) with odd parity, while the spin-singlet pair state ($S = 0$) corresponds to an orbital pair wave function $\psi(\mathbf{k})\ \alpha\ \Delta(\mathbf{k})$ with even parity, i.e., $\Delta(\mathbf{k}) = \Delta(-\mathbf{k})$. Because spin-orbit interaction is relatively small in cuprate superconductors, the spin-singlet and -triplet states are well defined. It can be shown from group-theoretical considerations that the gap function $\Delta(\mathbf{k})$ for each pair state can be expanded as a function of k_x, k_y, and k_z. Some examples follow:

$$\Delta_s(\mathbf{k}) = \Delta_s^0 + \Delta_s^1(\cos k_x + \cos k_y) + \Delta_s^2 \cos k_z + \cdots, \tag{14.133}$$

$$\Delta_{d_{x^2-y^2}}(\mathbf{k}) = \Delta^0_{d_{x^2-y^2}}(\cos k_x - \cos k_y) + \cdots, \tag{14.134}$$

$$\Delta_{d_{xy}}(\mathbf{k}) = \Delta^0_{d_{xy}}(\sin k_x \sin k_y) + \cdots. \tag{14.135}$$

Thus, the order parameters for both the possible d-wave pair states have node lines.

All cuprate superconductors are characterized by a relatively high ratio of c-axis to a-axis lattice constants. For example, the c/a ratio is 3.0 for YBCO, 5.7 for BI-2212, and 7.6 for Tl-2212. These ratios of c/a translate into a flattened Brillouin zone possessing the basic symmetric properties of the unit cell of a square/rectangular lattice. Recent studies of interplane dc and ac intrinsic Josephson effects has shown that high T_c superconductors such as Bi-2212 act as stacks of two-dimensional superconducting CuO_2-based layers coupled by Josephson interactions (Ref. 13). It has also been shown that the vortex state can be understood in terms of stacks of two-dimensional pancake vortices. The cores of these 2D vortices, localized in the CuO_2 layers, are connected by Josephson vortices with cores confined in the nonsuperconducting charge-reservoir layers (see Figure 14.16). Thus, the pairing symmetry should reflect the underlying CuO_2 square/rectangular lattices. It is more convenient to consider our study of pairing symmetry in a square lattice. A $\mathbf{k}$-space representation of allowed symmetry basis functions for the C_{4v} symmetry is shown in Figure 14.17.

Group-theoretic notation	A_{1g}	A_{2g}	B_{1g}	B_{2g}
Order parameter basis function	constant	$xy(x^2-y^2)$	x^2-y^2	xy
Wave function name	s-wave	g	$d_{x^2-y^2}$	d_{xy}
Schematic representation of $\Delta(\mathbf{k})$ in B.Z.	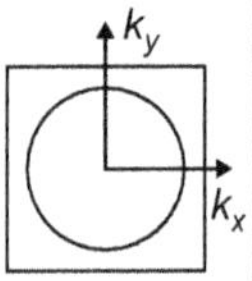			

FIGURE 14.17

A **k**-space representation of allowed symmetry basis functions for the C_{4v} symmetry.

Reproduced from Tsuei and Kirtley[27] with the permission of the American Physical Society.

A schematic presentation in **k** space is shown where black and white represent opposite signs of the order parameter.

14.9.4 Experimental Confirmation of *d*-wave Pairing

Angle-Resolved Photoemission

A number of both non-phase-sensitive and phase-sensitive experimental techniques confirm that pairing in cuprates is highly anisotropic with a line of nodes in the superconducting gap. We will first discuss the angle-resolved photoemission spectroscopy (ARPES), which has the advantage of directly investigating the momentum space of the gap. We will discuss $Bi_2Sr_2CaCu_2O_{8+x}$(Bi – 2212), of which the complex structure consists of a superlattice of orthorhombic units. However, the basic orthorhombic subunit of $Bi_2Sr_2CaCu_2O_{8+x}$(Bi – 2212) has essentially tetragonal symmetry with lattice parameters $a \approx b$. This pseudotetragonal subunit is closely approximated by a body-centered tetragonal structure of which the primitive cell is shown in Figure 14.18. The key elements of this structure are the presence of two Cu-O sheets similar to those in the 40° and 90° K materials, as well as a double layer of edge-sharing Bi-O octahedrals,

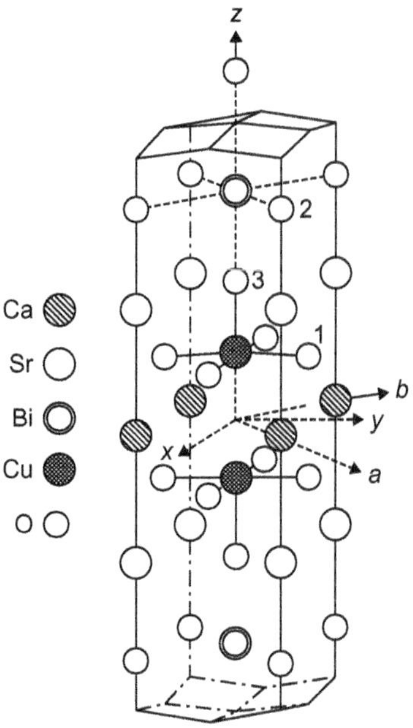

FIGURE 14.18

Primitive cell for body-centered tetragonal $Bi_2Sr_2CaCu_2O_{8+x}$. O(1), O(2), and O(3) denote oxygens in the Cu, Bi, and Sr planes, respectively. The *a* and the *b* axes at 45° to *x* and *y* are shown.

Reproduced from Stavola et al.[25] with the permission of the American Physical Society.

which fulfill a structural role that is analogous to the Cu-O chains in $YBa_2Cu_3O_7$. The main band features at the Fermi level include a pair of half-filled two-dimensional Cu-O $3d-2p$ bands similar to those found in other Cu-O planar superconductors, as well as slightly filled $6p$ bands, which provide additional carriers in the Bi-O planes.

The ARPES study shows that the gap in $Bi_2Sr_2CaCu_2O_{8+x}$(Bi−2212) (Ref. 10) is largest along the $\Gamma - M$ direction (parallel to a or b) and smallest among $\Gamma - Y$ (the diagonal line between them), as expected for a $d_{x^2-y^2}$ superconductor. Figure 14.19 shows the inferred value of the energy gap as a function of the angle in the Fermi surface (solid circles), compared to the prediction of a simple d-wave model (solid line). There is remarkable agreement between experimental and theoretical results. However, ARPES is not phase-sensitive and cannot distinguish between d-wave and highly anisotropic s-wave pairing.

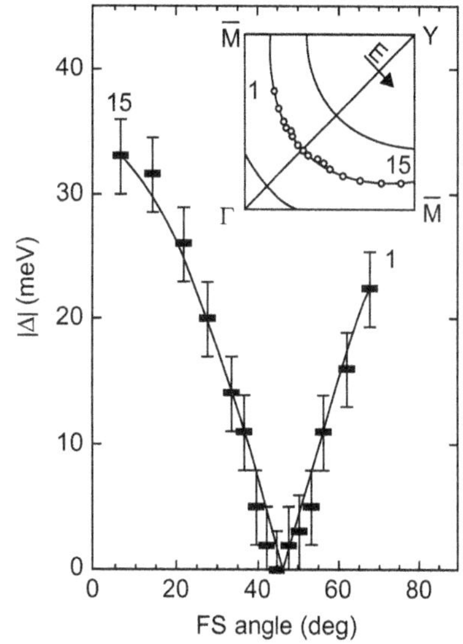

FIGURE 14.19

Energy gap in Bi − 2212, measured with ARPES as a function of an angle in the Fermi surface: solid curve, with fits to the data using a d-wave order parameter. Inset indicates the data points in the Brillouin zone.

Reproduced from Ding et al.[10] *with permission of the American Physical Society.*

Nuclear Magnetic Resonance

Nuclear-magnetic-resonance (NMR) measurements can probe the electronic properties of individual atomic sites on the CuO_2 sheets of the high-temperature superconductors. There is no Hebel-Slichter peak (found in normal superconductors due to density of states in the gap edge) and an increase in the nuclear relaxation rate T_1^{-1} near T_c, for both Cu and O in-plane sites. These properties can be explained by using a $d_{x^2-y^2}$ model with Coulomb correlations that yield (a) a weaker quasiparticle density-of-states singularity at the gap edge compared with an s-wave BCS gap, (b) the vanishing of the coherence factor for quasiparticle scattering for $\mathbf{q} \sim (\pi, \pi)$ for a $d_{x^2-y^2}$ gap, and (c) inelastic–scattering suppression of the peak, which is similar for both d-waves and s-waves. There has been excellent agreement between the experimental results for both the anisotropy ratio $(T_1^{-1})_{ab}/(T_1^{-1})_c$ and the transverse nuclear relaxation rate T_G^{-1} for $^{63}Cu(2)$ when a d-wave model is used, but absolutely no agreement when an s-wave model is used for the theoretical calculations.

Josephson Tunneling

As we discussed earlier, Josephson first pointed out that Cooper pairs can flow through a thin insulating barrier between two superconductors.[13] A schematic representation of a Josephson tunnel junction between a pure $d_{x^2-y^2}$ superconductor on the left and a superconductor with some admixture of s in a predominantly $d_{x^2-y^2}$ state on the right is shown in Figure 14.20. The gap states align with the crystal-line axes, which are rotated by angles θ_L and θ_R with respect to the junction normals $\mathbf{n}_L$ and $\mathbf{n}_R$.

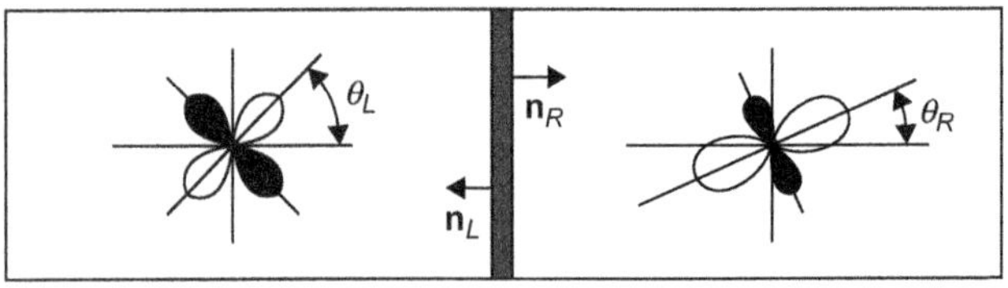

FIGURE 14.20

Schematic diagram of a Josephson junction showing the tunnel barrier sandwiched between two junction electrodes, with order parameters $\Delta_i(\mathbf{k}_i) = |\Delta_i|e^{i\phi_i}$.

Reproduced from Tsuei and Kirtley with the permission of the American Physical Society.[27]

The supercurrent I_s, proportional to the tunneling rate of Cooper pairs through the barrier, was given by Josephson as

$$I_s = I_c \sin \gamma, \tag{14.136}$$

where γ is the gauge-invariant phase difference at the junction,

$$\gamma = \phi_L - \phi_R + \frac{2\pi}{\Phi_0}\int_L^R \mathbf{A} \cdot d\mathbf{l}, \tag{14.137}$$

where $\mathbf{A}$ is the vector potential, and $d\mathbf{l}$ is the element of line integration from the left electrode (L) to the right electrode (R) across the barrier. It was shown by Cohen et al. (Ref. 6) that the supercurrent I_s at zero temperature is given by

$$I_s = \sum_{\mathbf{k},\mathbf{l}} |T_{\mathbf{k},\mathbf{l}}|^2 \frac{\Delta_L(\mathbf{k})\Delta_R(\mathbf{l})}{E_L(\mathbf{k})E_R(\mathbf{l})} \frac{1}{[E_L(\mathbf{k}) + E_R(\mathbf{l})]} \sin(\gamma_L - \gamma_R), \tag{14.138}$$

where $T_{\mathbf{k},\mathbf{l}}$ is the time-reversal-symmetry-invariant tunneling matrix element,

$$E_i(\mathbf{k}) = \sqrt{\in_i^2(\mathbf{k}) + \Delta_i^2(\mathbf{k})}, \tag{14.139}$$

and $\in(\mathbf{k})$ is the one-electron energy. Eq. (14.138) is used to determine the parity of superconductors. It was shown by Pals et al. (Ref. 24). that I_s vanishes up to the second order in $T_{\mathbf{k},\mathbf{l}}$, in tunnel junctions between spin-singlet (even parity) and -triplet (odd parity) superconductors. However, because pair tunneling exists for Josephson junctions made of a cuprate superconductor and low-T_c conventional superconductor such as Nb or Pb, this confirms that the superconducting state in cuprates, just as in low-T_c conventional superconductors, is that of even-parity spin-singlet pairing.

π–Rings

The sign changes in the pair critical current I_c are arbitrary for a particular junction because an arbitrary phase can always be added to either side of the junction. However, the signs of the critical currents in a closed ring of superconductors that is interrupted by Josephson weak links can be assigned self-consistently. One can determine if a particular geometry is frustrated by counting these sign changes. A frustrated geometry has a local maximum in its free energy with zero circulating supercurrent in the absence of an external magnetic field. A negative pair-tunneling critical current I_c can be considered as a phase shift of π at the junction interface, i.e.,

$$I_s = -|I_c| \sin \gamma = |I_c| \sin (\gamma + \pi). \tag{14.140}$$

A superconducting ring with an odd number of π shifts is frustrated.

In the Ginzburg–Landau formalism (Ref. 17), the order parameter is small near T_c, and hence, the free energy of the Josephson junction can be expanded as a power series of the order parameter. Further, the gap function can be expressed as a linear combination of the basis functions (χ_μ^j) of the irreducible representation (Γ^j), which corresponds to the highest T_c of the order parameter,

$$\Delta(\mathbf{k}) = \sum_{\mu=1}^{l_j} \eta_\mu \chi^j_\mu(\mathbf{k}), \tag{14.141}$$

where l_j is the dimensionality of Γ^j, and the expansion coefficient η_μ is invariant under all symmetry operations of the normal-state group G.

The free energy per unit area of Josephson coupling between two superconducting electrodes with order parameters ψ_L and ψ_R (Figure 14.20) can be written as

$$F_j = W \int ds \left[\psi_L \psi_R^* \exp\left[i(2\pi/\Phi_0) \int_L^R \mathbf{A} \cdot d\mathbf{l} \right] + c.c. \right], \tag{14.142}$$

where W is a measure of the Josephson coupling strength, and the integral is over the junction interface. By minimizing the total free energy with respect to ψ_L and ψ_R (Problem 14.15), we can obtain an expression for the Josephson current density J_s, flowing perpendicular to the junction interface from superconductor L to R,

$$J_s = t_{L,R}\chi_L(\mathbf{n})\chi_R(\mathbf{n})|\eta_L||\eta_R| \sin \gamma = J_c \sin \gamma. \tag{14.143}$$

Here, J_c is the critical current density; $\chi_{L,R}$, the basis function, is related to the gap function $\Delta(\mathbf{k})$ through Eq. (14.141); $\Delta_{L,R}(\mathbf{n}) = \eta_{L,R}(\mathbf{n})\chi_{L,R}(\mathbf{n})$, where $\mathbf{n}$ is the unit vector normal to the junction interface, $\eta_{L,R}(\mathbf{n}) = |\eta_{L,R}|e^{i\phi_{L,R}}$; and γ is defined in Eq. (14.137). $t_{L,R}$ is a constant characteristic of the junction. The basis functions $\chi(\mathbf{n})$ of a Josephson junction electrode with tetragonal symmetry (point group C_{4v}) are listed in Table 14.1. Here, n_x, n_y are the projections of the unit vector $\mathbf{n}$ onto the crystallographic axes $\mathbf{x}$ and $\mathbf{y}$, respectively.

It can be shown from Eq. (14.143), for Josephson junctions between two d-wave superconductors, $\chi(\mathbf{n}) = n_x^2 - n_y^2$, and of which the interface is clean and smooth,

$$J_s = A_s \cos(2\theta_L) \cos(2\theta_R) \sin \gamma, \tag{14.144}$$

Table 14.1 Basis Functions $\chi(\mathbf{n})$ for a Josephson Junction Electrode with Tetragonal Crystal Symmetry

Irreducible	A_{1g}	A_{2g}	B_{1g}	B_{2g}
Representation	s	g	$d_{x^2-y^2}$	d_{xy}
Basis function $\chi(\mathbf{n})$	1	$n_x n_y(n_x^2 - n_y^2)$	$n_x^2 - n_y^2$	$n_x n_y$

where θ_L and θ_R are the angles of the crystallographic axes with respect to the interface, and A_s is a constant characteristic of the junction. However, in real Josephson junctions made with cuprates, the electron wave vector normal to the junction interface can be significantly distorted by interface roughness, oxygen deficiency, strain, and so on. In this limit,

$$J_s = A_s \cos 2(\theta_L + \theta_R) \sin \gamma. \tag{14.145}$$

Flux Quantization of a Superconducting Ring

The flux quantization of a superconducting ring can be written as

$$\Phi_a + I_s L + \frac{\Phi_0}{2\pi} \sum_{ij} \gamma_{ij} = n\Phi_0, \tag{14.146}$$

where Φ_a is due to the flux of an external field, I_s is the supercurrent circulating in the ring

$$I_s = I_c^{ij}(\theta_i, \theta_j) \sin \gamma_{ij}, \tag{14.147}$$

γ_{ij} is defined in Eq. (14.137), and Φ_0 is a flux quantum. It can be shown easily (Problem 14.16) that the ground state of a superconducting ring containing an odd number of sign changes (π ring) has a spontaneous magnetization of a half-magnetic flux quantum, $I_s L \approx (1/2)\Phi_0$, when the external field is zero. Because $I_s = 0$ in the ground state for an even number of π shifts, the magnetic-flux state has an even number of quantization. To summarize,

$$\Phi = n\Phi_0 \quad \text{for } N \text{ even (0 ring)}, \tag{14.148}$$

and

$$\Phi = \left(n + \frac{1}{2}\right)\Phi_0 \quad \text{for } N \text{ odd } (\pi \text{ ring}), \tag{14.149}$$

where N is an integer.

Tricrystal Magnetometry

The multiple-junction ring consists of deliberately oriented cuprate crystals for defining the direction of the pair wave function. The presence or absence of the half-integer flux-quantum effect in such samples as a sample configuration differentiates between various pairing symmetries. In the first tricrystal experiment of Tsuei et al. (Ref. 24), an epitaxial YBCO film (1200 Å thick) was deposited using laser ablation on a tricrystal (100) $SrTiO_3$ substrate. In addition to the three-junction ring located at the tricrystal meeting point, two two-junction rings and one ring with no junction were also made as controls. The design was such that $I_c L \gg \Phi_0$ for observing the half-integer flux quantization. The magnetic flux threading through the superconducting cuprate rings in the tricrystal magnetometry experiments were directly measured by a high-resolution SQUID microscope. A series of tricrystal experiments with various geometrical configurations confirmed that only the $d_{x^2-y^2}$ frustrated configuration showed the half-integer flux-quantum effect.

d-Wave Pairing Symmetry

The evidence from both phase-sensitive symmetry as well as non-phase-sensitive symmetry techniques has conclusively proved that the cuprates have *d*-wave pairing symmetry. The identification of

d-wave symmetry is based on group theory and the macroscopic quantum coherence phenomena of pair tunneling and flux quantization. However, it does not necessarily specify a mechanism for high-temperature superconductivity.

14.9.5 Search for a Theoretical Mechanism of High T_c Superconductors

The identification of d-wave pairing symmetry does not necessarily specify a mechanism for high-temperature superconductors. The solution of this problem should include pairing symmetry, pairing interactions (mediated by phonons, spin fluctuations, or some other bosons) in the presence of strong correlations, and taking into consideration the anomalous normal state and charge segregation and the stripe phase. The observation of a pseudogap in the normal state, with a d-wave-like **k** dependence of many underdoped cuprate semiconductors, has generated renewed interest for a suitable theory. Recently, Kresin and Wolf proposed some experiments that will unambiguously resolve the issue.[16]

Anderson noted, "The consensus is that there is absolutely no consensus on the theory of high-T_c superconductivity." Bardeen et al.[2] proposed a microscopic theory of superconductivity in 1957, which was unanimously accepted 46 years after Kammerlingh Onnes discovered superconductivity in Hg in 1911.[15] We do not know how long we have to wait for a satisfactory theory for high T_c superconductivity, which was discovered by Bednroz and Muller in 1986.[3]

PROBLEMS

14.1. Show that Maxwell's equation (neglecting the displacement current **D** as well as replacing **H** with **B** because **j** is the mean microscopic current) can be written as

$$\vec{\nabla} \times \mathbf{B} = \frac{4\pi}{c}\mathbf{j}. \tag{1}$$

14.2. Minimizing Eq. (14.16) with respect to **A**, show that

$$\vec{\nabla} \times \mathbf{B} = \frac{4\pi}{c}\mathbf{j}, \tag{1}$$

where

$$\mathbf{j}(\mathbf{r}) = \frac{e^*\hbar}{2im^*}[\Psi^*\vec{\nabla}\Psi - \Psi\vec{\nabla}\Psi^*] - \frac{e^{*2}}{m^*c}\mathbf{A}\Psi^*\Psi. \tag{2}$$

14.3. Minimizing Eq. (14.16) with respect to Ψ^*, by first integrating such that all spatial derivatives act on Ψ and then taking functional derivatives with respect to Ψ^*, show that

$$\left[\alpha + \beta|\Psi|^2 + \frac{1}{2m^*}\left(\frac{\hbar}{i}\vec{\nabla} + \frac{e^*}{c}\mathbf{A}\right)^2\right]\Psi = 0. \tag{1}$$

14.4. Substituting Eq. (14.54) in Eqs. (14.52) and (14.53), we obtained

$$i\hbar\left(\frac{\dot{n}_1}{\sqrt{n_1}} + i\sqrt{n_1}\dot{\phi}_1\right)e^{i\phi_1} = (\varepsilon_1\sqrt{n_1}e^{i\phi_1} + \in\sqrt{n_2}e^{i\phi_2}) \tag{1}$$

and

$$i\hbar\left(\frac{\dot{n}_2}{\sqrt{n_2}}+i\sqrt{n_2}\dot{\phi}_2\right)e^{i\phi_2}=(\varepsilon_2\sqrt{n_2}e^{i\phi_2}+\epsilon\sqrt{n_1}e^{i\phi_1}). \tag{2}$$

Show from Eqs. (1) and (2) that

$$\dot{n}_1=2\frac{\epsilon n}{\hbar}\sin(\phi_2-\phi_1)=-\dot{n}_2=\frac{j}{e^*} \tag{3}$$

and

$$\dot{\phi}_2-\dot{\phi}_1=\frac{1}{\hbar}(\varepsilon_1-\varepsilon_2)=\frac{e^*}{\hbar}(V_2-V_1), \tag{4}$$

where $n=\sqrt{n_1n_2}$.

14.5. If we choose $\hat{S}$ such that

$$\hat{H}_1+[\hat{H}_0,\hat{S}]=0, \tag{1}$$

in which case, the electron–phonon interaction $\hat{H}_1$ is eliminated apart from a higher-order term. If we choose $\hat{S}$ of the form

$$\hat{S}=\sum_{\mathbf{k}\mathbf{q}s'}M_{\mathbf{q}}(\alpha\hat{a}^{\dagger}_{-\mathbf{q}}+\beta\hat{a}_{\mathbf{q}})\hat{c}^{\dagger}_{\mathbf{k}+\mathbf{q},s'}\hat{c}_{\mathbf{k},s'}. \tag{2}$$

From Eqs. (1) and (2), show that

$$\alpha^{-1}=E(\mathbf{k})-E(\mathbf{k}+\mathbf{q})-\hbar\omega_{\mathbf{q}} \tag{3}$$

and

$$\beta^{-1}=E(\mathbf{k})-E(\mathbf{k}+\mathbf{q})+\hbar\omega_{\mathbf{q}}. \tag{4}$$

14.6. From Eqs. (14.60) and (14.61), the next-order interaction term is $\frac{1}{2}[\hat{H}_1,\hat{S}]$, which is a sum of terms that contain operator products of the form

$$\hat{a}^{\dagger}_{\pm\mathbf{q}}\hat{a}_{\pm\mathbf{q}'}\hat{c}^{\dagger}_{\mathbf{k}'+\mathbf{q}',s'}\hat{c}_{\mathbf{k}',s'}\hat{c}^{\dagger}_{\mathbf{k}+\mathbf{q},s}\hat{c}_{\mathbf{k},s}. \tag{1}$$

Show that because $\mathbf{q}'=-\mathbf{q}$ (from momentum conservation), out of all the combinations, only one does not contain any phonon operators,

$$c^{\dagger}_{\mathbf{k}+\mathbf{q},s}c^{\dagger}_{\mathbf{k}'-\mathbf{q},s'}\hat{c}_{\mathbf{k}',s'}\hat{c}_{\mathbf{k},s}, \tag{2}$$

when

$$\mathbf{k}'\neq\mathbf{k},\ \mathbf{k}+\mathbf{q}. \tag{3}$$

14.7. Show from Eqs. (14.60), (14.62), and (14.66) that the proper form of Eq. (2) is

$$H_{eff}=\frac{1}{2}\sum_{\mathbf{k},\mathbf{k}',\mathbf{q},s,s'}|M_{\mathbf{q}}|^2(\alpha-\beta)\hat{c}^{\dagger}_{\mathbf{k}+\mathbf{q},s}\hat{c}^{\dagger}_{\mathbf{k}'-\mathbf{q},s'}\hat{c}_{\mathbf{k}',s'}\hat{c}_{\mathbf{k},s}. \tag{1}$$

14.8. Substituting the values of α and β from Eqs. (14.63) and (14.64) in Eq. (1) of Problem 14.7, show that

$$\langle f|H_{eff}|i\rangle = \frac{1}{2}\sum_{\mathbf{k},\mathbf{k}',\mathbf{q},s,s'} \langle f|V_{\mathbf{kq}}\hat{c}^{\dagger}_{\mathbf{k}+\mathbf{q},s}\hat{c}^{\dagger}_{\mathbf{k}-\mathbf{q},s'}\hat{c}_{\mathbf{k}'s'}\hat{c}_{\mathbf{k},s}|i\rangle, \tag{1}$$

where

$$V_{\mathbf{kq}} = \frac{2|M_{\mathbf{q}}|^2\hbar\omega_{\mathbf{q}}}{[E(\mathbf{k}+\mathbf{q}) - E(\mathbf{k})]^2 - (\hbar\omega_{\mathbf{q}})^2}. \tag{2}$$

14.9. Show that if we take the complex conjugate of Eq. (14.79), multiply by $b(\mathbf{k})$, and sum over $\mathbf{k}$, we obtain an equation that agrees with Eq. (14.76), provided $\beta = E$.

14.10. In Eq. (14.83), we derived

$$U\int_{E_F}^{E_F+\hbar\omega_{\mathbf{q}}} \frac{D(x)dx}{2x - E} = 1, \tag{1}$$

where $D(x)$ is the density of states. Show that if we approximate $D(x)$ by $D(E_F)$ in the narrow range of integration, we obtain

$$E = 2E_F - \frac{2\hbar\omega_{\mathbf{q}}\exp[-2/D(E_F)U]}{1-\exp[-2/D(E_F)U]} \approx 2E_F - 2\hbar\omega_{\mathbf{q}}\exp[-2/D(E_F)U]. \tag{2}$$

14.11. We introduced the $\hat{\alpha}-$ operators in Eq. (14.88):

$$\begin{aligned} \hat{\alpha}_{\mathbf{k}} &= u_{\mathbf{k}}\hat{c}_{\mathbf{k}} - v_{\mathbf{k}}\hat{c}^{\dagger}_{-\mathbf{k}} \\ \hat{\alpha}_{-\mathbf{k}} &= u_{\mathbf{k}}\hat{c}_{-\mathbf{k}} + v_{\mathbf{k}}\hat{c}^{\dagger}_{\mathbf{k}} \\ \hat{\alpha}^{\dagger}_{\mathbf{k}} &= u_{\mathbf{k}}\hat{c}^{\dagger}_{\mathbf{k}} - v_{\mathbf{k}}\hat{c}_{-\mathbf{k}} \\ \alpha^{\dagger}_{-\mathbf{k}} &= u_{\mathbf{k}}c^{\dagger}_{-\mathbf{k}} + v_{\mathbf{k}}\hat{c}_{\mathbf{k}}. \end{aligned} \tag{1}$$

In addition, we use the conditions $u_{\mathbf{k}}^2 + v_{\mathbf{k}}^2 = 1$, $u_{\mathbf{k}} = u_{-\mathbf{k}}$, and $v_{\mathbf{k}} = v_{-\mathbf{k}}$. Show that these conditions guarantee that the commutation relations for the c-operators also apply to the α-operators.

14.12. From Eqs. (14.91), (14.94), and (14.95), show that

$$\Delta = \frac{U}{2}\sum_{\mathbf{k}} \frac{\Delta}{\sqrt{\varepsilon^2(\mathbf{k}) + \Delta^2}}. \tag{1}$$

14.13. In Eq. (14.102), we obtained

$$E = D(E_F)\int_0^{\hbar\omega_{\mathbf{q}}} \left(\varepsilon - \frac{1}{2}\frac{2\varepsilon^2 + \Delta^2}{\sqrt{\varepsilon^2 + \Delta^2}}\right) d\varepsilon. \tag{1}$$

After integrating Eq. (1), show that

$$E = \frac{D(E_F)}{2}(\hbar\omega_{\mathbf{q}})^2\left[1-\sqrt{1+\left(\frac{\Delta}{\hbar\omega_{\mathbf{q}}}\right)^2}\right] \approx -\frac{D(E_F)\Delta^2}{4}. \tag{2}$$

14.14. Because $\Delta(0) \ll \hbar\omega_{\mathbf{q}}$, from Eqs. (14.98) and (14.114), show that

$$\ln\frac{\Delta(T)}{\Delta(0)} = -2\int_0^\infty \frac{dx}{\sqrt{x^2+1}} f\left(\sqrt{x^2+1}\frac{\Delta(T)}{\Delta(0)}\left[\frac{k_BT}{\Delta(0)}\right]^{-1}\right) = F\left(\frac{\Delta(T)}{\Delta(0)}, \frac{k_BT}{\Delta(0)}\right). \tag{1}$$

14.15. The free energy per unit area of Josephson coupling between two superconducting electrodes with order parameters ψ_L and ψ_R (Figure 14.20) can be written as

$$F_j = W\int ds\left[\psi_L\psi_R^* \exp\left[i(2\pi/\Phi_0)\int_L^R \mathbf{A}\cdot d\mathbf{l}\right] + c.c.\right], \tag{1}$$

where W is a measure of the Josephson coupling strength, and the integral is over the junction interface. By minimizing the total free energy with respect to ψ_L and ψ_R, show that the expression for the Josephson current density J_s, flowing perpendicular to the junction interface from superconductor L to R, is

$$J_s = t_{L,R}\chi_L(\mathbf{n})\chi_R(\mathbf{n})|\eta_L||\eta_R|\sin\gamma = J_c\sin\gamma. \tag{2}$$

Here, J_c is the critical current density; $\chi_{L,R}$, the basis function, is related to the gap function $\Delta(\mathbf{k})$ through Eq. (14.141); $\Delta_{L,R}(\mathbf{n}) = \eta_{L,R}(\mathbf{n})\chi_{L,R}(\mathbf{n})$, where $\mathbf{n}$ is the unit vector normal to the junction interface, $\eta_{L,R}(\mathbf{n}) = |\eta_{L,R}|e^{i\phi_{L,R}}$; and γ is defined in Eq. (14.137). $t_{L,R}$ is a constant characteristic of the junction.

14.16. The flux quantization of a superconducting ring can be written as

$$\Phi_a + I_sL + \frac{\Phi_0}{2\pi}\sum_{ij}\gamma_{ij} = n\Phi_0, \tag{1}$$

where Φ_a is due to the flux of an external field, I_s is the supercurrent circulating in the ring

$$I_s = I_c^{ij}(\theta_i, \theta_j)\sin\gamma_{ij}, \tag{2}$$

where γ_{ij} is defined in Eq. (14.137), and Φ_0 is a flux quantum. Show that the ground state of a superconducting ring containing an odd number of sign changes (π ring) has a spontaneous magnetization of a half-magnetic flux quantum, $I_sL \approx (1/2)\Phi_0$, when the external field is zero. Because $I_s = 0$ in the ground state for an even number of π shifts, the magnetic-flux state has an even number of quantization. To summarize,

$$\Phi = n\Phi_0 \quad \text{for } N \text{ even (0 ring)}, \tag{3}$$

and

$$\Phi = \left(n + \frac{1}{2}\right)\Phi_0 \quad \text{for } N \text{ odd } (\pi \text{ ring}), \tag{4}$$

where N is an integer.

References

1. Aschroft NW, Mermin ND. *Solid state physics.* New York: Brooks/Cole; 1976.
2. Bardeen J, Cooper LN, Schrieffer JR. Theory of Superconductivity. *Phys Rev* 1957;**108**:1175.
3. Bednroz G, Muller KA. Possible high Tc superconductivity in the Ba-Ca-Cu-O system. *Z Physik* 1986; **B64**:189.
4. Bednroz G, Muller KA. Perovskite-type oxides-The new approach to high-Tc superconductivity. *Rev Mod Phys* 1988;**60**:585.
5. Bogoliubov NN. A new method in the theory of superconductivity. *Soviet Phys JETP* 1958;**7**:41.
6. Cohen MH, Falicov LM, and Phillips JC. Superconducting Tunneling.
7. Cooper LN. Bound Electron Pairs in a Degenerate Fermi Gas. *Phys Rev* 1986;**104**:1189.
8. Deaver BS, Fairbank MS. Experimental Evidence for Quantized Flux in Superconducting Cylinders. *Phys Rev Lett* 1961;**7**:43.
9. de Gennes P-G. *Superconductivity of metals and alloys.* Reading, MA: Addison-Wesley; 1992.
10. Ding H, Norman MR, Campuzino JC, Randeria M, Bellmar AF, Yokoya T, et al. Angle-resolved photoemission spectroscopy study of the superconducting gap anisotropy in Bi2Sr2CaCu2O8+x. *Phys Rev B* 1996;**54**:R9678.
11. Dynes R. McMillan's equation and the Tc of superconductors. *Solid State Commun.* 1972;**10**:615.
12. Doll R, Nabauer M. Experimental Proof of Magnetic Flux Quantization in a Superconducting Ring. *Phys Rev Lett* 1961;**7**:51.
13. Josephson BD. Possible new effects in superconducting tunneling. *Phys Lett* 1962;**1**:251.
14. Josephson BD. The discovery of tunneling supercurrents. *Rev Mod Phys* 1974;**46**:251.
15. Kammerlingh Onnes H. The resistance of platinum at helium temperature. *Comm Phys Lab Univ Leiden* 1911;**119b**:19.
16. Kresin VJ, Wolf SA. Electron-lattice interaction and its impact in high-Tc superconductivity. *Rev Mod Phys* 2009;**81**:481.
17. Landau LD, Ginzburg VL. On the theory of superconductivity. *JETP (USSR)* 1950;**20**:1064.
18. London F, London H. The Electromagnetic Equations of the Superconductor. *Proc. Roy. Soc.* London 1935;**A149**:71.
19. Madelung O. *Introduction to solid state physics.* New York: Springer-Verlag; 1978.
20. Marder MP. *Condensed matter physics.* New York: John Wiley & Sons; 2000.
21. McMillan W. Transition Temperature of Strong-Coupled Superconductors. *Phys Rev* 1968;**167**:331.
22. Meissner W, Ochsenfeld R. A new effect in penetration of superconductors. *Die Naturwissenschaften* 1933;**21**:787.
23. Mercereau. Superconductivity vol. 1, editor. Parks RD. New York: Marcel Dekker; 1961, p. 393.
24. Pals JA, Haeringen W, van Maaren MH. *Phys. Rev. B* 1977;**15**:2592.
25. Stavola M, Krol DM, Schneemeyer LF, Sunshine SA, Fleming RN, Waszczak JV, et al. Raman scattering from single crystals of the 84-K superconductor Bi2.2La0.8Sr2Cu2O8+Δ. *Phys Rev B* 1988;**41**:R5110.
26. Tinkham M. *Introduction to superconductivity.* New York: McGraw-Hill; 1996.
27. Tsuei CS, Kirtley JR. Pairing Symmetry in Cuprate Superconductors. *Rev Mod Phys* 2000;**72**:969.

28. Tsuei CC, Kirtley JR, Chi CC, Yu-Jahnes LS, Gupta A, Shaw T, Sun J-Z, Ketchen MB. Pairing Symmetry and Flux Quantization in a Tricrystal Superconducting Ring of YBa2Cu3O7-Δ. *Phys. Rev. Lett.* 1994;**73**:593.
29. Wu MK, Ashburn JR, Torng CJ, Hor PH, Meng RL, Gao L, et al. Superconductivity at 93 K in a new mixed phase Y-Ba-Cu-O compound system. *Phys Rev Lett* 1987;**58**:908.

CHAPTER

Heavy Fermions

15

CHAPTER OUTLINE

15.1 Introduction 488
15.2 Kondo-Lattice, Mixed-Valence, and Heavy Fermions 490
15.2.1 Periodic Anderson and Kondo-Lattice Models 490
15.2.2 Mixed-Valence Compounds 492
15.2.3 Slave Boson Method 493
15.2.4 Cluster Calculations 494
15.3 Mean-Field Theories 498
15.3.1 The Local Impurity Self-Consistent Approximation 498
15.3.2 Application of LISA to Periodic Anderson Model 499
15.3.3 RKKY Interaction 500
15.3.4 Extended Dynamical Mean-Field Theory 501
15.4 Fermi-Liquid Models 502
15.4.1 Heavy Fermi Liquids 502
15.4.2 Fractionalized Fermi Liquids 505
15.5 Metamagnetism in Heavy Fermions 506
15.6 Ce- and U-Based Superconducting Compounds 508
15.6.1 Ce-Based Compounds 508
15.6.2 U-Based Superconducting Compounds 509
15.7 Other Heavy-Fermion Superconductors 513
15.7.1 $PrOs_4Sb_{12}$ 513
15.7.2 $PuCoGa_5$ 513
15.7.3 $PuRhGa_5$ 515
15.7.4 Comparison between Cu and Pu Containing High-T_c Superconductors 516
15.8 Theories of Heavy-Fermion Superconductivity 516
15.9 Kondo Insulators 516
15.9.1 Brief Review 516
15.9.2 Theory of Kondo Insulators 517
Problems 519
References 524

15.1 INTRODUCTION

Heavy fermions, which are also sometimes referred to as heavy electrons, are a loosely defined collection of intermetallic compounds containing lanthanide (mostly Ce, Yb) or actinide (mostly U, Np) elements. They also include other compounds such as quasi two-dimensional $CeCoIn_5$ and "Skutterdites" such as $PrOs_4Sb_{12}$. The common feature of the heavy fermions is that they have large effective mass m^* (50–1000 times greater than the mass of a free electron) below a coherence temperature T^*. The effective mass is estimated through the electronic specific heat. In general, for very low temperatures, the specific heat C of a metal can be expressed as

$$C/T = \gamma + \beta T^2, \tag{15.1}$$

where

$$\gamma = V_m k_F k_B^2 m^* / 3\hbar^2. \tag{15.2}$$

Here, V_m is the molar volume, k_F is the Fermi vector, m^* is the effective mass of the electron, T is the absolute temperature, γ is the electronic contribution, and β is the contribution of the phonons to the specific heat. There is an additional spin-fluctuation term $\delta T^3 \ln T$ in the specific heat of UPt_3 and UAl_2.

For normal metals such as copper or aluminum, γ is of the order 1 mJ/mol K^2 at low temperatures. A generally accepted definition of heavy fermions is those systems that have $\gamma > 400$ mJ/f atom mol K^2 below the coherence temperature T^*. γ is generally normalized to a mole of f atoms so that there can be a comparison between systems with different structure. Some of the other properties of heavy fermions include (a) an enhanced Pauli spin susceptibility indicating a large effective mass; (b) a Wilson ratio of approximately one; (c) a huge T^2 term in the electrical resistivity; and (d) highly temperature-dependent de Haas–van Alphen oscillation amplitudes at very low temperatures. The Wilson ratio (Ref. 35) R is defined as

$$R = \frac{\pi^2 k_B^2 \chi(0)}{g^2 \mu_B^2 J(J+1)\gamma(0)}. \tag{15.3}$$

Here, $\chi(0)$ and $\gamma(0)$ are the magnetic susceptibility and specific heat at zero temperature, J is the total angular momentum, g_J is the Landé g factor, and the other symbols have their usual meanings.

$CeAl_3$, which earlier had been considered a mixed-valence compound, was the first heavy-fermion system discovered by Andres et al. in 1975. They found that below 0.2° K, $\gamma = 1620$ mJ mole/K^2 and the coefficient of the T^2 term in $\rho = AT^2$, $A = 35$ μΩ cm/K^2. The intense interest in heavy-fermion systems started with the discovery of superconductivity in $CeCu_2Si_2$ by Steglich et al.[29] in 1979. Their results are shown in Figures 15.1 and 15.2.

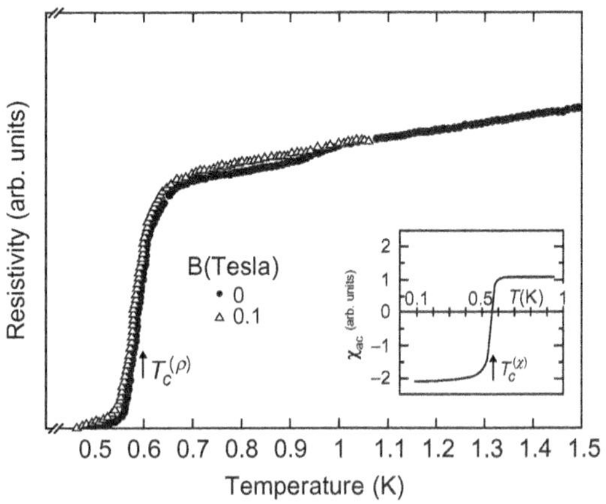

FIGURE 15.1

Resistivity (main part) and low-field ac susceptibility (inset) of $CeCu_2Si_2$ as a function of temperature. Arrows give transition temperatures $T_c^{(\rho)} = 0.60 \pm 0.03$° K and $T_c^{(\chi)} = 0.54 \pm 0.03$° K.

Reproduced from F. Steglich et al.[29] with permission of the American Physical Society.

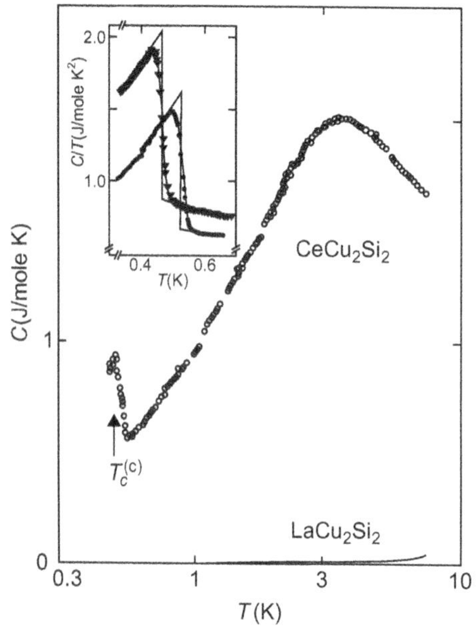

FIGURE 15.2

Molar specific heat of $CeCu_2Si_2$ as a function of temperature on a logarithmic scale. Inset shows the specific heat jumps of two other samples.

Reproduced from F. Steglich et al.[29] with the permission of the American Physical Society.

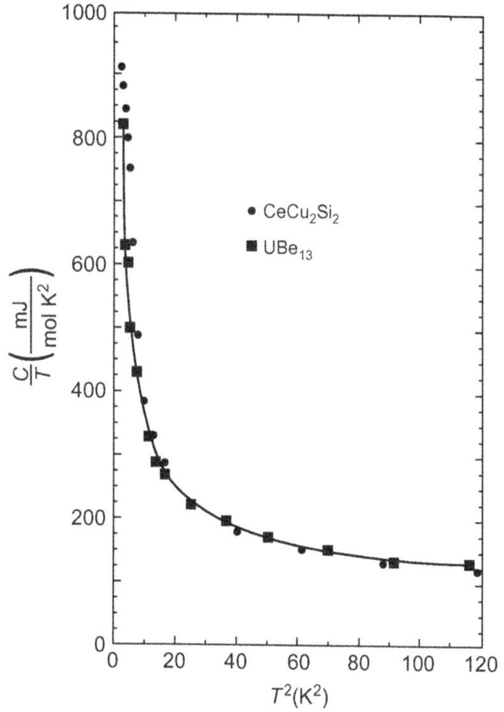

FIGURE 15.3

Specific heat of nonsuperconducting single crystals (•) and a piece of superconducting crystal (■) of UBe_{13}.

Reproduced from Stewart[30] with the permission of the American Physical Society.

The main part of Figure 15.2 shows, in a logarithmic scale, the molar specific heat of $CeCu_2Si_2$ at $B = 0$ as a function of temperature. The inset in Figure 15.2 shows in a C/T plot the specific heat-jumps of two other $CeCu_2Si_2$ samples that do not look very profound.

The specific heat-jumps below the coherence temperature T^*, which is characteristic of heavy-fermion systems, are elegantly displayed when one plots C/T versus T^2. Stewart[30] plotted C/T versus T^2 of nonsuperconducting single crystals of $CeCu_2Si_2$ and a piece of a superconducting single crystal of UBe_{13}. These results are reproduced in Figure 15.3, in which the line through UBe_{13} serves as a guide to the eye.

Since 1974, approximately 50 heavy-fermion compounds have been discovered, but there is no uniformity in their properties. For example, UBe_{13} is a superconductor in the ground state with non-Fermi-liquid properties in the normal state, whereas UPt_3 orders antiferromagnetically below the Néel temperature (T_N), exhibits a heavy Fermi-liquid state well below T_N, and has unconventional superconductivity with a multicomponent superconducting parameter. $CeAl_2$ and U_2Zn_{17} are antiferromagnets with weak moments at very low temperatures, and CeNiSn and $Ce_3Bi_4Pt_3$ are narrow-gap semiconductors with quasiparticles having large effective masses. Some heavy-fermion superconductors such as $CeCoIn_5$ are quasi two-dimensional. The only common feature is the large effective mass below the coherence temperature and the fact that all these are highly correlated electron systems. In addition, some heavy fermions such as $CeRu_2Si_2$ exhibit metamagnetism, which has a wide variety of technological applications.

There are many factors that lead to the conclusion that the large effective mass of heavy fermions below the coherence temperature is not due to band-structure renormalization. For example, the magnitude of the nuclear relaxation rate of UBe_{13} and the ultrasonic attenuation in UPt_3 in the normal state are the same as ordinary metals. The thermal conductivity measurements in $CeCu_2Si_2$, UBe_{13}, and UPt_3 yield results similar to ordinary metals.

There have been several powerful techniques applied to discuss the theory of these strongly correlated systems. However, the theory of these systems lags behind the experiment. In this chapter, we will discuss the properties of the wide variety of these correlated systems without going into the details of the complex theories.

15.2 KONDO-LATTICE, MIXED-VALENCE, AND HEAVY FERMIONS

15.2.1 Periodic Anderson and Kondo-Lattice Models

It has been noted that the majority of the rare-earth and actinide compounds have local moments and can be classified as systems in the magnetic regime. The f orbitals have no charge fluctuation in this region and have integral valence. Therefore, they can be considered to be in a Mott insulating stage. Weak residual spin polarization of the conduction electrons, Rudderman–Kittel–Kasuya–Yosida (RKKY) interactions between the local moments (Refs. 12, 21, 36), magnetic transition at low temperatures, and spin-wave excitations occur. The spin waves scatter the conduction electrons at low temperatures.

To correlate and to study their dependence on the various relevant parameters, the simplest Hamiltonian is the orbitally nondegenerate periodic Anderson model (Ref. 1). The periodic Anderson model for a system consisting of a set of N sites is denoted by sites i, j. On each site, there are two orthogonal nondegenerate orbitals that will be referred to as C and f. The Hamiltonian is assumed to have the form

$$H = t\sum_{i \neq j,\sigma} \hat{C}^{\dagger}_{i\sigma}\hat{C}_{j\sigma} + V\sum_{i \neq j,\sigma} (\hat{C}^{\dagger}_{i\sigma}\hat{f}_{j\sigma} + \hat{f}^{\dagger}_{j\sigma}\hat{C}_{i\sigma}) + \varepsilon_f \sum_{i,\sigma} \hat{f}^{\dagger}_{i,\sigma}\hat{f}_{i\sigma} + U\sum_{i} \hat{f}^{\dagger}_{i\uparrow}\hat{f}_{i\uparrow}\hat{f}^{\dagger}_{i\downarrow}\hat{f}_{i\downarrow}. \tag{15.4}$$

Here, t (which can be positive or negative) is the transfer (hopping) integral of the extended orthogonal orbitals between sites i and j (restricted to nearest neighbors in our model). $\hat{C}^{\dagger}_{i\sigma}$ and $\hat{C}_{j\sigma}$ are the creation and annihilation operators for these extended orbitals at sites i and j with spin σ. There is one extended orbital per site per spin with a mean energy that is the origin of the energy scale. $\hat{f}^{\dagger}_{i\sigma}$ and $\hat{f}_{i\sigma}$ are the creation and annihilation operators for the localized f orbitals (i denotes the site) with energy ε_f. V is a positive hybridization parameter between the localized and the band orbitals in neighboring sites. The third term represents the single-particle energy of the isolated f orbitals. The fourth term is an interaction of the Hubbard type between electrons of the f orbitals on the same site. U is the Coulomb repulsion between two electrons of opposite spin in the f orbital and describes a short-range interaction between them. U is positive, whereas t and ε_f can have either sign. When we consider the f orbital only on a single site, the model (Eq. 15.4) is reduced to a single-impurity Anderson model (Ref. 1), as discussed in Section 13.9, except that the localized magnetic moment for rare-earth metals is due to $s-f$ mixing instead of $s-d$ mixing as originally visualized in the single-impurity Anderson model.

The Hamiltonian (Eq. 15.4) can be augmented by additional terms such as second-neighbor hopping or Coulomb repulsion between extended orbitals and f electrons, or between electrons on different sites. Because the orbital degeneracy is neglected, there is no Hund's rule coupling between the f orbitals in this model. However important such terms are in applications to real systems, they are ignored here in the belief that they would contribute nothing really essential to the qualitative physics.

When each f orbital is occupied by a single electron (either up-spin or down-spin), the system is described as the Kondo regime. The empty sites and doubly occupied sites become virtual states. The low-energy physics of the periodic Anderson model (Eq. 15.4) can be described by an effective model where the f-electron degrees of freedom are represented by localized spins. Schrieffer and Wolff (Ref. 24) (Problem 15.1) used a second-order perturbation with respect to V to obtain an effective Hamiltonian

$$H = t\sum_{j\neq i,\sigma}(C_{i\sigma}^{\dagger}C_{j\sigma} + H.C.) + J\sum_{i} S_i \cdot S_i^c, \tag{15.5}$$

where

$$S_i = \frac{1}{2}\sum_{\sigma,\sigma'} \tau_{\sigma,\sigma'} f_{i\sigma}^{\dagger} f_{i\sigma'}, \tag{15.6}$$

$$S_i^c = \frac{1}{2}\sum_{\sigma,\sigma'} \tau_{\sigma,\sigma'} C_{i\sigma}^{\dagger} C_{i\sigma'}, \tag{15.7}$$

and τ are the Pauli spin matrices. Thus, S_i^c are the spin-density operators of the conduction electrons, and S_i are the localized spins. J is the exchange interaction, which is antiferromagnetic $(J > 0)$ and inversely proportional to U. Under symmetric conditions, $J = 8V^2/U$.

Thus, the rare-earth compounds that have either localized four f-electrons (Ce, Yb) or five f electrons (U, Np) can be considered as a Kondo-lattice (Ref. 9), where at each lattice site a local moment interacts via an exchange coupling J with the spin of any conduction electron sitting at the site. The Hamiltonian in Eq. (15.5) is also known as the Kondo-lattice model (Ref. 9). The exchange coupling is the source of interesting many-body effects in the Kondo-lattice model. The complexity of solving the Kondo-lattice model arises due to the complex correlation effect involving both the localized spin and the itinerant electron degrees of freedom. In fact, a conduction electron undergoes a spin-flip process with a localized spin if the spin is antiparallel. The conduction electron leaves a trace of its spin exchange processes with the localized spins while moving around the lattice. The direction of the localized spins is determined by the history of the electrons that passed through this site. Thus, the conduction electrons are no longer independent. There are similar correlation effects in the periodic Anderson model due to the dynamic aspects of the localized electrons. Because these systems are highly correlated, most of the theoretical models developed during the past 30 years are approximate treatments of the complex problem.

15.2.2 Mixed-Valence Compounds

The properties of rare-earth metals and their compounds as well as the actinide compounds have been the subject of a great deal of interest for the past 40 years. A subclass of these rare-earth and the actinide compounds is known as mixed-valence compounds, which are poor metals but have a fluctuating valence. Clear indication that two ionic valence states are present in these compounds is provided by X-ray photoelectron spectra in which the two valence states are seen side by side. They are also evident by both photoemission measurements as well as by isomer-shift measurements. In these compounds, near the Fermi energy, the s and d electrons as well as the much heavier f electrons are present. A simple explanation is that because in the ground state, both f^n and (f^{n-1} + conduction electron) configurations are present, their energies must be very close. The difference of energy is on the order of the hopping line width. The extra electron is assumed to go into an extended state, so its energy is equal to the Fermi energy. The extra available f orbital can be described as a localized state, with energy ϵ_f nearly equal to the Fermi energy E_F, and that can accept one electron but not two.

The mixed-valence compounds generally form with rare-earth elements only at the beginning, the middle, and the end of the rare-earth series. The reason that the beginning and the end of the rare-earth series are favored is that a closed shell screens the nuclear charge very effectively. Hence, the $4f$ electron in Ce and the $4f$ hole in Yb are loosely bound and not far off from the $5d$ configuration. The middle of the rare-earth series is favored because of the importance of Hund's rule coupling. The final occupied f-level, even for Sm, is not far below the d-level.

A typical example of a mixed-valence compound is SmS, which is a semiconductor at normal pressure. Sm has the electronic structure [Xe] $4f^5 5d^0 6s^2$, and S has the electronic structure [Ne] $3s^2 3p^4$. In compounds, the d-level broadens into a band and hybridizes with the $6s$ band, but the f-levels are essentially unaffected. In a schematic electronic structure, one can visualize a localized f^6-level in the gap between the $5d-6s$ band and the $s-p$ bands.

Under pressure, the lower of the crystal field split d-bands broadens and moves down in energy relative to the f-level and ultimately crosses it. When the $f-d$ gap goes to zero, a metal insulator transition occurs. The electronic structure and the density of states for the metallic state are shown schematically in Figure 15.5.

The f-levels hybridize with the d level on a neighboring atom because they cannot hybridize with the d-level of the same atom (the f^6 configuration has a total $J=0$). The "bandwidth" of the hybridized f-band is very narrow so that in the density of states, over the smooth $s-d$ background, there is a sharp-peak attributable to the f-like atomic character in a tight-binding representation. The wave functions near this peak are linear combinations of f-like and d-like wave functions and can be written as

$$\psi_{\mathbf{k}}(\mathbf{r}) = a_{\mathbf{k}}\phi_d(\mathbf{r}) + b_{\mathbf{k}}\phi_f(\mathbf{r}), \tag{15.8}$$

where the proportion $a_{\mathbf{k}}$ to $b_{\mathbf{k}}$ varies rapidly near the peak. The f^6-level, which is nondegenerate due to correlation energy, accommodates only one electron per atom. Because this peak is derived from the f^6-level, the integrated density

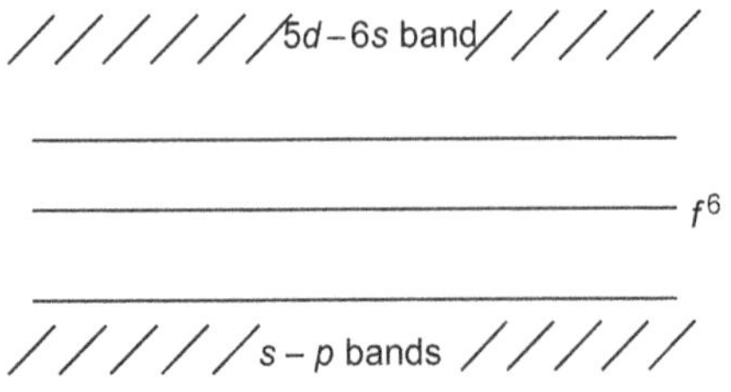

FIGURE 15.4

Schematic structure of SmS in the semiconducting phase.

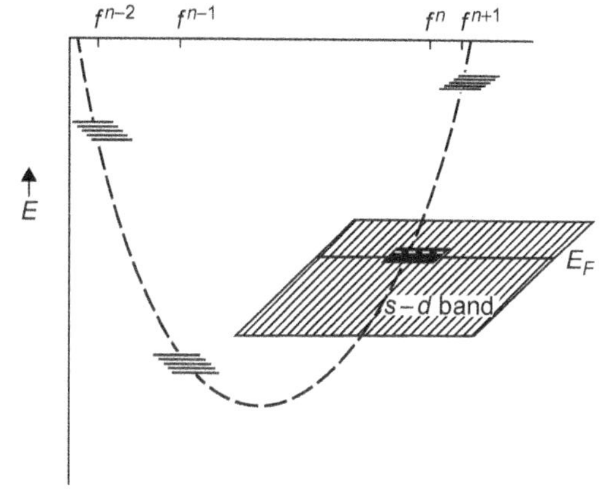

FIGURE 15.5

Electronic energy levels for mixed-valence materials. A wide *sd* band overlaps one of the configurations of the multiplet splitting of *f* electrons.

Reproduced from Varma[32] with the permission of the American Physical Society.

of states of f-like character in it is also one-electron per atom. Hence, at $T=0$, the Fermi level is pinned to lie in the f peak. This characteristic is in a sense the definition of mixed-valence compounds. The wave function in Eq. (15.8) represents the linear combination of the atomic orbital states, which is partly f^5d and partly f^6. The d electron is not affected by the local atomic exchange and correlations because it is relatively free. Thus, $\psi_{\mathbf{k}}(\mathbf{r})$ represents a linear combination of 3+ and 2+ valence states on the rare-earth ion, and the compound is known as a mixed-valence system. The average valence can be defined as

$$V_{av} = \sum_{\mathbf{k}} |a_{\mathbf{k}}|^2 / |\sum_{\mathbf{k}} |b_{\mathbf{k}}|^2, \tag{15.9}$$

where the sum is over the occupied part of the occupied band. However, the instantaneous valence V_{inst} will be different from V_{av} because the f electrons have a nonzero bandwidth.

15.2.3 Slave Boson Method

The slave boson method was developed for the $U=\infty$ periodic Anderson model that had the constraint $n_f \leq 1$ at each site. The essential feature of this method is that the localized electron operators are written as a composition of a fermion $\hat{f}$ and a boson $\hat{b}$, where we may consider the boson as an f vacancy. Every site is occupied either by an $\hat{f}$ fermion or a $\hat{b}$ boson. The localized electron operators are written as a composition of a boson $\hat{b}$ and a fermion $\hat{f}$. One defines

$$f_{i\sigma}^{\dagger} = \hat{f}_{i\sigma}^{\dagger}\hat{b}_i \quad \text{and} \quad f_{i\sigma} = \hat{b}_i^{\dagger}\hat{f}_{i\sigma}. \tag{15.10}$$

The operator equality,

$$\sum_{\sigma} \hat{f}_{i\sigma}^{\dagger}\hat{f}_{i\sigma} + \hat{b}_i^{\dagger}\hat{b}_i = 1, \tag{15.11}$$

satisfies the preceding condition. The Anderson lattice Hamiltonian can be written as

$$H = \sum_{\mathbf{k},\sigma} \varepsilon_{\mathbf{k}} c_{\mathbf{k}\sigma}^{\dagger} c_{\mathbf{k}\sigma} + \varepsilon_f \sum_{i,\sigma} \hat{f}_{i\sigma}^{\dagger}\hat{f}_{i\sigma} + V\sum_{i,\sigma}(c_{i\sigma}^{\dagger}\hat{b}_i^{\dagger}\hat{f}_{i\sigma} + \hat{f}_{i\sigma}^{\dagger}\hat{b}_i c_{i\sigma}) + \sum_i \lambda_i \left(\sum_{\sigma} \hat{f}_{i\sigma}^{\dagger} f_{i\sigma} + \hat{b}_i^{\dagger} b_i - 1\right). \tag{15.12}$$

Here, λ_i is a Lagrangian multiplier for the site i and is needed to impose the local constraints. The properties of the Hamiltonian (Eq. 15.12) are usually discussed in a mean-field approximation. It is assumed that the bosons have Bose condensations, $<\hat{b}_i> = b_0$, and the Lagrange multiplier $\lambda_i = \lambda_0$ for all sites. Thus, the constraint is obeyed only on the average over the whole system.

15.2.4 Cluster Calculations

To correlate the mixed-valence, Kondo, and heavy-fermion behavior and to study how they correspond to different regimes of one fundamental phenomenon (at least in Ce systems), Misra et al.[17] considered the application of the periodic Anderson model to finite clusters with periodic boundary conditions. Although the phrase "periodic Anderson model" is somewhat inappropriate when applied to a small system, they used it in reference to a four-atom cluster in which each site has a localized orbital and an extended orbital with appropriate Coulomb repulsion, hybridization, and transfer matrix elements. In a later paper, they extended the number of electrons to eight particles, but the results were similar. The value of small cluster calculations is that exact solutions of the Hamiltonian are obtained. However, it has to be recognized that in some respects, small clusters are not representative of bulk materials. For example, at sufficiently low temperatures the specific heat of a cluster model will vanish exponentially, and the magnetic susceptibility will either be infinity or zero. The large number of states obtained even for a small cluster suggests that statistical mechanics may give results that fairly represent a large system over a reasonable range of temperatures.

Misra et al. (Ref. 17) applied the periodic Anderson model (Eq. 15.4) to four-site tetrahedral clusters of equal length with periodic boundary conditions, thereby including the band structure effects. For example, their model Hamiltonian for a tetrahedron is identical to that of an fcc lattice if the Brillouin zone sampling is restricted to four reciprocal-lattice points, the zone center Γ, and the three square-face-center points X. They studied the region of crossover between the magnetic, Kondo, and mixed-valence regimes by varying the different parameters $U/|t|$, $V/|t|$, and $E_f/|t|$, and their results for the tetrahedron, which reflect the properties of cerium alloys, are presented in Figure 15.6.

One can distinguish the three regimes by considering n electrons per site. Consider the non-f electrons to constitute an electron reservoir. In Figure 15.6, E_d is the Fermi level when there are n non-f electrons per site, and $E_{d'}$, is the Fermi level when there are $n-1$ electrons per site (if there is no interaction with the f electrons). E_F is the chemical potential when the f electrons are in contact with the electron reservoir. Let E_f be the energy boundary such that at $T=0$, the ion will be in the f^0 state if $E_F < E_f$ and in the f^1 state if $E_F > E_f$. $E_f + U$ is another ionization boundary separating the f^1 state from the f^2 state. The crossover from one regime to another depends sensitively on the various parameters U,V, and E_f as well as on the geometry (band structure).

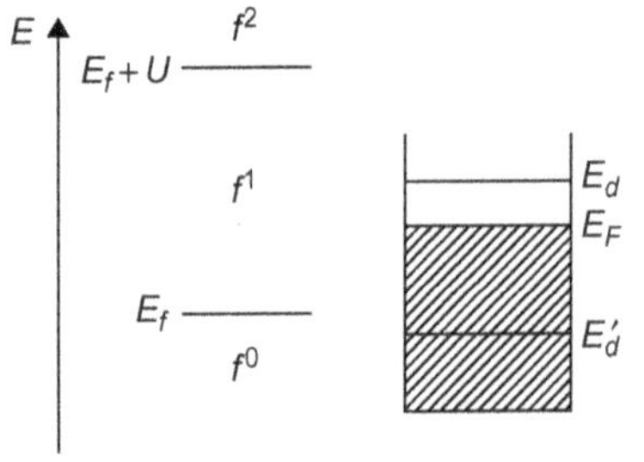

FIGURE 15.6

A schematic diagram illustrating the dependence of the three different regimes on the position of the f-level.

The Hamiltonian (Eq. 15.4) is conveniently considered on a basis of states diagonal in occupation numbers; Misra et al. (Ref. 17) calculated the many-body eigenstates and eigenvalues. Because spin is a good quantum number, the states can be classified as spin singlets, triplets, and quintets. For $n = 4$, there are 784 singlet, 896 triplet, and 140 quintet states. For $n = 8$ (Ref. 3), there are 12,870 states available for eight particles. These rather large numbers of states should tend to make the results somewhat representative of large systems except at extremely low temperatures (lower than the separation between the ground state and the first excited state). The significant results are projected through the diagram in Figure 15.7.

Misra et al. (Ref. 17) constructed a computer program to diagonalize the Hamiltonian within subspaces of fixed values of S_z. They calculated the f-state occupation (n_f), temperature dependence of specific heat (C_v), and the magnetic susceptibility (χ_f) of the f electrons (by using a canonical ensemble) for a large number of parameters. In Figure 15.8, a typical example is presented by plotting C_v/T against T for E_f ranging from -5.0 to -4.0 (n_f varies from 0.9943 to 0.9788).

We notice that for $E_f = -5.0$, C_v/T increases very rapidly at very low temperatures (which mimics the onset of heavy-fermion behavior) but gradually decreases as E_f is increased until the

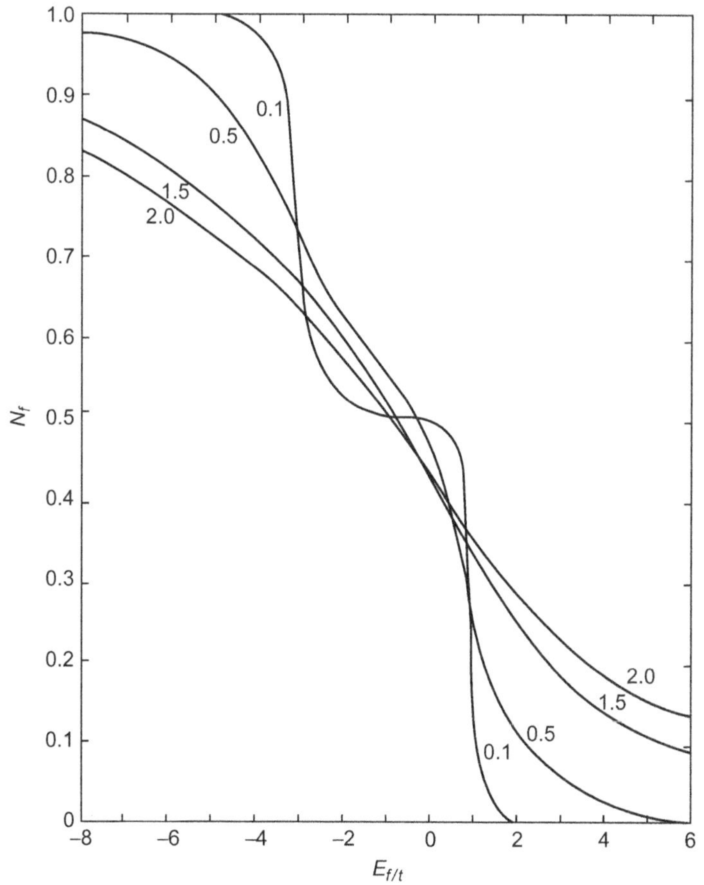

FIGURE 15.7

The f occupation number per site, n_f, in the four-electron ground state in terms of E_f and various hybridization energies V for $t = -1$, $U = 50$ for a tetrahedron.

Reproduced from Misra et al.[17] with the permission of the American Physical Society.

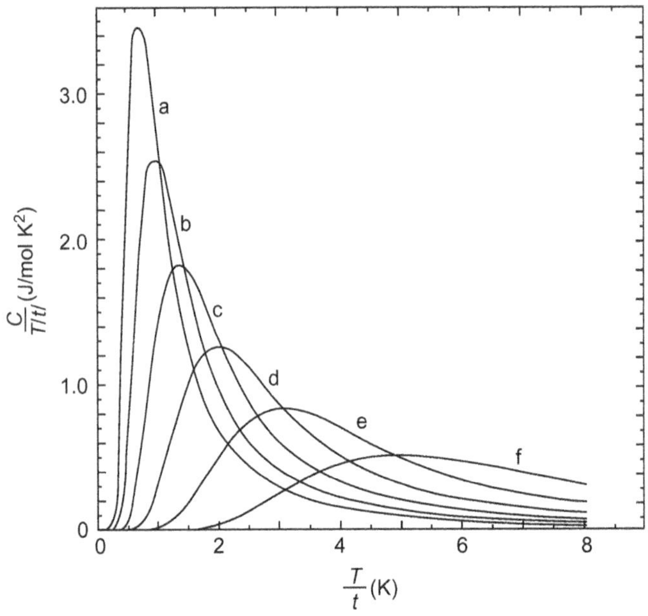

FIGURE 15.8

$C_V/T/|t|$ versus $T/|t|$ for various E_f for E negative, $U = 50$, and $V = 0.1$ for a tetrahedron. Curve (a), $E_f = -5.0$; curve (b), $E_f = -4.8$; curve (c), $E_f = -4.6$; curve (d), $E_f = -4.4$; curve (e), $E_f = -4.2$; curve (f), $E_f = -4.0$. (All parameters in units of $|t|$.)

Reproduced from Misra et al.[17] with the permission of the American Physical Society.

heavy-fermion feature has practically disappeared when $E_f = -4.2$. To explain the unusual increase in C_V/T, Misra et al. (Ref. 17) plotted the energy-level diagram (Figure 15.9) of the first few many-body states for each of these E_f as well as for $E_f = -3.0$.

In Figure 15.9, for $E_f = -5.0$, the ground state is a singlet, but the next two higher-energy states are a triplet and a quintet, which are nearly degenerate with the ground state. The low-temperature rise in C_v is determined by these three levels. As E_f increases, the separation between the lowest three levels increases, and the rise in C_v/T correspondingly decreases. Thus, the heavy-fermion behavior is obtained when the many-body ground state is a singlet but nearly degenerate to two other magnetically ordered states. The same pattern is repeated for a tetrahedron for $t = 1$, except that in some cases the ground state is a magnetically ordered triplet state. In such cases, the ground state of the heavy-fermion system would be magnetically ordered.

In Figure 15.10, Misra et al. plotted $k_B \chi_f T/(g\mu_B^2)(\equiv \chi_f T)$ versus $T/|t|$ to compare their results with the benchmark results for the single-impurity Anderson model. They defined a "frozen-impurity" regime ($\chi_f T = 0$), a free orbital regime ($\chi_f T \approx 0.125$), a valence-fluctuation regime ($\chi_f T \approx 0.167$), and a local moment regime ($\chi_f T \approx 0.25$). In addition, they defined an "intermediate regime" for which $0 < \chi_f T < 0.125$, but $\chi_f T$ essentially remains a constant in this regime.

We note from Figure 15.10 that when $E_f = -5.0$ ($n_f = 0.994$), there is a transition from the frozen-impurity to the local-moment regime. For $E_f = -3.0$ ($n_f = 0.725$), there is a transition from

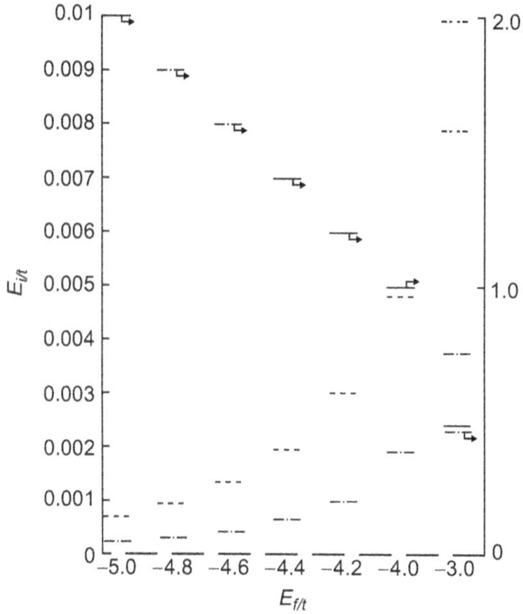

FIGURE 15.9

Energy-level diagram of the first few many-body states for various E_f for t negative, $U=50$, $V=0.1$ for a tetrahedron. (All parameters in units of $|t|$.)

Reproduced from Misra et al.[17] with the permission of the American Physical Society.

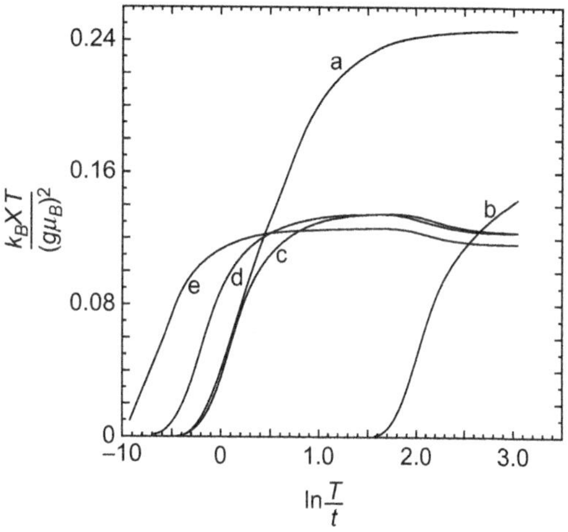

FIGURE 15.10

$k_B\chi_f T/(g\mu_B^2)$ versus ln $T/|t|$ for $t=-1$, $U=50$, and $V=0.1$ for various E_f for a tetrahedron. Curve (a), $E_f=-5.0$; curve (b), $E_f=-3.0$; curve (c), $E_f=-1.0$; curve (d), $E_f=-0.5$; curve (e), $E_f=0.5$. (All parameters in units of $|t|$.)

Reproduced from Misra et al.[17] with the permission of the American Physical Society.

the frozen-impurity to the valence-fluctuation regime. When E_f is further increased, the transition is from the frozen-impurity to the free orbital regime. The high-temperature results are in excellent agreement with the single-impurity "benchmark" results.

When we compare the specific heat curves with the χT curves for the same parameters, the specific heat maxima generally occur below the temperature at which χT reaches its high-temperature value (i.e., the crossover temperature from enhanced Pauli- to Curie-like susceptibility). The main reason is that at low temperatures where C_v is a maximum, the many-body states with magnetic moments are still just becoming thermally populated. The same broad features have also been observed experimentally.

It was generally believed that as E_f is increased from far below E_d (Figure 15.6), there would be a transition from the magnetic to the Kondo-lattice regime. However, Misra et al. (Ref. 17) found that for some choice of parameters, the system undergoes a transition from a Kondo-lattice to a magnetic regime as E_f is increased. Subsequently, it reenters a Kondo-lattice regime for higher values of E_f. This unusual feature of reentry to the Kondo-lattice regime is very sensitive to the hybridization parameter and occurs only for low $V/|t|$ values, which are the most important parameters in determining n_f as well as the thermodynamic properties.

15.3 MEAN-FIELD THEORIES

15.3.1 The Local Impurity Self-Consistent Approximation[17]

The dynamical mean-field theory is a very powerful tool for studying the strongly correlated system. In this approach, a lattice model is replaced by a single-site quantum impurity problem embedded in an effective medium determined self-consistently. This leads to an intuitive picture of the local dynamics of a quantum many-body problem. Because the impurity problem has been extensively studied, the self-consistency condition incorporates the translation invariance and the coherence effects of the lattice. This approach is now popularly known as the local impurity self-consistent approximation (LISA). The LISA freezes spatial fluctuations but includes local quantum fluctuations and is therefore characterized as a dynamical mean-field theory. The on-site quantum problem is still a many-body problem that can be addressed by using a variety of techniques. The dynamical mean-field theory becomes exact in the limit of large spatial dimensions $d \to \infty$ or in the limit of large lattice coordination.

In the mean-field theory, a lattice problem with many degrees of freedom is approximated by a single-site effective problem. The dynamics at a given site are the interaction of the degrees of freedom at this site with an external bath created by the degrees of freedom on the other sites. A simple example is an application to the Hubbard model in which the Hamiltonian is

$$H = -\sum_{<ij>,\sigma} t_{ij}(C_{i\sigma}^{\dagger}C_{j\sigma} + C_{j\sigma}^{\dagger}C_{i\sigma}) + U\sum n_{i\uparrow}n_{i\downarrow}. \tag{15.13}$$

An imaginary-time action (the local effective action) for the fermionic degrees of freedom $(C_{o\sigma}, C_{o\sigma}^{\dagger})$ at site o is

$$S_{eff} = -\int_0^{\beta} d\tau \int_0^{\beta} d\tau' \sum_{\sigma} C_{o\sigma}^{\dagger}(\tau) g_0^{-1}(\tau - \tau') C_{o\sigma}(\tau') + U\int_0^{\beta} d\tau n_{o\uparrow}(\tau) n_{o\downarrow}(\tau). \tag{15.14}$$

Here, $g_0(\tau - \tau')$, the generalized "Weiss function," is the effective amplitude for a fermion to be created on the isolated site at time τ (coming from the "external bath") and destroyed at time τ' (going back to the bath). Because g_0 is a function of time, it accounts for local quantum fluctuations. It can be shown that (Problem 15.2)

$$g_0(i\omega_n)^{-1} = i\omega_n + \mu + G(i\omega_n)^{-1} - R[G(i\omega_n)^{-1}]. \tag{15.15}$$

$G(i\omega_n)$, the on-site interacting Green's function, is calculated from

$$G(\tau - \tau') = -< TC(\tau)C^{\dagger}(\tau') >_{S_{eff}} \tag{15.16}$$

$$G(i\omega_n) = \int_0^{\beta} d\tau G(\tau) e^{iw_n\tau}, \omega_n \equiv \frac{(2n+1)\pi}{\beta}. \tag{15.17}$$

Here, $R(G)$ is the reciprocal function of the Hilbert transform of the density of states corresponding to the lattice. As an example, in the Hubbard model,

$$D(\varepsilon) = \sum_k \delta(\varepsilon - \varepsilon_k), \quad \varepsilon_k = \sum_{ij} t_{ij} e^{i\vec{k}\cdot(\vec{R_i} - \vec{R_j})}. \tag{15.18}$$

The Hilbert transform $\overline{D}(\xi)$ and its reciprocal function R are defined by

$$\overline{D}(\xi) \equiv \int_{-\infty}^{\infty} d\varepsilon \frac{D(\varepsilon)}{\xi - \varepsilon}, \quad R[\overline{D}(\xi)] = \xi. \tag{15.19}$$

Eqs. (15.14) through (15.16) are the basic equations of the LISA method. However, the major difficulty lies in the solution of S_{eff}. It can be shown that solving these equations yields the local quantities, and all the $\vec{k}$-dependent correlation functions of the original lattice Hubbard model can be obtained.

It may be noted that the LISA approach freezes spatial fluctuations but retains local quantum fluctuations. Each site undergoes transition between the four possible quantum states $|0\rangle, |\uparrow\rangle, |\downarrow\rangle, |\uparrow,\downarrow\rangle$ by exchanging electrons with the rest of the lattice or "the external bath." As an example, one can consider $(C_{o\sigma}, C_{o\sigma}^{\dagger})$ as an impurity orbital. The bath can be described as a "conduction band" described by the operators $(a_{l\sigma}, a_{l\sigma}^{\dagger})$, and the Hamiltonian is the well-known single-impurity Anderson Hamiltonian

$$H_{AM} = \sum_{l\sigma} \widetilde{\epsilon}_l a_{l\sigma}^{\dagger} a_{l\sigma} + \sum_{l\sigma} V_l (a_{l\sigma}^{\dagger} C_{o\sigma} + C_{o\sigma}^{\dagger} a_{l\sigma}) - \mu \sum_{\sigma} C_{o\sigma}^{\dagger} C_{o\sigma} + U n_{o\uparrow} n_{o\downarrow}. \tag{15.20}$$

Eq. (15.20) is quadratic in $a_{l\sigma}^{\dagger}, a_{l\sigma}$, and integrating these gives rise to S_{eff} of the form given in Eq. (15.14), provided

$$g_0^{-1}(i\omega_n)^{AM} = i\omega_n + \mu - \int_{-\infty}^{\infty} d\omega \frac{\Delta(\omega)}{i\omega_n - \omega} \tag{15.21}$$

and

$$\Delta(\omega) = \sum_{l\sigma} V_l^2 \delta(\omega - \widetilde{\varepsilon}_i). \tag{15.22}$$

If the parameters $V_l, \widetilde{\varepsilon}_l$ are chosen to obtain g_0, the solution of the mean-field equations, H_{AM} becomes the Hamiltonian representation of S_{eff}. Here, $\widetilde{\varepsilon}_l$'s are effective parameters and not $\varepsilon_{\mathbf{k}}$, the single-particle energy. In addition, $\Delta(\omega)$, the conduction bath density of states, is obtained when the self-consistent problem is solved.

Thus, by using the LISA approach, one obtains the Anderson impurity embedded in a self-consistent medium from the Hubbard model. The dynamical mean-field equations are solved such that the proper g_0 is obtained. When this g_0 is inserted into the Anderson model, the resulting Green's function should obey the self-consistency condition in Eq. (15.15). The mapping onto impurity models, which have been studied by a variety of analytical and numerical techniques, is used to study the strongly correlated lattice models in large dimensions. However, it is important to solve S_{eff} by using reliable methods.

15.3.2 Application of LISA to Periodic Anderson Model

We will now briefly describe the application of the LISA method to heavy-fermion systems and the Kondo insulators (Ref. 20). This is done by using the periodic Anderson model (PAM). This model describes a band of conduction electrons that hybridize with localized $f-$ electrons at each lattice site. The PAM Hamiltonian can be written as

$$H = \sum_{\mathbf{k}\sigma} \epsilon_{\mathbf{k}} C_{\mathbf{k}\sigma}^{\dagger} C_{\mathbf{k}\sigma} + V \sum_{i\sigma} (C_{i\sigma}^{\dagger} f_{i\sigma} + f_{i\sigma}^{\dagger} C_{i\sigma}) + \epsilon_f \sum_{i\sigma} f_{i\sigma}^{\dagger} f_{i\sigma} + U \sum_{i} (n_{fi\uparrow} - 1/2)(n_{fi\downarrow} - 1/2), \tag{15.23}$$

where the terms were defined in Section 15.2. In the $d \to \infty$ limit, the local interaction gives rise to **k**-independent self-energy, and the various Green's functions are obtained in the form

$$\begin{aligned} G_c(i\omega_n, \mathbf{k})^{-1} &= i\omega_n - \epsilon_{\mathbf{k}} - \frac{V^2}{i\omega_n - \epsilon_f - \Sigma_f(i\omega_n)}, \\ G_f(i\omega_n, \mathbf{k})^{-1} &= i\omega_n - \epsilon_f - \Sigma_f(i\omega_n) - \frac{V^2}{i\omega_n - \epsilon_{\mathbf{k}}}, \\ G_{cf}(i\omega_n, \mathbf{k})^{-1} &= \frac{1}{V}\{[(i\omega_n - \epsilon_{\mathbf{k}})(i\omega_n - \epsilon_f - \Sigma_f(i\omega_n)] - V^2\}, \end{aligned} \tag{15.24}$$

where $\Sigma_f(i\omega_n)$ is the self-energy of the f electrons, and μ, the chemical potential, is absorbed in the definitions of $\epsilon_{\mathbf{k}}$ and ϵ_f. It can be shown by reducing to a self-consistent single-site model that the effective action is

$$S_{eff} = -\int_0^\beta d\tau \int_0^\beta d\tau' \sum_\sigma f_\sigma^\dagger(\tau) g_0^{-1}(\tau - \tau') f_\sigma(\tau') + U\int_0^\beta d\tau [n_{f\uparrow}(\tau) - 1/2][n_{f\downarrow}(\tau) - 1/2]. \tag{15.25}$$

The f self-energy is obtained from

$$\Sigma_f = g_0 - G_f^{-1}, G_f \equiv -<Tff^\dagger>_{S_{eff}}. \tag{15.26}$$

Because the self-consistency condition requires that the Green's function of the impurity problem must be equal to the local f Green's function of the lattice model, we obtain

$$G_f(i\omega_n) = \int_{-\infty}^{\infty} \frac{d\epsilon\, D(\epsilon)}{i\omega_n - \epsilon_f - \Sigma_f(i\omega_n) - V^2/(i\omega_n - \epsilon)}. \tag{15.27}$$

Here, $D(\epsilon)$ is the density of states (noninteracting) of the conduction electrons.

The temperature dependence of the electronic transport of the heavy-fermion systems can be calculated by using a self-consistent second-order perturbation theory in terms of the Coulomb repulsions U.

15.3.3 RKKY Interaction

There are two competing interactions in the heavy-fermion system: the indirect exchange between the moments mediated by the RKKY interaction (Refs. 12, 21, 36) and the Kondo exchange between the conduction electrons and the moments. The conducting electrons and the moments retain their identities and interact weakly. The RKKY interaction is described in the following section.

Ruderman and Kittel[21] considered the problem of nuclear-spin ordering in a metal and used second-order perturbation theory to derive an expression for the indirect nuclear spin–spin interaction (Problem 15.5),

$$H_{\text{RKKY}} = -\frac{9\pi}{8} n_c^2 \frac{J^2}{\varepsilon_F} \sum_{<ij>} \frac{\mathbf{S}_i \cdot \mathbf{S}_j}{r_{ij}^3} \left[2k_F \cos(2k_F r_{ij}) - \frac{\sin(2k_F r_{ij})}{r_{ij}} \right], \tag{15.28}$$

where k_F is the Fermi wave vector, and n_c is the density of conduction electrons. The spin–spin interaction is long ranged and changes its sign depending on the distance between the pair of spins. Kasuya discussed the magnetic properties of rare-earth metals based on Eq. (15.28).[12] Yosida[36] showed that the oscillatory behavior originates from the Friedel oscillation of the spin polarization of conduction electrons induced by a localized spin. Therefore, Eq. (15.28) is known as the RKKY interaction.

For rare-earth metals, the Fourier transform of the RKKY interaction is given by $\chi(\mathbf{q})$, the susceptibility of the conduction electrons for wave number $\mathbf{q}$. The ground state is usually ferromagnetic if $\chi(\mathbf{q})$ is maximum at $\mathbf{q}=0$. If the maximum of $\chi(\mathbf{q})$ occurs at $\mathbf{q}=\mathbf{Q}$, the antiferromagnetic wave vector, the ground state becomes antiferromagnetic. The ground state may have a spiral spin ordering if $\chi(\mathbf{q})$ becomes maximum at a general wave vector. The Kondo effect is suppressed whenever there is any type of magnetic ordering. The low-energy physics of the Kondo effect is given by the Kondo temperature

$$T_K = \varepsilon_F e^{-1/J\rho(\varepsilon_F)}. \tag{15.29}$$

However, the characteristic energy of the RKKY interaction is given by J^2/ε_F. This energy dominates over the Kondo temperature in the weak-coupling regime.

In the strong-coupling regime, the local moments are quenched because of the formation of local singlets. The Kondo effect or the effect of singlet formation is not considered for the derivation of the RKKY interaction. The relation between RKKY interaction and the Kondo effect depends on the conduction electron density, dimensionality, and the exchange coupling. As an example, we consider two localized spins, $\mathbf{S}_1$ and $\mathbf{S}_2$. The direct exchange coupling between the two spins can be expressed as

$$H = J_{\text{RKKY}}\mathbf{S}_1 \cdot \mathbf{S}_2, \tag{15.30}$$

where J_{RKKY}, the intersite coupling constant, is arbitrary. For $J>0$, the Kondo coupling is antiferromagnetic, and the ground state is a singlet. When $J>J_{\text{RKKY}}$, each of the two localized spins forms a singlet with conduction electrons, and hence, the interaction between the singlets is weak. When $J_{\text{RKKY}} \gg J$, the two localized spins form a singlet by themselves, and J is no longer important. There is a difference among theorists as to whether the change between the two regimes is smooth or sharp.

15.3.4 Extended Dynamical Mean-Field Theory[16]

The extended dynamical mean-field theory (EDMFT), which is an extension of DMFT, is particularly suitable to solve problems such as the competition between the exchange interaction and kinetic energy. In the EDMFT, the local quantum fluctuations are treated on the same level as the intersite quantum fluctuations. This is achieved by reducing the correlated lattice problem to a novel effective impurity problem corresponding to an Anderson impurity model with additional self-consistent bosonic baths. These bosonic baths reflect the influence of the rest of the lattice on the impurity site. As an example, they represent the fluctuating magnetic fields induced by the intersite spin-exchange interactions in the magnetic case. The intersite quantum fluctuations are included through self-consistency.

Smith and Si[28] applied the EDMFT method to the two-band Kondo-lattice model

$$H = \sum_{<ij>,\sigma} t_{ij} C^{\dagger}_{i\sigma} C_{j\sigma} + \sum_i J_K \vec{S_i} \cdot \vec{s_{c_i}} - \sum_{<ij>} J_{ij} \vec{S_i} \cdot \vec{S_j}, \tag{15.31}$$

where $\vec{S_i}$ is the impurity spin at site i, and $\vec{s_{c_i}}$ is the spin of conduction ($c-$) electrons at site i, t_{ij} is the hopping integral, and J_{ij} is the spin-exchange interaction. In the large D (dimension) limit, with $t_0 = t_{ij}\sqrt{D}$ and $J_0 = J_{ij}\sqrt{D}$, they derived an expression for the impurity action

$$S^{MF} = S_{top} + \int_0^\beta d\tau J_K \vec{S_i} \cdot \vec{s_c} - \int_0^\beta d\tau \int_0^\beta d\tau' \left[\sum_\sigma C_\sigma^\dagger(\tau) G_0^{-1}(\tau - \tau') C_\sigma(\tau') + \vec{S}(\tau) \cdot \chi_{s,0}^{-1}(\tau - \tau') \vec{S}(\tau') \right], \tag{15.32}$$

where S_{top} is the Berry phase of the impurity spin. The Weiss fields G_0^{-1} and $\chi_{s,0}^{-1}$ are determined by the self-consistency equations,

$$G_0^{-1}(i\omega_n) = i\omega_n + \mu - \sum_{ij} t_{i0} t_{0j} [G_{ij}(i\omega_n) - G_{i0}(i\omega_n) G_{0j}(i\omega_n)/G_{loc}(i\omega_n)] \tag{15.33}$$

and

$$\chi_{s,0}^{-1} = \sum_{ij} J_{i0} J_{0j} (\chi_{s,ij} - \chi_{s,i0} \chi_{s,0j}/\chi_{s,loc}). \tag{15.34}$$

Here, χ_s is the spin susceptibility. Smith and Si (Ref. 28) also showed that the effective action can be written in terms of the impurity problem,

$$\begin{aligned} H_{imp} = {} & \sum_{k\sigma} E_k \eta_{k\sigma}^\dagger \eta_{k\sigma} + \sum_q w_q \vec{\phi}_q^\dagger \cdot \vec{\phi}_q - \mu \sum_\sigma C_\sigma^\dagger C_\sigma \\ & + t \sum_{k\sigma} (C_\sigma^\dagger \eta_{k\sigma} + H.C.) + J_K \vec{S} \cdot \vec{s_c} + g \sum_q \vec{S} \cdot (\vec{\phi}_q + \vec{\phi}_{-q}^\dagger), \end{aligned} \tag{15.35}$$

where E_k, t, w_q, and g are determined from the Weiss fields G_0^{-1} and $\chi_{s,0}^{-1}$ specified by

$$i\omega_n + \mu - t^2 \sum_k 1/(iw_n - E_k) = G_0^{-1}(i\omega_n) \tag{15.36}$$

and

$$g^2 \sum_q w_q / [(i\nu_n)^2 - w_q^2] = \chi_{s,0}^{-1}(i\nu_n). \tag{15.37}$$

15.4 FERMI-LIQUID MODELS

15.4.1 Heavy Fermi Liquids

A number of universal features are associated with the coherent Fermi-liquid state in heavy-fermion systems. They can be summarized as follows:

a. The dimensionless Wilson ratio (Ref. 35)

$$R = \frac{\chi(0)/g_J^2 J(J+1)\mu_B^2}{\gamma(0)/\pi^2 k_B^2}$$

is close to the value of unity.

b. The specific heat C_V has a rapid downturn with increasing temperature, which has been fit to a function of the form $T^3 \ln T$.

c. The resistivity ρ is proportional to T^2.

d. The low-T susceptibility χ_T also appears to vary as T^2.

In addition, the evidence of universal behavior of heavy fermions is the observation of Kadowaki and Woods that $\rho/T^2 \equiv A$ is the same multiple of γ^2 for essentially all materials.[11]

A large number of the heavy-fermion systems become heavy Fermi liquids at low temperatures in the sense that Landau's Fermi-liquid theory is still adequate to describe the physics, provided the large effective mass is included. In the Fermi-liquid theory, the specific-heat enhancement C_v/C_{v0} is related, at low temperatures, to the quasiparticle density of states at the Fermi surface. This is equivalent to an average Fermi velocity or to that of an average mass. In view of the above, the effective mass is defined as

$$\frac{C_v}{C_{v0}} = \frac{m^*}{m}. \tag{15.38}$$

Here, it is important to comment on the physical interpretation of the quasiparticles. The f electrons are supposed to be hopping from site to site. There are a large number of f electrons, and the Luttinger theorem (described later) requires the Fermi surface to contain the total number of states and not a volume containing the mobile holes in the f-band. The f electrons have very large mass due to the weak effective hybridization. Thus, the quasiparticles are essentially f electrons, and the quasiparticle bands are f-bands that have moved up to the Fermi energy and have been narrowed by correlation. The heavy Fermi liquid arises due to the Kondo screening of the localized moments at each lattice site. In a sense, the localized moments "dissolve" into the Fermi sea.

We will now summarize the concepts of the various heavy Fermi-liquid models by following the elegant but brief review of Senthil et al.[25] The Kondo-lattice model can be written as

$$H_K = \sum_{\mathbf{k}} \varepsilon_{\mathbf{k}} C^{\dagger}_{\mathbf{k}\alpha} C_{\mathbf{k}\alpha} + \frac{J_K}{2} \sum_{r} \mathbf{S}_r \cdot C^{\dagger}_{r\alpha} \sigma_{\alpha\alpha'} C_{r\alpha'}. \tag{15.39}$$

Here, n_c is the density of conduction electrons with dispersion $\varepsilon_{\mathbf{k}}$, $C^{\dagger}_{\mathbf{k}\alpha}$ and $C_{\mathbf{k}\alpha}$ are the creation and annihilation operators of conduction states, $\mathbf{k}$ is the momentum, and $\alpha = \uparrow, \downarrow$ is a spin index. The conduction electrons interact with f electron spins $\mathbf{S}_r$ via the antiferromagnetic Kondo exchange coupling constant J_K. Here, r is a lattice position, and $\boldsymbol{\sigma}$ are the Pauli spin matrices.

In the heavy-fermion liquid models, the charge of the $f_{r\alpha}$ electrons is fully localized on the rare-earth sites. These electrons occupy a flat dispersionless band, as shown in Figure 15.11a. Because this band is half filled, it is placed at the Fermi level. The $C_{r\alpha}$ electrons occupy their own conduction band. The Kondo exchange turns on a small hybridization between these two bands. The hybridization can be represented by a bosonic operator

$$b_r \sim \sum_{\alpha} C^{\dagger}_{r\alpha} f_{r\alpha}. \tag{15.40}$$

Because $\langle b_r \rangle$ is nonzero, renormalized bands are formed (Figure 15.11b) due to the mixture of the two bands.

Because the f-band was initially dispersionless, the renormalized bands do not overlap. One now applies the Fermi-surface sum rule by Luttinger (Ref. 13), also known as the Luttinger theorem (Ref. 14). According to Luttinger, the volume enclosed by the Fermi surface is entirely determined by only the electron density. The volume is independent of the type and strength of an interaction, if the system remains a Fermi liquid and no phase transition occurs. Using this theorem leads to the conclusion that the occupied states are entirely within the lower band, and a single Fermi surface is obtained within wave vector k_F. The volume within k_F is obtained by the total density of f and c electrons.

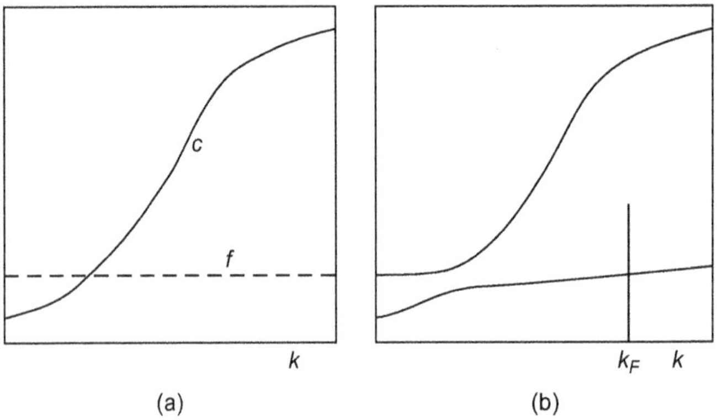

FIGURE 15.11

A completely flat f electron "band"—the dashed line in (a)—mixes with the conduction electrons to obtain the renormalized bands in (b). The single Fermi surface at k_F in the Fermi-liquid state contains states of which the wave number equals the sum of the c and f electrons.

Reproduced from Senthil, Sachdev, and Vojta[25] with the permission of Elsevier.

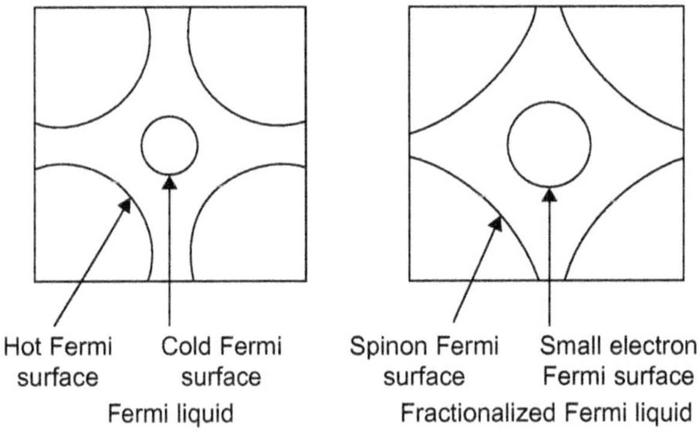

FIGURE 15.12

Fermi-surface evolution from FL to FL^*: The FL phase has two Fermi-surface sheets (the cold c and the hot f sheets) close to the transition. The f sheet becomes the spinon Fermi surface while the c sheet is the small conduction-electron Fermi surface on the FL^* side.

Reproduced from Senthil, Vojta, and Sachdev[27] with the permission of the American Physical Society.

The Fermi surface is in a region (Figure 15.11b) where the electrons primarily have an f character, and the band is flat. According to this model, this accounts for the large effective mass of the fermionic quasiparticles.

Because the charge fluctuations are quenched at the f electron sites, every rare-earth site has a constraint

$$\sum_{\alpha} f^{\dagger}_{r\alpha} f_{r\alpha} = 1, \tag{15.41}$$

which is obeyed at each rare-earth site. This implies that the theory is invariant under the space-time-dependent $U(1)$ gauge transformation

$$f_{ra}^{\dagger} \rightarrow f_{ra} e^{i\phi_r(\tau)}, \tag{15.42}$$

where τ is imaginary time.

15.4.2 Fractionalized Fermi Liquids

Senthil et al.[26,27] showed the existence of nonmagnetic translation-invariant small-Fermi-surface states, originally with a focus on two-dimensional Kondo lattices. These states are obtained when a local-moment system settles into a fractionalized spin liquid (FL^*) due to intermoment interactions. A weak Kondo coupling to conduction electrons leaves a sharp (but small) Fermi surface of quasiparticles (FL) of which the volume counts the conduction density, but the structure of the spin liquid is undisturbed. These states have fractionalized excitations that coexist with conventional Fermi-liquid-like quasiparticles.

In this paper, Senthil et al. considered a three-dimensional lattice by using $U(1)$ states. They focused on a three-dimensional $U(1)$ spin-liquid state with fermionic spinons that form a Fermi surface. The $U(1)$ spin-liquid state is stable to a weak Kondo coupling to conduction electrons. The $U(1)FL^*$ state consists of a spinon Fermi surface coexisting with a separate Fermi surface of conduction electrons. Senthil et al. used a mean-field theory to describe a $U(1)FL^*$ state and its transition to a heavy FL. They considered a three-dimensional Kondo–Heisenberg model on a cubic lattice,

$$H = \sum_{k} \epsilon_{\mathbf{k}} C_{\mathbf{k}\alpha}^{\dagger} C_{\mathbf{k}\alpha} + \frac{J_K}{2} \sum_{r} \vec{S}_r \cdot C_{r\alpha}^{\dagger} \vec{\sigma}_{\alpha\alpha'} C_{r\alpha'} + J_H \sum_{<rr'>} \vec{S}_r \cdot \vec{S}_{r'}. \tag{15.43}$$

Here, $C_{\mathbf{k}\alpha}$ is the conduction electron destruction operator, $\vec{S}_r$ are the spin-1/2 local moments, and summation over repeated spin indices α is implied. In a fermionic "slave-particle" representation of the local moments

$$\vec{S}_r = 1/2 f_{r\alpha}^{\dagger} \vec{\sigma}_{\alpha\alpha'} f_{r\alpha'}, \tag{15.44}$$

where $f_{r\alpha}$ is a spinful fermion destruction operator at site r. The decoupling of the Kondo and the Heisenberg exchange is made using two auxiliary fields by a saddle-point approximation, and the mean-field Hamiltonian is

$$H_{mf} = \sum_{\mathbf{k}} \epsilon_{\mathbf{k}} C_{\mathbf{k}\alpha}^{\dagger} C_{\mathbf{k}\alpha} - \chi_0 \sum_{<rr'>} (f_{r\alpha}^{\dagger} f_{r'\alpha} + H.C.) + \mu_f \sum_{r} f_{r\alpha}^{\dagger} f_{r\alpha} - b_0 \sum_{\mathbf{k}} (C_{\mathbf{k}\alpha}^{\dagger} f_{\mathbf{k}\alpha} + H.C.). \tag{15.45}$$

Here, b_0 and χ_0 are assumed to be real, and additional constants to H are dropped. The mean-field parameters b_0, χ_0, and μ_f are obtained from (Problem 15.3)

$$1 = \langle f_{r\alpha}^{\dagger} f_{r\alpha} \rangle, \tag{15.46}$$

$$b_0 = J_K/2 \langle C_{r\alpha}^{\dagger} f_{r\alpha} \rangle, \tag{15.47}$$

$$\chi_0 = J_H/2 \langle f_{r\alpha}^{\dagger} f_{r'\alpha} \rangle, \tag{15.48}$$

where r and r' are nearest neighbors. At zero temperature, in the Fermi-liquid (FL) phase, χ_0, b_0, and μ_0 are nonzero. In the FL^* phase, $b_0 = \mu_0 = 0$ but $\chi_0 \neq 0$. In this state, the conduction electrons are decoupled from the local moments and form a small Fermi surface. The local-moment system is described as a spin fluid with a Fermi surface of neutral spinons.

The mean-field is diagonalized by the transformation (Problem 15.4),

$$C_{\mathbf{k}\alpha} = u_{\mathbf{k}}\gamma_{\mathbf{k}\alpha+} + v_{\mathbf{k}}\gamma_{\mathbf{k}\alpha-} \tag{15.49}$$

and

$$f_{\mathbf{k}\alpha} = v_{\mathbf{k}}\gamma_{\mathbf{k}\alpha+} - u_{\mathbf{k}}\gamma_{\mathbf{k}\alpha-}. \tag{15.50}$$

The Hamiltonian can be written in terms of the new fermionic operators $\gamma_{\mathbf{k}\alpha\pm}$,

$$H_{mf} = \sum_{\mathbf{k}\alpha} E_{\mathbf{k}+}\gamma^{\dagger}_{\mathbf{k}\alpha+}\gamma_{\mathbf{k}\alpha+} + E_{\mathbf{k}-}\gamma^{\dagger}_{\mathbf{k}\alpha-}\gamma_{\mathbf{k}\alpha-}, \tag{15.51}$$

where

$$E_{\mathbf{k}\pm} = \frac{\epsilon_{\mathbf{k}} + \epsilon_{\mathbf{k}f}}{2} \pm \sqrt{\left(\frac{\epsilon_{\mathbf{k}} - \epsilon_{\mathbf{k}f}}{2}\right)^2 + b_0^2}. \tag{15.52}$$

Here, $\epsilon_{\mathbf{k}f} = \mu_f - \chi_0 \sum_{a=1,2,3} \cos(k_a)$. The $u_{\mathbf{k}}, v_{\mathbf{k}}$ are determined by

$$u_{\mathbf{k}} = -\frac{b_0 v_{\mathbf{k}}}{E_{\mathbf{k}+} - \epsilon_{\mathbf{k}}}, \quad u_{\mathbf{k}}^2 + v_{\mathbf{k}}^2 = 1. \tag{15.53}$$

For the FL* phase, $b_0 = \mu_0 = 0$ but $\chi_0 \neq 0$. The conduction-electron dispersion ϵ_k determines the electron Fermi surface and is small. The spinon Fermi surface encloses one spinon per site and has volume half that of the Brillouin zone. Senthil *et al.* assumed that the conduction–electron filling is less than half, and the electron Fermi surface does not intersect the spinon Fermi surface. In the FL phase near the transition (small b_0), there are two bands corresponding to $E_{\mathbf{k}\pm}$: one derives from the c electrons with f character (c-band), whereas the other derives from the f particles with weak c character (f-band).

As shown in Figure 15.12, for small b_0, the Fermi surface consists of two sheets because both bands intersect the Fermi energy. The total volume is large because it includes both local moments and conduction electrons. When b_0 decreases to zero, the transition moves to FL^*, the c-Fermi surface expands in size to match onto the small Fermi surface of FL^*, and the f-Fermi surface shrinks to match onto the spinon Fermi surface of FL^*.

15.5 METAMAGNETISM IN HEAVY FERMIONS

The name *metamagnetism* was originally introduced for antiferromagnetic (AF) materials where, at low temperatures, for a critical value of the magnetic field (H), the spin flips, which gives rise to a first-order phase transition. This was extended to paramagnetic (Pa) systems where field reentrant ferromagnetism (F) would appear in itinerant magnetism. Eventually, it was used to describe a crossover inside a persistent paramagnetic state between low-field Pa phase and an enhanced paramagnetic polarized (PP) phase.

In heavy-fermion systems, the f electrons are located near the border between itinerant and local moment behavior, as shown in the phase diagram in Figure 15.13. Doniach considered a one-dimensional analog of a system of conduction electrons exchange-coupled to a localized spin in each cell of a lattice.[6] He suggested that a second-order transition from an antiferromagnetic to a Kondo spin-compensated ground state would occur as the exchange coupling constant J increased to a critical value J_c. For J near to, and slightly smaller than J_c, there would exist antiferromagnets with very weak, "nearly quenched" moments, even though the f electrons are in a state with a well-defined local nonzero spin state. The existence of this transition can be understood by comparing the binding energy of the Kondo singlet

$$W_K \sim N(0)^{-1} e^{-1/N(0)J}, \tag{15.54}$$

with that of an RKKY antiferromagnetic state

$$W_{AF} \sim CJ^2 N(0), \tag{15.55}$$

where $N(0)$ is the density of conduction electron state, and C is a dimensionless constant that depends on the band structure. As shown in Figure 15.13, for $JN(0)$ less than a critical value, the RKKY state dominates, whereas above this, the Kondo singlet binding dominates. The RKKY binding again takes over at large J, but the weak coupling formula (Eq. 15.54) breaks down in this regime.

The heavy fermions that exhibit metamagnetism are $CeRu_2Si_2$, $Sr_3Ru_2O_7$, $CeCu_{6-x}Au_x$, UPt_3, UPd_2Al_3, URu_2Si_2, $CePd_2Si_2$, $YbRh_2Si_2$, and $CeIr_3Si_2$.

There have been many theories proposed for metamagnetism of heavy fermions, but no satisfactory model is yet available. A review of the various theoretical models was made by Misra.[16]

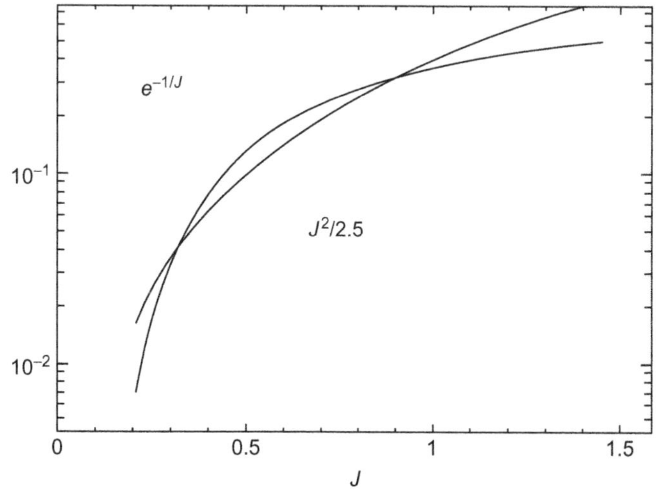

FIGURE 15.13

Comparison of *AF* with Kondo energies.

Reproduced from Doniach[6] with the permission of Elsevier.

15.6 CE- AND U-BASED SUPERCONDUCTING COMPOUNDS[16]

15.6.1 Ce-Based Compounds

Since 1979, approximately 25 unconventional superconductors have been discovered in heavy-fermion systems. Although most of these systems are Ce- and U-based compounds, a few others are quasi-two-dimensional in nature and filled skutterdites. The multiphase diagrams in UPt_3 and $U(Be_{1-x}Th_x)_{13}$ indicate unusual superconductivity with multicomponents. In fact, UPt_3 is the first odd-parity superconductor to be discovered in heavy-fermion systems. UPd_2Al_3 and UNi_2Al_3 are unconventional superconductors coexisting with the AF phase and are considered to have even- and odd-parity pairing states, respectively. There is coexistence of hidden-order and unconventional superconductivity in URu_2Si_2. UPt_3, URu_2Si_2, UNi_2Al_3, and UPd_2Al_3 have the following common features: (a) they order antiferromagnetically below T_N, ranging from 5 to 17° K; and (b) they exhibit, well below T_N and coexisting with AF order, a heavy Landau Fermi-Liquid (LFL) state that becomes unstable against a superconducting transition at T_c (ranging between 0.5 and 2° K).

Recently, a variety of heavy-fermion Ce-based superconductors were discovered due to progress in experiments under pressure. They include $CeCu_2Ge_2$, $CePd_2Si_2$, $CeRh_2Si_2$, $CeNi_2Ge_2$, and $CeIn_3$. These materials, which have the same $ThCr_2Si_2$-type crystal structure as $CeCu_2Si_2$ (except $CeIn_3$), are AF metals at ambient pressure, whereas under high pressures, the AF phases abruptly disappear accompanied by SC transitions.

The family of $CeTIn_5$ (T = Co, Rh, and Ir), which has a $HoCoGa_5$-type crystal structure (Ref. 16), has attracted a great deal of attention because they possess a relatively high transition temperature (T_c) such as $T_c = 2.3°$ K for $CeCoIn_5$, which is the highest among Ce- and U-based heavy-fermion superconductors. It has been proposed that valence fluctuations are responsible for the superconductivity in some Ce-based compounds. This is due to the fact that in metallic cerium, the phase diagram shows a first-order valence discontinuity line. This line separates the γ-Ce with a $4f$ shell occupation $n_f = 1.0$ from the α-Ce with $n_f \approx 0.9$. The valence transition is isostructural, and the line has a critical end point in the vicinity of $p_{cr} = 2$ GPa and $T_{cr} = 600°$ K. In cases in which p_{cr} is positive, either T_{cr} is very high or T_{cr} is negative, and only a crossover regime is accessible even at $T = 0$. The exceptions are $CeCu_2Si_2$ and $CeCu_2Ge_2$, for which T_{cr} is likely positive although small. In such a situation, the associated low-energy valence fluctuations can mediate superconductivity.

Holmes et al.[10] proposed that the superconducting phase diagram for $CeCu_2(Ge,Si)_2$, shown in Figure 15.14, exhibits a maximum in the transition temperature in close vicinity to a valence-changing critical point. Miyake (Ref. 16) has argued that superconductivity may develop around the region where the critical end point is suppressed to zero, to become a quantum critical point.

The heavy-fermion superconductor $CeCoIn_5$ is a quasi-two-dimensional (2D) system, and the de Haas–van Alphen effect data indicate a quasi-2D Fermi surface. These properties have led to the possibility of a Fulde–Ferrell–Larkin–Ovchinnikov (FFLO) (Refs. 7, 13) superconducting state in $CeCoIn_5$. These states result from the competition between superconducting condensate energy and the magnetic Zeeman energy that lowers the total energy of the electrons in the normal state. This competition is strong when the superconductivity is of a spin-singlet nature. In this case, the superconducting Cooper pairs form with opposite spins, and the electrons cannot lower the total energy of the system by preferentially aligning their spins along the magnetic field. This effect, called Pauli limiting, leads to suppression of superconductivity in the magnetic field. The characteristic

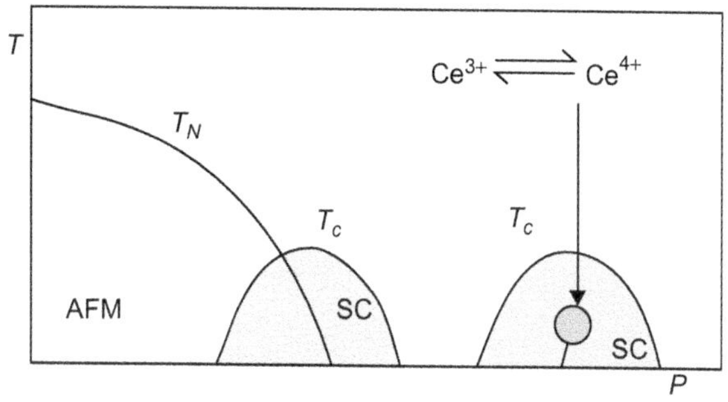

FIGURE 15.14

Schematic phase diagram for $CeCu_2(Ge,Si)_2$ illustrating a possible valence fluctuation critical point beneath the superconducting dome at high pressures.

Reproduced from P. Coleman, (Ref. 5), with the permission of Elsevier.

Pauli field H_P determines the upper limit of the superconducting upper critical field H_{c2}. When Pauli limiting is the dominant mechanism for suppression of superconductivity, a new inhomogeneous superconducting FFLO state would appear at high fields between the normal and the mixed, or the vortex, state below the critical temperature T_{FFLO} with planes of normal electrons that can take advantage of Pauli susceptibility. In the FFLO state, pair breaking due to the Pauli paramagnetic effect is reduced by the formation of a new pairing state $(\mathbf{k}_\uparrow, -\mathbf{k}+\mathbf{q}_\downarrow)$, with $|\mathbf{q}| \sim 2\mu_B H/\hbar v_F$ (v_F is the Fermi velocity) between the Zeeman split parts of the Fermi surface. One of the intriguing features is the T and H phase dependence of the phase boundary between the FFLO and non-FFLO superconducting state. $H^{\|}{}_{FFLO}$ ($H^{\|}ab$) exhibits an unusually large shift to higher fields at higher temperatures. The results of Bianchi et al.[2] are shown in Figure 15.15.

15.6.2 U-Based Superconducting Compounds

The first two U-based heavy-fermion superconductors, UBe_{13} ($T_c = 0.9°$ K) and UPt_3 ($T_c = 0.54°$ K), were discovered in 1983 by Ott et al. (Ref. 19) and in 1984 by Stewart et al.,[31] respectively. It was evident within a few years that UPt_3 had three superconducting phases, which created great impetus for further study of this unusual heavy-fermion superconductor.

UBe_{13} was the first actinide-based heavy-fermion compound that was found to be a bulk superconductor below approximately 0.9° K. The cubic UBe_{13} is also one of the most fascinating HF superconductors because superconductivity develops out of a highly unusual normal state characterized by a large and strongly T-dependent resistivity. In addition, upon substituting a small amount of Th for U in $U_{1-x}Th_xBe_{13}$, a nonmonotonic evolution of T_c and a second-phase transition of T_{c2} below T_{c1}, the superconducting one, is observed in a critical concentration range of x.

It was also shown that the superconducting state is formed by heavy-mass quasiparticles. This was demonstrated by plotting C_p/T versus T (at low temperatures), which is shown in Figure 15.16. The anomaly at T_c is compatible with the large γ parameter in the normal state at this temperature.

The temperature dependence of the specific heat of UBe_{13} well below T_c was the first indication of the unconventional superconductivity. Figure 15.17 shows the nonexponential but power-law-type

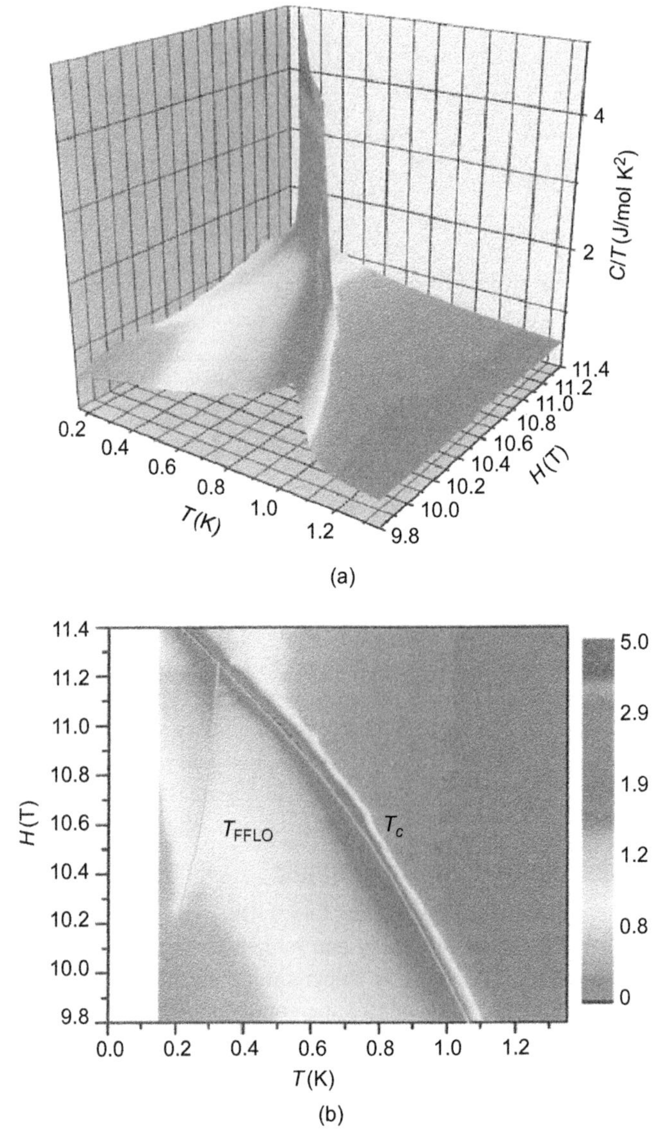

FIGURE 15.15

(a) Electronic specific heat of $CeCoIn_5$ divided by temperature with H‖ [110] collected with the temperature decay method, as a function of field and temperature. (b) Contour plot of the data in (a) in the *H*-*T* plane. Gray lines indicate the superconducting phase transition T_c and the FFLO-mixed state T_{FFLO} anomaly. The color scale is the same in (a) and (b).

Reproduced from Bianchi et al.[2] with the permission of the American Physical Society.

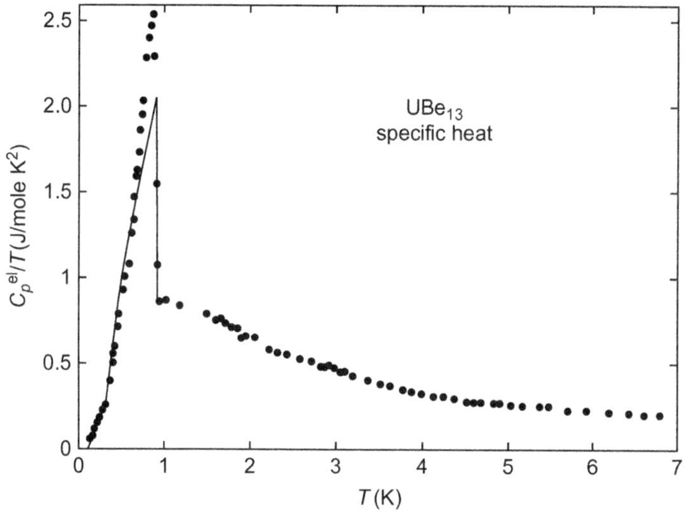

FIGURE 15.16

Electronic specific heat of UBe_{13} below 7° K. The solid line represents the BCS approximation of the anomaly at and below T_c.

Reproduced from Ott[18] with the permission of Elsevier.

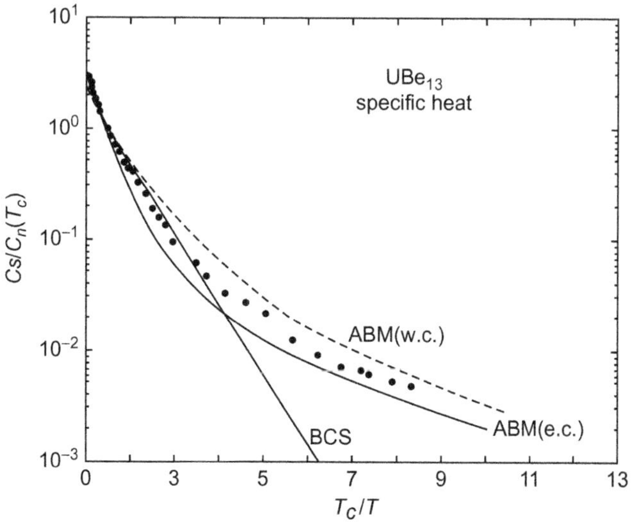

FIGURE 15.17

Normalized electronic specific heat of UBe_{13} below T_c, plotted versus T_c/T. The solid and broken lines represent calculations assuming point nodes in the gap.

Reproduced from Ott[18] with the permission of Elsevier.

decrease of $C_p(T)$ that was interpreted as being the consequence of nodes in the gap of the electronic excitation spectrum.

It was also found that when small amounts of U atoms in UBe_{13} were replaced with other elements, there was a substantial reduction of the critical temperature. T_c is also first substantially reduced with the alloys $U_{1-x}Th_xBe_{13}$ as x is increased. However, when $x > 0.018$, T_c increases again until it passes over a willow maximum at $x = 0.033$ and gradually decreases with a reduced slope when x is further increased. Further, in the range $0.019 < x < 0.05$, a second transition at T_{c2} below T_c was discovered by measuring the specific heat of these alloys at very low temperatures. Measurements of $\rho(T)$ and $\chi(T)$ confirmed that the phase at temperatures below the second anomaly of $C_p(T)$ was superconducting.

The phase diagram of superconductivity of $U_{1-x}Th_xBe_{13}$, from these observations as well as from thermodynamic arguments, is shown in Figure 15.18. One can identify three different superconducting phases: F, L, and U.

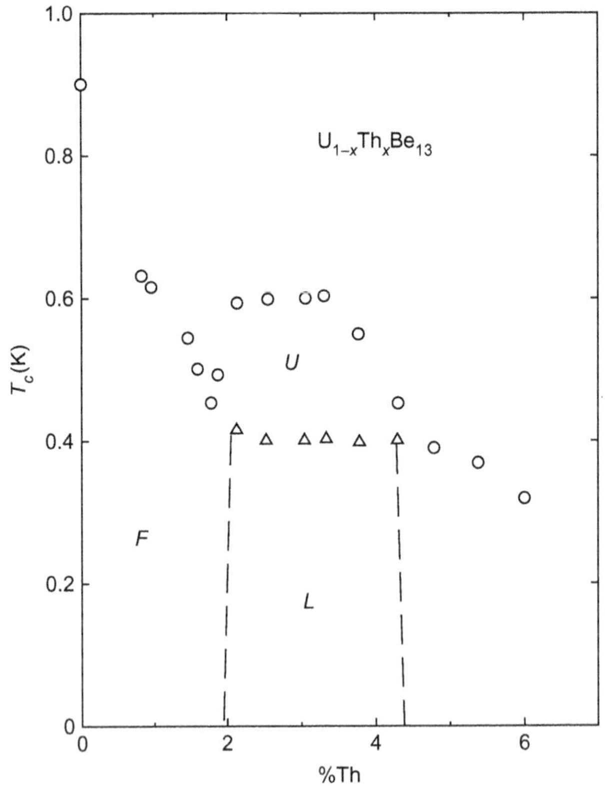

FIGURE 15.18

An x, T phase diagram for superconducting $U_{1-x}Th_xBe_{13}$ as derived from the measurements of the specific heat. The letters F, L, and U denote three superconducting phases.

Reproduced from Ott[18] with the permission of Elsevier.

Superconductivity has also been discovered in UPt_3, and its alloys, URu_2Si_2; and in UPd_2Al_3 and its alloys, UNi_2Al_3, UGe_2, URhGe, and UIr.

The discovery of superconductivity in UGe_2 in single crystals of UGe_2 under pressure below $P_c \sim 16$ kbar was very surprising. The sensational part of this discovery is that the pressure $P \sim 12$ kbar, where the superconducting temperature $T_S = 0.75°$ K is strongest, the Curie temperature $T_C \sim 35°$ K is two orders of magnitude higher than T_S; superconductivity occurs in a very highly polarized state ($\mu(T \to 0°\text{ K}) \sim \mu_B$).

The superconductivity in UGe_2 disappears above a pressure $P_c \approx 16$ kbar that coincides with the pressure at which the ferromagnetism is suppressed. The pressure-temperature phase diagram of UGe_2 is shown in Figure 15.19.

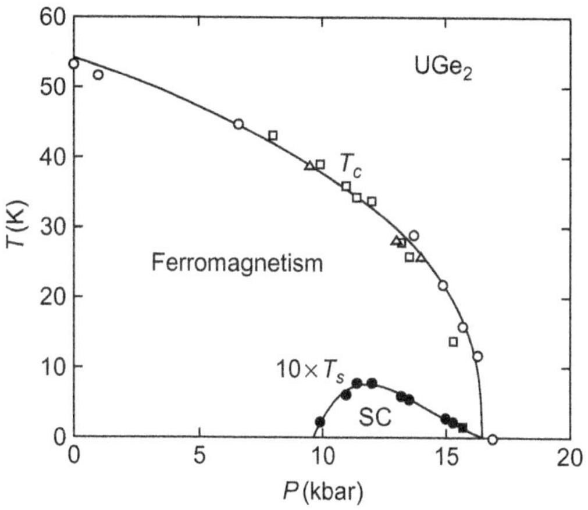

FIGURE 15.19

The pressure-temperature phase diagram of UGe_2.
Reproduced from Demuer et al.[4] with permission of Elsevier.

15.7 OTHER HEAVY-FERMION SUPERCONDUCTORS

15.7.1 $PrOs_4Sb_{12}$

The filled skutterdite $PrOs_4Sb_{12}$ becomes superconducting at $T_c = 1.85°$K. It appears to involve heavy-fermion quasiparticles with effective mass $m^* \sim 50\, m_e$. There is speculation that the quadrupolar fluctuations play a role in the heavy-fermion superconductivity of $PrOs_4Sb_{12}$. The ground state of Pr^{3+} ions in the cubic CEF appears to be the Γ_3 nonmagnetic doublet. Therefore, the heavy-fermion behavior possibly involves the interaction of the Pr^{3+} Γ_3 quadrupole moments and the charges of the conduction electrons. In such a case, the quadrupolar fluctuations would play a role in the heavy-fermion superconductivity of $PrOs_4Sb_{12}$.

The variation of C at low temperature and the magnetic phase diagram inferred from C, the resistivity and magnetization, show that there was a doublet ground state. The two distinct superconducting anomalies in C provide evidence of two superconducting critical temperatures at $T_{C1} = 1.75°$ K and $T_{C2} = 1.85°$ K. This could arise from a weak lifting from of the ground-state degeneracy, which supports the theory of quadrupolar pairing; i.e., superconductivity in $PrOs_4Sb_{12}$ is neither of electron–phonon nor of magnetically mediated origin.

The H-T superconducting phase diagram of $PrOs_4Sb_{12}$ determined by specific heat measurements is shown in Figure 15.20.

15.7.2 $PuCoGa_5$

The discovery of superconductivity in the transuranium compound $PuCoGa_5$ with $T_c \approx 18.5°$ K, which is by far the highest critical temperature for any heavy-fermion superconductor, has attracted considerable attention. $PuCoGa_5$ crystallizes in the $HoCoGa_5$ structure, the same type as the $CeMIn_5$ materials. The *H-T* phase diagram of $PuCoGa_5$, inferred from the heat capacity data as a function of temperature in a magnetic field applied along the three orthogonal directions, is shown in Figure 15.21.

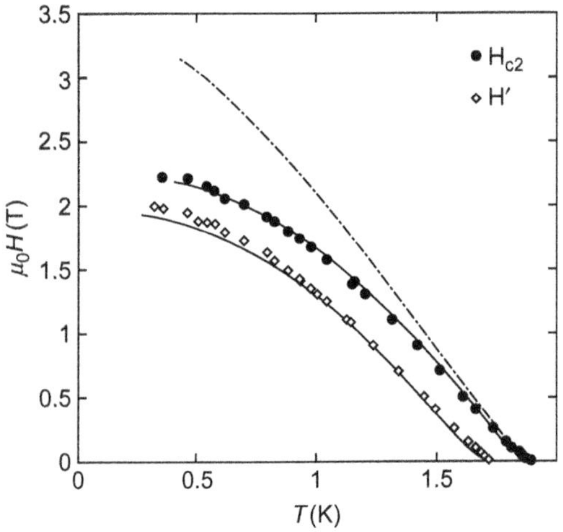

FIGURE 15.20

H-T superconducting phase diagram of $PrOs_4Sb_{12}$. The field dependences of T_{c1} and T_{c2} are identical. The dashed-dotted line is the same fit with the same parameters as the other lines but without paramagnetic limitation.

Reproduced from Measson et al.[15] *with the permission of Elsevier.*

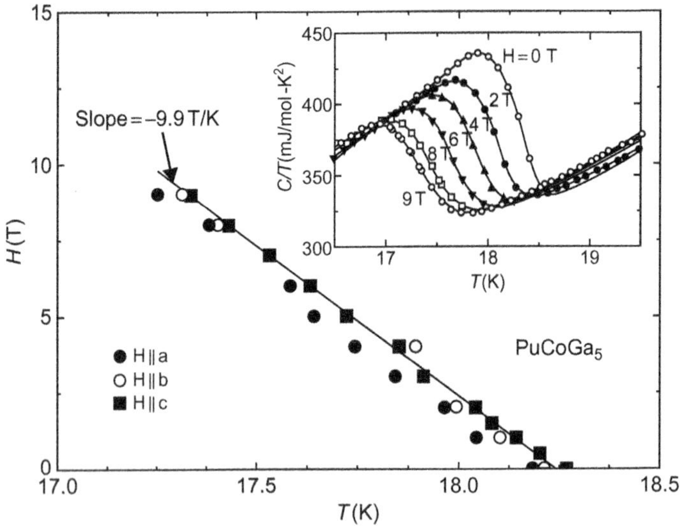

FIGURE 15.21

H-T diagram of $PuCoGa_5$ inferred from single-crystal heat capacity measurements with magnetic field applied along three orthogonal directions. The inset shows representative heat capacity data from which $T_c(H)$ was inferred.

Reproduced from J.L. Sarrao et al. (Ref. 23), with the permission of Elsevier.

15.7.3 $PuRhGa_5$

The discovery of superconductivity in $PuRhGa_5$ with $T_c \approx 9°$ K was reported by Wastin et al. (Ref. 34). $PuRhGa_5$ crystallizes in the tetragonal $HoCoGa_5$ structure with the lattice parameters $a =$ 4.2354 Å and $c = 6.7939$ Å. This structure has a two-dimensional feature, where alternating $PuGa_3$ and $RhGa_2$ layers are stacked along the c-axis. There are two crystallographically inequivalent Ga sites in this structure, which are denoted Ga(1) (the 1c site) and Ga(2) (the 4i site), respectively. The Ga(1) site is surrounded by four Pu atoms in the c plane, whereas the Ga(2) site is surrounded by two Pu and two Rh atoms in the a plane.

The high-pressure measurements on $PuRhGa_5$ are shown in Figure 15.22, in which the electrical resistance is plotted against temperature for pressures up to 18.7 GPa. This figure displays a metallic shape in the normal state, but an NFL behavior ($\rho(T) \sim T^{1.3}$) develops up to 50–60° K. The inset of Figure 15.22 shows the plot of T_c of both $PuRhGa_5$ and $PuCoGa_5$ against pressure.

The variation of T_c as a function of pressure (Figure 15.22) suggests that the pairing mechanism is differently affected by pressure for the two materials. The layered crystal structure associated with the quasi-2D Fermi surface calculated for these materials suggests that anisotropic properties might be the cause for this difference.

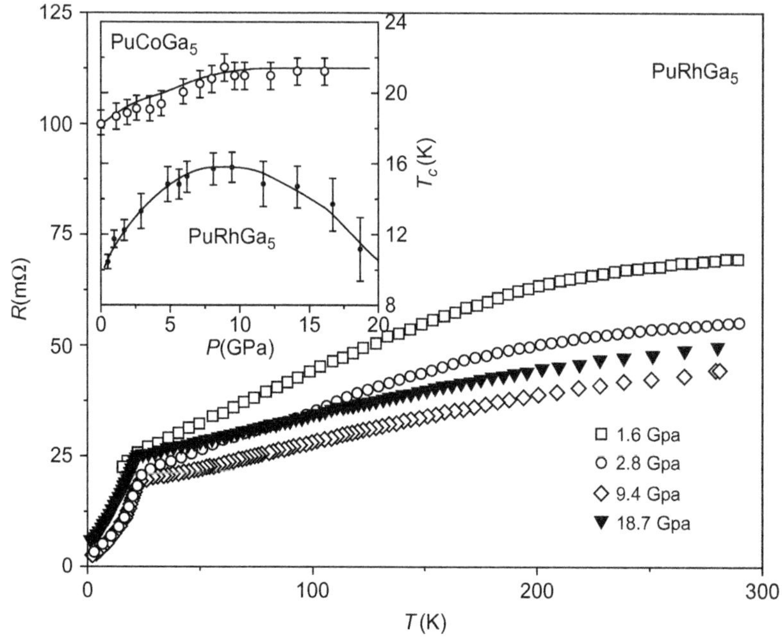

FIGURE 15.22

Evolution of the electrical resistance of $PuRhGa_5$ crystal up to 18.7 GPa. The inset shows T_c behavior of $PuCoGa_5$ as a function of the applied pressure.

Reproduced from Griveau et al.[8] with the permission of Elsevier.

The (*T*-*P*) phase diagram of $PuRhGa_5$ and $PuCoGa_5$ can be compared to that of $CeCoIn_5$. In these isostructural compounds, the superconducting transition temperature increases with increasing pressure and reaches a maximum before decreasing at higher pressure. The NFL behavior is also maintained over a large range of pressure.

15.7.4 Comparison between Cu and Pu Containing High-T_c Superconductors

Recently, Wachter[33] compared the Cu "high T_c superconductors" with equivalent measurements on "high T_c" $PuCoGa_5$ and $PuRhGa_5$. He observed the following common features. First, in all materials, spin pseudogaps were observed, which necessitates at least antiferromagnetic short-range order, i.e., in clusters. Second, all Cu and Pu superconductors are of mixed valence, as photoemission data have shown. The majority ions (Cu or Pu) are magnetic, and the minority ions are nonmagnetic and act as spin holes. Only short-range correlations remain because these spin holes have a concentration of 10% and hence dilute the antiferromagnetic order. According to Wachter, two dimensionality is not essential and *n*- or *p*-type conductivity is not important.

15.8 THEORIES OF HEAVY-FERMION SUPERCONDUCTIVITY

The superfluid ^3He, the physical properties of which were extensively studied prior to the discovery of heavy-fermion superconductivity, exhibited gap anisotropy and nodal structures like some heavy-fermion compounds. After the discovery of heavy-fermion superconductors, it was natural to compare them with superfluid ^3He to be able to understand the former. However, there are many differences between the two systems. For example, the presence of a crystal field and the fact that charged particles are paired in heavy fermions instead of pairing of the neutral atoms in ^3He are important. In addition, the strong correlation effects and the spin-orbit interaction in heavy-fermion systems are major factors to be considered.

In heavy-fermion compounds, the *f*-shell electrons are strongly correlated. These *f* electrons determine the properties of the quasiparticles at the Fermi level, which gives rise to a large effective mass. It is generally believed that superconductivity is mainly by the heavy quasiparticles. These quasiparticles with *f* characters would have difficulty forming ordinary *s*-wave Cooper pairs, characteristic of the BCS theory of superconductivity, due to the strong Coulomb repulsion. To avoid a large overlap of the wave functions of the paired particles, the system would rather choose an anisotropic channel, such as a *p*-wave spin triplet (as is done in superfluid ^3He) or a *d*-wave spin singlet state to form pairs.

We cannot review here in detail the theory of superconductivity of each heavy-fermion compound. In addition, heavy-fermion systems are one of the areas in physics where the experimentalists are well ahead of the theorists and superconductivity in various heavy-fermion compounds has a different origin.

15.9 KONDO INSULATORS

15.9.1 Brief Review

The strongly correlated *f*-electron materials called Kondo insulators have recently attracted much attention because of their unusual physical properties. At high temperatures, they behave like metals with a local magnetic moment, whereas at low temperatures, they behave as paramagnetic insulators with a small energy gap at the Fermi level. It appears that a gap in the conduction band opens at the Fermi energy as the temperature is reduced. Despite intensive theoretical and experimental studies, the mechanism of gap formation is still unclear, and there is considerable controversy on how to describe the physics of Kondo insulators. We will concentrate on the 4*f* and 5*f* compounds, i.e., those *f*-element compounds that are in a certain sense "valence" compounds. The general properties of these materials are characterized by a small gap. The *f* elements that are present in these compounds have unstable valence, with the valence corresponding to the nonmagnetic *f* state of the element satisfying the valence requirements of the other elements in the material. The Kondo insulator can be viewed as a limiting case of the correlated electron lattice: exactly one half-filled band interacting with one occupied *f*-level. This can also be viewed as the limiting case of the Kondo lattice with one conduction electron to screen one moment at each site. However, there has been no clear definition of Kondo insulators, and this situation stems from the confusion over how to understand various types of Kondo insulators consistently.

Following is a variety of Kondo insulators, some of which are semiconductors that become Kondo insulators with application of pressure: CeNiSn, $Ce_3Bi_4Pt_3$, CeRhAs, CeRhSb, CeNiSn, $CeRu_4Sn_6$, URu_2Sn, $CeFe_4P_{12}$, $CeRu_4P_{12}$, $CeOs_4Sb_{12}$, UFe_4P_{12}, TmSe, URu_2Sn, YbB_{12}, SmB_6, and SmS.

A detailed review of the experimental properties of each one of these Kondo insulators is available in Misra.[16]

15.9.2 Theory of Kondo Insulators

The Anderson Lattice Model

The Anderson lattice model provides a basic description of the electronic properties of the heavy-fermion materials. The solution of the model at half-filling is expected to exhibit an indirect gap in the density of states. The chemical potential lies directly in the gap making the system semiconducting. Thus, if there are four states per atom—two states per atom in the upper hybridized band and two states per atom in the lower hybridized band—then at half-filling, two electrons per atom completely fill the doubly degenerate lower hybridized band and the noninteracting system is semiconducting. According to Luttinger's theorem, if the interactions are turned on adiabatically so that perturbation theory converges, the ground state of the interacting system will remain insulating. The Hamiltonian can be written as

$$H = H_f + H_d + H_{fd}, \tag{15.56}$$

where H_f is the Hamiltonian of the lattice of localized f electrons, H_d is the Hamiltonian of the conduction electron states, and H_{fd} is the hybridization Hamiltonian,

$$H_f = \sum_{i,\alpha} E_f f^\dagger_{i,\alpha} f_{i,\alpha} + \sum_{i,\alpha,\beta} \frac{U_{ff}}{2} f^\dagger_{i,\alpha} f^\dagger_{i,\beta} f_{i,\beta} f_{i,\alpha}, \tag{15.57}$$

$$H_d = \sum_{\mathbf{k},\alpha} \varepsilon_d(\mathbf{k}) d^\dagger_{\mathbf{k},\alpha} d_{\mathbf{k},\alpha}, \tag{15.58}$$

and

$$H_{fd} = N_s^{-1/2} \sum_{i,\mathbf{k},\alpha} [V(\mathbf{k})\exp(-i\mathbf{k}\cdot\mathbf{R}_i) f^\dagger_{i,\alpha} d_{\mathbf{k},\alpha} + V^*(\mathbf{k})\exp(i\mathbf{k}\cdot\mathbf{R}_i) d^\dagger_{\mathbf{k},\alpha} f_{i,\alpha}]. \tag{15.59}$$

Here, E_f is the binding energy of a single f electron to a lattice site, and U_{ff} is the Coulomb repulsion between a pair of f electrons located on the same lattice site. Due to the spin and orbital degrees of freedom, the total degeneracy of each f orbital is 14. This degeneracy can be lifted by spin-orbit coupling and crystal field splitting. We will consider the degeneracy of the lowest f multiplet to be $N = 2$. The operators $f_{i\alpha}^{\dagger}(f_{i\alpha})$ create (destroy) an f electron at site i with a combined spin-orbit label α. The summation is over all lattice sites and all degeneracy labels. $\varepsilon_d(\mathbf{k})$ is the dispersion relation for the d-bands; the operators $d_{\mathbf{k}\alpha}^{\dagger}(d_{\mathbf{k}\alpha})$ create and annihilate an electron in the αth d sub-band state labeled by the Bloch wave vector $\mathbf{k}$. The hybridization between the f states and the states of the d-band is governed by H_{fd}. The first term represents a process in which a conduction electron in the Bloch state $\mathbf{k}$ hops into the f orbital located at site i. However, α is conserved in the process. The Hermitian conjugate term describes an electron in the f orbital at site i tunneling into the conduction band state labeled by the Bloch state $\mathbf{k}$. The summation runs over the total number of lattice sites N_s and over the $\mathbf{k}$ values of the first Brillouin zone.

Riseborough's Theory

Riseborough (Ref. 20) showed that the noninteracting Hamiltonian ($U_{ff} \to 0$) is exactly soluble and the electronic states fall into two quasiparticle bands of mixed f and conduction band character. He showed that in this limit, the binding energy of the f-levels falls within the width of the unhybridized conduction band, which has a width of $2\ W = 12t$ in the tight-binding approximation. The indirect gap is between the zone boundary of the upper branch and the $k = 0$ state of the upper branch. The direct gap occurs for k values halfway along the body diagonal and has a magnitude of $2V$. A sketch of the hybridized bands is shown in Figure 15.23. Each band can contain a maximum of $2N$ electrons.

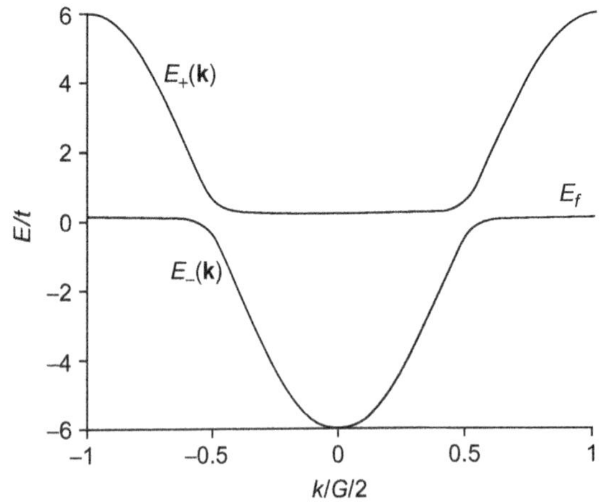

FIGURE 15.23

Sketch of the hybridized band structure, for k vectors along the body diagonal of the first Brillouin zone.

Reproduced from Riseborough[20] with the permission of Taylor & Francis Ltd.

Thus, the noninteracting system is a semiconductor. If the interactions are turned on adiabatically, Luttinger's theorem implies that the ground state of the system will be an insulator. In the mean-field approximation, one can use the slave boson technique described earlier. This approach to the $U_{ff} \to \infty$ limit of the Anderson lattice model projects the states of double f occupancy. The f electron operators are replaced by a product of an f quasiparticle operator and a slave boson field,

$$f_{i,\alpha}^{\dagger} = \tilde{f}_{i,\alpha}^{\dagger} b_i, \quad f_{i,\alpha} = b_i^{\dagger} \tilde{f}_{i,\alpha}, \tag{15.60}$$

where b_i and $b_i^{\dagger}$ are the annihilation and creation operators for the site i, and the f quasiparticle operators are $\tilde{f}_{i,\alpha}^{\dagger}$ and $\tilde{f}_{i,\alpha}$. These operators satisfy the constraints

$$\sum_{\alpha} \tilde{f}_{i,\alpha}^{\dagger} \tilde{f}_{i,\alpha} + b_i^{\dagger} b_i = Q_i = 1. \tag{15.61}$$

The slave boson field satisfies the equation of motion

$$i\hbar\frac{\partial}{\partial t}(b_i^\dagger) = \lambda_i b_i^\dagger + \frac{1}{N_s^{1/2}}\sum_{\mathbf{k},\alpha} V(\mathbf{k})\exp[i\mathbf{k}\cdot\mathbf{R}_i]\widetilde{f}_{i,\alpha}^\dagger d_{\mathbf{k},\alpha}. \tag{15.62}$$

The lowest-order approximation, the terms of zeroth order in the boson fluctuation operators b_i, is retained. If b_0 is finite, this corresponds to a time-independent macroscopic equation of the $k = 0$ state that is equivalent to assuming that the boson field has undergone Bose–Einstein condensation. In this approximation, Eq. (15.62) can be written as (Problem 15.6)

$$\lambda_i b_0^* = \frac{-1}{N_s^{1/2}}\sum_{\mathbf{k},\alpha} V(\mathbf{k})\exp[i\mathbf{k}\cdot\mathbf{R}_i]\langle\widetilde{f}_{i,\alpha}^\dagger d_{\mathbf{k},\alpha}\rangle. \tag{15.63}$$

Here, b_0 and λ_i can be determined self-consistently from Eqs. (15.62) and (15.63). The hybridization matrix element is renormalized through

$$\widetilde{V}(\mathbf{k}) = b_0 V(\mathbf{k}), \tag{15.64}$$

and the f-level energy is renormalized through

$$\widetilde{E}_f = E_f + \lambda. \tag{15.65}$$

This moves the quasiparticle component of the f structure from the incoherent bare f-level component of the density of states to a position near the chemical potential. It can be shown that the quasiparticle dispersion relations are obtained as (Problem 15.7)

$$E_\pm(\mathbf{k}) = \frac{1}{2}[\widetilde{E}_f + \varepsilon_d(\mathbf{k}) \pm ([\widetilde{E}_f - \varepsilon_d(\mathbf{k})]^2 + 4|\widetilde{V}(\mathbf{k})|^2)^{1/2}]. \tag{15.66}$$

In this formulation, the amplitude of the slave boson condensate b_i is temperature dependent and vanishes at a critical temperature, T_c, for the semiconductor system. It can be shown that

$$k_B T_c = 1.14W\exp\left[\frac{E_f - \mu}{N\Delta}\right], \tag{15.67}$$

where $\Delta = |V|^2/W$, W represents approximately half the width of the conduction band, and the direct gap has a magnitude of $2V$. It is interesting to note that this temperature dependence is related to the Kondo temperature, in which the effects of both band edges are taken into account.

It may be noted that the slave boson mean-field theory is exact only when the degeneracy of the f-level approaches infinity. In addition, it is valid only when the lower band is fully occupied, which is true only for some Kondo insulators but not others, such as SmB_6. The effect of the magnetic field gives rise to a Zeeman splitting of the quasiparticle bands, reducing the hybridization gap. It has been shown by using the periodic Anderson lattice (in the limit of infinite spatial dimensions, $d \to \infty$) that the semiconductor-to-metal transition associated with the high field closing the gap may be of the first order.

PROBLEMS

15.1. The Schrieffer-Wolff transformation[24] can be easily used to relate the Anderson model of a localized magnetic moment to that of Kondo. The two models can be shown to be equivalent in small $s-f$ mixing. The Anderson Hamiltonian for a single localized orbital f is

$$\hat{H} = \sum_{k\sigma}\epsilon_k n_{k\sigma} + \sum_\sigma \epsilon_f n_{f\sigma} + Un_{f\uparrow}n_{f\downarrow} + \sum_{k\sigma}[V_{kf}\hat{C}_{k\sigma}^\dagger\hat{C}_{f\sigma} + V_{kf}^*\hat{C}_{f\sigma}^\dagger\hat{C}_{k\sigma}] = \hat{H}_0 + \hat{H}_1, \tag{1}$$

where $\in_k$ and $\in_f$ are the one-electron energies of the conduction and localized orbitals, measured relative to the Fermi energy, and $\hat{H}_0$ is the sum of the first three terms in Eq. (1). The model can be characterized by two dimensionless ratios

$$r_\pm \equiv \Gamma_\pm / |\in_\pm|, \tag{2}$$

where

$$\begin{aligned} \in_f &= \in_f + U, \quad \alpha = +, \\ &= \in_f, \qquad \alpha = -, \end{aligned} \tag{3}$$

and

$$\Gamma_\alpha = \pi N(\in_\alpha) |V_{kf}|^2_{AVE}. \tag{4}$$

$N(\in_\alpha)$ is the density of band states in the perfect crystal at energy $\in_\alpha$, and the matrix elements are averaged over k states of this energy. If $\in_+ > 0$ and $\in_- < 0$, then for $V_{kf} \to 0$, the ground state is given by the filled Fermi sea and a single electron occupying the f orbital. A localized moment occurs even at zero temperature because the states with f-electron spin $\uparrow$ and $\downarrow$ are degenerate. For $r_\alpha \ll 1$, these two spin states are mixed by electrons hopping on and off the f orbital due to V. Because arbitrarily small energy denominators $\in_k - \in_{k'} \simeq 0$ occur in fourth and higher orders of V, V cannot be treated directly by perturbation theory. However, the interactions that dominate the dynamics of the system for $r_\alpha \ll 1$ can be isolated. Show that one can perform a canonical transformation,

$$\overline{H} \equiv e^S H e^{-S}, \tag{5}$$

by requiring that V_{kf} is eliminated to the first order, where

$$[H_0, S] = H_1. \tag{6}$$

Show from Eqs. (4) and (6) that

$$S = \sum_{k\sigma\alpha} \frac{V_{kf}}{\in_k - \in_\alpha} n^\alpha_{f,-\sigma} C^\dagger_{k\sigma} C_{f\sigma} - H.C., \tag{7}$$

where the projection operators $n^\alpha_{f,-\sigma}$ are defined by

$$\begin{aligned} n^\alpha_{f,-\sigma} &= n_{f,-\sigma}, \qquad \alpha = +, \\ &= 1 - n_{f,-\sigma}, \quad \alpha = -. \end{aligned} \tag{8}$$

Show also that in the limit $r_\alpha \ll 1$,

$$\overline{H} \approx H_0 + H_2, \tag{9}$$

where

$$H_2 = \frac{1}{2}[S, H_1] \approx H_{ex} = -\sum_{kk'} J_{k'k} (\Psi^\dagger_{k'} \mathbf{S} \Psi_k) \cdot (\Psi^\dagger_f \mathbf{S} \Psi_f), \tag{10}$$

where $2\mathbf{S} = \tau$ are the Pauli spin matrices, Ψ_k and Ψ_f are the field operators

$$\Psi_k = \begin{pmatrix} c_{k\uparrow} \\ c_{k\downarrow} \end{pmatrix}, \ \Psi_d = \begin{pmatrix} c_{f\uparrow} \\ c_{f\downarrow} \end{pmatrix}, \tag{11}$$

and

$$J_{k'k} = V_{k'f}V_{fk}\{(\in_k - \in_+)^{-1} + (\in_{k'} - \in_+)^{-1} - (\in_{k'} - \in_-)^{-1} - (\in_{k'} - \in_-)^{-1}\}. \tag{12}$$

For k and $k' \simeq k_F$, $J_{kk'}$ is given by

$$J_{k_F k_F} \equiv J_0 = 2|V_{k_F f}|^2 \frac{U}{\in_f(\in_f + U)}. \tag{13}$$

This coupling is antiferromagnetic. If there is an f electron at every site of the lattice, from Eq. (10), the f-electron degrees of freedom are represented by localized spins.

15.2. Show that

$$g_0(i\omega_n)^{-1} = i\omega_n + \mu + G(i\omega_n)^{-1} - R[G(i\omega_n)^{-1}]. \tag{1}$$

Here, $G(i\omega_n)$, the on-site interacting Green's function, is calculated from the effective action S_{eff} defined in Eq. (15.14),

$$G(\tau - \tau') = - < TC(\tau)C^\dagger(\tau') >_{S_{eff}} \tag{2}$$

$$G(i\omega_n) = \int_0^\beta d\tau G(\tau)e^{i\omega_n\tau}, \omega_n \equiv \frac{(2n+1)\pi}{\beta}. \tag{3}$$

Here, $R(G)$ is the reciprocal function of the Hilbert transform of the density of states corresponding to the lattice. The noninteracting density of states is

$$D(\in) = \sum_{\mathbf{k}} \delta(\in - \in_{\mathbf{k}}), \tag{4}$$

where

$$\in_{\mathbf{k}} \equiv \sum_{ij} t_{ij} e^{i\mathbf{k}\cdot(\mathbf{R}_i - \mathbf{R}_j)}. \tag{5}$$

The Hilbert transform $\overline{D}(\xi)$ and its reciprocal function R are defined by

$$\overline{D}(\xi) = \int_{-\infty}^{\infty} d\in \frac{D(\in)}{\xi - \in} \tag{6}$$

and

$$R\left[\overline{D}(\xi)\right] = \xi. \tag{7}$$

In principle, G can be computed as a functional of g_0, using the impurity action S_{eff}. Thus, Eqs. (15.14), (1), and (2) form a complete system of functional equations for the on-site Green's function G and the Weiss function g_0.

15.3. The Hamiltonian of a three-dimensional Kondo–Heisenberg model on a cubic lattice is

$$H = \sum_k \in_{\mathbf{k}} C^\dagger_{\mathbf{k}\alpha} C_{\mathbf{k}\alpha} + \frac{J_K}{2} \sum_r \vec{S}_r \cdot C^\dagger_{r\alpha} \vec{\sigma}_{\alpha\alpha'} C_{r\alpha'} + J_H \sum_{<rr'>} \vec{S}_r \cdot \vec{S}_{r'}. \tag{1}$$

Here, $C_{\mathbf{k}\alpha}$ is the conduction electron destruction operator, $\vec{S}_r$ are the spin-1/2 local moments, and summation over repeated spin indices α is implied. In a fermionic "slave-particle" representation of the local moments

$$\vec{S}_r = 1/2 f^{\dagger}_{r\alpha} \vec{\sigma}_{\alpha\alpha'} f_{r\alpha'}, \tag{2}$$

and $f_{r\alpha}$ is a spinful fermion destruction operator at site r. The decoupling of the Kondo and the Heisenberg exchange is made using two auxiliary fields by a saddle-point approximation, and the mean-field Hamiltonian is

$$H_{mf} = \sum_{\mathbf{k}} \epsilon_{\mathbf{k}} C^{\dagger}_{\mathbf{k}\alpha} C_{\mathbf{k}\alpha} - \chi_0 \sum_{<rr'>} (f^{\dagger}_{r\alpha} f_{r'\alpha} + H.c.) + \mu_f \sum_r f^{\dagger}_{r\alpha} f_{r\alpha} - b_0 \sum_{\mathbf{k}} (C^{\dagger}_{\mathbf{k}\alpha} f_{\mathbf{k}\alpha} + H.C.). \tag{3}$$

Here, b_0 and χ_0 are assumed to be real and additional constants to H are dropped. Show that the mean-field parameters b_0, χ_0, and μ_f are obtained from

$$1 = <f^{\dagger}_{r\alpha} f_{r\alpha}>, \tag{4}$$

$$b_0 = J_K/2 <C^{\dagger}_{r\alpha} f_{r\alpha}>, \tag{5}$$

$$\chi_0 = J_H/2 <f^{\dagger}_{r\alpha} f_{r'\alpha}>, \tag{6}$$

where r and r' are nearest neighbors. At zero temperature, in the Fermi-liquid (FL) phase, χ_0, b_0, and μ_0 are nonzero. In the FL^* phase, $b_0 = \mu_0 = 0$, but $\chi_0 \neq 0$. In this state, the conduction electrons are decoupled from the local moments and form a small Fermi surface. The local-moment system is described as a spin fluid with a Fermi surface of neutral spinons.

15.4. In Problem 15.3, the mean-field is diagonalized by the transformation (Senthil et al.[26]),

$$C_{\mathbf{k}\alpha} = u_{\mathbf{k}} \gamma_{\mathbf{k}\alpha+} + v_{\mathbf{k}} \gamma_{\mathbf{k}\alpha-} \tag{1}$$

and

$$f_{\mathbf{k}\alpha} = v_{\mathbf{k}} \gamma_{\mathbf{k}\alpha+} - u_{\mathbf{k}} \gamma_{\mathbf{k}\alpha-}. \tag{2}$$

Show that the Hamiltonian (Eq. 3 in Problem 15.3) can be written in terms of the new fermionic operators $\gamma_{\mathbf{k}\alpha\pm}$,

$$H_{mf} = \sum_{\mathbf{k}\alpha} E_{\mathbf{k}+} \gamma^{\dagger}_{\mathbf{k}\alpha+} \gamma_{\mathbf{k}\alpha+} + E_{\mathbf{k}-} \gamma^{\dagger}_{\mathbf{k}\alpha-} \gamma_{\mathbf{k}\alpha-}, \tag{3}$$

where

$$E_{\mathbf{k}\pm} = \frac{\epsilon_{\mathbf{k}} + \epsilon_{\mathbf{k}f}}{2} \pm \sqrt{\left(\frac{\epsilon_{\mathbf{k}} - \epsilon_{\mathbf{k}f}}{2}\right)^2 + b_0^2}. \tag{4}$$

Here, $\epsilon_{\mathbf{k}f} = \mu_f - \chi_0 \sum_{a=1,2,3} \cos(k_a)$. The $u_{\mathbf{k}}, v_{\mathbf{k}}$ are determined by

$$u_{\mathbf{k}} = -\frac{b_0 v_{\mathbf{k}}}{E_{\mathbf{k}+} - \epsilon_{\mathbf{k}}}, \quad u_{\mathbf{k}}^2 + v_{\mathbf{k}}^2 = 1. \tag{5}$$

15.5. The metals of the rare-earth (lanthanide) group have very small $4f^n$ magnetic cores immersed in a sea of conduction electrons from the $6s$–$6p$ bands. The magnetic properties of these metals can be understood in detail in terms of an indirect exchange interaction between the magnetic cores via the conduction electrons. If the spins of the local magnetic moments at $\mathbf{r} = \mathbf{r}_i$ are $\mathbf{S}_i$ and at $\mathbf{r} = \mathbf{r}_j$ are $\mathbf{S}_j$, the second-order interaction between the two spins is given by

$$H''(\mathbf{x}) = \sum_{\mathbf{k}\mathbf{k}'ss'}{}' \frac{<\mathbf{k}s\,|H|\,\mathbf{k}'s'><\mathbf{k}'s'\,|H|\,\mathbf{k}s>}{\varepsilon_{\mathbf{k}} - \varepsilon_{\mathbf{k}'}}. \tag{1}$$

Here,

$$H = \sum_{\mathbf{k}\mathbf{k}'ss'} \left[\int d^3x \phi^*_{\mathbf{k}'s'}(\mathbf{x}) A(\mathbf{x} - \mathbf{r}_i) \mathbf{S} \cdot \mathbf{S}_i \phi_{\mathbf{k}s}(\mathbf{x}) \right] c^\dagger_{\mathbf{k}'s'} c_{\mathbf{k}s}, \tag{2}$$

where $A(\mathbf{x} - \mathbf{r}_i)$ is the interaction (which is proportional to the delta function) between the spin of the electron $\mathbf{S}$ and the spin $\mathbf{S}_i$ of the local moment at site $\mathbf{r}_i$. Here, $\phi_{\mathbf{k}s}$ are the Bloch functions $\phi_{\mathbf{k}s} = \phi_{\mathbf{k}}|s>$, and $\mathbf{S}$ operates on the spin part of $\phi_{\mathbf{k}s}$. Show that Eq. (2) can be rewritten as

$$H = \frac{1}{2} \sum_{\mathbf{k},\mathbf{k}'} e^{i(\mathbf{k}-\mathbf{k}')\cdot\mathbf{R}_i} J(\mathbf{k}', \mathbf{k}) [S_i^+ c^\dagger_{\mathbf{k}'\downarrow} c_{\mathbf{k}\uparrow} + S_i^- c^\dagger_{\mathbf{k}'\uparrow} c_{\mathbf{k}\downarrow} + S_i^z (c^\dagger_{\mathbf{k}'\uparrow} c_{\mathbf{k}\uparrow} - c^\dagger_{\mathbf{k}'\downarrow} c_{\mathbf{k}\downarrow}], \tag{3}$$

where

$$J(\mathbf{k}, \mathbf{k}') = \int d^3x \, \phi^*_{\mathbf{k}'}(\mathbf{x}) A(\mathbf{x}) \phi_{\mathbf{k}}(\mathbf{x}). \tag{4}$$

If

$$A(\mathbf{x}) = J\delta(\mathbf{x}), \tag{5}$$

$$J(\mathbf{k}', \mathbf{k}) = J. \tag{6}$$

From Eqs. (1), (2), and (6), show that

$$H''(\mathbf{x}) = \sum_{\mathbf{s}} (\mathbf{S} \cdot \mathbf{S}_i)(\mathbf{S} \cdot \mathbf{S}_j) m J^2 \hbar^{-2} (2\pi)^{-6} P \int_o^{k_F} d^3k \int_{k_F}^{\infty} d^3k' \frac{e^{-i(\mathbf{k}-\mathbf{k}').\mathbf{x}}}{k^2 - k'^2} + cc. \tag{7}$$

The sum over electron spin states is done with the help of the standard relation between Pauli operators,

$$(\sigma \cdot \mathbf{S}_i)(\sigma \cdot \mathbf{S}_j) = \mathbf{S}_i \cdot \mathbf{S}_j + i\sigma \cdot \mathbf{S}_i \times \mathbf{S}_j. \tag{8}$$

Because the trace of any component of σ vanishes,

$$\sum_{\mathbf{s}} (\mathbf{S} \cdot \mathbf{S}_i)(\mathbf{S} \cdot \mathbf{S}_j) = \frac{1}{2} \mathbf{S}_i \cdot \mathbf{S}_j. \tag{9}$$

From Eqs. (7) and (9), by performing the integrations, show that

$$H''(\mathbf{x}) = \frac{4J^2 m k_F^4}{(2\pi)^3 \hbar^2 r_{ij}^4} [2k_F r_{ij} \cos(2k_F r_{ij}) - \sin(2k_F r_{ij})] \mathbf{S}_i \cdot \mathbf{S}_j. \tag{10}$$

The density of conduction electrons,

$$n_c = \frac{k_F^3}{3\pi^2}. \tag{11}$$

From Eqs. (10) and (11), we obtain the RKKY interaction,

$$H_{\mathrm{RKKY}} = -\frac{9\pi}{8} n_c^2 \frac{J^2}{\varepsilon_F} \sum_{<ij>} \frac{\mathbf{S}_i \cdot \mathbf{S}_j}{r_{ij}^3} \left[2k_F \cos(2k_F r_{ij}) - \frac{\sin(2k_F r_{ij})}{r_{ij}} \right], \tag{12}$$

where a factor of ½ has been multiplied to avoid double counting of i and j. The spin–spin interaction is long ranged and changes its sign depending on the distance between the pair of spins.

15.6. The slave boson field satisfies the equation of motion

$$i\hbar \frac{\partial}{\partial t}(b_i^\dagger) = \lambda_i b_i^\dagger + \frac{1}{N_s^{1/2}} \sum_{\mathbf{k},\alpha} V(\mathbf{k}) \exp[i\mathbf{k} \cdot \mathbf{R}_i] \widetilde{f}_{i,\alpha}^\dagger d_{\mathbf{k},\alpha}. \tag{1}$$

The lowest-order approximation, the terms of zeroth order in the boson fluctuation operators b_i, is retained. If b_0 is finite, this corresponds to a time-independent macroscopic equation of the $k = 0$ state, which is equivalent to assuming that the boson field has undergone Bose–Einstein condensation. In this approximation, show that Eq. (1) can be rewritten as

$$\lambda_i b_0^* = \frac{-1}{N_s^{1/2}} \sum_{\mathbf{k},\alpha} V(\mathbf{k}) \exp[i\mathbf{k} \cdot \mathbf{R}_i] \langle \widetilde{f}_{i,\alpha}^\dagger d_{\mathbf{k},\alpha} \rangle. \tag{2}$$

15.7. Show that in the slave boson mean-field theory, the quasiparticle dispersion relations are obtained as

$$E_\pm(\mathbf{k}) = \frac{1}{2} [\widetilde{E}_f + \varepsilon_d(\mathbf{k}) \pm ([\widetilde{E}_f - \varepsilon_d(\mathbf{k})]^2 + 4|\widetilde{V}(\mathbf{k})|^2)^{1/2}]. \tag{1}$$

References

1. Anderson PW. Localized Magnetic States in Metals. *Phys Rev B* 1961;**124**:41.
2. Bianchi A, Movshovich R, Caban C, Pagluso PG, Sarrao JL. A possible Fulde-Ferrel-Larkin-Ovchinnikov superconducting state in Ce Co In5. *Phys Rev Lett* 2003;**91**:187004.
3. Callaway J, Chen DP, Kanhere DG, Misra PK. Cluster Simulation of the Lattice Anderson Model. *Phys Rev B* 1988;**38**:2583.
4. Demuer A, Sheikin I, Braithwaite D, Fak B, Huxley A, Raymond S, Flouquet J. *J. Magn. Matter* 2001;**17**:226.
5. Coleman P. *Physica B* 2006;**378–380**:1160.
6. Doniach S. The Kondo lattice and weak antiferromagnetism. *Physica B* 1977;**91**:231.
7. Fulde P, Ferrell RA. Superconductivity in Strong Spin-Exchange Field. *Phys Rev A* 1964;**135**:550.
8. Griveau J-C, Boulet P, Collineau E, Wastin F, Rebizant J. Pressure effect on PuMGd5(M=Co, Rh, Zr). *Physica B* 2005;**359–361**:1093.
9. Hewson AC. *The Kondo problem in heavy fermions*. Cambridge: Cambridge University Press; 1993.
10. Holmes AT, Jaccard D, Miyake K. Signatures of Valence Fluctuations in $CeCu_2Si_2$ under high pressure. *Phys Rev B* 2004;**69**:024508.
11. Kadowaki K, Woods SB. A universal relationship of the resistivity and specific heat of heavy-Fermion compounds. *Solid State Commun* 1986;**58**:507.

12. Kasuya T. A Theory of Metallic Ferro- and Antiferromagnetism in Zener's model. *Prog Theor Phys (Kyoto)* 1956;**16**:45.
13. Larkin AI, Ovchinnikov YN. Inhomogeneous state of superconductor. *Sov Phys JETP* 1965;**20**:762.
14. Luttinger JW. Fermi Surface and Some Simple Equilibrium Properties of a system of Interacting Fermions. *Phys Rev* 1960;**119**:1153.
15. Measson M-A, Brison JP, Seyfarth G, Braithwaite D, Lapertat G, Salce B, et al. Superconductivity of the filled skutterdite-PrOs4Sb12. *Physica* 2005;**359–361**:827.
16. Misra PK. *Heavy-fermion systems.* Amsterdam: Elsevier; 2008.
17. Misra PK, Kanhere DG, Callaway J. Periodic Anderson Model for four-site clusters. *Phys Rev B* 1987;**35**:5013.
18. Ott HR. Heavy-electrons and non-Fermi liquids, the early times. *Physica B* 2006;**378–380**:1.
19. Ott HR, Rudigier H, Delsing P, Fisk Z. UBe3:An unconventional Actinide Superconductor. *Phys Rev Lett* 1983;**50**:1595.
20. Riseborough PS. Heavy fermion semiconductors. *Adv Phys* 2000;**49**:257.
21. Ruderman MA, Kittel C. Indirect Exchange Coupling of Nuclear Magnetic Moments by Conduction Electrons. *Phys Rev* 1954;**96**:99.
22. Sarrao JL, Morales LA, Thompson JD, Scole BL, Stewart GR, Wastin F, et al. Plutonium-based Superconductivity with a transition temperature above 18K. *Nature* 2002;**420**:297.
23. Sarrao JL, Brauer ED, Morales LA, Thompson JD. Structural tuning and anisotropy in PuCoGa5. *Physica B* 2005;**359–361**:1144.
24. Schrieffer JR, Wolff PA. Rekation between the Anderson and Kondo Hamiltonian. *Phys Rev* 1966;**149**:491.
25. Senthil T, Sachdev S, Vojta M. Quantum phase transition out of the heavy Fermi liquid. *Physica* 2005;**359–361**:9.
26. Senthil T, Sachdev S, Vojta M. Fractionalized Fermi liquids. *Phys Rev Lett* 2003;**90**:216403.
27. Senthil T, Vojta M, Sachdev S. Weak magnetism and non-Fermi liquids near heavy-fermion critical points. *Phys Rev B* 2004;**69**:035111.
28. Smith JL, Si Q. Spatial correlations in dynamical mean-field theory. *Phys Rev B* 2000;**61**:5184.
29. Steglich F, Aarts J, Bredl CD, Lieke W, Meschede D, Franz W, et al. Superconductivity in the Presence of Strong Pauli Paramagnetism: $CeCu_2Si_2$. *Phys Rev Lett* 1979;**43**:1892.
30. Stewart GR. Heavy-fermion systems. Possibility of Coexistence of Bulk Supercon. *Rev Mod Phys* 1984;**56**:755.
31. Stewart GR, Rudigier H, Delsing P, Fisk Z. Possibility of Coexistence of Bulk Supercon. *Phys Rev Lett* 1984;**52**:679.
32. Varma CM. Mixed-Valence Compounds. *Rev Mod Phys* 1976;**48**:219.
33. Wachter P. Similarities between Cu and Pu containing "high-Tc Superconductor." *Physica C* 2007;**453**:1.
34. Wasrtin F, Boule P, Rebizant J, Collineau E, Lander GH. Advances in the preparation and characterization of transuranium systems. *J. Phys. Condens. Matter* 2003;**15**:S2279.
35. Wilson KG. The renormalization group: Critical phenomena and the Kondo problem. *Rev Mod Phys* 1975;**47**:773.
36. Yosida K. Magnetic Properties of Cu-Mn Alloys. *Phys Rev* 1957;**106**:893.

CHAPTER

Metallic Nanoclusters

16

CHAPTER OUTLINE

16.1 Introduction ... 528
16.1.1 Nanoscience and Nanoclusters ... 528
16.1.2 Liquid Drop Model ... 528
16.1.3 Size and Surface/Volume Ratio ... 528
16.1.4 Geometric and Electronic Shell Structures ... 530
16.2 Electronic Shell Structure ... 531
16.2.1 Spherical Jellium Model (*Phenomenological*) ... 531
16.2.2 Self-Consistent Spherical Jellium Model ... 532
16.2.3 Ellipsoidal Shell Model ... 535
16.2.4 Nonalkali Clusters ... 535
16.2.5 Large Clusters ... 535
16.3 Geometric Shell Structure ... 537
16.3.1 Close-Packing ... 537
16.3.2 Wulff Construction ... 537
16.3.3 Polyhedra ... 538
16.3.4 Filling between Complete Shells ... 540
16.4 Cluster Growth on Surfaces ... 540
16.4.1 Monte Carlo Simulations ... 540
16.4.2 Mean-Field Rate Equations ... 541
16.5 Structure of Isolated Clusters ... 542
16.5.1 Theoretical Models ... 542
16.5.2 Structure of Some Isolated Clusters ... 546
16.6 Magnetism in Clusters ... 547
16.6.1 Magnetism in Isolated Clusters ... 547
16.6.2 Experimental Techniques for Studying Cluster Magnetism ... 549
16.6.3 Magnetism in Embedded Clusters ... 553
16.6.4 Graphite Surfaces ... 555
16.6.5 Study of Clusters by Scanning Tunneling Microscope ... 555
16.6.6 Clusters Embedded in a Matrix ... 557
16.7 Superconducting State of Nanoclusters ... 558
16.7.1 Qualitative Analysis ... 558
16.7.2 Thermodynamic Green's Function Formalism for Nanoclusters ... 559
Problems ... 562
References ... 565

16.1 INTRODUCTION

16.1.1 Nanoscience and Nanoclusters

Nano objects have a size that is intermediate between atoms or molecules and bulk matter. The recent invention of a variety of tools for studying systems at the atomic level, coupled with the development of techniques for producing nanoclusters, has led to the use of nanoscience as a new field of study. The scanning probe microscopes (the first of which was the scanning tunneling microscope, or STM), make it possible to "see" individual atoms and molecules on surfaces of materials as well as to move them on the nanoscale.

New sources to produce clusters in the gas phase were developed in the 1960s and 1970s, but it was in 1980s that Knight et al.[16] first produced clusters of alkali metals with approximately 100 atoms and systematically studied their properties. Nanoclusters can now be formed from most elements of the periodic table. They can be classified as metallic, semiconductor, ionic, rare gas, or molecular, according to their constituents. Clusters are classified as homogeneous if they contain a single type of atom or heterogeneous if they comprise more than one constituent. They may be neutral or charged (anions or cations). There have been two main approaches in creating nanostructures: top-down and bottom-up. In the top-down method, starting from a large piece of material, a nanostructure is formed by removing material from it through etching or machining by using an electron beam or focused ion beam lithography. In the bottom-up approach, nanoparticles or molecules are produced by chemical synthesis followed by ordered structures by physical or chemical interactions between the units. In this chapter, we will concentrate on the study of metallic nanoclusters.

16.1.2 Liquid Drop Model

The simplest description of a metallic nanocluster is the liquid drop model (LDM). The cluster is represented as a sphere of radius R, which is related to the number of atoms N through the Wigner–Seitz radius r_s,

$$R = N^{1/3} r_s. \tag{16.1}$$

Here, the Wigner–Seitz radius r_s, originally defined as the radius of the volume occupied by each valence electron, is equivalent to the volume occupied by each atom in a monovalent nanometal. The internal structure of the cluster is ignored in this model. This model is equivalent to the free electron theory of solids. However, the solid box has macroscopic dimensions with a continuum of energy levels, whereas the cluster box is a nanoscale entity and the energy levels are discrete. The first few energy levels for noninteracting electrons in a spherical box and the number of electrons required for complete filling of the shells are shown in Figure 16.1.

16.1.3 Size and Surface/Volume Ratio

The fraction of atoms that are on the surface of a cluster distinguishes the difference of the properties of the cluster from the bulk. One simple way to analyze this is to cut a regularly shaped object

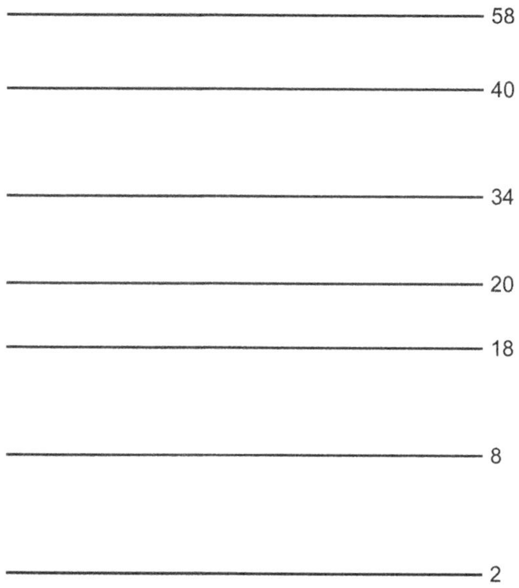

FIGURE 16.1

First few energy levels for noninteracting electrons in a spherical box.

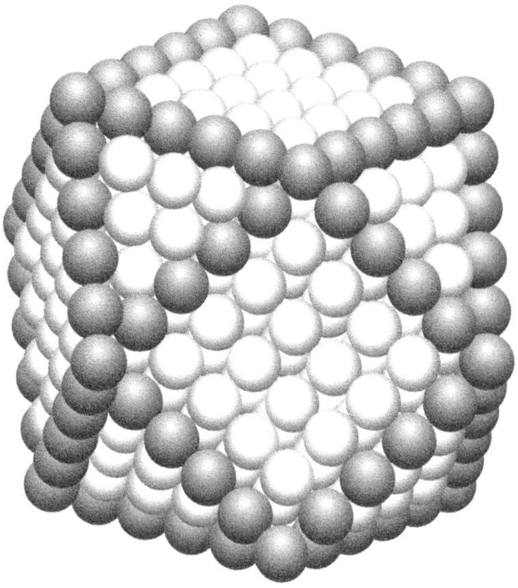

FIGURE 16.2

A 561-atom cut from a cuboctahedral cluster cut from a bulk fcc crystal.

Reproduced from Misra[28] with the permission of Elsevier.

from a fcc lattice and count the surface atoms for different size clusters. Such clusters have eight triangular and six square faces, as shown in Figure 16.2. The fraction of atoms that are on the surfaces of the clusters is shown in Figure 16.3.

Clusters represent a state of matter that is intermediate between atoms and the solid or liquid state, with properties that depend on the size, shape, and material of the particle. The arbitrary property per atom, $x(N)$, can be expressed as

$$x(N) = a + bN^{-1/3}, \tag{16.2}$$

where the first term is the "bulk" contribution, and the second term is the "surface" contribution.

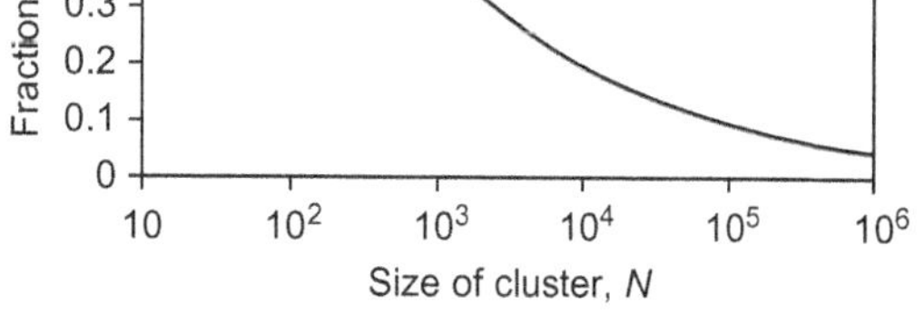

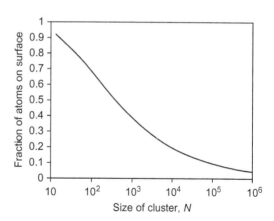

FIGURE 16.3

Fraction of atoms on cuboctohedral clusters of N atoms.

Reproduced from Misra[28] with the permission of Elsevier.

16.1.4 Geometric and Electronic Shell Structures

Knight et al.[16] produced clusters through the supersonic expansion of a metal/carrier gas mixture. The alkali metal is first vaporized and then seeded in the inert carrier gas (typically argon). The supersonic expansion of the mixture results in adiabatic cooling. The cluster is then ionized and passed through a mass spectrometer. Knight et al.[16] observed marked peaks in the mass spectrum of the clusters indicating high stability at particular sizes. Their experiment results are reproduced in Figure 16.4.

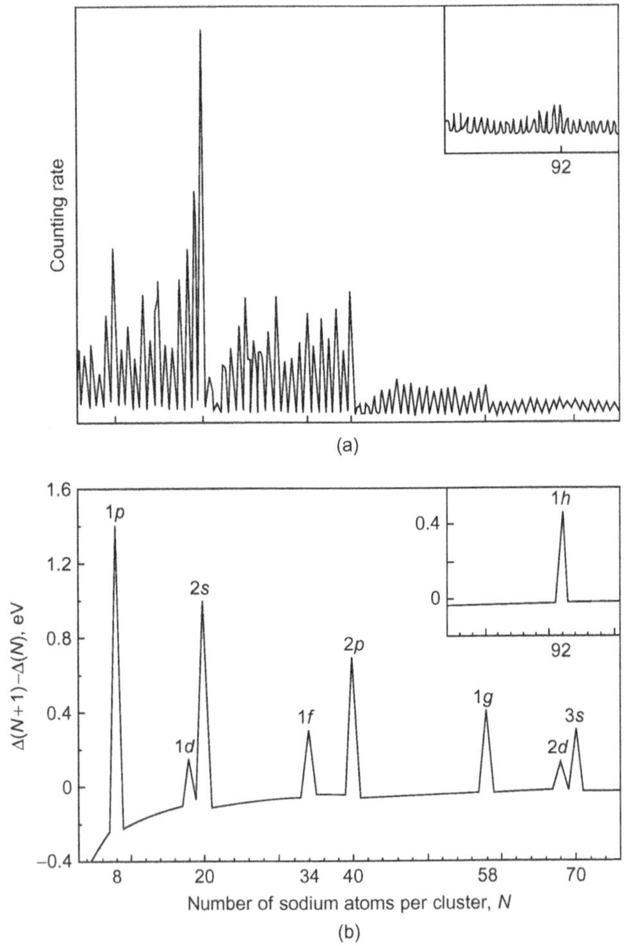

FIGURE 16.4

(a) The mass spectrum of sodium clusters with $N = 4$–75 and 75–100 (inset). (b) The calculated change in the electronic energy difference, $\Delta(N+1) - \Delta(N)$ versus N. The peak labeling corresponds to closed-shell orbitals.

Reproduced from Knight et al.[16] with the permission of the American Physical Society.

The marked peaks or steps at special numbers in the sodium-cluster mass spectra are called "magic numbers." These magic numbers were associated with complete filling of electronic shells, a concept of the electronic shell model borrowed from nuclear physics. From Figure 16.4, the magic number peaks are seen at cluster sizes $N = 8$, 20, 40, 58, and 92.

16.2 ELECTRONIC SHELL STRUCTURE

16.2.1 Spherical Jellium Model (*Phenomenological*)

In the spherical jellium model, the ionic cores are replaced by a uniform positive charge background of radius R. The electrons are treated as independent particles moving in a parametrized phenomenological potential. The basic parameter of the model is the Wigner–Seitz radius r_s defined in Eq. (16.1). Further, the wave function for a spherically symmetric potential can be written as

$$\psi_{nlm}(\mathbf{r}) = R_{nl}(r)Y_{lm}(\theta, \phi). \tag{16.3}$$

There are three empirical potentials used to describe the nanoclusters, all borrowed from basic ideas of nuclear physics. The simplest model is the harmonic oscillator potential.

Harmonic Oscillator Potential

The harmonic oscillator potential is described by

$$V(r) = \frac{1}{2} m\omega_0 r^2. \tag{16.4}$$

The energy of the harmonic oscillator potential is given by

$$E_\nu = \left(\frac{3}{2} + \nu\right)\hbar\omega_0. \tag{16.5}$$

The quantum numbers (n, l) can be used for any spherically symmetric potential. Thus, all orbitals with the same value of $(2n + l)$ are degenerate and the energies are written in terms of the single quantum number $(2\nu + l - 2)$. The potential energy due to the background charge is

$$V(r) = \begin{cases} \dfrac{3e^2N}{8\pi\varepsilon_0 R^3}\left(\dfrac{r^2}{3} - R^2\right) & \text{if } r < R, \\ e^2 Z & \\ \dfrac{e^2N}{4\pi\varepsilon_0 r} & \text{if } r > R. \end{cases} \tag{16.6}$$

So the harmonic oscillator potential mimics the potential felt by the electrons inside the cluster, provided the electron–electron interaction is ignored.

Spherical Square-Well Potential

The spherical square-well potential is described by

$$\begin{aligned} V(r) &= C \quad \text{for } r < R, \\ &= \infty \quad \text{otherwise.} \end{aligned} \tag{16.7}$$

Here, C is a constant. The radial wave function $R_{nl}(r)$ for the square-well potential is written in terms of the spherical Bessel function $j_l(\kappa_{nl}R)$, where

$$\kappa_{nl} = \left[\frac{2m|E|}{\hbar^2}\right]^{1/2}. \tag{16.8}$$

The energy levels are determined by the boundary conditions $j_l(\kappa_{nl}R) = 0$ and, for each l, the first zero of j_l is given the quantum number $n = 1$, the second $n = 2$, and so on. The order of the energy levels (which are borrowed from nuclear physics and are different from atomic physics) are 1s, 1p, 1d, 2s, 1f, 2p, 1g, 2d, and so on. If two solutions have the same number of radial nodes, the one with higher l has higher energy. In this notation, the principal quantum number in atomic physics is equal to $n+l$. The interior of the jellium cluster will be electrically neutral if one includes the exchange-correlation contribution to the electrostatic potential of the electrons, and the effective potential will be nearly constant. The square-well potential essentially represents this phenomenon.

Woods–Saxon Potential

The Woods–Saxon potential R yields a better phenomenological representation of a potential that is flat in the middle of the cluster and rounded at the edges. This potential is described by

$$U(r) = \frac{-U_0}{\exp\left[(r-R)/\varepsilon\right]+1}. \tag{16.9}$$

U_0 is the sum of the Fermi energy and the work function of the bulk metal. R is determined by Eq. (16.4) with r_s, the Wigner-Seitz radius of the bulk. The parameter ε is taken to match the variation in the potential at the surface. This potential is flat in the middle of the cluster but rounded at the edges. The three potentials are shown in Figure 16.5.

In Figure 16.5, the degeneracy of the states of the Woods–Saxon potential is similar to that of the square-well potential, but the ordering of the energy levels is different. Knight et al.[16] used the Woods–Saxon potential (Ref. 35) in the analysis of the mass spectra shown in Figure 16.3. The ordering of the energy levels is shown, and the cumulative totals of electrons are indicated above the energy levels. At the bottom of the figures are sketches of the potentials as a function of r.

16.2.2 Self-Consistent Spherical Jellium Model

For an interacting electron gas, the density functional theory (DFT) based on the Hohenberg–Kohn (HK) theorem (discussed in detail in Section 7.8.1) is used. The basic assumption of DFT is that the total energy of the system is a functional of the electron density $n(\mathbf{r})$. The HK theorem states that the exact ground-state energy of a correlated electron system is a functional of the density, and the minimum is the ground-state density. One can apply the HK theorem to clusters. The density $n_I(r)$ of the smeared-out positive charge of the ions can be expressed as

$$n_I(r) = n_0\theta(r-R), \tag{16.10}$$

where R is the cluster radius given in Eq. (16.1), θ is a step function, and n_0 is the constant bulk density of the metal

$$n_0 = \left(\frac{4\pi r_s}{3}\right)^{-1}. \tag{16.11}$$

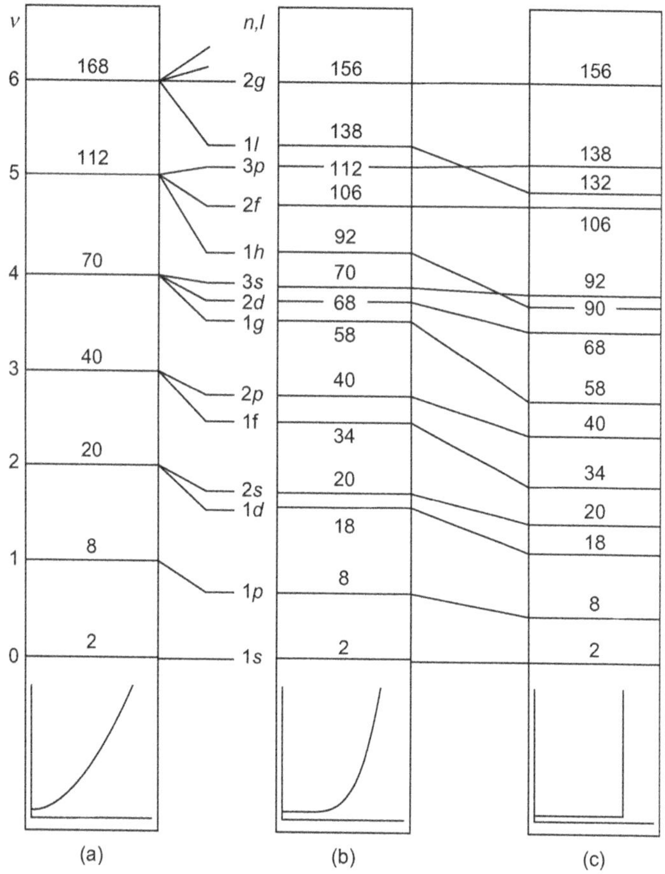

FIGURE 16.5

Comparison of the (a) harmonic, (b) Woods–Saxon, and (c) square-well potentials.

Reproduced from Misra[28] with the permission of Elsevier.

The total energy can be written as

$$E[n] = E_{es}[n] + E_{ex}[n], \tag{16.12}$$

where $E_k[n]$ is the kinetic energy of a system of independent particles of density n,

$$E_k[n] = \sum_i \frac{\hbar^2}{2m} |\nabla \psi_i|^2. \tag{16.13}$$

where $E_{es}[n]$ is the electrostatic energy,

$$E_{es}[n] = \frac{e^2}{2} \iint d\mathbf{r}'\, d\mathbf{r}\, \frac{[n(\mathbf{r}) - n_I(\mathbf{r})][n(\mathbf{r}') - n_I(\mathbf{r}')]}{|\mathbf{r} - \mathbf{r}'|}. \tag{16.14}$$

Following the procedure outlined in Section 7.8.2 (Ref. 17) (subject to the modification for a metallic cluster), we obtain (Problem 16.1)

$$-\frac{\hbar^2}{2m}\nabla^2\psi_i(\mathbf{r}) + V_{KS}(\mathbf{r})\psi_i(\mathbf{r}) = \varepsilon_i\psi_i(\mathbf{r}), \tag{16.15}$$

where

$$V_{KS}(\mathbf{r}) = V_H(\mathbf{r}) + V_{xc}(\mathbf{r}) \tag{16.16}$$

and

$$V_H(\mathbf{r}) = 2e^2 \int d\mathbf{r}' \frac{n(\mathbf{r}') - n(\mathbf{r}')}{|\mathbf{r} - \mathbf{r}'|}. \tag{16.17}$$

The local density approximation assumes that $n(\mathbf{r})$ is slowly varying. The exchange-correlation part of the total energy can be written as

$$E_{xc}[n] = \int d\mathbf{r} n(\mathbf{r})\varepsilon_{xc}(n(\mathbf{r})), \tag{16.18}$$

and

$$V_{xc}(\mathbf{r}) = \frac{\partial\varepsilon_{xc}(n)}{\partial n} + \varepsilon_{xc}(n). \tag{16.19}$$

Here, $\varepsilon_{xc}(n)$ is the exchange and correlation energy of a uniform electron gas of density n.

A number of functional forms are used for ε_{xc}. The form used most often is given by Ekardt,[7]

$$\varepsilon_{xc}(n(\mathbf{r})) = -\frac{0.916}{r_s(\mathbf{r})} - 0.0666G(x(\mathbf{r})). \tag{16.20}$$

The first term is the exchange part, where

$$r_s(\mathbf{r}) = [3/4\pi n(\mathbf{r})]^{1/3}, \tag{16.21}$$

and the second term is the correlation part, where

$$G(x) = (1 + x^3)\ln\left[1 + \frac{1}{x}\right] - x^2 + \frac{x}{2} - \frac{1}{3} \tag{16.22}$$

and

$$x(\mathbf{r}) = r_s(\mathbf{r})/11.4. \tag{16.23}$$

The wave functions and hence the energy are obtained self-consistently from the Kohn–Sham equations (Ref. 17). The Kohn–Sham potentials are fairly flat in the center of the cluster and analogous to the Woods–Saxon potential discussed earlier.

Due to the absence of the ionic structure, the simple jellium model, which works reasonably well for alkali metals, does not work out for higher-density materials such as aluminum. Recently, in the "stabilized jellium model," the ionic pseudopotentials have been introduced through a modified exchange-correlation potential.

16.2.3 Ellipsoidal Shell Model

The use of spheres is a justified approximation only for closed-shell structures. However, for open-shell structures, the ellipsoidal shell model (borrowed from nuclear physics) is used. The model is based on the harmonic oscillator Hamiltonian

$$H = \frac{p^2}{2m} + \frac{1}{2}m\omega_0^2\left[\Omega_\perp^2(x^2+y^2)+\Omega_z^2z^2\right] - U\hbar\omega_0[l^2 - <l^2>_n]. \quad (16.24)$$

It is essentially a spheroid that has two equal axes (x and y) and one unequal axis (z). Constant volume is maintained by imposing the condition $\Omega_\perp^2\Omega_z = 1$, and the distortion is expressed by a parameter δ,

$$\delta = \frac{2(\Omega_\perp - \Omega_z)}{\Omega_\perp + \Omega_z}. \quad (16.25)$$

The last term in Eq. (16.24) is an empirical addition that splits the states of different angular momenta and gives the same ordering as the Woods–Saxon potential. The potential is elongated in the x and y directions, and the cluster is elongated in the z-axis (prolate distortion), but when δ is negative, there is oblate distortion as well as expansion in the x-y plane. In Eq. (16.24), $<l^2>_n = (\frac{1}{2})n(n+3)$ (Problem 16.2).

One can obtain the wave function self-consistently from the Kohn–Sham equation outlined earlier. The wave functions are characterized by the quantum numbers (n, n_3, Λ), where Λ is the component of **l** along z, and n, n_1, n_2, n_3 are defined in Problem 16.3. Shell filling occurs with a prolate distortion until half-filling and then reverts to oblate. Closed-shell clusters are spherical, but there are subshell closings that are ellipsoidal.

16.2.4 Nonalkali Clusters

The noble metals (Cu, Ag, and Au) lie at the end of the $3d, 4d$, and $5d$ periods, respectively, with a filled d-shell of 10 electrons and a single valence electron. In bulk, the d-band falls well below the Fermi level, and the valence electrons are expected to behave similarly to that observed in alkali metal clusters. In the experiments of Katakuse et al.,[15] the clusters were created as positively charged ions, so the number of electrons in a cluster was $N-1$, and the magic numbers corresponding to electronic shell charges had fair agreement with experimental results. The electronic shell model also applies to divalent metals where $2N$ corresponds to the shell-filling numbers shown in Figure 16.4, leading to magic numbers at $N = 4$, 9, 10, 17, 20, 29, 34, 35, 46, 53, 56, ... This is also experimentally confirmed by another experiment by Katakuse et al.[15]

16.2.5 Large Clusters

The mass spectra of sodium clusters have been recorded up to sizes of about 25,000 atoms, in which the magic numbers are evident. However, the electronic energy levels of large clusters tend to bunch together in groups so that there is a series of approximately degenerate levels (shells). From experimental results, it is apparent that these shells fill on a scale that is proportional to $N^{1/3}$. One can gain useful insight by considering the electronic shells that occur for potentials for which exact degeneracy occurs and for which there is an analytic solution.

We consider the harmonic oscillator potential and the energy-level scheme shown in Figure 16.5 and define a shell index K ($= \nu$ in Figure 16.5). If the states are labeled as (n, l), all states with $K = 2n + l$ are degenerate. If $N(K)$ denotes the number of atoms in a cluster with complete levels filling up to and including the shell $K (K = 0, 1, 2,...)$,

$$N(K) = \frac{1}{3}(K+1)(K+2)(K+3) \rightarrow \frac{K^3}{3}. \tag{16.26}$$

The shell index associated with the magic numbers is plotted against $N^{1/3}$ in Figure 16.6.

If all states are degenerate ($1/r$ potential), $K = n + l$ are degenerate. (The notation n_{at} in atomic physics is $n_{at} = n + l$). The number of states for a complete filling up to shell K is

$$N(K) = \frac{2}{3}K\left(K+\frac{1}{2}\right)(K+1) \rightarrow \frac{2}{3}K^3. \tag{16.27}$$

Thus, the general filling scale is $K \propto N^{1/3}$. Figure 16.6 shows the shell index K plotted against $N^{1/3}$.

Figure 16.6 distinctly shows that there is a break in behavior at about $N \approx 1500$. For $N < 1500$, the electronic shell model yields satisfactory results. However, for $N > 1500$, one has to use a different model. The model used is known as the geometric shell model.

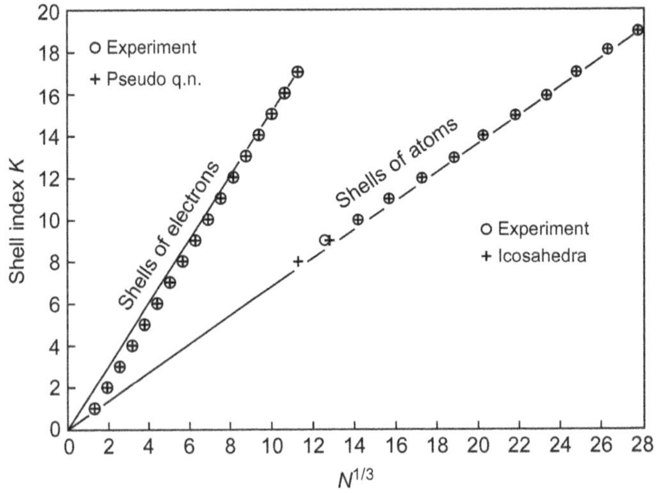

FIGURE 16.6

Shell index K plotted against $N^{1/3}$ for Na_N clusters.

Reproduced from Martin[25] with the permission of Elsevier.

16.3 GEOMETRIC SHELL STRUCTURE

16.3.1 Close-Packing

We have noted that for sodium clusters, for $N > 1500$, a new criterion for stability sets is based on the close-packing of atoms (the fcc or hcp structure in bulk) suitably modified by the presence of a surface. This is known as the geometric shell model. In fact, for other metallic clusters, the geometric shell model is the starting point for studies of stability and associated magic numbers. The bonding is dominated by the short-ranged $d-d$ interaction in metals such as nickel, and the configuration tends to maximize the number of nearest neighbors. The transition metals with unfilled $d-$ shells have directional dependence in their bonding. An example is the bcc and hcp structure in bulk iron and cobalt atoms.

16.3.2 Wulff Construction

Wulff proposed that at equilibrium, the polyhedron is such that the perpendicular distance from the center of the particle to a face of the polyhedron is proportional to the surface energy of the face.[36] When one uses this procedure for a fcc metal, a polyhedron comprising eight (111) and six (100) faces, shown in Figure 16.7, is obtained. If one defines the perpendicular distance of the two faces as p_{111} and p_{100}, it can be shown that (Problem 16.4) the lengths of the sides of the faces are β and $(1-2\beta)$ in units of $\sqrt{6}p_{111}$, as shown in Figure 16.7. The scale factor β is given by

$$\beta = 1 - p_{100}/\sqrt{3}p_{111}. \tag{16.28}$$

If γ_{100} and γ_{111} are the surface energies of the two faces, we have from Eq. (16.28)

$$\beta = 1 - \frac{\gamma_{100}}{\sqrt{3}\gamma_{111}}. \tag{16.29}$$

Assuming pairwise interaction (ϕ) between nearest neighbors, the specific surface energy $\gamma = n_d\phi/A$, where n_d is the number of neighbors a surface atom is deficient, and A is the surface area per atom. An atom on the (100) surface has a deficiency of four neighbors compared with an atom in the bulk. Therefore,

$$\gamma_{100} = 4\phi/d^2, \tag{16.30}$$

where d is the nearest-neighbor distance. An atom on the (111) surface has nine neighbors and has a deficiency of three compared with an atom in the bulk. We thus have

$$\gamma_{111} = 2\sqrt{3}\phi/d^2. \tag{16.31}$$

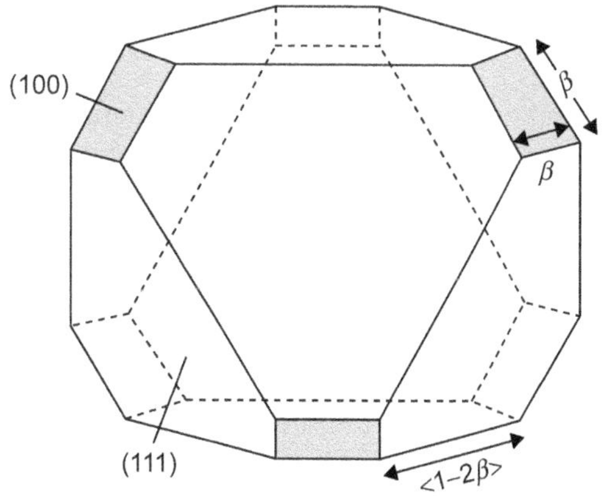

FIGURE 16.7

Polyhedral structure based on Wulff construction.

Reproduced from Marks (Ref. 24) with the permission of Elsevier.

From Eqs. (16.29) through (16.31), we obtain

$$\beta = \frac{1}{3}. \qquad (16.32)$$

Thus, $\beta = 1 - 2\beta$ and, hence, the lengths of the sides of the hexagonal and square faces are equal. This is shown in Figure 16.8 for a 586-atom polyhedron. It may be noted that the Wulff criterion actually applies for large sizes when vertex and edge effects can be neglected.

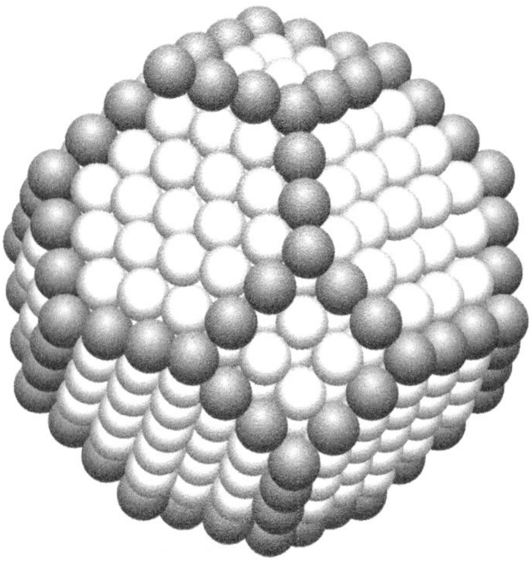

FIGURE 16.8

A 586-atom truncated octahedron with square and hexagonal faces.

Reproduced from Misra[28] with the permission of Elsevier.

16.3.3 Polyhedra

fcc

A basic shape that can be constructed from a symmetric cluster is the octahedron (see Figure 16.9) with eight triangular faces that are close-packed (111) planes with low surface energy.

In Figure 16.9, we defined a shell number K as the number of atoms along the edge of a face. A cluster with an even value of K is built around an elementary octahedron of six atoms while a cluster with an odd value of K has a single atom at its center. The total number of atoms N in a cluster containing K octahedral shells is

$$N = \frac{1}{3}(2K^3 + K). \qquad (16.33)$$

A cuboctahedron, which is an octahedron truncated by a cube, was shown earlier in Figure 16.2. It has eight (111) and six square (100) faces. The polyhedron has a central atom and can be considered as built of successive shells covering interior shells. The Kth shell contains $(10K^2 + 2)$ atoms. It can be shown that the total number of atoms in a cluster with K shells (Problem 16.5) is

$$N = \frac{1}{3}(10K^3 - 15K^2 + 11K - 3). \qquad (16.34)$$

From Eq. (16.34), the magic numbers associated with geometric shell filling are 1, 13, 55, 147, 309, 561, ...

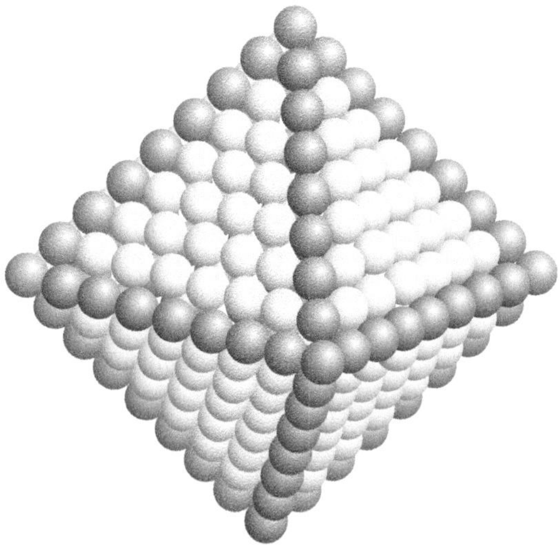

FIGURE 16.9

A 489-atom octahedron.

Reproduced from Misra[28] with the permission of Elsevier.

Mackay Icosahedron

It can be shown that the total surface energy can be minimized if all the faces of the polyhedron are (111) planes and the surface area can be kept at a minimum. The icosohedral structure shown in Figure 16.10a has 20 triangular faces and has this feature.

The Mackay icosahedron is made up of 20 distorted tetrahedra. The distortion is shown in Figure 16.10b. The vertex O is at the center. If the three edges meeting at the center (OA, OB, OC) are of unit length, the three sides of the equilateral triangle ABC are extended to 1.05146. The angles subtended at O are 63°26′. The icosohedra can be constructed by arranging 12 neighbors of a central atom at the corners of an icosohedron. One can build larger clusters by covering the 13-atom core with a second layer of 42 atoms, a third of 92 atoms, and so on. The total number of atoms is identical to that of the cuboctohedron with triangular faces, as in Eq. (16.34). The reason for assignment of magic numbers to geometrical shell structures for large sodium clusters is that experimental results support the values from Eq. (16.34) with $K = 10$–19, thereby confirming an icosohedral or cuboctahedral configuration. In addition, in the K against $N^{1/3}$ plots of Figure 16.6, there is a good fit to Eq. (16.36) in the large K limit (for $N > 1500$), whereas $N \approx 0.21\ K^3$ fits for small clusters. One can show that the later equation can be obtained by using the Woods–Saxon potential.

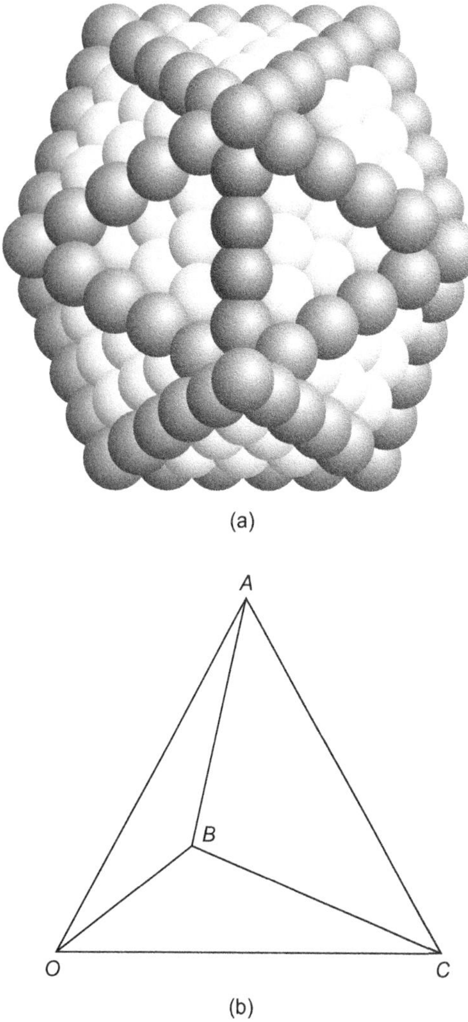

FIGURE 16.10

(a) An icosahedron of 561 atoms; (b) the Mackay icosahedron with vertex O at the center.

Reproduced from Misra[28] with the permission of Elsevier.

bcc

The 369-atom rhombic dodecahedron shown in Figure 16.11 is one of the main cluster shapes based on the bcc lattice.

There are 12 identical faces of which the diagonals have lengths with the ratio $1 : \sqrt{2}$, and this shape is similar to the cuboctahedron. The number of atoms, N, for a cluster of K shells is

$$N = 4K^3 - 6K^2 + 4K - 1. \tag{16.35}$$

Thus, the cluster sizes at successive shell fillings are 1, 15, 65, 369, ...

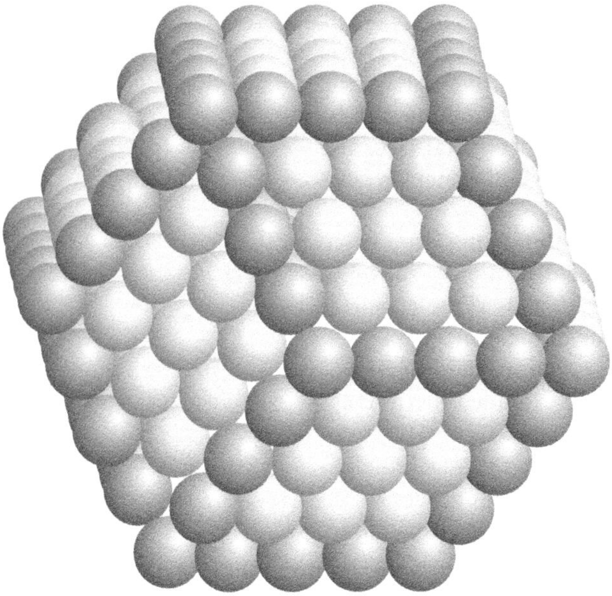

FIGURE 16.11

The 360-atom rhombic dodecahedron.

Reproduced from Misra[28] with the permission of Elsevier.

16.3.4 Filling between Complete Shells

There are two different models for filling of sub-shells in between the completion of the main shells of the Mackay (Ref. 21) icosohedra. In one model, one starts from a complete icosohedron and adds atoms that preferentially sit on the three-fold hollow formed by the three atoms on the (111) face, thereby maximizing their coordinate number. The atoms rearrange themselves to the Mackay packing arrangement when the shell is half filled. This model has been applied to rare-earth clusters. In the alternate model, also known as the umbrella model, the atoms are added to icosohedral sites on the surface of the core icosohedron. The covering of a complete face achieves enhanced stability.

16.4 CLUSTER GROWTH ON SURFACES

16.4.1 Monte Carlo Simulations

The easiest way to understand the key ideas of cluster growth on surfaces is through Monte Carlo simulation of cluster (or island) nucleation and growth during vapor deposition. Usually, through the use of a comprehensive set of simulations, an empty lattice is made to represent a substrate. The monomers are deposited at random into this lattice at a constant deposition rate. They can diffuse by random nearest-neighbor hops, and when they collide, they nucleate new point islands. The immobile islands "grow" whenever new monomers diffuse onto them. They remain as single points on the lattice while their recorded size increments by the number of adsorbed monomers. The density of islands nucleated in the simulation depends on the ratio, R, of the monomer deposition rate to the monolayer deposition rate. R dictates the Monte Carlo simulation procedure, by changing the probabilities of the next simulation step being a deposition event or a monomer diffusion event. Typically, $R \sim 10^5 - 10^{10}$ in experimental growth conditions.

The Monte Carlo simulations throw up many questions relevant to experiment, such as (1) how does the island density depend on growth rate and temperature, (2) how does the critical island size i change with these conditions, (3) how does the island size distribution (ISD) depend on this critical island size, and (4) why it displays scale invariance?

16.4.2 Mean-Field Rate Equations

The following assumptions are made in deriving the mean-field rate equations:

a. Only monomers are mobile.
b. Islands of size $s \geq 2$ grow by capturing the diffusing monomers.
c. Monomers are produced on the substrate by random deposition from vapor or by release from an existing island.
d. The direct impingement from vapor onto existing monomers and islands can be neglected in the early stages of nucleation and growth.

The time evolution of monomer density N_1 and density N_s of islands of size s is

$$\frac{dN_1}{dt} = F - 2D\sigma_1 N_1^2 - DN_1 \sum_{s>2} \sigma_s N_s + 2\frac{N_2}{\tau_2} + \sum_{s\geq 3} \frac{N_s}{\tau_s} \tag{16.36}$$

and

$$\frac{dN_s}{dt} = DN_1(\sigma_{s-1}N_{s-1} - \sigma_s N_s) + \frac{N_{s+1}}{\tau_{s+1}} - \frac{N_s}{\tau_s}. \tag{16.37}$$

Here, F is the monomer deposition rate (monolayer per second), D is the monomer diffusion rate, σ_s is the so-called "capture number" of monomers by an island of size s, and $1/\tau_s$ is the dissociation rate (rate of monomer release) of an island of size s.

The preceding equations can be numerically integrated up to some finite time and some maximum possible island size at that time, provided one knows σ_s, the capture numbers, and $1/\tau_s$, the dissociation rate of an island of size s. σ_s can be obtained by considering the monomer density $n_1(r,t)$ in the vicinity of a circular island of radius $r_s = \sqrt{s/\pi}$ at the origin and by using cylindrical symmetry. The monomer density is zero at the adsorbing island edge and rises to the global average N_1 a long way from the island. The diffusive flux into the island is given by

$$DN_1\sigma_s = D2\pi r_s \frac{\partial n_1}{\partial r}\Big|_{r_s}, \tag{16.38}$$

from which the capture number σ_s can be calculated. The monomer density field can be obtained from the diffusion equation by first assuming that the monomers can evaporate from the substrate at the rate $1/\tau_a$ and that $N_1 \rightarrow F\tau_a$, a long way from the island. The later assumption implies that the density of islands is sufficiently low. The diffusion equation can be written as

$$\frac{\partial n_1(r,t)}{\partial t} = D\nabla^2 n_1(r,t) - \frac{n_1(r,t)}{\tau_a} + F. \tag{16.39}$$

In the steady state $t > \tau_a$, it can be shown from Eqs. (16.38) and (16.39) that the capture number σ_s can be written as (Problem 16.6)

$$\sigma_s = 2\pi X_s \frac{K_1(X_s)}{K_0(X_s)}, \tag{16.40}$$

where K_0 and K_1 are the Bessel functions and

$$X_s = \frac{r_s}{\sqrt{D\tau_a}}. \tag{16.41}$$

If we assume that the islands around the central one (being considered) are randomly distributed and each one has the same size <s> and the same capture number σ_x, τ_s in Eq. (16.40) can be replaced by τ such that

$$\frac{1}{\tau} = \frac{1}{\tau_a} + D\sigma_x N, \tag{16.42}$$

where $N = \sum_{s>i} N_s$ is the density of stable islands. The evaporation of monomers can be switched off ($1/\tau_a = 0$) in Eq. (16.42) to model the "complete condensation" regime of the film growth.

In an improved method using the results from kinetic Monte Carlo simulations, it has been shown (Bales and Chrazan[1]) that a self-consistent solution for all σ_s can be used to reproduce the evolution of the monomer and total island density. In their work, Eq. (16.42) for the average capture number is replaced by

$$\frac{1}{\tau} = 2D\sigma_1 N_1 + \sum_{s\geq 2} D\sigma_s N_s + F\kappa_1, \tag{16.43}$$

which allows for monomer trappings by other monomers and direct hits from vapor deposition on existing monomers as well as capture by islands.

16.5 STRUCTURE OF ISOLATED CLUSTERS

16.5.1 Theoretical Models

Hartree–Fock Methods

The earliest method employed was the Hartree–Fock method (discussed in Chapter 4), but because it uses a single-determinant wave function, it does not describe electrons with different spins. The neglect of correlations produces energy values that are larger than the actual ones. There have been various methods used to include the effect of correlations in post-Hartree–Fock theories.

In the configuration interaction (CI) methods, one or more occupied orbitals in the Hartree–Fock determinant are replaced by virtual orbitals where each replacement is equivalent to an electron excitation. In the full CI method, the wave function is formed as a linear combination of the Hartree–Fock determinant and all possible substitute determinants. However, it is necessary to truncate the expansion at a fairly low level of substitution (single and double excitations) except for the smallest systems. There are refinements that use Rayleigh–Schrodinger perturbation theory and add the higher excitations as noniterative corrections. Usually, these methods are limited to small clusters of about 20 atoms.

Density Functional Theory

The local density approximation (LDA) in density functional theory (DFT) can be used by first writing the exchange-correlation energy as a sum of exchange and correlation part

$$E_{xc} = E_x + E_c. \tag{16.44}$$

If $\rho_\alpha(\mathbf{r})$ and $\rho_\beta(\mathbf{r})$ are the electron densities of the two spin components, $(\alpha = \uparrow)$ and $(\beta = \downarrow)$, the local exchange functional is given by the standard Dirac–Slater form

$$E_x^{LDA} = -\frac{3}{2}\left(\frac{3}{4\pi}\right)^{1/3} \int d\mathbf{r}[\rho_\alpha^{4/3} + \rho_\beta^{4/3}], \tag{16.45}$$

and the Kohn–Sham (Ref. 17) wave functions are atom-centered. The local spin-density correlation energy functional (Vosko et al.[34]) can be written as

$$E_c^{LDA}[\rho_\alpha, \rho_\beta] = \int d\mathbf{r}\rho\varepsilon_c(\rho_\alpha, \rho_\beta), \tag{16.46}$$

where ε_c is a complicated function of ρ and several constants. However, the correlation energy is overestimated by 100% in this approximation.

A more widely used exchange-energy functional based on the generalized-gradient approximation (GGA) of Becke (Ref. 2) is given by

$$E_x^{GGA} = E_x^{LDA} - \beta \sum_{\sigma=\alpha,\beta} \int d\mathbf{r}\, \rho_\sigma^{4/3} \frac{x_\sigma^2}{1 + 6\beta x_\sigma \sinh^{-1}(x_\sigma)}, \tag{16.47}$$

where x_σ is the dimensionless ratio

$$x_\sigma = \rho_\sigma^{-4/3}|\nabla\rho_\sigma|. \tag{16.48}$$

In general, the GGA exchange-correlation energy functionals can be expressed as

$$E_{xc}^{GGA}[\rho_\uparrow, \rho_\downarrow] = \int d\mathbf{r}\, f(\rho_\uparrow, \rho_\downarrow, \nabla\rho_\uparrow, \nabla\rho_\downarrow). \tag{16.49}$$

An interesting technique to go beyond GGA is to use hybrid functionals that are formulated as a mixture of Hartree–Fock and DFT exchange coupled with DFT correlation. However, a number of authors have suggested that hybrid methods should be avoided for metal clusters because these techniques do not yield reliable results.

Tight-Binding Methods

The tight-binding model is used mainly in the study of transition metal clusters. The Hamiltonian is expressed in terms of matrix elements in an orthogonal basis set composed from the $s, p(x, y, z), d(xy, yz, zx, x^2 - y^2, 3z^2 - r^2)$ valence atomic orbitals. The intersite matrix elements are determined by the Slater–Koster hopping integrals (Slater and Koster,[31]) $ss\sigma$, $sp\sigma$, $sd\sigma$, $pp\sigma$, $pd\pi$, $dd\delta$, which decay exponentially with distance, but the three-center integrals are ignored. In the model proposed by Mehl and Papaconstantopoulos,[27] the total energy is obtained by summing up the occupied energy levels,

$$\varepsilon_i = a_i + b_\lambda \rho_i^{2/3} + c_\lambda \rho_i^{4/3} + d_\lambda \rho_i^2, \tag{16.50}$$

where

$$\rho_i = \sum_{j \neq i} \exp\left[-\gamma\left(\frac{r_{ij}}{r_0 - 1}\right)\right]. \tag{16.51}$$

The three on-site terms depend on the orbital angular momentum $\lambda = s, p, d$. In Eq. (16.51), r_0 is the interaction term in the bulk, and a cutoff is included in the distance in the sum. The repulsive energy is incorporated by varying it as a function of the total density of each function.

The Hamiltonian for the model is written as

$$H = \sum_{i\lambda\sigma} \varepsilon_{i\lambda} c_{i\lambda\sigma}^{\dagger} c_{i\lambda\sigma} + \sum_{\substack{i \neq j \\ \lambda\mu\sigma}} \beta_{ij}^{\lambda\mu} c_{i\lambda\sigma}^{\dagger} c_{j\mu\sigma}. \tag{16.52}$$

In Eq. (16.52), λ and μ are orbital labels, σ is the spin label, $\beta_{ij}^{\lambda\mu}$ are thc hopping integrals between sites i and j, and the other symbols have their usual meanings. The different parameters are obtained by fitting to bulk properties. There are a large number of variations of the tight-binding model. Specifically, spin-polarized systems, a Hubbard-like term,

$$H_{\text{int}} = -\bar{J}_\lambda \sum_{i\lambda\sigma} \sigma c_{i\lambda\sigma}^{\dagger} c_{i\lambda\sigma}, \tag{16.53}$$

is added to Eq. (16.52). In Eq. (16.53),

$$\bar{J}_\lambda = \frac{1}{2} \sum_{\mu} J_{\lambda\mu} (\bar{n}_{i\mu\uparrow} - \bar{n}_{i\mu\downarrow}), \tag{16.54}$$

where $J_{\lambda\mu}$ is an exchange integral, and $\bar{n}_{i\mu\sigma}$ is the component of electron density at site i associated with orbitals μ and spin σ. A variety of additional terms are added to this basic model to obtain the computational results that can relate better to the cluster properties.

Semi-Empirical Potentials

To study large clusters, several semi-empirical potentials have been developed that contain many-body contributions and can be used with Monte Carlo or molecular dynamics simulations. The various models are known as embedded atom model (EAM), effective medium theory (EMT), glue model, and second-moment methods. All these methods start with a common functional form for the total energy,

$$E = \sum_i \left[\frac{1}{2} \sum_{j \neq i} V(r_{ij}) + F(\overline{\rho}_i) \right], \tag{16.55}$$

where $\frac{1}{2}\sum_{j \neq i} V(r_{ij})$ is a repulsive pair potential between atoms separated by a distance r_{ij}. In the EAM (Daw and Baskes[5]; Foiles et al.[10]), each term of a metal is considered as an impurity embedded in a host provided by the rest of the electrons. $F(\overline{\rho}_i)$ is the energy required to embed atom i in an electron gas of density $\overline{\rho}_i$, where

$$\overline{\rho}_i = \sum_{j \neq i} \rho(r_{ij}). \tag{16.56}$$

One has to use empirical fits to bulk properties to be able to generate the embedding functions and pair interactions.

In the second-moment approximations (Finnis and Sinclair,[9]; Sutton,[33]) the total energy is

$$E = \varepsilon \sum_i [\sum_{j \neq i} V(r_{ij}) - c\sqrt{\rho_i}]. \tag{16.57}$$

One can see the similarity as well as the contrast between Eqs. (16.55) and (16.57). In Eq. (16.57), ε is a parameter that has dimensions of energy, and c is a dimensionless parameter. ρ_i is expressed as a sum of pair potentials

$$\rho_i = \sum_{j \neq i} \phi(r_{ij}). \tag{16.58}$$

ρ_i can be described as the bond energy. We consider just one orbital site and local density of states $d_i(E)$ at site i with a center of gravity ε_i. The bond energy is given by

$$E_{bond} = 2\sum_i \int_{-\infty}^{E_F} (E - \varepsilon_i) d_i(E) dE. \tag{16.59}$$

The second moment of the local density of states can be written as

$$\mu_i^{(2)} = \int_{-\infty}^{\infty} (E - \varepsilon_i)^2 d_i(E) dE. \tag{16.60}$$

From Eqs. (16.59) and (16.60), we obtain an approximation in which moments no higher than the second are used (Problem 16.7),

$$E_{bond}^{(i)} = A\sqrt{\mu_i^{(2)}}, \tag{16.61}$$

where A is a constant. It can be shown that the local density of states can be written as

$$d_i(E) = <i|\delta(E-H|i>, \tag{16.62}$$

where H is the Hamiltonian for the system, and $|i>$ is the orbital on-site i. From Eqs. (16.60) and (16.62), we obtain an expression for the second moment,

$$\begin{aligned}\mu_i^{(2)} &= \int_{-\infty}^{\infty} (E-\varepsilon_i)^2 <i|\delta(E-H)|i> dE \\ &= <i|(\varepsilon_i - H)^2|i>.\end{aligned} \tag{16.63}$$

We can write the Hamiltonian H as

$$H = \sum_i \varepsilon_i |i><i| + \sum_{i \neq j} \beta_{ij} |i><j|, \tag{16.64}$$

where ε_i is the energy at site i, and β_{ij} is the hopping integral from site i to j. From Eqs. (16.63) and (16.64), we obtain

$$\mu_i^{(2)} = \sum_{j \neq i} \beta_{ij}^2. \tag{16.65}$$

The second-moment approximation is very effective for metals that tend to form close-packed structures. However, one needs the fourth moment, which controls the stability of the bcc structure, and the fifth and sixth moments if hcp and fcc packing arc to be differentiated. There are several other semi-empirical potentials that we will not discuss in this chapter.

There are several alternate models to choose the potential in Eqs. (16.57) and (16.58). Once a potential is chosen, there are many global optimization techniques that can be used to find the lowest energy of a cluster. The standard Monte Carlo or molecular dynamics procedure is used after incorporating some algorithms. The computation involved is very massive, but fairly good global minima have been obtained for clusters of more than 100 atoms.

16.5.2 Structure of Some Isolated Clusters

A schematic configuration of a number of commonly occurring small clusters is shown in Figure 16.12. There is usually some departure from the symmetric configurations shown in the figure. The linear and planar clusters usually show some departure from one or two dimensionality, and triangular configuration in 3b is isosceles instead of equilateral. The three-dimensional clusters occur at various sizes. One extreme is nickel, which has a tetrahedral structure at $N = 4$, whereas the other extreme is gold clusters, which are planar until $N > 10$. The alkali metals are between these two extremes. An interesting aspect is that anions and cations have different structures than neutral clusters.

A large number of experimental research is focused on probing clusters of special sizes to verify whether magic number clusters, such as icosahedral or cuboctahedral (13, 55, 147), truncated octahedral (38), or Marks decahedral (75, 146, 192), are more stable than clusters of other sizes as suggested by the geometrical shell model.

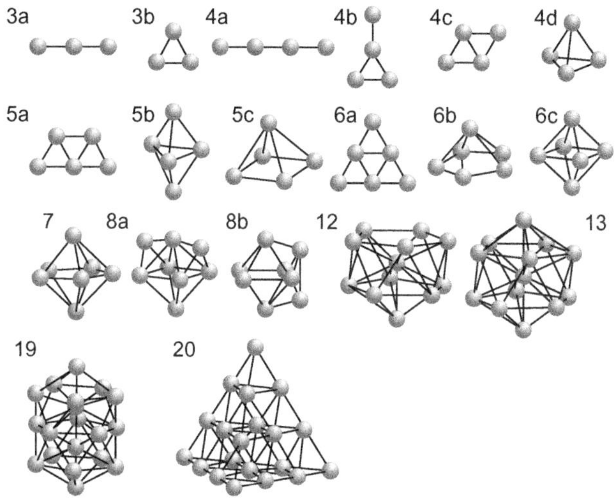

FIGURE 16.12

Schematic representation of some small clusters. The 3D figures are (4d) tetrahedron, (5b) trigonal bipyramid, (5c) square pyramid, (6b) pentagonal pyramid, (6c) octahedron, (7) pentagonal bipyramid, (8a) bicapped octahedron, (8b) bisphenoid, (12) incomplete icosahedron, (13) icosahedron, (19) double icosahedron, and (20) large tetrahedron.

Reproduced from Misra[28] with the permission of Elsevier.

16.6 MAGNETISM IN CLUSTERS

16.6.1 Magnetism in Isolated Clusters

Superparamagnetism and Blocking Temperature

There has been a rapidly growing interest in the nanostructures formed from magnetic clusters because of their enormous potential in the development of high-performance magnetic materials and devices. There has been considerable development of technologies in the manufacture of quantum dots, monolayers, self-organized islands, quantum wires, and deposited nanoclusters. However, magnetism in small clusters, which is in the mesoscopic regime, is not well understood in either the atomic or bulk states.

It has been shown (Skomski,[30]) that nanoclusters are single-domain particles of which the size is well below the critical radius R_{SD}, above which it is favorable to form domain walls. For a sphere,

$$R_{SD} = \frac{36\sqrt{AK}}{\mu_0 M_s^2}, \tag{16.66}$$

where M_s is the spontaneous magnetization, K is an anisotropy constant, and A is related to the exchange stiffness.

At low temperatures (well below the equivalent of the Curie temperature), the nanoparticles can be considered as giant magnetic moments of ferromagnetically coupled spins. This phenomenon is known as superparamagnetism. In an external field, the reorientation of this giant moment takes place coherently and the individual moments remain aligned with each other. The different magnetic alignments are separated by an anisotropy boundary KV, where V is the volume of the nanoparticle and K is the anisotropic constant. At high temperatures, because $k_BT \gg KV$, the anisotropic boundary becomes unimportant, but there is a competition between the external field, which tends to orient the particles, and the thermal fluctuations of the magnetic moments, which tend to magnetically disorder the array.

The volume V of the transition metal clusters is $V \leq 10^{-25} m^3$, and at room temperature, $k_BT \gg KV$. Because all magnetization directions have equal probability, the magnetization M of a cluster along the direction of an applied field H can be obtained from the classical Langevin function L,

$$M = \mu_c \mathscr{L}\left(\frac{\mu_c H}{k_B T}\right), \tag{16.67}$$

where μ_c is the giant magnetic moment of each cluster and $\mathscr{L}(x) = \coth(x) - 1/x$.

An assembly of nanoclusters, each of which has several hundred atomic spins, can be easily saturated, unlike an assembly of atoms. At temperatures less than 50° K, $k_BT \sim KV$, and the magnetization in a given field deviates from superparamagnetism. At very low temperatures, the moment in each particle becomes static. The blocking T_B is the temperature at which half the cluster moments have relaxed during the time of a measurement. Because a narrow temperature region around T_B separates the frozen moments from superparamagnetic behavior, these can be used in magnetic recording technology provided one can obtain deposited small clusters with T_B greater than the room temperature. In fact, the deposition of clusters or their embedding in a matrix is essential for practical applications.

Cluster Magnetism

The magnetism in small clusters is usually significantly different from bulk magnetism of the same atoms because a much larger percentage of atoms in a cluster lies at the surface. For example, 162 of the atoms in a 309-atom cuboctahedral cluster lie at the surface, and these atoms have a reduced coordination and are in a lower symmetry environment. In the second-moment approximation discussed earlier, the bandwidth W is proportional to $z^{1/2}$, where z is the coordination number. For simplicity, if we consider a rectangular band for the d electrons, the density of states scales as $1/W$ (Problem 16.8).

The reduced coordination of the surface atoms results in an enhanced density of states $D(\varepsilon_F)$. We have derived in Eq. (13.111),

$$M \approx \frac{2\mu_B^2 BD(\varepsilon_F)}{1 - UD(\varepsilon_F)}. \tag{16.68}$$

Thus, due to the Stoner criterion for ferromagnetic instability, $UD(\varepsilon_F) > 1$, there is possibility of ferromagnetic instability in clusters that was absent in the bulk material.

The reduced z_i, the coordination number of the surface atoms, will result in an increase of the local moments μ_i on the atom i on the surface compared with the bulk value μ_{bulk}.

A crude approximation is

$$\mu_i = \left(\frac{z_{bulk}}{z_i}\right)^{1/2} \mu_{bulk}. \tag{16.69}$$

However, a reduced coordination in the second-moment approximation causes a contraction in the interatomic distances, which reduces the magnetic moment. A simple expression proposed for the average moment $\overline{\mu}_N$ of a cluster of N atoms (Jensen and Bennemann,[14]) is given by

$$\mu_N = \mu_{bulk} + (\mu_{surf} - \mu_{bulk})N^{-1/3}. \tag{16.70}$$

The experiments indicate a more complex size dependence but confirm the trend of the decrease in moment toward the bulk value with increasing size.

16.6.2 Experimental Techniques for Studying Cluster Magnetism

Chemical Probe Methods

The chemical probe method is usually used to investigate the geometric structure of clusters of Fe, Co, and Ni (and other transition metals) up to the sizes of 200 atoms. Atoms and small molecules react with transition metal clusters in ways that are analogous to the physisorption and chemisorption processes that occur on metal surfaces. The geometric structure of the cluster can be inferred from the fact that the reactivity of the cluster is dependent on the environment of the atoms. The main probe molecules that have been used are NH_3, N_2, H_2O, and H_2 or D_2. The determination of the number of binding sites of a cluster for a particular molecule yields important clues as to possible structures. Because the *d*-orbitals have significant spatial extent and are involved in the transition metal bonding, the chemical properties are sensitive to both the number and the configuration of the *d* electrons.

The experimental arrangement is as follows. The clusters are produced by a pulsed laser vaporization of a metal target in a flow tube upstream of a flow-tube reactor (FTR) using helium gas. To ensure that the cluster growth is finished and clusters have cooled to ambient temperature before they enter the FTR, one uses a narrow flow tube and low helium pressure to ensure a rapid decrease in metal atom density. The reagent gas is introduced when the clusters enter the FTR. The reaction between the cluster, M_N, and the reagent molecule, A, produces an internally excited complex,

$$M_N + A \rightleftarrows M_N A^*. \tag{16.71}$$

The excited complex requires a collision with a third body (a carrier gas atom, C) to stabilize it,

$$M_N A^* + C \rightarrow M_N A + C. \tag{16.72}$$

The clusters and reaction products expand out of the nozzle and form into a molecular beam. The clusters are then pulsed laser ionized and mass analyzed in a time-of-flight mass spectrometer. There are different binding rules for the various reagent molecules. Surface studies indicate that the NH_3 molecule binds to the surface through the donation of the N-localized lone pair to a metal atom. However, in the case of clusters, the binding preferentially occurs on the low-coordination metal atoms, which means that saturation with ammonia counts the number of vertex atoms on a cluster. Clusters of cobalt and nickel yield more straightforward clues than iron about their structure. The reason is iron and other elements in the middle of the transition metal series have fewer filled *d*-orbitals, and the binding of the atoms in the cluster is much more directional. It is interesting to note that, although cobalt and nickel are structurally similar, there are significant differences for small clusters. This is shown in Figure 16.13.

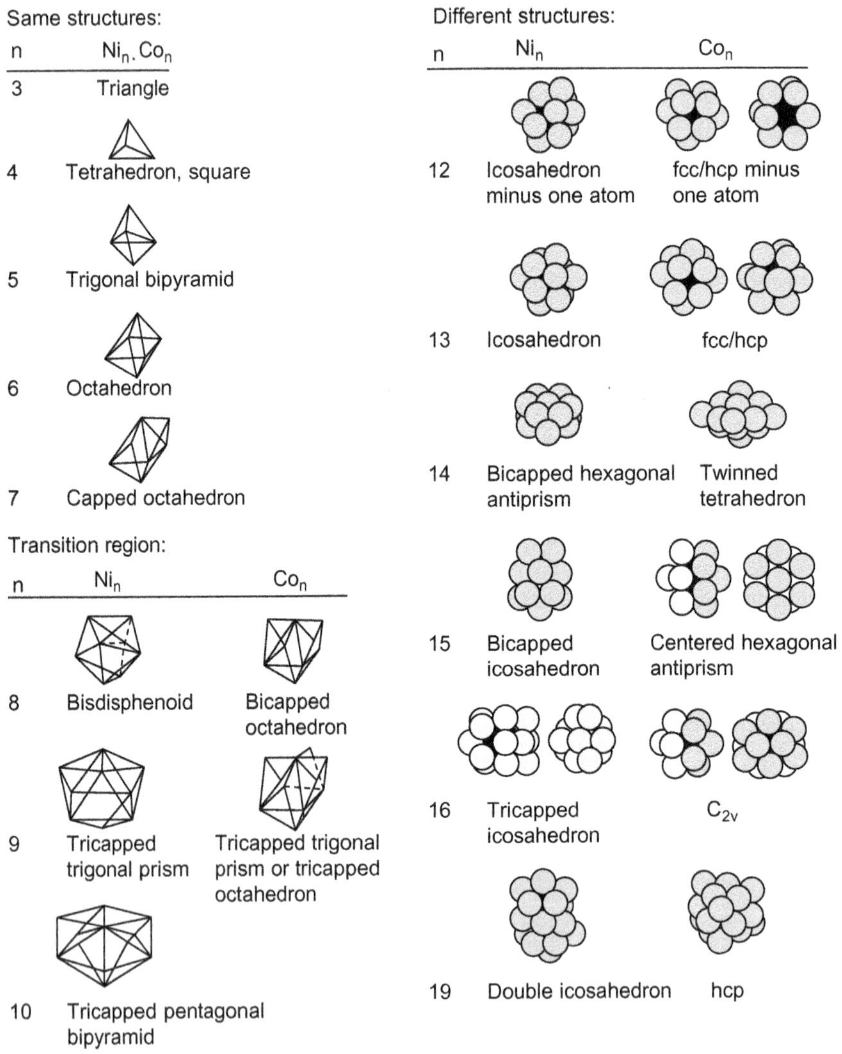

FIGURE 16.13

Comparison of structures of 3–19 atom nickel and cobalt structures.

Reproduced from Riley[29] with the permission of Elsevier.

Gradient-Field Deflection (Stern–Gerlach)

The gradient-field deflection method is based on the classic Stern-Gerlach experiment that first detected the electron spin. A schematic diagram of the experiment for measuring the magnetic moment of free clusters is shown in Figure 16.14.

A collimated cluster beam that is generated by a pulsed laser evaporation source with a variable temperature nozzle is guided into a magnetic field gradient dB/dz, which deflects a particle of magnetic moment M vertically by

$$d = M\frac{dB}{dz}L^2\frac{(2D/L+1)}{2mv_x^2}. \tag{16.73}$$

Here, m is the mass of the cluster, and v_x is its velocity when it enters the magnet. D is the distance from the end of the magnet of length L to the detector. v_x is measured by a controlled delay between the evaporation laser and the ionizing laser. A mechanical chopper is used in front of the source to define the pulse start time. Figure 16.15 shows the low-temperature total moments of Fe_N, Co_N, and Ni_N (in μ_B/atom) as a function of cluster size. The right scale in Figure 16.15 indicates the spin imbalance obtained from the equation

$$\frac{M_{spin}}{M_{tot}} = \frac{2}{g}, \tag{16.74}$$

where g is the gyromagnetic ratio.

One can see in Figure 16.15 that clusters with different sizes show locked-moment behavior. The moment of the Ni cluster drops rapidly to the bulk value at $N = 160$ and then increases slowly until it reaches the bulk value again at $N \approx 350$. The moments of Fe and Co clusters fall less rapidly. They reach the bulk value between $N \approx 400$–500.

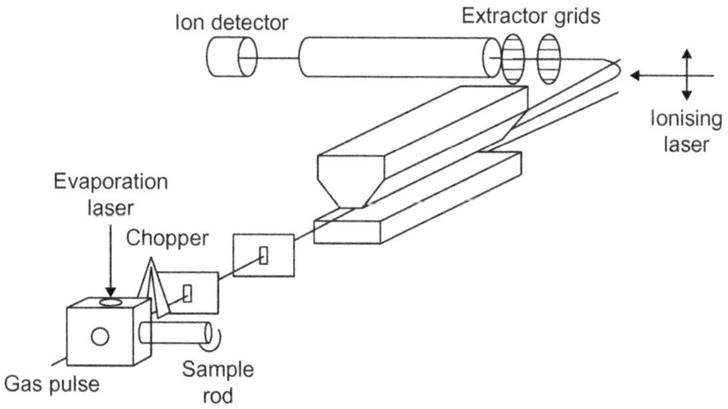

FIGURE 16.14

Schematic of the experiment for measuring the magnetic moment of free clusters.

Reproduced from Misra[28] with the permission of Elsevier.

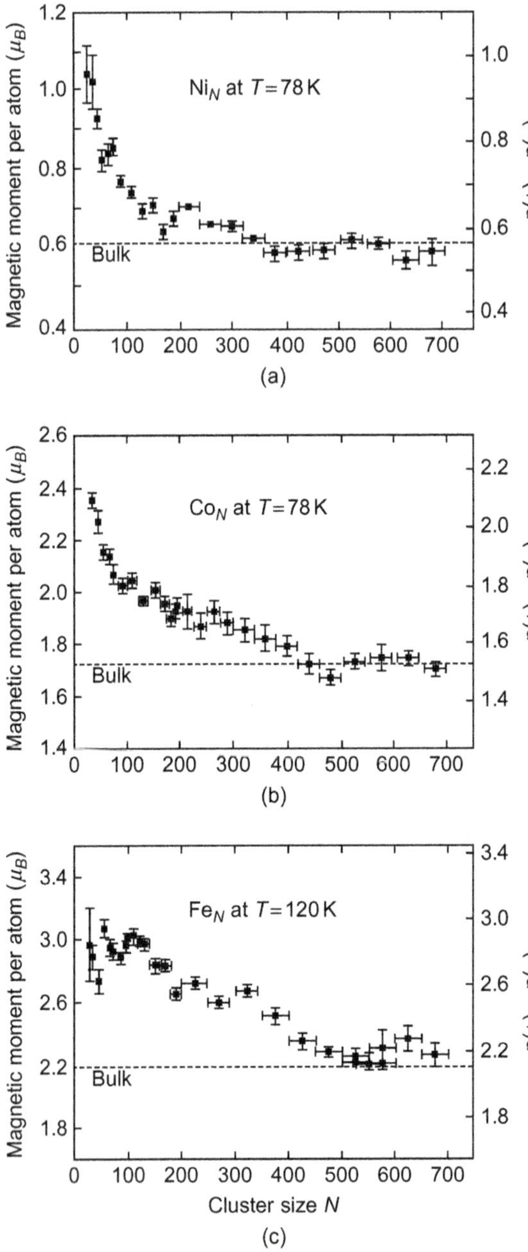

FIGURE 16.15

Low-temperature total moments (in μ_B/atom) of (a) Ni_N, (b) Co_N, and (c) Fe_N. The right scale indicates the spin imbalance defined in Eq. (16.74).

Reproduced from Isabella et al.[13] with the permission of Elsevier.

The magnetic behavior of clusters primarily depends on their structure. Most of the theoretical work has been done using density functional theory and geometry optimization; however, the details of the theoretical calculations are beyond the scope of this book.

16.6.3 Magnetism in Embedded Clusters

Switching and Blocking Temperature

The interest in magnetic nanoparticles has recently increased considerably because of the device applications such as in ultra-high-density recording as well as in spintronics. For device applications, the nanoclusters are either deposited on a surface or embedded in another material. The embedding material can modify the properties of the magnetic clusters to a great extent.

Switching (from magnetism to superparamagnetism) is a major factor in miniaturization of devices. When the energy barrier due to anisotropy, KV, becomes very small, the probability of switching due to the thermal energy becomes significant, which creates problems in such devices. The probability, P, of not switching in time t is given by the Boltzmann statistics

$$P = e^{-\frac{t}{\tau}}, \tag{16.75}$$

and the Arrhenius relation

$$\frac{1}{\tau} = f_0 \, e^{-\left(\frac{KV}{k_B T}\right)}, \tag{16.76}$$

where f_0 is the attempt frequency. The blocking temperature, which marks the onset of instability during an observation time τ, is obtained from Eq. (16.76) as

$$T_B = \frac{KV}{k_B \ln(f_0 \tau)}. \tag{16.77}$$

For a spherical particle of diameter D, $V = 1/6\, \pi D^3$, and from Eq. (16.77), we obtain

$$T_B = \frac{\pi K D^3}{6 k_B \ln(f_0 \tau)}. \tag{16.78}$$

For data storage, the time constant $\tau \approx 10$ years, which yields $\ln(f_0\tau) \approx 40$. For a typical uniaxial anisotropy $K \approx 0.2$ MJ/m^3, and Eq. (16.77) yields for a nanocluster with $D = 14$ nm, $T \approx 520°$ K for a 10-year blocking temperature, which is well above the operating temperature of a memory device. However, for a nanocluster for which $D = 7$ nm, $T \approx 65°$ K, which is far lower than the room temperature.

One option to increase the blocking temperature is to increase the anisotropy K by depositing ferromagnetic particles on a platinum surface (which has large spin-orbit interaction) or the formation of alloys such as CoPt or FePt. An alternate approach to increase the effective anisotropy is through relying on the exchange bias phenomenon. This is based on the enhanced stability of the ferromagnetic component due to the exchange coupling at the interface of a ferromagnetic and antiferromagnetic material as well as the high anisotropy in the antiferromagnet.

Small Clusters and 2D Nanostructures

In this section, we first study the introduction of small 3*d* metal clusters on surfaces of a noble metal (Cu(001)). The results of the calculations are based on density functional theory and the KKR method in which a Green's function technique is used to treat the perturbation of a surface into the ideal crystal and then placing an adatom or cluster on the surface. Figure 16.16 shows the spin magnetic moment of 3*d* metal clusters and monolayer on Cu(001).

One can note that the moments on the Co and Fe atoms plotted in Figure 16.16 are relatively insensitive to the island size.

The deposition of atoms of the 3*d* elements onto a ferromagnetic surface shows that the magnetic moments of the adatoms may align parallel or antiparallel to the direction of magnetization of the substrate. There is further complexity with 2*d* clusters or nanolayers if there is a tendency for antiferromagnetic ordering within the nanostructure itself. The spin magnetic moments obtained from using KKR Green's function method with exchange and correlation effects on monolayers on bcc Fe(001) are shown in Figure 16.17.

Similar results have been obtained for fcc Ni(001) substrate except for Mn. The total energy calculations for fcc Ni(001) substrate on Mn indicate ferromagnetic coupling for all positions. Further, the enhancement of the impurity moments is larger with Ni(001) than with Fe(001) because the hybridization is weaker due to the fact that the Ni wave function is less extended. The dimers on next-nearest-neighbor sites on top of Fe(001) couple to the surface with the same configuration as the adatoms except for Mn. The moments of the dimer atoms are parallel to each other. However, for the Mn dimer, the two states are energetically degenerate, one with the moments on both atoms ferromagnetically coupled to the substrate, whereas the other has the moments of the dimer atoms antiparallel.

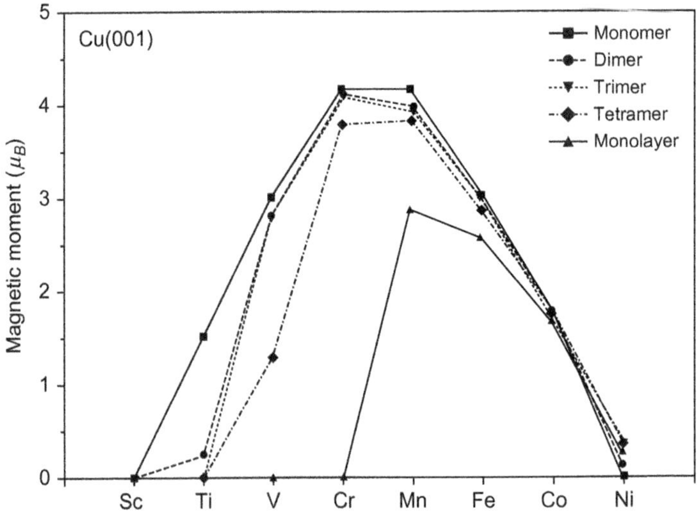

FIGURE 16.16

Spin magnetic moment of 3*d* metal clusters and monolayer on Cu(001).

Reproduced from Stepanyuk et al.[32] with the permission of Elsevier.

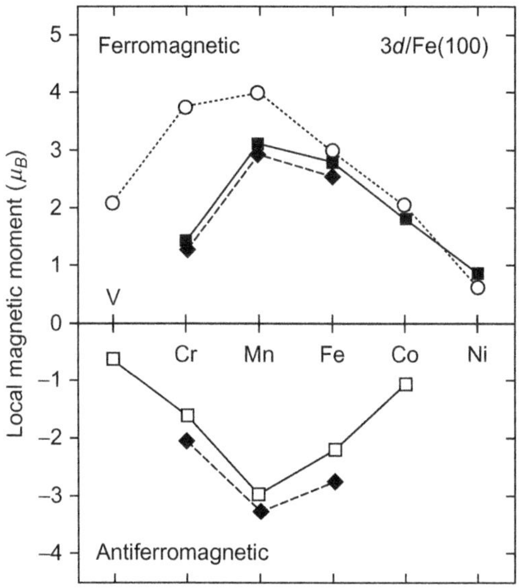

FIGURE 16.17

Magnetic moments of the monolayer on the surface: p(1 x 1) ferromagnetic coupling (solid squares/solid line); layered antiferromagnetically (open squares); c(2 x 2) ferromagnetic (diamonds/dashed line); 3*d* monolayers on Ag(001) for comparison.

Reproduced from Handschuh and Blugel[11] with the permission of Elsevier.

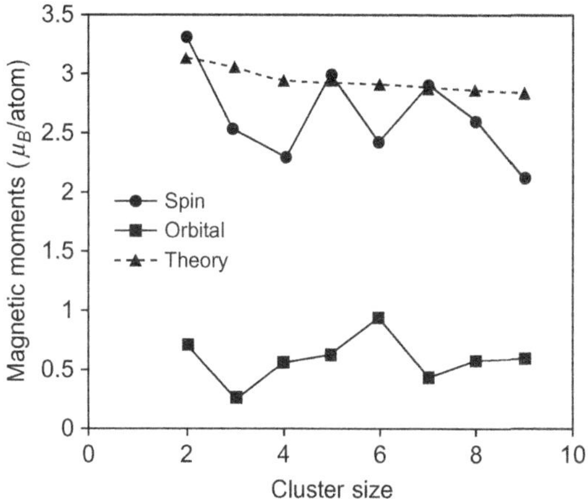

FIGURE 16.18

Experimental results from Lau et al.[15] of spin and orbital moments (in μ_B/atom) and theoretical results from Martinez et al.[26] for spin moments for Fe clusters deposited on Ni/Cu (001) clusters.

Reproduced from Misra[28] with the permission of Elsevier.

It is interesting to note that the spin moments of Fe clusters deposited on the Ni(001) surface are enhanced compared with bulk bcc Fe. The results of theoretical calculation, using a modified embedded atom model, show that the spin moments of $N = 2$–9 Fe clusters on Ni(001), which adopt a planar geometry on top of the Ni(001) surface, have spin moments that vary smoothly from 3.15 μ_B to 2.85 μ_B for $N = 2$ to $N = 9$ (Martinez et al.[26]). In contrast, the experimental results (Lau et al.[20]) indicate oscillations. These contradictory results are shown in Figure 16.18.

16.6.4 Graphite Surfaces

The spin magnetic moments of the 3*d* transition metal adatoms and dimers on a graphite surface from the calculations of Duffy and Blackman[6] and of 3*d* monolayers by Kruger et al. (Ref. 19) are shown in Figure 16.19.

16.6.5 Study of Clusters by Scanning Tunneling Microscope

The ability to image single impurity atoms on a surface by scanning tunneling microscope (STM) has become very convenient to study the properties of clusters on the surface of a metal. The images of magnetic atoms on the surface of a metal and the Kondo resonance can be obtained from the current versus voltage characteristics. STM experiments measure the bias voltage dependence of

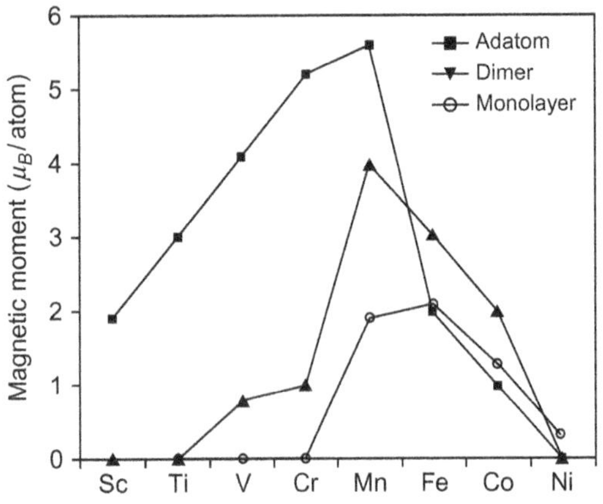

FIGURE 16.19

Composite showing the results of calculations of spin moments of 3*d* transition metal adatoms, dimers (Duffy and Blackman,[6] and monolayers (Kruger et al.[19]) on graphite. The Mn monolayer has antiferromagnetic and ferrimagnetic ground states that are degenerate.

Reproduced from Misra[28] with the permission of Elsevier.

the differential conductance dI/dV. The electrons tunnel between states at ε_F on the tip and states at $\varepsilon_F + eV$ on the surface, so the structure in the measured differential conductance reflects the surface density of states at that energy. Because the STM probes a localize state immersed in a Fermi sea, the spectroscopy is that of a discrete auto-ionized state. We discuss the case of noninteracting impurity resonances that have no Coulomb interactions. This is the situation when the *s*-orbital of an adatom hybridizes with a metal surface. Such a system can be described by a resonant-level model, where the electron is allowed to hop between the discrete atomic orbital and the continuum of the electronic band states. Thus, the Kondo resonance can be analyzed as a type of Fano resonance (Fano,[8]). Fano showed that the Hamiltonian for a discrete state and a continuum, ignoring the spin, can be written as

$$\hat{H}_0 = \sum \varepsilon_k \hat{c}^\dagger_{\mathbf{k}} \hat{c}_{\mathbf{k}} + \varepsilon_a \hat{a}^\dagger \hat{a} + \sum_{\mathbf{k}} (V_{a\mathbf{k}} \hat{a}^\dagger \hat{c}_{\mathbf{k}} + V^*_{a\mathbf{k}} \hat{c}^\dagger_{\mathbf{k}} \hat{a}), \tag{16.79}$$

where ε_a, $\hat{a}^\dagger$, and $\hat{a}$ are, respectively, the energy, creation, and annihilation operators of an electron residing in the discrete atomic state; and $V_{a\mathbf{k}}$ is the hybridization matrix element connecting the atomic state to the *k*th band state. The advanced atomic-state Green's function can be written as (Madhavan et al.[23])

$$G^0_{aa}(\varepsilon) = \frac{1}{\varepsilon - [\varepsilon_a + [\mathrm{Re}\,\Sigma_0(\varepsilon) + i\,\mathrm{Im}\,\Sigma_0(\varepsilon)]]}, \tag{16.80}$$

where the real and imaginary parts of the self-energy are

$$\mathrm{Re}\,\Sigma_0(\varepsilon) = \sum_{\mathbf{k}} |V_{a\mathbf{k}}|^2 P\left(\frac{1}{\varepsilon - \varepsilon_k}\right) \tag{16.81}$$

and

$$\mathrm{Im}\,\Sigma_0(\varepsilon) = \pi \sum_{\mathbf{k}} |V_{a\mathbf{k}}|^2 \delta(\varepsilon - \varepsilon_k). \tag{16.82}$$

Here, P denotes the Cauchy principal value. In the context of STM, the energy $\varepsilon = \mathrm{eV}$.

In a tunneling experiment, a second electrode (the tip) is added to the system, and electrons tunnel between the electrodes. The tip is modeled by a single state with energy ε_t, and $\hat{t}$ removes an electron from that state. $\hat{M}$ is a transfer Hamiltonian term (treated as a perturbation) that induces electrons to tunnel from one electrode to the other and can be expressed as

$$\hat{M} = (M_{at} \hat{a}^\dagger \hat{t} + H.c.) + \sum_{\mathbf{k}} (M_{\mathbf{k}t} \hat{c}^\dagger_{\mathbf{k}} \hat{t} + H.c.). \tag{16.83}$$

M_{at} and $M_{\mathbf{k}t}$ are the tunnel matrix elements that connect the STM tip to the discrete atomic state and continuum states, respectively. From Eqs. (16.79) and (16.83), we obtain

$$\hat{H} = \hat{H}_o + \hat{\varepsilon}_t \hat{t}^\dagger \hat{t} + \hat{M}. \tag{16.84}$$

It can be shown that the differential conductance can be written as (Problem 16.9)

$$\frac{dI}{dV} = \frac{2\pi e^2}{\hbar} \frac{(\varepsilon' + q)^2}{1 + \varepsilon'^2} \rho_{\text{tip}} \sum_{\mathbf{k}} |M_{\mathbf{k}t}|^2 \delta(eV - \varepsilon_k) + \text{constant}, \tag{16.85}$$

where

$$q = A/B, \tag{16.86}$$

$$A = M_{at} + \sum_{\mathbf{k}} M_{\mathbf{k}t} V_{a\mathbf{k}} P\left(\frac{1}{eV - \varepsilon_{\mathbf{k}}}\right), \tag{16.87}$$

$$B = \pi \sum_{\mathbf{k}} M_{\mathbf{k}t} V_{a\mathbf{k}} \delta(eV - \varepsilon_{\mathbf{k}}), \tag{16.88}$$

$$\varepsilon' = \frac{eV - \varepsilon_a - \text{Re}\,\Sigma(eV)}{\text{Im}\,\Sigma(eV)}, \tag{16.89}$$

and ρ_{tip} is the tip density of states. ρ_{tip} is usually treated as a constant to reflect the broadening of the tip state by contact with the remainder of the tip electrode. Here, the quantity A in Eq. (16.87) is an amplitude for tunneling to the discrete state modified by hybridization with the continuum (the second term in Eq. 16.87). The quantity B in Eq. (16.88) is an amplitude for tunneling into a set of continuum states contained in the range of energies on the order of the width of the resonance. If $q \gg 1$, tunneling via the resonance dominates, but if $q \ll 1$, the continuum states dominate the tunneling.

The experimental results for Co atoms on the Au(111) substrate are reproduced in Figure 16.20. This figure shows how the dI/dV spectra vary as the STM tip is moved with a clear feature of the intermediate q type appearing when the tip is in the vicinity of the Co atom, indicating Kondo resonance.

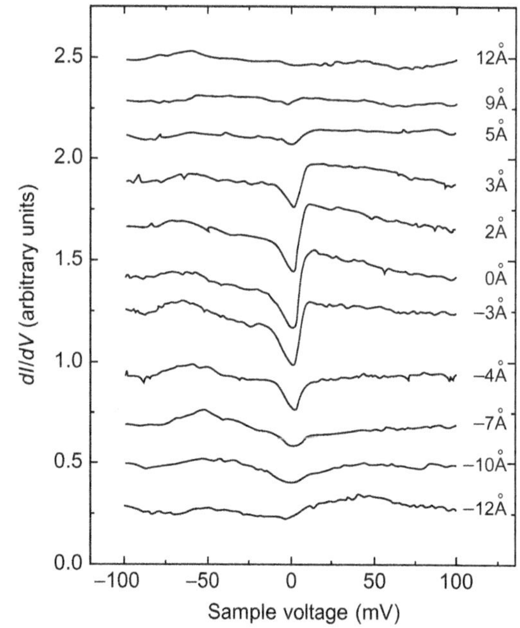

FIGURE 16.20

dI/dV spectra at various lateral distances away from the center of a single Co atom on Au(111) taken with the STM tip.

Reproduced from Madhavan et al.[22] with the permission of AAAS.

16.6.6 Clusters Embedded in a Matrix

The magnetic properties of particles are modified when they are embedded in a matrix of another material. The magnetic moment of a free particle is enhanced over the bulk value because the coordination of the atoms at the surface of the particle is reduced. In contrast, the embedded particle has the coordination of its surface atoms restored. The probability of the magnetic moment reverting to its bulk value depends on the electronic band structures of the particle and the host matrix. Further, if the particle and the matrix have the same bulk crystal structure with similar lattice spacings, there is the possibility of embedding with fairly good epitaxy. In contrast, if the two materials have different structures, the particle structure could either remain robust or adapt to that of the host matrix.

The overall picture for the full concentration range can be visualized by considering the magnetic phase diagram for films of deposited 3 nm diameter Fe nanoparticles embedded in Ag matrices, as shown in Figure 16.21.

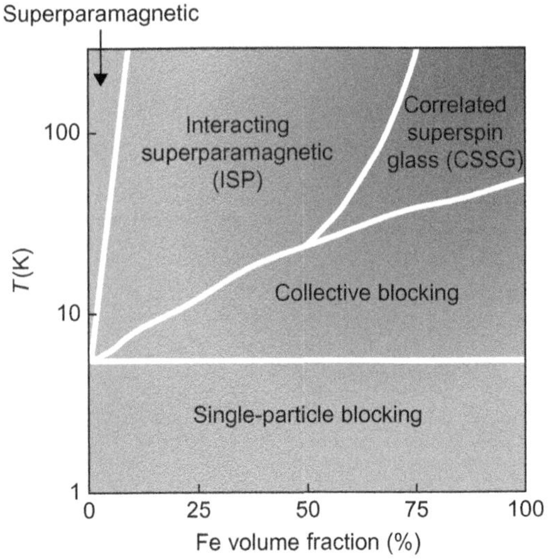

FIGURE 16.21

Magnetic phase diagram for films of deposited 3 nm diameter Fe nanoparticles embedded in Ag matrices as a function of volume fraction and temperature.

Reproduced from Binns et al.[3] with the permission of the Institute of Physics.

Ideal superparamagnetism occurs at the lowest concentrations above a blocking temperature. When the volume filling fraction (VFF) increases, the magnetic behavior is determined by the dipolar interactions and the aggregation. A correlated superspin glass (CSSG) state occurs at higher concentrations. Blocking occurs at low temperatures, and the particles are aligned to their anisotropy axes. There is a significant difference between single-particle blocking and collective blocking. The single-particle blocking occurs in isolated clusters because the intraparticle anisotropy is the only characteristic to stabilize the magnetic moment. Strong interparticle interactions lead to collective blocking, which occurs at higher temperatures. Collective blocking occurs due to the fact that a particle, which would be thermally activated if it was isolated, would have its magnetization stabilized by a neighbor that is slightly larger or more anisotropic.

16.7 SUPERCONDUCTING STATE OF NANOCLUSTERS

16.7.1 Qualitative Analysis

In Chapter 14, expressions for the critical temperature T_c in both the weak coupling (BCS) and the strong coupling (McMillan) theories were derived. The expression for the critical temperature T_c in the weak-coupling limit (Eq. 14.116) is

$$T_c \approx 1.13\,\theta_D \exp\left[-\frac{2}{D(\varepsilon_F)U}\right], \tag{16.90}$$

and in the strong coupling limit (Eq. 14.127), it is

$$T_c \approx \frac{\theta_D}{1.2} \exp\left[-\frac{1.04(1+\lambda)}{\lambda-\mu^*(1+0.62\lambda)}\right]. \tag{16.91}$$

In Eq. (16.91), λ is the coupling constant (Eq. 14.124),

$$\lambda = \nu < I^2 > / M\overline{\Omega}^2, \tag{16.92}$$

and ν is the bulk density of states (Eq. 14.125),

$$\nu = m^* p_F / 2\pi^2. \tag{16.93}$$

It is obvious from Eqs. (16.90) and (16.91) that T_c, the critical temperature of a superconductor, increases if the density of states increases at the Fermi level. In fact, it has been observed that the T_c of Al films (~2.1° K) is nearly double the value for bulk samples. Granular Al has $T_c \sim 3°$ K, which is almost three times the critical temperature of Al. These increases were explained by size quantization and corresponding increase in the effective density of states in films and isolated granules.

The metallic nanoclusters contain delocalized electrons of which the states organize into shells, similar to those in atoms or nuclei. In some clusters, the shells are completely filled all the way up to the highest occupied shell. These are known as the "magic numbers," $N = N_m = 20, 40, 58, 92, 138, 168, \ldots$, and the clusters are spherical. These magic clusters have similarity with atoms in the sense that their electronic states are labeled by radial quantum number n and orbital momentum l. The Cooper pairs are formed by electrons with opposite projections of orbital momentum, similar to the pairing in atomic nuclei. The degeneracy of the shell, which is $2(2l+1)$, is large if l is large, and the effective density of states correspondingly increases. Further, the energy spacing ΔE between neighboring shells varies. Some of them are separated by only a small ΔE.

The combination of high degeneracy and a small energy spacing ΔE between the highest occupied shell (HOS) and the lowest unoccupied shell (LUS) results in a large increase in the strength of the superconducting pairing interaction in the clusters. One can understand this result qualitatively from the fact that if the HOS is highly degenerate, the shell has a large number of electrons, which is equivalent to having a sharp peak in the density of states at the Fermi level. It is obvious from Eqs. (16.90) and (16.91) that the critical temperature of a superconductor increases if the density of states increases at the Fermi level.

16.7.2 Thermodynamic Green's Function Formalism for Nanoclusters

In this section, we will derive an expression for the critical temperature for the nanoclusters. We will show how it is possible that the critical temperature for some of these nanoclusters can be of the order of room temperature. We follow a method outlined by Kresin and Wolf.[18] However, we will not discuss in detail the strong coupling theory based on Green's function method of the many-body theory.

In the standard thermodynamic Green's function formalism of Abrikosov et al. (1963), the equation for the pairing order parameter, $\Delta(\omega)$, can be written as

$$\Delta(\omega_n)Z \approx \lambda T \sum_{\omega_{n'}} D(\omega_n - \omega_{n'}; \widetilde{\Omega}) F^+(\omega_{n'}), \tag{16.94}$$

where Z is the renormalization function that describes the "dressing" of electrons moving through the lattice, λ is the electron–phonon coupling constant,

$$\lambda = 2\int \alpha^2(\Omega)F(\Omega)\Omega^{-1}d\Omega, \tag{16.95}$$

$F(\Omega)$ is the phonon density of states, Ω is the phonon frequency, and $\alpha^2(\Omega)$ is a measure of the phonon-frequency-dependent electron–phonon interaction. Here, D is the phonon Green's function,

$$D = \Omega^2[(\omega_n - \omega_{n'})^2 + \Omega^2]^{-1}, \tag{16.96}$$

where

$$\omega_n = (2n+1)\pi T \tag{16.97}$$

and

$$\tilde{\Omega} = <\Omega^2>^{1/2}. \tag{16.98}$$

The average $<f(\Omega)>$ is determined by

$$<f(\Omega)> = \frac{2}{\lambda}\int d\Omega\, f(\Omega)\alpha^2(\Omega)F(\Omega)\Omega^{-1}, \tag{16.99}$$

and the pairing Green's function, introduced by Gorkov, is given by

$$F^+ = \Delta(\omega_n)/[\omega_n^2 + \xi^2 + \Delta^2(\omega_n)]. \tag{16.100}$$

In Eq. (16.100), ξ is the electron energy relative to the chemical potential. The metallic clusters contain delocalized electrons of which the states organize into shells. For some clusters that are spherical, the shells are filled up to the highest occupied shell, i.e., those with the magic numbers $N = N_m = 20, 40, 58, 92, 138, 168, \ldots$. The electronic states in the magic clusters are labeled by radial quantum number n and orbital momentum l. The Cooper pairs are formed by electrons with opposite projections of orbital momentum, and if the orbital momentum is large, the shell has a degeneracy of $2(2l+1)$.

For nanoclusters, of volume V, Eqs. (16.92), (16.93), (16.94), (16.96), and (16.100) are modified as

$$\Delta(\omega_n)Z = \eta\frac{T}{2V}\sum_{\omega_{n'}}\sum_{s} D(\omega_n - \omega_{n'}, \tilde{\Omega})\, F_s^+(\omega_{n'}), \tag{16.101}$$

where

$$\eta = <I^2>/M\tilde{\Omega}^2, \tag{16.102}$$

$$D(\omega_n - \omega_{n'}, \tilde{\Omega}) = \tilde{\Omega}^2[(\omega_n - \omega_{n'})^2 + \tilde{\Omega}^2]^{-1}, \tag{16.103}$$

$$F_s^+(\omega_{n'}) = \Delta(\omega_{n'})[\omega_{n'}^2 + \xi_s^2 + \Delta^2(\omega_{n'})]^{-1}, \tag{16.104}$$

and

$$\xi_s = E_s - \mu. \tag{16.105}$$

Here D and $F_s^+(\omega_n)$ are the vibrational propagator and the pairing function, $\Delta(\omega_n)$ is the order parameter, Z is the renormalization function, E_s is the energy of the sth electronic state, and μ is the chemical potential. The summation is over all the discrete electronic states. There are two aspects in which these equations differ from the corresponding equations for the bulk. First, they contain a summation over discrete energy levels E_S instead of integration over a continuous energy spectrum, as in a bulk superconductor. Second, because the number of electrons is fixed, the position of the chemical potential μ is determined by N and T and hence different from the value of Fermi level E_F.

Kresin and Wolf (Ref. 18) also showed that because "magic clusters" have a spherical shape, one can replace the summation over states by summation over shells $\Sigma_s \rightarrow \sum_j G_j$, where G_j is the shell degeneracy

$$G_j = 2(2l_j + 1), \tag{16.106}$$

and l_j is the orbital momentum. They showed that for such clusters, Eqs. (16.101) through (16.104) can be written in the form

$$\Delta(\omega_n)Z = \lambda \frac{2E_F}{3N} \sum_{\omega_{n'}} \sum_j G_j \frac{\tilde{\Omega}^2}{\tilde{\Omega}^2 + (\omega_n - \omega_{n'})^2} \frac{\Delta^2(\omega_{n'})}{\omega_{n'}^2 + \xi_l^2} |T_c. \tag{16.107}$$

In Eq. (16.107), the bulk coupling constant λ (Eq. 16.92) and the Fermi energy E_F are used because the characteristic vibrational frequency is close to the bulk value due to the fact that the pairing is mediated mostly by the short-wave part of the vibrational spectrum.

However, if the shell is incomplete, there is a Jahn–Teller deformation in the cluster. Because the shape becomes ellipsoidal, the s states are classified by their projection of the orbital momentum $|m| \leq l$, and each level contains up to four electrons for $|m| \geq 1$. In the weak coupling case, $\eta/V \ll 1$, which leads to $\pi T_c \ll \tilde{\Omega}$. In Eqs. (16.101), (16.103), and (16.104), one should substitute $Z = 1$ and $D = 1$. This leads to the BCS (weak coupling) theory.

It may be noted that Eqs. (16.101), (16.103), and (16.104) are different from the strong-coupling theory of bulk superconductors in two aspects. First, they contain a summation over discrete energy levels E_S, whereas one integrates over a continuous energy spectrum (over ξ) in bulk superconductors. Second, because the clusters have a finite Fermi system, the number of electrons N is fixed. Hence, the position of the chemical potential μ is different from the Fermi level E_F and is determined by the value of N and T.

The value of the critical temperature T_c in a nanocluster depends on the number of valence electrons N; the energy spacing $\Delta E = E_L - E_H$; and on the values of λ_b, E_F, and $\tilde{\Omega}$. One can obtain a high value of T_c (as high as 100° K) by using realistic values of these parameters.

The pairing in an isolated cluster can be observed from the strong temperature dependence of the excitation spectrum. When $T \sim 0°$ K, the excitation energy is strongly modified by the gap parameter and significantly exceeds the gap parameter when $T > T_c$. For example, the minimum absorption energy for Gd_{83} clusters at $T > T_c$ corresponds to $\hbar\omega \approx 6$ meV, whereas for $T \ll T_c$, its value is much larger: $\hbar\omega \approx 34$ meV.

Recently, Cao et al.[4] measured the heat capacity of an isolated cluster. They observed a jump in heat capacity for selected Al clusters (i.e., for Al_{35}^- ions) at $T \approx 200°$ K. The value of T_c as well as the amplitude of the jump and its width are in agreement with this theory.

PROBLEMS

16.1. We have shown that for a metallic cluster, the total energy can be written as

$$E[n] = E_k[n] + E_{es}[n] + E_{xc}[n], \tag{1}$$

where $E_k[n]$ is the kinetic energy of a system of independent particles of density n,

$$E_k[n] = \sum_i \frac{\hbar^2}{2m} |\nabla \psi_i|^2. \tag{2}$$

$E_{es}[n]$ is the electrostatic energy,

$$E_{es}[n] = \frac{e^2}{2} \iint d\mathbf{r}'\, d\mathbf{r}\, \frac{[n(\mathbf{r}) - n_I(\mathbf{r})][n(\mathbf{r}') - n_I(\mathbf{r}')]}{|\mathbf{r} - \mathbf{r}'|}, \tag{3}$$

and $E_{xc}[n]$ is the exchange-correlation term of the energy. Following the procedure outlined in Section 7.8.2 (subject to the modification for a metallic cluster), show that

$$-\frac{\hbar^2}{2m} \nabla^2 \psi_i(\mathbf{r}) + V_{KS}(\mathbf{r}) \psi_i(\mathbf{r}) = \varepsilon_i \psi_i(\mathbf{r}), \tag{4}$$

where

$$V_{KS}(\mathbf{r}) = V_H(\mathbf{r}) + V_{xc}(\mathbf{r}), \tag{5}$$

$$V_H(\mathbf{r}) = 2e^2 \int d\mathbf{r}'\, \frac{n(\mathbf{r}') - n_I(\mathbf{r}')}{|\mathbf{r} - \mathbf{r}'|}, \tag{6}$$

and

$$V_{xc}(\mathbf{r}) = \frac{\partial \varepsilon_{xc}(n)}{\partial n} \varepsilon_{xc}(n) \tag{7}$$

Here, $\varepsilon_{xc}(n)$ is the exchange-correlation energy of the electron gas.

16.2. Show that in Eq. (16.24),

$$< l^2 >_n = \left(\frac{1}{2}\right) n(n+3). \tag{1}$$

16.3. The ellipsoidal model is based on the harmonic oscillator Hamiltonian

$$H = \frac{p^2}{2m} + \frac{1}{2} m\omega_0^2 [\Omega_\perp^2 (x^2 + y^2) + \Omega_z^2 z^2] - U\hbar\omega_0 [l^2 - < l^2 >_n]. \tag{1}$$

Ignoring the anharmonic terms in Eq. (1), show that

$$E(n_1, n_2, n_3) = \hbar\omega_0 \left[\Omega_\perp (n_1 + n_2 + 1) + \Omega_z \left(n_3 + \frac{1}{2}\right)\right]. \tag{2}$$

By using the definition of the distortion parameter defined in the text,

$$\delta = \frac{2(\Omega_\perp - \Omega_z)}{\Omega_\perp + \Omega_z}, \tag{3}$$

show that Eq. (2) can be written as

$$E(n_1, n_2, n_3) = \hbar\omega_0 \left[n + \frac{3}{2} + \frac{\delta}{3}(n - 3n_3) + \frac{\delta^2}{18}(n + 3n_3 + 3) \right], \tag{4}$$

where

$$n = n_1 + n_2 + n_3. \tag{5}$$

16.4. If one defines the perpendicular distance of the two faces as p_{111} and p_{100}, show that the lengths of the sides of the faces of the Wulff polyhedron (Figure 16.7) are β and $(1-2\beta)$ in units of $\sqrt{6}p_{111}$. The scale factor β is given by $\beta = 1 - p_{100}/\sqrt{3}p_{111}$.

16.5. The polyhedron has a central atom and can be considered as built of successive shells covering interior shells. The Kth shell contains $(10K^2 + 2)$ atoms. Show that the total number of atoms in a cluster with K shells is

$$N = \frac{1}{3}(10K^3 - 15K^2 + 11K - 3). \tag{1}$$

16.6. The diffusive flux into the island is given by

$$DN_1\sigma_s = D2\pi r_s \frac{\partial n_1}{\partial r}\Big|_{r_s}, \tag{1}$$

from which the capture number σ_s can be calculated. The monomer density field can be obtained from the diffusion equation by first assuming that the monomers can evaporate from the substrate at the rate $1/\tau_a$ and that $N_1 \to F\tau_a$, a long way from the island. The latter assumption implies that the density of islands is sufficiently low. The diffusion equation can be written as

$$\frac{\partial n_1(r,t)}{\partial t} = D\nabla^2 n_1(r,t) - \frac{n_1(r,t)}{\tau_a} + F. \tag{2}$$

In the steady state $t > \tau_a$, which can be shown from Eqs. (1) and (2), show that the capture number σ_s can be written as

$$\sigma_s = 2\pi X_s \frac{K_1(X_s)}{K_0(X_s)}, \tag{3}$$

where K_0 and K_1 are the Bessel functions and

$$X_s = \frac{r_s}{\sqrt{D\tau_a}}. \tag{4}$$

16.7. We considered the simple case of one orbital site and local density of states $d_i(E)$ at site i with center of gravity $d_i(E)$,

$$E_{bond} = 2\sum_i \int_{-\infty}^{E_F} (E - \varepsilon_i) d_i(E) dE. \tag{1}$$

The second moment of the local density of states can be written as

$$\mu_i^{(2)} = \int_{-\infty}^{\infty} (E - \varepsilon_i)^2 d_i(E) dE. \tag{2}$$

We represent $d_i(E)$ by a Gaussian centered at ε_i and at width $\sqrt{\mu_i^{(2)}}$. By invoking the condition of charge neutrality to confirm that other factors arising out of Gaussian integration of Eq. (1) are site-independent, show that, in an approximation in which moments no higher than the second are used, the bond energy $E_{bond}^{(i)}$ is given by

$$E_{bond}^{(i)} = A\sqrt{\mu_i^{(2)}}, \tag{3}$$

where A is a constant.

16.8. In the second-moment approximation discussed earlier, show that the bandwidth W is proportional to $z^{1/2}$, where z is the coordination number. For simplicity, if we consider a rectangular band for the d electrons, show that the density of states scales is $1/W$.

16.9. Show that the low-temperature STM differential conductivity can be written as

$$\frac{dI}{dV} = \frac{2\pi e^2}{\hbar} \frac{(\varepsilon' + q)^2}{1 + \varepsilon'^2} \rho_{tip} \sum_{\mathbf{k}} |M_{\mathbf{k}t}|^2 \delta(eV - \varepsilon_k) + \text{constant}, \tag{1}$$

where

$$q = A/B, \tag{2}$$

$$A = M_{at} + \sum_{\mathbf{k}} M_{\mathbf{k}t} V_{a\mathbf{k}} P\left(\frac{1}{eV - \varepsilon_{\mathbf{k}}}\right), \tag{3}$$

$$B = \pi \sum_{\mathbf{k}} M_{\mathbf{k}t} V_{a\mathbf{k}} \delta(eV - \varepsilon_{\mathbf{k}}), \tag{4}$$

$$\varepsilon' = \frac{eV - \varepsilon_a - \text{Re}\,\Sigma(eV)}{\text{Im}\,\Sigma(eV)}, \tag{5}$$

and ρ_{tip} is the tip density of states. ρ_{tip} is usually treated as a constant to reflect the broadening of the tip state by contact with the remainder of the tip electrode. (Hint: Most of the derivation is outlined in the appendix of the research paper by Madhavan et al.[23]).

References

1. Bales GL, Chrazan DC. Dynamics of irreversible growth during submonolayer epitaxy. *Phys Rev B* 1994;**50**:6057.
2. Becke AD. Density-functional exchange energy approximation with correct asymptotic behavior. *Phys. Rev. A* 1988;**38**:3098.
3. Binns C, Trohidou KN, Bansmann J, Baker SH, Blackman J, Bucher J-P, et al. The behaviour of nonstructured magnetic materials produced by depositing gas-phase nanoparticles. *J Phys D: Appl Phys* 2005;**38**:R357.
4. Cao B, Neal CN, Starace AK, Ovchinnnikov YN, Kresin V, et al. Evidence of of High-Tc Superconducting Transition in Isolated Al 45- and Al 47- isolated nanoclusters. *J Supercond Novel Magn* 2008;**21**:163.
5. Daw MS, Baskes MI. Embedded-atom method: Derivation and application to impurities, surfaces and other defects in metals. *Phys Rev B* 1984;**29**:6443.
6. Duffy D, Blackman JA. The energy of Ag adatoms and dimers on graphite. *Surf Sci* 1998;**415**:L1016.
7. Ekardt W. Work function of small metal particles: Self-consistent spherical jellium background model. *Phys Rev B* 1984;**29**:1558.
8. Fano U. Effects of Configuration Interaction on Intersites and Phase Shifts. *Phys Rev* 1961;**124**:1866.
9. Finnis MW, Sinclair JE. A simple empirical potential for transition metals. *Philos Mag A* 1984;**50**:45.
10. Foiles SM, Baskes MI, Daw MS. Embedded-atom-method functions for the fcc metals Cu, Ag, Au, Ni, Pd, Pt, and their alloys. *Phys Rev B* 1986;**33**:7983.
11. Handschuh S, Blugel S. Magnetic Exchange Coupling of 3d metal monolayers on Fe(001). *Solid State Commun.* 1998;**108**:633.
12. Hohenberg P, Kohn W. Inhomogeneous electron gas. *Phys Rev B* 1970;**1**:4555.
13. Isabelle M, Billas L, de Heer WA. Magnetism of Fe, Co, Ni clusters in molecular beams. *J. Magn. Magn. Mater.* 1997;**168**:64.
14. Jensen PD, Bennemann KH. A simple empirical N-body potential for transition metals. *Z Phys D* 1995;**35**:273.
15. Katakuse IT, Ichihara Y, Fujita Y, Matsuo T, Sakurai T, Matsuda H. Correaltion between mass distribution of zinc, cadmium cluster and electronic shell structure. *Int J Mass Spectr Ion Processes* 1986; **69**:109.
16. Knight WD, Clemenger K, de Heer WA, Saunders WA, Chou MY, Cohen ML. Electroanic Shell Structures and Abundance of Sodium Cluster. *Phys Rev Lett* 1984;**52**:2141.
17. Kohn W, Sham LJ. Self-consistent Equations Including Exchange and Correlation effects. *Phys Rev* 1965;**140**:A 1133.
18. Kresin VZ, Wolff SA. Electron-lattice interaction and its impact on high Tc superconductivity. *Rev Mod Phys* 2009;**81**:481.
19. Kruger P, Rakoatomahevitra A, Parbelas JC, Demangeat C. Magnetism of eptaxial 3d-transition-metal monolayers on graphite. *Phys Rev B* 1998;**57**:5276.
20. Lau JT, Fohlisch A, Martins M, Nietubyc, Reif M, Wurth W. Spin and orbital magnetic moments of deposited small iron clusters studied by X-ray magnetic circular dichroism spectroscopy. *New J Phys* 2002;**4**:98.
21. Mackay AL. A dense non-crystallograohic packing of equal spheres. *Acta Crystallogr* 1962;**15**:916.
22. Madhavan V, Chen W, Jameneala J, Crommie MF, Wingreen NS. Tunneling into a single Magnetic Atom: Spectroscopic Evidence of the Kondo Resonance. *Science* 1998;**280**:567.
23. Madhavan V, Chen W, Jameneala J, Crommie MF, Wingreen NS. Local Spectroscopy of a Kondo impurity: Co on Au(111). *Phys Rev B* 2001;**64**:165412.
24. Marks LD. Modified Wulff construction for twinned particles. *J. Cryst. Growth* 1983;**61**:556.
25. Martin TP. Shells of Atoms. *Phys Rep* 1984;**273**:199.
26. Martinez E, Lange RC, Robles R, Vega A, Gallgo LJ. Structure and magnetic properties of small Fe clusters supported on the Ni (001) surface. *Phys Rev* 2005;**71**:165424.

27. Mehl MJ, Papaconstantopoulous DA. Applications of a tight-binding total-energy method for transition and noble metals: Elastic constants, vacancies and surfaces of monatomic metals. *Phys Rev B* 1996;**54**:4519.
28. Misra PK, series editor, Blackman J, editor. *Metallic Nanoparticles.* Amsterdam: Elsevier; 2009.
29. Riley SJ. The atomic structure of transition metal clusters. *J Non-Cryst Solids* 1996;**205–207**:781.
30. Skomski R. Nanomagnetics. *J Phys Condens Matter* 2003;**15**:R 841.
31. Slater JC, Koster GF. Simplified LCAO Method for the Periodic Potential Problem. *Phys Rev* 1954;**94**:1498.
32. Stepanyuk VS, Hergert A, Rennert P, Wildberger K, Zeller R, Dedelichs PA. Transition metal magnetic nanostructures on metal surfaces. *Surface Sci* 1997;**377–379**:495.
33. Sutton AP. *Electronic Structure of Materials.* Oxford: Oxford University Press; 1993.
34. Vosko SJ, Wilk L, Nusair M. Accurate spin-dependent electron liquid correlation energies for local spin density calculations: a critical analysis. *Can J Phys* 1980;**58**:1200.
35. Woods BD, Saxon DS. Diffuse Surface Optical Mode for Nucleon-Nucleon Scattering. *Phys Rev* 1954;**95**:577.
36. Wulff G. *Z Kirst* 1901;**34**:449.

CHAPTER

Complex Structures

17

CHAPTER OUTLINE

17.1 Liquids 568
17.1.1 Introduction 568
17.1.2 Phase Diagram 568
17.1.3 Van Hove Pair Correlation Function 569
17.1.4 Correlation Function for Liquids 570
17.2 Superfluid ^{4}He 570
17.2.1 Introduction 570
17.2.2 Phase Transition in ^{4}He 570
17.2.3 Two-Fluid Model for Liquid ^{4}He 571
17.2.4 Theory of Superfluidity in Liquid ^{4}He 571
17.3 Liquid ^{3}He 573
17.3.1 Introduction 573
17.3.2 Possibility of Superfluidity in Liquid ^{3}He 574
17.3.3 Fermi Liquid Theory 574
17.3.4 Experimental Results of Superfluidity in Liquid ^{3}He 575
17.3.5 Theoretical Model for the A and A_1 Phases 575
17.3.6 Theoretical Model for the B Phase 577
17.4 Liquid Crystals 578
17.4.1 Introduction 578
17.4.2 Three Classes of Liquid Crystals 578
17.4.3 The Order Parameter 580
17.4.4 Curvature Strains 581
17.4.5 Optical Properties of Cholesteric Liquid Crystals 581
17.5 Quasicrystals 583
17.5.1 Introduction 583
17.5.2 Penrose Tiles 583
17.5.3 Discovery of Quasicrystals 584
17.5.4 Quasiperiodic Lattice 584
17.5.5 Phonon and Phason Degrees of Freedom 586
17.5.6 Dislocation in the Penrose Lattice 589
17.5.7 Icosahedral Quasicrystals 589

17.6 Amorphous Solids 590
17.6.1 Introduction 590
17.6.2 Energy Bands in One-Dimensional Aperiodic Potentials 591
17.6.3 Density of States 593
17.6.4 Amorphous Semiconductors 593
Problems 594
References 597

17.1 LIQUIDS

17.1.1 Introduction

It is well known that matter exists in three different (solid, liquid, and gaseous) phases. Solids have an ordered arrangement of atoms or molecules as evidenced by rigid and sharp Bragg reflections in a diffraction experiment. Liquids and gases are fluids and will flow even under a small shear stress. In diffraction experiments, they yield only diffuse rings, showing that there is no long arrangement of molecules. In addition, there are glasses and amorphous solids that blur the distinction between solids and fluids. The atoms or molecules in glasses are arranged at random, whereas those in amorphous solids have short-range order.

Van der Waals first pointed out the continuity of liquid and gaseous states. At low temperatures below a critical temperature, two fluid phases can coexist in equilibrium: the dense phase is called liquid, and the less-dense phase is called gas. By heating above the critical temperature, compressing, and cooling, one can pass continuously from low-temperature gas to low-temperature liquid. The difference between liquid and gas is essentially a difference in density.

For roughly spherical molecules (rare gases), there is disorder in only translational motion. There is the possibility of rotational disorder in molecules that are far from spherical. In plastic crystals, both kinds of disorder occur, whereas in "liquid crystals," there can be translational order but rotational disorder.

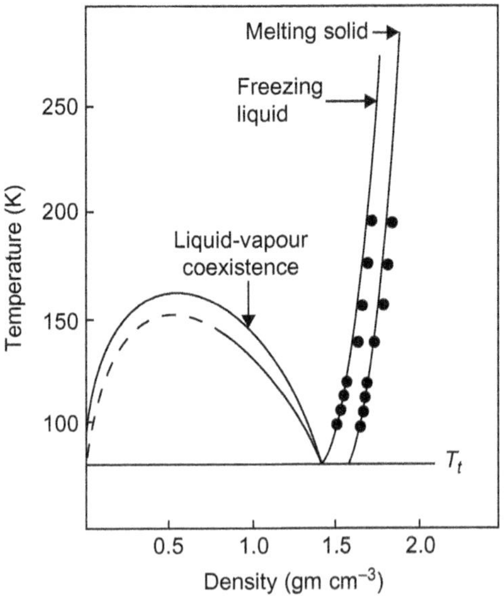

FIGURE 17.1

Phase diagram for the 6-12 fluid (solid lines) and for argon (dashed lines and circles). Here, $\epsilon/k = 119°$ K, $\sigma = 3.405$ Å.

Reproduced from Barker and Henderson[5] with the permission of the American Physical Society.

17.1.2 Phase Diagram

The study of the physics of liquids is mainly to understand why particular phases are stable in particular ranges of density, as shown in Figure 17.1. One has to relate the stability, structure, and dynamical properties of fluid phases to the size and shape of the molecules, atoms, or ions, and the nature of forces between them. Because the interactions that determine the bulk properties of matter are basically electrostatic in character (apart from small relativistic and retardation effects), and arise from the Coulomb interactions between nuclei and electrons, one way to predict the properties would be to solve the many-body Schrodinger equation describing the motion of nuclei and electrons. However, we will restrict the discussion to the concept of order parameters, a function that is present when order is present (solids) and vanishes when the desired order is absent (liquids).

17.1.3 Van Hove Pair Correlation Function

The best way to distinguish between liquids and solids is to first introduce the van Hove pair correlation function $p(\mathbf{r}_1, \mathbf{r}_2, t)$ defined as

$$p(\mathbf{r}_1, \mathbf{r}_2, t) = \left\langle \sum_{i \neq j} \delta(\mathbf{r}_1 - \mathbf{X}_i(0))\delta(\mathbf{r}_2 - \mathbf{X}_j(t)) \right\rangle. \tag{17.1}$$

Here, $p(\mathbf{r}_1, \mathbf{r}_2, t)$ is the probability that if a particle is found at position $\mathbf{r}_1$ at time t_1, some other particle is to be found at position $\mathbf{r}_2$ at time $t_1 + t$. The brackets mean a thermal average, and $\mathbf{X}_i(t)$ tracks the location of the particles. Here, $\mathbf{X}_i(t)$ is the Heisenberg operator defined for all i and all t by the equation

$$\mathbf{X}_i(t) = e^{iHt/\hbar}\, \mathbf{X}_i e^{-iHt/\hbar}, \tag{17.2}$$

where H is the Hamiltonian of the system.

The static structure factor, the dimensionless measure of scattering in a scattering experiment that measures the thermal average (because such experiments last much longer than the time scale of atomic motions), is defined as

$$S(\mathbf{q}) = \frac{1}{N} \sum_{i,j} \langle e^{i\mathbf{q}\cdot(\mathbf{X}_i - \mathbf{X}_j)} \rangle. \tag{17.3}$$

Eq. (17.3) can be rewritten in the alternate form

$$S(\mathbf{q}) = \frac{1}{N} \sum_{i,j} \int d\mathbf{r}_1 d\mathbf{r}_2\, e^{i\mathbf{q}\cdot(\mathbf{r}_1 - \mathbf{r}_2)} \langle \delta(\mathbf{r}_1 - \mathbf{X}_i)\delta(\mathbf{r}_2 - \mathbf{X}_j) \rangle. \tag{17.4}$$

From Eqs. (17.1) and (17.4), we obtain

$$S(\mathbf{q}) = 1 + \frac{1}{N} \int d\mathbf{r}_1\, d\mathbf{r}_2\, p(\mathbf{r}_1, \mathbf{r}_2, 0) e^{i\mathbf{q}\cdot(\mathbf{r}_1 - \mathbf{r}_2)}. \tag{17.5}$$

We define

$$p(\mathbf{q}) = \frac{1}{V} \iint d\mathbf{r}\, d\mathbf{r}'\, p(\mathbf{r} + \mathbf{r}', \mathbf{r}, 0) e^{i\mathbf{q}\cdot\mathbf{r}'}, \tag{17.6}$$

where V is the volume of the system. From Eqs. (17.5) and (17.6), we obtain

$$S(\mathbf{q}) = 1 + \frac{V}{N} p(\mathbf{q}). \tag{17.7}$$

17.1.4 Correlation Function for Liquids

In general, liquids are homogeneous and isotropic. Therefore, one can assume that the pair correlation function $p(\mathbf{r}_1, \mathbf{r}_2)$ will depend on the distance $r = |\mathbf{r}_1 - \mathbf{r}_2|$. If the density of the liquid is $\rho = N/V$, a dimensionless correlation function $g(r)$ can be defined as

$$g(r) = \frac{p(r)}{\rho^2}. \tag{17.8}$$

One obtains from Eqs. (17.5) and (17.8)

$$S(\mathbf{q}) = 1 + \rho \int d\mathbf{r}\, g(r) e^{i\mathbf{q}\cdot\mathbf{r}}. \tag{17.9}$$

Eq. (17.9) can be rewritten in the alternate form

$$S(\mathbf{q}) = 1 + \rho \int d\mathbf{r}\, e^{i\mathbf{q}\cdot\mathbf{r}} + \rho \int d\mathbf{r}\, [g(r) - 1] e^{i\mathbf{q}\cdot\mathbf{r}}. \tag{17.10}$$

The second term in Eq. (17.10) is a delta function, which is zero except when directly along the scattering beam. Thus, it can be dropped while analyzing the experimental results of scattering, and one obtains from Eq. (17.10)

$$S(\mathbf{q}) \approx 1 + \rho \int d\mathbf{r}\, [g(\mathbf{r}) - 1] e^{i\mathbf{q}\cdot\mathbf{r}}. \tag{17.11}$$

17.2 SUPERFLUID ^{4}He

17.2.1 Introduction

Kammerlingh Onnes discovered in 1908 that liquid helium never solidified under its own vapor pressure. The interaction between the helium atoms is very weak because helium is an inert gas. The liquid phase is very weakly bound, and the normal boiling point is very low (4.2° K). The large-amplitude quantum mechanical zero-point vibrations due to the small atomic masses and the weak interactions do not permit the liquid to freeze into the crystalline state. Liquid ^{4}He solidifies only when a pressure of at least 25 atmospheres is applied. Therefore, it is possible to study liquid ^{4}He all the way down to the neighborhood of absolute zero.

17.2.2 Phase Transition in ^{4}He

At 2.17° K, a remarkable phase transition was discovered in liquid ^{4}He under saturated vapor pressure. When the liquid was cooled through this temperature, all boiling ceased, and the liquid became perfectly quiescent. This effect occurs because liquid helium becomes an enormously good heat conductor. The thermal inhomogeneities that give rise to bubble nucleation are absent. The specific heat versus temperature curve of liquid ^{4}He was shaped like a Greek letter λ, characteristic of a second-order phase transition. This temperature is called the lambda point, and the experimental results are shown in Figure 17.2.

The temperature at the lambda point is usually denoted by T_λ. Below this temperature, liquid ^{4}He has remarkable flow as well as the "superheat" transport properties. If a small test tube containing the liquid were raised above the surrounding helium bath, a mobile film of the liquid would be transported up the inner walls. Eventually, it would drip back into the bath, and the test tube would be emptied. Kapitza (Ref. 10) showed that liquid ^{4}He could flow through the tiniest pores and cracks. Allen and Jones (Ref. 10) found that if a glass tube packed tightly with a powder was partially immersed in a ^{4}He bath and then heated, a fountain of helium rising high above the level of the surrounding helium bath was produced.

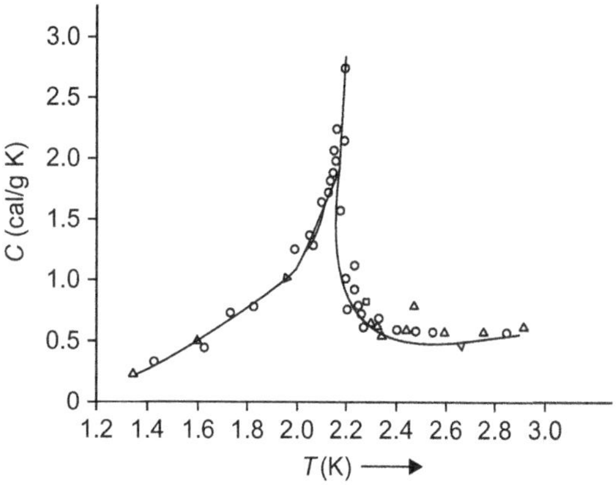

FIGURE 17.2

Schematic diagram of the specific heat of liquid helium versus temperature.

17.2.3 Two-Fluid Model for Liquid ^{4}He

Landau (Ref. 10) and Tisza independently developed a two-fluid model to describe these phenomena. According to the two-fluid model, below T_λ, liquid ^{4}He can be thought of as two interpenetrating fluids known as the normal and superfluid components. The superfluid component does not carry entropy and is involved in superflow through pores and cracks. In addition, it does not interact with the walls of a vessel containing the fluid in a dissipative fashion. In contrast, the normal component transports heat and exhibits viscosity, which allows transfer of energy between the liquid and the walls. The normal fluid density decreased with decreasing temperature, whereas the superfluid density increased, becoming dominant at the lowest temperature. The superfluid component replaces the normal fluid, which carries heat away from the heat source. The flow of the superfluid component toward a source of heat manifests in the fountain effect mentioned previously. The normal fluid consists of a gas of quantized thermal excitations that include the phonons (longitudinal sound waves) and rotons (short-wavelength compact excitations). It was predicted that heat transport would obey a wave equation that describes the compression and rarefactions in the proton/roton "gas," which is known as second sound.

17.2.4 Theory of Superfluidity in Liquid ^{4}He

London (Ref. 15) noted that as the temperature of liquid ^{4}He is reduced through the transition temperature, the occupancy of the one-particle ground state becomes macroscopic and can be thought of as a Bose–Einstein (BE) condensate, which is the superfluid component of the two-fluid picture, although strong interactions between the atoms in the liquid modify this picture. According to London, the superfluid atoms are governed by a wave-function-like entity called the order parameter that was originally introduced by Ginzburg and Landau to explain the phenomenon of superconductivity (described in detail in Section 14.4.1). The order parameter Ψ for a superfluid ^{4}He is given by

$$\Psi = \Psi_0 e^{i\phi}, \tag{17.12}$$

where Ψ_0 is roughly thought of as the square root of the density of the superfluid component, and ϕ is a phase factor. The fact that the macroscopic order parameter is also described by a definite phase is known as broken gauge symmetry. It has been shown that the superfluid velocity is proportional to the gradient of the phase. The macroscopic order parameter picture describes how the helium atoms march in "lock step" during superfluid flow. The existence of quantized vortices in superfluid ^{4}He is also a consequence of this model. In fact, this phenomena is also seen in superfluid liquid ^{3}He (to be discussed in the next section) and in superconductivity in solids, where a quantized current vortex must enclose a quantum of flux (discussed in Chapter 14).

The fundamental assumption that underlies the modern theory of superfluidity in a Bose system such as liquid ^{4}He is that the superfluid phase is characterized by a generalized Bose–Einstein Condensation (BEC). We assume that at any given time t, it is possible to find a complete orthonormal basis (which may itself depend on time) of single-particle states such that one and only one of these states is occupied by a finite fraction of all the particles, while the number of particles in any other single-particle state is of order 1 or less. The corresponding single-particle wave function $\chi_0(r, t)$ is called a condensate wave function, and the N_0 particles occupying it, the condensate. The $T = 0$ condensate fraction $N_0/N \sim 0.1$, where N is the total number of particles in the system. The macroscopic occupation occurs only in a single-particle state because according to the Hartree–Fock approximation, the macroscopic occupation of more than one state is always energetically unfavorable provided the effective low-energy interaction is repulsive, as is the case for ^{4}He.

Because the BEC occurs in the sense defined earlier, at any given time there exists one and only one single-particle state $\chi_0(r, t)$ that is macroscopically occupied, and the conceptual basis for superfluidity is quite simple. We can write

$$\chi_0(r, t) = |\chi_0(r, t)| e^{i\phi(r,t)} \tag{17.13}$$

and define the superfluid velocity $v_s(r, t)$ by

$$v_s(r, t) \equiv \frac{\hbar}{m} \nabla \phi(r, t), \tag{17.14}$$

from which we obtain

$$\nabla \times v_s = 0. \tag{17.15}$$

Thus, the superfluid flow is irrotational. In addition, because no "ignorance" is associated with the single state χ_0, the entropy must be carried entirely by the "normal" component, i.e., the particles occupying single-particle states other than χ_0. These two observations provide the basis for Landau's phenomenological two-fluid hydrodynamics. However, the superfluid density ρ_s, which occurs in the latter, is given by $(T \to 0)$ $\rho_s \to N/V$ while $N_0 \to 0.1\, N$.

In a region where $|\chi_0|$ is everywhere nonzero, the application of the Stokes theorem to the curl of Eq. (17.14) leads to the conclusion that the integral of v_s around any closed curve is zero (Problem 17.1). However, we consider a line or a region infinite in one dimension on which $|\chi_0(r, t)|$

vanishes. This would happen if the liquid is physically excluded from this region, or while atoms are present in this region, the single-particle state into which BEC has taken place happens to have a nodal line there. When we integrate Eq. (17.14) around a circuit that encloses the one-dimensional region, the fact that the phase of the wave function χ_0 must be single-valued modulo 2π leads to the Onsager–Feynman quantization condition (Problem 17.2)

$$\oint v_s \cdot dl = nh/m. \tag{17.16}$$

In a simply connected region of space, Eq. (17.16) can be satisfied by a "vortex," which is a pattern of flow in which $v_s \sim 1/r$, where r is the perpendicular distance from the "core." Because $|\chi_0|$ vanishes at $r = 0$, v_s is not defined, and hence, the singularity that appears at the core is physically irrelevant.

Thus, in a superfluid system, the circulation is quantized according to Eq. (17.16). In practice, the values $n = \pm 1$ are of interest because vortices with higher values of n are unstable against decay into these. However, vortices can be metastable for astronomical times even under equilibrium conditions.

We note that the superfluid velocity $v_s(r,t)$ is not a directly observable quantity, whereas the mass current $J(r,t)$ is observable. Landau (Refs. 10-13) showed in his phenomenological theory that in stable or metastable equilibrium, this quantity is given by

$$J(r,t) = \rho_s v_s(r,t) + \rho_n v_n(r,t), \tag{17.17}$$

where the "superfluid" and "normal" densities ρ_s and $\rho_n \equiv \rho - \rho_s$ are functions of only the temperature. The normal velocity $v_n(r,t)$ is assumed to behave like the velocity of a normal (nonsuperfluid) liquid. In equilibrium, $v_n(r,t)$ should be zero in the frame of reference in which the walls of the vessel are at rest.

In the 70 years since Landau's original proposal, although there has been almost universal belief that the key to superfluidity in liquid ^{4}He is the onset of BEC at the lambda temperature, it has proved very difficult to verify the latter phenomenon directly. The main evidence comes from the high-energy neutron scattering and from the spectrum of atoms evaporated from the surface of the liquid. Although both are consistent with the existence of a condensate fraction of approximately 10%, neither can be said to establish it beyond all possible doubt.

17.3 LIQUID ^{3}He

17.3.1 Introduction

^{3}He is a rare isotope of helium that occurs in both solid and liquid states. Because the nuclear moments are very small, solid ^{3}He undergoes nuclear magnetic ordering at a temperature of about 1 mK. Therefore, in the range of temperatures above 0.01° K, the nuclear spins of the ^{3}He atoms comprised of the solid are almost fully disordered. Hence, for spin 1/2 nuclei, the entropy $S_{solid} = R \ln 2$ per mole (Problem 17.3). In contrast, liquid ^{3}He obeys Fermi–Dirac statistics. Well below the Fermi degeneracy temperature (T_F), which is less than 1° K, both the specific heat and the entropy will be linear functions of the temperature, $S_{liquid} = \gamma T$. The liquid–solid phase equilibrium is determined by the Clausius–Clapeyron equation, according to which the melting curve is given by

$$\frac{dP}{dT} = \frac{S_{liquid} - S_{solid}}{V_{liquid} - V_{solid}} = \frac{Latent\ Heat}{T(V_{liquid} - V_{solid})}. \tag{17.18}$$

For ^{3}He because $V_{liquid} > V_{solid}$, the denominator is always positive. In the Fermi degenerate region (at the lowest temperatures), $S_{liquid} < S_{solid}$ and hence the slope of the melting curve becomes negative; i.e., *it takes heat to freeze liquid* ^{3}He.

17.3.2 Possibility of Superfluidity in Liquid ^{3}He

In 1972, Osheroff, Richardson, and Lee (Ref. 20) and Osheroff, Gully, Richardson, and Lee[21] discovered that liquid ^{3}He possesses three anomalous phases below 5 mK, each of which appears to display most of the properties expected of a superfluid. Because liquid ^{3}He obeys Fermi rather than Bose statistics, the mechanism of superfluidity cannot be simple BEC as in liquid ^{4}He. It is believed that just as in metallic superconductors, the fermions pair up to form "Cooper pairs," which are sort of giant diatomic quasimolecules of which the characteristic "radius" is much larger than the typical interatomic distance. These molecules, which are composed of two fermions, effectively obey Bose statistics and hence can undergo BEC. We will discuss superfluidity in liquid ^{3}He in the next section.

The strong short-range repulsion of the quasiparticles in liquid ^{3}He prevents pairing through Cooper pairs that have zero angular momentum $(1 = 0)$ so that the members of a pair do not rotate around one another. However, over the years, a number of higher orbital angular momentum pairing states for a hypothetical superfluid state of liquid ^{3}He were proposed. It is interesting to note that both p-wave $(l = 1)$ and d-wave $(l = 2)$ states of relative orbital angular momentum were proposed. However, the proposals for p-wave pairing by Anderson and Morel (Ref. 1) and Bailan and Werthamer (Ref. 4) were later identified as the actual superfluid phases of liquid ^{3}He. The basic characteristics of the hypothetical superfluid ^{3}He were that (1) there would be an intrinsic pairing mechanism not mediated by an ionic lattice (phonons), and (2) the resulting Cooper pairs would have internal degrees of freedom. These properties would distinguish superfluid ^{3}He from superfluid ^{4}He and superconducting electrons.

17.3.3 Fermi Liquid Theory

Liquid ^{3}He is composed of neutral atoms with nuclear spin angular momentum of $\hbar/2$ and a nuclear magnetic moment. The ^{3}He atom has an odd number of elementary particles, so it obeys Fermi–Dirac statistics and the Pauli exclusion principle. Because the atoms in the liquid interact strongly, Landau developed the Fermi liquid theory (Section 7.9) to explain the properties of liquid ^{3}He. The basic idea of the Fermi liquid theory is to consider the excitations of the strongly interacting system instead of concentrating on the nature of the ground state. The scattering rate of the fermions is considerably reduced due to the Pauli exclusion principle. Landau termed the excitations, which act like particles, as quasiparticles. The various properties of normal liquid ^{3}He qualitatively resembled the properties of ideal Fermi gas, but the numerical factors were obtained from the Fermi-liquid parameters (Section 7.9.3). In a Fermi liquid at low temperatures, the thermally excited quasiparticles occur in a narrow band near the Fermi surface with energy width on the order of $k_B T$. Only the quasiparticles in this narrow band participate in scattering or in thermal excitations. The width of the band shrinks as T is lowered, and fewer quasiparticles can participate in such events. Consequently, the specific heat C and the entropy S depend linearly on the temperature $(C = \alpha T)$, and the mean free path is proportional to T^{-2}. The thermal conductivity has a $1/T$ dependence, and the viscosity has a $1/T^2$ dependence. At the lowest temperatures in a Fermi liquid, the collisions are absent, and ordinary sound, which is produced as a result of propagation of waves of compression and rarefaction brought about by collisions of the molecules, dies away. This new mode of sound propagation is known as the *zero sound* and arises at the lowest temperatures due to self-consistent rearrangements of quasiparticles under the influence of Fermi-liquid interactions.

17.3.4 Experimental Results of Superfluidity in Liquid ^{3}He

The classical experiment to find superfluidity in ^{3}He was performed by Osheroff, Richardson, and Lee (Ref. 20) who used a Pomerchunk cell to observe two phase transitions (denoted as A and B) at temperatures $T_A \approx 2.7$ mK and $T_B \approx 2.1$ mK, respectively. They originally misinterpreted the results as a second-order magnetic phase transition in solid ^{3}He. However, subsequent nuclear magnetic resonance experiments made at the suggestion of Gully (Osheroff, Gully, Richardson, and Lee[21]) showed that the A and B phases were both superfluid phases of liquid ^{3}He. In addition, it was observed that the A phase split into two phases (A and A_1) in a magnetic field, whereas at about 0.6 Tesla, the B phase no longer existed. The early specific-heat measurements of liquid ^{3}He near the superfluid transition were done by Webb et al.[26] and are shown in Figure 17.3. The shape is characteristic of a BCS pairing transition.

The phase diagram of liquid ^{3}He in a magnetic field was investigated by Paulson et al.[22] at pressures below melting pressures; they studied the static magnetization of the liquid via SQUID interferometry. The A phase narrowed and finally vanished at a point called the polycritical point (PCP) at about 22 bar. In a larger magnetic field, the B phase is suppressed in favor of the A phase even at the lowest pressure, and the PCP disappears, as shown in Figure 17.4.

A schematic P-T-H diagram showing the general topology of the superfluid phases, A, A_1, and B of liquid ^{3}He, is shown in Figure 17.5. The A_1 phase occurs between the surfaces labeled A_1 and A_2. The A phase occurs at temperatures below the boundary labeled A_2. The boundary between phases A and B is labeled B. The surface labeled S corresponds to the melting curve.

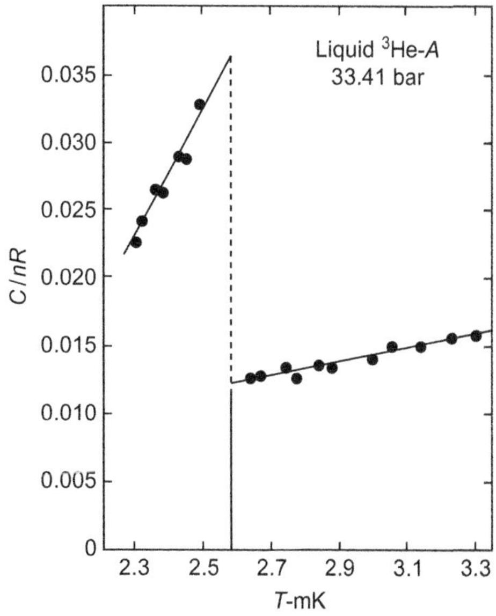

FIGURE 17.3

Specific-heat measurements of liquid ^{3}He near the superfluid transition.

Reproduced from Webb et al.[26] with the permission of the American Physical Society.

17.3.5 Theoretical Model for the A and A_1 Phases

The ^{3}He A phase corresponds to the p-wave equal spin pairing state first considered by Anderson and Morel.[1] It is a p-wave pairing state with total $L = 1$ and $S = 1$. It is an orbital

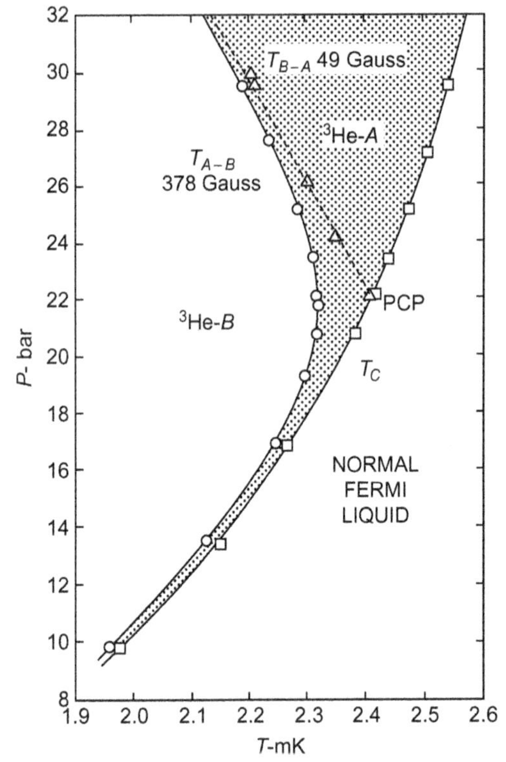

FIGURE 17.4

Experimental data for the phase diagram in a magnetic field.

Reproduced from Paulson et al.[22] with the permission of the American Physical Society.

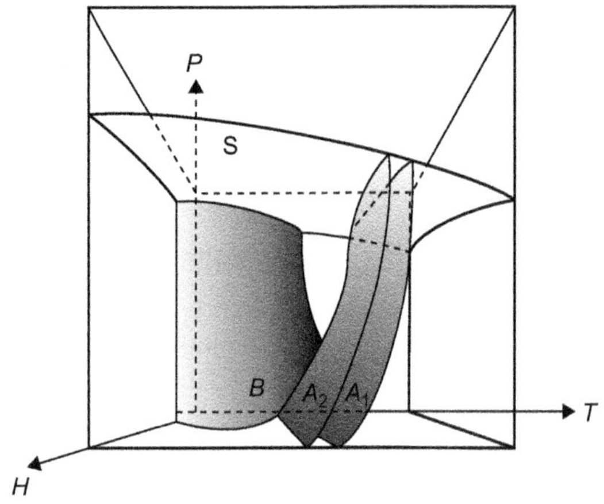

FIGURE 17.5

A schematic P-T-H diagram showing the superfluid phases, A, A_1, and B of liquid ^{3}He.

Reproduced from Lee[14] with the permission of the American Physical Society.

m state along some direction $\hat{l}$ and spin $m = 0$ state along direction $\hat{d}$, where $\hat{d}$ is the direction of zero spin projection. The Anderson–Morel order parameter (Ref. 1) can be expressed as

$$\psi_{AM} = (orbital\ part) \times (spin\ part). \qquad (17.19)$$

Here, the orbital part is in momentum space, and the spherical harmonic $Y_{11} \sim e^{i\varphi} \sin\theta$ defines a polar axis corresponding to the direction of the pair orbital angular momentum. Thus, the Anderson–Morel order parameter (Ref. 1) can be defined as

$$\psi_{AM} \sim e^{i\varphi} \sin\theta \left[\frac{1}{\sqrt{2}}(\downarrow\uparrow + \uparrow\downarrow)\right], \qquad (17.20)$$

where the spherical harmonic $Y_{11} \sim e^{i\varphi} \sin\theta$ defines a polar axis $\hat{l}$ corresponding to the direction of the pair orbital angular momentum. Because in the spin-triplet pair-wave function, the spin part appears along the $\hat{d}$-axis, only the $(\downarrow\uparrow + \uparrow\downarrow)$ component occurs. The three-dimensional representation of the Anderson–Morel order parameter is shown in Figure 17.6a. The vector $\hat{l}$ defines the axes of the order parameter. The amplitude is zero along this axis, which corresponds to $\sin\theta$ dependence where θ is the polar angle with respect to $\hat{l}$. The vector $\vec{d}$ has the same direction for all points on the Fermi surface. The shaded region in Figure 17.17b shows the anisotropic energy gap and the two nodes along $\hat{l}$.

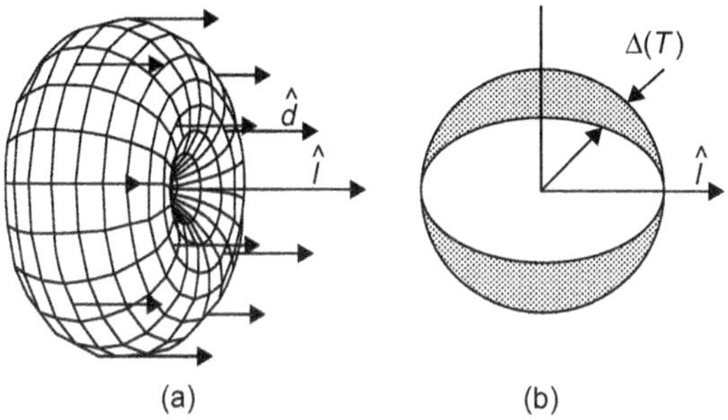

FIGURE 17.6

(a) A three-dimensional representation of the Anderson–Morel order parameter; (b) the anisotropic energy gap is indicated by the shaded region.

Reproduced from Lee[14] with the permission of the American Physical Society.

It can be shown by using a classical argument that the dipolar interaction combined with spontaneously broken symmetry would favor the state for which $\hat{l} \| \hat{d}$ (Problem 17.4). The order parameter has nodes at $\theta = 0$ and $\theta = \pi$. Because the behavior of the BCS energy gap follows the order parameter, the gap nodes also appear at $\theta = 0$ and $\theta = \pi$, which is shown in Figure 17.6b. The three-dimensional picture is obtained by a revolution about the $\hat{l}$-axis. The direction of $\hat{l}$, which is perpendicular to the walls of the container, is also sensitive to flow and to the magnetic field.

The spin state in Eq. (17.20) can be rotated in spin space, which results in the equal spin pairing version of the Anderson–Morel order parameter,

$$\psi_{AM} \sim e^{i\varphi} \sin\theta[(|\uparrow\uparrow\rangle + e^{i\Phi}|\downarrow\downarrow\rangle)], \quad (17.21)$$

where Φ is a phase factor. The A_1 phase has the orbital properties described by the Anderson–Morel state but has only $|\uparrow\uparrow\rangle$ spin pairs.

The concept of spin fluctuation effect was introduced earlier by Layzer and Fay, who noted that because the nuclear magnetic susceptibility of liquid ^{3}He was much higher than that of an ideal Fermi gas of comparable density, there was some tendency for the liquid to be ferromagnetic. When a ^{3}He quasiparticle passed through the liquid, it would polarize spins of neighboring quasiparticles parallel to its own spin because of this ferromagnetic tendency. Anderson and Brinkman[2] showed that the spin fluctuation feedback effect could indeed lead to a stable Anderson–Morel phase in a zero magnetic field. After this paper, the Anderson-Morel phase was known as the Anderson–Brinkman–Morel (ABM) phase.

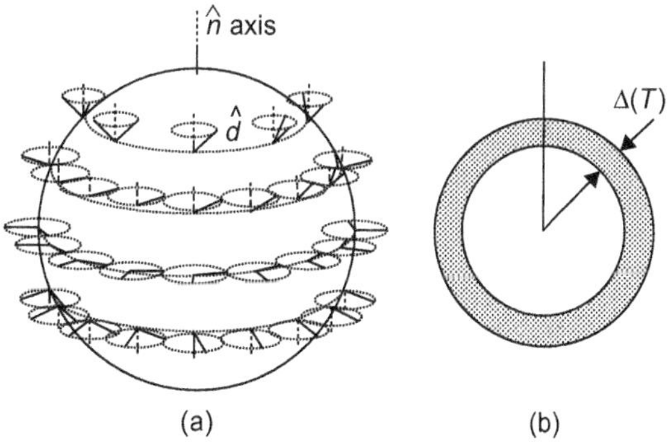

FIGURE 17.7

(a) The order parameter for superfluid ^{3}He B; (b) the isotropic energy gap of the BW order parameter.

Reproduced from Lee[14] with the permission of the American Physical Society.

17.3.6 Theoretical Model for the *B* Phase

The simplest Balian–Werthamer (BW) state (Ref. 4) is the $^3P_0$ state, represented by

$$\psi_{BW} \sim Y_{1,-1}|\uparrow\uparrow\rangle + Y_{10}|\uparrow\downarrow + \downarrow\uparrow\rangle + Y_{11}|\downarrow\downarrow\rangle. \quad (17.22)$$

Because $J=0$ for the $^3P_0$ state, the BW state is specified in terms of the vector $\hat{d}(k)=a\hat{k}$, where a is a constant. The order parameter for superfluid ^{3}He B showing $\hat{d}$ vectors (thick lines) rotated by 104° (Problem 17.4) about a vector along the radial directions (thin lines) for all points on the Fermi sphere is shown in Figure 17.7a. Figure 17.7b shows the isotropic energy gap of the BW order parameter.

17.4 LIQUID CRYSTALS

17.4.1 Introduction

Liquid crystal is a state of matter intermediate between that of an isotropic liquid and a crystalline solid. Liquid crystals have many of the properties of a liquid, such as high fluidity, formation and coalescence of droplets, and inability to support shear. They are also similar to crystals in the sense that they exhibit anisotropy in electric, magnetic, and optical properties. There are two broad types of liquid crystals. Liquid crystals that are obtained by melting a crystalline solid are called thermotropic where temperature and (secondarily) the pressure are the controllable parameters. Liquid crystalline behavior is also found in certain colloidal solutions and certain polymers. This type of liquid crystal is called lyotropic, for which concentration and (secondarily) the temperature are the controllable parameters.

Liquid crystals are found among certain organic compounds that may be of a variety of chemical types. However, the molecules forming liquid crystal phases have certain structural features that can be summarized as follows:

a. The molecules are elongated and have flat segments, as in benzene rings.
b. The long axis of the molecule is defined by a fairly rigid backbone containing double bonds.
c. The molecule should have strong dipoles and easily polarizable groups.
d. The extremities of the molecules are not very important.

17.4.2 Three Classes of Liquid Crystals

Para-azoxyanisole (PAA) and 2-*p*-methoxybenzylidene *n*-butylaniline (MBBA), the two liquid crystals that have been extensively studied, are shown in Figure 17.8.

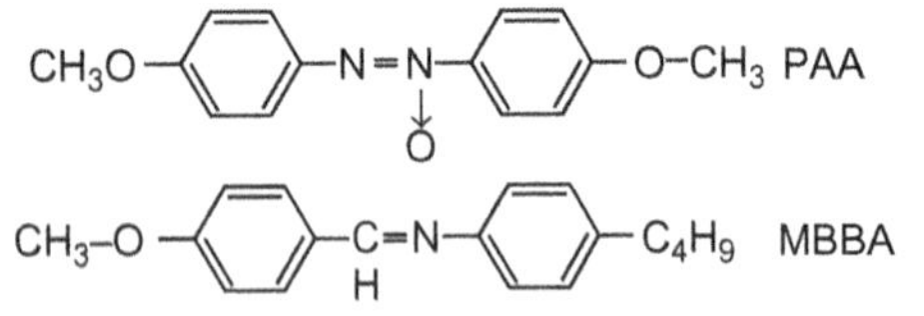

FIGURE 17.8

Molecular structure of para-azoxyanisole (PAA) and 2-*p*-methoxybenzylidene *n*-butylaniline (MBBA).

Reproduced from Stephen and Straley[25] with the permission of American Physical Society.

Liquid crystals are divided into three main classes: (a) nematic, (b) cholesteric, and (c) smectic phases. They are shown in Figure 17.9.

(a) Nematics: In the nematic phase, the long planar molecules are symbolized by ellipses. The long axes of the molecules align along a preferred direction, which indicates that there is a long-range orientational order. The locally preferred direction usually varies throughout the medium in the strained nematic. One defines a vector field $\mathbf{n}(\mathbf{r})$, known as the director, which gives its local orientation. Its magnitude is taken

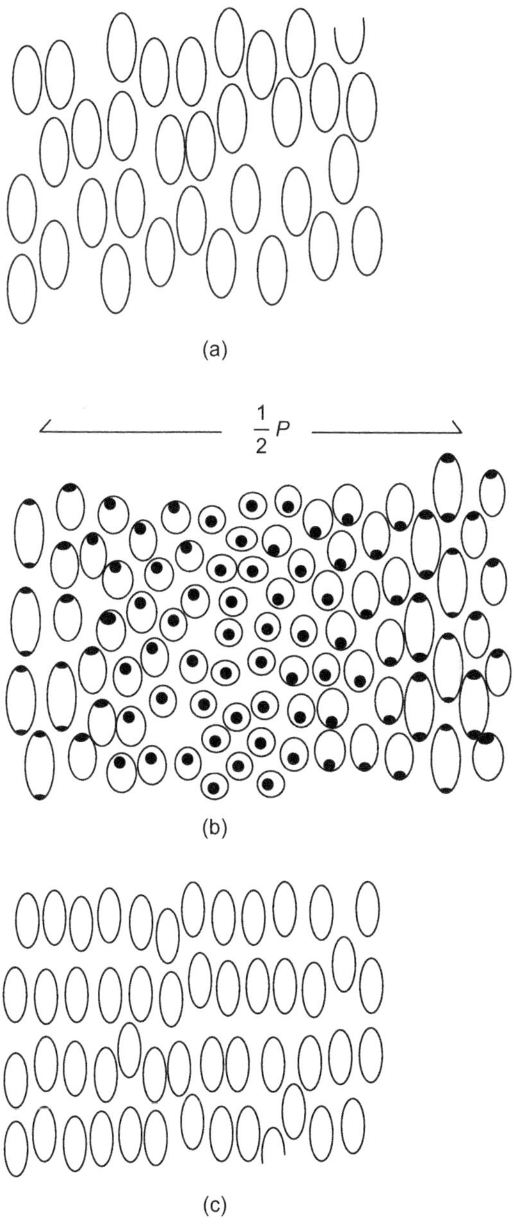

FIGURE 17.9

The arrangement of molecules in the (a) nematic, (b) cholersteric, and (c) smectic A phase.

Reproduced from Stephen and Straley[25] with the permission of the American Physical Society.

as unity. The director field can be aligned by electric and magnetic fields and by surfaces that have been properly prepared. Some structural perturbations appear as threads on optical examinations, which indicate that there is no idealized equilibrium configuration.

There is short-range order in nematics as in ordinary liquids, but no long-range order in the centers of mass of the molecules. There is no preferential arrangement of the two ends of the molecules if they differ, but they can rotate about their long axes. Thus, the sign of the director has no physical significance, and optically, a nematic behaves as a uniaxial material with a center of symmetry, which is confirmed by the absence of ferroelectric phenomena. It has been suggested on the basis of X-ray and optical data that there exists another type of nematic phase, known as the cybotactic phase. In this phase, the molecules are arranged in groups such that the centers of mass of the molecules in each group lie in a plane.

(b) Cholesterics: The cholesteric phase differs from the nematic phase in the sense that the director **n** varies in direction in the medium in a regular way. If a nematic initially aligned along the *y*-axis is twisted about the *x*-axis, a cholesteric configuration would be obtained. The director and the Fresnel ellipsoid rotate as one proceeds along the twist axis. The long axes of the molecules tend to align along a single preferred direction in any plane perpendicular to the twist axis. However, in a series of equidistant parallel planes, the preferred direction rotates as shown in Figure 17.9b. The distance measured along the twist axis over which the director rotates through a full circle is known as the pitch (P) of the cholesteric. Because **n** and **–n** are indistinguishable, the periodicity length of the cholesteric is $P/2$. The pitch of cholesterics, which is sensitive to temperature flow, chemical composition, and applied magnetic or electric fields, is comparable with visible light because it is of the order of several thousand angstroms.

The characteristic colors of the cholesterics in reflection (through Bragg reflection by the periodic structure) and their very large optical rotatory power are due to the spiral arrangement.

(c) Smectics: The molecules in the smectics are arranged in layers, and in addition to orientational layering, they exhibit orientational ordering. There are a number of different classes of smectics, which are briefly listed here:

Smectic A: In this phase, the molecules are aligned perpendicular to the layers, but there is no long-range crystalline order within a layer (see Figure 17.9c). The layers can slide freely over one another.

Smectic B: In this phase, there is a hexagonal crystalline order within the layers. The layers can slip on each other but cannot rotate on each other.

Smectic C: In this phase, the preferred axis is not perpendicular to the layers so that the phase has biaxial symmetry.

Smectic D: Optically, the D phase appears to have a cubic structure, and the X-ray patterns are consistent with a cubic packing.

Smectic E: The X-ray patterns obtained from the smectic E phase show the presence of a layered structure and a high degree of order arrangement within the planes.

17.4.3 The Order Parameter

If we assume that the molecules of a nematic or cholesteric liquid crystal are rigid and rodlike in shape, then we can describe the orientation of the *i*th molecule by introducing a unit vector $\vec{v}^{(i)}$ along its axis. $\vec{v}^{(i)}$ is different from the director **n**, which gives the average preferred direction of the molecules. It is not possible to introduce a vector order parameter for liquid crystals, which possess a center of symmetry due to which the average of $\vec{v}^{(i)}$ vanishes. Hence, the order parameter can be expressed only as a second-rank tensor

$$S_{\alpha\beta}(\mathbf{r}) = \frac{1}{N}\sum_i \left(v_\alpha^{(i)} v_\beta^{(i)} - \frac{1}{3}\delta_{\alpha\beta}\right), \tag{17.23}$$

where the sum is over all the N molecules in a macroscopic volume located at **r**, and the v_α are the components of $\vec{v}$ referred to by a set of laboratory-fixed axes. $S_{\alpha\beta}$ is a symmetric traceless tensor of rank two and has five independent components. In the isotropic case, where the molecules have random orientation, $S_{\alpha\beta} = 0$ (Problem 17.5).

To express the order parameter for nonlinear rigid molecules, one can introduce a Cartesian coordinate system $x'y'z'$ fixed in the molecules. In the case of a uniaxial liquid crystal, the order parameter tensor is defined by

$$S_{\alpha'\beta'}(\mathbf{r}) = <\cos\theta_{\alpha'}\cos\theta_{\beta'} - \frac{1}{3}\delta_{\alpha'\beta'}>, \tag{17.24}$$

where $\cos\theta_{\alpha'}$ is the angle between the α' molecular axis, and the preferred direction or the optic axis. The angle brackets indicate an average over the molecules in a small but macroscopic volume. It can be shown that Eq. (17.24) is equivalent to Eq. (17.23) in the case of linear molecules or molecules with a well-defined long axis about which they rotate rapidly (Problem 17.6).

In real liquid crystals, different parts of the molecules might have to be described by different $S_{\alpha\beta}$ tensors. It is preferable to define the order parameter through a macroscopic property such as the anisotropy in the diamagnetic susceptibility,

$$Q_{\alpha\beta} = \chi_{\alpha\beta} - \frac{1}{3}\delta_{\alpha\beta}\chi_{\gamma\gamma}, \tag{17.25}$$

where $\chi_{\alpha\beta}$ is the magnetic susceptibility tensor per unit volume. By convention, repeated indices are to be summed over. Here, $Q_{\alpha\beta}$ is a symmetric traceless tensor of rank two and has five independent components.

The diamagnetic susceptibility is approximately the sum of the susceptibilities of individual molecules. We choose the principal susceptibilities of a rigid molecule to be $\chi_1^{(0)}$, $\chi_2^{(0)}$, and $\chi_3^{(0)}$, and choose the fixed axes x', y', and z' of the molecule to coincide with the principal axes of the susceptibility. It can be easily shown for a uniaxial liquid crystal (Problem 17.7),

$$Q_{xx} = Q_{yy} = 2N[(S_{y'y'} + S_{z'z'})\chi_1^{(0)} + (S_{z'z'} + S_{x'x'})\chi_2^{(0)} + (S_{x'x'} + S_{y'y'})\chi_3^{(0)}] \tag{17.26}$$

and

$$Q_{zz} = N(S_{x'x'}\chi_1^{(0)} + S_{y'y'}\chi_2^{(0)} + S_{z'z'}\chi_3^{(0)}), \tag{17.27}$$

where N is the number of molecules per unit volume. Further, it can be easily shown that

$$S_{x'x'} + S_{y'y'} + S_{z'z'} = 0. \tag{17.28}$$

Hence, there are only two independent parameters on the right side of Eqs. (17.26) and (17.27).

17.4.4 Curvature Strains

There is a preferred axis along which the molecules orient themselves in the microscopic region of a liquid crystal. The direction of this axis varies from place to place in equilibrium, and it can also be forced to vary by external forces. The relative orientations away from the equilibrium position are known as curvature strain, and the restoring forces are known as curvature stresses. The free energy density is a quadratic function of the curvature strains. The theory of the curvature elastic energy is based on the symmetry properties of the liquid crystal.

The three distinct curvature strains of a liquid crystal are shown in Figure 17.10 by assuming that $\hat{n}(\mathbf{r})$ is a unit vector giving the direction of the preferred orientation at the point $\mathbf{r}$ and varies slowly from point to point in molecules with permanent dipole moments. At $\mathbf{r}$, we introducc a coordinate system with z parallel to $\mathbf{n}$. The curvature has six components, resulting in three curvature strains: splay, twist, and bend.

17.4.5 Optical Properties of Cholesteric Liquid Crystals

The unusual optical properties of cholesteric liquid crystals include the color effects seen in reflection under white light. These effects are due to the interactions of the light with the twisted arrangement of the molecules as well as the spatial variation of the dielectric constant. The variation of the dielectric constant through the medium is small ($\Delta n/n \sim 0.03$), and we will assume that the normal form of the waves propagating through the medium is approximately that of ordinary circularly

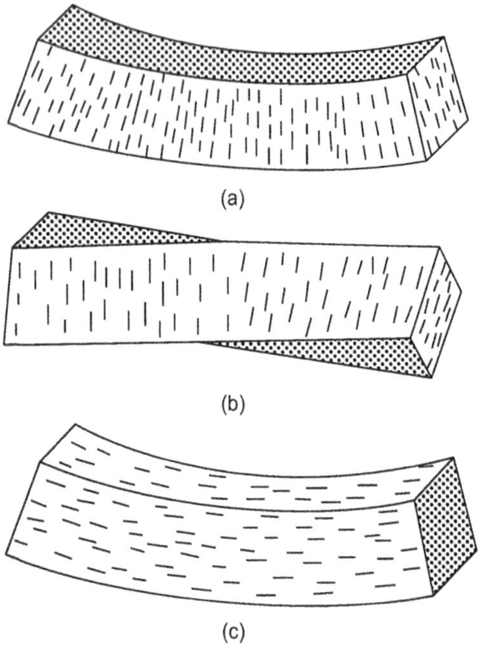

FIGURE 17.10

The three curvature strains of liquid crystals: (a) splay: $s_1 = \partial n_x/\partial x$, $s_2 = \partial n_y/\partial y$; (b) twist: $t_1 = -\partial n_y/\partial x$, $t_2 = \partial n_x/\partial y$; (c) bend: $b_1 = \partial n_x/\partial z$, $b_2 = \partial n_y/\partial z$.

Reproduced from Stephen and Straley[25] with the permission of the American Physical Society.

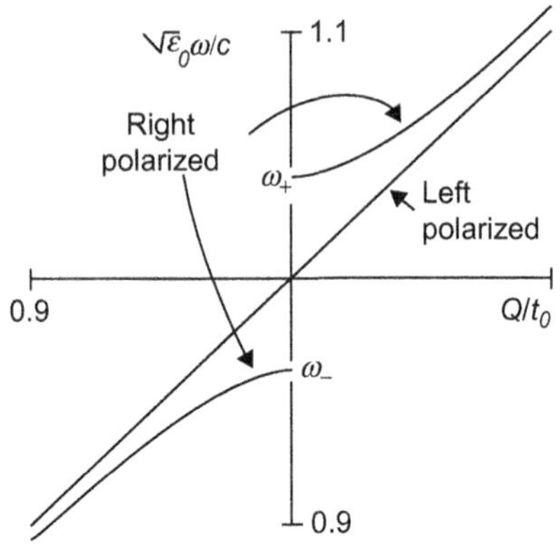

FIGURE 17.11

Dispersion of light in a right-hand twisting cholesteric.

Reproduced from Stephen and Straley[25] with the permission of the American Physical Society.

polarized waves. These waves would be strongly affected only if the half-wavelength nearly matches the periodicity length (or its projection on the wave vector) and if the sense of its rotation is the same as the twisted structure. Otherwise, the effect on the dispersion relation of the wave would be the same as in other periodic structures. Because **n** and −**n** are indistinguishable, the periodicity is one-half the pitch, and according to the Bragg formula, a "band gap" would appear at a matching wavelength

$$\lambda_0 = p \sin \theta, \tag{17.29}$$

where p is the periodicity length. The periodic perturbation determines the size of the band gap. The form of the dispersion relation is shown in Figure 17.11. The right-hand circularly polarized wave is strongly affected when its half-wavelength nearly matches the periodicity length. The left-hand circularly polarized wave of the same wavelength is unaffected.

The extrema of the band gaps are obtained from the fact that light of wavelength λ_0 can travel at two distinct speeds with its electric vector aligned with the principal axes of the dielectric constant. Thus, the extrema of the band gap are

$$\begin{aligned} \omega_+ &= 2\pi c/n_-\lambda_0, \\ \omega_- &= 2\pi c/n_+\lambda_0. \end{aligned} \tag{17.30}$$

Here, n_+ and n_- are the refraction indices. If light of frequencies between ω_+ and ω_- and of the appropriate circular polarization to match the twist is directed at the liquid crystal, it is totally reflected, and the polarization of the reflected light matches the sense of the twist. Thus, a right-hand cholesteric reflects right-polarized light, whereas left-polarized light suffers only weak reflection. Because the reflected band is narrow ($\Delta\omega/\omega \sim \Delta n/n \sim 0.03$), a pure color is reflected; this color depends on the pitch as well as the angle of incidence.

Liquid crystals are extensively used in fiber-optic devices applied in telecommunication circuits. The optics of liquid-crystal devices (LCD) has evolved extensively in the past decade.

17.5 QUASICRYSTALS

17.5.1 Introduction

It had been generally accepted among solid state physicists that crystals can have translational periodicity as well as one-, two-, three-, four-, and six-fold rotational symmetries. However, five-fold rotational symmetry could not exist in equilibrium-condensed phases. An icosahedron, which is the most locally densely packed arrangement, had been observed in liquids and amorphous solids. It was thought that an icosahedral rotational symmetry contradicts the translational periodicity and is unlikely to be found.

17.5.2 Penrose Tiles

Penrose introduced the concept of two-dimensional tiles in 1974 (the concept was published in 1977)[23]; he proposed that it is possible to cover (tile) any flat two-dimensional space with only two different tile shapes (known as fat and skinny rhombus) in an infinite number of aperiodic ways. The tiles, which are rhombi, must be placed such that the matching arrows are always adjacent. The Penrose tiles are shown in Figure 17.12. The smaller angle of the fat rhombus is $2\pi/5$, and the smaller angle of the skinny rhombus is $\pi/5$. If the length of each side is 1, the long diagonal of the fat rhombus (dashed line) has length $\tau = (\sqrt{5}+1)/2$, and the short diagonal of the skinny rhombus (dashed line) has length $1/\tau$.

The Penrose lattice is obtained by tiling a plane using a collection of the two Penrose tiles. The Penrose lattice, shown in Figure 17.13, shows the local regions of five-fold symmetry.

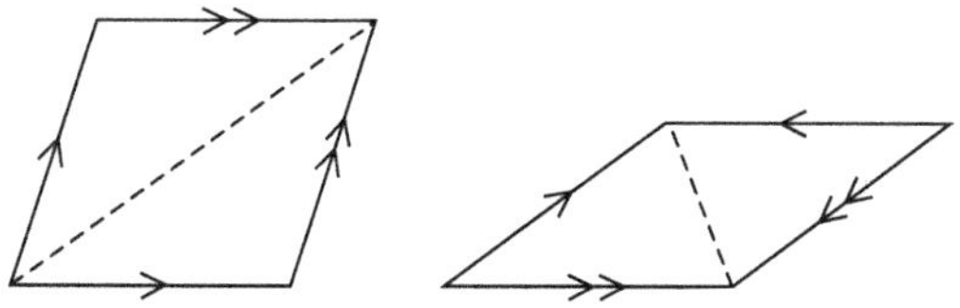

FIGURE 17.12

Penrose tiles.

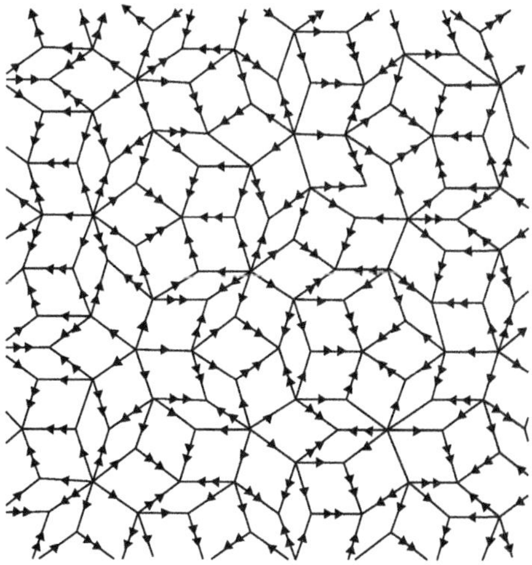

FIGURE 17.13

The Penrose lattice.

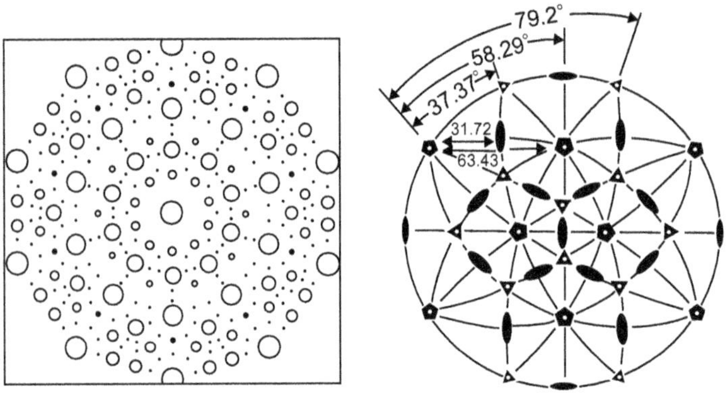

FIGURE 17.14

Diffraction pattern of five-fold symmetry (area proportional to the intensity) and stereogram of the icosahedral group.

Reproduced from Misra[18] with the permission of Elsevier.

17.5.3 Discovery of Quasicrystals

In 1984, Shectman et al. (Ref. 24) discovered sharp diffraction patterns of icosahedral symmetry in Al-Mn alloy rapidly quenched from the melt. The icosahedral phase in rapidly solidified Al-Mn alloy is resistant to crystallization up to 350° C and is metastable. The results of Shectman et al. are shown schematically in Figure 17.14.

In view of the above, Levine and Steinhardt generalized the notion of crystal to include a quasiperiodic translational order and named it "quasicrystal." A quasicrystal has long-range quasiperiodic order as well as long-range crystallographically forbidden orientation symmetry. The classes of quasicrystal include Al-based transition metal alloys (TM alloys: AlMn, AlMnSi, AlCuFe, and AlPdMn); alloys having similar composition to that of the Frank–Kasper phase with tetrahedrally close-packed structure (FK alloys: Al-Cu-Li, Zn-Mg-RE; RE = rare earth); and stable binary alloys (Cd alloys: $Cd_{5.7}Yb$ and $Cd_{17}Ca_3$). One can observe a decagonal phase with a diffraction pattern of 10-fold rotational symmetry in AlMn, AlFe, AlCuCo, and AlCoNi. There is a one-dimensional translational periodicity along the 10-fold rotational axis in these alloys. CrNiSi and VNiSi are two-dimensional octagonal quasicrystals. The two-dimensional dodecagonal quasicrystals are TaTe and a polymer alloy.

17.5.4 Quasiperiodic Lattice

The various methods used to mathematically generate the structures of quasicrystals include the inflation–deflation operation, utilization of matching rules, grid method, strip projection method, cut projection method, or generalized dual method. We will discuss the strip projection method and note that the other methods are closely related.

In the strip projection method, the lattice points in the hypercubic lattice in the n-dimensional space $\mathbf{E}''$ are projected on vertices in the d-dimensional quasiperiodic lattice. $\mathbf{E}''$ is decomposed into two subspaces,

$$\mathbf{E}'' = \mathbf{E}^{\parallel} + \mathbf{E}^{\perp}, \tag{17.31}$$

where the subspace $\mathbf{E}^{\parallel}$ is the d-dimensional physical space, and $\mathbf{E}^{\perp}$ is orthogonal to $\mathbf{E}^{\parallel}$. $\mathbf{E}^{\perp}$ is called the "perpendicular space" (perp-space). The projection of a unit "hypercube" onto $\mathbf{E}^{\perp}$ is known as the "window" $\mathbf{W}$. For any $\mathbf{x}_0$ in $\mathbf{E}^{\parallel}$, we write $\mathbf{W}_0$ for $\mathbf{W}+\mathbf{x}_0$. When a lattice point in $\mathbf{E}''$ is located inside $\mathbf{W}_0$, its projection onto $\mathbf{E}^{\parallel}$ is selected as a vertex in the physical space $\mathbf{E}^{\parallel}$. If the "slope" of the hyperplane $\mathbf{E}^{\parallel}$ is rational to the hypercubic lattice in $\mathbf{E}''$, the projected lattice is periodic. However, if the "slope" of the hyperplane $\mathbf{E}^{\parallel}$ is irrational, the projected lattice becomes quasiperiodic. The hypercubic lattice, the directions of the two orthogonal subspaces $\mathbf{E}^{\parallel}$ and $\mathbf{E}^{\perp}$, and $\mathbf{W}$ must be invariant under operations of the noncrystalline group of the quasicrystal.

In Figure 17.15, we show a one-dimensional quasiperiodic lattice obtained by projection from a two-dimensional square lattice. The Fibonacci lattice is generated when the $\mathbf{E}^{\parallel}$-axis is generated at $\theta = \arctan 1/\tau$ to the x-axis. Here, the golden mean $\tau = (\sqrt{5}+1)/2$.

In Figure 17.16, we show a Fibonacci lattice, obtained in this manner, that can also be described as a one-dimensional section ($d = 1$) of a two-dimensional periodic function ($N = 2$). The two-dimensional periodic structure consists of a periodic arrangement of a line segment extending in the direction of $E_{\perp}$. The line segment is called an atomic surface. A point sequence is obtained on the $E_{\parallel}$ section comprised of two spacings L and S. The irrational shape indicates the lack of periodicity in the arrangement of L and S.

The distribution function of lattice points $\rho(\mathbf{x})$ of a two-dimensional square lattice is given by

$$\rho(\mathbf{x}) = \rho(x^{\parallel}, x^{\perp}) = \sum_{jl} \delta(x^{\parallel} - j\cos\theta - l\sin\theta) \times \delta(x^{\perp} + j\sin\theta - l\cos\theta), \tag{17.32}$$

and the distribution function on the projected lattice on $\mathbf{E}^{\parallel}$ is (Problem 17.8)

$$\rho_0(x^{\parallel}) = \int_{-\infty}^{\infty} dx^{\perp} \rho(x^{\parallel}, x^{\perp}) W(x^{\perp}). \tag{17.33}$$

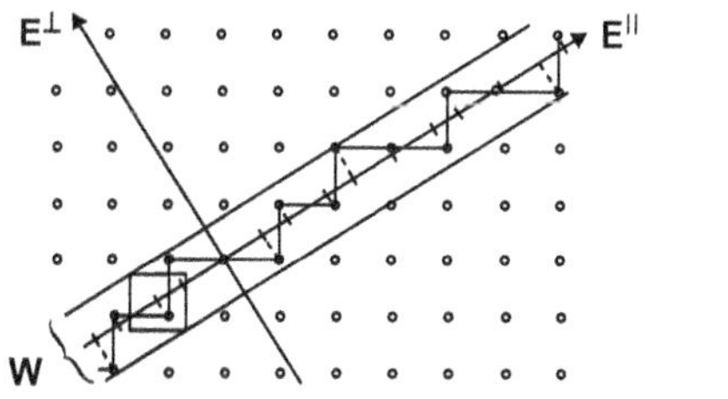

FIGURE 17.15

One-dimensional quasiperiodic lattice (crossmarks on the $\mathbf{E}^{\parallel}$-axis) generated by projection.

Reproduced from Misra[18] with the permission of Elsevier.

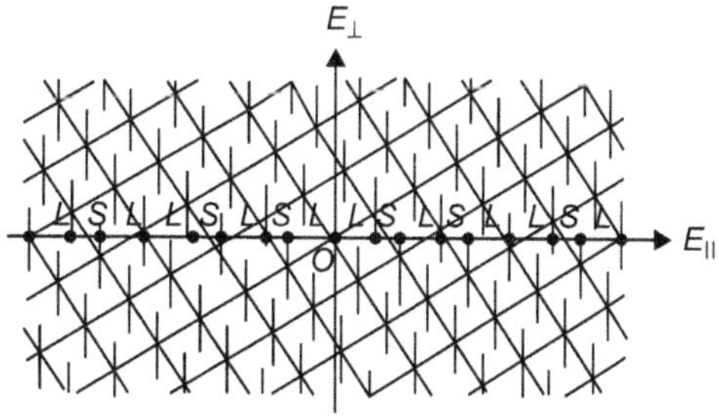

FIGURE 17.16

A Fibonacci lattice.

Reproduced from Misra[18] with the permission of Elsevier.

Here,

$$W(x^{\perp}) = \begin{cases} 1, & \text{if } 2|x^{\perp}| < (\cos\theta + \sin\theta) \\ 0, & \text{otherwise.} \end{cases} \tag{17.34}$$

The diffraction pattern of the one-dimensional Fibonacci lattice is expressed by a structure factor

$$\begin{aligned} f(k) &= \int_{-\infty}^{\infty} dx^{\parallel} e^{-ikx^{\parallel}} \rho_0(x^{\parallel}) \\ &\approx \sum_{m,n=-\infty}^{\infty} \frac{\sin k^{\perp}\omega}{k^{\perp}\omega} \delta(k - 2\pi(m \sin\theta + n\cos\theta)). \end{aligned} \tag{17.35}$$

Here,

$$\omega = \frac{1}{2}(\cos\theta + \sin\theta) \tag{17.36}$$

and

$$k^{\perp} = 2\pi(m\cos\theta + n\cos\theta). \tag{17.37}$$

This procedure of projection and calculation of structure factor is generalized to higher-dimensional cases. For example, the pentagonal quasicrystal is obtained by a projection of the hypercubic lattice to a two-dimensional space and the icosahedral quasicrystal by that of the six-dimensional one to a three-dimensional space. The three-dimensional icosahedral quasicrystals are constructed by two rhombohedral units: prolate and oblate.

17.5.5 Phonon and Phason Degrees of Freedom

The diffraction intensity pattern $I(\mathbf{q})$ of a solid is, in general, given by

$$I(\mathbf{q}) \equiv |S(\mathbf{q})|^2, \tag{17.38}$$

where

$$S(\mathbf{q}) = \int \rho(\mathbf{r}) e^{-2\pi i \mathbf{q}\cdot\mathbf{r}} d\mathbf{r}, \tag{17.39}$$

$\mathbf{q}$ is the wave vector, and $\rho(\mathbf{r})$ is the atomic-density function in real space. For a quasicrystal, the characteristics of the function $I(\mathbf{q})$ observed experimentally are as follows:

a. It consists of $\delta-$ functions.
b. The number of basis vectors necessary for indexing the positions of the $\delta-$ functions exceeds the number of dimensions.
c. It shows a rotational symmetry forbidden in the conventional crystallography.

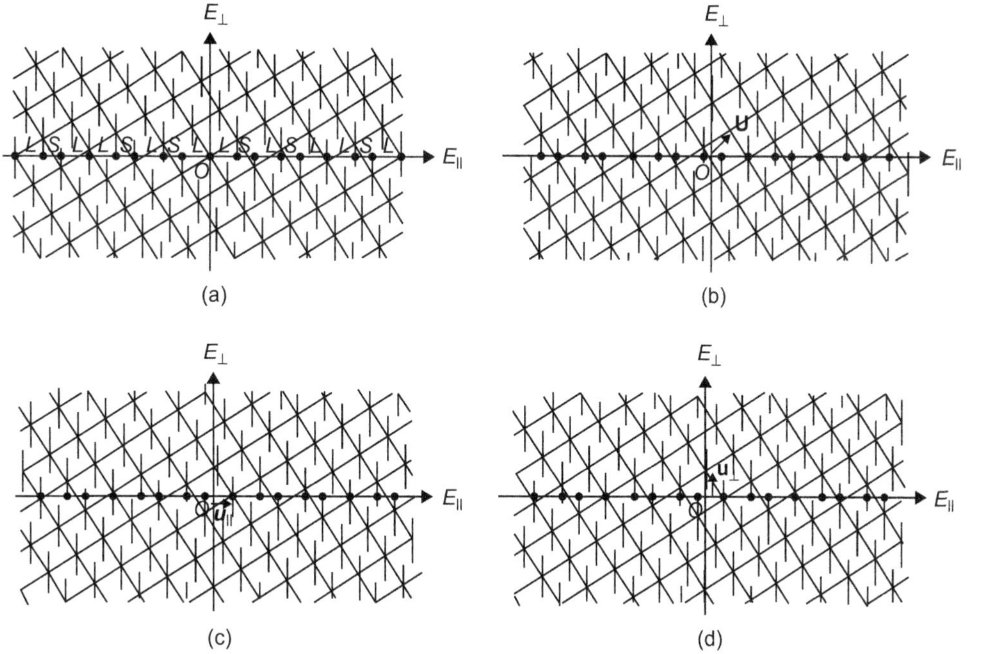

FIGURE 17.17

(a) A Fibonacci lattice, the structure resulting from a displacement of (b) **U**, (c) **u**, and (d) **w**.

Reproduced from Misra[18] with the permission of Elsevier.

As an example, we consider the Fibonacci lattice in Figure 17.17a. We consider the translation of the two-dimensional periodic structure by a vector **U** with respect to the origin of the physical space $E_{\parallel}$ (Figure 17.17b). It can be shown that the two structures on $E_{\parallel}$ before and after the displacement **U** can be overlapped out to large finite distances by a finite translation in $E_{\parallel}$. These two structures are said to belong to the same local isomorphic class (LI class). Because these structures are physically indistinguishable, they give the same diffraction intensity function I(**q**) and have the same energy. The vector **U** can be written as

$$\mathbf{U} = \mathbf{u} + \mathbf{w}, \tag{17.40}$$

where **u** represents the degrees of freedom of d-dimensional translation in physical space that crystals also possess, and **w** represents $(N-d)$ degrees of freedom characteristic of a quasiperiodic system. Figure 17.17c shows **u** results in a translation of the Fibonacci lattice in $E_{\parallel}$, and Figure 17.17d shows how **w** generates a rearrangement of L and S.

Here, **u** and **w** are called phonon and phason displacements. When these displacements vary spatially, their gradients yield a strain. The gradient of **u** yields the conventional elastic strain (phonon strain), whereas the gradient of **w** yields the phason strain. A phonon-strained Fibonacci lattice is shown in Figure 17.18. A uniform phonon strain is introduced by a compression deformation of the two-dimensional structure.

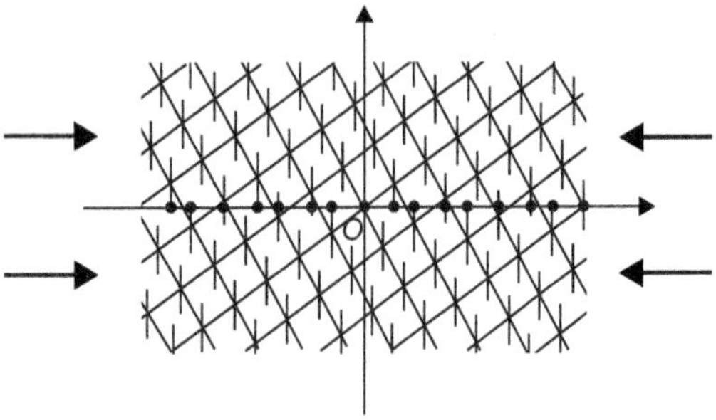

FIGURE 17.18

Phonon-strained Fibonacci lattice.

Reproduced from Misra[18] with the permission of Elsevier.

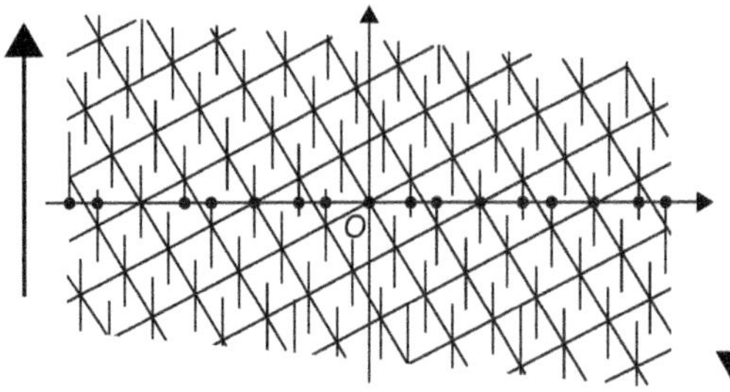

FIGURE 17.19

Phason-strained Fibonacci lattice.

Reproduced from Misra[18] with the permission of Elsevier.

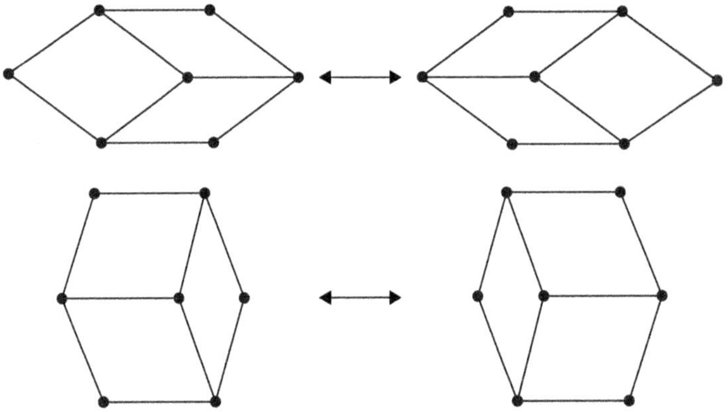

FIGURE 17.20

Examples of phason flips in the two-dimensional Penrose lattice.

Reproduced from Misra[18] with the permission of Elsevier.

A uniform phason strain is introduced by a shear deformation of the Fibonacci lattice, as shown in Figure 17.19.

A phason displacement results in a local rearrangement of points (atoms) such as $LS \leftrightarrow SL$, which is known as the phason flip. Examples of the phason flip in two-dimensional Penrose lattice, which is known as a two-dimensional decagonal quasicrystal, are shown in Figure 17.20.

To decompose properly the total N degrees of freedom into d phonon and $(N-d)$ phason degrees of freedom, i.e., to embed properly a given d-dimensional quasiperiodic structure into an N-dimensional hypercrystal, one needs to know the point group symmetry of the system.

17.5.6 Dislocation in the Penrose Lattice

The perfect edge dislocation introduced in the two-dimensional Penrose lattice (or Penrose tiling) is shown in Figure 17.21. The phason strain field cannot be easily recognized around the dislocation. However, the tiling pattern changes when the dislocation position is translated to the left by a distance represented by the arrow.

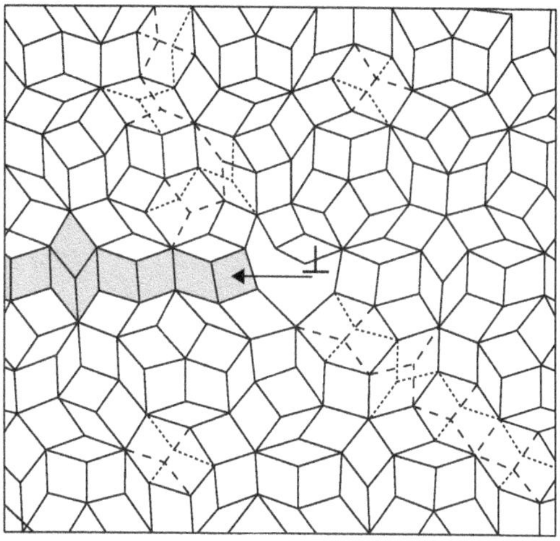

FIGURE 17.21

A perfect edge dislocation in the Penrose lattice. The shaded tiles are destroyed to produce intratile phason defects after gliding the dislocation to the left.

Reproduced from Misra[18] with the permission of Elsevier.

17.5.7 Icosahedral Quasicrystals

An icosahedron, shown in Figure 17.22, is a regular polyhedron with 20 identical equilateral triangular sides. The main characteristics of a quasicrystal structure are that it is a combination of a quasiperiodic lattice and a cluster decorating it. Three types of icosahedral quasilattices, P-, F-, and I-types, are known theoretically, out of which two types, P and F, have been observed experimentally. These two types are distinguished by reflection conditions.

The reflection vector of **g** of a icosahedral quasicrystal can be written as

$$\mathbf{g} = \frac{1}{a_{6D}} \sum_{1}^{6} \mathbf{m}_i \mathbf{e}_{i//}, \qquad (17.41)$$

where the six $\mathbf{e}_{i//}$ vectors with length 1/2 are parallel to the lines connecting to the center and vertices of an icosahedron, and a_{6D} is the lattice parameter of the six-dimensional hypercubic lattice in the framework of the section. There are no restrictions for the indices in the case of P-type reflections, whereas either all odd or all even indices appear in the F-type. The reflection condition in the F-type exhibits a τ-scaling rule (τ: golden mean), whereas the P-type has a τ^3-scaling rule. However, in real space, a P-type quasilattice can be decomposed into two F-type sublattices. Thus, the F-type quasicrystal can be interpreted as an ordered phase, in which two kinds of atomic clusters with different atomic configurations are arranged regularly.

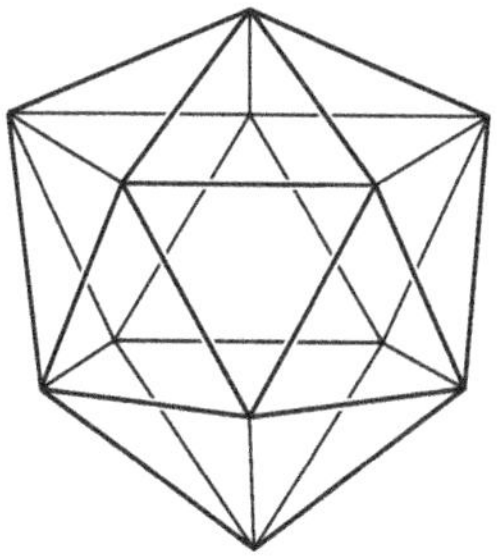

FIGURE 17.22

The icosahedron.

In the case of Al-transition metal quasicrystals, the Mackay-type cluster is considered as a basic structural unit. The 54 atoms form the triple shells as presented in Figure 17.23a. The first and the

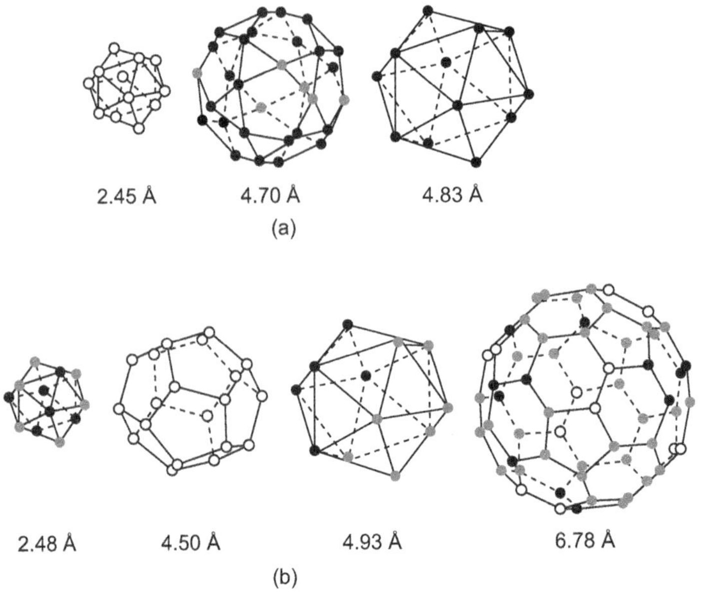

FIGURE 17.23

Icosahedral atomic clusters. (a) Mackay type in Al-Mn-Si approximant and (b) Bergman type in Zn-Mg-Al approximant. The radius in each shell is inserted.

Reproduced from Misra[18] with the permission of Elsevier.

second shell, which are composed of Al, are an icosahedron and an icosidodecahedron, respectively. The third shell is a larger icosahedron of the transition metal. The five ternary alloys, which are of the stable Mackay type, are Al-Cu-(Fe, Ru, Os) and Al-Pd-(Mn, Re).

The Zn-based quasicrystals are the Bergman clusters shown in Figure 17.23b. They include 104 atoms. This cluster has four concentric shells: an icosahedron, a dodecahedron, a larger icosahedron, and a truncated icosahedron (soccer ball). The 10 ternary alloys, which are of the Bergman type, are Al-Li-Cu, Zn-Mg-Ga, Ti-Zr-Ni, Mg-Al-Pd, and Zn-Mn-(Y, Dy, Gd, Ho, Tb, and Er).

17.6 AMORPHOUS SOLIDS

17.6.1 Introduction

Amorphous solids have attracted a great deal of attention in recent years. There are an infinite number of ways in which the geometrical arrangement of atoms can be visualized. Sometimes, amorphous solids have been visualized as frozen-liquid structures, but there are some significant differences between the two types of states. For example, there are distinct differences between the radial distribution of functions for a true liquid and an amorphous solid. Another interesting difference in their properties is that whereas liquid Ge and Si are metals, amorphous Si and Ge are semiconductors. The bonds are broken in liquid Ge and Si, where presumably the sp^3 configuration in the solid is changed to an s^2p^2 configuration in the liquid, which makes it metallic. However, in the amorphous state, the sp^3 bonds are still present between nearest neighbors. These bonds are imperfect because the amorphous films are produced at relatively low temperature. Due to this bonding, amorphous Ge and Si are semiconductors like their crystalline counterpart, but there are significant differences arising out of the fact that (1) the bonding is imperfect because there is no true tetrahedral symmetry in spite of four-fold coordination over small regions; (2) different regions are not linked because there is no long-range order; and (3) many atoms could have only three nearest neighbors due to which the bond angles are severely distorted from the ideal value.

Amorphous solids can be obtained in a variety of ways. In some cases, they are obtained in the frozen-liquid amorphous state either by evaporation into a cooled substrate or by extremely rapid cooling from the melt. The more popular method is by a process known as "sputtering" the components onto a cooled substrate. In this process, the atoms in a solid are knocked out by energetic ions of inert gas such as argon. The inert gas is at a reduced pressure. It is ionized by an electrical discharge. A substrate is placed above the solid, which is in contact with an electrode, and the atoms are discharged by the impingement of the energetic argon ionized by the electric discharge. The atoms condense above the substrate and form a thin film. The sputtering process has the advantage that a thin layer of an amorphous film is deposited on the substrate. The disadvantage of this method is that both argon and oxygen are invariably present as impurities.

17.6.2 Energy Bands in One-Dimensional Aperiodic Potentials

To consider the localized states, we consider the energy bands in one-dimensional aperiodic potentials.[8] We consider a finite segment of a line $a<x<b$, where $b-a=L$. The potential energy of the segment is periodic but can be derived from a periodic potential by a disordering process. The two linearly independent real solutions of the Schrodinger equation are

$$\int_a^b \psi_1^2 dx = 1, \quad \text{where } \psi_1(a) = \psi_1(b), \tag{17.42}$$

and

$$\int_a^b \psi_1(x)\psi_2(x)dx = 0, \quad \int_a^b \psi_2^2(x)dx = 1. \tag{17.43}$$

Economou and Cohen (Ref. 8) postulated that the solutions of the aperiodic case that most resemble the Bloch functions of the periodic case are those combinations of φ that satisfy an extremal condition in relation to the average value of momentum $<\hat{p}>$. φ is so chosen that the real part of $<\hat{p}>$ is as extremum. One can write φ as

$$\varphi = \psi_1 + (\alpha + i\beta)\psi_2/[1+\alpha^2+\beta^2]^{\frac{1}{2}}, \tag{17.44}$$

where x and y are real. The average value of $\hat{p}$ (Problem 17.9) is

$$<\hat{p}> = -i\hbar \int \varphi^*(d\varphi/dx)dx = \hbar \frac{\beta(\pi_{12} - \pi_{21}) - i[\pi_{11} + (\alpha^2 + \beta^2)\pi_{22} + \alpha(\pi_{12} + \pi_{21})]}{1 + \alpha^2 + \beta^2}, \tag{17.45}$$

where

$$\pi_{ij} = \int_a^b \psi_i d\psi_j, \quad i,j = 1,2. \tag{17.46}$$

The extremum requirement on Re $<\hat{p}>$ gives

$$\alpha = 0,\ \beta = 1 \text{ or } \alpha = 0,\ \beta = -1. \tag{17.47}$$

To avoid obtaining the nonzero imaginary part of the expectation value of the extremum value of $<\hat{p}>$ from the preceding solutions, Economou and Cohen imposed periodic boundary conditions,

$$\psi_2(a) = \psi_2(b), \tag{17.48}$$

so that the energy eigenvalues originating from Eq. (17.48) are exactly the same as the first, third, fifth, and so on, obtained by setting $V = \infty$ outside (a, b). Thus, there are half the number of values in this problem as in the infinite-barrier problem.

From Eqs. (17.44) and (17.47), there are two functions that make $<\hat{p}>$ an extremum (Problem 17.10),

$$\varphi = (\psi_1 + i\psi_2)/\sqrt{2} \tag{17.49}$$

and

$$\varphi^* = (\psi_1 - i\psi_2)/\sqrt{2}. \tag{17.50}$$

The corresponding values of $<\hat{p}>$ are

$$p_{ext} = \pm\frac{1}{2}\hbar(\pi_{12} - \pi_{21}) = \pm\hbar\pi_{12}. \tag{17.51}$$

One can easily show that the set of functions $\varphi_n(x), \varphi_n^*(x)$, $n = 1, 2, 3, \ldots$ is complete and orthonormal and makes each momentum expectation value extremal. This set is the closest one can come to a set of Bloch functions in the aperiodic case. A schematic representation of these results is shown in Figure 17.24. Here, E_c is the energy at which the states change from localized to extended. E_B is the energy at which the density of states vanishes, the bound of the spectrum. They coincide in the periodic case and move in the opposite direction as the aperiodicity is increased.

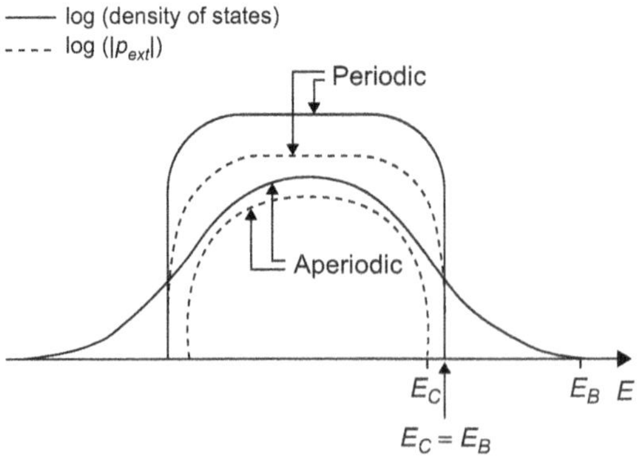

FIGURE 17.24

Density of states (solid curve) and distribution of the extended states (dashed curve) for an energy band of an ordered and disordered lattice.

Reproduced from Economou and Cohen (Ref. 8) with the permission of the American Physical Society.

17.6.3 Density of States

Concepts such as crystal momentum, effective mass, or band structure function E(**k**) cannot be used for localized one-electron states. However, the concept that is used and still valid is the density of states, $g(E)$, which can be defined as

$$g(E) = \frac{1}{V_g} \sum_i \delta(E - E_i). \tag{17.52}$$

In delocalized states of disordered solids, bands can exist in which tails with localized states are attached to their edges. The region without states between the tails of two adjacent bands is called a gap, or if the tails overlap, the region is known as a pseudogap.

17.6.4 Amorphous Semiconductors

We first assume that each Ge or Si atom in the amorphous state has four nearest neighbors as in the crystalline state. However, the covalent bonds in the crystalline phase are distorted in the amorphous phase. The band edges of the conduction and valence band contain localized states. They are separated from the extended states by the mobility edges. The abrupt band edges are shown in Figure 17.25.

The electrons occupying the tail states cannot take part in the conduction because they are localized due to the disorder in the potential.

Normally, each Si or Ge atom, which has four valence electrons, shares one electron in the covalent bond with its neighbor. However, in the amorphous state, the three-fold-coordinated atoms produce a dangling bond because one bond remains uncompensated. A dangling bond essentially means an electron and an empty state. The electron in the dangling bond is localized at 0° K because there are no adjacent sites available for it to move in the amorphous state. The concentration of these dangling bonds is very high (on the order of $10^{25} m^{-3}$), and they control the position of the Fermi level.

However, at finite temperatures, the process of electrical conduction occurs in three different regimes: the propagating regime, the jumping regime, and the hopping regime. The propagating regime, which is commonplace in the crystalline solid state theory, is dominant in disordered systems only at very high temperatures. At intermediate temperatures, electronic conduction in such systems takes place by diffusion or Brownian motion. This type of conduction, encountered near the mobility edges of amorphous solids, is known as jumping conduction. At low temperatures, the hopping regime is prevalent. In this regime, the electrons can move only through phonon-assisted hopping. At relatively higher temperatures in the hopping regime of conduction, the largest tunnel contribution arises from jumps to unoccupied levels of nearest-neighbor centers. At lower temperatures, the number and

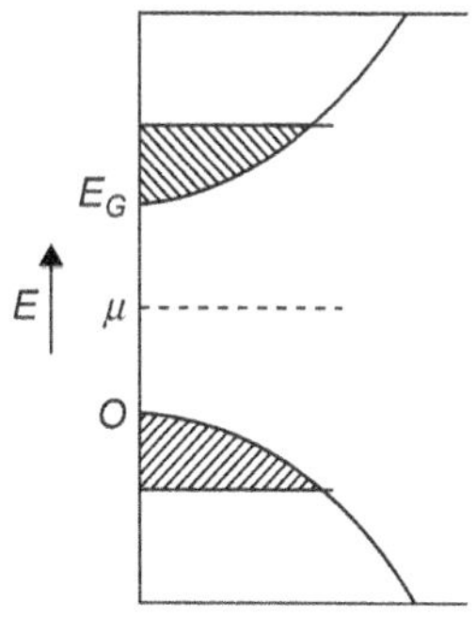

FIGURE 17.25

Localized states at the edges of the valence and conduction bands in amorphous semiconductors.

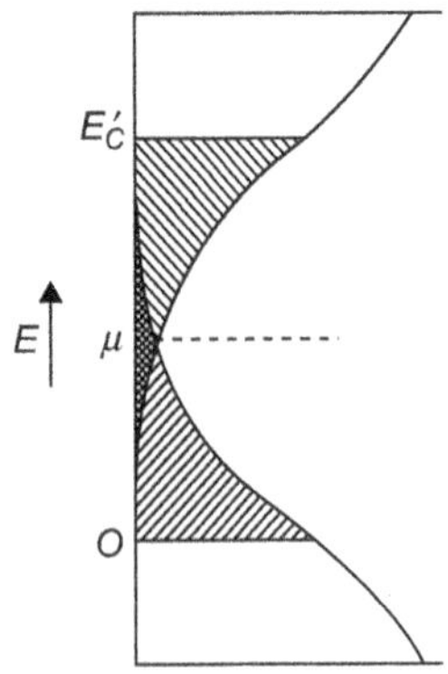

FIGURE 17.26

The overlap of the tails of the localized states in amorphous tetrahedral semiconductors.

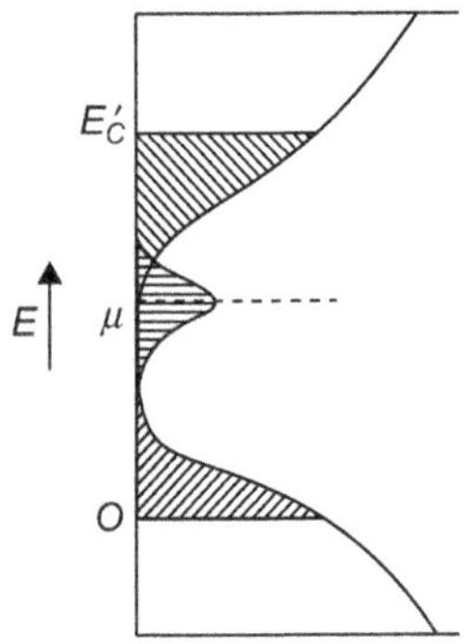

FIGURE 17.27

Impurity band of local defects in the gap or the pseudogap.

energy of phonons available for absorption decrease so that tunneling is restricted to seek centers that are not nearest neighbors but lie energetically closer. This type of conduction is known as variable-range hopping.

The overlap of tails of the localized states in amorphous tetrahedral semiconductors is shown in Figure 17.26.

The regime of variable-range hopping is subdivided into two ranges. At relatively higher temperatures of this regime, variable-range hopping transport is done by tunneling conduction in a band tail with somewhat large activation energy. At the lower temperature range, hopping transport by tunneling conduction near the Fermi level is favored.

The local defects, such as dangling bonds, can cause impurity bands to appear in the gap or pseudogap. This is shown in Figure 17.27.

The extended states above the mobility gap are important in deciding the optical properties. However, at low temperatures, the transport properties of amorphous semiconductors are mainly determined by the states of the local defects rather than the localized states of the tail states.

PROBLEMS

17.1. Show that in a region where $|\chi_0|$ is everywhere nonzero, the application of the Stokes theorem to the curl of Eq. (17.14) leads to the conclusion that the integral of v_s around any closed curve is zero.

17.2. Show that when we integrate Eq. (17.14) around a circuit that encloses the one-dimensional region, the fact that the phase of the wave function χ_0 must be single-valued modulo 2π leads to the Onsager–Feynman quantization condition

$$\oint v_s \cdot dl = nh/m. \tag{1}$$

17.3. Show that because, in the range of temperatures above 0.01° K, the nuclear spins of the ^{3}He atoms comprising the solid are almost fully disordered, the entropy $S_{solid} = R \ln 2$ (per mole).

17.4. The dipolar interaction is calculated by taking a quantum mechanical average of the dipolar Hamiltonian over the order parameter (pair-wave function). It can be shown that the dipolar free energies are given by

$$\Delta F_D = \begin{cases} -\frac{3}{5} g_D(T)[1 - (\vec{d} \cdot \vec{l})^2], & A \text{ phase}, \\ \frac{4}{5} g_D(T)\left\{\cos\theta + 2\cos^2\theta + \frac{3}{4}\right\}, & B \text{ phase}, \end{cases} \tag{1}$$

where

$$g_D \approx \left(1 - \frac{T}{T_c}\right) 10^{-3} \text{ergs/cm}^3. \tag{2}$$

Show that to minimize the free energy, $\vec{l}$ and $\vec{d}$ must be parallel in the A phase, and the dipolar energy is minimized for $\theta = \cos^{-1}(-\frac{1}{4}) = 104°$ in the B phase.

17.5. For a nematic or cholesteric liquid crystal where the molecules are rigid and rodlike in shape, the order parameter can be expressed as a second-rank tensor

$$S_{\alpha\beta}(\mathbf{r}) = \frac{1}{N}\sum_i \left(v_\alpha^{(i)} v_\beta^{(i)} - \frac{1}{3}\delta_{\alpha\beta}\right), \tag{1}$$

where the sum is over all the N molecules in a macroscopic volume located at $\mathbf{r}$, and the v_α are the components of $\vec{v}$ referred to by a set of laboratory-fixed axes. Show that in the isotropic case, where the molecules have random orientation, $S_{\alpha\beta} = 0$.

17.6. In the order parameter for nonlinear rigid molecules, one can introduce a Cartesian coordinate system $x'y'z'$ fixed in the molecules. In the case of a uniaxial liquid crystal, the order parameter tensor is defined by

$$S_{\alpha'\beta'}(\mathbf{r}) = <\cos\theta_{\alpha'}\cos\theta_{\beta'} - \frac{1}{3}\delta_{\alpha'\beta'}>, \tag{1}$$

where $\cos\theta_{\alpha'}$ is the angle between the α' molecular axis, and the preferred direction or the optic axis. The angle brackets indicate an average over the molecules in a small but macroscopic volume. Show that Eq. (1) in Problem (17.6) is equivalent to Eq. (1) in Problem (17.5) in the case of linear molecules or molecules with a well-defined long axis about which they rotate rapidly.

17.7. Show that for a uniaxial liquid crystal,

$$Q_{xx} = Q_{yy} = 2N[(S_{y'y'} + S_{z'z'})\chi_1^{(0)} + (S_{z'z'} + S_{x'x'})\chi_2^{(0)} + (S_{x'x'} + S_{y'y'})\chi_3^{(0)}] \tag{1}$$

and

$$Q_{zz} = N(S_{x'x'}\chi_1^{(0)} + S_{y'y'}\chi_2^{(0)} + S_{z'z'}\chi_3^{(0)}), \tag{2}$$

where N is the number of molecules per unit volume where the symbols are defined in the text.

17.8. In a quasicrystal, the distribution function of lattice points $\rho(\mathbf{x})$ of a two-dimensional square lattice is given by

$$\rho(\mathbf{x}) = \rho(x^{\parallel}, x^{\perp}) = \sum_{jl} \delta(x^{\parallel} - j\cos\theta - l\sin\theta) \times \delta(x^{\perp} + j\sin\theta - l\cos\theta). \tag{1}$$

Show that the distribution function on the projected lattice on $\mathbf{E}^{\parallel}$ is

$$\rho_0(x^{\parallel}) = \int_{-\infty}^{\infty} dx^{\perp} \rho(x^{\parallel}, x^{\perp}) W(x^{\perp}). \tag{2}$$

Here,

$$W(x^{\perp}) = \begin{cases} 1, & \text{if } 2|x^{\perp}| < (\cos\theta + \sin\theta) \\ 0, & \text{otherwise.} \end{cases} \tag{3}$$

17.9. Show that if one writes φ as

$$\varphi = \psi_1 + (\alpha + i\beta)\psi_2 / [1 + \alpha^2 + \beta^2]^{\frac{1}{2}}, \tag{1}$$

where x and y are real, the average value of the momentum $\hat{p}$ is

$$<\hat{p}> = -i\hbar \int \varphi^* (d\varphi/dx) dx = \hbar \frac{\beta(\pi_{12} - \pi_{21}) - i[\pi_{11} + (\alpha^2 + \beta^2)\pi_{22} + \alpha(\pi_{12} + \pi_{21})]}{1 + \alpha^2 + \beta^2}, \tag{2}$$

where

$$\pi_{ij} = \int_a^b \psi_i d\psi_j, \quad i, j = 1, 2. \tag{3}$$

17.10. Show from Eqs. (17.44) and (17.47) that the two functions that make $<\hat{p}>$ an extremum are

$$\varphi = (\psi_1 + i\psi_2)/\sqrt{2} \tag{1}$$

and

$$\varphi^* = (\psi_1 - i\psi_2)/\sqrt{2}. \tag{2}$$

References

1. Anderson PW, Morel P. Generalized Bardeen-Cooper-Schrieffer States and the proposed low-temperature Phase of Liquid He3. *Phys Rev* 1961;**123**:1911.
2. Anderson PW, Brinkman WF. Anisotropic Superfluidity in He3: A Possible Interpretation of its Stability as a Spin-Fluctuation Effect. *Phys Rev Lett* 1973;**30**:1108.
3. Ashcroft NW, Mermin ND. *Solid state physics*. New York: Brooks/Cole; 1976.
4. Balian R, Werthamer NR. *Phys. Rev.* 1963;**131**:1553. Superconductivity with Pairs in a Relative p Wave.
5. Barker JA, Henderson DH. What is "liquid"? Understanding the states of matter. *Rev Mod Phys* 1976;**48**:587.
6. Chandrasekhar S. *Liquid Crystals*. Cambridge: Cambridge University Press; 1992.
7. de Gennes P-G, Prost J. *The physics of liquid crystals*. Oxford: Clarendon Press; 1993.
8. Economou EN, Cohen MH. Energy Bands in One-Dimensional Aperiodic Potential. *Phys Rev Lett* 1970;**24**:218.
9. Kramer P. Non-periodic Central space filling with icosahedral symmetry using copies of seven elementary cells. *Acta Crystallogr* 1982;**A38**:257.
10. Landau LD. *J Phys.* USSR 1941;**5**:71.
11. Landau LD. *JETP* 1957;**3**:920.
12. Landau LD. *JETP* 1957;**5**:101.
13. Landau LD, Lifshitz EM. *Statistical physics, part 1*. Oxford: Pergamon Press; 1980.
14. Lee DM. The Extraordinary phases of liquid He3. *Rev Mod Phys* 1997;**69**:645.
15. London F. On the Bose-Einstein Condensation. *Phys Rev* 1938;**54**:947.
16. Mackay AL. Cryatallography and the Penrose patterns. *Physica* 1982;**114A**:609.
17. Marder MP. *Condensed matter physics*. New York: John Wiley & Sons; 2000.
18. Misra PK, Fujiwara T, Ishii Y. *Quasicrystals*. Amsterdam: Elsevier; 2008.
19. Mott NF, Davis EA. *Electronic properties of noncrystalline materials*. Oxford: Clarendon Press; 1979.
20. Osheroff DD, Richardson RC, Lee DM. Evidence for a New Phase of Solid He3. *Phys Rev Lett* 1972;**28**:885.
21. Osheroff DD, Gully WJ, Richardson RC, Lee DM. New magnetic Phenomena in Liquid He3 below 300 K. *Phys Rev Lett* 1972;**29**:920.
22. Paulson DN, Kojima H, Wheatley JC. Preferred Effect of a Magnetic Field in the Phase Diagram of Superfluid 3He. *Phys Rev Lett* 1974;**32**:1098.
23. Penrose R. *Bull Inst Math Appl* 1974;**10**:266.
24. Shechtman D, Blech L, Gratias D, Cahn JW. Metallic Phase with Long Range Orientational Order and no Translational Symmetry. *Phys Rev Lett* 1984;**53**:1951.
25. Stephen MA, Straley JP. Physics of Liquid Crystals. *Rev Mod Phys* 1974;**46**:617.
26. Webb RA, Greytak JT, Johnson RT, Wheatley. Observation of a second-order Phase Transition and its Associated P-T phase Diagram in liquid He3. *Phys Rev Lett* 1973;**30**:210.

CHAPTER

Novel Materials

18

CHAPTER OUTLINE

18.1 Graphene 600
18.1.1 Introduction 600
18.1.2 Graphene Lattice 601
18.1.3 Tight-Binding Approximation 602
18.1.4 Dirac Fermions 606
18.1.5 Comprehensive View of Graphene 608
18.2 Fullerenes 608
18.2.1 Introduction 608
18.2.2 Discovery of C_{60} 609
18.3 Fullerenes and Tubules 613
18.3.1 Introduction 613
18.3.2 Carbon Nanotubeles 614
18.3.3 Three Types of Carbon Nanotubes 614
18.3.4 Symmetry Properties of Carbon Nanotubes 616
18.3.5 Band Structure of a Fullerene Nanotube 617
18.4 Polymers 617
18.4.1 Introduction 617
18.4.2 Saturated and Conjugated Polymers 618
18.4.3 Transparent Metallic Polymers 621
18.4.4 Electronic Polymers 621
18.5 Solitons in Conducting Polymers 622
18.5.1 Introduction 622
18.5.2 Electronic Structure 623
18.5.3 Tight-Binding Model 623
18.5.4 Soliton Excitations 624
18.5.5 Solitons, Polarons, and Polaron Excitations 626
18.6.6 Polarons and Bipolarons 626
18.6 Photoinduced Electron Transfer 627
Problems 627
References 630

18.1 GRAPHENE

18.1.1 Introduction

Carbon has four perfect crystalline forms: graphite, diamond, "Buckminsterfullerene" and a fullerene nanotube. In addition, graphene is a one-atom-thick allotrope of carbon, which is a honeycomb lattice of carbon atoms. Graphene also has two-dimensional Dirac-like excitations. We discussed the properties of graphene as well as its possible applications in electronics in Sections 10.7 and 10.8. In the following sections, we will discuss graphene as a building block for all novel materials of carbon as well as derive the theory of Dirac fermions discussed in Section 10.7.

One can view graphite as a stack of graphene layers, and carbon nanotubes can be considered as rolled cylinders of graphene. "Buckminsterfullerene" (C_{60}) can be viewed as molecules obtained by introducing pentagons on the hexagonal lattice of wrapped graphene. These are shown in Figure 18.1.

Diamond is not shown in the diagram because it is primarily used in making jewelry due to its beauty and elegance, and it does not have any major applications in materials science, presumably because of its cost. In addition, each atom in diamond is surrounded in all three directions in space by a full coordination. Because all directions are taken up, it would be nearly impossible for an atom in a diamond lattice to have any bonding with any other atom in the outside 3D space.

Graphene is a two-dimensional (2D) allotrope of carbon that can be imagined to be benzene rings stripped out from the hydrogen atoms. Fullerenes are molecules where carbon atoms are arranged spherically and are zero-dimensional (0D) objects that have discrete energy states. Fullerenes can be thought of as wrapped-up graphene because they are obtained from graphene with the introduction of pentagons, which create positive curvature defects. Carbon nanotubes, which have only hexagons and can be thought of as one-dimensional (1D) objects, are obtained by rolling graphene along a definite direction and reconnecting the carbon bonds. Graphite, which is a three-dimensional (3D) allotrope of carbon, is made out of stacks of graphene layers that are weakly coupled by van der Waals forces. Two-dimensional materials like graphene were presumed not to exist until 2004, when it was obtained in liquid suspension.[2] Graphene could also be obtained on top of noncrystalline substrates[3–5] and was eventually spotted in optical microscopes due to the subtle optical effects made by it on top of an SiO_2 substrate. Graphene exhibits high crystal quality, in which charge carriers can travel thousands of interatomic distances without scattering.

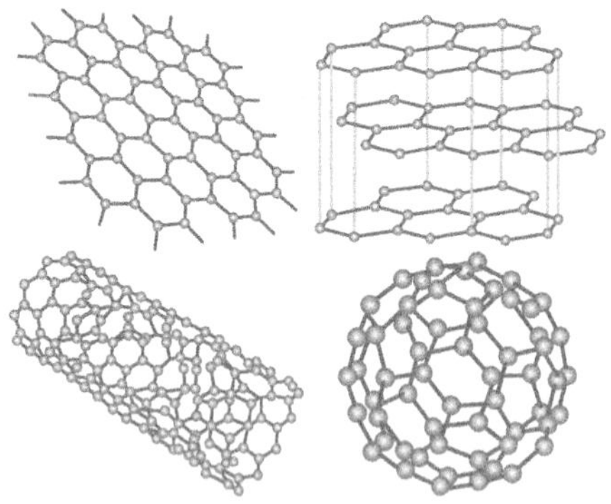

FIGURE 18.1

Clockwise: graphene (2D), graphite (3D), "Buckminsterfullerene" (0D), and carbon nanotubes (1D).

Reproduced from Castro Neto et al.[1] with the permission of the American Physical Society.

The Coulomb interactions are considerably enhanced in small geometries such as graphene quantum dots that lead to Coulomb blockade effects. The transport properties of graphene lead to a variety of applications, which range from single molecule detection to spin injection. Because graphene has unusual structural and electronic flexibility, it can be tailored: deposition of metal atoms or molecules on top; incorporation of boron and/or nitrogen in its structure; and using different substrates that modify the electronic structure. The control of graphene properties can be extended in new directions that would allow for the creation of graphene-based systems with magnetic and superconducting properties.

18.1.2 Graphene Lattice

Carbon has four valence electrons, three of which form tight bonds with neighboring atoms in the plane. Their wave functions are of the form

$$\frac{1}{\sqrt{3}}\left(\psi_e(2s)+\sqrt{2}\psi_e(\tau_i 2p)\right), \quad (i=1,2,3), \tag{18.1}$$

where $\psi_e(2s)$ is the $(2s)$ wave function for carbon, and $\psi_e(\tau_i 2p)$ are the $(2p)$ wave functions of which the axes are in the directions τ_i joining the graphite atom to its three neighbors in the plane. The fourth electron is in the $2p_z$ state. Its nodal plane is the lattice plane and its axis of symmetry perpendicular to it. Because the three electrons forming coplanar bonds do not play any part in the conductivity, graphene can be considered to have one conduction electron in the $2p_z$ state.

The unit cell of the hexagonal layer, designated as $PQRS$ in Figure 18.2, contains two carbon atoms A and B. The distance $AB \approx a = 1.42$ Å. The fundamental lattice displacements are $a_1 = AA'$ and $a_2 = AA''$, and their magnitude is $a_1 = \sqrt{3}\times 1.42$ Å $= 2.46$ Å. The reciprocal lattice vectors have magnitude $8\pi/3a$ and are in the directions AB and AS, respectively. Hence, the first Brillouin zone is a hexagon (see Figure 18.3) of which the sides are at a distance $4\pi/3a$ from its center. The density of electron states in **k** space is $2A$, where A is the area of the crystal. The zone has exactly one electron per atom. Therefore, the first Brillouin zone of graphene has $2N$ electron states, and the second Brillouin zone is empty. As we discussed in Sections 10.7 and 10.8, it becomes a semiconductor at finite temperatures.

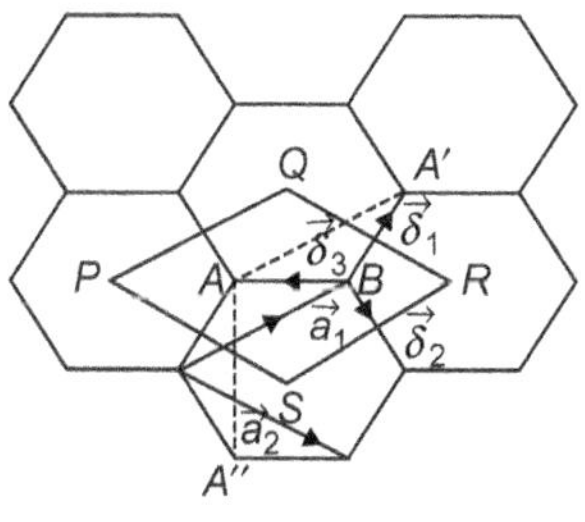

FIGURE 18.2

Honeycomb lattice structure of graphene, made out of two interpenetrating triangular lattices. The lattice unit vectors $\vec{a}_1$ and $\vec{a}_2$ and the nearest-neighbor vectors $\vec{\delta}_1, \vec{\delta}_2$, and $\vec{\delta}_3$. $AB = a$.

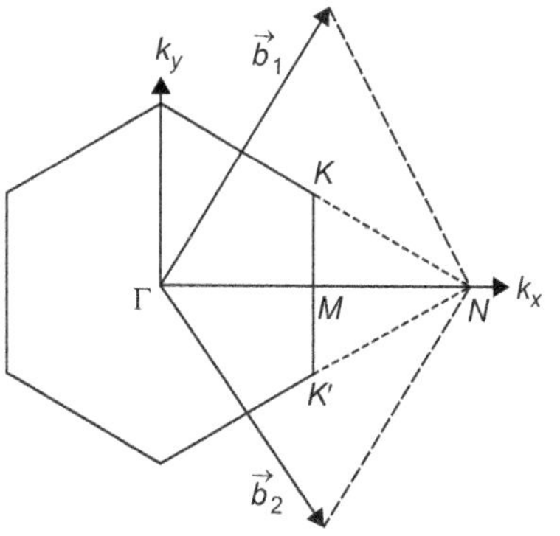

FIGURE 18.3

First Brillouin zone of the honeycomb lattice and the Dirac points K and K' at the corners.

The lattice vectors of graphene can be written as (Figure 18.2)

$$\mathbf{a}_1 = \frac{3a}{2}\left(\hat{x} + \frac{1}{\sqrt{3}}\hat{y}\right); \quad \mathbf{a}_2 = \frac{3a}{2}\left(\hat{x} - \frac{1}{\sqrt{3}}\hat{y}\right), \tag{18.2}$$

where the carbon–carbon distance is $a \approx 1.42$ Å. The reciprocal lattice vectors (shown in Figure 18.3) are given by (Problem 18.1)

$$\mathbf{b}_1 = \frac{2\pi}{3a}\left(\hat{x} + \sqrt{3}\hat{y}\right); \quad \mathbf{b}_2 = \frac{2\pi}{3a}\left(\hat{x} - \sqrt{3}\hat{y}\right). \tag{18.3}$$

The positions of the two Dirac points K and K', located at the corners of the Brillouin zone (of which the significance is to be explained later), are (Problem 18.2)

$$\mathbf{K} = \frac{2\pi}{3a}\left(\hat{x} + \frac{1}{\sqrt{3}}\hat{y}\right); \quad \mathbf{K}' = \frac{2\pi}{3a}\left(\hat{x} - \frac{1}{\sqrt{3}}\hat{y}\right). \tag{18.4}$$

The three nearest-neighbor vectors in real space are given by

$$\vec{\delta}_1 = \frac{a}{2}\left(\hat{x} + \sqrt{3}\hat{y}\right); \quad \vec{\delta}_2 = \frac{a}{2}\left(\hat{x} - \sqrt{3}\hat{y}\right); \quad \vec{\delta}_3 = -a\hat{x}. \tag{18.5}$$

The six second-nearest neighbors are located at $\vec{\delta}_1' = \pm\mathbf{a}_1$, $\vec{\delta}_2' = \pm\mathbf{a}_2$, and $\delta_3' = \pm(\mathbf{a}_2 - \mathbf{a}_1)$.

18.1.3 Tight-Binding Approximation

Wallace (Ref. 19) developed a "tight-binding" method for the band theory of graphite. Because the spacing of the lattice planes of graphite is large (3.37 Å) compared with the hexagonal spacing of the layer (1.42 Å), he neglected, as a first approximation, the interactions of the planes and assumed that conduction takes place in the layers. This is precisely graphene, which at that time was merely a concept. We note that some the notations have different values in that paper, presumably because it was published in 1947, but these have been modernized in the present derivation.

If $\chi(r)$ is the normalized orbital $2p_z$ wave function for an isolated atom, the wave function in the tight-binding approximation has the form

$$\psi = \phi_1 + \lambda\phi_2, \tag{18.6}$$

where

$$\phi_1 = \frac{1}{\sqrt{N}}\sum_A e^{i\mathbf{k}\cdot\mathbf{r}_A}\,\chi(\mathbf{r} - \mathbf{r}_A) \tag{18.7}$$

and

$$\phi_2 = \frac{1}{\sqrt{N}}\sum_B e^{i\mathbf{k}\cdot\mathbf{r}_B}\,\chi(\mathbf{r} - \mathbf{r}_B). \tag{18.8}$$

Here, the first sum is taken over A and all the lattice points generated from it by primitive lattice translations, and the second sum is similarly over the points generated from B (Figure 18.2). Neglecting the overlap integrals,

$$\int \chi(\mathbf{r} - \mathbf{r}_A)\chi(\mathbf{r} - \mathbf{r}_B)\, d\mathbf{r} = 0, \tag{18.9}$$

and substituting in Eq. (18.6),

$$H\psi = E\psi, \tag{18.10}$$

we obtain (Problem 18.3)

$$H_{11} + \lambda H_{12} = E \tag{18.11}$$

and

$$H_{12} + \lambda H_{22} = \lambda E. \tag{18.12}$$

Here,

$$H_{11} = H_{22} = \int \phi_1{}^* H\phi_2 dv; \qquad H_{12} = H_{21}^* = \int \phi_1{}^* H\phi_2 dv \tag{18.13}$$

and

$$\int \phi_1{}^* \phi_1 dv = \int \phi_2{}^* \phi_2 dv = 1. \tag{18.14}$$

Eliminating λ from Eqs. (18.11) and (18.12), we obtain the secular equation

$$\begin{vmatrix} H_{11} - E & H_{12} \\ H_{21} & H_{22} - E \end{vmatrix} = 0. \tag{18.15}$$

From Eq. (18.15), it is easy to show that

$$E = \frac{1}{2}\left\{ H_{11} + H_{22} \pm \left((H_{11} - H_{22})^2 + 4|H_{12}|^2 \right)^{\frac{1}{2}} \right\}. \tag{18.16}$$

Because $H_{11} = H_{22}$, Eq. (18.16) can be rewritten in the alternate form

$$E_{\pm} = H_{11} \pm |H_{12}|. \tag{18.17}$$

The positive sign in Eq. (18.17) will apply to the outside of the hexagonal zone and the negative sign to the inside. The discontinuity of energy across the zone boundary is

$$\Delta E = 2|H_{12}|. \tag{18.18}$$

From Eqs. (18.7), (18.13), and (18.17), we obtain

$$H_{11} = \frac{1}{N}\sum_{A,A'} e^{-i\mathbf{k}\cdot(\mathbf{r}_A - \mathbf{r}_{A'})} \int \chi^* (\mathbf{r} - \mathbf{r}_A) H \chi(\mathbf{r} - \mathbf{r}_{A'})\, dv. \tag{18.19}$$

Keeping only the nearest-neighbor integrals among the atoms A and writing

$$E_0 = \int \chi^* (\mathbf{r}) H \chi(\mathbf{r})\, dv \tag{18.20}$$

and

$$\gamma_0' = -\int \chi^* (\mathbf{r} - \vec{\rho}\,') H \chi(\mathbf{r})\, dv, \tag{18.21}$$

where $\vec{\rho}' = \mathbf{a}_1(say)$ is a vector joining the nearest neighbor among atoms A, we can show that (Problem 18.4)

$$H_{11} = E_0 - 2\gamma_0'\left[\cos\left(\sqrt{3}k_y a\right) + 2\cos\left(\frac{3}{2}k_x a\right)\cos\left(\frac{\sqrt{3}}{2}k_y a\right)\right]. \tag{18.22}$$

Writing

$$H = H_0 + (H - H_0), \tag{18.23}$$

where H_0 is the Hamiltonian of an isolated carbon atom, and using

$$H - H_0 = V - U < 0, \tag{18.24}$$

where U is the potential field of an isolated atom and V is the periodic potential of the lattice because

$$H_0\chi = \overline{E}\chi, \tag{18.25}$$

($\overline{E}$ is the energy of an electron in the $2p_z$ state in carbon), from Eqs. (18.20), (18.21), (18.24), and (18.25), we obtain

$$E_0 = \overline{E} - \int \chi^*(\mathbf{r})(U - V)\chi(\mathbf{r})\,dv \tag{18.26}$$

and

$$\gamma_0' = \int \chi^*(\mathbf{r} - \vec{\rho}')(U - V)\chi(\mathbf{r})\,dv > 0. \tag{18.27}$$

Similarly, we obtain the expression for H_{12},

$$H_{12} = \frac{1}{N}\sum_{A,B} e^{-i\mathbf{k}\cdot(\mathbf{r}_A - \mathbf{r}_B)} \int \chi^*(\mathbf{r} - \mathbf{r}_A)H\chi(\mathbf{r} - \mathbf{r}_B). \tag{18.28}$$

Considering only the nearest-neighbor interactions in the lattice (between atoms of type A and type B and vice versa), we write (in analogy with Eq. 18.27)

$$\gamma_0 = \int \chi^*(\mathbf{r} - \vec{\rho})(U - V)\chi(\mathbf{r})\,dv > 0, \tag{18.29}$$

where

$$\vec{\rho} = \mathbf{AB}. \tag{18.30}$$

It can be shown that (Problem 18.5)

$$H_{12} = -\gamma_0\left[e^{-ik_x a} + 2\cos\left(\frac{\sqrt{3}}{2}k_y a\right) e^{i\left(\frac{3}{2}k_x a\right)}\right] \tag{18.31}$$

and

$$|H_{12}|^2 = \gamma_0^2\left[1 + 4\cos^2\left(\frac{\sqrt{3}}{2}k_y a\right) + 4\cos\left(\frac{3}{2}k_x a\right)\cos\left(\frac{\sqrt{3}}{2}k_y a\right)\right]. \tag{18.32}$$

From Eqs. (18.22) and (18.32), we can write

$$E_{\pm}(\mathbf{k}) = H_{11} \pm H_{12} = E_0 - \gamma_0' f(\mathbf{k}) \pm \gamma_0[3 + f(\mathbf{k})]^{1/2}, \tag{18.33a}$$

where

$$f(\mathbf{k}) = 2\left[\cos(\sqrt{3}k_y a) + 2\cos\left(\frac{3}{2}k_x a\right)\cos\left(\frac{\sqrt{3}}{2}k_y a\right)\right]. \tag{18.33b}$$

The energies at the various points in the Brillouin zone can be written as

$$\begin{aligned} \Gamma&: E = E_0 - 3\gamma_0 - 6\gamma_0', \\ N&: E = E_0 + 3\gamma_0 - 6\gamma_0', \\ K&: E = E_0 + 3\gamma_0', \\ M(inside)&: E = E_0 - \gamma_0 + 2\gamma_0', \\ M(outside)&: E = E_0 + \gamma_0 + 2\gamma_0'. \end{aligned} \tag{18.34a}$$

Across the boundary at any point over a side of the zone (Figure 18.3), there is a discontinuity of energy of amount

$$2\gamma_0\left[2\cos\left(\frac{\sqrt{3}}{2}k_y a\right) - 1\right], \tag{18.34b}$$

which is a maximum at the center and decreases to zero at the corners. The degeneracy at K and similar points (called Dirac points) and the zero-energy gap at these points are consequences of the symmetry of the lattice and are independent of any approximation.

The energy contours are given by

$$E = E_0 - 3\gamma_0 - 6\gamma_0' + \frac{3}{4}(\gamma_0 + 6\gamma_0')(k_x^2 + k_y^2)\, a^2. \tag{18.35}$$

The curves of constant energy are shown in Figure 18.4.

It may be noted that near the corners K or K' (Dirac points),

$$|E - E_K| = 3\gamma_0' \pm \frac{3}{2}\gamma_0|\mathbf{k} - \mathbf{K}|a - \frac{9}{4}\gamma_0'|\mathbf{k} - \mathbf{K}|^2 a^2. \tag{18.36}$$

The surfaces of constant energy are circular. If one neglects γ_0' relative to γ, Eq. (18.36) can be rewritten near the corners of the zone,

$$\begin{aligned} |E - E_K| &\approx \frac{3}{2}\gamma_0 a|\mathbf{k} - \mathbf{K}| + O[(q/K^2)] \\ &\approx \frac{3}{2}\gamma_0 a|\mathbf{q}| + O[(q/K^2)], \end{aligned} \tag{18.37}$$

which can be rewritten in the alternate form

$$E_{\pm}(\mathbf{q}) \approx v_F|\mathbf{q}| + O[(q/K)^2], \tag{18.38}$$

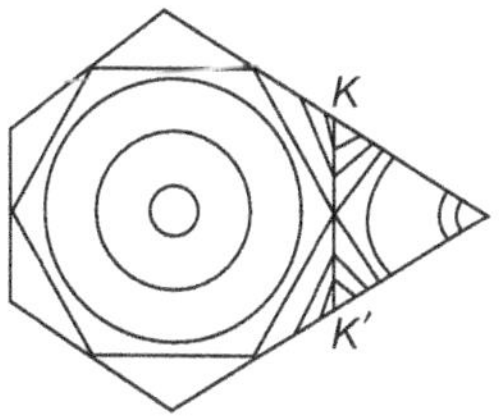

FIGURE 18.4

Curves of constant energy.

Reproduced from Wallace[19] with the permission of the American Physical Society.

where $\mathbf{q}$ is the momentum (in units $\hbar = 1$) measured relatively to the Dirac points, and v_F is the Fermi velocity,

$$v_F = 3\gamma_0 a/2. \tag{18.39}$$

The numerical value of $v_F \approx 1 \times 10^6$ m/s. From Eqs. (18.36) and (18.39), we obtain

$$E_{\pm}(\mathbf{q}) = 3\gamma_0' \pm v_F|\mathbf{q}| + O[(q/K)^2]. \tag{18.40}$$

Conduction in the graphene layer will take place through the electrons excited into the upper band and through the equal number of positive holes created in the lower band, as shown in Figure 18.5.

For moderate temperatures, $N(E)$ is even in $\in = |E - E_K|$ over the whole range in which the Fermi distribution $f(E)$ is different from its value at absolute zero. One can write $\xi = E_K$ and express

$$f(E) = f(\in) = 1/(e^{\in/k_B T} + 1). \tag{18.41}$$

It is interesting to note that the original tight-binding method, used by Wallace[19] in 1947 as a first approximation for the calculation of band structure of a single layer of graphite, is now being widely used to study the energy bands of graphene.

The tight-binding Hamiltonian for electrons in graphene can be written in the second-quantization form[1] (in units such that $\hbar = 1$),

$$\hat{H} = -t \sum_{<i,j>,\sigma} (\hat{a}^{\dagger}_{i,\sigma}\hat{b}_{j,\sigma} + H.c.) - t' \sum_{<<i,j>>,\sigma} (\hat{a}^{\dagger}_{i,\sigma}\hat{a}_{j,\sigma} + \hat{b}^{\dagger}_{i,\sigma}\hat{b}_{j,\sigma} + H.C.), \tag{18.42}$$

where $\hat{a}^{\dagger}_{i,\sigma}(\hat{a}_{i,\sigma})$ are the creation and annihilation operators with spin σ ($\sigma = \uparrow, \downarrow$) on site $\mathbf{R}_i$ on sublattice A, and $\hat{b}^{\dagger}_{j,\sigma}(b_{j,\sigma})$ are the corresponding operators on site $\mathbf{R}_j$ on sublattice B.

Here, $t(\approx 2.7$ eV) is the nearest-neighbor hopping energy (between A and B), and $t' = -0.2\,t$ is the next nearest-neighbor hopping energy (between two A's or two B's). We note that $t = \gamma_0$ and $t' = \gamma_0'$ in Wallace's theory.

The electronic dispersion in the honeycomb lattice is shown in Figure 18.6, for finite values of $t = 2.7$ eV and $t' = -0.2\,t$. We also note the most striking difference between the results of Eq. (18.39) and the usual case in which $\in(\mathbf{q}) = \hbar^2 q^2/2m$, where m is the electron mass. In Eq. (18.39), the Fermi velocity v_F does not depend on the energy and momentum while in the usual case, $v = \hbar k/m = \sqrt{2E/m}$, and hence the velocity changes substantially with energy. We also note that the presence of the second-order terms (arising due to t' in Eq. 18.40) shifts in energy the position of the Dirac point and breaks the electron-hole symmetry.

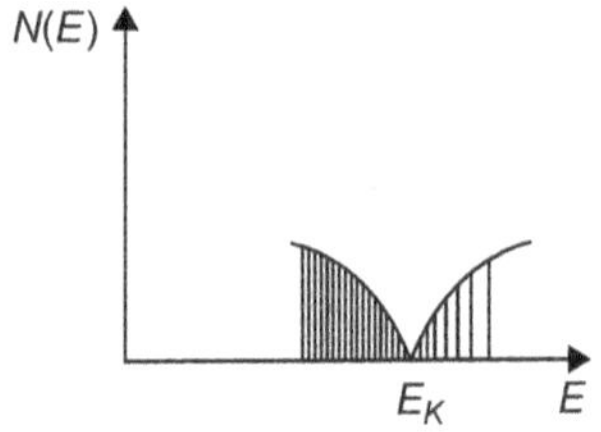

FIGURE 18.5

The form of the electronic energy states, $N(E)$, near E_K.

18.1.4 Dirac Fermions

Graphene's charge carriers have a particularly unique nature. Its charge carriers mimic relativistic particles and are described starting with the Dirac equation rather than the Schrodinger equation. The interaction of the electrons with the graphene's honeycomb lattice gives rise to new

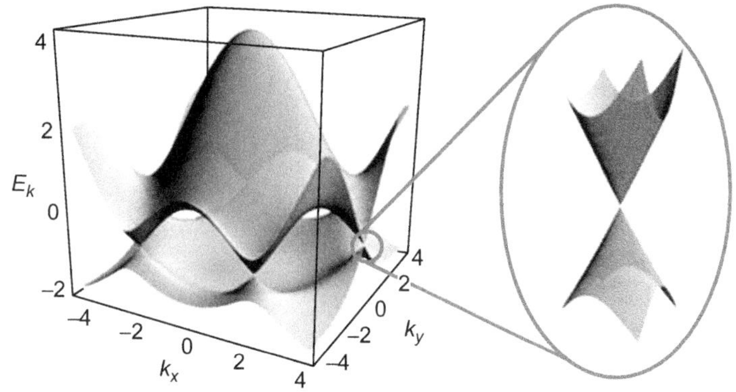

FIGURE 18.6

Left: Energy spectrum (in units of t) for $t = 2.7$ eV and $t' = -0.2\,t$. Right: Close-up of the energy bands near to one of the Dirac points.

Reproduced from Castro Neto et al.[1] *with the permission of the American Physical Society.*

quasiparticles, which, at low energies E, are accurately described by the (2 + 1)-dimensional Dirac equation with an effective speed of light $v_F \approx 10^6\ \mathrm{m}^{-1}\mathrm{s}^{-1}$. These quasiparticles are called massless Dirac fermions. They can be viewed as electrons that have lost their rest mass m_e or as neutrinos that have acquired the electron charge e. The reason the quasiparticles are known as Dirac fermions is as follows.

The Dirac equation for an electron in a periodic potential V can be written as

$$\hat{H}\phi_i = (c\vec{\alpha}\cdot\vec{p} + \beta m_e c^2 + IV)\phi_i = \varepsilon_i\phi_i, \tag{18.43}$$

where

$$\vec{\alpha} = \begin{bmatrix} 0 & \vec{\sigma} \\ \vec{\sigma} & 0 \end{bmatrix}, \quad \beta = \begin{bmatrix} E & 0 \\ 0 & -E \end{bmatrix}, \quad I = \begin{bmatrix} E & 0 \\ 0 & E \end{bmatrix}, \tag{18.44}$$

$\vec{\sigma}$ is the Pauli spin matrix vector, E is a 2×2 unit matrix, $\vec{p}$ is the momentum operator, m_e is the rest mass of the electron, and ϕ_i is a four-component Bloch function with an energy ε_i. The suffix i signifies a set of the wave vector, band index, and spin direction and is limited to positive energy states.

Graphene is a zero-gap semiconductor, in which low-E quasiparticles within each valley can be described by the Dirac-like Hamiltonian

$$\hat{H} = \hbar v_F \begin{pmatrix} 0 & k_x - ik_y \\ k_x + ik_y & 0 \end{pmatrix} = \hbar v_F \vec{\sigma}\cdot\vec{k}. \tag{18.45}$$

Eq. (18.43) can be approximated by Eq. (18.45) when the k-independent Fermi velocity v_F plays the role of the velocity of light c, $\vec{p} = \hbar\vec{k}$, and because the electrons are fermions, they are called Dirac fermions. The honeycomb lattice is made up of two equivalent carbon sublattices A and B, and the cosine-like energy bands associated with the sublattices intersect at zero E near the edges of the Brillouin zone, giving rise to conical sections of the energy spectrum. The electronic states at the intersection of the bands are composed of states belonging to the different sublattices, and their relative

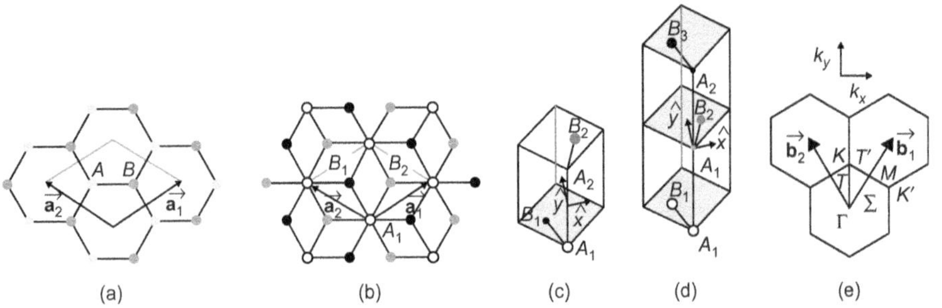

FIGURE 18.7

A top view of (a) unit cell of monolayer graphene, showing the inequivalent atoms A and B and unit vectors $\mathbf{a}_1$ and $\mathbf{a}_2$; (b) real-space bilayer graphene in which the light-/dark-gray dots and black circles/black dots represent the carbon atoms in the upper and lower layers; (c) the unit cell and the $\hat{x}$ and $\hat{y}$ unit vectors of bilayer graphene; (d) the same as (c) for trilayer graphene; and (e) the reciprocal space unit cell showing the first Brillouin zone with its high symmetry points.

Reproduced from Malard et al.[12] with the permission of Elsevier.

contributions are taken into account by using two-component wave functions (spinors). The index to indicate sublattices A and B is known as pseudospin $\vec{\sigma}$ because it is similar to the spin index (up and down) in quantum electrodynamics (QED). The real spin of the electrons must be described by additional terms in the Hamiltonian. Because the QED-specific phenomena are often inversely proportional to c and because $c/v_F \approx 300$, the pseudospin effects usually dominate over those due to the real spin.

One can introduce the concept of chirality, which is formally a projection of $\vec{\sigma}$ on the direction of motion $\vec{k}$, and is positive (negative) for electrons (holes). Chirality in graphene signifies the fact that $\vec{k}$ electrons and $-\vec{k}$ hole states originate from the same sublattice. The concepts of chirality and pseudospin are important because they are conserved quantities.

18.1.5 Comprehensive View of Graphene

A comprehensive view of the unit cell of monolayer, bilayer, and trilayer graphene and the first Brillouin zone with its high symmetry points are shown in Figure 18.7.

18.2 FULLERENES

18.2.1 Introduction

If one forms a vapor of carbon atoms and lets them condense slowly while keeping the temperature high, as the intermediate species grow, there is a path where the bulk of all reactive kinetics follows that make spheroidal fullerenes. There are two types of fullerenes that are famous for different reasons. The "Buckminsterfullerene" (C_{60}) is the most symmetric of all possible molecules. In addition, it is possible by adding a few percent of other atoms (nickel and cobalt) to trick the carbon into making tubes. The (10,10) fullerene nanotube is the most famous nanotube. The propensity for bonding that causes C_{60} to be the end point of 30–40% of all the reactive kinetics leads to the (10,10) nanotube. The metal atoms (nickel and cobalt) prevent the addition of the seventh, eighth, and ninth pentagons, and ultimately, the growing tubelet can anneal to its most energetically favored form.

The idea that C_{60} would form a stable molecule originated from Euler's rule stating that a solid figure with any even number n of 24 or more vertices could be constructed with 12 pentagons and $(n-20)/2$ hexagons. The spheroidal carbon–cage carbon molecules consisting only of pentagons and hexagons were given the generic name "fullerenes."

18.2.2 Discovery of C_{60}

The truncated icosahedron form of C_{60} is shown in Figure 18.8. It was discovered by Kroto et al.[9] by using a supersonic laser-vaporization nozzle source, as shown in Figure 18.9a.

C_{60} is chemically a very stable structure. Cluster "cooking" reactions in the "integrating cup" were responsible for the C_{60} cluster's becoming over 50 times more intense than any other cluster in the nearby size range. The up-clustering reactions with small carbon chains and rings reacted away nearly all clusters except for C_{60}, which survived because of its perfect symmetry. C_{60} does not have any dangling bonds because the valences of every carbon atom are satisfied. There is no specific point of chemical attack because every atom is equivalent by symmetry. While curving the intrinsically planar system of double bonds into a spherical shape, strain is introduced. However, this strain is uniformly and symmetrically distributed over the molecule. No other structure has this high degree of symmetry, and hence, the experimental observation that carbon-vapor condensation conditions could be found where the intensity of the mass spectrum peak of the C_{60} in the carbon cluster beam was many times the intensity of any of its near neighbors in mass is shown in Figure 18.9b.

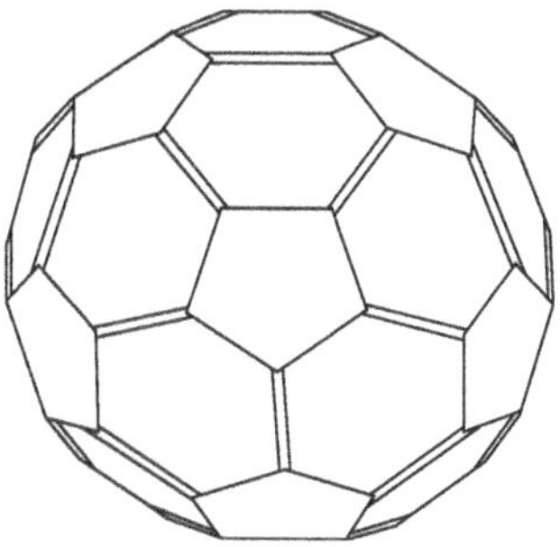

FIGURE 18.8

Truncated icosahedron C_{60}, popularly known as "Buckminsterfullerene."

Reproduced from Curl[2] with the permission of the American Physical Society.

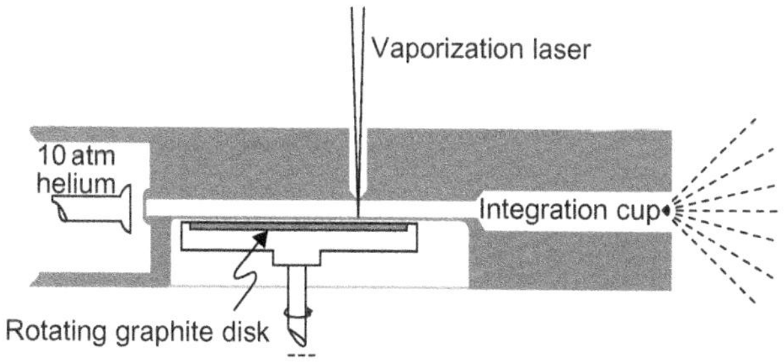

FIGURE 18.9a

Schematic cross-sectional drawing of the supersonic laser-vaporization nozzle source used in the discovery of fullerenes.

Reproduced from Smalley[16] with the permission of the American Physical Society.

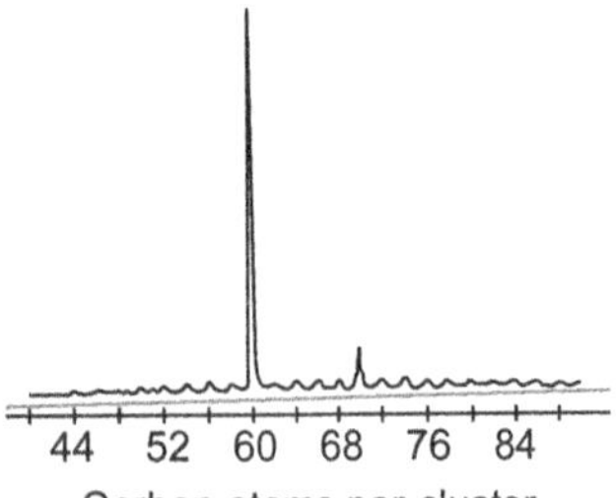

FIGURE 18.9b

Intensity of the mass spectrum peak of the C_{60} in the carbon cluster beam relative to its neighbors in mass.

Reproduced from Curl[2] with the permission of the American Physical Society.

To confirm the existence of C_{60}, Kroto and coworkers[9] made two pioneering NMR experiments. The first experiment was done on a solution of C_{60} with benzene, which yielded a very strong resonance line at 128 ppm (for benzene) and a very tiny NMR trace in which C_{60} resonance was identified at 143 ppm. However, a second experiment (Ref. 18) in which C-NMR spectrum obtained from chromatographically purified samples of soluble material extracted from arc-processed graphite, yielded a spectrum of purified C_{60}, in which a strong resonance was obtained at 143 ppm. This result is shown in Figure 18.10.

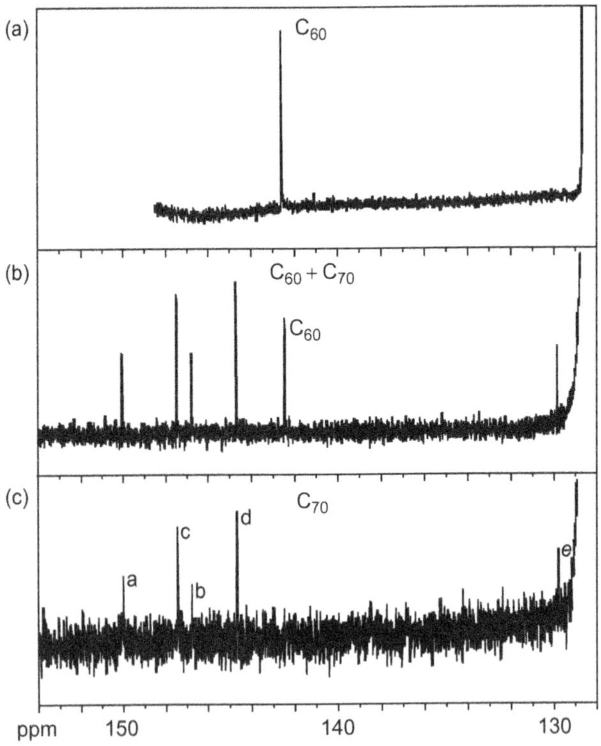

FIGURE 18.10

C-NMR spectrum of (a) purified C_{60} (143 ppm); (b) mixed sample; (c) purified C_{70} (five lines).

Reproduced from Kroto[8] with the permission of the American Physical Society.

The fullerenes have a wide variety of technological applications. An example of a fullerene-based *n*-channel FET is shown in Figure 18.11. A highly doped *n*-type silicon wafer takes the place of the gate metal, a ~30–300 nm thick layer of SiO_2 serves as the oxide, and the fullerene film serves as the semiconductor.

When an appropriate positive gate voltage V_G is applied, the drain current I_D increases, which indicates that a conduction channel is formed near the fullerene-insulator interface.

Another application of fullerene is in C_{60} photolithography. The sequence of steps (deposition, exposure, development, and pattern transfer) used in photolithography, in which C_{60} acts as a negative photoresist, is shown in Figure 18.12.

One of the many important potential applications of fullerenes is the nature of the fullerenes and metallic and semiconductor substrates. Direct rectification between solid C_{60} and *p*-type crystalline Si has been shown in Nb/Co_{60}/p-Si and Ti/Co_{60}/p-Si heterojunctions, which are strongly rectifying. Because the potential barriers at the Nb-C_{60} and Ti-C_{60} interfaces are close to

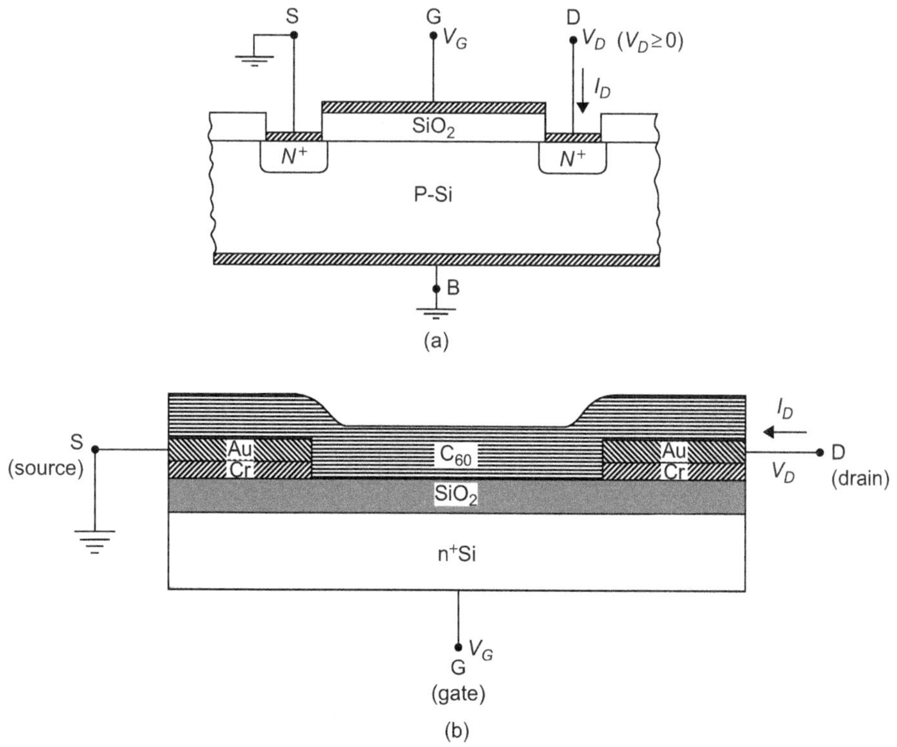

FIGURE 18.11

(a) The terminal designations and blasting conditions for Si-based MOSFETs. G, D, B, and S, respectively, denote the ground, drain, base, and source. (b) The corresponding structure for the fullerene C_{60} device.

Reproduced from Dresselhaus et al.[3] with the permission of Elsevier.

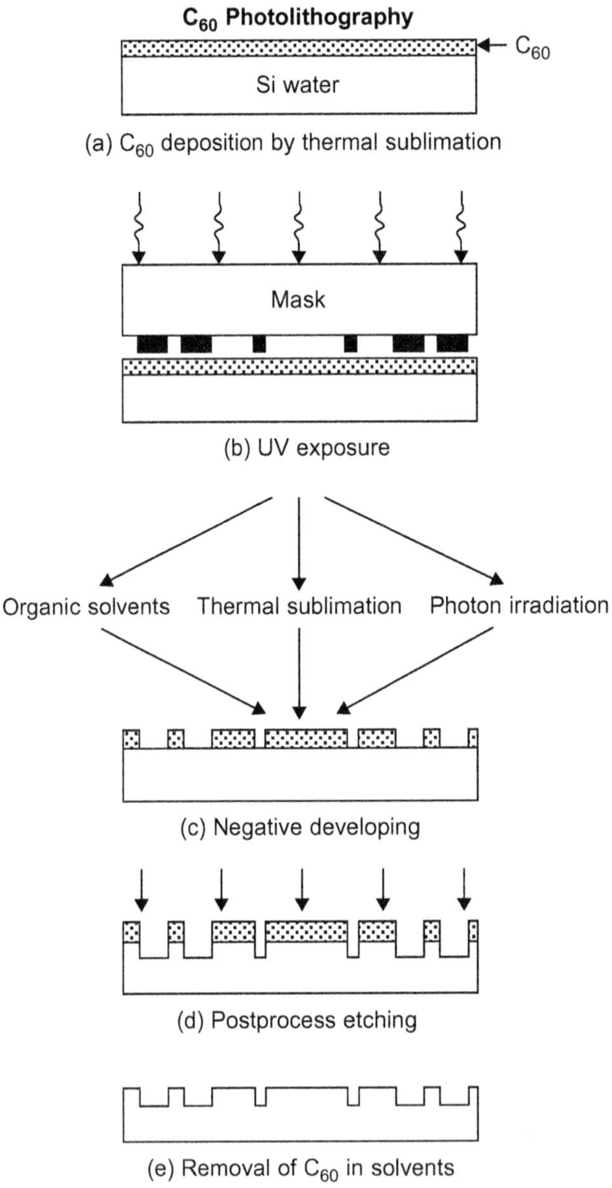

FIGURE 18.12

Sequence of steps of C_{60} photolithography.

Reproduced from Dresselhaus et al.[3] with the permission of Elsevier.

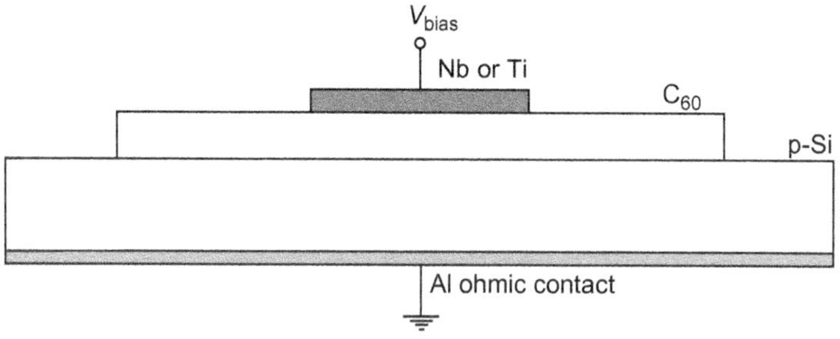

FIGURE 18.13

Schematic cross-section of an Nb/C_{60}/p-Si structure used as a heterojunction diode.

Reproduced from Dresselhaus et al.[3] with the permission of Elsevier.

zero, it is the C_{60}/p-Si interface that is responsible for the strong rectifying properties of the heterostructure. A schematic cross-section of the Nb/C_{60}/p-Si interface is shown in Figure 18.13.

18.3 FULLERENES AND TUBULES

18.3.1 Introduction

The fullerene nanotube (10,10) mentioned in the introduction of the previous section, with one end open, is shown in Fig. 18.14. The (10,10) tube is formed because the metal atoms frustrate the ability of the open edge to curve in and close. The addition of the seventh, eighth, and ninth pentagons is prevented, and by appropriate choice of temperature and reaction rate, the growing tubelet can anneal to its most energetically favored form.

The closed end is a hemifullerene dome (one half of C_{240}), whereas the other end is left open. These ends are directly amenable to the formation of excellent C-O, C-N, or C-C covalent bonds to attach any molecule, enzyme, membrane, or surface to the end of the tube. If two objects A and B are attached to the two ends, they will communicate with each other by metallic transport along the tube. Thus, the (10,10) tube is a metallic wave guide for electrons.

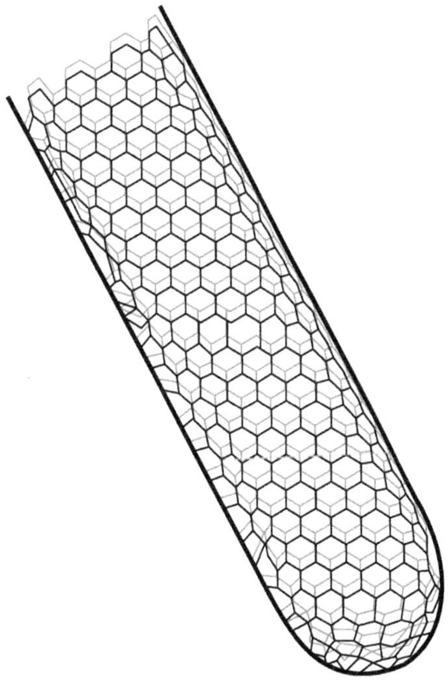

FIGURE 18.14

Section of a (10,10) fullerene nanotube with one end open.

18.3.2 Carbon Nanotubeles

It is convenient to specify a general carbon nanotubule in terms of the tubule diameter d_t and the chiral angle θ, which are shown in Figure 18.15 as the rectangle bounded by the chiral vector **OA** or $\mathbf{C}_h$. The chiral vector

$$\mathbf{C}_h = n\mathbf{a}_1 + m\mathbf{a}_2 \tag{18.46}$$

is defined on the honeycomb lattice by unit vectors $\mathbf{a}_1$ and $\mathbf{a}_2$. $\mathbf{C}_h$ connects two crystallographically equivalent sites O and A on a two-dimensional graphene sheet where a carbon atom is located at each vertex of the honeycomb structure. Figure 18.15 shows the chiral angle θ of the nanotube with respect to the zigzag direction $(\theta = 0)$ and the unit vectors $\mathbf{a}_1$ and $\mathbf{a}_2$ of the hexagonal honeycomb lattice. The armchair tubule (Figure 18.16a) corresponds to $\theta = 30°$ on this construction. An ensemble of chiral vectors can be specified by Eq. (18.46) in terms of pairs of integers (n, m), and this ensemble is shown in Figure 18.17. Each pair of integers (n, m) defines a different set of rolling the graphene sheet to form a carbon nanotube.

Along the zigzag axis $\theta = 0°$. Also shown in the figure is the basic translation vector $\mathbf{OB} = \mathbf{T}$ of the 1D tubule unit cell, and the rotation angle ψ and the translation τ, which constitute the basic symmetry operation $R = (\psi|\tau)$. The integers (n, m) uniquely determine the tubular diameter d_t and θ. The diagram is constructed for $(n, m) = (4, 2)$.

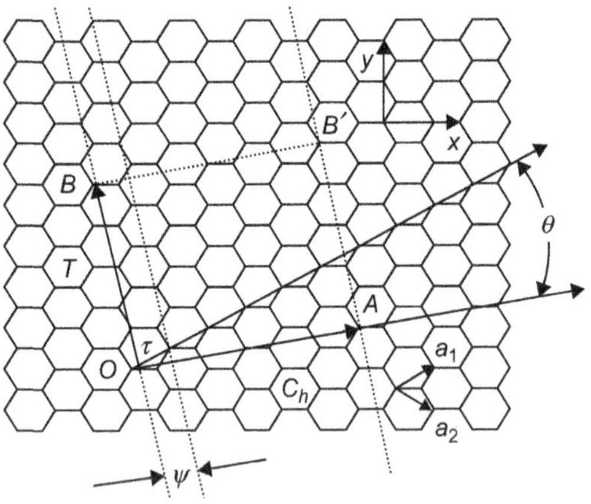

FIGURE 18.15

The 2D graphene sheet is shown along with the vector that specifies the chiral nanotube.

Reproduced from Dresselhouse et al.[4] with the permission of Elsevier.

18.3.3 Three Types of Carbon Nanotubes

When the two ends of the vector $\mathbf{C}_h$ are superimposed, the cylinder connecting the two hemispherical caps of Figure 18.16 is formed. The line AB' (in Figure 18.15) is joined to the parallel line OB, where the lines OB and AB' are perpendicular to the vectors $\mathbf{C}_h$ at each end. There are no distortions of the bond angles in the chiral tubule except the distortions caused by the cylindrical curvature of the tubule. Differences in the tubular diameter d_t give rise to the differences in the various properties of carbon nanotubes. The vectors $(n, 0)$ denote zigzag tubules, and the vectors (n, n) denote armchair tubules. The larger the value of n, the larger the tubule diameter. The $(n, 0)$ and (n, n) have high symmetry and exhibit a mirror symmetry plane normal to the tubular axis. The other vectors (n, m) correspond to chiral nanotubes. Because both right- and left-handed chirality are

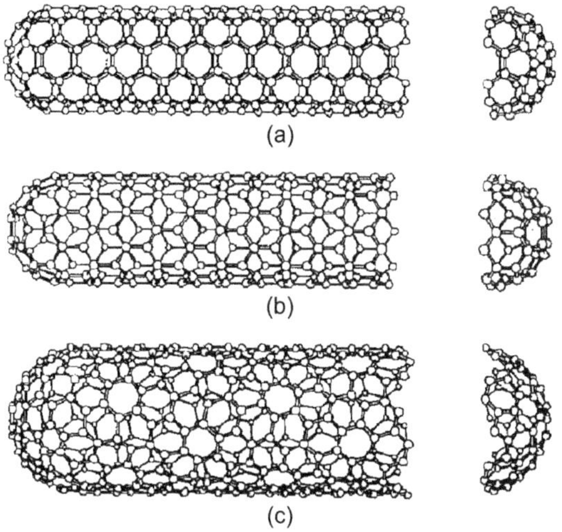

FIGURE 18.16

Three types of nanotubes obtained by rolling a graphene sheet into a cylinder and capping each end of the cylinder with half of a fullerene molecule; a "fullerene-derived tubule" that is one atomic layer in thickness is formed: (a) $\theta = 30°$ (an armchair tubule); (b) $\theta = 0°$ (a) zigzag tubule; and (c) chiral tubule.

Reproduced from Dresselhaus et al.[4] with the permission of Elsevier.

possible for chiral nanotubes, the chiral tubules are optically active to either right- or left-circularly polarized light propagating along the tubule axis.

The tubular diameter d_t is given by

$$d_t = C_h/\pi = \sqrt{3}a_{C-C}(m^2 + mn + n^2)^{1/2}/\pi, \tag{18.47}$$

where a_{C-C} is the nearest-neighbor C–C distance, C_h is the length of the chiral vector $\vec{C_h}$, and the chiral angle θ is given by

$$\theta = \tan^{-1}[\sqrt{3}m/(m + 2n)]. \tag{18.48}$$

The three types of carbon nanotubes are shown in Figure 18.16.

Figure 18.17 shows the number of distinct caps that can be formed theorctically from pentagons and hexagons, such that each cap fits continuously onto the cylinder of the tubule, specified by a given (n, m) pair. It shows that the hemispheres of C_{60} are the smallest caps that satisfy these requirements, so that the smallest carbon nanotube is expected to be 7 Å, which is in agreement with the experiment. Figure 18.17 also shows that the number of possible caps increases rapidly with increasing tubular diameter. Below each pair of integers (n, m) is listed the number of distinct caps that can be joined continuously to the cylindrical carbon tubule denoted by (n, m).

Due to the point group symmetry of the honeycomb lattice, several values of (n, m) will give rise to equivalent nanotubes. Therefore, one restricts consideration to the nanotubes arising from the 30° wedge of the 3D Bravais lattice shown in Figure 18.17. Because the length-to-diameter ratio of

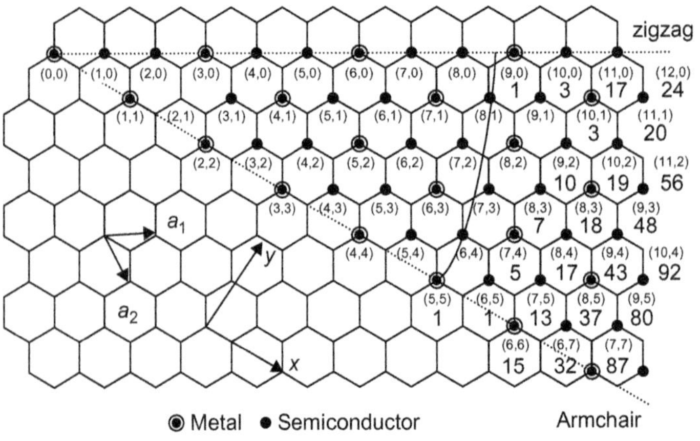

FIGURE 18.17

The 2D graphene sheet that specifies the chiral nanotube.

Reproduced from Dresselhaus et al.[4] with the permission of Elsevier.

carbon nanotubes is $>10^3$ while the diameter is only ~ 10 Å, carbon nanotubes are an important system for studying one-dimensional physics.

18.3.4 Symmetry Properties of Carbon Nanotubes

To study the properties of carbon nanotubes as 1D systems, we define the lattice vector $\mathbf{T}$ along the tubule axis normal to the chiral vector $\mathbf{C}_h$ defined in Eq. (18.46) and Figure 18.15. The vector $\mathbf{T}$ defines the unit cell of the 1D carbon nanotube. The length T of the translation vector $\mathbf{T}$ corresponds to the first lattice point of the 2D graphene sheet through which the vector $\mathbf{T}$ passes. Thus, we obtain from Figure 18.15 and these definitions

$$\mathbf{T} = [(2m+n)\mathbf{a}_1 - (2n+m)\mathbf{a}_2]/d_R, \tag{18.49}$$

with a length

$$T = \sqrt{3}C_h/d_R. \tag{18.50}$$

The length C_h is defined in Eq. (18.47). Defining d as the highest common divisor of (n,m), we have

$$d_R = \begin{cases} d \text{ if } n-m \text{ is not a multiple of } 3d \\ 3d \text{ if } n-m \text{ is a multiple of } 3d. \end{cases} \tag{18.51}$$

The relation between the fundamental symmetry vector $\mathbf{R} = p\mathbf{a}_1 + q\mathbf{a}_2$ of the 1D unit cell and the two vectors that specify the carbon nanotube (n,m), the chiral vector $\mathbf{C}_h$, and translation vector $\mathbf{T}$ are shown in Figure 18.18.

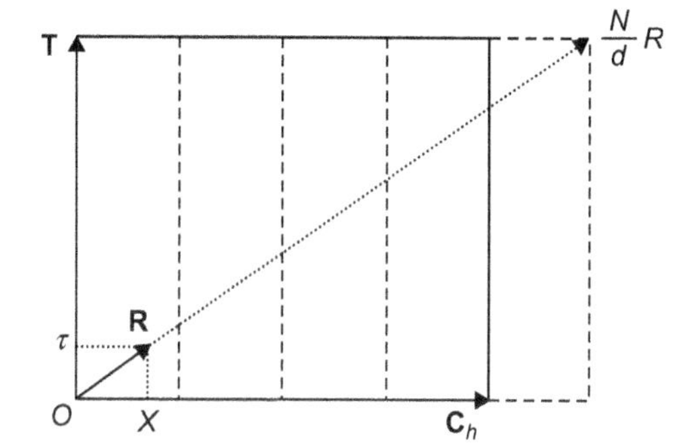

FIGURE 18.18

Relation between **R**, $\mathbf{C}_h$, and **T**.

Reproduced from Dresselhaus et al.[3] with the permission of Elsevier.

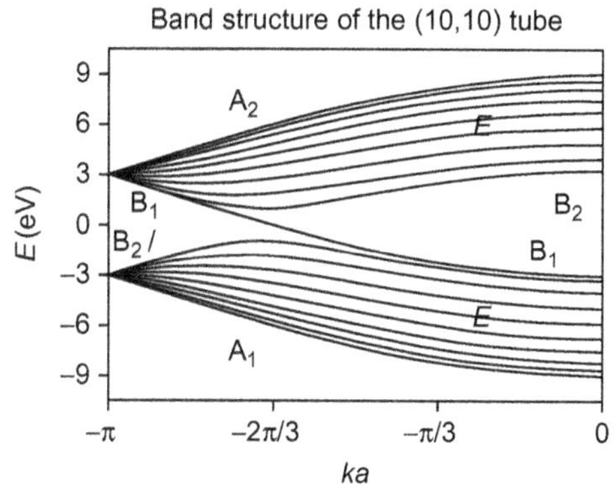

FIGURE 18.19

Band structure of a (10,10) fullerene nanotube, calculated by Dresselhaus et al., using zone folding from the band structure of an infinite 2D graphene sheet.

Reproduced from Dresselhaus et al. (Ref. 3) with the permission of Elsevier.

The projection of **R** on the $\mathbf{C}_h$ and **T** axes yields ψ and τ. **X** in the figure is ψ scaled by

$$\mathbf{C}_h/2\pi.$$

18.3.5 Band Structure of a Fullerene Nanotube

The electronic band structure of a (10,10) fullerene nanotube was first calculated by Dresselhaus et al.[3] by using tight-binding methods and by using zone folding from the band structure of an infinite 2D graphene sheet. Their results are shown in Figure 18.19.

As one can see in Figure 18.19, the two bands that cross the Fermi energy at $ka = -2\pi/3$ have different symmetry and guarantee that the tube will be a metallic conductor.

18.4 POLYMERS

18.4.1 Introduction

A long chain of molecules that has a backbone of carbon atoms is known as a polymer. The basic building block, which usually but not necessarily consists of one carbon atom and two hydrogen atoms, is known as a monomer. A polymer is formed by repeating the structure of the monomer over and over again. In fact, a single polymer can be constituted from several thousand monomers.

Materials composed from polymers, such as bone, wood, skin, and fibers, have been used by man since prehistoric times. However, polymer science was developed in the twentieth century by Hermann Staudinger, who developed the concept of macromolecules in the 1920s. Wallace Carothers showed the great industrial potential of synthetic polymers and invented nylon in 1935. Synthetic polymers are now used in large quantities in a variety of applications. In the 1950s, Ziegler and Natta discovered polymerization catalysts, which led to the development of the modern plastics industry. Some of the most popularly known polymers are rubber, plastic, and Teflon. They do not have any common property other than the fact that they are lightweight, flexible, resistant to corrosion, and easy to mold or cut into any desired shape. Some polymers such as rubber are very elastic and can be deformed very easily. Other polymers that have similar elastic property to rubber are known as elastomers.

Paul Flory created modern polymer science through both experimental and theoretical studies of "macromolecules." A polymer is essentially a giant molecule. There are a variety of ways in which giant molecules can be obtained. Other examples of such giant molecules are branched polymers, in which hydrogen atoms are replaced by any of the halogen elements that need one electron to have a filled subshell. When the hydrogen atoms are replaced by molecules, more complex polymers are obtained. A third alternative is that the carbon chain can be replaced by silicon atoms. In view of the fact that there is a huge variety of ways in which polymers can be obtained, and the fact that they can be easily and cheaply produced, there is a great deal of excitement in finding new ways of obtaining and using complex polymers.

18.4.2 Saturated and Conjugated Polymers

In the saturated polymers studied by Staudinger, Flory, Ziegler, and Natta, all four valence electrons of carbon are used up in covalent bonds, and hence, they are insulators. Therefore, they are viewed as unsuitable for use as electronic materials. In contrast, in conjugated polymers, the chemical bonding leads to one unpaired electron (the π electron) per carbon atom. The carbon orbitals are in the sp^2p_z configuration in the π bonding. Because the orbitals of the successive carbon atoms overlap, the electrons are delocalized along the backbone of the polymer, which provides the path for charge mobility along the polymer chain. The molecular structure of some conjugated polymers is shown in Figure 18.20. The bond-alternated structure of polyacetylene is characteristic of conjugated polymers, which are typically semiconductors.

The electronic structure of conducting polymers can be determined by the symmetry of the chain. These polymers can exhibit either metallic or semiconducting properties. These electrically conducting polymers are known as the "fourth generation of polymeric materials."

The electronic structure in conducting polymers is determined by the number and kinds of atoms within the repeat unit (chain symmetry). The classic example is *trans*- and *cis*-polyacetylene, $(\text{-CH})_n$, which is shown in Figure 18.21.

In polyacetylene, if the carbon–carbon bond lengths were equal, the chemical formula, $(\text{-CH})_n$ with one unpaired electron per formula unit, would yield a metallic state. If the electron–electron interactions were too strong, $(\text{-CH})_n$ would be an antiferromagnetic Mott insulator, a possibility that has been eliminated through a variety of studies together with the fact that these are easily converted to a metallic state on doping.

The structure in polyacetylene is dimerized due to Peierls instability with two carbon atoms in the repeat unit, $(-\text{CH}=\text{CH})_n$. Thus, the π band is divided into π and π^* bands, each of which can hold two electrons per atom (spin-up and spin-down). The $\pi-\pi^*$ energy gap E_g implies that there are no partially filled bands and polyacetylene is a semiconductor. However, the electrical conductivity of polyacetylene can be increased by more than a factor of 10^7 to a level approaching that of a metal by using a dopant. The electrical conductivity of *trans*-(CH), as a function of (AsF_5) dopant concentration, is shown in Figure 18.22.

However, until 1990, there were no known examples of stable metallic polymers. It was shown that polyaniline (PANI) could be rendered, conducting either through oxidation of the

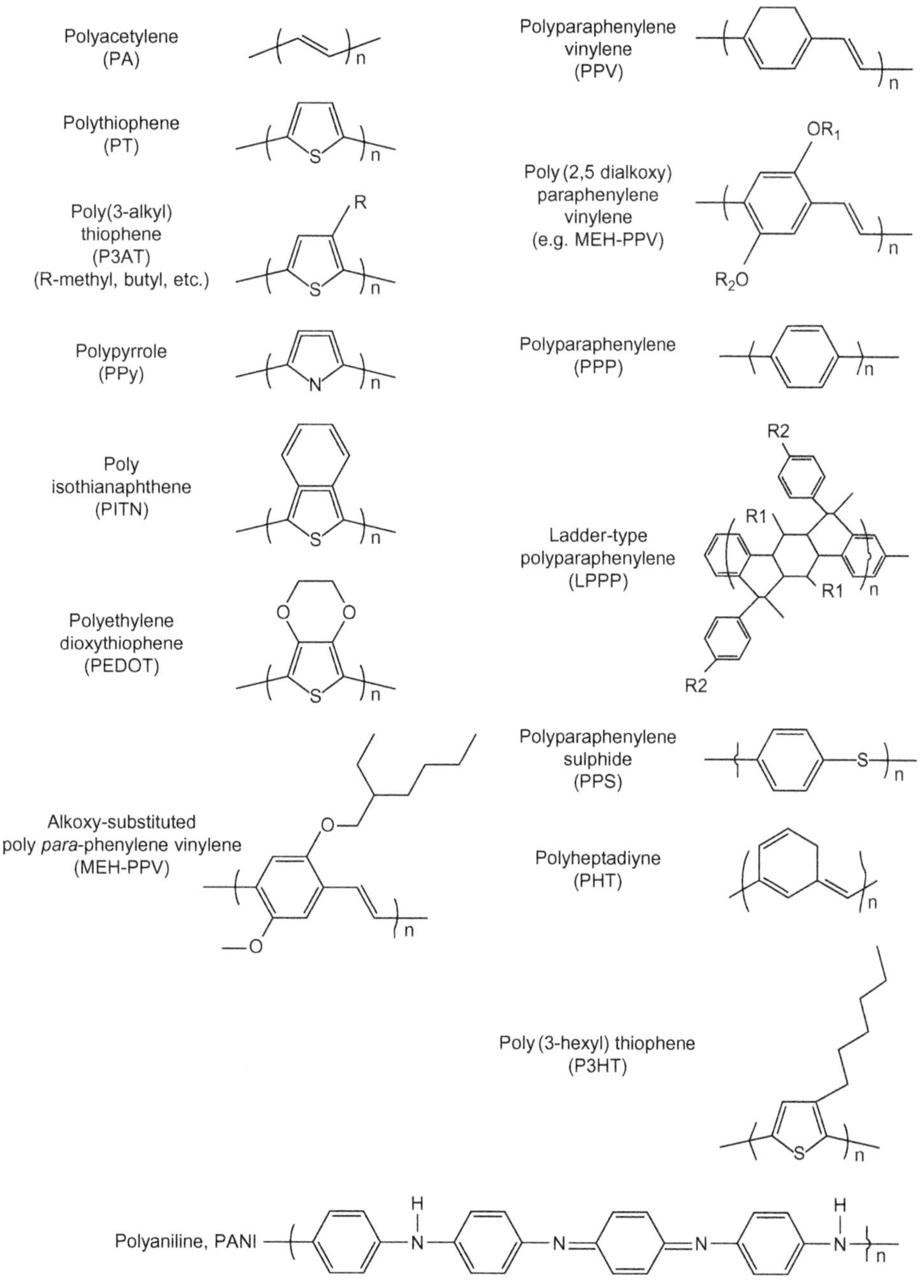

FIGURE 18.20

Molecular structures of some conjugated polymers (bond-alternated structures).

Reproduced from Heeger[6] with the permission of the American Physical Society.

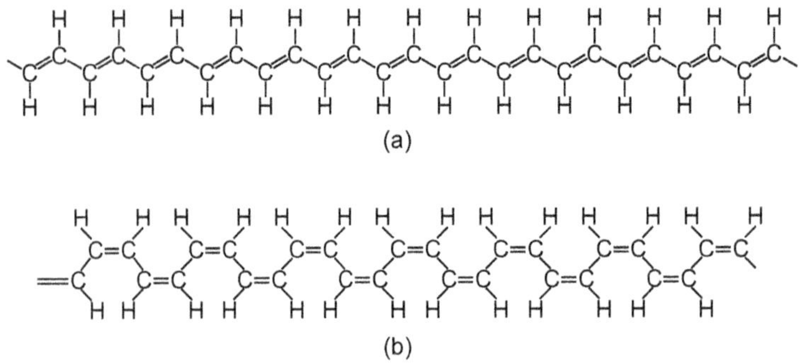

FIGURE 18.21

(a) All *trans*- and (b) all *cis*-polyacetylene.

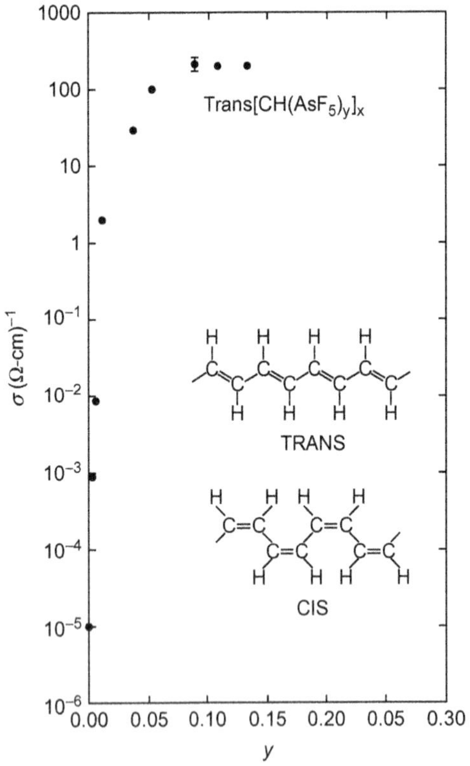

FIGURE 18.22

Electrical conductivity of *trans*-(CH) as a function of (AsF_5) dopant concentration.

Reproduced from Heeger[6] with the permission of the American Physical Society.

leucoemeraldine base or protonation of the emeraldine base. Eventually, protonic acids were used to both convert PANI to metallic form and render the resulting PANI complex soluble in common organic solvents. The processibility of PANI induced by the "surfactant" counterions has made possible the fabrication of conducting polymer blends with a variety of host polymers. The "surfactant" counterions led to the formation of a self-assembled network, due to which the PANI network remains connected and conducting even after the removal of the host polymer. This led to the fabrication of novel electrodes for use in electronic devices.

The chemistry and physics of these polymers in the semiconducting state are also of great interest since their application to "plastic electronic" devices. The polymer diodes were fabricated in the 1980s and the light-emitting diodes (LEDs) in 1990. The other "plastic" optoelectronic devices include lasers, high-sensitivity plastic photodiodes, photovoltaic cells, ultrafast image processors, thin-film transistors, and all-polymer integrated circuits. All these are fabricated semiconducting and metallic polymers that are thin-film devices in which the active layers are fabricated by casting the semiconducting and/or metallic polymers from solution.

18.4.3 Transparent Metallic Polymers

In conventional metals, the length of the interchain spacing and of the repeat unit are large compared to the interatomic distances. However, in metallic polymers, N, the number of electrons per unit volume is on the order of $N \sim 2-5 \times 10^{21}\,\mathrm{cm}^{-3}$. The plasma frequency (the frequency below which the metals reflect light) is given by

$$\omega_p^2 = 4\pi N e^2 / m^*, \tag{18.52}$$

where m^* is the effective mass of the electrons. Thus, the plasma frequency of metallic polymers is on the order of 1 eV. Hence, they are semitransparent in the visible part of the spectrum but exhibit high reflectance in the infrared. Therefore, optical-quality thin films of metallic polymers are used as transparent electrodes. Transparent conducting films are used as antistatic coatings, as electrodes in liquid-crystal display cells or in polymer LEDs, or for fabricating electrochromic windows.

18.4.4 Electronic Polymers

Electronic polymers are those polymers of which the conductivity can be increased by several orders of magnitude, which can be obtained by doping. The increase of conductivity of electronic polymers by doping is shown in Figure 18.23.

In Figure 18.23, *trans*-$(CH)_x$ and the emeraldine base form of polyamine are shown to give an example of the increase in electrical conductivity by doping. Electronic polymers are extensively used in light-emitting diodes. In addition, superconductivity has been discovered in regioregular

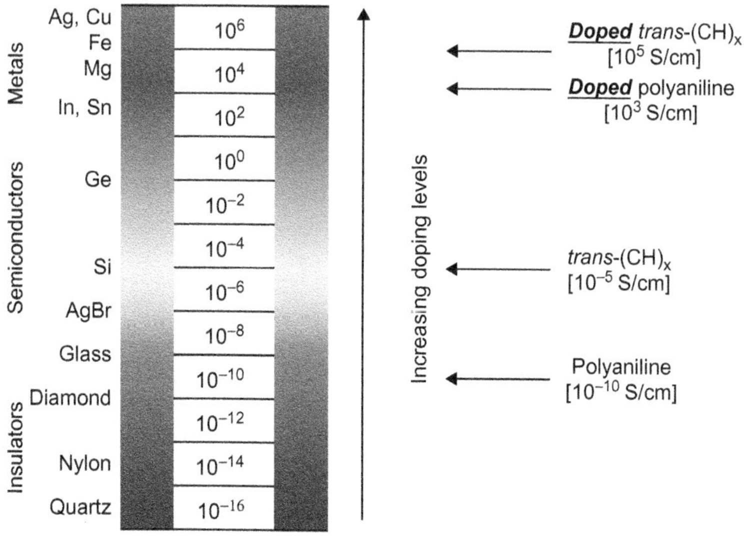

FIGURE 18.23

Increase of conductivity of electronic polymers by doping.

Reproduced from MacDiarmid[11] with the permission of the American Physical Society.

poly (3-hexylthiophene) around a critical temperature of 2° K. The future application of this amazing discovery is still being speculated.

18.5 SOLITONS IN CONDUCTING POLYMERS

18.5.1 Introduction

Trans-polyacetylene (*trans*-$(CH)_x$) was the first highly conducting organic polymer. The simple molecular structure –CH– units are repeated, which implies that each carbon contributes a single p_z electron to the π band, and as a result, the π band would be half-filled. Thus, polyacetylene would be a one-dimensional metal. However, in 1955, Peierls showed that 1D metals are unstable with respect to a structural distortion that opens up an energy gap at the Fermi level; thereby, 1D metals end up as semiconductors. The periodicity of the Peierls distortion is $\Lambda = \pi/k_F$, where the Fermi wave vector $k_F = 2\pi/a$ for the half-filled band of *trans*-$(CH)_x$. Hence, it converts *trans*-polyacetylene into *trans*-$(-HC = CH-)_x$, which is essentially alternating single and double bonds, as shown in Figure 18.24a.

However, a chain of monomers can be dimerized in two distinct patterns, both of which have the same energy (the degenerate *A* and *B* phases), as shown in Figure 18.24b.

Thus, in addition to electron and hole excitations in a dimerized semiconductor, a domain wall separates regions of different bonding structures or different vacua, which would be a new type of excitation. The large width of the domain walls leads to a small effective mass for the excitations, on the order of electron mass instead of the ionic mass. The domain-wall excitation, which propagates freely, has been called a "soliton." Because a moving soliton converts *A*– phase material into *B*– phase material (or vice versa), these objects can only be created or destroyed in pairs. The creation of a soliton is shown in Figure 18.24c.

As we will see, because the midgap state is a solution to the Schrodinger equation in the presence of the structural kink, it can be occupied with zero, one, or two electrons. However, a charged soliton, which has either zero or two electrons in the gap state, carries a charge $\pm e$ and has spin zero, rather than the spin $\frac{1}{2}$ as for an electron or a hole. These reversed spin-charge relations are a fundamental feature of the soliton model of polyacetylene and are supported by experiment.

The π band of $(CH)_x$ is split into two subbands: a fully occupied π band (valence band) and an empty π^* band (the conduction band), each with a wide bandwidth (~5 eV) and significant distortion. The resulting band structure results from the opening of the band gap that originates from the doubling of the unit cell. This is shown in Figure 18.25.

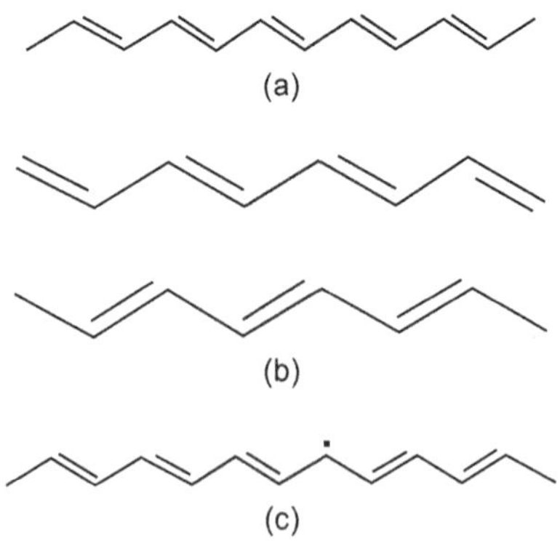

FIGURE 18.24

(a) Dimerized structure due to Peierls instability. (b) Degenerate *A* and *B* phases of dimerized structure. (c) Soliton in *trans*-polyacetylene.

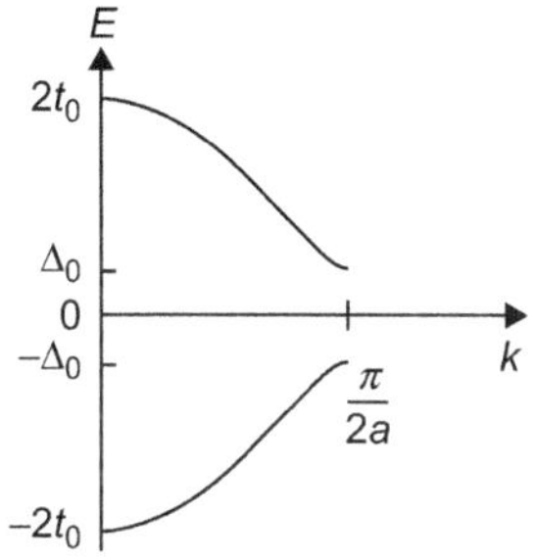

FIGURE 18.25

Band structure of polyacetylene. Energy opening at $k = 2\pi/a$ due to Peierls distortion.

18.5.2 Electronic Structure

The electronic structure of conjugated polymers (SSH) was described by Su et al.[17] by using a tight-binding model in which the π electrons are coupled to distortions in the polymer backbone by the electron–phonon interaction. In the SSH model, photoexcitation across the $\pi - \pi^*$ band gap creates the excitations of conducting polymers: solitons (in degenerate ground-state systems), polarons, and bipolarons. Direct photogeneration occurs due to the overlap between the uniform chain in the ground state and the distorted chain in the excited state.

18.5.3 Tight-Binding Model

Each (CH) group (Figure 18.26) has six degrees of freedom for nuclear translation. Su et al.[17] postulated that only the dimerization coordinate u_n, which specifies the displacement of the nth group along the molecular symmetry group, is important. The tight-binding Hamiltonian can be written as

$$\begin{aligned} H &= H_\pi + H_{\pi-ph} + H_{ph} \\ &= \sum_{n,s}[-t_0 + \alpha(u_{n+1} - u_n)](c^\dagger_{n+1,s}c_{n\cdot s} + c^\dagger_{n,s}c_{n+1,s}) + \sum_n \left[\frac{p_n^2}{2M} + \frac{1}{2}K(u_{n+1} - u_n)^2\right], \end{aligned} \tag{18.53}$$

where p_n are the nuclear momenta, u_n are the displacements from equilibrium, M is the carbon mass, and K is an effective spring constant. The $c^\dagger_{n,s}$, $c_{n,s}$ are the fermion creation and annihilation operators for site n. The first term describes the hopping of $\pi(p_z)$ electrons along the chain without spin flip. The second term describes the $\pi-$ electron–phonon interaction where the terms linear in u_n dominate higher-order terms for the weak-coupling systems. The last two terms are, respectively, a harmonic "spring constant" term, which represents the increase in potential energy that results from displacement from the uniform bond lengths in $(CH)_x$, and a kinetic energy term, where M is the mass of the (CH) group and p_n is the momentum conjugate to u_n.

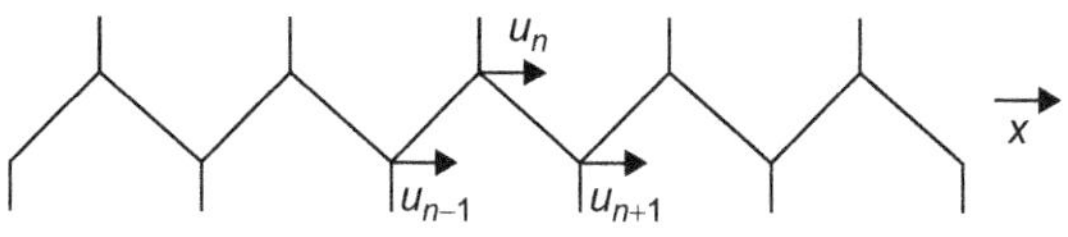

FIGURE 18.26

Dimerization coordinate u_n defined for *trans*-$(CH)_x$.

Reproduced from Heeger et al. (Ref. 7), with the permission of the American Physical Society.

18.5.4 Soliton Excitations

The ground state of a one-dimensional metal is spontaneously distorted to form a charge-density wave $<u_n> \neq 0$, as per Peierls theorem. The strongest instability occurs for a charge-density wave of wave number $Q = k_F = \pi/a$, and hence, we consider the adiabatic ground-state energy E_0 as a function of the mean amplitude of distortion u, where the u_n's are considered to be

$$u_n \to <u_n> = (-1)^n u. \tag{18.54}$$

For u_n given by Eq. (18.54), $H_{\pi-ph}$ is invariant under spatial translation $2ma$, $m = \pm 1, \pm 2, \ldots,$ and H can be diagonalized in k space in the reduced zone, $-\pi/2a < k < \pi/2a$, for the valence ($-$) and conduction ($+$) bands. For a chain of monomers of ring geometry, we obtain (Problem 18.6)

$$\hat{H}(u) = -\sum_{n,s} [t_0 + (-1)^n 2\alpha u](\hat{c}^\dagger_{n+1,s}\hat{c}_{n,s} + \hat{c}^\dagger_{n,s}\hat{c}_{n+1,s}) + 2NKu^2. \tag{18.55}$$

For $\alpha = 0$, $H(u)$ can be made diagonal by using the Bloch operators,

$$\hat{c}_{ks} = (N^{-1/2}) \sum_{n,s} e^{-ikna}\hat{c}_{ns} \tag{18.56}$$

in the extended zone, where $-\pi/a < k \leq \pi/a$.

For $\alpha \neq 0$, the big zone can be folded into the little zone, as shown in Figure 18.27. The valence- and conduction-band operators are defined as

$$\hat{c}_{ks-} = (N)^{-1/2} \sum_{n,s} \mathrm{e}^{-ikna}\, \hat{c}_{ns} \tag{18.57}$$

and

$$\hat{c}_{ks+} = -i(N)^{-1/2} \sum_{n,s} e^{-ikna}(-1)^n \hat{c}_{ns}. \tag{18.58}$$

From Eqs. (18.55), (18.57), and (18.58), we obtain (Problem 18.8)

$$\hat{H}(u) = \sum_{ks} \left[\varepsilon_k(\hat{c}^\dagger_{ks+}\hat{c}_{ks-} - \hat{c}^\dagger_{ks-}\hat{c}_{ks-}) + \Delta_k(\hat{c}^\dagger_{ks+}\hat{c}_{ks-} + \hat{c}^\dagger_{ks-}\hat{c}_{ks+})\right] + 2NKu^2, \tag{18.59}$$

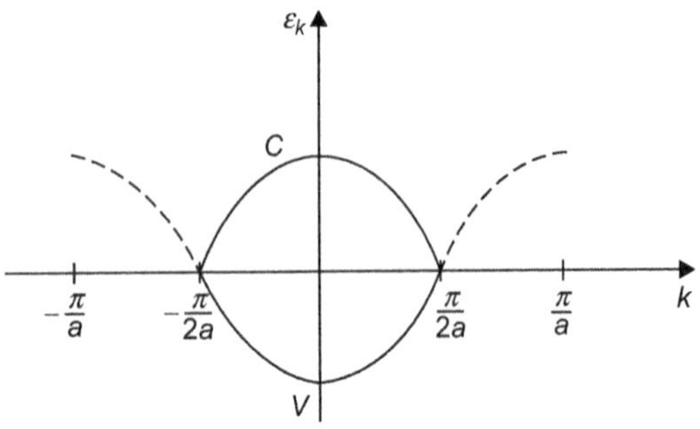

FIGURE 18.27

The reduced and extended zone schemes.

where Δ_k, the energy gap parameter, is defined as

$$\Delta_k = 4\alpha u \sin ka. \tag{18.60}$$

In the reduced zone, the unperturbed band energy is defined as

$$\varepsilon_k = 2t_0 \cos ka, \quad (-\pi)/(2a) < k < (\pi)/(2a), \tag{18.61}$$

where ε_k describes particles for ($+$) and holes for ($-$). We now convert H in Eq. (18.59) to a diagonal form by making the transformation

$$\hat{a}_{ks-} = \alpha_k\hat{c}_{ks-} - \beta_k\hat{c}_{ks+}$$

and

$$\hat{a}_{ks+} = \beta_k \hat{c}_{ks-} + \alpha_k \hat{c}_{ks+}. \tag{18.62}$$

We have, in analogy with BCS theory of superconductivity,

$$|\alpha_k|^2 + |\beta_k|^2 = 1, \tag{18.63}$$

so that the a's satisfy the Fermi anticommutation relations. Because H has no term that mixes the $(+)$ and $(-)$ operators, we obtain from Eq. (18.59) through (18.62)

$$H = \sum_k E_k (n_{ks+} - n_{ks-}) + 2NKu^2. \tag{18.64}$$

The quasiparticle energy is given by (in analogy with BCS superconductivity theory)

$$E_k = (\epsilon_k^2 + \Delta_k)^{1/2}, \tag{18.65}$$

$$\alpha_k = [(1 + \varepsilon_k/E_k)/2)]^{1/2}, \tag{18.66}$$

$$\beta_k = [(1 - \varepsilon_k/E_k)/2)]^{1/2} \text{ sgn } \Delta_k, \tag{18.67}$$

and

$$\alpha_k \beta_k = \Delta_k/(2E_k). \tag{18.68}$$

If one were to clamp u_n at the ground-state mean-field value, $(-1)^n u$, the chain would behave as a conventional semiconductor with electron and hole excitations. However, because of the two-fold-degenerate ground-state $E(u_0) = E(-u_0)$, the system supports nonlinear excitations, which act as moving walls separating the A phase $(+u_0)$ and B phase $(-u_0)$.This is shown in Figure 18.28, which implies that the nonlinear excitations, solitons, will bef important. To determine the soliton, one introduces

$$\varphi_n = (-1)^n u_n, \tag{18.69}$$

so that in the A phase, $\varphi_n = u_0$, and in the B phase, $\varphi_n = -u_0$. However, numerical calculations have shown that the form of φ_n, for the preferred width ξ of the soliton that minimizes the total energy, is

$$\varphi_n = u_0 \tanh [(n - n_0)a/\xi)]. \tag{18.70}$$

It has been calculated that $\xi \approx 7a$, and the energy to create a soliton at rest is $E_s \approx 0.42$ eV. Thus, $E_s < 0.5\ \Delta$, where Δ is the single-particle gap. Thus, it is less costly to create a soliton than an electron or hole, and they are spontaneously generated by photoexcitation, by thermal generation, or by injection of electrons and/or holes. It

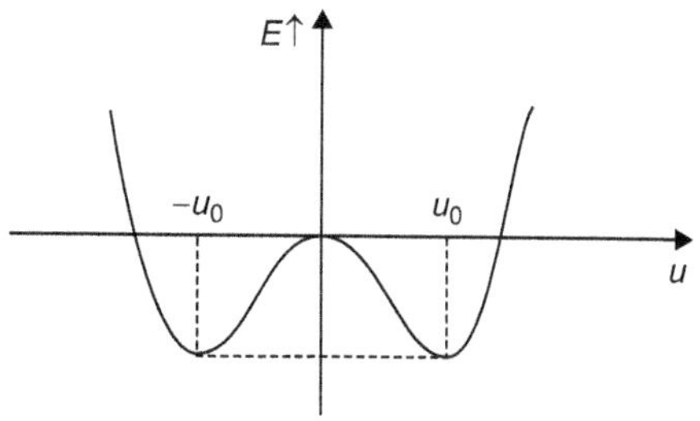

FIGURE 18.28

The total energy of the dimerized polyacetylene chain.

can also be shown that for each widely spaced soliton or antisoliton (a reverse boundary from the *B* phase back to the *A* phase), there exists a normalized single-electron state in the gap center that can accommodate zero, one, or two electrons.

18.5.5 Solitons, Polarons, and Polaron Excitations

It can be seen from Figure 18.29c that because the nonbonding or atomic state formed by the chain relaxation can be mapped to a specific atomic site, the resulting distribution of charge and spin can be understood. If the state is unoccupied (doubly occupied), the carbon atom at the boundary is left with a positive (negative) charge, but there are no unpaired spins. Therefore, the charged soliton is positively (negatively) charged but spinless. Single occupation of the soliton state neutralizes the electronic charge of the carbon nucleus, while introducing an unpaired spin onto the chain. The localized electronic state associated with the soliton is a nonbonding state at an energy that lies at the middle of the $\pi - \pi^*$ gap, between the bonding and antibonding levels of the perfect chain.

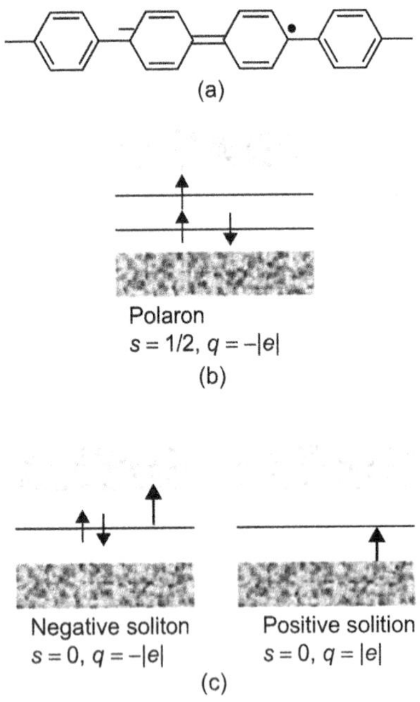

FIGURE 18.29

(a) Schematic picture of a polaron in polyparaphenylene (PPP); (b) band diagram of an electron polaron—the lower gap state is single occupied and the upper gap state is empty for a hole polaron; (c) band diagrams for positive and negative solitons.

Reproduced from Heeger[6] with the permission of the American Physical Society.

18.6.6 Polarons and Bipolarons

A polaron can be thought of as a bound state of a charged soliton and a neutral soliton of which the midgap energy states hybridize to form bonding and antibonding levels. The neutral soliton contributes no charge and a single spin. The charged soliton carries a charge of $\pm e$ and no spin. The resulting polaron is a fermion with the charge-spin relationship $q = \pm e$ and $s = 1/2$. The positive polaron is a radical cation; the negative polaron is a negative anion. Both are quasiparticles consisting of a single electronic charge dressed with a local geometrical relaxation of the bond lengths. The schematic picture of polarons in PPP is shown in Figures 18.29a and b.

A bipolaron is a bound state of two charged solitons of like charges or two polarons of which the neutral solitons annihilate each other with two corresponding midgap levels. The bipolarons are shown in Figure 18.30. Because each charged soliton has a single electronic charge and no spin, the bipolaron has charge $\pm 2e$ and zero spin. Both positive and negative bipolarons are doubly charged bound states of two polarons bound together by enhanced geometrical relaxation of the bond lengths or the overlap of a common lattice distortion.

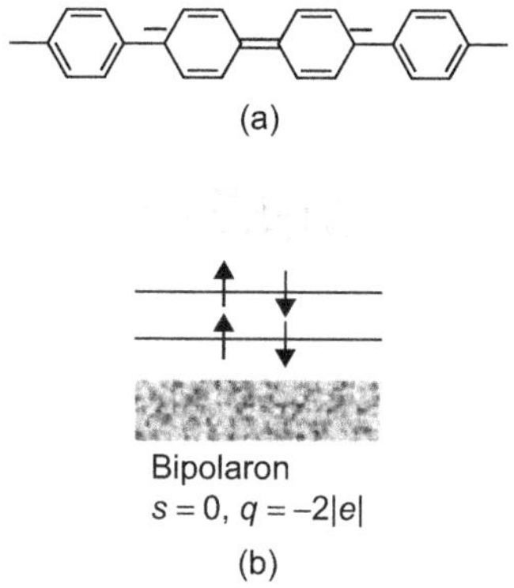

FIGURE 18.30

Bipolarons in polymer with nondegenerate ground state. (a) Negative bipolaron in PPP; (b) band diagram for a negative bipolaron. For a positive bipolaron, both gap states are unoccupied.

Reproduced from Heeger[6] with the permission of the American Physical Society.

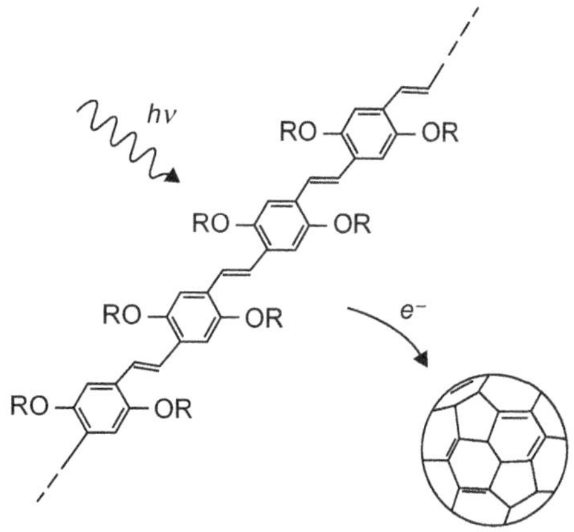

FIGURE 18.31

Photoinduced electron transfer from a conjugated semiconducting polymer to C_{60}.

Reproduced from Heeger[6] with the permission of the American Physical Society.

18.6 PHOTOINDUCED ELECTRON TRANSFER

The electrons are promoted to the antibonding π^* band of semiconducting polymers when they are photoexcited. The photoinduced nonlinear excitations (such as polarons) on the conjugated polymer backbone are quite stable. Thus, semiconducting polymers are electron donors when they are photoexcited. On the other hand, Buckministerfullerene, C_{60}, is an excellent electron acceptor capable of accepting up to six electrons. It forms charge-transfer salts with several types of donors. The photoinduced transfer process from a conjugated semiconducting polymer to C_{60} is shown schematically in Figure 18.31.

A study of the dynamics of the photoinduced electron transfer from semiconducting polymers to C_{60} by Lanzani et al. (Ref. 10) demonstrated that the charge transfer occurs within 50 fs after photoexcitation. This charge transfer rate is more than 1000 times faster than any other process. The quantum efficiency for charge separation approaches one. In addition, the charge separation state is metastable.

PROBLEMS

18.1. Show that the reciprocal lattice vectors of honeycomb lattice of graphene (Figure 18.3) are

$$\mathbf{b}_1 = \frac{2\pi}{3a}(\hat{i} + \sqrt{3}\hat{j}); \quad \mathbf{b}_2 = \frac{2\pi}{3a}(\hat{i} - \sqrt{3}\hat{j}). \tag{1}$$

18.2. Show that the positions of the two Dirac points K and K', located at the corners of the Brillouin zone (Figure 18.3), are

$$\mathbf{K} = \frac{2\pi}{3a}\left(\hat{i} + \frac{1}{\sqrt{3}}\hat{j}\right); \quad \mathbf{K}' = \frac{2\pi}{3a}\left(\hat{i} - \frac{1}{\sqrt{3}}\hat{j}\right). \tag{1}$$

18.3. Show that by substituting in Eq. (18.6),

$$H\psi = E\psi, \tag{1}$$

we obtain

$$H_{11} + \lambda H_{12} = E \tag{2}$$

and

$$H_{12} + \lambda H_{22} = \lambda E. \tag{3}$$

Here,

$$H_{11} = \int \phi_1{}^* \, H\phi_1 \, d\nu, \; H_{12} = H_{21}{}^* = \int \phi_1{}^* \, H\phi_1 \, d\nu; \; H_{22} = \int \phi_2{}^* \, H\phi_2 \, d\nu \tag{4}$$

and

$$\int \phi_1{}^* \phi_1 d\nu = \int \phi_2{}^* \phi_2 d\nu = 1. \tag{5}$$

18.4. It has been shown in Eq. (18.19) that

$$H_{11} = \frac{1}{N} \sum_{A,A'} e^{-i\mathbf{k}\cdot(\mathbf{r}_A - \mathbf{r}_{A'})} \int \chi^* (\mathbf{r} - \mathbf{r}_A) H \chi(\mathbf{r} - \mathbf{r}_{A'}) \, d\nu. \tag{1}$$

Keeping only the nearest-neighbor integrals among the atoms A, and writing

$$E_0 = \int \chi^* (\mathbf{r}) H \chi(\mathbf{r}) \, d\nu \tag{2}$$

and

$$\gamma_0' = -\int \chi^* (\mathbf{r} - \vec{\rho}\,') H \chi(\mathbf{r}) \, d\nu, \tag{3}$$

where $\vec{\rho}\,' = \mathbf{a}_1 (say)$ is a vector joining the nearest neighbor among atoms A, show that

$$H_{11} = E_0 - 2\gamma_0' \left[\cos(2\pi k_y a) + 2\cos(\pi k_x a\sqrt{3}) \cos(\pi k_y a) \right]. \tag{4}$$

18.5. It has been shown in Eq. (18.28) that

$$H_{12} = \frac{1}{N} \sum_{A,B} e^{-i\mathbf{k}\cdot(\mathbf{r}_A - \mathbf{r}_B)} \int \chi^* (\mathbf{r} - \mathbf{r}_A) H \chi(\mathbf{r} - \mathbf{r}_B). \tag{1}$$

Considering only the nearest-neighbor interactions in the lattice (between atoms of type A and type B and vice versa), we write (in analogy with Eq. 18.27)

$$\gamma_0 = \int \chi^* (\mathbf{r} - \vec{\rho}) (U - V) \chi(\mathbf{r}) \, d\nu > 0, \tag{2}$$

where

$$\vec{\rho} = \mathbf{AB}. \tag{3}$$

Show that

$$H_{12} = -\gamma_0\left[e^{-ik_xa} + 2\cos(\sqrt{3}k_ya)e^{3ik_xa/2}\right] \tag{4}$$

and

$$|H_{12}|^2 = \gamma_0^2\left[1 + 4\cos^2\left(\frac{\sqrt{3}}{2}k_ya\right) + 4\cos\left(\frac{\sqrt{3}}{2}k_ya\right)\cos\left(\frac{3}{2}k_xa\right)\right]. \tag{5}$$

18.6. The translation vector of a general chiral tubule can be written as

$$\mathbf{T} = [(2m+n)\mathbf{a}_1 - (2n+m)\mathbf{a}_2]/d_R. \tag{1}$$

The length T of the translation vector $\mathbf{T}$ corresponds to the first lattice point of the graphene sheet. Show that

$$T = \sqrt{3}C_h/d_R, \tag{2}$$

where the length C_h is defined in Eq. (18.47). Defining d as the highest common divisor of (n, m), we have

$$d_R = \begin{cases} d \text{ if } n-m \text{ is not a multiple of } 3d \\ 3d \text{ if } n-m \text{ is a multiple of } 3d. \end{cases} \tag{3}$$

18.7. The tight-binding Hamiltonian for the polymer *trans*-(CH)$_x$ is given by (Eq. 18.53)

$$\begin{aligned} H &= H_\pi + H_{\pi-ph} + H_{ph} \\ &= \sum_{n,s}[-t_0 + \alpha(u_{n+1} - u_n)](\hat{c}^\dagger_{n+1,s}\hat{c}_{n.s} + \hat{c}^\dagger_{n,s}\hat{c}_{n+1,s}) + \sum_n\left[\frac{p_n^2}{2M} + \frac{1}{2}K(u_{n+1} - u_n)^2\right]. \end{aligned} \tag{1}$$

Show that if

$$u_n \to \langle u_n\rangle = (-1)^n u, \tag{2}$$

one can write

$$H(u) = -\sum_{n,s}[t_0 + (-1)^n 2\alpha u](\hat{c}^\dagger_{n+1,s}\hat{c}_{n,s} + \hat{c}^\dagger_{n,s}\hat{c}_{n+1,s}) + 2NKu^2 \tag{3}$$

for a chain of monomers in a ring geometry.

18.8. Show that for $\alpha = 0$, $H(u)$ in Eq. (3) of Problem 18.7 can be made diagonal by using the Bloch operators,

$$\hat{c}_{ks} = (N^{-1/2})\sum_{n,s} e^{-ikna}\hat{c}_{ns}, \tag{1}$$

in the extended zone, where $-\pi/a < k \le \pi/a$.

18.9. Show that by using the operators

$$\hat{c}_{ks-} = (N)^{-1/2} \sum_{n,s} e^{-ikna} \hat{c}_{ns} \tag{1}$$

and

$$\hat{c}_{ks+} = -i(N)^{-1/2} \sum_{n,s} e^{-ikna} (-1)^n \hat{c}_{ns}, \tag{2}$$

Eq. (3) in Problem 18.7 can be rewritten as

$$H(u) = \sum_{ks} \left[\varepsilon_k (\hat{c}^\dagger_{ks+} \hat{c}_{ks-} - \hat{c}^\dagger_{ks-} \hat{c}_{ks-}) + \Delta_k (\hat{c}^\dagger_{ks+} \hat{c}_{ks-} + \hat{c}^\dagger_{ks-} \hat{c}_{ks+}) \right] + 2NKu^2, \tag{3}$$

where Δ_k, the energy gap parameter, is defined as

$$\Delta_k = 4\alpha u \sin ka. \tag{4}$$

18.10. From Eqs. (18.59) through (18.62), show that

$$H = \sum_k E_k (n_{ks+} - n_{ks-}) + 2NKu^2. \tag{1}$$

References

1. Castro Neto AH, Guinea F, Peres NMR, Noboselov KS, Geim AK. The electronic properties of graphene. *Rev Mod Phys* 2009;**81**:109.
2. Curl RF. Dawn of the fullerenes: experiment and conjecture. *Rev Mod Phys* 1997;**69**:691.
3. Dresselhaus MS, Dresselhaus G, Ekland PC. *Science of fullerenes and nanoparticles.* Amsterdam: Elsevier Science; 1996.
4. Dresselhaus MS, Dresselhaus G, Saito R. Physics of Carbon Nanotubes. *Carbon* 1995;**33**:883.
5. Geim AK, Novoselov KS. The rise of Graphene. *Nature materials* 2007;**6**:183.
6. Heeger AJ. Semiconducting and metallic polymers: The fourth generation of polymeric materials. *Rev Mod Phys* 2001;**73**:681.
7. Heeger AJ, Kivelson S, Schrieffer JR, Su W-P. Solitons in conducting Polymers. *Rev Mod Phys* 1988;**60**:781.
8. Kroto HW. Symmetry, space, stars and C60. *Rev Mod Phys* 1997;**69**:703.
9. Kroto HW, Heath JR, O'Brien SC, Curl RF, Smalley RE. C60: Buckminsterfullerene. *Nature* 1985; **318**:162.
10. Lanzani G, Zenz C, Cerullo G, Graupner W, Leising G, Scherf U, De Silvestri S. *Synth Met* 2000; **111–112**:493.
11. MacDiarmid AG. "Synthetic Metals." A novel role for organic polymers. *Rev Mod Phys* 2001;**73**:701.
12. Malard LM, Pimenta MA, Dresselhaus G, Dresselhaus MS. Raman Spectroscopy in Graphene. *Phys Rep* 2009;**473**:51.
13. Novoselov KS, Geim AK, Morozov SV, Jiang D, Zhang Y, Grigoneva IV, et al. Electric field effect in atomically thin carbon films. *Science* 2004;**306**:666.

14. Novoselov KS, Geim AK, Morozov SV, Jiang D, Katsnelson MI, Grigoneva IV, et al. Two-dimensional gas of massless Dirac fermions in graphene. *Nature* 2005;**438**:197.
15. Novoselov KS, Jiang D, Schedin F, Booth TJ, Khotkevich VV, Morozov SV, et al. Two-dimensional atomic crystals. *Proc Natl Acad Sci* 2005;**102**:10451.
16. Smalley RE. Discovering the fullerenes. *Rev Mod Phys* 1997;**69**:723.
17. Su WP, Schrieffer JR, Heeger AJ. Solitons in Polyacetylene. *Phys Rev Lett* 1979;**42**:1698.
18. Taylor R, Hare JP, Abdul-Sada AK, Kroto J. *Chem Soc Chemical Commu* 1990;1423.
19. Wallace PR. The Band Theory of Graphite. *Phys Rev* 1947;**71**:622.

APPENDIX

Elements of Group Theory

A

CHAPTER OUTLINE

A.1 Symmetry and Its Consequences 633
A.1.1 Symmetry of Crystals 633
A.1.2 Definition of a Group 633
A.1.3 Symmetry Operations in Crystal Lattices 634
A.2 Space Groups 634
A.2.1 Introduction 634
A.2.2 Space Group Operations 634
A.3 Point Group Operations 636
A.3.1 Introduction 636
A.3.2 Description of Point Groups 636
A.3.3 The Cubic Group O_h 638

A.1 SYMMETRY AND ITS CONSEQUENCES

A.1.1 Symmetry of Crystals

Crystalline solids possess translational symmetry as well as symmetries involving rotations and reflections. The fundamental principle is that the quantum mechanical operators representing a symmetry operation of a crystal must commute with the Hamiltonian of the crystal. This facilitates the derivations of restrictions limiting possible Hamiltonians, the classification of eigenstates according to symmetry, as well as determining the selection rules for transitions between states. Often, one has to use group theory to consider these symmetry considerations.

A.1.2 Definition of a Group

A group has to satisfy the following three basic conditions:

a. The group must contain the unit element, generally denoted by E.
b. The product of any two elements in the group must be another element in the group.
c. The inverse of every element in the group must belong to the group.

If the members commute with each other, the group is Abelian.

A.1.3 Symmetry Operations in Crystal Lattices

For the purpose of considering the symmetry operations of crystals as well as their reciprocal lattices, we consider a set of symmetry operations (matrices) that form a group. We denote this group as $\{G\}$. There are two types of such groups known as space groups and point groups.

A.2 SPACE GROUPS

A.2.1 Introduction

The elements of symmetry groups relevant to solids include translations as well as proper and improper rotations. These are also known as *space groups.*

A Bravais lattice is generally characterized by the specification of all rigid operations that take the lattice into itself. This set of operations is known as the space group or symmetry group of the Bravais lattice. All translations through the lattice vectors of a Bravais lattice are included in the operations of a symmetry group. In addition, rotations, reflections, and inversions can also take the Bravais lattice into itself.

For example, a cubic Bravais lattice can be taken into itself by a rotation through 90° about a line of lattice points in a <100> direction, a rotation through 120° about a line of lattice points in a <111> direction, and reflection of all points in a $\{100\}$ lattice plane. In contrast, a simple hexagonal Bravais lattice can be taken into itself by a reflection in a lattice plane perpendicular to the c-axis, or rotation through 60° about a line of points parallel to the c-axis.

The symmetry operation of a Bravais lattice can be compounded out of a translation $T_{\mathbf{R}}$ through a lattice vector $\mathbf{R}$ and a rigid operation leaving at least one point fixed. A simple example of two such consecutive operations (symmetry operation S and translation $T_{-\mathbf{R}}$ through $-\mathbf{R}$) is also known as a composite operation $T_{-\mathbf{R}}S$. In this example, because S transports the origin of the lattice $\mathbf{O}$ to $\mathbf{R}$ while $T_{-\mathbf{R}}$ carries $\mathbf{R}$ back to the origin $\mathbf{O}$, $T_{-\mathbf{R}}S$ also leaves at least one lattice point ($\mathbf{O}$) fixed. In contrast, if the operation $T_{\mathbf{R}}$ is performed after performing the operation $T_{-\mathbf{R}}S$, the result is the operation S alone. In the process, $T_{\mathbf{R}}$ has reversed the previous application of $T_{-\mathbf{R}}$. Hence, S can be compounded out of operations that leave a point fixed ($T_{-\mathbf{R}}S$), translations through Bravais lattice vectors ($T_{\mathbf{R}}$), or through successive applications of operations of both types.

A.2.2 Space Group Operations

An operator of a space group contains a part that is denoted by the symbol $\{\widetilde{\alpha}|\mathbf{t}\}$, where $\widetilde{\alpha}$ denotes a proper or improper rotation and $\mathbf{t}$ denotes a translation part. The coordinate transformation from $\mathbf{x}$ to $\mathbf{x}'$ can be expressed as

$$\mathbf{x}' = \widetilde{\alpha}\mathbf{x} + \mathbf{t}. \tag{A.1}$$

Here, $\widetilde{\alpha}$ can be represented by a 3×3 orthogonal matrix. One can multiply two such operators by

$$\{\widetilde{\beta}|\mathbf{t}'\}\{\widetilde{\alpha}|\mathbf{t}\} = \{\widetilde{\beta\alpha}|\widetilde{\beta}\mathbf{t} + \mathbf{t}'\}. \tag{A.2}$$

It is easily verified that the unit operator is $\{\varepsilon|0\}$. It can be shown from Eq. (A.2) that the inverse of the operator $\{\tilde{\alpha}|\mathbf{t}\}$ is

$$\{\tilde{\alpha}|\mathbf{t}\}^{-1} = \{\tilde{\alpha}^{-1}| -\tilde{\alpha}^{-1}\mathbf{t}\}. \tag{A.3}$$

In the block notation, we can write Eq. (A.1) in the alternate form, where the position vector $\mathbf{x}$ is denoted by

$$\mathbf{x} = \begin{pmatrix} x_1 \\ x_2 \\ x_3 \end{pmatrix}, \mathbf{t} = \begin{pmatrix} t_1 \\ t_2 \\ t_3 \end{pmatrix}, \text{ and } \tilde{\alpha} = \begin{pmatrix} \tilde{\alpha}_{11} & \tilde{\alpha}_{12} & \tilde{\alpha}_{13} \\ \tilde{\alpha}_{21} & \tilde{\alpha}_{22} & \tilde{\alpha}_{23} \\ \tilde{\alpha}_{31} & \tilde{\alpha}_{32} & \tilde{\alpha}_{33} \end{pmatrix}. \tag{A.4}$$

Eq. (A.1) can be written as

$$\begin{pmatrix} 1 \\ x_1' \\ x_2' \\ x_3' \end{pmatrix} = \begin{pmatrix} 1 & 0 & 0 & 0 \\ t_1 & \tilde{\alpha}_{11} & \tilde{\alpha}_{12} & \tilde{\alpha}_{13} \\ t_2 & \tilde{\alpha}_{21} & \tilde{\alpha}_{22} & \tilde{\alpha}_{23} \\ t_3 & \tilde{\alpha}_{31} & \tilde{\alpha}_{32} & \tilde{\alpha}_{33} \end{pmatrix} \begin{pmatrix} 1 \\ x_1 \\ x_2 \\ x_3 \end{pmatrix} \tag{A.5}$$

or in the block notation

$$\begin{pmatrix} 1 \\ \mathbf{x}' \end{pmatrix} = \begin{pmatrix} 1 & 0 \\ \mathbf{t} & \tilde{\alpha} \end{pmatrix} \begin{pmatrix} 1 \\ \mathbf{x} \end{pmatrix}. \tag{A.6}$$

From Eqs. (A.2) and (A.6), we obtain

$$\begin{pmatrix} 1 & 0 \\ \mathbf{t}' & \tilde{\beta} \end{pmatrix} \begin{pmatrix} 1 & 0 \\ \mathbf{t} & \tilde{\alpha} \end{pmatrix} = \begin{pmatrix} 1 & 0 \\ \mathbf{t}' + \tilde{\beta}\mathbf{t} & \tilde{\beta}\tilde{\alpha} \end{pmatrix}, \tag{A.7}$$

and the inverse is

$$\begin{pmatrix} 1 & 0 \\ \mathbf{t} & \tilde{\alpha} \end{pmatrix}^{-1} = \begin{pmatrix} 1 & 0 \\ -\tilde{\alpha}^{-1}\mathbf{t} & \tilde{\alpha}^{-1} \end{pmatrix}. \tag{A.8}$$

$\{\varepsilon|\mathbf{R}_i\}$ is an operator that represents a lattice translation through $\mathbf{R}_i$, each $\{\varepsilon|\mathbf{R}_i\}$ has an inverse $\{\varepsilon|-\mathbf{R}_i\}$, and because the sum $\mathbf{R}_i + \mathbf{R}_{i'} = \mathbf{R}_j$, the lattice translations form a group. We will now define a few common terms used in group theory.

Abelian: If the members commute with each other, the group is Abelian. The operators of the full space group, which contains both rotations and translations, are not necessarily Abelian.

Invariant subgroup: If A is a member of a subgroup and X is a member of the full group and if $B = XAX^{-1}$ is in the subgroup concerned for all A and all X, the subgroup is said to be invariant. This can be proved as follows.

Proof: Let $\mathbf{R}_i$ be any direct lattice vector and $\tilde{\alpha}$ be the rotational part of any space group operation, then $\tilde{\alpha}\mathbf{R}_i$ is also a lattice vector. It is easy to show that

$$\{\tilde{\alpha}|\mathbf{t}\}\{\tilde{\varepsilon}|\mathbf{R}_i\}\{\tilde{\alpha}|\mathbf{t}\}^{-1} = \{\tilde{\alpha}|\tilde{\alpha}\mathbf{R}_i + \mathbf{t}\}\{\tilde{\alpha}^{-1}|-\tilde{\alpha}^{-1}\mathbf{t}\} = \{\tilde{\varepsilon}|\tilde{\alpha}\mathbf{R}_i\}. \tag{A.9}$$

Thus, if $\{\varepsilon|\mathbf{R}_i\}$ is a lattice translation, then $\{\varepsilon|\tilde{\alpha}\mathbf{R}_i\}$ is also a lattice translation and Eq. (A.9) proves that the subgroup is invariant. Thus, a space group is defined as a group of operators of the form $\{\tilde{\alpha},\mathbf{t}\}$, which possesses an invariant subgroup of pure translations.

A.3 POINT GROUP OPERATIONS

A.3.1 Introduction

It can be easily shown that a point group can contain rotations $\tilde{\alpha}$ through $60°, 90°$, and (or) multiples of these, but it forbids five-fold rotational symmetry.

Proof: If $\mathbf{R}$ is a direct lattice vector,

$$\mathbf{R} = \sum_{j=1}^{3} n_j \mathbf{a}_j, \tag{A.10}$$

then $\tilde{\alpha}\mathbf{R}$ is also a direct lattice vector. Here, $\mathbf{a}_j$ are the primitive translation vectors and the n_j's are integers. Thus, $\tilde{\alpha}\mathbf{R}$ can be expressed in terms of $\mathbf{a}_j$ with integer coefficients.

Because $\tilde{\alpha}$ can be a rotation through an angle θ about some axis, in considering the operation of $\tilde{\alpha}$ on $\mathbf{R}$, one has to include the possibility that $\mathbf{a}_j$ may not be orthogonal and assume that the $\mathbf{a}_j$ are related to a Cartesian system by a nonsingular matrix $\tilde{A}$. If the original vector is represented by $\mathbf{n}$ (with components n_j) and the rotated vector is denoted by $\mathbf{n}'$, we have

$$\mathbf{n}\tilde{A}\tilde{\alpha} = \mathbf{n}'\tilde{A}, \tag{A.11}$$

which can be written in the alternate form

$$\mathbf{n}\tilde{A}\tilde{\alpha}\tilde{A}^{-1} = \mathbf{n}'. \tag{A.12}$$

The elements of $\tilde{A}\tilde{\alpha}\tilde{A}^{-1}$ as well as the trace of $\tilde{A}\tilde{\alpha}\tilde{A}^{-1}$ must be integers because the elements of both $\mathbf{n}$ and $\mathbf{n}'$ are arbitrary integers. Because the trace of a matrix is invariant under a similarity transformation,

$$tr(\tilde{A}\tilde{\alpha}\tilde{A}^{-1}) = tr(\tilde{\alpha}) = 1 + 2\cos\theta = \text{integer}. \tag{A.13}$$

Eq. (A.13) implies that $\theta = 60°, 90°$, or an integral multiple of these angles.

A.3.2 Description of Point Groups

There are 32 possible crystal point groups. There are two widely used but different notations for point groups: the Schoenflies system and the international system. We will first discuss the Schoenflies notation (see Table A.1).

Table A.1 Shoenflies Symbol

Operation	Shoenflies Symbol
Identity	E
Rotation through $2\pi/n$	C_n
Inversion	i
Improper rotation through $2\pi/n$	S_n
Reflection in a plane	σ
Reflection in a plane perpendicular to highest symmetry axis	σ_h
Reflection in a plane containing the highest symmetry axis	σ_v
Reflection in a plane containing the highest symmetry axis and bisecting the angle between two-fold axes perpendicular to symmetry axis	σ_d

Reproduced from Quantum Theory of the Solid State *by J. Callaway (Academic Press, New York, 1976) with the permission of Elsevier.*

1. C_n: These point groups contain only a single axis of symmetry, around which rotations through angles of $2\pi/n$ are permitted. As we have seen, $n = 1, 2, 3, 4, 6$.
2. C_{nv}: In addition to the operation C_n, these groups have a mirror plane that contains the axis of rotation, plus as many mirror planes as the existence of the n-fold axis requires. The reflection symmetry must be consistent with the rotational symmetry, and n-fold rotational symmetry about some axis demands the existence of n reflection planes at angles of π/n if there are any reflection planes containing the axis. The allowed values of n are 2, 3, 4, 6.
3. C_{nh}: These groups contain, in addition to the n-fold axis, a single mirror plane that is perpendicular to the axis. Thus, there is a "horizontal" reflection plane (operation σ_h), which is a reflection in a plane to the origin perpendicular to the axis of highest symmetry. The group also contains the inversion operation i if n is even. In this case, $n = 1, 2, 3, 4, 6$.
4. S_n: These groups have n-fold improper rotation (rotation combined with improper reflection in a plane perpendicular to the axis of rotation). Because the group S_3 is identical with C_{3h}, it is not counted. Thus, the distinct groups are S_2 (simple inversion), S_4, and S_6.
5. D_n: In addition to the n-fold rotation axis, these groups contain a two-fold axes perpendicular to the highest symmetry (C_n) axis plus as many additional two-fold axes as are required by the existence of the n-fold axis. Thus, one can have $n = 2, 3, 4, 6$.
6. D_{nh}: These (the most symmetric of the groups) contain all the elements of D_n plus mirror planes containing the n-fold axes, which bisect the angles between the two-fold axes. Therefore, there are twice as many elements in D_{nh} as in D_n.
7. D_{nd}: These contain the elements of D_n plus reflections in a "diagonal" plane (σ_d) containing the symmetry axis and bisecting the angle between the two-fold axis. The diagonal planes are special cases of vertical planes. There are two cases: D_{2d} and D_{3d}.

The international notation for point groups and the three Schoenflies equivalents are given in Table A.2.

Table A.2 International Notation for Point Group Symbols

Axis	Symbol*	Schoenflies Equivalent
n-fold rotation axis	n	$n \leftrightarrow C_n$
Improper *n*-fold rotation axis	$\bar{n}$	
Rotation axis with reflection plane perpendicular to it	$\frac{n}{m}$	
n-fold rotation axis with two-fold axis perpendicular to it	$n2$	$n22 \leftrightarrow D_n$
n-fold rotation axis with reflection plane containing the axis	nm	$nmm \leftrightarrow C_{nv}$
Improper rotation axis with two-fold axis perpendicular to it	$\bar{n}2$	
Improper rotation axis with reflection planes containing the axis	$\bar{n}m$	
Rotation axis with a perpendicular reflection and a set of reflection planes containing it	$\frac{n}{mm}$ or $\frac{n}{m}m$	

* *n is an integer 1, 2, 4, 6.*
Reproduced from Quantum Theory of the Solid State *by J. Callaway (Academic Press, New York, 1976) with the permission of Elsevier.*

A.3.3 The Cubic Group O_h

The operations of the full cubic group O_h, which is the point group of highest symmetry, may be described as follows:

a. The identity E.
b. Rotations by $\pm 90°$ about a coordinate (four-fold) axis. Class C_4, six operations.
c. Rotations by 180° about the same axis. There are six such axes. Class C_2, six operations.
d. Rotations by $\pm 120°$ about a three-fold axis (body diagonal of the cube). There are four such axes. Class C_3, eight operations.
e. The inversion with respect to the origin, class J, one operation.
f. X Classes JC_4, JC_4^2, JC_2, and JC_3. Combinations of the preceding operations with the inversion. These four classes contain 23 operations.

The cubic group O_h is described in Table A.3. The operations on a position vector **r** (with components x, y, z) can be specified as possible permutations of x, y, z with change of signs.

The operations contained in the classes E, C_4^2, C_4, C_2, and C_3 form the subgroup O. The subgroup T_d are composed of E, C_4^2, JC_4, and C_3.

A class is a set of elements of a group that are conjugate to each other. For example, if X and A are members of a group, the element $B = XAX^{-1}$ is conjugate to A. If two elements A and C are conjugate to a third member D, they are conjugate to each other.

Table A.3 The Cubic Group O_h

Class	Operation	Class	Operation
E	$x \quad y \quad z$	J	$-x \quad -y \quad -z$
C_4^2	$-x \quad -y \quad z$	JC_4^2	$x \quad y \quad -z$
	$x \quad -y \quad -z$		$-x \quad y \quad z$
	$-x \quad y \quad -z$		$x \quad -y \quad z$
C_4	$-y \quad x \quad z$	JC_4	$y \quad -x \quad -z$
	$y \quad -x \quad z$		$-y \quad x \quad -z$
	$x \quad -z \quad y$		$-x \quad z \quad -y$
	$x \quad z \quad -y$		$-x \quad -z \quad y$
	$z \quad y \quad -x$		$-z \quad -y \quad x$
	$-z \quad y \quad x$		$z \quad -y \quad -x$
C_2	$y \quad x \quad -z$	JC_2	$-y \quad -x \quad z$
	$z \quad -y \quad x$		$-z \quad y \quad -x$
	$-x \quad z \quad y$		$x \quad -z \quad -y$
	$-y \quad -x \quad -z$		$y \quad x \quad z$
	$-z \quad -y \quad -x$		$z \quad y \quad x$
	$-x \quad -z \quad -y$		$x \quad z \quad y$
C_3	$z \quad x \quad y$	JC_3	$-z \quad -x \quad -y$
	$y \quad z \quad x$		$-y \quad -z \quad -x$
	$z \quad -x \quad -y$		$-z \quad x \quad y$
	$-y \quad -z \quad x$		$y \quad z \quad -x$
	$-z \quad -x \quad y$		$z \quad x \quad -y$
	$-y \quad z \quad -x$		$y \quad -z \quad x$
	$-z \quad x \quad -y$		$z \quad -x \quad y$
	$y \quad -z \quad -x$		$-y \quad z \quad x$

Reproduced from Quantum Theory of the Solid State *by J. Callaway (Academic Press, New York, 1976) with the permission of Elsevier.*

Reference

1. Callaway J. *Quantum theory of the solid state*. New York: Academic Press; 1986.

APPENDIX

Mossbauer Effect

B

CHAPTER OUTLINE

B.1 Introduction 641
B.2 Recoilless Fraction 642
B.3 Average Transferred Energy 643
Reference 644

B.1 INTRODUCTION

We will first consider the case of free nucleus at rest of mass M. We assume that the nucleus is initially in an excited state of energy E_x above the ground state. When it emits a γ ray of energy E_γ (and momentum $p = E_\gamma/c$), some recoil energy R is transferred to the nucleus. The recoil energy is given by

$$R = p^2/2M = E_\gamma{}^2/2Mc^2. \tag{B.1}$$

Because R is much smaller than E_x, the energy of the γ ray is

$$E_y = E_x - R = E_x - (E_\gamma{}^2/2Mc^2) \approx E_x[1 - (E_x/2Mc^2)]. \tag{B.2}$$

The typical values are 10 keV$< E_x <$100 keV while $2 \times 10^{-4} < R < 5 \times 10^{-2}$ eV. For example, for ^{57}Fe, $E_x \approx 14$ keV and $R \approx 0.002$ eV. However, the natural linewidth $\Gamma \approx 4.6 \times 10^{-9}$ eV. An energy $E_\gamma + R$ must be supplied to a nucleus in its ground state to absorb a γ ray and make a transition to an excited state. However, because $R \gg \Gamma$, the emitted γ rays will not be reabsorbed with appreciable probability.

If the radioactive nucleus is rigidly bound in a solid, the macroscopic mass of the solid, instead of the nuclear mass, appears in Eq. (B.1) and hence $R = 0$. This is an ideal case, because the binding of an atom to a solid is not rigid, and the lattice possesses vibrational degrees of freedom at finite temperature, which can be excited by a displaced nucleus. In general, a fraction of the emission of γ rays takes place with no perceptible recoil energy, and a width close to the natural width. This is known as the Mossbauer effect because Mossbauer (Ref. 1) had observed in 1958 that when radioactive nuclei, which emit low-energy γ rays, are bound in a lattice (solid), a considerable portion of the decay occurs without any transfer of energy to the lattice. Thus, the γ ray spectrum is extremely sharp and the resonant absorption of the γ rays by unexcited atoms can be easily observed. The Mossbauer effect is a significant tool for determining the hyperfine interactions, i.e., the interactions between a nucleus and the surrounding electrons. The hyperfine interactions involve the product of a nuclear quantity, such as the nuclear magnetic dipole moment and the electric quadrupole moment, as well as an atomic quantity, such as the electron density at the nucleus or the electric field gradient at the nuclear site.

This larger fraction f (known as the recoilless fraction and originally observed by Mossbauer) will be calculated in Section B.2. A smaller fraction R, the average energy transferred to the lattice in the emission process, was calculated by Lipkin by using a sum rule. Thus, in a solid, one can find both a sharp γ ray line unshifted by frequency and a broad shifted background. The relative proportions are temperature dependent, and the width of the sharp line is independent of temperature. The natural width is usually dominant.

B.2 RECOILLESS FRACTION

The interaction of a single radioactive nucleus bound in a crystal with a radiation field can be written (using a semiclassical theory) as

$$H_I = -(e/m_p)\sum_i \mathbf{A}(\mathbf{x}_i)\cdot\mathbf{p}_i, \tag{B.3}$$

where m_p is the proton mass, $\mathbf{x}_i$ and $\mathbf{p}_i$ are the proton coordinate and momentum, $\mathbf{A}(\mathbf{x}_i)$ is the vector potential, and the sum includes all the protons in the nucleus. Here, the interaction with the magnetic moment has been neglected.

We consider a monatomic lattice and introduce the following notations to simplify the problem. Consider the center of mass of the nucleus located at the position

$$\mathbf{X}_\nu = \mathbf{R}_\nu + \mathbf{u}_\nu, \tag{B.4}$$

and relative coordinates for the protons in a given nucleus

$$\mathbf{x}_i = \mathbf{X}_\nu + \mathbf{r}_i. \tag{B.5}$$

The vector potential $\mathbf{A}(\mathbf{x}_i)$ is proportional to $e^{i\mathbf{k}_\mu\mathbf{x}_i}$ where the plane electromagnetic wave vector has a wave vector. When the nucleus makes a transition $i \to f$, the matrix element for the transition is

$$M_T = <\{n_f(\mathbf{q})\}|e^{i\mathbf{k}_\mu\mathbf{X}_\nu}|\{n_i(\mathbf{q})>\} <f|a(\mathbf{k}_\mu)|i>, \tag{B.6}$$

where

$$a(\mathbf{k}_\mu) = K\sum_j e^{i\mathbf{k}_\mu\mathbf{r}_j}\mathbf{A}_0\cdot\mathbf{p}_j, \tag{B.7}$$

K is a proportionality constant, and $\mathbf{A}_0$ is the amplitude of the radiation field. $<f|a(\mathbf{k}_\mu)|i>$ is independent of the lattice position because $a(\mathbf{k}_\mu)$ involves all the nuclear physics of the problem.

The matrix element for a transition between phonon states is such that the transition rate is proportional to

$$P(n_f, n_i) = |<n_f|e^{i\mathbf{k}_\mu\mathbf{X}_\nu}|n_i>|^2. \tag{B.8}$$

The recoilless fraction f is the probability that a transition would occur between phonon states $(n_i \to n_f)$ without a change in energy of the phonon distribution and can be written as

$$f = \sum_f |<n_i|e^{i\mathbf{k}_\mu \cdot \mathbf{X}_\nu}|n_f>|^2 \delta(E_f - E_i). \tag{B.9}$$

However, Eq. (B.9) should be multiplied by the probability $P_{ni}(T)$ of a phonon distribution at temperature T and summing over all states n_i. In addition, it can be assumed that the equilibrium position of the radioactive atom is at the origin, i.e., $\mathbf{X}_0 = \mathbf{u}_0$, the displacement of this atom. Incorporating these, we rewrite Eq. (B.9) as

$$f = \sum_{if} P_{ni}(T)|<n_i|e^{i\mathbf{k}_\mu \cdot \mathbf{u}_0}|n_f>|^2 \delta(E_f - E_i). \tag{B.10}$$

Because the natural width of the γ ray line is negligible compared to phonon energies, the transition can be considered to be sharp. If the final state in emission differs from the initial state by energy $\hbar\omega$, Eq. (B.10) can be generalized as

$$f(\omega) = \sum_{if} P_{ni}(T)|<n_i|e^{i\mathbf{k}_\mu \cdot \mathbf{u}_0}|n_f>|^2 \delta(E_f - E_i - \hbar\omega). \tag{B.11}$$

B.3 AVERAGE TRANSFERRED ENERGY

The eigenstates of the Hamiltonian H contain a definite number of phonons. Because the kinetic energy operator for the radioactive nucleus, $\mathbf{p}^2/2M$, is the only portion of H that does not commute with $e^{i\mathbf{k}_\gamma \cdot \mathbf{u}_0}$, we obtain

$$[H, e^{i\mathbf{k}_\mu \cdot \mathbf{u}_0}] = [p^2/2M, e^{i\mathbf{k}_\mu \cdot \mathbf{u}_0}] = e^{i\mathbf{k}_\mu \cdot \mathbf{u}_0}[(\hbar^2\mathbf{k}_\mu{}^2/2M) + (\hbar/M)\mathbf{k}_\mu \cdot \mathbf{p}]. \tag{B.12}$$

Similarly,

$$[[H, e^{i\mathbf{k}_\mu \cdot \mathbf{u}_0}], e^{-i\mathbf{k}_\mu \cdot \mathbf{u}_0}] = 2H - e^{i\mathbf{k}_\mu \cdot \mathbf{u}_0} H e^{-i\mathbf{k}_\mu \cdot \mathbf{u}_0} - e^{-i\mathbf{k}_\mu \cdot \mathbf{u}_0} H e^{i\mathbf{k}_\mu \cdot \mathbf{u}_0} = -\hbar^2\mathbf{k}_\mu{}^2/M. \tag{B.13}$$

From Eq. (B.13), we obtain

$$<n_i|[[H, e^{i\mathbf{k}_\mu \cdot \mathbf{u}_0}], e^{-i\mathbf{k}_\mu \cdot \mathbf{u}_0}]|n_i> = 2E_i - <n_i|e^{i\mathbf{k}_\mu \cdot \mathbf{u}_0} H e^{-i\mathbf{k}_\mu \cdot \mathbf{u}_0}|n_i> - <n_i|e^{-i\mathbf{k}_\mu \cdot \mathbf{u}_0} H e^{i\mathbf{k}_\mu \cdot \mathbf{u}_0}|n_i>. \tag{B.14}$$

By inserting a complete set of final states, $|n_f><n_f|$, Eq. (B.14) can be written in the alternate form

$$\begin{aligned} <n_i|[[H, e^{i\mathbf{k}_\mu \cdot \mathbf{u}_0}], e^{-i\mathbf{k}_\mu \cdot \mathbf{u}_0}]|n_i> = 2E_i &- \sum_f <n_i|e^{i\mathbf{k}_\mu \cdot \mathbf{u}_0}|n_f><n_f|H|e^{-i\mathbf{k}_\mu \cdot \mathbf{u}_0}|n_i> \\ &+ \sum_f <n_i|e^{-i\mathbf{k}_\mu \cdot \mathbf{u}_0}|n_f><n_f|He^{i\mathbf{k}_\mu \cdot \mathbf{u}_0}|n_i> \end{aligned} \tag{B.15}$$

$$= 2E_i - 2\sum_f E_f|<n_i|e^{i\mathbf{k}_\mu \cdot \mathbf{u}_0}|n_f>|^2. \tag{B.16}$$

We use the identity

$$\sum_{n_f} |<n_i|e^{i\mathbf{k}_\mu \cdot \mathbf{u}_0}|n_f>|^2 = \sum_{n_f} <n_i|e^{i\mathbf{k}_\mu \cdot \mathbf{u}_0}|n_f><n_f|e^{-i\mathbf{k}_\mu \cdot \mathbf{u}_0}|n_i> = <n_i|n_i> = 1. \tag{B.17}$$

From Eqs. (B.8), (B.13), (B.16), and (B.17), we obtain

$$\sum_f (E_f - E_i)P(n_f, n_i) = \hbar^2 \mathbf{k}_\mu{}^2/2M. \tag{B.18}$$

Thus, the average energy transferred to the phonon system is equal to the recoil energy $\frac{\hbar^2 \mathbf{k}_\mu{}^2}{2M}$, which was the result originally derived by Lipkin.

The advantage of the Mossbauer effect is that one can measure the magnetic fields at the nuclear sites through the Zeeman effect. If a nucleus has spin I, the magnetic moment $\vec{\mu}_n = \gamma_n \hbar \mathbf{I}$ (γ_n is the nuclear gyromagnetic ratio) would interact with an effective magnetic field H_{eff}. The interaction is given by $H_z = -\vec{\mu}_n \cdot \mathbf{H}_{eff}$, which removes the (2I + 1) degeneracy of the spin orientation. This spin splitting is easily observable.

Reference

1. Mossbauer R. Nuclear resonance fluroscence of gamma rays in Ir 191. *Zeitschrift. fur Physik* 1958;**181**:124.

APPENDIX

C

Introduction to Renormalization Group Approach

CHAPTER OUTLINE

C.1 Critical Behavior 645
C.2 Theory for Scaling 646
C.3 Renormalization Group Approach 648
References 649

C.1 CRITICAL BEHAVIOR

We consider a ferromagnet in equilibrium at temperature T and under the action of a uniform magnetic field H. The reduced temperature variable is defined as

$$t = (T - T_c)/T_c, \tag{C.1}$$

where T_c is the critical temperature, and we consider the properties as $t \to 0$ with $H = 0$. The initial susceptibility in this regime diverges as

$$\chi_0(T) \approx C/t^{\gamma}, \tag{C.2}$$

where γ is known as the critical exponent. For ferromagnets like Fe and Ni, the critical exponent has values near 1.36, which implies that the mean-field, or "classical," prediction $\gamma = 1$ is incorrect. The zero-field specific heat can be written as

$$C_{H=0}(T) \approx A/t^{\alpha}, \tag{C.3}$$

or in the alternate form

$$C_{H=0}(T) \approx \overline{A}(t^{-\alpha} - 1)/\alpha, \tag{C.4}$$

where $\alpha \simeq -0.1$ for Ni and other isotropic magnets. The scaling theory of critical behavior asserts that the singular part of the free energy $F(T, H)$ varies asymptotically as

$$f(T,H) = -(k_B T)^{-1} F_{\text{sing}}(T,H) \approx t^{2-\alpha} Y(H/t^{\Delta}), \tag{C.5}$$

where the gap exponent

$$\Delta = \frac{1}{2}(2 - \alpha + \gamma). \tag{C.6}$$

Physics of Condensed Matter

$Y(y)$ is known as the scaling function and depends only on a single variable but is not explicitly given by the theory. From Eq. (C.4), we find that the spontaneous magnetization vanishes when $t \to 0$ as

$$M_0(T) \approx B|t|^{\beta}, \tag{C.7}$$

where β is predicted by the *exponent relation*

$$\beta = \frac{1}{2}(2-\alpha-\gamma). \tag{C.8}$$

$\beta \approx 0.36$ for magnets. Further, the equation of state $M = \mathfrak{M}(T,H)$ can be written in the scaled form

$$M/t^{\beta} \approx W(H/t^{\Delta}), \tag{C.9}$$

where $W(y)$ is a single-variable scaling function.

The basic two-point correlation function is defined as

$$G(\mathbf{x},T) = <\vec{S}_0 \cdot \vec{S}_{\mathbf{x}}>, \tag{C.10}$$

where $\vec{S}_{\mathbf{x}}$ denotes a localized spin at site $\mathbf{x}$. The variation of the scattering intensity in the critical region is given by

$$\hat{G}(\mathbf{q},T) = \sum_{\mathbf{x}} e^{i\mathbf{q}\cdot\mathbf{x}}\, G(\mathbf{x},T). \tag{C.11}$$

At the critical point,

$$\begin{aligned} G_c(\mathbf{x}) &\approx D_c/x^{d-2+\eta} \quad \text{as} \quad x \to \infty, \\ \hat{G}_c(\mathbf{q}) &\approx \hat{D}_c/q^{2-\eta} \quad \text{as} \quad q \to 0. \end{aligned} \tag{C.12}$$

Here, the critical exponent η lies in the range 0.03–0.1.

In zero field, as $t \to 0$, the form predicted by scaling is

$$G(\mathbf{x},T) \approx x^{-d+2-\eta} D(x/\xi), \qquad \xi \sim t^{-\nu}, \tag{C.13}$$

where ξ is the correlation length. Eq. (C.13) can be written in the equivalent form

$$\hat{G}(\mathbf{q},T) \approx C t^{-\gamma} \hat{D}(q^2/t^{2\nu}). \tag{C.14}$$

The correlation length exponent is given by

$$\nu = \gamma/(2-\eta). \tag{C.15}$$

The scaling function $\hat{D}(z^2)$ represents the scattering "line shape" near T_c.

C.2 THEORY FOR SCALING

Any adequate theory for scaling should (a) show how to calculate the exponents $\alpha, \gamma,$ and η; (b) predict or justify scaling; (c) lead to explicit calculations of scaling functions $Y(y), W(y), \hat{D}(z^2)$, etc; and

(d) describe the corrections of the asymptotic scaling laws and give a concrete estimate of their magnitude.

The first step in building a theory is to describe a lattice structure of spacing a, generated by a set of nearest-neighbor vectors $\vec{\delta}$. In usual treatments of such problems, we consider the lattice sites with coordinate vectors $\mathbf{x} = (x_i)$ with $i = 1, 2, d$, where $d = 3$ and is known as the *spatial dimensionality*. However, because d enters theoretical calculations in an essential way only through space and momentum (or wave number integrals), the definition of dimensionality can be extended to continuous values of d beyond 3. It turns out that the difference

$$\in = 4 - d \tag{C.16}$$

is a small but important parameter. The lattice, as constructed above with sites $\mathbf{x}$, is populated with spins $\vec{S_x}$. $\vec{S_x}$ has n components, i.e., $S_x = (S_x{}^{\mu})$ with $\mu = 1, 2, \ldots, n$, which enter equally into interactions. Some known examples are as follows:

a. $n = 3$, Heisenberg spins: $\overline{S} = (S^x, S^y, S^z)$.
b. $n = 2$, XY or "planar" spins: $\overline{S} = (S^x, S^y)$.
c. $n = 1$, uniaxial or Ising spins: $\overline{S} = S^z$.

One can show that for systems with *short range, isotropic* coupling is represented by an interaction Hamiltonian of the form

$$\mathcal{H}\{\vec{s_x}\} = \mathcal{H}_{iso.exch.} = -\frac{1}{2}\sum_{\mathbf{x},\mathbf{x}'} J(\mathbf{x}-\mathbf{x}')\,\vec{s_{\mathbf{x}}}\cdot\vec{s_{x'}}. \tag{C.17}$$

The parameters n and d are the only ones that apparently determine the critical exponents. In Eq. (C.17), $\vec{s}$ is the normalized spins or local variables

$$\vec{s} = \vec{S}/[S(S+1)]^{1/2}. \tag{C.18}$$

It is often convenient to allow the spin length $|\vec{S}|$ to vary continuously but introduce a *spin weighting function* in order to restrict the fluctuations in spin length. The partition function for a system of N spins is

$$Z_N[\mathcal{H}] = Tr_N\{\exp \mathscr{H}\}, \tag{C.19}$$

where the reduced Hamiltonian, $\mathscr{H}$, is defined as

$$\mathscr{H} = -(\mathscr{H}/k_B T) + \mathcal{W}. \tag{C.20}$$

$\mathcal{W}$ is a spin weighting function, which is a sum of identical terms $-w(\vec{s_j})$ for each spin. $\mathcal{W}$ becomes very large and negative as $|\vec{S}|$ becomes large. The specification of $\mathscr{H}$ determines the temperature, the external magnetic field, pressure, and so on, as well as all the interactions that are translationally invariant. The thermodynamics follow from the free energy per spin (the thermodynamic limit $N \to \infty$ is essential if the critical behavior is to be investigated),

$$f[\mathscr{H}] = -F(T,H)/k_B T = \lim_{N\to\infty} N^{-1} \ln Z_N[\mathscr{H}]. \tag{C.21}$$

C.3 RENORMALIZATION GROUP APPROACH

The exact construction of a renormalization group is in general a very difficult task. The main ideas can be summarized as follows:

a. The initial Hamiltonian $\mathcal{H}$ is renormalized to obtain a new Hamiltonian

$$\mathcal{H} \to \mathcal{H}' = \mathbf{R}[\mathcal{H}]. \tag{C.22}$$

b. The renormalization group operator $\mathbf{R}$ reduces the spin variables (the degrees of freedom) from N to

$$N' = N/b^d, \tag{C.23}$$

where b ($b > 1$) is known as the spatial rescaling factor and d is the dimensionality.

c. Sometimes $\mathbf{R}$ is defined via a partial trace over $(N - N')$ of the spin variables so that

$$\exp \mathcal{H}' = Tr_{N-N'}\{\exp \mathcal{H}\}. \tag{C.24}$$

d. The partition function must be preserved by $\mathbf{R}$,

$$Z_{N'}[\mathcal{H}'] = Z_N\{\mathcal{H}\}. \tag{C.25}$$

e. It is implicitly assumed that the renormalized Hamiltonian $\mathcal{H}'$ displays translational invariance.

f. All spatial vectors entering into correlation functions are rescaled by the factor b according to

$$\mathbf{x} \Rightarrow \mathbf{x}' = \mathbf{x}/b, \tag{C.26}$$

so that the spatial density of degrees of freedom (of spins) is preserved. Thus, the momenta are rescaled by

$$\mathbf{q} \Rightarrow \mathbf{q}' = b\mathbf{q}. \tag{C.27}$$

g. The basic spin fluctuation magnitude is preserved by rescaling the renormalized spin vectors as

$$\vec{s}_x \Rightarrow \vec{s}_{x'}' = \vec{s}_x/c, \tag{C.28}$$

where the spin rescaling factor c depends on $\mathcal{H}$, i.e., $c = c[\mathcal{H}]$. This step is an essential feature of *linear* renormalization groups where the local variable is transformed into an equivalent renormalized variable to which it is linearly related.

The renormalization group operator depends on b and c. From Eqs. (C.20) and (C.25), we obtain that the free energy is transformed according to

$$f[\mathcal{H}'] = b^d f[\mathcal{H}], \tag{C.29}$$

and from Eq. (C.28), we find that the basic spin–spin correlation function transforms as (for a linear renormalization group)

$$G[\mathbf{x}; \mathcal{H}] = c^2 G[\mathbf{x}/b; \mathcal{H}']. \tag{C.30}$$

Eqs. (C.29) and (C.30) lead to scaling properties. The steps followed in completing the renormalization process are as follows:

a. The transformation of the renormalization group operator **R** (defined in Eq. C.22) is iterated until one obtains a fixed-point Hamiltonian $\mathcal{H}^*$ by varying the parameters of the initial Hamiltonian, such that

$$\mathbf{R}[\mathcal{H}^*] = \mathcal{H}^*. \tag{C.31}$$

b. It can be shown from Eqs. (C.30) and (C.31) that one can obtain a functional equation with the unique solution

$$G[\mathbf{x}; \mathcal{H}^*] \sim 1/x^{2\omega} \tag{C.32}$$

and

$$c^* = c[\mathcal{H}^*] = b^{-\omega}. \tag{C.33}$$

Comparison of Eqs. (C.10), (C.12), and (C.33) shows that η is determined by the equation

$$c^* = b^{-(d-2+\eta)/2}. \tag{C.34}$$

The preceding results are brief outlines only to introduce the reader to the fascinating field of the magnetic phenomena arising from different symmetry, spatial form, and magnitude of the Hamiltonians of magnetic materials. The field of renormalization group, which offers a systematic approach for studying the effect of such terms on critical behavior and distinguishing between them, is much more complex and requires much further reading and solving a host of existing problems before one can use them in practice.

References

1. Fisher ME. The renormalization group in the theory of critical behavior. *Rev Mod Phys* 1974;**46**:597.2.
2. Wilson KG. *Phys Rev B* 1972;**4**:3174,3184.

Index

Page numbers in *italics* indicate figures and tables

A

Abelian, 633, 635
AC Josephson effect, 462
Acoustic mode in diatomic linear chain, *46*, 46
Aluminum atom in silicon crystal, *287*
Amorphous semiconductors, 296–299
  physical properties of, 299
Amorphous solids
  density of states, 593
  energy bands in one-dimensional aperiodic potentials, 591–592
  introduction, 590–594
  semiconductors, 593–594
    Ge and Si, 590, 593
Amorphous systems, electron energy levels of, 297
Anderson model, 439–440
  lattice model, 517–518
Anderson–Morel order parameter, 576–577, *577*
Angle-resolved photoemission spectroscopy (ARPES), *476*, 476–477
Antiferromagnetic (AF) alignment, 344
Antiferromagnetic (AF) materials, 506
Antiferromagnetic configuration, 341
Antiferromagnetic Mott insulator, 618
Antiferromagnetism, spin waves in, 421–422
Antiparallel (AP) configuration, 342
APW method, *see* Augmented plane-wave method
ARPES, *see* Angle-resolved photoemission spectroscopy
Atomic magnetic susceptibility, 371–378
  Curie's law, 377–378
  formulation, 371–372
  Hund's rules, 373–374
  Landé g factor, 375–376
  Larmor diamagnetism, 372–373
  Van Vleck paramagnetism, 374–375
Atomic scattering factor, 33–34
Augmented plane-wave (APW) method, 150–152, 156
Average transferred energy, 643–644

B

Balian–Werthamer (BW) state, 577
Band gaps, extrema of, 582
Band index, 101
Band model, exchange self-energy in, 234–235, 393–394
Band picture, 166
  of Ge, *170*
  of InSb, *171*
  of Si, *170*
Band structure
  of fullerene nanotube, *617*, 617
  of *trans*-polyacetylene, 622, *623*
Basic two-point correlation function, 646
bcc lattice, *see* Body-centered cubic lattice
BCS theory
  microscopic theory of superconductivity, 466
  strong coupling theory, 472
Bergman clusters, 590
Bipolar transistors, *318*, 318–319
  amplification of current in, *319*
Bipolarons, 626
Bloch electrons, 233–234
  effective mass, 247–248
  electron–phonon interaction, 264–271
  in external fields
    applied DC field, 253–254
    derivation for, 252–253
    Lorentz force equation, 248
    time evolution of, 250–252
  holes
    electrons and, *257*, 257
    Fermi surface, *258*, 258
    negative effective mass, 257
    positive effective mass, *256*, 256–257
  k · p perturbation theory, 245–246
  quasiclassical dynamics, 246–247
  semiclassical model, 243–244
  velocity operator, 244–245
  Zener breakdown, 258–260
  Zener tunneling, calculation of, 261–264
Bloch functions, 99, 132, 591–592
  in crystalline lattice, *139*
Bloch oscillations, *254*, 254–255
Bloch representation, equation of motion in, 388–389
Bloch theorem, 99
  effective Hamiltonian, 103
  translational symmetry, 103–105
Body-centered cubic (bcc) lattice, 5–7
  application to, 32–33
  Brillouin zones, 29–30, *30*, *119*, 119
  lattice constants of, *6*
  polyhedra, 539, *540*
  primitive cell of, 5, *6*
  Wigner–Seitz cell of, *6*
Boltzmann equation, 183–184
  for semiconductors, 313

Bond picture, 167
of GaAs, *171*
of InSb, 170
Born–Oppenheimer approximation, 200
Born–von Karman boundary conditions, 77
plane wave solutions, 97–98
Bose operators, 417
Bose–Einstein (BE) condensation, 571–572
Bosons, creation and annihilation operators, 54–58
Boundary conditions, Ginzburg–Landau theory, 457
Bragg diffraction, 19–20
Bragg formula, 582
Bragg law, 19
Bragg planes, *26*, 26, 28, *107*, 107
Brillouin zones, 117
Bravais lattice, 2, 9–11, 137, 634
with basis, 13
and crystal structures, 13–14
of diamond, 276
in three dimensions, 4–11
in two dimensions, 2, *3*, 4–11
Brillouin function, 423
Brillouin zones, 45, 134, *137*, 276
bcc lattice, 29–30, *30*, *119*, 119
Bragg plane, 117
definition, 27
energies in, 605
fcc lattice, 30–31, *31*, *119*, 119
of graphene, 601
Harrison's method of construction, 121–124, *122*, *123*, *124*
hcp structure, *119*, 119
of one-dimensional lattice, *28*, 28
simple cubic lattice, *118*, 118
two-dimensional centered rectangular lattice, 117, *118*
two-dimensional square lattice, 28–29, *29*, 117
Buckminsterfullerene (C_{60}), 600
discovery of, 609–613

C

C-NMR spectrum of C_{60}, *610*, 610
C_{60} photolithography, 611, *612*
Camley–Barnas model, *345*, 345–348, *347*, *348*
Carbon
nanotubes, 600, 614
symmetry properties of, 616–617
types of, 614–616
wave functions of, 601
Carrier-mediated magnetism, 291
Ce-based superconducting compounds, 508–509
$CeCu_2Si_2$, molar specific heat of, *489*
Cellular method, 142–145
Cesium chloride structure, *15*, 15–16, *173*
Chemical probe methods, 549–550, *550*
Chirality in graphene, 608
Cholesteric liquid crystals, 579
optical properties of, 581–583
CIP, *see* Current in plane
Circuits, integrated, 325
cis polyacetylene, *620*
Clusters
calculations, 494–497
embedded in matrix, 557–558, *558*
growth on surfaces
mean-field rate equations, 541–542
Monte Carlo simulations, 540–541
isolated structure
DFT, 543
Hartree–Fock method, 542–543
schematic representation of, 546, *547*
semi-empirical potentials, 544–546
tight-binding methods, 544
magnetism, 548–549
in embedded clusters, 553–555, *554–555*
experimental techniques for, 549–553, *550–552*
graphite surfaces, 555, *556*
in isolated clusters, 547–549
STM, 555–557, *557*
Coherence length, Ginzburg–Landau theory, 457–458
Cohesion of solids, 174–179
Condensate wave function, 572
Conducting polymers
electronic structure of, 618
solitons in
bipolarons, 626
electronic structure, 623
excitations, 624–626
introduction, 622
polarons and polaron excitations, 626
tight-binding model, 623
Conduction bands, 275, *276*
Conductivity
electrical, 186–187
thermal, 187–188
weak scattering theory of, 188–192
Conjugated polymers, 618–620
electronic structure of, 623
Constant energy curves, *605*, 605
Cooper pairs, 574
microscopic theory of superconductivity, *464*, 464–466
Coordination numbers, 5
Correlated super-spin glass (CSSG), 558
Correlation function for liquids, 570
Correlation length exponent, 646
Coulomb blockade system, *324*

Coulomb interactions, 204–205, 324
graphenes, 600
Covalent crystals, 177–178
CPP, *see* Current perpendicular to plane
Crystal defects, 14
Crystal lattices, 2–3
basis, diffraction, 31–34
symmetry operations in, 634
Crystal structures, 15–18
Bravais lattice and, 13–14
packing fraction of, 14
Crystalline solids, categories of, 275
Crystals
covalent, 177–178
ionic, 176–177
molecular, 174–176
CSSG, *see* Correlated super-spin glass
Cu superconductor, 516
Cubic group, 638, *639*
Cuboctahedral cluster, 528–529, *529*
Curie temperature, 410, 422–424
Curie's law, 377–378, 422
Curie–Weiss law, 441
Current in plane (CIP), 348
Current perpendicular to plane (CPP) geometry, 348
Current perpendicular to plane-GMR, *349*, 348–352
of multilayered nanowires, 350–352, *351*, *353*
theory of, 350–352
Curvature strains, liquid crystals, 581, *582*
Cybotactic phase, 579

D

2D graphene sheet, *614*, *616*
2D nanostructures, 554–555, *554*, *555*
d-wave pairing, 474–476
experimental confirmation of
ARPES, *476*, 476–477
flux quantization of superconducting ring, 480
high T_c superconductors, theoretical mechanism of, 481
Josephson tunneling, 477–478, *478*
NMR, 477
π–rings, 478–480, *478*, *479*
tricrystal magnetometry, 480
symmetry, 480–481
Dangling bonds, 299
amorphous semiconductors, 593
DC electrical conductivity, 86–87
DC Josephson effect, *462*, 462
de Haas–van Alphen effect, 378, 380, 383–387
Debye frequency, 50
Debye model of specific heat, 49–51, *51*
Debye temperature, 49, 51
Degenerate semiconductors, 281
Degrees of freedom, phonon and phason, 586–588
Delocalized states of disordered solids, 593
Density functional theory (DFT), 532
LDA, 224–225
Schrodinger's equation, 223
structure of isolated clusters, 543
universal function, 223
Density of electron states
amorphous solids, *592*, 593
comparison of, *82*, 82
concept of, 81
vs. energy, 81, *82*
Density operator
Fermi distribution function, 76
grand canonical ensemble, 75
DFT, *see* Density functional theory
Diamagnetic susceptibility, 378, 581
Diamond, 600
Bravais lattice of, 168, 276
structure, *16*, 16, *168*, 168
Diluted magnetic semiconductors (DMS), 290–296
Dimerization coordinate, *623*
Dimerized polyacetylene chain, total energy, *625*
Diode laser, *327*
efficiency of, 327
Dirac delta function, 192
Dirac equation, 607
Dirac Fermions, graphene, 606–608
Dirac points, 602
Direct exchange model, 416–417, *417*
DMS, *see* Diluted magnetic semiconductors
Doped semiconductors, 285–287
Doping in electronic polymers, *621*, 621
Double-barrier channel, *323*
Double-heterostructure laser, *327*, 327
Drude model
electrical and thermal conductivity, 73
electron density concept, 72
kinetic theory of gas, 73
3s electrons energy levels, *72*, 72
scattering of electrons, 71, *72*
sodium atoms, *72*, 72
valence electrons, 71
Dulong and Petit law, 49, 51

E

EAM, *see* Embedded atom model
EDMFT, *see* Extended Dynamical Mean-Field Theory
Effective Hamiltonian, 102–103
Effective mass, 247–248
tensor, 278

Effective medium theory (EMT), 544
Einstein frequency, 37
Einstein model of specific heat, 52–53
Einstein temperature, 52
Elastomers, 618
Electric breakdown, *see* Zener breakdown
Electrical conductivity, 186–187
  and thermal conductivity, 73
  of *trans*-(CH), 618, *620*
Electron
  and holes, 278–279
    densities in equilibrium, 279–282
  in phosphorus impurity, *286*
  polarization of tunneling, 354
  trajectories, *342*
Electron paramagnetic resonance (EPR) shift, 291–295
Electron–electron interaction
  Born–Oppenheimer approximation, 200
  density functional theory
    LDA, 224–225
    Schrodinger's equation, 223
    universal function, 223
  Fermi liquid theory
    energy functional, 227–230
    Fermi liquid parameters, 230–231
    quasiparticles, 225–227
  frequency and wave-number-dependent dielectric constant
    Fermi distribution function, 218
    Fourier component, 218
    plasma mode, 221
    time-independent perturbation theory, 220
  Friedel sum rule and oscillations
    Legendre polynomial, 215
    Plane wave, *215*
    wave function, 214, 216
  Green's function method
    Bloch electrons, 233–234
    exchange self-energy, band model, 234–235
    Fourier transformation, 232
  Hartree approximation
    Lindhard theory, 202
    Pauli principle, 202
    Schrodinger equation, 200
    self-consistent field approximation, 202
    Thomas–Fermi theory, 202
  Hartree–Fock approximation
    jellium model, 204–207
    Pauli principle, 203
    spin-up functions, 203
  Mott transition, 222
  Schrodinger equation, 200
  screening effects
    Fourier transform, 207
    Lindhard theory, 209–214
    Poisson's equations, 207
    Thomas–Fermi approximation, 208–209
Electronic energy levels for mixed-valence materials, *493*
Electronic energy states, graphene, *606*
Electronic polymers, 621
Electronic shell structures, *530*, 530–531
  ellipsoidal shell model, 535
  large clusters, *533*, 535–536, *536*
  nonalkali clusters, 535
  spherical jellium model
    harmonic oscillator potential, 531
    self-consistent, 532–535
    spherical square-well potential, 531–532
    Woods–Saxon potential, 532, *533*
Electronic structure of conjugated polymers, 623
Electronics, graphene-based, 332
Electron–phonon interaction, 264–271
  superconductors, 452
Electrons
  density concept, 72
  in weak periodic potential
    plane wave solutions, 97–99
    two-dimensional rectangular lattice, *96*, 96
Elementary band theory of solids, 111–112
Ellipsoidal shell model, 535
Embedded atom model (EAM), 544
Empirical pseudopotentials, 157–158
EMT, *see* Effective medium theory
Energy bands, *259*, 259
  in one-dimensional aperiodic potentials, 591–592
Energy gap, 275–276
EPR shift, *see* Electron paramagnetic resonance shift
Equilibrium, p-n junction in, *308*, 307–310
Equivalent Hamiltonian operator, 181
Extended Dynamical Mean-Field Theory (EDMFT), 501–502
Extended zone scheme, *101*, 101, *110*, 110
External electric fields, Bloch electrons in
  applied DC field, 253–254
  derivation for, 252–253
  Lorentz force equation, 248
  time evolution of, 250–252
Extrinsic semiconductors, 284–285

F

Face-centered cubic (fcc) lattice, 7–8
  application to, 33
  Brillouin zones, *31*, 31, *119*, 119

lattice constants of, *8*
polyhedra, *538*, 538
primitive cell of, *7*, 7
Wigner–Seitz cell of, *7*, 8
Faraday's law of induction, 400
fcc lattice, *see* Face-centered cubic lattice
Fermi distribution, 606
function, 218, 435–436
grand canonical ensemble, 74
Fermi energy, 79, 438, 617
and chemical potential, 82–84
Fermi function, 184
Fermi level, 383–384
Fermi liquid theory, 503, 574–575
energy functional
effective mass, 228, 230
Fermi wave vector, 228
Pauli exclusion principle, 228
Fermi liquid parameters, 230–231
quasiparticles
Green's function, 227
Hartree–Fock approximation, 225
Landau's argument, 225, 227
law of conservation of energy, 226
N-electron system, 226
Thomas–Fermi screened potential, 226
Fermi liquids
fractionalized, 505–506
heavy, 502–505
models, 502–506
parameters, 230–231
Fermi sphere, *79*, 79
displacement of, *86*, 87
Fermi surfaces, 503, *504*
Harrison's method of construction, 121–124, *122, 123, 124*
of metal, 383
in three dimensions, 121, *121, 122*
in two dimensions, 119–120, *120*
Fermi wave vector, 79, 228
Fermi–Dirac distribution function
density operator, 76
ground-state energy of electron gas, 80
specific heat of electron gas, 84–86
variation of, *84*, 84
Fermi–Dirac distributions, 428
Fermi–Dirac statistics, 573–574
Fermions
creation and destruction operators for, 56
energy levels for, *60*
Ferromagnetic domains, 425–426
Ferromagnetic (F) configuration, 341
Ferromagnetic ground state, 437–439
Ferromagnetism
direct, indirect, and super exchange, 416–417
Heisenberg Hamiltonian, application, 418–420
Heisenberg model, 416
Heitler–London approximation, 412–414
magnons, 417
metallic ions, 412
Schwinger representation, 417–418
in solids
Curie temperature, 422–424
ferromagnetic domains, 425–426
hysteresis, 426–427
Ising model, 427
spin Hamiltonian, 414–416
in transition metals
Fe, Co, and Ni, 430–431
free electron gas model, 431–432
Hubbard model, 433
magnetic moments of, 428
Stoner model, 428–429
FET, *see* Field-effect transistor
Fibonacci lattice, *585*, 585, *587*
diffraction pattern of, 586
Field-effect transistor (FET), 319–321
First-order perturbation theory, 209
First-principles pseudopotentials, 158–160
Five-fold symmetry, diffraction pattern of, *584*
Flow-tube reactor (FTR), 549
Flux quantization, 459–460, *460*
of superconducting ring, 480
FM-T-N junction, theory of, 358–361
Form factor, *see* Atomic scattering factor
Fourier component, 218
Fourier transform, 207, 232
Fourth generation of polymeric materials, 618
Fractionalized Fermi liquids, 505–506
Free-electron model
failures of, 89–90
Fermi gas
Born–von Karman boundary conditions, 77
definition, 77
Fermi energy, 79
Fermi sphere, *79*, 79
Fermi wave vector, 79
Schrodinger equation, 77
two-dimensional k-space, points, *78*, 78
gas model, 431–432
Frequency dependent dielectric constant
Fermi distribution function, 218
Fourier component, 218

Frequency dependent dielectric constant (*Cont.*)
plasma mode, 221
time-independent perturbation theory, 220
Friedel sum rule and oscillations
Legendre polynomial, 215
plane wave, *215*
wave function, 214, 216
FTR, *see* Flow-tube reactor
Fulde–Ferrell–Larkin–Ovchinnikov (FFLO) superconducting state, 508
Fullerene derived tubule, *615*
Fullerenes
carbon nanotubes, 614
symmetry properties of, 616–617
types of, 614–616
nanotube, *613*, 613
band structure of, *617*, 617
see also Buckminsterfullerene (C_{60})

G

GaAs-AlGaAs interface, *323*
advantage of, 322
Gauss theorem, 155
Ge, 168–169
band structure of, 168–169, *277*, 277
Geometric shell structure
close-packing of atoms, 537
polyhedra
bcc, 539, *540*
fcc, *538*, 538
Mackay icosahedron, 539, *539*
Wulff construction, *537*, 537, *538*
Geometrical structure factor, 32
Giant magnetic resonance (GMR), 339, 340–342
effect, 341, 352
metallic multilayers, 340–342, *341–342*
Ginzburg–Landau theory
boundary conditions, 457
coherence length, 457–458
London penetration depth, 458–459
order parameter, 456–457
GMR, *see* Giant magnetic resonance
Gradient-field deflection method, *551*, 551–553, *552*
Grand canonical ensemble, 74
density operator, 75–77
Fermi distribution function, 74
Graphene, 329, *329*, *330*
ambipolar electric field effect in, *331*
Dirac Fermions, 606–608
introduction, 600–601
lattice, 601–602
tight-binding approximation, 602–606
Graphene-based electronics, 332
Graphite, 600
surfaces, magnetism clusters, 555, *556*
Green's function (KKR) method, 152–156, *154*, 554
Bloch electrons, 233–234
exchange self-energy, band model, 234–235
Fourier transformation, 232
Ground-state energy of electron gas, 79–81
Group, definition of, 633
Group theory
point group operations, 636–639
space groups, 634–636
symmetry of crystals, 633–634

H

Hall effect, geometry of, *87*, 87
Hall field, 89
Hamiltonian, renormalized group, 648–649
Hamiltonian function, 180, 246
Harmonic approximation, 39
Harmonic oscillator potential, 531
Harmonic spring constant, 623
Harrison's method of construction
Brillouin zones, 121–124, *122*, *123*, *124*
Fermi surface, 121–124, *122*, *123*, *124*
Hartree approximation
Lindhard theory, 202
Pauli principle, 202
Schrodinger equation, 200
self-consistent field approximation, 202
Thomas–Fermi theory, 202
Hartree–Fock approximation, 572
exchange term, 203
jellium model
Coulomb interaction, 204–205
exchange term, 205
kinetic energy, 204
Kronecker delta function, 206
Lindhard dielectric function, 206
N-electron system, 206
Thomas–Fermi theory, 207
Pauli principle, 203
spin-up functions, 203
Hartree–Fock method, 542–543
hcp structure, *see* Hexagonal close-packed structure
Heat capacity of solid, Einstein theory, 37
Heavy Fermi liquids, 502–505
Heavy-fermions, 488
properties of, 488
superconductivity theories, 516

superconductors
Ce-based compounds, 508–509
$PrOs_4Sb_{12}$, 513, *514*
$PuCoGa_5$, 513, *514*
$PuRhGa_5$, *515*, 515–516
U-based compounds, *511*, 509–513
Heisenberg Hamiltonian, application of, 418–420
Heisenberg model, 416
Heitler–London model, 414
Helmholtz free energy, 377
Heterojunction diode, *613*
Hexagonal close-packed (hcp) structure, 17–19, *18*
Brillouin zones, *119*, 119
lattice constants of elements, *19–18*
unit cell of, 17
Hexagonal systems, *10*, 10
Highest occupied shell (HOS), 559
Hilbert transform, 498
Hohenberg–Kohn (HK) theorem, 532
Holstein–Primakoff transformation, 418
Homogeneous semiconductor, *see* Intrinsic semiconductor
Honeycomb lattice structure of graphene, *601*, 607
electronic dispersion in, 606, *607*
HOS, *see* Highest occupied shell
Houston functions, 252
Hubbard model, 433, 498
Hund's rules, 373–374
Hydrogen-bonded structures, 174
Hydrogen molecule, covalent bond, *167*
Hyperfine interactions, 641
Hysteresis, 426–427, *427*

I

Icosahedral atomic clusters, *590*
Icosahedral group, stereogram of, *584*
Icosahedral quasicrystals, 589–590
Impurity band of local defects, *594*
Impurity scattering, 189–192
Indirect exchange model, 416–417, *417*
Indirect semiconductors, 276
InGaAsP laser, structure of, *327*
Injection coefficient, 361–364
Insulators, 112–117, 276, *354*
Integrated circuits, 325
Intrinsic semiconductor, 276, 278, 283–284
chemical potential for, *283*
types of, 276
Invariant subgroup, 635
Ionic crystals, 176–177
Ionic magnetic susceptibility, *see* Atomic magnetic susceptibility
Ionic solids, 172
Ising model, 427
Isotope effect, superconductors, 454

J

Jellium model
Coulomb interaction, 204–205
exchange term, 205
kinetic energy, 204
Kronecker delta function, 206
Lindhard dielectric function, 206
N-electron system, 206
Thomas–Fermi theory, 207
Josephson effect
AC, 462
DC, *462*, 462
two superconductors, oxide layer, *461*, 460–462
Josephson tunneling, 477–478, *478*

K

$\vec{k} \cdot \vec{\pi}$ model, 295–296
k · p perturbation theory, 245–246
Kinetic theory of gas, 73
KKR, *see* Green's function method
Kohn–Sham equations, 158–159, 534
Kondo effect, 439, 501
Kondo insulators, 516–519
theory of, 517–519
Kondo-lattice models, 490–491, 501
Kronecker delta function, 206

L

Lagrangian function, 180
Lambda point, 571
Landau diamagnetism, 380–383
Landau Fermi-Liquid (LFL) state, 508
Landau gauge, 379
Landau levels, 384, *384*, *386*
Landau theory, second-order phase transitions, 441–442
Landau's argument, 225, 227
Landé g factor, 375–376
Langevin susceptibility, *see* Larmor diamagnetic susceptibility
Large clusters, *533*, 535–536, *536*
Larmor diamagnetic susceptibility, 373
Larmor diamagnetism, 372–373
Laser, 326
diode, *327*, 327
double-heterostructure, *327*, 327
InGaAsP, structure of, *327*
master, 327–328
semiconducting, 327
slave, 327–328

Lattice
planes, 11–12, *12*, *13*
specific heat
Debye model of, 49–51
Einstein model of, 52–53
theory, 48–49
structure, graphene, 601–602
translation, 636
Lattice constants, 4–5, 44, 61
of bcc lattices, *6*
of elements with hcp structure, *18–19*
of fcc lattices, *8*
Lattice dynamics, 37–46
normal modes
of one-dimensional chain with basis, 44–46
of one-dimensional monoatomic lattice, 41–44
theory, 37–41
Lattice points, 41, *41*, *42*
group, 3
Lattice waves, quantization of, 61–65
Laue condition, alternative formulation of, 25–27
Laue method, 20–21
Law of conservation of energy, 226
Law of mass action, 282
LCAO, *see* Linear combination of atomic orbitals method
LCH model, *see* Linear combination of hybrids model
LDA, *see* Local density approximation
LDM, *see* Liquid drop model
LED, *see* Light-emitting diode
Legendre polynomial, 215
Leonard–Jones potential, *175*, 175
Light dispersion in cholesteric, *582*
Light-emitting diode (LED), 326
Lindhard dielectric function, 206
Lindhard theory, 202
Fermi sphere, *213*, 213
first-order perturbation theory, 209
Fourier transform, 214
Friedel oscillations, 214
Thomas–Fermi approximation, 211
Linear combination of atomic orbitals (LCAO) method, 134–140
Linear combination of hybrids (LCH) model for tetrahedral semiconductors, 297–299, *299*, *300*
Liouville's theorem, 182
Liquid crystals, 568
classes of, 578–580
curvature strains, 581, *582*
optical properties of, 581–583
order parameter, 580–581
Liquid drop model (LDM), 528, *529*
Liquid ^{3}He
Fermi liquid theory, 574–575
liquid–solid phase equilibrium, 573
superfluidity in, 574
experimental results of, 575
theoretical model
A and A_1 phases, 575–577
B phase, 577–578
Liquids
correlation function, 570
phase diagram, 568–569
specific heat *vs.* temperature, *571*, 571
Van Hove pair correlation function, 569
Local density approximation (LDA), 158, 224–225
Local impurity self-consistent approximation (LISA), 498–499
application to periodic Anderson model, 499–500
Localized states
in amorphous semiconductors, *593*, *594*
overlap of tails, *594*
London equation, 455–456
London penetration depth, Ginzburg–Landau theory, 458–459
Longitudinal acoustic phonons, 270
Lorentz force equation, 253
Lowest unoccupied shell (LUS), 559
Luttinger theorem, 503, 518
Luttinger–Kohn ($\vec{k}\cdot\vec{p}$) model, 295
Lyotropic liquid crystal, 578

M

Mackay icosahedron, *539*, 539
filling between complete shells, 540
Macroscopic order parameter, 572
Magnetic breakdown, 252
Magnetic dipole moments, 411
Magnetic ordering
Anderson model, 439–440
antiferromagnetism, spin waves in, 421–422
ferrimagnetism, 410
ferromagnetism
direct, indirect, and super exchange, 416–417
Heisenberg Hamiltonian, application, 418–420
Heisenberg model, 416
Heitler–London approximation, 412–414
magnons, 417
metallic ions, 412
Schwinger representation, 417–418
in solids, 422–427
spin Hamiltonian, 414–416
in transition metals, 412, 427–433
Kondo effect, 439
magnetic dipole moments, 411
spontaneous magnetization, 410

Magnetic permeability, 370
Magnetic phase transition
Landau theory, second-order phase transitions of, 441–442
order parameter, 441
Magnetic random access memory (MRAM), 355
principles of, *355*
Magnetic susceptibility
of atoms, *see* Atomic magnetic susceptibility
of Bloch electrons in solids, many-body theory, 388–396
definition, 370
of free electrons in metals, 378–387
de Haas–van Alphen effect, 378, 382–387
formulation, 378–380
Landau diamagnetism, 380–383
Pauli paramagnetism, 380–383
general formula for, 390–393
of nonferromagnetic solid, 388
orbital contribution to exchange and correlation effects, 395–396
spin contribution to exchange enhancement, 394–395
spin-orbit contribution to, 393
exchange and correlation effects on, 396
Magnetic switching, 356–357
Magnetic tunnel junctions (MTJ), 352–356
Magnetism
in embedded clusters, 553–555, *554*, *555*
2D nanostructures, *554*, 554–555, *555*
small clusters, *554*, 554–555, *555*
switching and blocking temperature, 553
experimental techniques for cluster
chemical probe methods, 549–550, *550*
gradient-field deflection method, *551*, *552*, 551–553
graphite surfaces, 555, *556*
in isolated clusters, 547–549
cluster magnetism, 548–549
superparamagnetism and blocking temperature, 547–548
Magnetization
of Bloch electrons
correlations contribution to, 436–437
quasiparticle contribution to, 435–436
theory of magnetization, 434–435
in zero external magnetic field, 291
Magnetoresistance, 340
effect, 348
Mass spectrum peak of C_{60}, *610*
Massless Dirac fermions, 331, 607
Master laser, 327–328
Matrix element, transition in phonon states, 642
Maxwell–Boltzmann distribution function, 80
MBE, *see* Molecular-beam epitaxy
Mean-field approximation, 423
Mean-field rate equations, 541–542
Mean-field theories, 498–502
Meissner–Ochsenfeld effect, *455*, 455
Metals, 112–117, 275
cohesion in, 178–179
Metamagnetism in heavy fermions, 506–507
Miller indices, 11–12, *13*
Mixed-valence compounds, 492–493
Model pseudopotentials, 156–157
for KKR matrix elements, *157*
Molecular-beam epitaxy (MBE), 322
Molecular crystals, 174–176
Molecular solids, 174
Molecular structures
of conjugated polymers, *619*
MBBA, *578*
PAA, *578*
Monoatomic lattice, 41–44
Monoclinic systems, 10–11, *11*
Monomer, 617
Monte Carlo simulations, 540–541
MOSFET, 320, *611*
inversion layer in, *320*, 320
n-type, structure of, *319*, 319
Mossbauer effect
advantage of, 644
average transferred energy, 643–644
definition of, 641
recoilless fraction, 642–643
Mott transition, 140, 222
Mott's theory of spin-dependent electron scattering, 342–344
MRAM, *see* Magnetic random access memory
MTJ, *see* Magnetic tunnel junctions
Muffin-tin potential, 149–150, *150*

N

n-doped semiconductor, 285–286
charge movement to metal, *306*
donor and acceptor levels for, *287*
n-p-n bipolar transistor, *318*, 318
n-type MOSFET, structure of, *319*, 319
Nanoclusters
nanoscience and, 528
superconducting state of
qualitative analysis, 558–559
thermodynamic Green's function, 559–561
Negative effective mass, 257
Negative soliton, *626*
Nematic liquid crystals, 578
molecule arrangement in, *579*
Newton's law, 247
NMR, *see* Nuclear-magnetic-resonance
Noble gases, 174–176

Nonalkali clusters, 535
Nondegenerate semiconductors, 281
Nuclear-magnetic-resonance (NMR), 477

O

One-dimensional aperiodic potentials, energy bands, 591–592
One-dimensional electron gas (1DES), *323*
One-dimensional quasiperiodic lattice, *585*
Optical modes in diatomic linear chain, *46*, 46
Optoelectronic devices, 325–329
OPW method, *see* Orthogonalized plane-wave method
Order parameter, 441
  Ginzburg–Landau theory, 456–457
  liquid crystals, 580–581
  for superfluid ^{3}He, *577*, 578
Orthogonalized plane-wave (OPW) method, 145–146, *147*
Orthorhombic systems, 9, *10*

P

3*p* energy bands, *113*, 114
p-n junction, 306–310, *307*
  in equilibrium, 307–310, *308*
  rectification by, 311–318
    equilibrium case, 311–313
    nonequilibrium case, 313–318
P-T-H diagram, 575, *576*
p-type semiconductor, 285
  donor and acceptor levels for, *287*
p-wave pairing, 474–476
PAM, *see* Periodic Anderson model
Para-azoxyanisole (PAA), molecular structure, *578*
Paramagnetism
  susceptibility, 378
  Van Vleck, 374–375
Pauli exclusion principle, 73, 228, 575
Pauli limiting, 508
Pauli paramagnetism, 380–383
Peierls instability, *622*, 622
Peierls theorem, 624
Penrose lattice, *583*, 583
  perfect edge dislocation in, *589*, 589
Penrose tiles, 583
Pentagonal quasicrystal, 586
Periodic Anderson model (PAM), 490–491
  application of LISA to, 499–500
Periodic zone scheme, 100–101, *101*, *111*, 111, 243
Phase diagram
  of liquid ^{3}He, 575
  of liquids, *568*, 568–569
  in magnetic field, *576*
  superconductors, *454*, 454
Phase transition, liquid ^{4}He, 570–571
Phason degrees of freedom, 586–588
Phason flips in two-dimensional Penrose lattice, *588*, 588
Phason-strained Fibonacci lattice, *588*
Phonon-assisted hopping, 593
Phonon-strained Fibonacci lattice, 587, *588*
Phonons, 38–39, 48
  absorption, 265
  average energy transfer in, 643–644
  degrees of freedom, 586–588
  emission, 265
  resistivity due to scattering by, 192–194
Phosphorus atom in silicon crystal, *286*
Photoconductivity, 325
Photodiode, 326
Photoexcitation, SSH model, 623
Photoinduced electron transfer, semiconducting polymer to C_{60}, *627*, 627
Photons, 48
  absorption of, 276
Plane wave solutions
  Born–von Karman boundary conditions, 97–98
  Schrodinger equation, 97
Plasma frequency of metallic polymers, 621
Plasma mode, 221
Plastic crystals, 568
Point groups
  cubic group, 638, *639*
  description of, 636–637
  symbols, international notation for, *638*
Poisson's equations, 207
Polarons, 626
  excitations, 626
Polyacetylene structure, 618
Polyaniline (PANI) as conducting polymer, 618
Polycritical point (PCP), 575
Polyhedra
  bcc, 539, *540*
  fcc, *538*, 538
  Mackay icosahedron, *539*, 539
Polymers, 617–618
  conducting polymers, solitons in
    bipolarons, 626
    electronic structure, 623
    excitations, 624–626
    introduction, 622
    polarons and polaron excitations, 626
Polyparaphenylene (PPP)
  bipolaron in, *627*
  polaron in, *626*
Positive effective mass, 257

Positive soliton, *626*
Primitive cell, 2, *3*, 3
of bcc lattice, 5, *6*
of fcc lattice, *7*, 7
of rectangular lattice, *3*
of two-dimensional square Bravais lattice, *2*
π-rings, *478*, 478–480, *479*
$PrOs_4Sb_{12}$, 513, *514*
Pseudogap, 593
Pseudopotentials, 147–149
empirical, 157–158
first-principles, 158–160
model, 156–157
Pseudospin, 608
Pu superconductor, 516
$PuCoGa_5$, 513, *514*
$PuRhGa_5$, *515*, 515–516

Q

QED, *see* Quantum electrodynamics
QHE, *see* Quantum Hall effect
Quantization of lattice waves, 65–66
formulation, 61–65
Quantized Hall resistance, 397, *400*, 400
Quantum electrodynamics (QED), 331, 608
chirality, 332
Quantum Hall effect (QHE)
fractional, *400*, 400
from gauge invariance, 400
in two-dimensional electron gas, 396–397
in strong magnetic field, 397–399
Quasi-electrons, microscopic theory of superconductivity, 463–464
Quasiclassical dynamics, 246–247
Quasicrystals
discovery of, 584
icosahedral quasicrystals, 589–590
Penrose tiles, 583
perfect edge dislocation in the Penrose lattice, *589*, 589
phonon and phason degrees of freedom, 586–588
quasiperiodic lattice, 584–586
see also Liquid crystals
Quasiparticles
Green's function, 227
Hartree–Fock approximation, 225
Landau's argument, 225, 227
law of conservation of energy, 226
N-electron system, 226
Thomas–Fermi screened potential, 226
Quasiperiodic lattice, 584–586

R

Reciprocal lattice
definition, 21–22
properties of, 22–25
Recoil energy, 641
Recoilless fraction, 642–643
Rectification
in heterojunctions, 611
by p-n junction, 311–318
Reduced zone scheme, *100*, 100, 109–110, *110*
Relaxation time, 188
approximation, 184–185
Renormalization group approach, 648–649
Repeated zone scheme, 100–101, *101*, *111*, 111
Rhombohedral systems, *10*, 10
Riseborough's theory, 518–519
Rudderman–Kittel–Kasuya–Yosida (RKKY) interactions, 490, 500–501
Russsell–Saunders Coupling, 373

S

3s electrons energy levels, *72*, 72
3s energy bands, *113*, 114
s-wave pairing, 474–476
Saturated polymers, 618–620
Saturation magnetization, *425*
sc lattice, *see* Simple cubic lattice
Scaling theory, 646–647
of critical behavior, 645
Scanning tunneling microscope (STM), 555–557, *557*
Scattering
of electrons, 71, *72*
impurity, 189–192
probability, 188
resistivity due to, 192–194
Schoenflies notation, point groups, 636, *637*
Schottky diode, 305
rectifying effect of, *306*
Schrieffer-Heeger (SSH) model, 623
Schrodinger equations, 77, 132, 135, 200, 223, 244
effective Hamiltonian, 102
Josephson effect, 461
plane wave solutions, 98
Schwinger representation, 417–418
SDL, *see* Spin diffusion length
Second-order perturbation theory, 245
Second-order phase transition, superconductors, *454*, 454
Second quantization, 53–60
creation and annihilation operators, 54–58
field operators and Hamiltonian, 58–60
occupation number representation, 53–54

Semi-empirical potentials, 544–546
Semiclassical model, 179–182, 243–244
Semiclassical theory, 642
Semiconductors, 112–117
  amorphous, 296–299
    physical properties of, 299
  Boltzmann equation for, 313
  degenerate, 281
  DMS, 290–296
  doped, 285–287
  elements, 276
  extrinsic, 284–285
  III-V zincblende, 277
  III–V semiconductors, 277
  indirect, 276
  intrinsic, *see* Intrinsic semiconductor
  n-type, 285–286
    donor and acceptor levels for, *287*
  p-type, 285
    donor and acceptor levels for, *287*
  spintronics with, 357–364
  tetrahedral, LCH model, 297–299, *299*, *300*
  work function of, 305
  zinc-blende, 170
Semimetals, 115
SET, *see* Single-electron transistor
SHE, *see* Spin Hall effect
Short circuit effect, 342, *342*
Si, 168–169
  band structure of, *170*
  tetrahedral bonding, *167*
Simple cubic (sc) lattice, 4, *13*
  Brillouin zones, *118*, 118
Single-electron transistor (SET), 321–325
Slater–Koster hopping integrals, 544
Slave boson method, 493
Slave laser, 327–328
Small clusters, *554*, 554–555, *555*
Smetic liquid crystals, 579
  molecule arrangement, *579*
Sodium chloride, *172*
  structure, *15*, 15
Sodium metal in Wigner–Seitz method, *143*
Solids
  cohesion of, 174–179
  ionic, 172
  molecular, 174
Solitons in conducting polymers
  bipolarons, 626
  electronic structure, 623
  excitations, 624–626
  introduction, 622
  polarons and polaron excitations, 626
  tight-binding model, 623
Sommerfeld model
  density of electron states
    comparison of, *82*, 82
    concept of, 81
    vs. energy, 81–82, *82*
  density operator, 75–77
  Fermi distribution function, 74
  free-electron Fermi gas
    Born–von Karman boundary conditions, 77
    definition, 77
    Fermi energy, 79
    Fermi sphere, *79*, 79
    Fermi wave vector, 79
    Schrodinger equation, 77
    two-dimensional k-space, points, *78*, 78
  ground-state energy of electron gas, 79–81
  Pauli exclusion principle, 73
Space groups
  operations, 634–636
  see also Point groups
Spatial dimensionality, 647
Specific heat
  Debye model of, 49–51, *51*
  Einstein model of, 52–53
  of electron gas, 84–86
  measurements of liquid ^{3}He, *575*
Spherical jellium model
  harmonic oscillator potential, 531
  self-consistent, 532–535
  spherical square-well potential, 531–532
  Woods–Saxon potential, 532, *533*
Spherical square-well potential, 531–532
Spin accumulation in ferromagnetic and nonmagnetic layer, *353*
Spin-dependent electron scattering, *341*
  Mott's theory of, 342–344
Spin diffusion length (SDL), nonmagnetic and ferromagnetic layers, 350, 352
Spin fluctuation effect, 577
Spin Hall effect (SHE), 362
Spin Hamiltonian, 414–416
Spin-orbit interaction, 362
Spin-orbit splitting, modulation of, 358
Spin states, 421
Spin-transfer, *357*
  torques, 356–357, *357*
Spin transistor, 358
Spin-triplet pair-wave function, 576
Spin-up functions, 203
Spin-valve system, 344

Spin weighting function, 647
Spintronics, 339
with semiconductors, 357–364
Squeezed-vacuum-state generation, *328*
Stabilized jellium model, 535
Step function, 192
Stern-Gerlach experiment, *551*, 551–553, *552*
STM, *see* Scanning tunneling microscope
Stoner model, 428–429
Strong coupling theory, 472–473
Super exchange model, 416–417, *417*
Superconducting compounds
Ce-based, 508–509
U-based, 509–513, *511*
Superconductivity, microscopic theory
BCS theory, 466
Cooper pairs, *464*, 464–466
excited states at $T = 0$, 469–470, *470*
excited states at $T \neq 0$, 470–472, *471*
quasi-electrons, 463–464
superconducting electron gas, ground state, 466–469
Superconductors
high-temperature
ARPES, *476*, 476–477
atomic structure of, *473*, 473
d-wave pairing, 474–476, 480–481
flux quantization of superconducting ring, 480
Josephson tunneling, 477–478, *478*
NMR, 477
p-wave pairing, 474–476
π–rings, 478–480, *478*, *479*
properties of, *474*, 474
s-wave pairing, 474–476
theoretical mechanism of, 481
tricrystal magnetometry, 480
properties of
electron–phonon interaction, 452
isotope effect, 454
phase diagram, *454*, 454
resistivity, *452*, 452–453, *453*
second-order phase transition, *454*, 454
type I superconductors, *453*, 453
type II superconductors, *454*, 454
Superfluid component, 571
Superfluid ^{4}He
phase transition, 570–571
theory of superfluidity, 571–573
two-fluid model, 571
Superfluidity in liquid ^{3}He, 574
Superparamagnetism, 547–548
Surface effects, 14–15
Symmetry of crystals, 633, 634

T

Temperature variable, reduced, 645
Tetragonal systems, *9*, 9
Tetrahedral bonding, *167*
of Si, *169*
Tetrahedral semiconductors for LCH model, 297–299, *299*, *300*
Theory of magnetization, 434–435
Theory of superfluidity, liquid ^{4}He, 571–573
Thermal conductivity, 187–188
Thermal equilibrium
acceptor levels, 288–289
donor levels, 288
impurity levels in, 288–290
Thermodynamic Green's function, 559–561
Thermodynamic potential, 390
Thermotropic liquid crystal, 578
Thomas–Fermi approximation, 208–209, 323
Tight-binding approximation, 131–134
graphene, 602–606
Tight-binding methods, 544
Tight-binding model, 623
Time-independent perturbation theory, 220
TMR, *see* Tunneling magnetoresistance
trans-polyacetylene, 618, *620*, 622
Transistors
bipolar, *318*, 318–319
amplification of current in, *319*
FET, 319–321
SET, 321–325
Transparent metallic polymers, 621
Transverse acoustical (TA) mode
for diatomic linear lattice, *47*
displacement direction of ions in, *47*
Transverse optical (TO) mode
for diatomic linear lattice, *47*
displacement direction of ions in, *47*
Transverse phonons, 270
Triclinic systems, 10–11, *11*
Tricrystal magnetometry, 480
Trigonal systems, *10*, 10
Truncated icosahedron, Buckminsterfullerene C_{60}, *609*, 609
Tunneling magnetoresistance (TMR), 352–356
Two-fluid model, liquid ^{4}He, 571
Type I superconductors, *453*, 453
Type II superconductors, *454*, 454

U

U-based superconducting compounds, 509–513, *511*
Umklapp process, 250, 270

Unit cell, *3*, 3
  of cubic lattices, *4*, 5
  of hcp structure, 17
  of monolayer graphene, *608*
  of rectangular lattice, *3*
Universal function, 223

V

Valence bands, 276, *276*
Valence electrons, 71
van der Waals interaction, 174
Van Hove pair correlation function, 569
Van Vleck paramagnetism, 374–375
Variable-range hopping, 594
Velocity operator, 244–245
V–F theory, 350
Volume filling fraction (VFF), 558
Vortex, 573

W

Wannier functions, 140–141, *142*
Wave functions, 214, 216, 413
  carbon, 601
Wave-number-dependent dielectric constant
  Fermi distribution function, 218
  Fourier component, 218
  plasma mode, 221
  time-independent perturbation theory, 220
Weak scattering theory of conductivity, 188–192
Weiss Field Model and Spin-wave theory, 424–425
Wiedemann–Franz law, 188
Wigner–Eckart theorem, 375
Wigner–Seitz cell, 3–4
  of bcc Bravais lattice, *6*
  of fcc Bravais lattice, *7*, 8
  of reciprocal lattice, 27
  of sodium metal, *142*
  of two-dimensional Bravais lattice, *3*
Wilson ratio, 488
WKB approximation, 259–260
Woods–Saxon potential, 532, *533*
Wulff construction, 537, *537*, *538*

Z

Zeeman effect, 644
Zener breakdown, 252, 258–260
Zener tunneling, calculation of, 261–264
Zero-field specific heat, reduced, 645
Zero-gap semiconductor, graphene as, 607
Zinc-blende semiconductors, 170
  structure, *17*, 17
Zinc oxide (ZnO), 296
Zinc sulphide bond, *171*
Zone boundary, 105–109

Printed and bound by CPI Group (UK) Ltd, Croydon, CR0 4YY
03/10/2024
01040314-0002